AF527631

Air Pollution:

Prevention and Control Technologies

Second Edition

Air Pollution:
Prevention and Control Technologies
Second Edition

Prof. Anjaneyulu Yerramilli
Ph.D., FRSC (London)
Professor & Director, Center for Energy Studies,
K L University, Green Fields, Vaddeswaram,
Guntur Dist., India.

BS Publications
A unit of **BSP Books Pvt. Ltd.**
4-4-309/316, Giriraj Lane,
Sultan Bazar, Hyderabad - 500 095.

Air Pollution: Prevention and Control Technologies, ***Second Edition***
by ***Anjaneyulu Yerramilli***

Published by:

BS Publications

A unit of **BSP Books Pvt., Ltd.**
4-4-309/316, Giriraj Lane, Sultan Bazar,
Hyderabad - 500 095, T.S.
Phone : 040 - 23445600, 23445688
e-mail : info@bspbooks.net
Website : www.bspbooks.net

ISBN: 978-93-87593-42-8 (Hardbound)

Dedication

This book is dedicated to my Late Parents Sri. Y. Satchidanandam and Smt. Y. Seetamma who are my inspiring guidance and for their great love and affection.

Koneru Lakshmaiah Education Foundation
(Deemed to be University estd u/s 3 of the UGC Act, 1956)
Campus: Green Fields, Vaddeswaram, Guntur District, Andhra Pradesh, India. PIN 522 502
Phone No. 0863 - 2399999; www.kluniversity.in

FOREWORD

Due to increasing air quality & climate change legislations and rigorous implementation of government policies towards sustainable future, air pollution career opportunities are tremendously growing. For example the demand for scientifically trained power in air quality management, climate changes forecasting & regulation, industrial emission monitoring, implementation of new cost effective control technologies, environmental and human health impact assessment is expected to increase rapidly along with industrial development.

To meet the fast growing demands for skilled man power/scientists/researchers/ engineers in this emerging area, courses on air pollution, preventive measure and control technologies are being offered by a number of institutes in Science and Technology in India which need Resource/Reference material with well updated information.

I am happy to write Foreword for this book on "Air Pollution: *Prevention and Control Technologies* Second Edition" (which serves the above requirement), authored by Prof. Anjaneyulu Yerramilli, Distinguished Professor and Director for our Center for Advanced Energy Studies who has very rich experience in both Teaching & Research in the area of Air pollution and Control both at National and International level.

I understand this book covers in addition to all fundamental concepts of Air Pollution extensively, latest information on scientific and engineering aspects of air pollution monitoring, assessment predictive modelling, state of art air pollution control technologies, air pollution from automobiles noise pollution and methodologies for impact assessment on air quality which can be useful as a Text book/Reference material to under graduate/graduate students in Science and Engineering, Air Quality Management Professionals/Researchers.

I congratulate Prof. Anjaneyulu Yerramilli for his great efforts in writing this very useful book on a very important topic Air Pollution & Control and wish him to bring some more on various emerging technologies addressing the needs of students in Science/ Engineering /Research.

Koneru Satyanarayana

GANDHI INSTITUTE OF TECHNOLOGY AND MANAGEMENT (GITAM)
Visakhapatnam | Hyderabad | Bengaluru
(Declared as Deemed to be University u/s 3 of UGC Act. 1956)
Accredited by NAAC with A+ Grade

Prof. M. S. Prasada Rao
Vice-Chancellor

Prologue

Day by day, the air quality in the environment is getting depleted because of the increase of particulates, biological molecules and other harmful gasses, mainly oxides of carbon, sulphur and nitrogen in the atmosphere. Industrial growth and rapid increase in population have necessitated the production and use of energy that triggered deterioration of Air Quality throughout the world damaging human health, ecosystems and vegetation. The air pollution in many cities has spiked far beyond acceptable limits due to rapid urbanization, increase in vehicular emissions and lack of strict implementation of environmental regulations. As per WHO report, out of the top twenty most polluted cities in the world, thirteen exist in India.

Sustainable development of any country is possible only with the adoption of innovative control technologies and rigorous implementation of air quality/climate change legislations by the government. This needs programs for development of skilled work force, competent technologists and committed researchers in this area.

Prevention of air pollution and adoption of control technologies need support from experts in Air quality management trained in a number of Science and Engineering disciplines. Many Universities/Institutes are now offering courses on Air Pollution & Control Technologies at basic and advanced levels, both in under-graduate and post-graduate programs. Students successfully completing these courses are finding lucrative employment opportunities in environmental consultancy firms, industry or local government departments.

This book on "Air Pollution Prevention and Control Technologies" authored by Prof. Anjaneyulu Yerramilli will serve as a Reference / Textbook to under-graduate/post-graduate students of Science and Engineering. Air Quality Management Professionals and Researchers also find it useful as a reference book, as it provides comprehensive understanding of the consequences of air pollution, management techniques, modeling and control technologies explained through illustrations and worked out problems.

This book authored by a person of Prof. Anjaneyulu's eminence with rich exposure to academics and research, will definitely be a useful source of information to the students, faculty and researchers in the field.

I greatly appreciate the efforts of Prof. Anjaneyulu in bringing out this book on Air Pollution & Control Technologies, a subject of contemporary relevance and importance. I wish him great success in all his future endeavors.

(M.S. PRASADA RAO)

26th April, 2018

Gandhinagar Campus, Rushikonda, Visakhapatnam-530 045, A.P., India. www.gitam.edu
Ph.: +91-891 - 2840202, 2795313, EPABX: 2790101, Fax: 2795312, E-mail: vicechancellor@gitam.edu

PREFACE TO SECOND EDITION

This second edition on **Air Pollution: Prevention and Control Technologies, Second Edition,** covering in **14 Chapters**, provides a strong interdisciplinary approach in presenting topics on Air Pollution, by integrating chemistry, physics, meteorology, engineering, health effects, policy and regulatory aspects.

To address the requirements of **undergraduate/graduate students** in **Science/Engineering** taking **Air Pollution** courses at a basic level, the fundamental aspects of air pollution are presented with suitable examples and worked out problems in **Chapters 1-4**.

To provide technological perspectives, the scientific and engineering aspects of air pollution monitoring, assessment, predictive modelling and control technologies are discussed exhaustively in separate **chapters 5-14**. Methodologies for air quality Impact assessment for new projects are also discussed in these chapters.

The speciality of this book is that it extensively deals on various aspects of air pollution, providing up-to-date information on global air pollution trends, preventive measures adopted by different countries. The book exhaustively presents latest modeling techniques for air quality assessment/prediction/management and state-of-art air pollution control technologies with suitable examples, illustrations and worked out problems.

A Stronger systems approach is adopted in presenting the nature of air pollutants, their dispersion under different meteorological conditions, modelling and control technologies at various scales – starting at the individual/niche level and expanding to local /regional/global scale.

The present book addresses the needs of undergraduate/graduate students in **Science and Engineering** pursuing **Air Pollution Course** both as a basic introductory or as an advanced Core course.

This book also serves as a resource material for researchers/practicing professionals in this important area.

Anjaneyulu Yerramilli

ACKNOWLEDGEMENTS

The permission provided by several authors and publishers for using some generic figures, flow charts and data etc. shown as sources at appropriate places in the text, is gratefully acknowledged.

I am greatly indebted to our Honble, President K. L. University Er. Koneru Satyanarayana garu for his constant support and encouragement in writing this book.

I gratefully appreciate Dr. L. S. S Reddy Vice Chancellor, K. L. University for all his administrative support in my efforts.

My Special thanks are for my junior Colleagues at KLU, Dr. Srinivas Desamsetty; Dr. Naga Mahesh K; Mr. Nitin Ralph. P; and Mr. Sandeep Yechury for their help in preparing question bank and manuscript preparation.

Anjaneyulu Yerramilli

CONTENTS

Chapter 4: Air Pollution and Global Climatic Changes

Chapter 5: Meteorological aspects of Dispersion of Air Pollutants

Chapter 6: Air Pollution Sampling and Analysis

Chapter 7: Air Pollution Control: General Source Correction Methods and Particulate Control Technologies

Chapter 8: Air Pollution Control Technologies for Gaseous Pollutants

Chapter 9: Air Pollution Control Technologies for Oxides of Sulphur

Chapter 10: Air Pollution Control Technologies for Oxides of Nitrogen

Chapter 11: Control Technologies for Volatile Organics, Hydrocarbons and Hydrogen Sulphide

Chapter 12: Air Pollution by Automobiles and its Control Technologies

Chapter 13: Noise Pollution

Chapter 14: Prediction and Assessment of Impacts on the Air Environment

CHAPTER 1
Air Pollution: Sources and Effects

1.1 Introduction

Air pollution is a gas (or a liquid or solid dispersed through ordinary air) released into atmosphere in a large quantity to harm the health of people or other animals, kill plants or stop them growing properly, damage or disrupt some other aspect of the environment (such as making buildings crumble), or cause some other kind of nuisance (reduced visibility, perhaps, or an unpleasant odor).

As with water pollution and land contamination, it's the quantity (or concentration) of a chemical in the air that makes the difference between "harmless" and "pollution." Carbon dioxide (CO_2), for example, is present in the air around you at a typical concentration of less than 0.05 percent and breathing it in usually does no harm (you breathe it out all day long); but air with an extremely high concentration of carbon dioxide (say, 5–10 percent) is toxic and could kill anybody in a matter of minutes. Forest fires, erupting volcanoes, and gases released from radioactive decay of rocks inside Earth are just three examples of natural air pollution that can have hugely disruptive effects on people and the planet.

1.1.1 Definition of air pollution

Air pollution may be described as "the imbalance in quality of air so as to cause adverse effects on the living organisms existing on earth".

The World Health Organization (WHO) defines **air pollution** as "limited to situations in which the outer ambient atmosphere contains materials in concentrations which are harmful to man and his environment".

Air pollution is broadly defined as "the presence in the atmosphere, of one or more contaminants such as fumes, dust, gases, mist, grit, odour, smoke, smog or vapours in considerable quantities and of duration which is injurious to human, animal or plant life or which unreasonably interferes with the comfortable enjoyment of life and property".

Thus, **air pollution** is generally a disequilibrium condition of air caused by the introduction of foreign elements from natural and man-made sources into the air such that the air becomes injurious to biological communities.

1.1.2 Air pollution due to natural sources

Forest fires (which often start naturally) can produce huge swathes of smoke that drift for miles over neighboring cities, countries, or continents. Giant volcanic eruptions can spew so much dust into the atmosphere that they block out significant amounts of sunlight and cause the entire planet to cool down for a year or more. Radioactive rocks can release a gas called radon when they decay, which can build up in the basements of buildings with serious effects on people's health.

All these things are examples of serious air pollution that happen without any human activity; although we can adapt to natural air pollution, and try to reduce the disruption it causes, we can never stop it happening completely.

However, since Earth's atmosphere is very turbulent the natural air pollution will often disperse relatively quickly.

1.1.3 Air pollution due to anthropogenic activities

Today most of the natural resources are polluted due to increased pace of anthropogenic activities. The increase in population and transportation and industrial development has resulted in the release of all kinds of air pollutants into the environment. The natural capacity of the environment to tolerate and sustain development has dwindled in the face of the ever-increasing discharge of air pollutants.

The assessment of air quality and control of air pollution are difficult tasks facing an environmental manager. Both soil and water can be confined and gathered at one place while air cannot be gathered and confined in one place. Hence air quality assessment, air pollution prevention strategies and control technologies are of great importance in air quality management

1.1.4 Top-ten gases in anthropogenic air pollution

Any gas could qualify as pollutant if it reaches a high enough concentration to do harm. Theoretically, that means there are dozens of different pollution gases. In practice, about ten different substances cause most concern:

Sulfur dioxide: Coal, petroleum, and other fuels are often impure and contain sulfur as well as organic (carbon-based) compounds. When sulfur (spelled "sulphur" in some countries) burns with oxygen from the air, sulfur dioxide (SO_2) is produced. Coal-fired power plants are the world's biggest source of sulfur-dioxide air pollution, which contributes to smog, acid rain, and health problems that include lung disease.

Carbon monoxide: This highly dangerous gas forms when fuels have too little oxygen to burn completely. It spews out in car exhausts and it can also build up to dangerous levels inside a home if you have a poorly maintained gas boiler, stove, or fuel-burning appliance.

Carbon dioxide: This gas is central to everyday life and isn't normally considered a pollutant: we all produce it when we breathe out and plants such as crops and trees need to "breathe" it in to grow. However, carbon dioxide is also a greenhouse gas released by engines and power plants. Since the beginning of the Industrial Revolution, it's been building up in Earth's atmosphere and contributing to the problem of global warming and climate change.

Nitrogen oxides: Nitrogen dioxide (NO_2) and nitrogen oxide (NO) are pollutants produced as an indirect result of combustion, when nitrogen and oxygen from the air react together. Nitrogen oxide pollution comes from vehicle engines and power plants, and plays an important role in the formation of acid rain, ozone and smog. Nitrogen oxides are also "indirect greenhouse gases" (they contribute to global warming by producing ozone, which is a greenhouse gas).

Volatile Organic Compounds (VOCs): These carbon-based (organic) chemicals evaporate easily at ordinary temperatures and pressures, so they readily become gases. That's precisely why they're used as solvents in many different household chemicals such as paints, waxes, and varnishes. Unfortunately, they're also a form of air pollution: they're believed to have long-term (chronic) effects on people's health and they also play a role in the formation of ozone and smog.

Particulates: These are the sooty deposits in air pollution that blacken buildings and cause breathing difficulties. Particulates of different sizes are often referred to by the letters PM followed by a number, so PM10 means soot particles of less than 10 microns (10 millionths of a meter or 10μm in diameter). In cities, most particulates come from traffic fumes.

Ozone: Also called trioxygen, this is a type of oxygen gas whose molecules are made from three oxygen atoms joined together (so it has the chemical formula O_3), instead of just the two atoms in conventional oxygen (O_2). In the stratosphere (upper atmosphere), a band of ozone ("the ozone layer") protects us by screening out harmful ultraviolet radiation (high-energy blue light) beaming down from the Sun. At ground level, it's a toxic pollutant that can damage health. It forms when sunlight strikes a cocktail of other pollution and is a key ingredient of smog (see box below).

Chlorofluorocarbons (CFCs): Once thought to be harmless, these gases were widely used in refrigerators and aerosol cans until it was discovered that they damaged Earth's ozone layer. We discuss this in more detail down below.

Unburned hydrocarbons: Petroleum and other fuels are made of organic compounds based on chains of carbon and hydrogen atoms. When they burn properly, they're completely converted into harmless carbon dioxide and water; when they burn incompletely, they can release carbon monoxide or float into the air in their unburned form, contributing to smog.

Lead and heavy metals: Lead and other toxic "heavy metals" can be spread into the air either as toxic compounds or as aerosols (when solids or liquids are dispersed through gases and carried through the air by them) in such things as exhaust fumes and the fly ash (contaminated waste dust) from incinerator smokestacks.

1.1.5 Composition of air

Air constitutes 80% of man's daily intake of materials by weight. We breathe 22,000times a day on the average, inhaling 16 kg of air. The earth's atmosphere is an envelope of gases extending to a height of 200 km from the earth's surface. The atmosphere is a protective cover which sustains life and protects it from the unfriendly environment of outer space. The lower atmosphere, i.e, the troposphere contains about 70% of the mass of the atmosphere. It has mainly three categories of gaseous components: major, minor, and trace (Table 1.1). For major gases, like N_2 and O_2, the concentrations vary but to a very little extent with time, while for trace gases, the values tend to converge as the averaging time lengthens. Oceans as well as vegetation are the most important **sinks** for most atmospheric gases. A **sink** is a medium which is capable of retaining and interacting with along-lived pollutant, though not necessarily indefinitely.

Table 1.1 *Average composition of clean dry air near sea level (ppm by volume)*

Components	Average conc. (ppm)	Concentration in volume percent	Estimated residence time
Major			
Nitrogen, N_2	78.09×10^4	78.09	Continuous
Oxygen, O_2	20.94×10^4	20.94	Continuous
Minor			
Argon, Ar	93×10^2	0.93	Continuous
Carbon dioxide, CO_2	32×10^2	0.0318	2 to 4 years
Trace			
Neon, Ne	18	0.0018	Continuous
Helium, He	5.2	0.00052	~ 2 million years
Methane, CH_4	1.3	0.00013	4 to 7 years
Krypton, Kr	1.0	0.0001	Continuous
Hydrogen, H_2	0.5	0.00005	Little is known about residence time
Carbon monoxide, CO	0.1	0.00001	0.5 year
Ozone, O_3	0.02	0.000002	~ 60 days
Ammonia, NH_3	0.01	0.000001	7 days
Nitrogen dioxide, NO_2	0.001	0.0000001	5 days
Sulphur dioxide, SO_2	0.0002	0.0000002	4 days
Hydrogen sulphide, H_2S	0.002	0.0000002	2 days
Xenon, Xe	0.081	-	Continuous

(**Source:** *Engineering Tool Box (www.engineeringtoolbox.com)*)

1.1.6 Concentration units of air pollutants

The concentration of any pollutant can be expressed in a number of ways involving units of mass, or volume per unit mass or volume per volume of air. There are four ways in which the concentrations of pollutants can be expressed.

1. The ratio of the mass of pollutant m_p to the mass of pure air m_a plus the mass of the pollutant m_p in a given volume is called the **mass concentration ratio, ω_p**.

$$\omega_p = m_p/(m_a + m_p) \qquad .....(1.1)$$

2. The ratio y_p of the volume of pollutant v_p to the volume of pure air v_a is called the **volume concentration ratio.**

$$y_p = v_p/(v_a + v_p) \qquad .....(1.2)$$

3. The more common way of expression of the concentrations is **ppm** which **is volume concentration in parts per million.**

$$y_{ppm} = y_p \times 10^6 \qquad .....(1.3)$$

   1 ppm of SO_2 = 1part of the gas pollutant/10^6 parts of air

   1 m^3 of gas/10^6 m^3 of air = 1mL of gas/1 litre air.

4. The fourth way is to express the concentration as the ratio of the mass of pollutant m_p to the volume of the pollutant v_p plus the volume of air v_a

$$\rho_p = m_p/(v_a + v_p) \qquad .....(1.4)$$

The concentration of the pollutant (gas or particle), p_p is generally given in micrograms per cubic meter or $\mu g/m^3$.

While all the above concentration ratios are interconvertible, the most important is the interconversion of y_{ppm} and p_p in $\mu g/m^3$ for gaseous pollutants.

Substituting Equations 1.2 and 1.3 in 1.4 we have

$$\rho_p = (m_p/v_p)\, y_{ppm}\, 10^{-6} \qquad .....(1.5)$$

Assuming that the perfect gas law holds for pollutants that are usually formed in the atmosphere, we can express the mass density of the pure component pollutant, m_p/v_p in units of $\mu g/m^3$ as,

$$\frac{m_p}{v_p} = \frac{PM_p\,(10^9)}{(8.314 \times 10^{-2})\,T} \qquad .....(1.6)$$

where P is the total pressure in bars, M_p is the molecular weight of the pollutant, T is the absolute temperature in K, and 8.314×10^{-2} is the value of the universal gas constant expressed in units of m^3 bar/kg mol –K. At the standard temperature (25 °C) and pressure (1.0133 bars) the above equation reduces to

$$\frac{m_p}{v_p} = \frac{M_p\,(10^9)}{24.45} \qquad .....(1.7)$$

Substituting Equation 1.7 in Equation 1.5, we get the relation between ρ_p $\left(\mu g/m^3\right)$ and y_{ppm}

$$\rho_p = \frac{M_p\, y_{ppm}\left(10^3\right)}{24.45} \qquad .....(1.8)$$

The constant in the denominator becomes 22.4 at 0 °C and 1 atmo pressure.

Thus

$$1\text{ ppm} = \frac{\left(22.4 \times 10^{-3}\right) M_p \text{ g}}{10^6 \text{ m}^3} = \frac{M_p}{22.4} \times 1000\ \mu\text{g / m}^3 \text{ at } 0\,^\circ C \qquad .....(1.9)$$

$$= \frac{M_p}{24.45} \times 1000\ \mu\text{g / m}^3 \text{ at } 25\ ^\circ C \qquad .....(1.10)$$

Any mass per unit volume measurement must take into account changing temperature as well as pressure while the mass per unit volume of an ideal gas is independent of temperature and pressure. 1 g mole of any gas at STP conditions occupies 22.4 litres (64 g of SO_2, or 30 g of NO or 46 g of NO_2 or 48 g of O_3, all of them occupy 22.4 litres at 0°C and 760 mm pressure while the volume occupied is 24.45 litres at 25°C and 760 mm pressure.).

In the volume scale the following relations hold good

$$\mu\text{g / m}^3 = \frac{n l\, 1^{-1} \times M_p \times 10^{-3} \times T_0 \times p}{V_0 \times T \times P_0} \qquad .....(1.11)$$

$$n l\, 1^{-1} = \frac{\mu\text{g / m}^3 \times V_0 \times 10^{-3} \times T \times P_0}{M_p \times T_0 \times P} \qquad .....(1.12)$$

where M_p is the gram molecular weight and V_0 is the molar volume of an ideal gas which is equivalent to $22.4 \times 10^{-3} m^3$ $mole^{-1}$ at standard temperature (T_0 in Kelvin) and pressure (P_0 = 101.3 k Pa).

The important point is that mass per unit volume (i.e., $\mu g/m^3$) expressions do not allow an immediate comparison of one pollutant with another in terms of numbers of molecules. In contrast, an atmosphere containing 1 $\mu\, l^{-1}$ SO_2 contains the same number of pollutant molecules as one containing 1 μl 1^{-1} O_3 or NO_2 because the molecular weight and the volume of gas containing 1 gram molecule have already been taken into account. Interconversions from ppm to $\mu g/m^3$ and vice versa at different temperatures for common air pollutants are given in Table 1.2 and Table 1.3.

Table 1.2 *Amounts of different pollutants in $\mu g\ m^{-3}$ equivalent to 1 ppm or 1 $\mu l\ l^{-1}$ (1000 nl l^{-1}) at different temperatures but standard atmospheric pressure*

T°C	SO_2	NO_2	NO	NH_3	O_3	CS_2	CO	H_2S	HF
–5	2914	2092	1365	774	2183	2001	1274	1550	910
0	2860	2054	1340	760	2143	1965	1250	1521	893
5	2809	2017	1316	747	2104	1929	1228	1494	877
10	2759	1981	1292	733	2067	1895	1206	1468	862
15	2711	1947	1270	721	2031	1862	1185	1442	847
20	2665	1914	1248	708	1997	1831	1165	1417	832
25	2620	1882	1227	696	1963	1800	1146	1394	818
30	2577	1850	1207	685	1931	1770	1127	1371	805
35	2535	1821	1187	674	1899	1742	1108	1348	792

Table 1.3 *Amounts of different pollutants in ppb or nl l^{-1} equivalent to 1000 mg m^{-3} at different temperature but standard atmospheric pressure*

T°C	SO_2	NO_2	NO	NH_3	O_3	CS_2	CO	H_2S	HF
–5	343	478	733	1291	458	500	785	645	1099
0	349	487	746	1315	467	509	800	657	1120
5	356	496	760	1339	475	518	814	669	1140
10	362	505	774	1364	484	528	829	681	1161
15	368	514	787	1388	492	537	844	693	1181
20	375	523	801	1412	501	546	858	705	1202
25	382	531	815	1436	509	556	873	718	1222
30	388	540	828	1460	518	565	888	730	1243
35	394	549	842	1484	527	574	902	742	1263

Problem 1.1

The Central Pollution Control Board standards for NO_2, SO_2, and CO for sensitive, industrial and residential areas (exposure limits at atmospheric pressure and 25 °C) are given in Table 1.4. Express their concentrations in ppm.

Table 1.4 *CPCB standards ($\mu g/m^3$) at 1 atm pressure at 25°C*

Parameter	Sensitive areas	Industrial areas	Residential areas
SO_2	30	120	80
NO_2	30	120	80
CO	0.75	1.5	1.0

Solution: One mole of an ideal gas at 1atm and 25°C occupies a volume of 24.45 l ($24.45 \times 10^{-3}\ m^3$).

SO_2 standards

The molecular weight of SO_2= 32.06 + 2 × 16.0 = 64.60

In sensitive areas:

$$\text{ppm of } SO_2 = \frac{24.45 \times 10^{-3}\ \text{m}^3/\text{mol} \times 30 \times 10^{-6}\text{g}/\text{m}^3}{64.6\ \text{g/mol}} = 11.35 \times 10^{-9} \text{ or } 0.01135$$

In industrial areas:

$$\text{ppm of } SO_2 = \frac{24.45 \times 10^{-3}\ \text{m}^3/\text{mol} \times 120 \times 10^{-6}\text{g}/\text{m}^3}{64.6\ \text{g/mol}} = 0.045$$

In residential areas:

$$\text{ppm of } SO_2 = \frac{24.45 \times 10^{-3}\ \text{m}^3/\text{mol} \times 180 \times 10^{-6}\text{g}/\text{m}^3}{64.6\ \text{g/mol}} = 0.03$$

NO_2 standards

The molecular weight of NO_2 = 14 + 2 × 16 = 46.0

In sensitive areas:

$$\text{ppm of } NO_2 = \frac{24.45 \times 10^{-3}\ \text{m}^3/\text{mol} \times 30 \times 10^{-6}\text{g}/\text{m}^3}{46\text{g}/\text{mol}} = 0.0159 \times 10^{-6} = 0.016$$

In industrial areas:

$$\text{ppm of } NO_2 = \frac{24.45 \times 10^{-3}\ \text{m}^3/\text{mol} \times 120 \times 10^{-6}\text{g}/\text{m}^3}{46\text{g}/\text{mol}} = 0.0159 \times 10^{-6} = 0.016$$

In residential areas:

$$\text{ppm of } NO_2 = \frac{24.45 \times 10^{-3}\ \text{m}^3/\text{mol} \times 180 \times 10^{-6}\text{g}/\text{m}^3}{46\text{g}/\text{mol}} = 0.0424$$

CO standards

The molecular weight of CO = 12 + 16 = 28.0

In sensitive areas:

$$\text{ppm of CO} = \frac{24.45 \times 10^{-3}\ \text{m}^3/\text{mol} \times 1000 \times 10^{-6}\text{g}/\text{m}^3}{28\text{g}/\text{mol}} = 0.873$$

In industrial areas:

$$\text{ppm of CO} = \frac{24.45 \times 10^{-3}\ \text{m}^3/\text{mol} \times 1000 \times 10^{-6}\text{g}/\text{m}^3}{28\text{g}/\text{mol}} = 0.873$$

In residential areas:

$$\text{ppm of CO} = \frac{24.45 \times 10^{-3}\ \text{m}^3/\text{ml} \times 2000 \times 10^{-6}\ \text{g/m}^3}{28\,\text{g/mol}} = 17.500$$

Problem 1.2

A car emits CO as exhaust gas at 2% by volume. Calculate the concentration of CO in μg/m^3 at 0 °C and 25 °C at 1 atmospheric pressure.

Solution: 2% by volume of CO = 20 ml CO in 1 litre air = 20 ppm.

By Equations 1.9 and 1.10,

$$\text{At } 0\ ^\circ\text{C, 1 ppm of CO} = \frac{28}{22.4} \times 1000\ \mu\text{g/m}^3 = 1250\ \mu\text{g/m}^3$$

$$20\ \text{ppm of CO} = 1250 \times 20 = 25000\ \mu\text{g/m}^3$$

$$\text{At } 25\ ^\circ\text{C, 1 ppm of CO} = \frac{28 \times 1000\ \mu\text{g/m}^3}{24.45} = 1145.20\ \mu\text{g/m}^3$$

$$20\ \text{ppm of CO} = 1145\ \mu\text{g/m}^3 \times 20 = 22900\ \mu\text{g/m}^3$$

Various units used for air pollutants are summarized in Table 1.5.

Table 1.5 *Units of measurement in air pollution control*

Parameter	Unit	Description
Emission		
Mass concentration c	mg/m^3 (at 0 °C 1013 mbar)	Pollutant mass per cubic metre exhaust gas in normal condition, usually dry
Volume concentration c_v	cm^3/m^3, also ppm v 1 cm^3/m^3 = 1 ppm v	Pollutant volume per exhaust gas volume, parts per million
Emission mass flow	kg/h t/a	Emitted pollutant mass per unit of time
Emission factor	mg/kg kg/t	Emitted pollutant mass per mass of fired fuel or produced quantity
Emission factor	kg/TJ	Kilograms per terajoule; emitted pollutant mass per thermal power (in furnaces)
Emission factor	g/km	Emitted pollutant mass per kilometer driven (in motor vehicles)
Air quality concentrations		
Mass concentration c	mg/m^3 μg/m^3 1000 μg = 1 mg	Pollutant mass per cubic metre of air
Volume concentration c_v (mixing ratio)	cm^3/m^3 also ppm v ppb (μl/m^3)	Pollutant volume per air volume; Parts per million; Parts per billion

Table 1.5 ***Contd...***

Parameter	Unit	Description
Air quality concentrations		
Pollutant concentration in rain water	mg/*l* µg/*l*	Pollutant mass per litre of rain water
Pollutant concentration (e.g., deposited particulate matter, but also wet precipitation and gases)	mg/g^2/day µg/g^2/day	Deposited pollutant mass per surface area and time
Pollutant dose	mg/m^2 h	Concentration times time (c – t)
Dose of effect	µg/kg	Received (effective) pollutant mass per acceptor (recipient) mass

1.2 Classification of Air-pollutants, their Sources of Emission and Air Quality Standards

1.2.1 Classification

As clean air in the troposphere moves across the earth's surface, it collects the products of both natural events (dust storms and volcanic eruptions) and human activities (emissions from sources like transportation, fuel combustion, industrial operations, solid waste disposal and various other activities). These potential pollutants, called, **primary pollutants** which are emitted directly from the source, mix with the churning air in the troposphere. Some primary pollutants may react with one another or with the basic components of the atmosphere in chemical and photochemical reactions to form new pollutants which are called **secondary pollutants**. The reaction mechanisms and various steps involved in the process are influenced by many factors such as concentrations of reactants, the amount of moisture in the atmosphere, degree of photo-activation, meteorological forces, and local topography (Fig. 1.1). Long-lived pollutants travel far before they return to the earth as particles, droplets, or chemicals dissolved in precipitation.

Pollutants are also classified into the following categories: gaseous pollutants, particulate matter pollutants, aerosol pollutants, pesticides, metallic contaminants, carcinogenic pollutants, radioactive pollutants and biological contaminants. Gaseous pollutants include oxides of nitrogen, oxides of sulphur, CO and CO_2, hydrocarbons, etc.

Table 1.6(a) lists the major classes of pollutants found in ambient air, Carbon oxides (CO and CO_2), nitrogen oxides (mostly NO and NO_2 or NO_x), sulphur oxides (SO_2 and SO_3), volatile organic compounds (VOCs, mostly hydrocarbons), and suspended particles – all produced primarily by the combustion of fossil fuels.

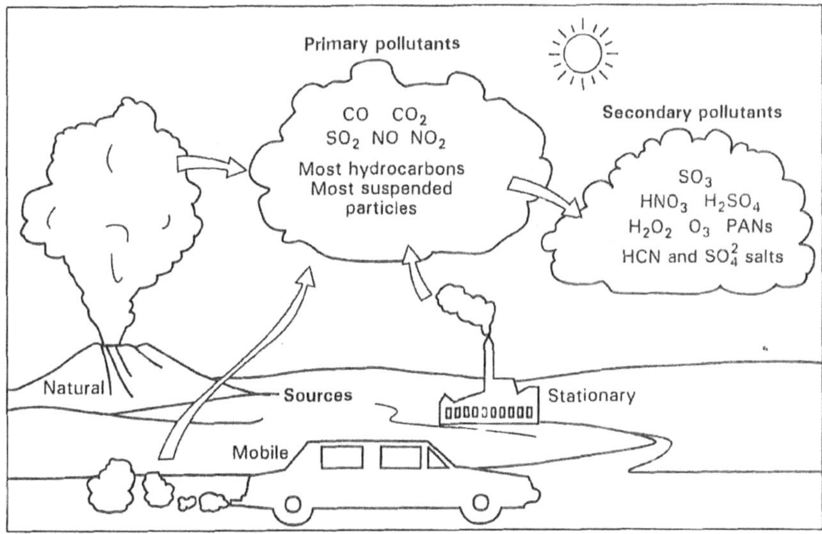

Fig. 1.1 *Primary and secondary air pollutants*

1.2.2 Emission sources of air pollutants

The various emission sources of air pollutants can be classified into natural and man-made sources as presented in Fig. 1.2.

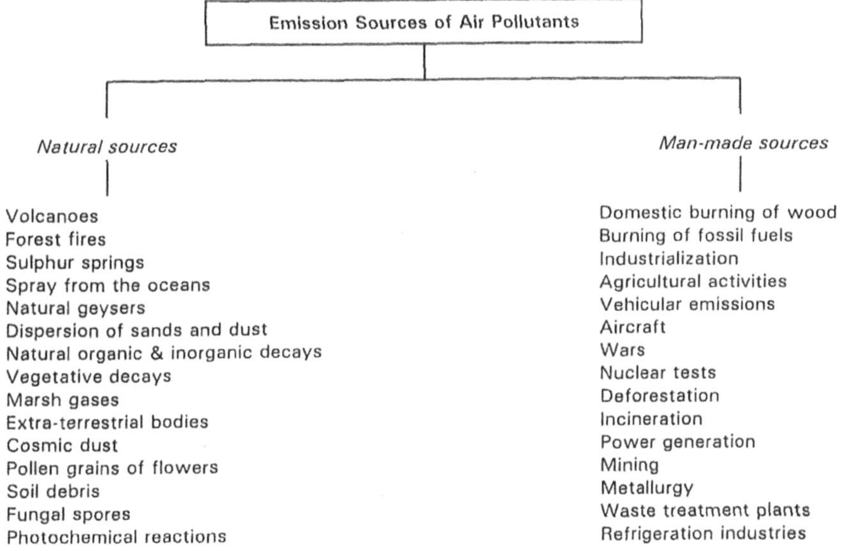

Fig. 1.2 *Emission sources of air pollutants*

Table 1.6 (a) *Summary of sources, concentrations, and scavenging process of atmospheric trace gases*

Contaminant	Major pollutant sources	Natural sources	Estimated emission pollutants	Annual Tg/yr* Natural	Atmospheric background concentrations	Estimated atmospheric residence time	Removal reactions and sinks	Remarks
SO_2	Fossil fuel combustion	Volcanoes , reactions of biogenics emissions	212[1]	20[2]	About 0.1 ppb[3]	1-4 days	Oxidation to sulphate photo-chemical reactions or in liquid droplets	High reaction rates in summer due to photochemical processes
H_2S and organic sulphides[4]	Chemical processes, sewage	Volcanoes, biogenic processes in soil and water	3	84[5]	H_2S: 0.05-0.1 ppb COS: 0.5 ppb[6]	H_2S: 1-2 days COS: 1-2 yr[6]	Oxidation to SO_2 and SO_4	Atmospheric data are incomplete COS residence time can be 20 yr[2]
CO	Auto exhaust, general combustion	Forest fires, Photochemical reactions	700[7]	2100[7]	0.1-0.2 ppm (N. Hemisphere) 0.04-0.06 ppm (S. Hemishphere)	1-3 months	Photo-chemical reactions with	No long-term changes in the atmosphere have been detected
NO, NO_2	Combustion	Biogenic processing in soil, lightniug	75[8] (as NO_2)	180[9] (as NO_2)	About 0.1 ppb[10]	2-5 days	Oxidation to nitrate	Natural processes mostly estimated; background concentrations are in doubt but may be as low as 0.01 ppb

Table 1.6(a) ***Contd...***

Contaminant	Major pollutant sources	Natural sources	Estimated emission pollutants	Annual Tg/yr* Natural	Atmospheric background concentrations	Estimated atmospheric residence time	Removal reactions and sinks	Remarks
NH_3	Waste treatment, combustion	Biogenic processes in soil	6[11]	260[9]	About 10 ppm[9]	1-7 days	Reaction with SO_2 to form $(NH_4)_2$ SO_4	Atmospheric measurements are sparse
N_2O	Small amounts from combustions	Biogenic processes in soil and water	3[12]	340[13]	330 ppb	20-100 yr	Photochemical in stratosphere	Some estimates place natural source at 100 Tg or less[12]
CH_4	Combustion, natural gas	Biogenic process in soil and water	160[14]	1050[14]	1.5 ppm	8 yr[15]	Reaction with OH to form CO	Pollutant source includes 60 Tg yr^{-1}
Isoprene and terpenes	None	Biogenic plant Emission	None	830[7]	0	1-2 h	Photochemical reactions with OH and	Not found ambient atmosphere away from source regions O_2
Total non CH_4 hydro-carbons	Combustion	Biogenic process in soil and vegetation	10[17]	2×10^4 [16]	0-1 $\mu g\ m^{-3}$ for C_2S	Hours to a few days	Photochemical reactions with NO and O_3	Concentration given for C_2S in rural atmosphere
CO_2	Combustion	Biological processes	22.000[16]	10^6 [13]	345 ppm (1981)	2-4 yr	Biogenic processes, photosynthesis absorption in oceans	Forest destruction and changes in earth's biomass may add 20-30 × 10^3 Tg CO_2/yr atmosphere

Table 1.6(a) ***Contd...***

Contaminant	Major pollutant sources	Natural sources	Estimated emission pollutants	Annual Tg/yr* Natural	Atmospheric background concentrations	Estimated atmospheric residence time	Removal reactions and sinks	Remarks
CH_3Cl	Combustion	Oceanic biological processes	2[19]	4-6[16,19]	660 ppt[19,20]	1-2 yr[19]	Stratospheric reactions	Photochemical reactions in stratosphere may impact on O_3 layer
HCl, CL_2	Combustion, Cl manufacturing	Atmospheric reactions of NaCl, volcanoes	4[21]	100-200[23]	About 0.5 ppb[21]	About 1 wk	Precipitation	Volcanoes can release 10-20 Tg Cl yr^{-1} [22]

Source: *Elmer Robinson. (**Pullman,** Wash.: Washington State University)*

Notes: *Tg/yr = 10^{12} g/yr or 10^6 metric ton/yr; 1-Based on 1978 global fuel usage and estimated sulphur contents. 2-Major reference is R .D. Cadle, 1980, Rev. Geophys. Space Phys. 18,746-752.3-P. J. Maroulis, A. L., Torres, A. B. Goldberg, and A. R. Bandy, 1980, J. Geophys. Res.85. 7345-7349 4-Includes COS, CS_2, $(CH_3)_2$,S_2, CH_3, and SH. 5-Adapated from D. F. Adams, S. O. Farwell, E. Robinson, and M. R Pack, 1980, Biogenic sulphur emissions in the SURE region. Final report by Washington State University for Electric Power Research Institute, EPRI Report No.EA-1516, 6-A. L. Torres, P. J. Maroulis, A. B. Goldberg, and A. R. Bandy, 1980. J. Geophys. Res. 85,7357-7360.7-P. R. Zimmerman, R. B. Chat field, J. Fishman, P. J. Crutzen, and P. L. Hanst, 1978, Geophys. Res. Lett. 5, 679-682. 8-Based on 1978 global combustion estimates. 9-1. E. Galbally, Tellus 27, 67-70. 10-Approximate value combining values given in several reference. 11-R. Soderfund, and B. H. Svensson, 1976. The global nitrogen cycle, in SCOPE Report 7, Swedish National Science Research Council, Stockholm. 12-1978 fuel usage figures apply to the following references: R. F. Weiss, and H. Craig, Geophys. Res. Lett.3, 751-753; and D. Pierotti, and R. A. Rasmussen, 1976.Geophys. Res Lett. 3,263-267.13-E. Robinson, and R. C. Robbins, Emissions, concentrations and fate of gaseous atmospheric pollutants, in Air Pollution Control, edited by W. Strauss, 1-93, Part 2 of New York: Wiley. 14-J. C. Sheppard, H. West berg, J. F. Hooper, and K. Ganesan, 1982, J. Geophys. Res. 87,1305-1312.15-L. E. Heidt, J. P. Krasnec, R. A. Lueb, W. H. Pollock, B. E. Henry, and P. J. Crutzen, 1980. J. Geophys. Res. 85, 7329-7336. 16-R. E. Graedel. 1979. J. Geophys. Res. 84, 273-286.17-Reference 13 tabulation updated to approximate 1978 emissions. 18-G.M. Woodwell, R. H. Whittaker, W. A. Reiners, G. E. Likens, C. C. Delwiche, and D. B. Botkin, 1978, Science 199, 141-146. 19-R. A. Rasmussen, L. E. Rasmussen, M. A. K. Khalil, and R. W. Dalluge 1980. J. Geophys. Res. 85, 7350-7356. 20-E. Robinson, R. A. Rasmussen, J. Krasnec, D. Pierotti, and M. Jakubovic, 1977. Atm. Environ. 11, 213-215. 21-J. A.Ryan, and N. R. Mukherjee, 1975. Rev. Geophys. Space Phys. 13, 650-658. 22-R. D. Cadle, 1980. Rev. Geophys. 18,746-752.23-Based on estimated reaction of NaCl to form Cl_2.

Natural sources produce considerable pollutants but most of them are essential components of a balanced ecosystem. None of these natural pollutants normally accumulate to a level that is dangerous for life. However, many industrial activities (Table 1.6(b)) produce air pollutants in levels that exceed the normal natural assimilation processes. Air pollution problems are concentrated in and around urban areas because industries are concentrated in or around urban areas spewing poisonous gases, and particulates and increasing ambient temperature. Vehicular traffic is also dense in urban areas causing pockets of intense air pollution. All recorded air pollution episodes are around developed cities only. There is a qualitative difference between air pollution due to industries and air pollution due to vehicles, as can be seen by an analysis of the air polluted by them.

Table 1.6 (b) *Classification of anthropogenic or man-made air pollution sources*

Source type	Category	Important sources	Typical pollutants
Combustion	Stationary	Power plants, industrial boilers, diesel generators, municipal or industrial incineration	SO_x, NO_x, CO, smoke fly ash
	Refuse burning Mobile	Trace metal oxides Motor vehicles, aircraft	CO, HC, NO_x, SO_x, particulates
Roasting and heating processes	Non-ferrous metallurgical Ferrous metallurgical	Roasting, smelting and refining operations Material handling, ore sintering and pelletising, coke ovens, blast furnaces, steel furnaces	Dust, smoke, metal fumes (Cu, Zn, and Pb), oxides of sulphur Smoke, fumes, CO, odours, H_2S, organic vapour, fluorides
	Non-metallic minerals	Crushed stone, cement, glass, refractories, ceramic manufacture, coal cleaning	Mineral and organic particulates
Food and agriculture	Food processing crop spraying and dusting. Field burning	Drying, preserving, packaging Pest and weed control Refuse burning	Vapour, odour, and dust Organic phosphates, chlorinated HC, organic, lead Smoke, fly ash and soot
petroleum, chemicals, pulp and paper	Petroleum refining	Boilers, process heaters, catalyst regenerators, flares, storage tanks, compressor engines	So_x, HC, NO_x, particulate matter, CO, aldehyde, ammonia, odours
	Inorganic chemicals	Sulphuric acid plants, fertilizer manufacture, nitric acid and ammonia plants, phosphoric acid manufacture	So_x, HF, H_2S, NO_x, NH_3, particulate matter, H_3PO_4, etc.
	Organic chemicals	Plastics, paint and varnish manufacture, synthetic rubber, rayon, insecticides, soap and detergent manufacture, methanol, phenol, etc.	Particulate matter, odours, SO_2, CO, organic intermediates, solvent vapours
	Pulp and paper	Digester blow oxidation towers	Mercaptans, dimethyl suphide, SO_2

Among the emission sources, some are stationary point sources while others are moving point sources. The pollution from industries is almost continuous. The vehicular pollution waxes and wanes according to the peak hour traffic during the day and night.

The common air pollutants, their sources and pathogenic effects are given in Table 1.7 while their typical effects on vegetation are shown in Fig. 1.3. Of the large number of primary pollutants emitted into the atmosphere, only a few are present in sufficiently high concentrations to be of immediate concern. The relative weighting factors and the tolerance limits of the major air pollutants are given in Table 1.8.

Table 1.7 *Common air pollutants, their sources and pathological effects on man*

Pollutant	Source	Pathological effects on man
Sulphur dioxide	Colourless gas produced by coal and oil combustion and certain industrial sources	Respiratory irritant, aggravates asthma and other lung and heart diseases, reduces lung function.
Nitrogen oxides	Brownish orange gas produced by motor vehicles and combustion at major industrial sources	Inhibits cilia action so that soot and dust penetrate far into the lungs.
Hydrogen sulphide	Refineries, chemical industries and bituminous fuels	Causes nausea, irritates eyes and throat.
Carbon monoxide	Burning of coal, gasoline, motor exhausts	Reduces oxygen carrying capacity of blood.
Hydrogen cyanide	Blast furnace, fumigation, chemical manufacturing, metal plating, etc.	Interferes with nerve cells, produces dry throat, indistinct vision, headache, etc.
Ammonia	Explosives, dye making fertilizer plants and lacquers	Inflames upper respiratory passages.
Phosgene or carbonyl chloride	Chemical and dye making	Induces coughing, irritation and fatal pulmonary edema.
Aldehydes	Thermal decomposition of oils, fats, or glycerols	Irritate nasal and respiratory tracts.
Arsines	Processes involving metal or acids containing arsenic, soldering	Damage red cells in blood, kidneys and cause jaundice.
Suspended particles (ash, soot, smoke, etc.)	Solid or liquid particles produced by combustion and other processes at major industrial sources (e.g. steel mills, power plants, chemical plants, incinerators and almost every manufacturing process)	Respiratory irritants, aggravate asthma and other lung and heart diseases (especially in combination with sulphur dioxide); many are known as carcinogens. Toxic gases and heavy metals absorb onto these particulates and are commonly carried deep into the lungs. Cause emphysema, eye irritation and possibly cancer.

Table 1.7 ***Contd...***

Pollutant	Source	Pathological effects on man
Lead	Very small particles emitted from motor vehicles and smelters	Toxic to nervous and blood-forming systems, in high concentrations can cause brain and organ damage.
Ozone	A colourless gas formed from reactions between motor vehicle emissions and sunlight. It is the major component of smog.	Respiratory irritant, aggravates asthma and other lung and heart diseases, impairs lung functions, ozone is toxic to plants and corrodes materials.

Table 1.8 *Major air pollutants, their weighting factors and tolerance limits*

Pollutant	Relative toxicity (weighting factor)	Tolerance levels	
		ppm	μg/m^3
CO	1.00	32.0	40,000
hydrocarbons	2.08	-	19,300
SO_x	28.0	0.25	1430
NO_x	77.8	0.50	515
particulates	106.7	-	375

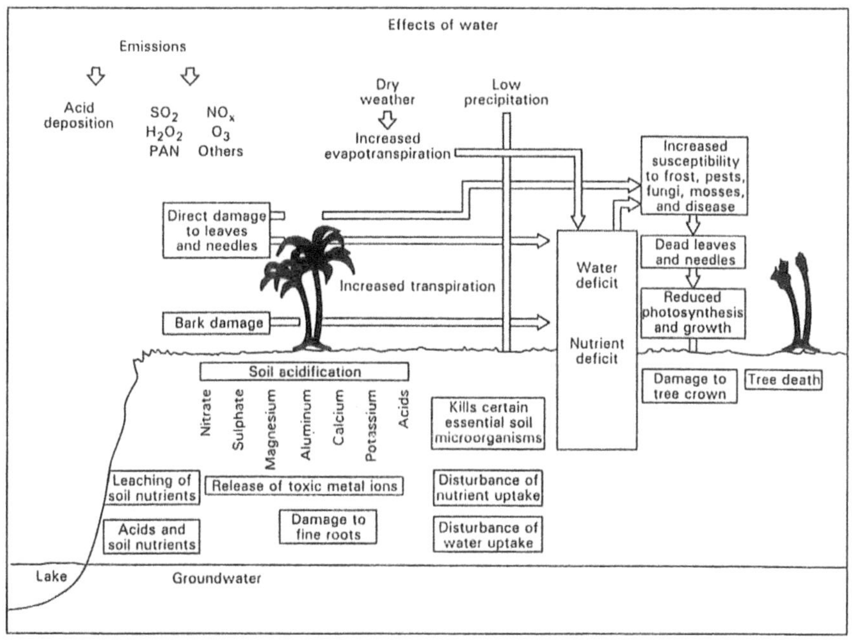

Fig. 1.3 *Possible or suspected harmful effects of air pollutants on trees*

1.2.3 Cycling of air pollutants

Air pollution emitted from various sources are removed from the atmosphere by different types of sinks. A typical cycle is shown in Fig. 1.4. Most pollutants have both natural and man-made sources: although the natural sources are often of sizeable magnitude at the global level, man-made sources predominate at local levels.

In the atmosphere, pollutants may move from a dry, gaseous phase into a liquid phase before falling to the earth's surface. If this is acidic in nature, it is popularly known as acid rain. Wet and dry deposition are the two main pathways by which atmospheric pollutants are returned to the surface of the earth. Other terms like rain-out and run off are also used (Fig. 1.5).

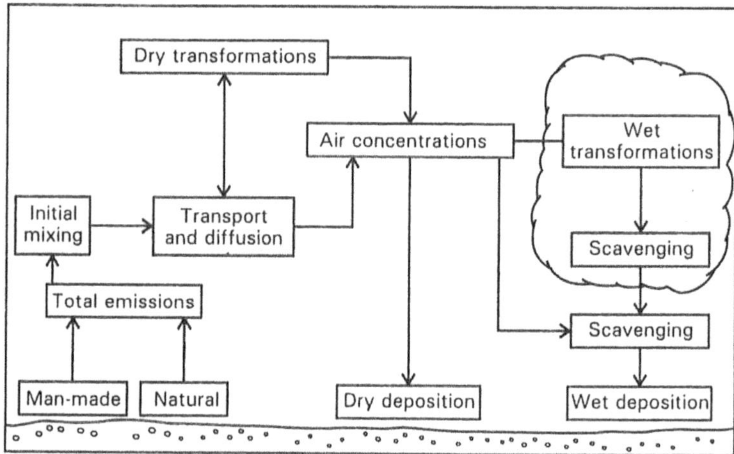

Fig. 1.4 *Typical atmospheric cycle of a pollutant*

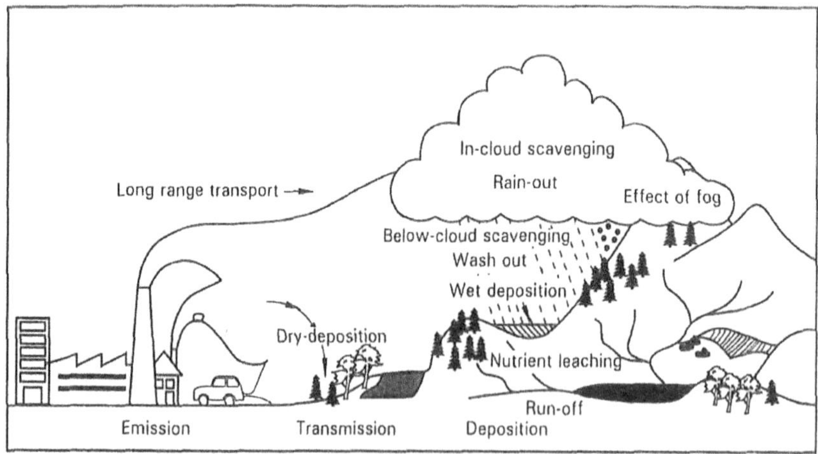

Fig. 1.5 *Emissions and deposition of atmospheric pollutants*

Dry deposition involves the transfer and removal of gases and particles at land and sea surfaces without the intervention of rain or snow. For gases removed at the surface, dry deposition is driven by a concentration gradient caused by surface depletion; for particles this mechanism operates in parallel with gravitational settling of the large particles. The efficiency of dry deposition is described by the deposition velocity, V_g, defined as

$$V_g \left(\text{ms}^{-1}\right) = \frac{\text{Flux to surface } \left(\mu\text{g m}^{-2}\, s^{-1}\right)}{\text{Atmospheric concentration } \left(\mu\text{g m}^{-3}\right)} \quad \text{.....(1.13)}$$

Some typical values of deposition velocity are given in Table 1.9. For gases, such as sulphur dioxide, which have a fairly high V_g, dry deposition has little influence upon near surface concentrations, but may appreciably influence ambient levels at large downwind distances.

Table 1.9 *Some typical values of deposition velocity*

Pollutant	Surface	Deposition velocity (cm s^{-1})
SO_2	Grass	1.0
SO_2	Ocean	0.5
SO_2	Soil	0.7
SO_2	Forest	2.0
O_3	Dry grass	0.5
O_3	Wet grass	0.2
O_3	Snow	0.1
HNO_3	Grass	2.0
CO	Soil	0.05
Aerosol (< 2.5 mm)	Grass	0.15

1.3 Physico-Chemical Characteristics, Sources and Effects of Primary Air Pollutants

1.3.1 Particulate pollutants

1.3.1.1 Physico-chemical characteristics of particulate pollutants

Airborne small, solid particles and liquid droplets are commonly known as particulates. When present in air in excess, they pose a serious pollution threat. The life period of particulates varies from a few seconds to several months; it depends on the settling rate, size, and density of particles and turbulence.

Particulates can be inert or extremely reactive materials ranging in size from 100 μm down to 0.1 μm and less. The inert materials do not react readily with the environment nor do they exhibit any morphological changes as a result of combustion or any other process, whereas reactive materials could be further oxidized or may react chemically with the environment.

1.3.1.2 Classification of particulates

Dust: Particles of size 1-200 μm belong to this category and are formed by the natural disintegration of rocks and soil or by mechanical processes like grinding and spraying. They are removed from the air by gravity and other inertial processes by large settling velocities and also act as centres of catalysis for many of the chemical reactions taking place in the atmosphere.

Smoke: Particles of size 0.01-1 μm constitute smoke which can be either in the liquid or solid form and is formed by combustion or other chemical processes. Smoke may have different colours depending on the nature of materials burnt.

Fumes: Solid particles of size 0.1-1 μm which are normally released from chemical or metallurgical processes belong to this category.

Mists: Liquid droplets generally smaller than 10 μm which are formed by condensation in the atmosphere or released from industrial operations represent mist.

Fog: It is the mist in which the liquid is water and is sufficiently dense to obscure vision.

Aerosols: All airborne suspensions, either solid or liquid belong to this category and these are generally smaller than 1 μm.

Particles of size 1-10 μm have measurable settling velocities but are readily stirred by air movements, whereas particles of size 0.1-1 μm have small settling velocities. Particles below 0.1 μm, a sub microscopic size found in urban air, undergo random Brownian motion resulting from collision among individual molecules. The respective sizes of particulate matter and their sources of origin are given in Fig. 1.6.

Figure 1.7 gives the particle size distribution of a typical atmospheric particulate sample.

Urban air contains particulates, mainly of size 0.1-10 μm. The finest and the smallest particles are the ones which cause significant damage to health. Particulate matter larger than 10 μm which enters the respiratory system is filtered to a maximum extent by the horizontal lining of the nose. Particles of size 5-10 μm enter into the tracheal bronchial system, will be absorbed by mucus and sent back to the throat by small hair-like cilia and removed by spitting or swallowing. Particles smaller than 5 μm will be able to pass through many turns and bends into the upper respiratory system and finally enter the lungs and may be deposited there. Thus, particles of size 0.5-5 μm are sufficiently small to settle in the lungs by sedimentation; they may release slowly absorbed toxic gases causing serious health problems.

1.3.1.3 Chemical composition of particulate pollutants

As the chemical composition of particulate pollutants depends upon the origin of the particulates, it varies widely from place to place.

(a) Calcium, aluminium, and silicon compounds will be present in particulates originating from soils and minerals.

(b) A number of organic compounds constitute smoke from coal, oil, wood, and solid waste.

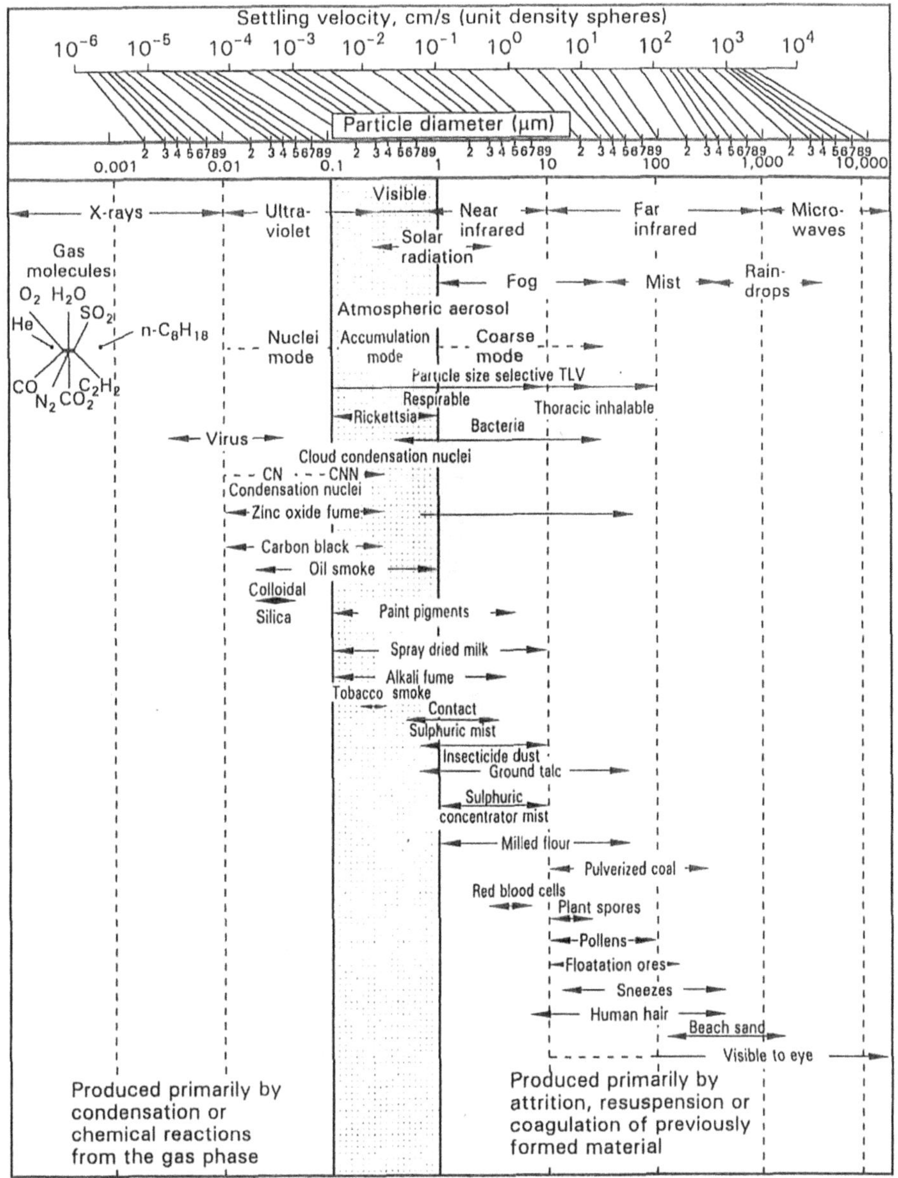

Fig. 1.6 *Molecular and aerosol particle diameters, © P. C. Reist. Molecular diameters calculated from viscosity data. See Bird, Stewart, and Lightfoot, 1960, Transport phenomena, Wiley. (Adapted from Lapple, 1961, Stanford Research Institute Journal, 3rd quarter; and J. S. Eckert and R. F. Strigle, Jr., 1974 JAPCA 24: 961–965.)*

(c) Insecticide dust and certain fumes released from chemical plants also contain organic compounds. Hydrocarbons themselves can coalesce into aerosol droplets which may contain the highly harmful components of incomplete combustion known as Polycyclic Organic Matter (POM) which are derivative of benz-α-pyrene, a potential carcinogen.

Certain, toxic heavy metals like cadmium, lead, nickel and mercury are frequently observed in different types of particulates in a variety of combined forms such as oxides, hydroxides, sulphates and nitrates. The typical composition of particulates in a sample of the atmosphere in Hyderabad is given in Table 1.10.

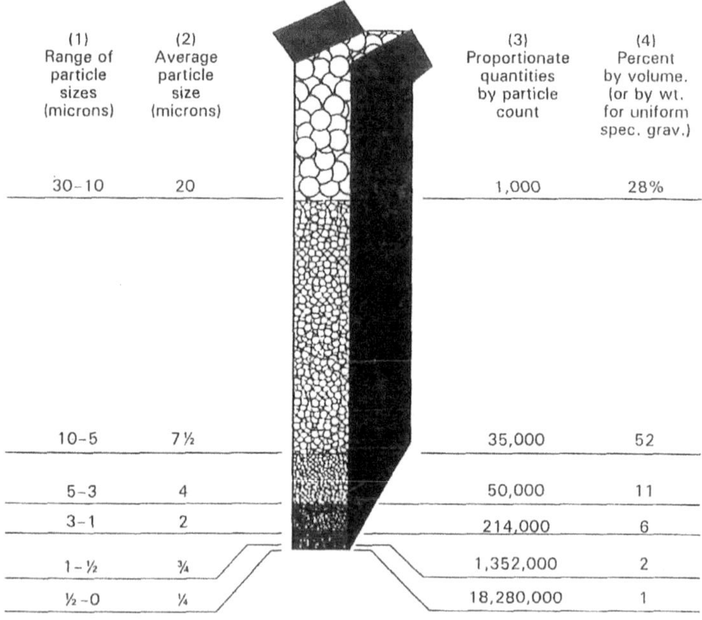

Fig. 1.7 *Particle size distribution of a typical atmospheric particulate sample*

(**Source:** *The American Air Filter Co.*)

Table 1.10 *Composition of Suspended Particulate Matter (SPM) in a sample of the atmosphere in Hyderabad*

Particulate	Composition (μg/m^3)	Particulate	Composition (μg/m^3)
Nitrate ions	2.8	Zinc	0.91
Ammonium ions	1.5	Antimony	0.02
Sulphate ions	9.3	Organic solvents	8.1
Lead	0.61	Arsenic	0.03
Iron	1.90	Molybdenum	0.006
Manganese	0.21	Beryllium	0.008

Table 1.10 ***Contd...***

Particulate	Composition (μg/m^3)	Particulate	Composition (μg/m^3)
Nickel	0.17	Bismuth	0.007
Tin	0.02	Cobalt	0.001
Titanium	0.08	Sodium	1.20
Copper	0.06	Silicon	11.10
Chromium	0.02	Calcium	9.2
Vanadium	0.08		

Source: *Y. Anjaneyulu, 1997. Air pollution and its effects. Proceedings of international conference on air pollution and control technologies, ICIPACT-1997.*

1.3.1.4 Sources of emission of particulate pollutants

In arid regions like deserts, storms frequently stir up fine dust and sand reducing visibility and causing irritation of the respiratory tract. Autopsies carried out on dwellers of the Sahara Desert in North Africa, where sandstorms are frequent, showed that a significant number of particles was present in the lungs. The presence of sulphur dioxide in a dust laden atmosphere is dangerous to the respiratory tract, the lungs and, indirectly, the heart.

Table 1.11 *Particulate emission rates*

Emission source	Emission factor
Natural gas combustion	
Power plants	240 mg/m^3 of gas burned
Industrial boilers	290 mg/m^3 of gas burned
Domestic and commercial furnaces	310 mg/m^3 of gas burned
Distillate oil combustion	
Industrial and commercial furnaces	1.70kg/m^3 of oil burned
Domestic furnaces	I kg/m^3 of oil burned
Residual oil combustion	
Power plants	1.10 kg/m^3 of oil burned
Industrial and commercial furnaces	1 kg/m^{3}oil burned
Coal combustion	
Cyclone furnaces Other pulverized coal furnaces Spreader stokers Other stokers	
Incineration	
Municipal, multiple chamber	8.00 kg/ton of refuse burned
Commercial, multiple chamber	1.40 kg/ton of refuse burned
Flue-fed incinerator	12.8 kg/ton of refuse burned
Domestic, gas fired	7.00 kg/ton of refuse burned
Open burning of refuse	7.20 kg/ton or refuse burned
Motor vehicles	
Petrol-powered engines	1.36 kg/m^3 of petrol burned

Table 1.11 ***Contd...***

Emission source	**Emission factor**
Incineration	
Diesel-powered engines	12.5 kg/m^3 of fuel burned
Cement manufacturing	17.2 kg/barrel of cement produced
Kraft pulp mills	
Lime kilns	43 kg/ton of dried pulp produced
Recovery furnaces (with scrubbers)	70 kg/ton of dried pulp produced
Steel manufacturing	
Open-hearth furnace	10 kg/ton of steel produced
Electric are furnace	7 kg/ton of metal charged
Sulphuric acid manufacturing	3 kg/ton of acid produced

Many industries emit fine particles with different emission rates (Table 1.11). Lead asbestos, manganese, beryllium, arsenic, copper, zinc are some of the many particulate pollutants that may be encountered. Since leaded petrol contains about 3 mL of tetraethyl lead per 3.785 litres, automobiles are a source of lead pollution. However, paint and industrial operations are the major sources of lead pollution. The Environmental Protection Agency (EPA) of USA determined that a lead level above 2 μg/m^3 involves enough risk of adverse physiological effects to constitute endangerment to public health. In 1972, 27 cities of the United States had composite lead levels above the EPA limit [1].

Particulate emissions in USA have decreased substantially in the past few decades due to the Clean Air Regulation Act, 1970 (Fig. 1.8).

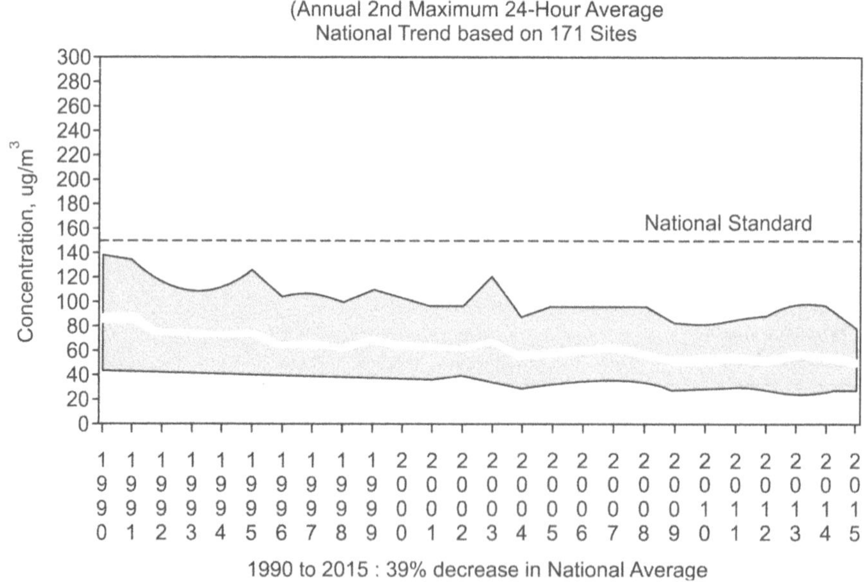

Fig. 1.8 *Emission pattern of particulates in USA, 1990-2015*

(**Source:** *National ambient air pollutant emission estimates, USEPA1990-2015.)*

1.3.1.5 National trends in PM 10 levels

Using a nation wide network of monitoring sites, US - EPA has developed ambient air quality trends for particle pollution, also called Particulate Matter (PM). PM10 describes inhalable particles, with diameters that are generally 10 micrometers and smaller. Under the Clean Air Act, EPA sets and reviews national air quality standards for PM. Air quality monitors measure concentrations of PM throughout the country. EPA, state, tribal and local agencies use that data to ensure that PM in the air is at levels that protect public health and the environment. Nationally, average PM10 concentrations have decreased over the years.

1.3.1.6 Effect of particulate pollutants on visibility

One of the main problems of particulate air pollution is the reduction in visibility caused by light scattering due to the particles. Scattering efficiency per unit mass of aerosol is critically dependent upon the size distribution of the aerosol, but aerosol mass loading is a good predictor of visibility impairment, as indicated in Fig. 1.9.

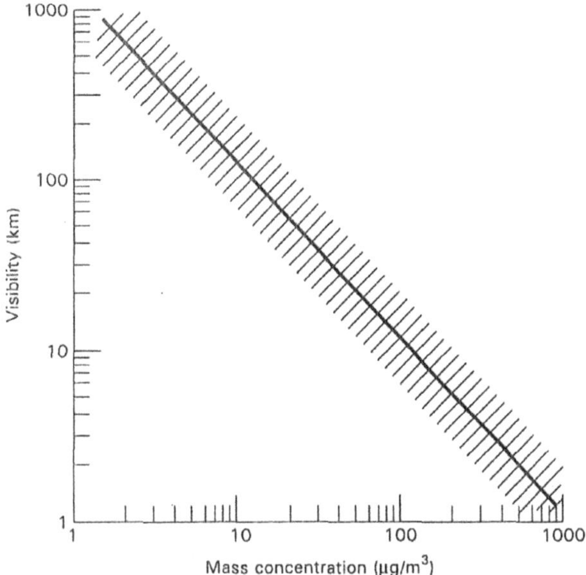

Fig. 1.9 *Visibility of atmosphere as a function of SPM levels. Though this is approximate, most data fit within the shaded area*

Problem 1.3

The most widely used equation for estimating visibility is the Koschmeider equation:

$$L_v = \frac{1200 \text{ km } \mu\text{g/m}^3}{\text{Particle concentration } \mu\text{g/m}^3} \qquad \text{.....(1.14)}$$

where L_v is the visual range, the distance at which an average person can barely distinguish a dark object (such as a mountain or skyscraper) against the sky. This equation is an approximation, based on an average set of atmospheric particles.

(a) Use Equation 1.14 to estimate the visual range when the particulate concentrations equal to 2.00 μg/m³.

(b) If the particle concentration in the atmosphere is increased by 1 μg/m³, what is the percentage decrease in the visual range if the initial visual range is 20 km? If the initial visual range is 200 km, what is the percentage decrease?

Solution:

(a) Visibility range when particulate concentration is 2.00 μg/m³:

$$L_v = \frac{1200 \text{ km } \mu\text{g/m}^3}{2 \ \mu\text{g/m}^3} = 600 \text{ km}$$

(b) When the initial visibility range is 20 km:

$$20 \text{ km} = \frac{1200 \text{ km } \mu\text{g/m}^3}{\text{x } \mu\text{g/m}^3}$$

$$x = 60 \ \mu\text{g/m}^3$$

If the concentration of SPM is increased by 1 μg/m³,

$$L_v = \frac{1200 \text{ km } \mu\text{g/m}^3}{60 + 1 \ \mu\text{g/m}^3} = 19.67 \text{ km}$$

So the percentage decrease in the visibility range is 1.65.

When the initial visual range is 200 km:

$$L_v = 200 \text{ km} = \frac{1200 \text{ km } \mu\text{g/m}^3}{x \ \mu\text{g/m}^3}$$

$$x = 6 \ \mu\text{g/m}^3$$

When the SPM concentration is increased by 1μg/m³,

$$L_v = \frac{1200 \text{ km } \mu\text{g/m}^3}{6+1 \ \mu\text{g/m}^3} = 171.4 \text{ km}$$

So the decrease in the visibility range is 14.3%.

1.3.1.7 *Effects of particulate pollutants on plants*

Plants are adversely affected by gaseous pollutants and deposition of particulates on soil. The deposition of toxic metals on soil makes the soil unsuitable for plant growth. Several particulate pollutants fall on the soil by acid rain; this tends to lower the pH of the soil making it more acidic and infertile. Deposited particulates on leaves restrict the absorption of CO_2, thereby reducing the rate of photosynthesis. Dust mixed with mist or light rain forms a thick crust on the upper surfaces of leaves; this blocks the sunlight necessary for carbon assimilation. Some plants are very sensitive to the traces of toxic metals, which inhibit the action of the plant enzyme system. Arsenic is a cumulative, potent, protoplasmic poison which inhibits the SH-group in enzymes. This is present in almost all types of soils in minute quantities and affects plant growth.

1.3.1.8 *Effects of particulate pollutants on human health*

The effects of particulate pollutants are largely dependent on the particle size. Air borne particles, i.e., dust, soot, fumes, and mists are potentially dangerous to human health. The nasal system prevents coarser particulates bigger than 5 microns from entering the respiratory system. Soluble aerosols will be absorbed into the blood from the alveoli while the insoluble aerosols are carried to the lymphatic stream and get deposited in pulmonary lymphatic depot points or in the lymph glands, where they create toxicity in the respiratory system. Lead interferes with the development and maturation of red blood cells. It is reported that a smoker can easily develop symptoms of asthma which his also due to a concentration of lead greater than in non-smokers.

1.3.1.9 *Effects of particulate pollutants on materials*

Particulates affect a variety of materials in various ways. They cause damage to buildings, paints, furniture, etc. Painted surfaces are very susceptible to damage in wet conditions.

1.3.2 Sulphur oxides

Sulphur dioxide is one of the major air pollutants discharged by various pollutant sources. Further, it reacts photo chemically or catalytically with other pollutants or normal atmospheric constituents to form sulphur trioxide, sulphuric acid and salts of sulphuric acid.

It is estimated that sulphur dioxide remains in the air for an average of two to four days; during this time it may be transported a distance of 1,000 km before it is deposited on the ground. Thus, the harmful effects of SO_2 and its deposition as sulphuric acid may be felt far away from the source; the pollution problem becomes an international one affecting countries which may not have pollutant sources discharging sulphur dioxide.

SO_2 is an unpleasant and highly irritating gas, when it is present in concentrations greater than 1 ppm and adversely affects men, animals, plants and materials. It is perhaps the most damaging among the various gaseous air pollutants. Along with SO_2, SO_3, is discharged, at about 1-5 percent of the SO_2 concentration, and it combines rapidly with moisture in the atmosphere to form sulphuric acid which has a low dew point. Both these oxides are rapidly removed from the atmosphere by rain or settle out as aerosol due to which their concentration is less compared to their emissions from human activities.

1.3.2.1 *Sources of emission of sulphur oxides*

The global sulphur fluxes per year into the atmosphere by anthropogenic and natural sources are shown in Fig. 1.10 while their distribution in different forms between various global compartments is shown in Fig. 1.11.

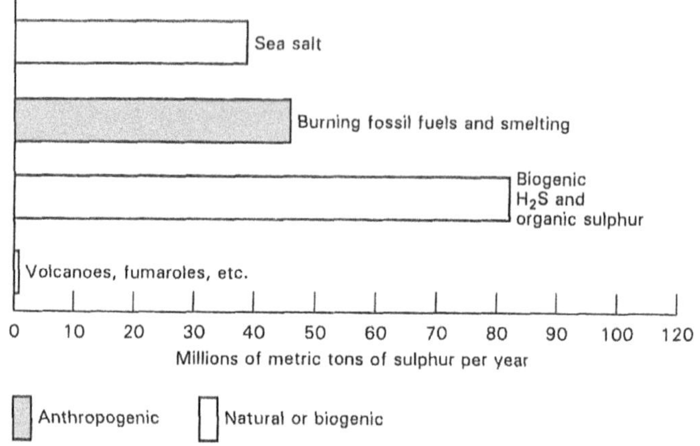

Fig. 1.10 *Sulphur fluxes into the atmosphere*
(**Source:** *R.J. Charlson, 1968. J. Air Pollut. Contr. Assoc. 18, 652.*)

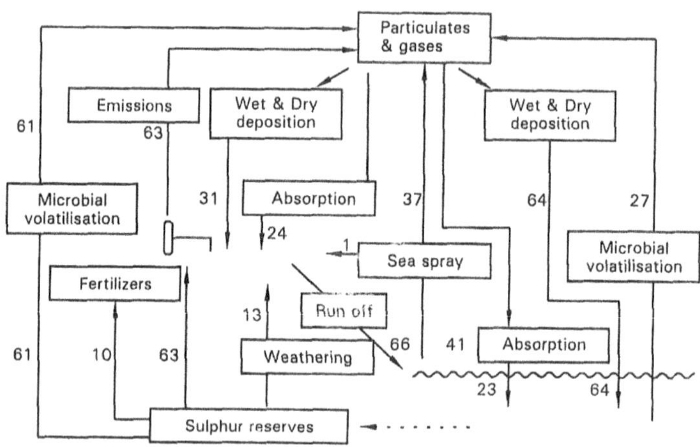

Fig. 1.11 *Distribution of sulphur between various environmental phases (expressed in gig amoles of sulphur per annum)*

(**Source:** *Economic commission for Europe (OECD). Airborne sulphur pollution. Air pollution study no. 1 U.N. 1984. New York.*)

Global emission of SO_2 from natural sources due to microbial activity, volcanoes, sulphur springs, volatilization from various surfaces including vegetation, sea spray, and weathering processes (OECD, 1984) amount to 128 Mt of sulphur (S) per annum (a^{-1}). These are only 50% greater than those caused by human activities (currently 70 Mt a^{-1}), although they are more widely distributed over the surface of the planet and provide natural background levels of SO_2 of about 0.1 ppm. However, biogenic emissions of reduced S-containing compounds from vegetation, wetlands, and oceans are larger.

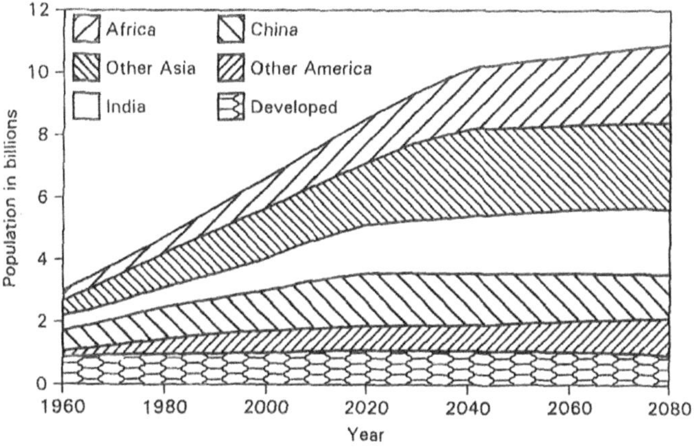

Fig. 1.12(a) *Projections of population increase (in billions) in different parts of the world (data from the UN population division)*

More than 90% of the anthropogenic emissions of SO_2 are over Europe, North America, India and the Far East. Emissions of SO_2 were highest (77 Mt S a^{-1}) during the late 1970s but have fallen over the last two decades as a result of emission controls, changes in the patterns of fuel consumption, and economic recession. Unfortunately, forecasts of increased coal consumption beyond year 2000 (Fig. 1.12) predict that global anthropogenic emissions of SO_2 will probably rise again by 30% over the present day rates to 85 Mt S a^{-1}.Typical urban levels in developed countries currently range from0.1 to 0.5 ppm but concentrations in excess of these levels are still common despite legislation

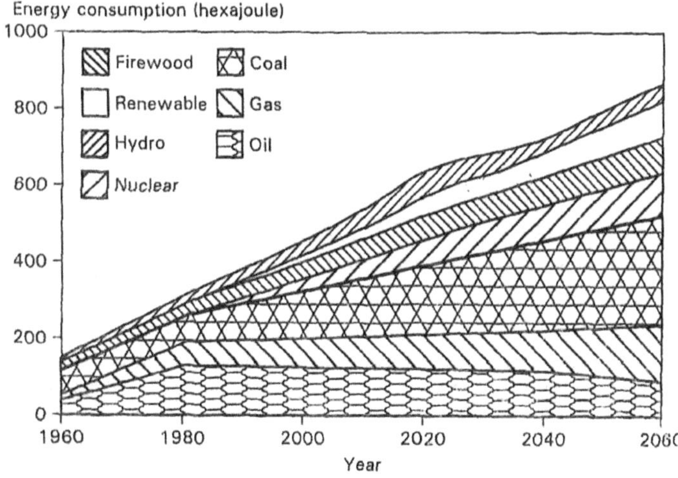

Fig. 1.12(b) *Projected worldwide energy consumption for different types of fuels*
(**Source:** *Proceedings of the World Energy Conference, Cannes, 1986.*)

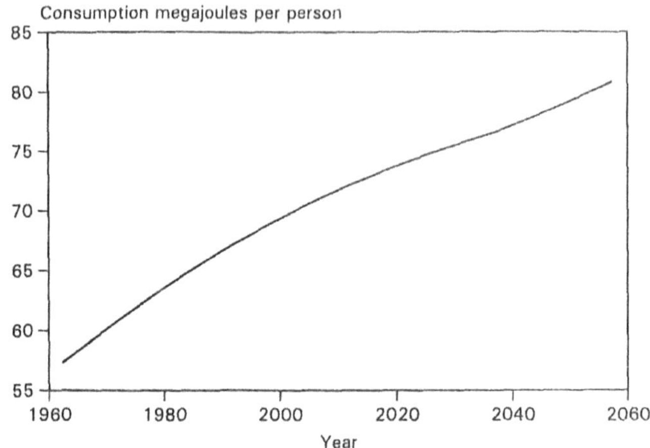

Fig.1.12(c) *Future likely trend of global energy consumption per capita derived from data in Figs. 1.12(a) and 1.12(b)*

In USA it is observed that fossil fuel concentration in stationary sources contribute more than 80% of SO_2 emission from anthropogenic sources. Out of these anthropogenic emissions nearly 85% are mainly from thermal power plants while only 2% come from highway vehicles. Petroleum refining, copper smelters and cement manufacturing are only the significant on combustion sources. As a result of use of fuels with lower sulphur content, flue gas scrubbing, and more extensive controls on sulphuric acid plants and smelters in USA, SO_2 emissions are fairly stabilized at the 1984 levels (Fig. 1.13(a) and Fig. 1.13(b)) with a 25% overall reduction in total SO_2 emission.

SO_2 emissions can be reduced by using fuels with the lowest possible S content: considerable advances have been made in the removal of S from fuels (including pulverized coal) before combustion. Consequently, the balance between the higher costs of such fuels and those associated with the installation of additional pollution controls is now becoming important.

Fluidized-bed combustion involves the burning of small particles of solid or liquid fuel at 500-700 °C. These are held in a state of suspension by upwardly directed jets of air injected under pressure. Steam and water pipes immersed in the 'bed' then transfer the heat to turbines very efficiently. The potential amount of SO_2 emitted by combustion varies according to the S content of the fuel. However, limestone or dolomite particles can be added to a fluidized bed to reduce SO_2 emissions by up to 90% because the S is trapped in the form of sulphates of calcium or magnesium within the ash. This technique therefore adds to the amount of 'involuntary' S that may be recovered and allows coal of higher S content to be used in the presence of stringent emission controls. The lower temperature of combustion also reduces the emissions of nitrogen oxides and CO because of the air saturating conditions. Of all the combustion technologies likely to be used for future power generation, coal still appears to have many advantages (Table 1.12).

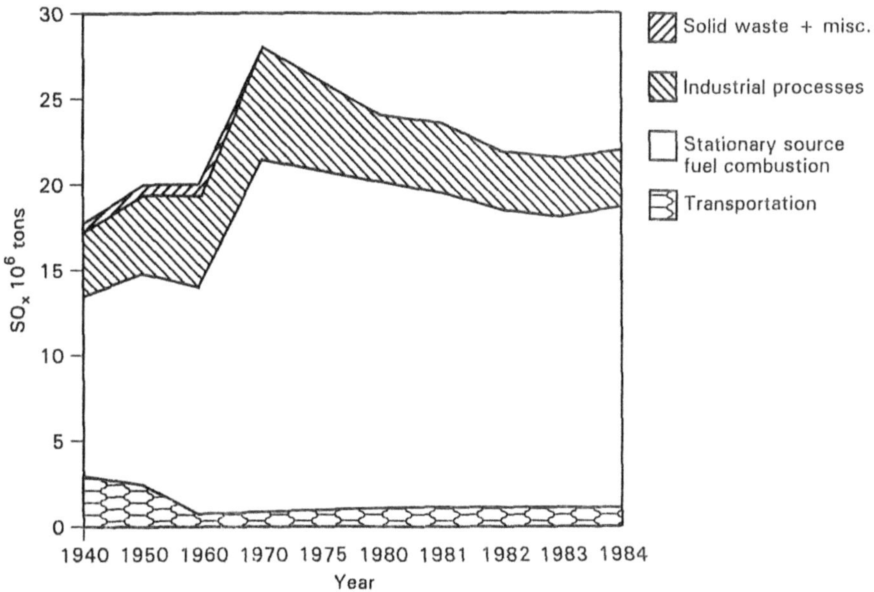

Fig. 1.13(a) *Trends in sulphur oxide emissions in USA, 1940-1984 (USEPA, 1940-1984)*

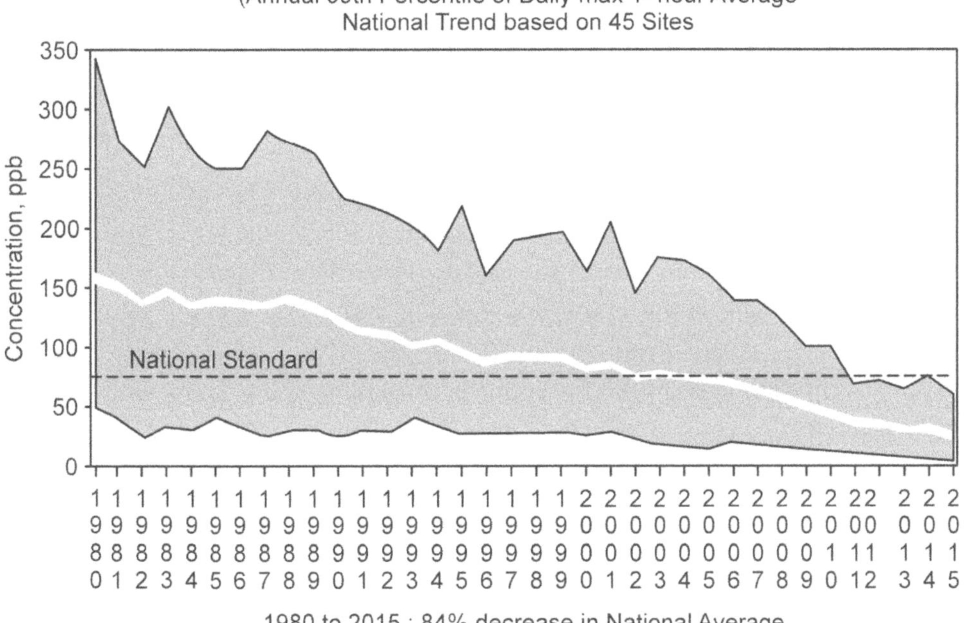

Fig. 1.13(b) *Trends in sulphur dioxide emissions in USA, 1980-2015*

Table 1.12 *Efficiencies and emission problems (without flue-gas desulphurization) of power generation by different combustion processes*

Type of power plant	Maximum thermal efficiency	Emission		
		CO	SO_2	$NO_2 + NO$
Oil-fired	35-37%	Trace	High	V. high
Natural gas-fired	35-37%	Trace	Low	V. high
Conventional coal-fired	35-40%	Low	High	V. high
Gas turbine	18-28%	High	V. high	V. high
Fluidized bed	40-50%	Trace	low	Low

1.3.2.2 *National trends in sulfur dioxide levels*

Using a nation wide network of monitoring sites, US-EPA has developed ambient air quality trends for sulfur dioxide (SO_2). Trends starting in 1980, 1990, and 2000 are shown here. Under the Clean Air Act, EPA sets and reviews national air quality standards for SO_2. Air quality monitors measure concentrations of SO_2 throughout the country. EPA, state, tribal and local agencies use that data to ensure that SO_2 in the air is at levels that protect public health and the environment. Nationally, average SO_2 concentrations have decreased substantially over the years.

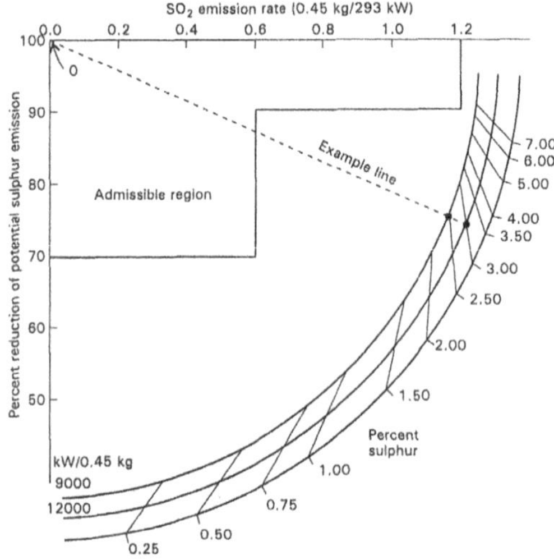

Fig. 1.14 *Graphical representation of new source performance standards for SO_2 emissions from coal-fired power plants*

(**Source:** *J. Mol burg, 1980. A graphical representation of the NSPS for sulphur dioxide. Journal of air pollution control association. 30(1): 172.*)

Most anthropogenic SO_2 emissions are mainly through tall stacks injecting the gases into the troposphere to heights at which the temperatures of the surrounding air and the rapidly cooling stack gases match. A discrete plume is then often carried away at that level. The behavior of this plume is dependent upon the prevailing weather conditions and the contents may be dry-deposited close by or carried many hundreds of kilometers away from the source. However, when the plume encounters wet conditions, considerable amounts of gaseous SO_2 (and nitrogen oxides) are removed by wet deposition.

For coal-fired, steam-electric power plants the emission standard for SO_2 is complex and is best illustrated in Fig. 1.14 where a region of allowable SO_2 emissions is identified. The sulphur content (% sulphur) of the fuel and the fuel's energy content (kW/kg) are parameters used to determine a point on one of the indicated arcs. A line drawn from the origin to the sulphur/heat point crosses the "admissible region" boundary at one point, corresponding to the allowable SO_2 emission rate.

Problem 1.4

Estimate the emission rate of SO_2 in a 5000 megawatt (MW) coal-fired power plant which is 60% efficient. The coal to be burned contains 2.5% sulphur and coal has a heat content of 9495 kW/0.45 kg. What is the permissible emission rate of SO_2 and if a control system has to be installed, what percent reduction in SO_2 emission should it achieve?

Solution: Since the power plant is working with 60% efficiency, and it delivers 5000 MW of electricity (5.0×10^6kW), the input heat rate is

$$Q_{in} = \frac{Q_{out}}{\eta} = \frac{5.0 \times 10^6 \text{ kW}}{0.60} = 8.3 \times 10^6 \text{ kW}$$

Converting it in terms of kW/day we have

$$Q_{in} = 8.3 \times 10^6 \text{ kW} \times \frac{24 \text{ h}}{day} = 1.992 \times 10^8 \text{ kW/day}$$

From Fig. 1.14 at 2.5% sulphur and 9495 kW/0.45 kg of coal, the maximum allowable emission rate is 0.45 kg/293 kW and controls must achieve an SO_2 reduction of about 87%. Totally daily emissions would be:

$$SO_2 \text{ emission rate} = \frac{1.992 \times 10^8 \text{kw}}{day} \times \frac{0.45\text{kg}}{293\text{kw}} = \frac{0.8964 \times 10^8}{293}$$

$$= 3.06 \times 10^5 \text{ kg/day} = 306000 \text{ kg/day}$$

This is also a significant amount.

1.3.3 Removal mechanisms of sulphur oxides

The transformation from SO_2 gas to sulphate particles (SO_4) is gradual, taking a matter of days. During that time, sulphur may be deposited back onto the land or into water, either in the form of SO_2 or sulphate. In either form, sulphur can be deposited by precipitation (wet deposition), or by slow, continuous removal processes that occur without precipitation (dry deposition). Figure 1.15 suggests the effects of time and distance on the conversion and deposition of sulphur.

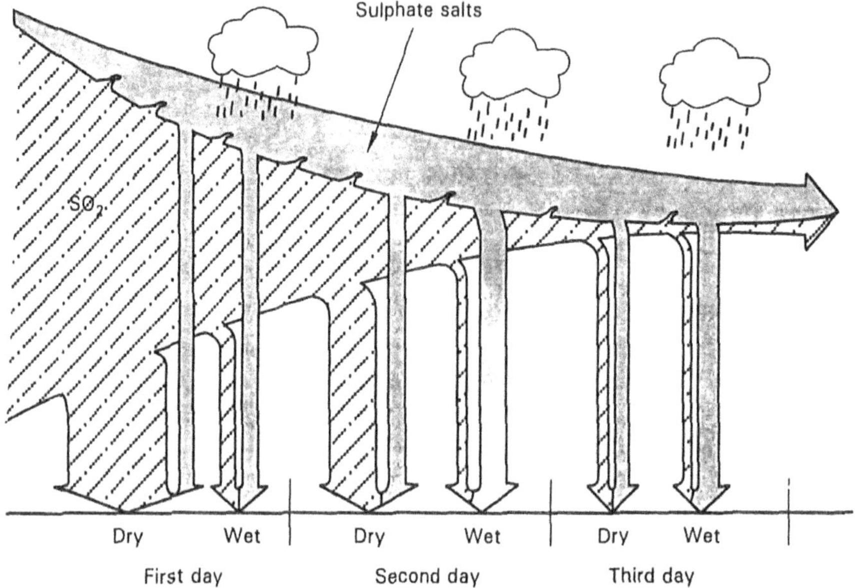

Fig. 1.15 *The effects of time and distance on conversion and deposition of sulphur pollutants*

(**Source:** *Office of Technology Assessment (OTA) 1984. Acid rain and transported air pollutants. Implications of public policy. Washington DC.*)

SO_2 in the discharged plumes may react with atomic oxygen formed within the stratosphere to form SO_3, and finally sulphuric acid as shown below.

$$SO_2 + O + M \longleftrightarrow SO_3 + M$$

$$SO_3 + H_2O \longleftrightarrow H_2SO_4$$

However, a second mechanism based upon one reaction which relies upon the extensive presence of free radical hydrocarbons, may also contribute to a smaller extent to the removal of SO_3.

$$O_3 + C_2H_4 \longleftrightarrow CH_2O_2 + HCHO$$

A third photochemical reaction may produce an energized form of SO_2 which then combines with O_2 to form SO_4. However, as this chemical compound has never been positively identified in the atmosphere, this light-driven acidification process is highly unlikely.

M^1 is a third entity, such as a surface or another gas molecule, which is capable of carrying away excess energy associated with the formation of the new bond.

$$2SO_2 + 2H_2O + O_2 \longleftrightarrow 2H_2SO_4$$

This process has been shown to be catalysed by metal salts such as iron and manganese, commonly found in flyash. Flyash particles serve as nucleation sites for droplet formation and the sulphuric acid droplet may in turn react with metal salts (such as NaCl from sea salt particles), metal oxides such as MgO, Fe_2O_3, ZnO, and Mn_2O_3, or ammonia to produce corresponding sulphates.

$$2NaCl + H_2SO_4 \longrightarrow Na_2\,SO_4 + 2HCl$$

$$MgO + H_2SO_4 \longrightarrow MgSO_4 + H_2O$$

$$2NH_3 + H_2SO_4 \longrightarrow (NH_4)_2\,SO_4$$

$$CH_2O_{2*} + SO_2 \longrightarrow SO_3 + HCHO$$

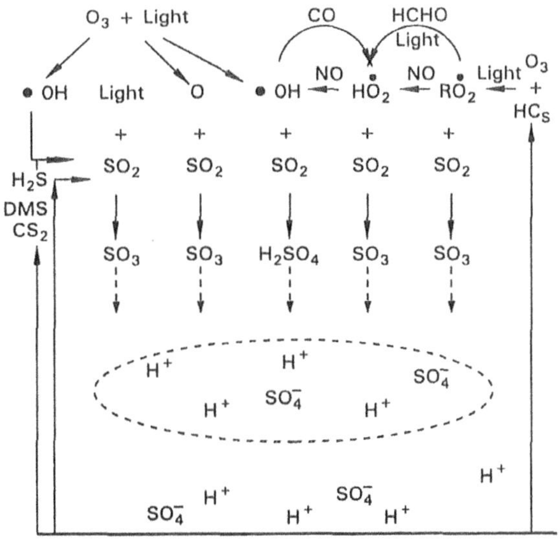

Fig. 1.16 *Some inter relationships between the wet and dry mechanisms of SO_2 removal*

(**Source:** *Dis Con and Penkett, AERE, U. K. Harwell and D. Reidel. Publ. Co., The Netherlands.*)

As shown in Fig. 1.16 the reaction of SO_2 with highly reactive hydroxyl (•OH) free radicals formed from O_3 or monoatomic oxygen are responsible for sulphuric acid formation in the atmosphere. The predominance of this mechanism of acidification involving OH radicals indicates that warm, bright summer days, which favour O_3, and monoatomic oxygen formation (and hence OH radicals) promote dry-phase acidification, while wet-phase reactions are predominant in winter.

$$O_3 + \text{light} \longleftrightarrow O + O_2$$

$$O + H_2O \longleftrightarrow 2{\bullet}OH$$

$$\bullet OH + SO_2 + M^1 \longleftrightarrow HSO_3\bullet + M$$

$$HSO_3\bullet + O_2 \longleftrightarrow HO_2\bullet + SO_3$$

Figure 1.16 shows a number of interrelationships between the wet and dry mechanisms of SO_2 removal.

1.3.3.1 Effects of sulphur oxides on plants

Cell and leaf anatomy: Since the leaf is the primary indicator of air pollution effects on plants, the functions of the leaf are given below. Atypical plant cell has three main components (Fig. 1.17): the cell wall, the protoplast, and the inclusions. Much like the human skin, the cell wall is thin in young plants and gradually thickens with age. Protoplast is the term used to describe the protoplasm of one cell. It consists primarily of water, but it also includes protein, fat, and carbohydrates. The nucleus is the dense bit of protoplasm that controls the operation of the cell. The protoplasm located outside the nucleus is called cytoplasm. Within the cytoplasm are three types of tiny bodies or plastids: chloroplasts, leucoplasts and chromoplasts. Chloroplasts contain the chlorophyll that manufactures the plant's food through photosynthesis. Leucoplasts convert starch into starch grains. Chromoplasts are responsible for the red, yellow and orange colour of the fruit and flowers.

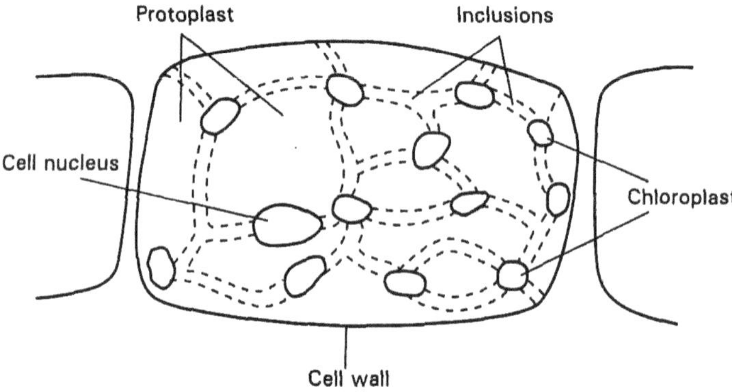

Fig. 1.17 *Typical plant cell*

(**Source:** *I. Hindawi, 1970. Air pollution injury to vegetation. U.S.Dept. of Health, Education and Welfare, National Air Pollution Control Administration, Washington DC. US Govt. Printing Office.*)

A cross-section through a typical mature leaf reveals (Fig. 1.18) three primary tissue systems: the epidermis, the mesophyll, and the vascular bundle. Chloroplasts are usually not present in epidermal cells. The opening in the underside of the leaf is called the stoma. (The plural of stoma is stomata.) The mesophyll, which includes both the palisade parenchyma and the spongy bundles carries water, minerals and food throughout the leaf and to and from the main stem of the plant. The guard cells regulate the passage of gases and water vapour in and out of the leaf. When it is hot, sunny and windy, the processes of photosynthesis and respiration are increased. The guard cells open and allow for increased transpiration of water and minerals from the roots.

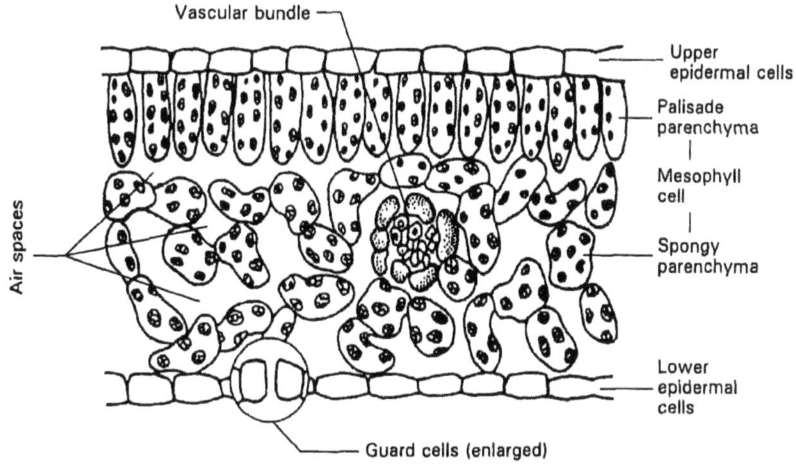

Fig. 1.18 *Cross-section of intact leaf*

(**Source:** *I. Hindawi, 1970. Air pollution injury to vegetation. U.S. Dept. of Health, Education and Welfare, National Air Pollution Control A dministration, Washington DC. US Govt. Printing Office.*)

The layer of still air above the leaf will be the first barrier encountered by a gaseous pollutant before reaching a leaf. The imaginary still air on both sides of the plant imposes a 'boundary layer resistance' to the entry of atmospheric pollutants (Fig. 1.19). Their movement across such a layer is achieved by diffusion of gas molecules in response to differences in concentration. The boundary layer resistance also varies with wind speed and leaf properties such as size, shape, and orientation. As wind speed increases, resistance to the inward movement of pollutant molecules falls and uptake increases; in this way, increased wind speed can facilitate the uptake of pollutants. Moreover, the boundary layer is generally thinner at the edges of a leaf than at the centre, which accounts for increased pollutant damage often seen at margins of leaves, especially on grasses and cereals.

A greater proportion of a leaf surface is occupied by epidermal cells than stomatal pores; the waxy cuticle covering them offers a greater penetration barrier to most pollutant gases. SO_2 deposited on wet leaf and stem surfaces may also dissociate and react with cuticular waxes, a certain amount of SO_2 may then enter leaves by penetrating the damaged cuticle even though most atmospheric SO_2 still enters through the stomata. When the cuticle remains damp for long periods, especially at night or in the winter, the deposition rate of SO_2 onto a leaf surface increases despite the fact that the stomata are closed. This causes additional damage for evergreens which retain their leaves and needles over winter.

For stomata to open, guard cells of the stomata have to be turgid which causes the gap between them to widen. In most plants, this occurs during the day as normal exchanges of water and CO_2 take place. As a result, uptake of atmospheric gases like SO_2 into plants takes place during the day. In very dull conditions, the stomatal opening is reduced and therefore, the rates of SO_2 uptake are also reduced. At night, however, when the stomata are closed, the rates of removal of SO_2 (or O_3) by plants approach those of other inert surfaces. Consequently those changes in stomatal aperture which provide a plant with mechanisms to control the movement of CO_2 into leaves and H_2O outwards, also influence pollutant gas uptake. Stomatal behaviour, frequency and distribution are therefore important factors which affect the amount of pollutant entering a plant.

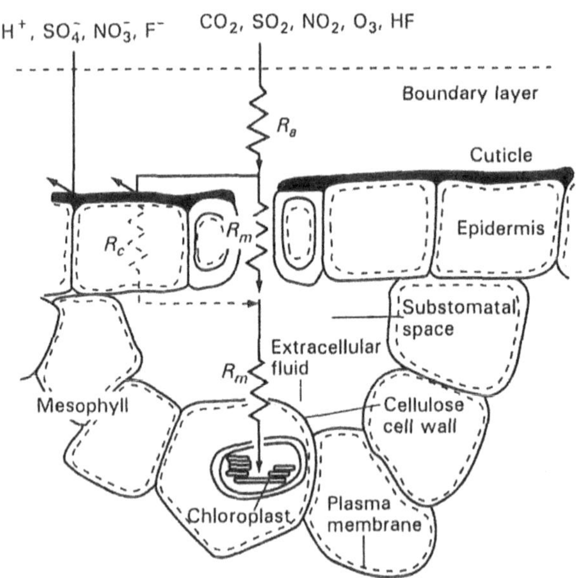

Fig. 1.19 *Likely access routes for SO_2 into a leaf*

(**Source:** *Alan Wellburn, 1998. Airpollution and climate change-the biological impact. Longman, U.K.*)

SO_2 is nearly 30 times as soluble as CO_2 in aqueous fluid. This means that SO_2 uptake takes place mainly on the lower inner surfaces of the stomatal guard cells, where most (77-90%) of the transpired H_2O is lost, rather than from the mesophyll extracellular fluid which contributes only 10-23% of H_2O loss but 85% of CO_2 uptake. This is largely because removal mechanisms of CO_2 are associated with photosynthetic carbon fixation. Uptake of SO_2 close to the stomata also means that the immediate effects of SO_2 are focused in this area. Indeed, SO_2 injuries both to the plasma membranes of guard cells and to the transport cells loading the vascular elements of leaves (companion cells in angiosperms, Strasburger cells in gymnosperms) are especially important. This means that increased levels of acidity and other ions (e.g., sulphite and sulphate) occurat these damaged locations and together they can have a detrimental effect upon the ability of a plant to move and use H_2O and photosynthetic products.

Inhibition of photosynthesis due to decrease in pH or increase in acidity is frequently recognized as one of the immediate effects of SO_2 pollution on plants.

Thus, the increase in acidity due to SO_2 pollution is likely to be significant. For example, direct effects of SO_2 have been demonstrated upon the enzyme of CO_2 fixation, ribulose-1,5 bi-phosphate carboxylase/oxygenase, which requires a pH change in the stroma from pH 7.5 to pH 9.0 to become active. It has been calculated that a reduction of chloroplast stroma by 0.5 of a pH unit will cause a 50% decrease in net photosynthesis. So even a small change could significantly affect the growth of polluted plants over a long period. The injurious levels of SO_2 for different types of trees are given in Table 1.13.

Table 1.13 *Injurious levels of SO_2 for different types of trees*

Tree	Concentration of SO_2 (ppm)	Time to produce injury (hours)
Apple	0.48	6
Pear	0.48	6
Mango	0.50	7
Mountain ash	0.54	3
Ponderosa pine	0.50	8

1.3.3.2 Effects of sulphur oxides on human health

Irritant properties: Sulphur dioxide and its derivatives produce strong irritation in the eyes and also within the nasal passageways. It causes intense irritation, even at 2.5 ppm levels, to the eyes and the respiratory tract. It is absorbed by the nasal system, leading to swelling and stimulated mucus secretion. Sulphur dioxide damages lung tissue and promotes respiratory diseases. Raised levels of SO_2 in the atmosphere may also cause lung cancer. Higher concentrations of SO_2 induce desquamation or peeling off of the surface epithelium in the mucosa. Cilia which protect the respiratory tract are also affected by SO_2. SO_3 on a molar basis produces more than four times the irritant response of SO_2 because it combines immediately with water to form sulphuric acid. Sulphates in particulates can also cause similar irritation and the intensity of the effect is dependent upon the concentration of gaseous SO_2. SO_2 also induces an involuntary coughing reflex. The eye irritation combined with the choking cough immediately draws the attention of those affected by the hazards of the surrounding atmosphere, although there is a wide variation in response susceptibility between individuals. The taste threshold limit is 0.3 ppm, and SO_2 produces an unpleasant smell at 0.5 ppm. SO_2 can also be detected by smell by some individuals better than others, but at very high concentrations (>3 ppm) this sense of smell is quickly paralysed in all.

The threshold concentration of SO_2 producing injury to plants, man and animals is presented in Table 1.14. Table 1.15 shows the human response to different levels of SO_2 for different periods.

Table 1.14 *The threshold concentrations of SO_2 producing injury*

Category	Short term exposure (one hour/day)	Long term exposure (one hour/day)
Plants	0.5-2.0	0.01-0.05
Man	0.5-3.0	0.02-0.10
Animals	1.0-4.0	0.05-0.20

Table 1.15 *Human response to different levels of SO_2 for different periods*

Concentration ($\mu l\ l^{-1}$)	Period	Effect
0.03-0.05	Continuous	Condition of bronchitic patients worsened
0.3-1	20 seconds	Brain activity changes
0.5-1.4	1 minute	Odour perceived
0.3-1.5	15 minutes	Increased eye sensitivity
1-5	30 minutes	Increased lung air way resistance, sense of smell lost
1.6-5	>6 hours	Constriction of nasal and lung passageways
5-20	>6 hours	Lung damage reversible if exposure ceases
20 upwards	>6 hours	Water logging of lung passageways and tissues, eventually leading to paralysis and/or death.

The primary effect of most air pollutants on human health seems to be the injury of delicate tissues, by causing damage to the cellular membranes. This often sets in motion an inflammatory response–a complex series of interactions between damaged cells, surrounding tissues, and the immune system. One of the first symptoms of inflammation is leakage of fluid (plasma) from blood vessels. Exposure of respiratory tissues to severe irritants can result in so much edema (fluid accumulation) in the lungs that one effectively drowns. Bronchitis is a persistent inflammation of bronchi and alveoli (large and small airways in the lung) that causes a cough, copious production of sputum (mucus and dead cells) and involuntary muscle spasms that constrict airways. Acute bronchitis can obstruct airways so severely that death results. Smoking is undoubtedly the largest cause of chronic bronchitis in most countries. Persistent smog and acid aerosols can also cause this disease.

Severe bronchitis can lead to emphysema, an irreversible obstructive lung disease in which airways become permanently constricted and alveoli are damaged or even destroyed. Stagnant air trapped in blocked airways swells the tiny air sacs in the lung (alveoli), blocking blood circulation. As cells die from lack of oxygen and nutrients, the walls of the alveoli break down, creating large empty spaces incapable of gas exchange (Fig. 1.20). Thickened walls of the bronchioles lose elasticity and breathing becomes more difficult. Victims of emphysema make a characteristic whistling sound when they breathe. Often they need supplementary oxygen to make up for reduced respiratory capacity.

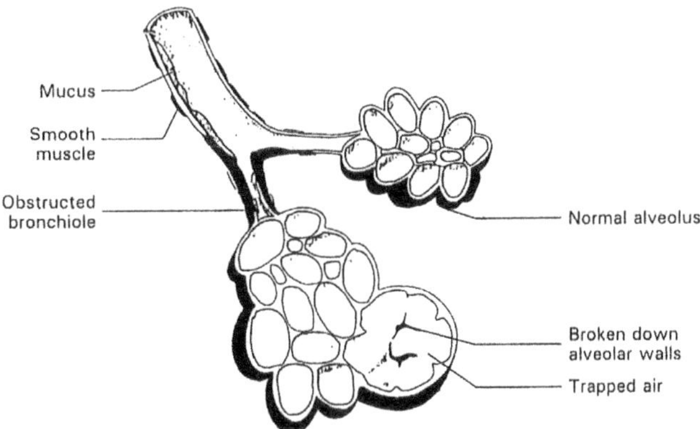

Fig. 1.20 *Emphysema results from chronic irritation and obstruction of airways and alveoli. In extreme cases, alveolar walls break down and enlarged air sacs full of trapped stagnant air fill the lungs*

Figure 1.21 shows the major air passageways into the lungs which terminate in the alveoli (fluid-lined air sacs) across which O_2 and CO_2 exchanges take place with the bloodstream. The mode of entry of SO_2 into this system, however, differs from those associated with nitrogen oxides, O_3 and photochemical oxidants because more than 95% of inhaled SO_2 is absorbed in the airways above the larynx (voice box) and less than 1% gets down into the lung alveoli when an individual is resting. Exercise increases the exposure of lung alveoli to SO_2 but at the same time, the upper airways are constricted and the mucus lining dries up. This bronchoconstriction and dryness may induce chronic bronchitis in some individuals, especially those predisposed to asanma. The same group often also show a sensitivity to sodium sulphite used commercially as sterilizing agent or as a preservative.

During severe short term exposures to SO_2, sulphate and sulphite anions formed on the cell surfaces of the nasal linings penetrate the surface mucosal cells and bind to granules within mast cells beneath. This causes local release of histamine which then acts as a local modulator to cause constriction of the airways and initiates local inflammation. Ciliary cells lining parts of the airways often release extra mucus to carry away some of the harmful anions as the nose is blown or one coughs. Similarly, tears running from the eyes provide relief from the local irritation.

Cardiovascular stress from lack of oxygen in the blood is a common complication of all obstructive lung diseases. About twice as many people die of heart failure associated with smoking as die of lung cancer. Irritants in the air are so widespread that about half of all lungs examined at autopsy in the United States have some degree of alveolar deterioration.

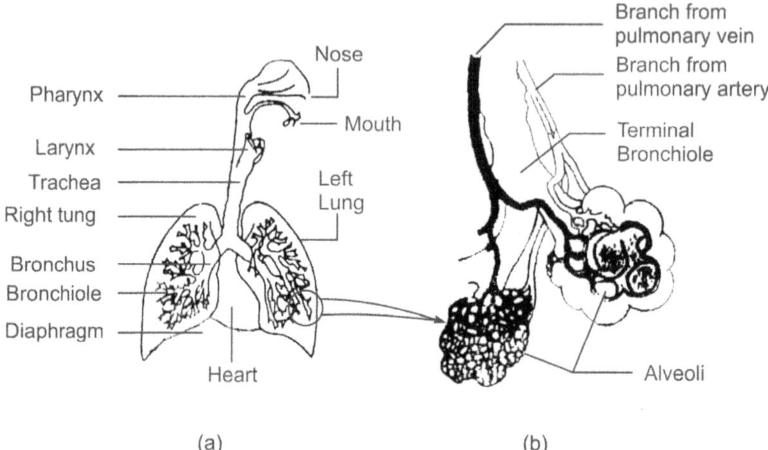

Fig. 1.21 *Cut away diagrams of the mammalian respiratory system at different scales of resolution*

Figure 1.22 summarizes the effect of various concentrations of SO_2 and exposure time on human health while Fig. 1.23 gives daily mortality as a function of SO_2 concentration observed in London.

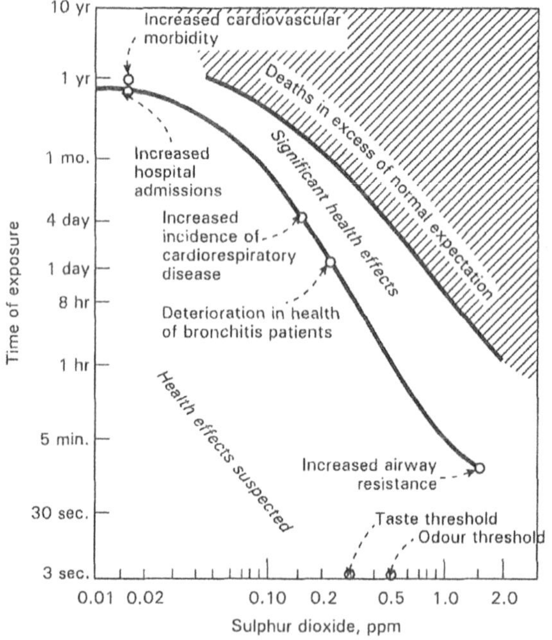

Fig. 1.22 *Hazards to health from SO_2 pollution*

(**Source:** *A. M. Squires, Chemical Engineering, Vol. 74, No., 23, Nov. 6, 1967*)

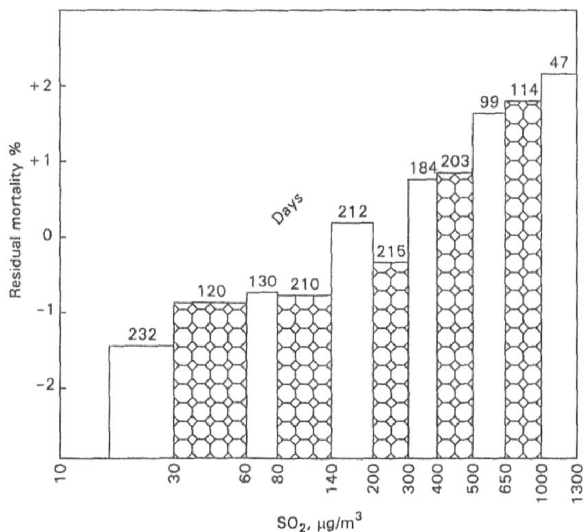

Fig. 1.23 *Daily mortality as a function of SO_2 concentration*

(**Source:** *Helen Dwight Reid Educational Foundation, Heldref Publications, 1319 Eighteenth St., N. W. Washington, D.C. 20036-1802.) [2]*

During accidental non-fatal exposure, victims experience inflammation of the eyes, nausea, vomiting, abdominal pain, a sore throat, bronchitis, and often pneumonia. Normally, the lungs are (and have to be) highly sterile, but this weakening of natural antibacterial mechanisms points the way to emergency treatment of SO_2-exposure cases. Prompt use of oxygen and bronchodilators along with heavy doses of antibiotics prevents permanent damage and may save life.

The possibilities of chronic poisoning by SO_2 over longer periods at low levels are greater for the general population than accidental acute cases. A number of theories confirm the high correlation between chronic chest disease and levels of SO_2 in urban air and indicate that chronic exposure to SO_2 along with particulate matter plays a significant role in the cause and development of chronic respiratory disease.

Thus, sulphur dioxide is found to be more harmful in a dusty atmosphere. This effect may be explained as follows: The respiratory tract is lined with hair-like cilia, which by their regular sweeping action force out foreign substances entering the respiratory tract through the mouth. However, SO_2 and H_2SO_4 molecules paralyse the cilia, rendering them ineffective in rejecting particulates, which thus penetrate deeper into the lungs. By themselves these molecules are too small to remain in the lungs, but some SO_2 molecules are also absorbed on larger particles, which penetrate into the lungs and settle there, bringing concentrated amounts of the irritant SO_2 into prolonged contact with the fine lung tissues.

The British Committee on air pollution estimated that a total of 4,000 more deaths than usual were caused by killer pollutants in the 1952 air pollution episode. The relationship between death rate and smoke and SO_2 concentrations observed is illustrated in Fig. 1.24. During this air pollution episode over a period of two months, there were 8,000 more deaths than usual. Symptoms of the illness were cough with little sputum, sore throat, nasal discharge, and sudden vomiting. Further, those with a previous history of chest trouble (such as chronic bronchitis, bronchiectasis, asthma, or one of the forms of pulmonary fibrosis) were among the most severely ill. The combined action of sulphuroxides and particulate matter was the suspected cause of irritation and disaster. During the episode, the concentration of sulphur oxides reached 0.7ppm, and at one location, the level was 1.3 ppm for a short period of time. Such heavy concentrations came from industry and home heating by coal and coke, which contain, on the average, 1.5% sulphur. It was estimated that 190 million tons of coal was used in England in 1948, and this sent into the atmosphere 4.7 million tons of deadly SO_2. In the winter of 1956, another episode took place which caused 1,000 additional deaths.

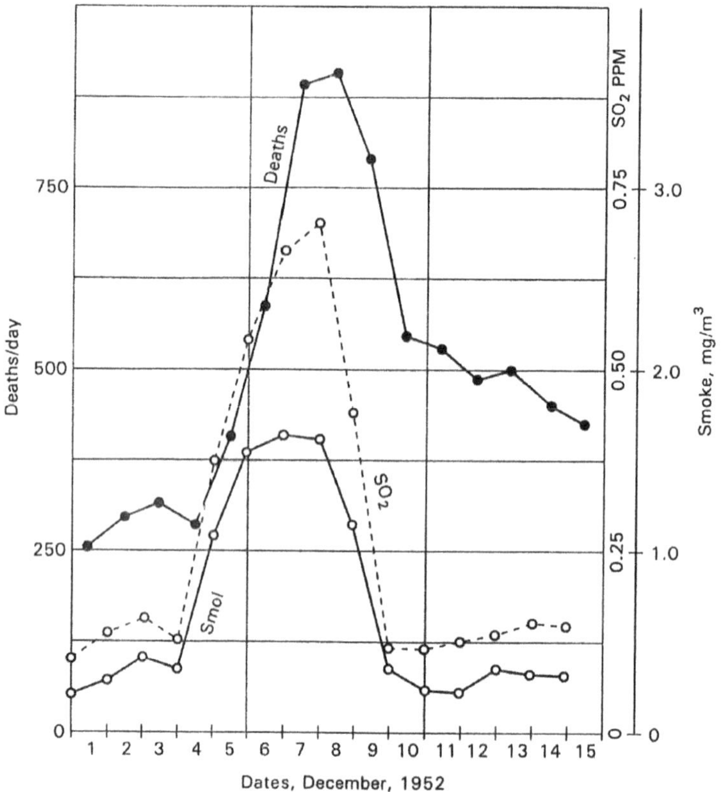

Fig, 1.24 *Daily air pollution and deaths in London, 1952*

(**Source:** *American Industrial Hygiene Association Air Pollution Manual, 1960.*)

Each year air pollutants cause a lot of damage to various materials. Expensive cleaning methods are required to remove soot and grit deposited on buildings, cars and clothes. Air pollutants break down exterior paint on cars and houses, and they deteriorate roofing materials. Irreplaceable marble statues, historic buildings, and stained-glass windows throughout the world have been pitted and discoloured by air pollutants. Damage to buildings in the United States from acid depositions alone was estimated at $5 billion per year. The various harmful effects of SO_x on materials are summarized in Table 1.16.The relationship between the corrosion of mild steel and mean SO_2 concentrations for various exposure times observed in Chicago are summarized in Fig. 1.25.

Table 1.16 *Harmful effects of air pollutants on materials*

Material	Effects	Principal air pollutants
Stone and concrete	Surface erosion, discoloration, soiling	Sulphur dioxide, sulphuric acid, nitric acid, particulate matter
Metals	Corrosion, tarnishing, loss of strength	Sulphur dioxide, sulphuric acid, nitric acid and particulate matter
Ceramics and glass	Surface erosion	Hydrogen fluoride, particulate matter
Paints	Surface erosion, discoloration, soiling	Sulphur dioxide, hydrogen sulphide, ozone, particulate matter
Paper	Embrittlement, discoloration	Sulphur dioxide
Rubber	Cracking, loss of strength	Ozone
Leather	Surface deterioration, loss of strength	Sulphur dioxide
Textile	Deterioration, fading, soiling	Sulphur dioxide, nitrogen dioxide, ozone, particulate matter

1.3.3.3 Effects of sulphur oxides on materials

SO_2 is injurious not only to man and plants, but it also attacks rapidly marble, limestone, roofing slate, electrical contacts, paper, textiles, and buildings. It can even dissolve nylon. Paper absorbs sulphur dioxide which will be oxidized to H_2SO_4 causing the paper to become brittle and fragile. Air with SO_2 accelerates the corrosion rates of metals such as Fe, steel, Zn, Cu.

1.3.3.3.1 Factors that influence deterioration of materials

Moisture, temperature, sunlight, and position of the exposed material are among the more important factors that influence its rate of deterioration.

Moisture, in the form of humidity, is essential for most of the mechanisms of deterioration to occur. Metal corrosion does not appear to occur even at relatively high SO_2 pollution levels until the relative humidity exceeds 60 percent. On the other hand humidities above 70 to 90 percent will promote corrosion without air pollutants. Rain reduces the effects of pollutant-induced corrosion by dilution and washing away of the pollutant.

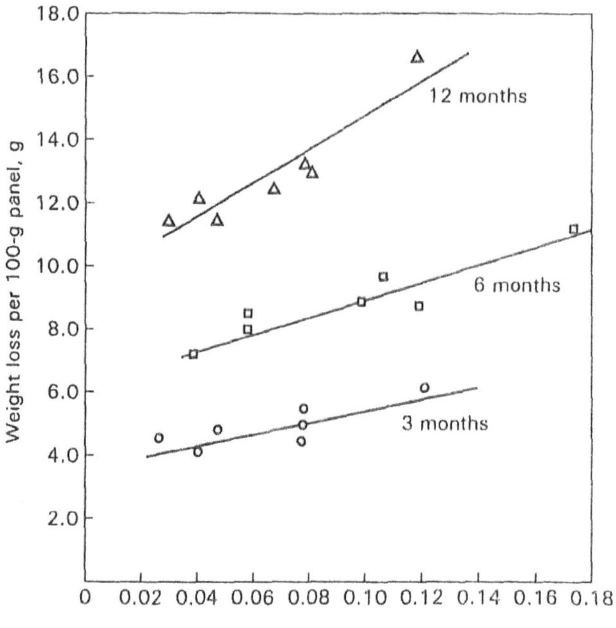

Fig, 1.25 *Relationship between corrosion of mild steel and corresponding mean sulphur dioxide concentrations for various exposure times at seven sites in Chicago (September, 1963-64)*

(**Source:** *Air and Waste Management Association Publication.) [3]*

Solubilization and oxidation/reduction reactions typify direct chemical attack. Frequently, water must be present as a medium for these reactions to take place. Sulphur dioxide and SO_3 in the presence of water react with limestone ($CaCO_3$) to form calcium sulphate ($CaSO_4$) and gypsum ($CaSO_4$ $2H_2O$). Both $CaSO_4$ and $CaSO_4$ $2H_2O$ are more soluble in water than $CaCO_3$, and both are leached away when it rains. The tarnishing of silver by H_2S is a classic example of an oxidation/reduction reaction.

Indirect chemical attack occurs when pollutants are absorbed and then react with some component of the absorbent to form a destructive compound. The compound may be destructive because it forms an oxidant, reductant, or solvent. Further, a compound can be destructive by removing an active bond in some lattice structure. Leather becomes brittle after it absorbs SO_2, which reacts to form sulphuric acid because of the presence of minute quantities of iron. The iron acts as a catalyst for the formation of the acid. A similar result has been noted for paper.

Oxidation/reduction reactions cause local chemical and physical differences on metal surfaces. These differences, in turn, result in the formation of microscopic anodes and cathodes. Electrochemical corrosion results from the potential that develops in these microscopic batteries.

Higher air temperatures generally result in higher reaction rates. However, when low air temperatures are accompanied by cooling of surfaces to the point where moisture condenses, then the rates may be accelerated.

1.3.4 Nitrogen oxides

Natural stratospheric oxides of nitrogen are produced by the action of cosmic rays in the upper atmosphere. Emissions of oxides of nitrogen from man-made sources vary in different areas. Nitrogen oxides are 10 to 100 times greater in the urban atmosphere as compared to rural areas.

Oxides of nitrogen which include N_2O, NO, NO_2, N_2O_3, and N_2O_5 are usually represented by the symbol NO_x. The two major pollutants among them are nitric oxide, NO, and nitrogen dioxide, NO_2.

As N_2 circulates, between plants and animals and microbes in the soil, wide differences exist between different reduced and oxidized components. The electronic configuration about an N atom may vary widely; this permits a number of different oxidation-reduction states of N to exist (Fig. 1.26). This, in turn, gives rise to a wide range of N-based compounds and together these features dominate the biological nitrogen cycle (Fig. 1.27).

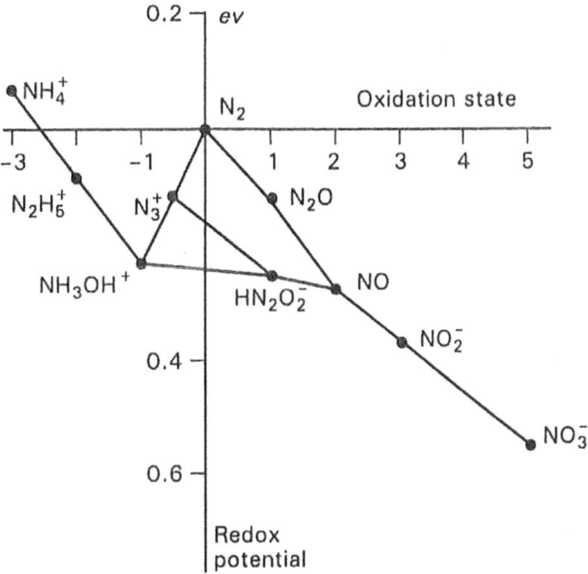

Fig. 1.26 *Oxidation states of nitrogen at pH 7.0 in relation to redox potentials*

(**Source:** *Air Pollution Climatic Change, Alan Well burn, Longman, 1998.*)

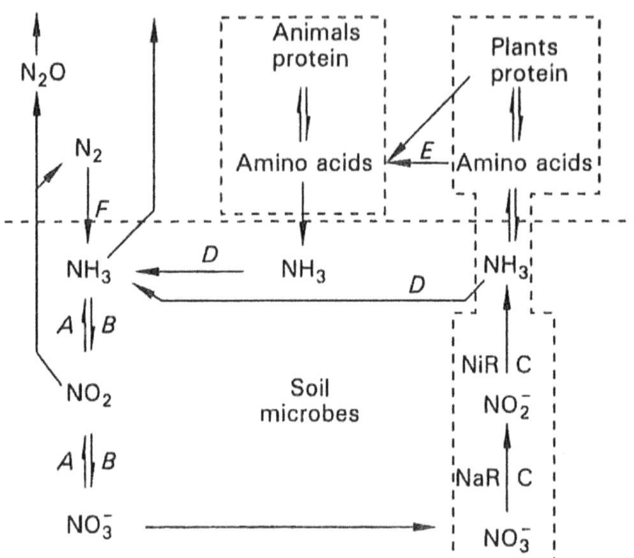

Fig. 1.27 *The nitrogen cycle as it relates to plants, animals, the soil and the atmosphere. The various processes are as follows: A = denitrification or dissimilatory nitrogen metabolism which releases N_2O; B = microbial nitrification; C = assimilatory nitrogen metabolism; NaR = nitrate reductase: NiR = nitrite reductase; D = decay after death or putrefaction; E = digestion; and F = nitrogen fixation*

(**Source:** *Air Pollution Climatic Change, Alan Wellbum, Longman, 1998.*)

Any excess artificial fertilizer remaining after application, which has not drained away, is usually removed by denitrification. Increasing global use of artificial fertilizers therefore contributes to higher NO_x levels. Table 1.17 shows the global transfer rates of nitrogen by various processes.

Table 1.17 *Global transfer rates of nitrogen by various processes*

Process	10^{12} g N per year
Denitrification/nitrification	160
Natural nitrogen fixation	150
Fires and combustion	70
Industrial fixation	40
Ionic exchange by rain, etc.	80
Run off to oceans	35

1.3.4.1 *Sources of emission of nitrogen oxides*

Major man-made activities like combustion of coal, oil, natural gas and gasoline produce up to 50 ppm of oxides of nitrogen. Atmospheric nuclear explosions can be another potential serious source of man-made stratospheric oxides of nitrogen. There are two ways of formation of oxides of nitrogen during the combustion of fossil fuels. Thermal NO_x are created when nitrogen and oxygen in the combustion air are heated to a high enough temperature (above about 1000 K) to oxidize the nitrogen. Fuel NO_x result from the oxidation of nitrogen compounds that are chemically bound in the fuel molecules themselves. Different fuels have different amounts of nitrogen in them, with natural gas having almost none and some coal having as much as 3% by weight. Both thermal NO_x and fuel NO_x can be significant contributors to the total NO_x emissions, but fuel NO_x often is the dominant source.

The global fluxes of reactive nitrogen gases into the atmosphere per year are shown in Fig. 1.28.

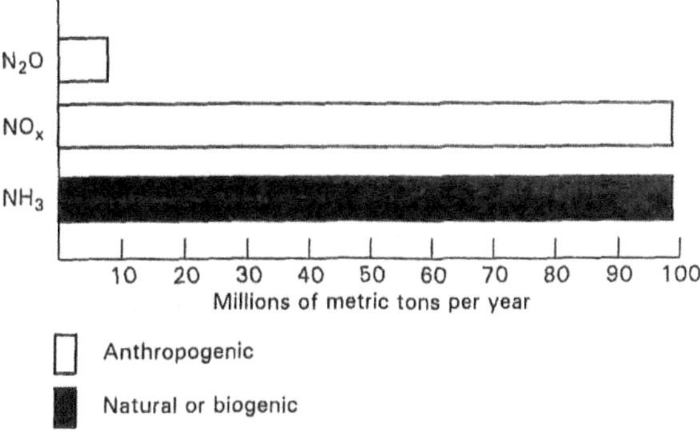

Fig. 1.28 *Fluxes of reactive nitrogen gases into the atmosphere*
(**Source:** *American Industrial Hygiene Association Air Pollution Manual, 1960.*)

Anthropogenic impacts upon the global exchange of N are considerable but the bulk of the global N cycling remains microbial.

1.3.4.2 *US -National trends in nitrogen dioxide levels*

Using a nationwide network of monitoring sites, US EPA has developed ambient air quality trends for nitrogen dioxide (NO_2). Under the Clean Air Act, EPA sets and reviews national air quality standards for NO_2. Air quality monitors measure concentrations of NO_2 throughout the country. EPA, state, tribal and local agencies use that data to ensure that NO_2 in the air is at levels that protect public health and the environment. Nationally, average NO_2 concentrations have decreased substantially over the years. Figure 1.29 shows the trends in nitrogen oxide emissions from 1980-2015.

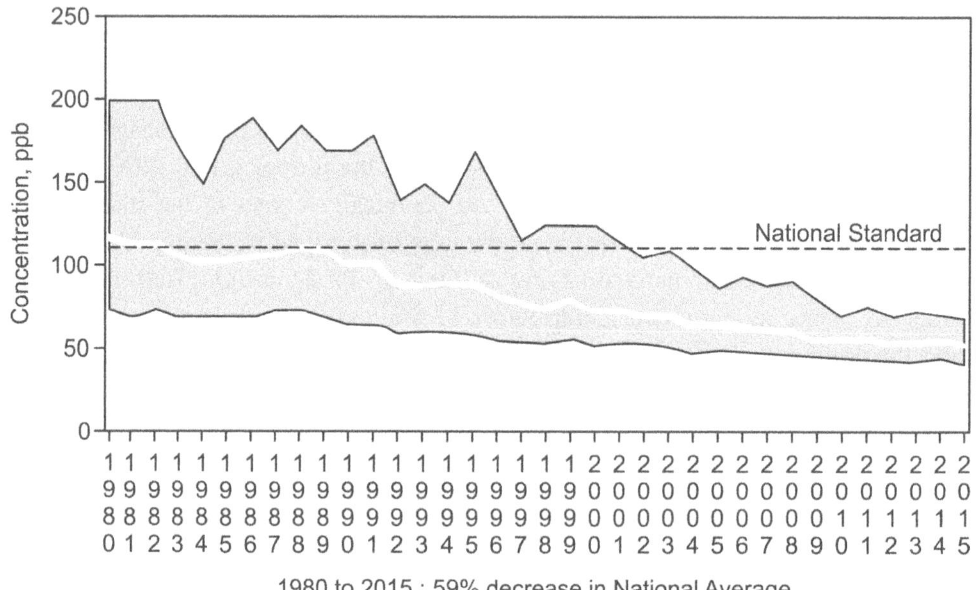

Fig. 1.29 *Trends in nitrogen oxide emissions in USA, 1980-2015*

1.3.4.3 Nitric oxide

Nitric oxide is colourless, but it is photo chemically converted to NO_2, which is one of the major NO is mainly formed by the combination of atmospheric N_2 and O_2 at high temperatures with a lesser contribution from N-containing components of fuels.

$$N_2 + O_2 \Rightarrow 2NO$$

The above reaction is often quoted as the major process, but actually it takes place in separate reactions (see below) involving highly reactive atoms created within a flame – the second being much faster than the first.

$$N_2 + O \Rightarrow NO + N$$

$$N + O_2 \Rightarrow NO + O$$

In the fuel-rich regions of a flame, highly reactive hydroxyl (OH*) radicals form NO but HCN, NH_3, or amines have also been identified as precursors of NO.

$$N + OH \Rightarrow NO + H$$

Odd as it may first appear, the amount of NO produced decreases as the N content of the fuel rises. By contrast, removal of S from fuels increases the amount of NO produced. Therefore, burning of heavy oil, which usually has a low N content, leads to NO emissions as great as those from power plants using coal with much higher N content.

1.3.4.4 Nitrous acid

Nitrous acid forms rapidly in the gas phase by reactions between NO, NO_2 and H_2O but decomposes to NO and nitric acid in aqueous solution by a series of inter conversions.

$$CH_3CO.O_2 + NO_2 + M \Rightarrow CH_3COO_2NO_2 + M$$

$$NO + NO_2 + H_2O \Rightarrow 2HNO_2$$

$$2HNO_2 \Rightarrow N_2O_3 + H_2O$$

$$2N_2O_3 \Rightarrow N_2O_4 + 2NO$$

$$N_2O_4 + H_2O \Rightarrow HNO_2 + HNO_3$$

1.3.4.5 Nitrous oxide

The most abundant nitrogen oxide in the atmosphere is N_2O. This is mainly released by denitrification which is an undesirable agricultural process carried out by certain soil microbes using nitrate instead of O_2 to respire. Soil conditions where O_2 levels are low favour denitrification. Consequently, stagnant, waterlogged or compacted soils are major sources of N_2O, and one of the main purposes of ploughing and draining is to discourage this anaerobic process.

Global production of N_2O by denitrification each year amounts to about 5% of the total global pool of atmospheric N_2O but, because N_2O is a relatively unreactive molecule, it has a long residence time of 20 years or more in the atmosphere. Any subsequent atmospheric increase in N_2O levels due to enhanced denitrification will therefore develop slowly but persist for a long time.

The composition of nitrogen oxides in the atmosphere is therefore very complex. Amultitude of different reactions are influenced to a greater or lesser extent by temperature, light flux, and different levels of dilution because all these molecules, atoms or radicals have to collide to react. The chances of this are reduced as pressure falls with increasing altitude but, as molecules rise upwards, some of the intervening collisions with unreactive molecules are reduced and radiation levels increase. Consequently, half-lives of reacting species vary with altitude and some intermediates, which exist for very short periods at ground level, persist for longer in the stratosphere. On the other hand, those species which are relatively inert at ground level have a chance to be slowly transported upwards into the stratosphere where they subsequently react. N_2O is a good example of this possibility.

1.3.4.6 Removal mechanisms of nitrogen oxides

Apart from photolytic breakdown (Fig. 1.30), a number of reactions of alternative phenomena like dry and wet deposition (Fig. 1.31) may remove NO_x.

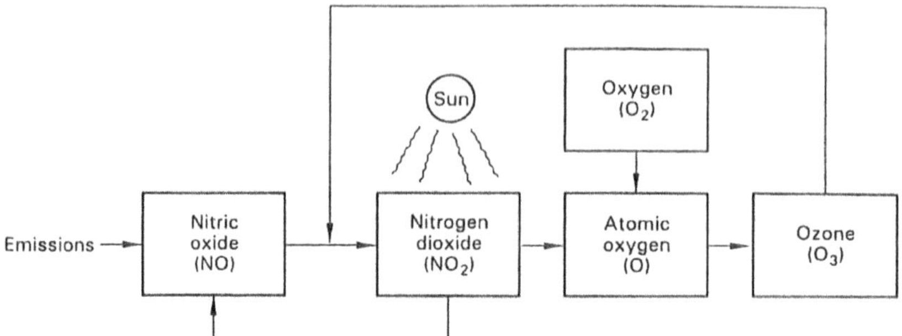

Fig. 1.30 *Simplified atmospheric nitrogen photolytic cycle*

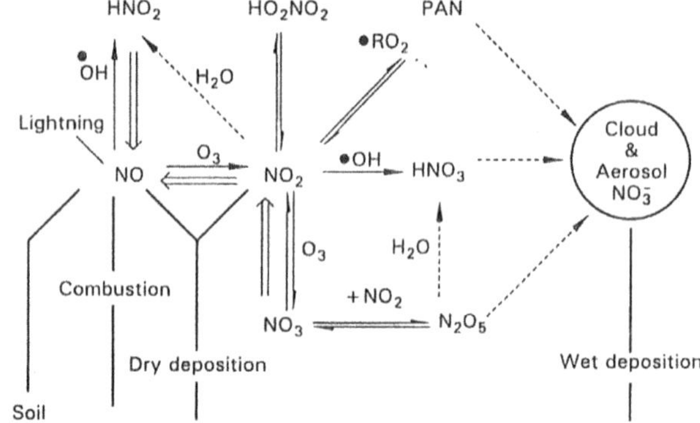

Fig. 1.31 *Atmospheric interconversions of oxides of nitrogen*

(**Source:** *Dis Con and Penkett, AERE, Harewell, UK, and D. Reidel Publ. Co. Dordrecht, The Netherlands.*)

1.3.4.7 Dry deposition

All oxides of nitrogen are removed from the atmosphere by absorption at ground level by land or ocean surfaces. The velocity of deposition of each atmospheric gas differs widely and depends upon the nature of the uptake surfaces. Rates for chemical species like nitric acid are very high because they are so reactive, while the deposition velocity of NO_2 is lower than that of SO_2 or O_3 but much higher than that of NO or N_2O. Estimates of global wet and dry deposition rates of both nitrate and nitric acid amount to about 25 million tons per annum, while dry deposition of NO_2 accounts for an additional third of this amount.

Measurements of deposition rates are complicated because the gases may interact in many different ways. For example, undissolved nitric acid forms ammonium nitrate aerosol – a salt which has a high a dissociation constant so that significant equilibrium levels of both NH_3 and nitric acid are present. The relative deposition rates of different nitrogen-containing compounds are as follows:

$$HNO_3 = NO_3 > NH_3 > NH^{+}_{4} + > NO_2 > NO > N_2O$$

1.3.4.8 *Effects of nitrogen oxides on plants*

Plant growth may be inhibited by continuous exposure to 0.5 ppm of NO_x. Levels of NO_x in excess of 2.5 ppm for periods of four hours or more are required to produce necrosis (surface spotting due to plasmolysis or loss of protoplasm).

Higher concentrations of NO_x affect the leaves of plants, retard the photosynthetic activity and cause chlorosis. NO_2 is highly injurious to plants. Vegetative growth is suppressed when plants are exposed to 0.3-0.5 ppm of NO_2 for 10 to 15 days. Exposure to 10 ppm of NO_2 checks the metabolic activities in plant tissues, e.g. bean and tomato plants, on fumigation showed decreased activity of CO_2 absorption and photosynthetic rate.

Generally, the movement of nitrogen oxides into leaves is subject to similar diffusive resistances as found for SO_2 but the stomata are not the only path of entry for nitrogen oxides. Cuticular resistance against NO_x entry is lower than that for SO_2 or O_3. This means that, even when stomata are closed, some oxides of nitrogen possibly enter the leaf through the epidermal layers even though this is a small fraction of the whole.

Inside a leaf, the exposed surface area of the moist cell wall below is considerably larger than the external surface area of a leaf and this provides a large surface area for the absorption of nitrogen oxides. The solubility of nitrogen oxides in the extra cellular fluid is therefore an important factor in determining their rate of uptake. Both gaseous NO_2 and NO are only slightly soluble in H_2O, but NO_2 vastly increases its apparent solubility by reacting with H_2O to produce nitric acid. Gaseous NO, however, does undergo a comparatively slow reaction with H_2O to form both nitric acid and nitrous acid. Normally, nitrous acid is oxidized in solution to nitric acid by O_3, H_2O_2 and other oxidizing agents, but it may be converted back to NO when reducing agents like ferrous iron are present.

1.3.4.9 *Effects of nitrogen oxides on human health*

NO is biochemically inert and not extremely toxic. The health effects vary with the degree of exposure. NO_2 has irritating effects on the mucous membrane. An exposure to 50 to 100 ppm of NO_2 for 30 to 50 minutes for a period of 5to 8 weeks causes inflammation of lung tissues. NO_2 poisoning results in Silo filter's disease when the NO_2 is produced by the fermentation of ensilage containing nitrates. The threshold limit value for NO_2 is 5 ppm (Table 1.18). Breathing it for 30 min to 1 h could affect lung tissues, while breathing it for 30 minutes to 1 h at 100 ppm-150 ppm could produce serious pulmonary edema, or swelling of lung tissues. A few breaths at 200 ppm-700 ppm may cause fatal pulmonary edema.

Table 1.18 *Likely effects of nitrogen dioxide upon humans*

Pollutant range, ppm	Symptoms
0-0.21	No effects
0.11-0.21	Slight odour detected
0.22-1.1	Some metabolic effects associated with either toxicity, adaptation or repair of lung tissues (e.g. inhibited metabolism of prostaglandin E_2)
1.1-2	Significant changes to respiratory rate and lung volume. Enhanced susceptibility to infection and evidence of tissue repair
2.1-5.3	Deterioration of lung tissue not balanced by repair mechanisms
Above 5.3	Gross distortion of lung tissues and emphysema, possible death if prolonged

The WHO has recommended a value of 0.23 ppm of NO_2, not to be exceeded for a period of 1 h. Over a period of 24 h, this should be reduced below 0.08 ppm to safeguard human health.

The lack of immediate warning symptoms from oxides of nitrogen means that exposure symptoms such as coughing, headaches and chest tightness develop a day later. If particularly severe, they are followed by either sudden circulatory collapse or congestion and water accumulation in the lungs a short while after this. Treatment with rest, oxygen, steroids and antibiotics is often recommended. However, once initial recovery is made, affected individuals are always likely to have a subsequent history of chronic chest complaints.

Damage is not just confined to the lungs. NO causes the of red blood cells (RBC) to form a type of methaemoglobin. In addition, excess blood nitrate may reduce blood pressure. The destruction of RBCs will be enhanced, inducing liver and kidney defects as increased efforts are made by the body to eliminate the products of blood break down through the bile duct or urine. The deterioration of the liver and kidneys then often gives rise to jaundice and other related conditions. Unfortunately, in workers who do not experience sudden massive doses but have steady inhalations over a period of time, the chronic illness they later suffer is often unrecognized and not directly linked with their occupation.

Penetration and retention of NO_2 depends on many factors. In tissue fluids, for example, reactivity of NO_2, its dilution in inspired air, the length of exposure, and the depth and frequency of breathing are all important. As NO_2 is soluble, it readily enters tissue fluids much as it enters the extracellular fluid of plants. By the same measure, NO has the potential to enter with rather more difficulty as a mixture of nitrate, nitrite and undissociated nitrous acid. Consequently, the airways passing through the nasal regions, the terminal bronchioles of the lungs, and finally, the alveoli all have their surfaces exposed to the products of nitrogen oxides in solution.

During light breathing, about 40% of the oxides of nitrogen are taken up in the nose and throat regions but, as when the body is exercised and deeper breathing takes place through the mouth, this balance shifts towards the lung tissues which then become the most important uptake surfaces. At low concentrations of NO_2, the most sensitive areas are the junctions between the bronchioles and the alveolar regions. With higher concentrations, the injury extends to the larger airways and into deeper tissues.

Other changes due to exposure to oxides of nitrogen include accumulation of tissue fluid and cellular debris or mucus. There is also a clumping of certain white blood cells which normally accumulate as one of the first lines of defence against infection. Animal lungs have a remarkable ability to rid themselves of inhaled microbes, some of which are pathogenic, and this is essential to keep the alveoli sterile. Low concentrations of NO_2, however, disturb these natural defences of the lungs and cause them to become more susceptible to bacterial attack.

1.3.4.10 Risk from NO_2 indoors

The levels of indoor NO_2 range from 20 to 250 ppb with at least as much NO. These levels of NO_2 are around one-eighth of the industrial Threshold Limit Value (TLV) but are experienced for a duration longer than an industrial working day by those who do not venture out frequently – young children, pregnant mothers, and the old are at a greater risk medically.

Smoking adds to indoor pollution. Cigarette smoke contains oxides of nitrogen, especially NO, as well as a multitude of other potential irritants.

The levels of NO_2 in the home during the winter can be considerable, especially in kitchens and living rooms. Table 1.19 shows the concentrations of NO_2 detected in different rooms of domestic homes.

Table 1.19 *Range of NO_2 concentrations detected in different rooms of domestic homes*

Concentration, ppb	Bedrooms (%)	Kitchens (%)	Living rooms (%)
0-21	48	11	40
22-42	33.5	29	49
43-63	13.5	19	18
64-84	2.5	13	5
85-105	2	10	2
106-126	0	7	1
127-147	0.5	3	1
148-168	0	1	1
> 168	0	5	0

1.3.4.11 *Effects of nitrogen oxides on materials*

NO_x affects a number of materials like cotton, rayon, acetate and viscose rayon. It was found that NO_x levels reaches 1 to 2 ppm, during the combustion of natural gas which is used to heat dryers. Higher concentrations of particulate nitrates in airborne dust accumulate adjacent to cracked areas of metals like nickel and brass.

1.3.5 Carbon monoxide

Carbon monoxide is the most abundant gaseous pollutant emitted through anthropogenic sources into the troposphere – 0.075 Gt per annum which is a very high rate when compared to other gaseous pollutants and this rate is still rising.

Carbon monoxide is a colourless, odourless, tasteless gas which is present in small concentrations (0.1 ppm) in the natural atmosphere and has a residence time of about six months. CO is produced markedly when carbonaceous fuels are burned under less than ideal conditions. Incomplete combustion, yielding CO instead of CO_2, results when any of the following four variables are not kept sufficiently high:

1. oxygen supply
2. flame temperature
3. gas residence time at high temperature, and
4. combustion chamber turbulence.

1.3.5.1 *Sources of emission of carbon monoxide*

Carbon monoxide in urban areas is produced mainly by automobiles, burning coal, oil, etc. (Table 1.20). About 70% of CO emission is from the transportation sector, with almost all of that coming from motor vehicles, Personal exposure to CO is very much determined by the proximity of heavy motor vehicular traffic.

Table 1.20 *Main sources of CO emission*

Contributor	**CO emission (%)**
Automobiles	
Transportation	64
Motor vehicles	59.5
Aircraft	2.5
Railroads	0.1
Industrial processes	
Iron, steel, paper and petroleum industries, etc.	9.6
Miscellaneous sources	
Forest fires	6.9
Agricultural burning (forest debris, crop residues, brush and weeds, etc.)	8.3

Carbon monoxide pollution is of basic concern in most large cities of the world, where traffic is usually congested and heavy. In New York City alone, automobile traffic produces 3.76×10^6 kg of CO daily; and in Los Angeles, the estimate is 9 million kg per day. A car emits about 11.69 mg of CO/km at 25 kmph and about 96 mg of CO/km at 10 kmph. The trends in CO emissions of USA between 1980 and 2015 are shown in Fig. 1.32.

Biogenic processes, by contrast, produce a tenth of the CO produced by human activities. Half of this comes from the oceans and the other half from a variety of natural processes such as volcanoes, electrical storms, and 'marsh gas' from rotting vegetation, etc.

1.3.5.2 US-National trends in CO levels

Using a nationwide network of monitoring sites, EPA has developed ambient air quality trends for carbon monoxide (CO). Under the Clean Air Act, EPA sets and reviews national air quality standards for CO. Air quality monitors measure concentrations of CO throughout the country. EPA, state, tribal and local agencies use that data to ensure that CO remains at levels that protect public health and the environment. Nationally, average CO concentrations have decreased substantially over the years.

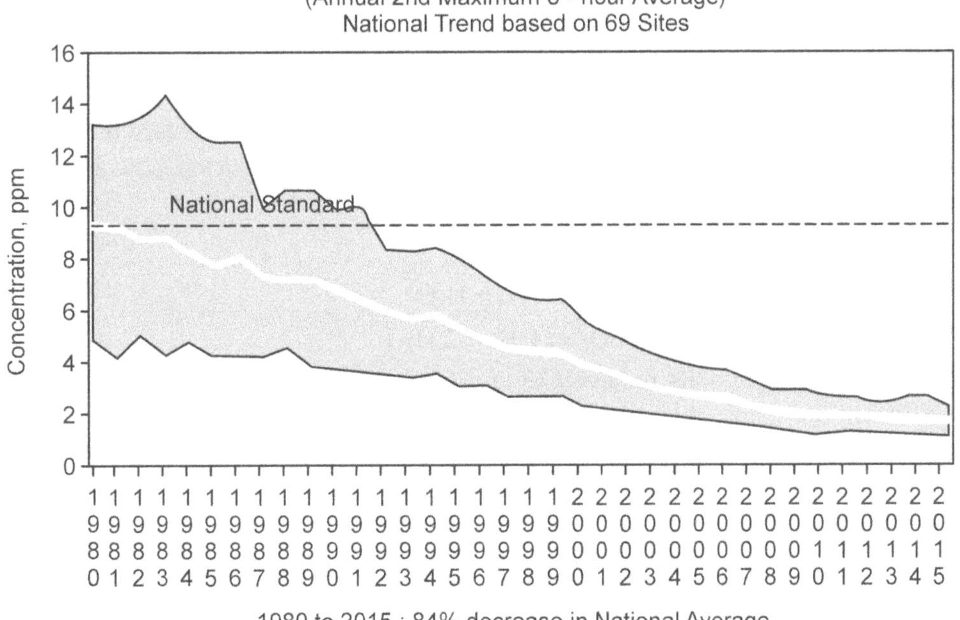

Fig. 1.32 *Trends in carbon monoxide emission in USA. 1980-2015*

(Source: *National ambient air pollutant emission estimates. USEPA.)*

1.3.5.3 Removal mechanisms of carbon monoxide

Over land areas and away from major sources, atmospheric levels of CO are relatively constant which means that natural sinks exist for the removal of excess CO. Various removal mechanisms include oxidation of CO to CO_2 in the troposphere by OH radicals, oxidation in the stratosphere, or absorption by soil microbes or plants.

CO is produced by incomplete combustion and adds to global warming because it is rapidly oxidized to CO_2 as in the reaction below.

$$OH + CO \Rightarrow CO_2 + H$$

Removal of CO, therefore, depends on the presence of sufficient OH radicals which, in turn, requires atomic O and H_2O. Consequently, sunny but wet atmospheric conditions are most favourable for OH production and CO removal. Overcast and cold but dry winter days are far from ideal. Any CO that rises above the troposphere into the stratosphere is oxidized to CO_2 by atomic O or O_3.

$$CO + O + M' = CO_2 + M$$

But this is counter balanced by photolysis of CO_2 back to CO.

$$CO_2 + \text{light} \Rightarrow CO + O$$

The oceans are a major sink for CO_2 but not for CO. A function of partial pressure, the solubility of CO in sea water is much less than that of CO_2 or SO_2. However, the surface waters of oceans contain many times the theoretical CO concentrations that would be expected from gaseous uptake from the atmosphere. In other words, marine biological activity contributes substantial quantities of CO to surface waters and transfers more CO into the atmosphere.

Soils have a variety of soil microbes which oxidize CO to CO_2 or reduce it to CH_4. For example, an anaerobic CH_4 producer like *Methanobacterium for micum* generates H_2:

$$CO + H_2O \Rightarrow CO_2 + H_2$$

This H_2 then reduces CO or CO_2.

$$3H_2 + CO \Rightarrow CH_4 + H_2O$$

$$4H_2 + CO_2 \Rightarrow CH_4 + 2\,H_2O$$

Generally, different soils remove CO at different rates but cultivated soils are much less efficient in doing this than natural soils high in organic matter. Agricultural processes, therefore, select against those microbes most efficient at CO removal. Consequently, the health of non-cultivated soils is vital, not just for protection against acid deposition, but also to reduce CO levels.

Some global estimates of the total potential capacity for soil microbial uptake of CO have been put as high as 14 Gt per annum, well in excess of what is discharged by human activities. By contrast, other calculations have estimated global rates of removal to be 0.45 Gt per annum, much lower than the annual rate of production. Various assumptions account for these discrepancies. Soil microbial uptake has been established as the major global mechanism for CO removal, but this is a slow temperature-dependent process which explains the rhythmical seasonal changes in levels of global atmospheric CO. As northern forest soils warm in summer they supplement the steady uptake of CO by tropical forest soils. Furthermore, as more tropical and temperate areas are brought into cultivation, the cyclical seasonal changes in global CO concentrations become more pronounced. Should northern forest soils also suffer damage by other types of pollution then overall levels of CO and, thereby, CO_2 will rise and enhance global warming.

1.3.5.4 CO Plant uptake

Oxidation of CO to CO_2 by certain crops has been detected using ^{14}CO. However, if respiration is artificially inhibited with dinitrophenol, then this process is much reduced. The implication of this is that the major oxidation of CO by plants is linked to the cytochrome oxides activity of mitochondria. Some trees, however, do not take up CO at all while green algae have their nitrogen metabolism inhibited by CO. Some seaweeds are even capable of emitting CO. Consequently, any judgment on the relative significance of vegetation to remove CO is difficult to make. Current estimates indicate global uptake of CO by vegetation to be 25% of that of soils.

1.3.5.5 Effects of carbon monoxide

CO is a colourless, odourless gas which is lethal to humans within a few minutes at concentrations exceeding 1,000 ppm (Table 1.21). All the gaseous pollutants of carbon mainly cause severe damage to the respiratory system. But CO affects the human body in a unique manner. It affects human health by forming carboxyhaemoglobin, which restricts transport of oxygen from lungs to cells.

Table 1.21 *Effects of CO on humans and the accompanying carboxyhaemoglobin blood concentrations*

Exposure[a] ($\mu l\ l^{-1}$)	Effects and symptoms	COHb/Hb[b] (%)
0-10	No discomfort or effect	0-2
10-50	Some tiredness, impaired vigilance and reduction in manual dexterity	2-10
50-100	Slight headache, tiredness and irritability	10-20
100- 200	Mild headache	20-30
200-400	Severe headache, visual impairment, nausea, general weakness and vomiting	30-40
400-600	As above, but with greater possibility of collapse	40-50
600-800	Fainting, increased pulse rate and convulsions	50-60
800-1600	Coma, weak pulse and possibility of death	60-70
1600 +	Death within a short period	70+

(**Source:** *American Industrial Hygiene Association Air Pollution Manual, 1960.*)

(a) After 2 h exposure. (b) Likely percentage of carboxyhaemoglobin (COHb) to total haemoglobin (Hb) content of blood sometime afterwards in a resting individual but which may be achieved three times faster during heavy work.

1.3.5.5.1 Blood biochemistry and pollution

Haemoglobin, the red pigment of RBCs, consists of the protein globin and a *tetrapyrrole* ring system called haem which contains ferrous iron. If this iron is oxidized to the ferric form, the haemoglobin turns into methaemoglobin which is unable to bind O_2. Oxidation of haemoglobin to methaemoglobin occurs naturally as RBCs age and is accelerated by the additional presence of H_2O_2. Normally, free radical scavenging ensures that this does not occur prematurely but free radical generation by other pollutants such as O_3 or PAN promotes early formation of methaemoglobin. Ultimately, these older RBCs high in methaemoglobin are broken up by certain white blood cells while they pass through the spleen.

During normal respiration, O_2 combines with haemoglobin to form oxyhaemoglobin, an association-dissociation reaction which is dependent upon the partial pressure of O_2, the temperature and pH. Changes of blood acidity are largely due to alteration in blood levels of dissolved CO_2 and the equilibrium between the different forms of CO_2.

$$CO_2 + H_2O\ H_2CO_3 \Leftrightarrow H^+ + HCO_3$$

The enzyme carbonic anhydrate brings about a faster interconversion between dissolved CO_2 and carbonic acid while H ions from the dissociation of carbonic acid cause the dissociation of oxyhaemoglobin.

Lactate, which is produced by contracting muscles working at low O_2 levels, also produces acidity which then causes more O_2 to be released from oxyhaemoglobin to alleviate the local shortage. The final factor that encourages the release of O_2 from haemoglobin is 2, 3-diphospho-glycerate. This substance combines reversibly with haemoglobin to change its structure in favour of O_2 release. Consequently, 2, 3 diphospho-glycerate is released by tissues which are especially short of O_2 to ensure that O_2 is delivered to those cells most in need.

About 7% of CO_2 produced by body tissues is dissolved in the blood plasma and carried to the lungs as a dissolved gas. The bulk is transported as bicarbonate and the remainder reacts with haemoglobin to form carbaminohaemoglobin. At the high partial pressures of CO_2 in the tissue capillaries, the formation of this complex is encouraged, but in the lungs where partial pressures of CO_2 are low it splits apart again.

CO combines with haemoglobin with an affinity which is between 200 and 300 times greater than the affinity of haemoglobin for O_2. The resultant carboxyhaemoglobin is extremely stable. CO at very low concentrations in the blood will combine with over half of the haemoglobin and immediately reduce the O_2-carrying capacity by a similar proportion. Only high O_2 partial pressures such as those achieved when treating CO poisoning cases with pure O_2 can reverse the combination of haemoglobin with CO. Other O_2-releasing mechanisms may also be affected adversely by CO. For example, levels of 2, 3-diphospho-glycerate in plasma circulating through O_2-deficient tissues are lower in the presence of CO.

With the bloodstream carrying less oxygen, brain functions are affected and the heart rate increases in an attempt to offset the oxygen deficit.

The amount of carboxyhaemoglobin in the blood is normally expressed as a percentage of the saturation level, % CO Hb. Three factors: the CO concentration, the length of time exposed, and the breathing rate control the extent of CO Hb formation in blood.

They are related by the following equation

$$\% \text{ COHb} = 0.005\ [\text{CO}]^{0.85}\ (\alpha t)^{0.63} \qquad ....(1.15)$$

where % COHb = carboxyhaemoglobin as a percentage of saturation

[CO] = carbon monoxide concentration in ppm

α = a physical activity level coefficient

t = exposure time in minutes.

For low activity, α = 1, while for heavy work, a factor of 3 is sometimes used. Equation 1.15 breaks down for large values of time, corresponding to the bloodstream reaching saturation.

Problem 1.5

CPCB standard for CO: Calculate the % COHb expected to form for a 1 h exposure to 40 ppm (the CPCB standard) for low activities, and for heavy work.

Solution: From Equation 1.15, for low activities with α = 1

$$\% \text{ COHb} = 0.005 \times (40)^{0.85} \times (60)^{0.63} = 1.52$$

Assuming α = 3 for heavy activity

$$\% \text{ CO Hb} = 0.005 \times (40)^{0.85} \times (3 \times 60)^{0.63} = 4.56$$

At small percentages of COHb, certain physiological effects can be observed whose severity increases with increasing concentration. Heart patients will be the worst sufferers since to offset the reduction in oxygen, the heart has to work under more stress.

The brain functions to perceive and react will also be greatly affected by the reduction of oxygen in the bloodstream. Impairment in time-interval discrimination was observed at 2.5% COHb while at 5%, psychomotor response times were affected. Patients with heart disease experienced increased physiological stress. Many people will experience dizziness and headache at 10% COHb while concentrations above 50% can be lethal.

In urban areas CO levels vary from 5 to 50 ppm; it is estimated that in congested highways drivers may be exposed to CO concentrations upto 100 ppm. Cigarette smoke contains more than 400 ppm CO. Smokers frequently have COHb levels between 5 to 10%. Further cigarette smoke in bars and restaurants often raises indoor CO levels to 20-30 ppm, which is close to the 1 h ambient standard [4].

It can be seen from Fig. 1.33 which depicts the relative accumulation of COHb with different exposure times of CO that within 3–4 h COHb reaches 50% of its saturation value. CO will be removed from the bloodstream when clean air is breathed and about half of the CO from the blood will be removed in the breathing process within 3-4 h. So the physiological effects caused by CO exposure will be temporary.

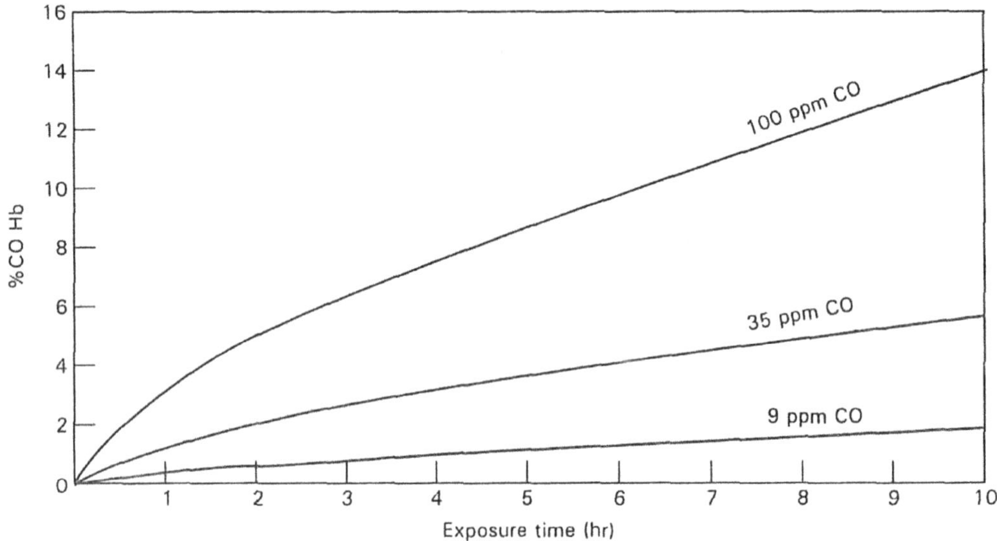

Fig. 1.33 *Percent CO Hb plotted for different exposure times with varying concentrations of CO*

Exposure of humans to CO may cause a variety of clinical effects (Table 1.21). Generally the onset of responses is hastened by heat, humidity and exertion. At work, it is the younger rather than the older workers who are more susceptible to CO, but alcoholism, obesity, old age, heart conditions, and lung diseases all exacerbate the problem. However, some physiological compensation or conditioning occurs with repeated exposures in steelworkers and smokers.

As COHb levels or duration of exposure increase, health effects become more serious

COHb level, percent	Demonstrated effects
Less than 1.0	No apparent effect
1.0 to 2.0	Some evidence of effect on behaviourial performance
2.0 to 5.0	Central nervous system effects impairment of time interval discrimination, visual acuity, brightness discrimination and certain other psychomotor functions
Greater than 5.0	Cardiac and pulmonary functional changes
10.0 to 80.0	Headaches, fatigue, drowsiness, coma, respiratory failure, death

There are clinical differences between O_2 deprivation and CO poisoning. In the former, respiratory symptoms precede nervous symptoms, but in the latter this order is reversed. This has led to the suggestion that the driving out of O_2 from RBCs is not the only physiological effect of CO.

The progressive symptoms listed in Table 1.21 approach without warning because CO is non-irritating and odourless. Only the presence of other contaminants such as hydrocarbons, mercaptans, ammonia or traces of other materials gives an advance warning of any hazard within fumes. It was a long time before CO poisoning became universally accepted as a specific disease in the industrial and urban environments because many of the early symptoms caused by CO are also associated with repetitive work or stressful situations. It is now recognized that misjudgments due to enhanced blood CO levels lead to more accidents at work, on the road, and in the home.

Unfortunately, the affinity of CO for human fetal haemoglobin is higher than that for normal haemoglobin. This means that unborn babies in the womb are especially sensitive to CO poisoning. Undoubtedly, the single most important factor which exposes an unborn child to greater than average CO concentrations is smoking by the mother. The relationship between maternal smoking and low birth weight is undeniable. Most of the number of developmental and clinical defects in the newborn ascribed to deprivation of O_2 are mostly neurological. However, as with adults, it is still not possible to distinguish in these babies the direct effects of CO caused by O_2 deprivation from those caused by other factors in tobacco smoke.

Related to CO poisoning is the disease Shinshu myocardiosis. This occurs in those exposed to high levels of CO over long periods and is typified by defective heart valves as well as by angina pectoris and arteriosclerosis. It was first detected in 1955 in the village of Kinasa, Japan where, during cold winters, it had been the custom for silk production to continue in tightly sealed, heated rooms. As a result, many of the workers experienced severe circulatory problems due to CO poisoning.

The effects of carboxyhaemoglobin levels and duration of exposure on human health are shown in Fig. 1.34. Carbon monoxide concentrations of 50-100 ppm are commonly encountered in the atmosphere of crowded cities, especially at heavy traffic rush hours. For example, in parts of New York city, the 8 AM and 4 PM concentrations of CO are respectively 86 and 76 ppm. In tunnels, CO concentrations may exceed I00 ppm. Such concentrations adversely affect people's ability to drive and can cause accidents.

The WHO recommends that levels of CO be reduced so that the level of carboxyhaemoglobin in the blood of the general population does not exceed 3%. This means that exposures should not exceed 50 ppm over 30 minutes, 25 ppm over 1 h, and10 ppm over 8-24 h.

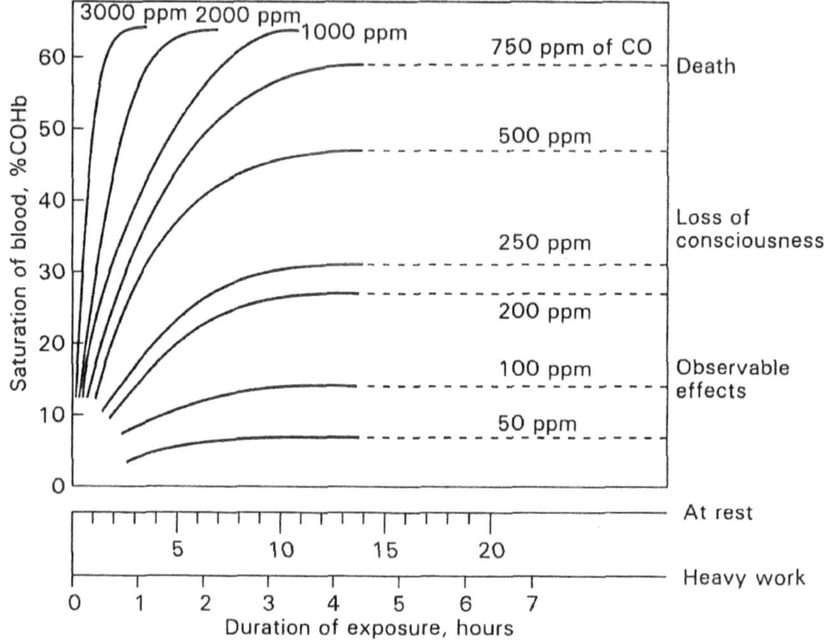

Fig. 1.34 *COHb levels in the blood depend upon the amount of CO in the atmosphere, duration of exposure, and type of physical activity*

1.3.6 Hydrocarbons

As their name indicates, the components of hydrocarbons are hydrogen and carbon; hydrocarbons constitute the major chemicals in petrol and other petroleum products. They are a nuisance and a health hazard, mainly because of their photochemical reactions. In the presence of sunlight, nitrogen dioxide reacts with hydrocarbons to give a series of extremely complex reactions; chief among the products of these reactions are peroxyacetyl nitrate (PAN) and ozone:

$$\text{Hydrocarbons} + NO_2 + O_2 + \text{sunlight} \rightarrow \text{PAN} + O_3 + \text{etc.}$$

Historically, the hydrocarbons have been regarded as important to human health only through their role in the formation of photochemical oxidants. With the exception of the aromatics, hydrocarbons have been considered toxicologically inert. In the 1980s, the USEPA began a major effort to identify and regulate toxic organic compounds that fall into the hydrocarbon category. Many of the polynuclear aromatic hydrocarbons (PAHs) have been recognized as carcinogens for many years. Other air toxics, including pesticides, polychlorinated biphenyls (PCBs), and dioxins, have also been identified.

1.3.6.1 *Sources of emission of hydrocarbons*

The various sources and their contribution to the emission of global hydrocarbons are given in Table 1.22. Natural sources, particularly trees emit huge quantities of hydrocarbons into the air. Plants, mostly emitting terpenes belong to the families *Coniferae* and *Mystaceae* and the *genus Uterus*. Automobile exhausts emit the maximum amount of hydrocarbons into the atmosphere. Some industrial sources, especially refineries, emit hydrocarbons, but the major source of this type of pollution is automotive emissions. The emissions from incomplete combustion in car engines, along with evaporative emissions from fuel tanks, carburettors, and crank cases, amount to approximately 12 million tons per year. Human activities contribute nearly 20% of the hydrocarbons emitted to the atmosphere every year; animals contribute about 80-85million tones of methane to the atmosphere every year.

Table 1.22 *Global emissions of hydrocarbons*

Source	Emission (%)
Petroleum	55.05
Gasoline	38.5
Refining	7.2
Oil	0.4
Evaporation of solvents in storage	8.8
Coal	3.3
Wood (fuel and forest fire)	2.2
Industrial uses	0.8
Organic solvent evaporation	11.3
Incinerators and refuse burning	28.3
Power generation	0.2

Among the various hydrocarbons involved in air pollution, 565 have been clearly identified by gas chromatography. The emission trends of the volatile organics in a typical highly industrialized country like USA with a massive transportation network is shown in Fig. 1.35. It indicates declining trends after the implementation of the clean air regulations of 1990 in USA. From different source categories.

1.3.6.2 *Effects of hydrocarbons on plants*

Exposure to hydrocarbons and photochemical oxidants is injurious to plants and causes chlorosis. Ozone enhances plant injury creating light flecks or stipples on the upper leaf surface inhibiting photosynthetic activity of leaves. Ethylene even at 1 ppm concentration shows adverse effects on vegetation. Acetylene and propylene at 50-500 ppm show extreme toxicity towards plants, damaging growth of vegetation.

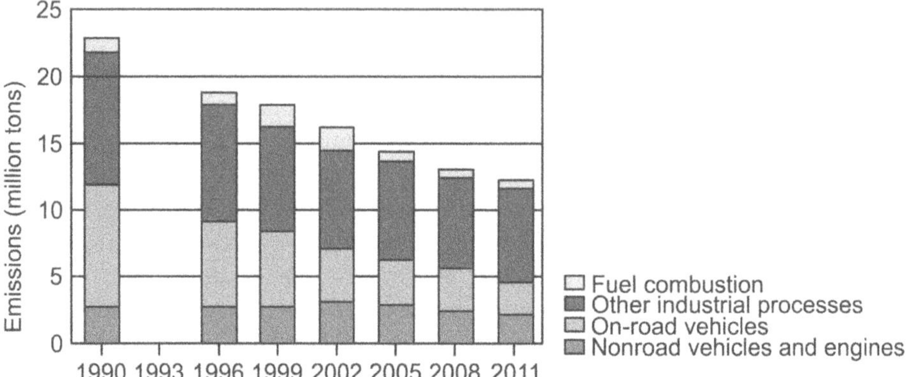

Fig. 1.35 *Trends in volatile organic compound emissions in USA, 1990-2011 by Source Category (USEPA)*

1.3.6.3 Effects of hydrocarbons on human health

Hydrocarbons at high concentrations (500-1000 ppm) have carcinogenic effects on lungs. They mainly cause swelling when they enter the lungs. Aromatic hydrocarbons like benzene, toluene etc., are more dangerous than acyclic and alicyclic hydrocarbons. The inhalation of their vapours irritates the mucous membrane.

Secondary pollutants (PAN) produced by hydrocarbons and NO_x result in the formation of photochemical smog which causes irritation of eyes, nose, throat and respiratory distress. Excess of hydrocarbons increases mucous secretion as a result of which the respiratory tracts are blocked causing coughs. Continuous cough causes pressure in the trachea of the lungs due to which the lining membrane of alveoli bursts.

1.3.6.4 Carcinogens

Carcinogens such as polycyclic hydrocarbons are present in the exhaust emissions of internal combustion engines, diesel or petrol. The two major carcinogens are benzopyrene, which is a strong cancer-inducing agent, and benzanthracene, which is a weak one. They are essentially non-volatile organic compounds associated with solids or polymeric substances in the air. Benzopyrene is present in trace amounts into bacco, charcoal, boiled stacks and gasoline exhausts, etc.

These compounds are not very stable, and they are destroyed at varying rates by other air pollutants and by sunlight. However, as a result of industrialization and urbanization, these substances are discharged into the atmosphere in significant quantities, causing a steady increase in the frequency of human lung cancer in the world. The concentrations of carcinogens in the New York air which has a large transport network are given in Table 1.23.

Table 1.23 *Concentrations of carcinogens in New York air*

Location	Benzopyrene Annual Avg.	μg/1000 m^3 Max.	Benzanthracene Annual Avg.	μg/1000 m^3Max.
Herald square	3.9	9.4	5.6	16.0
Scarsdale, Westchester Country (suburb)	0.3	0.6	0.6	1.2

Source: *Air pollutants and their effects on human health animal and plants. Esber. I. Shaheen Tech. of Air. Poll. Control, Pennwell Books, 1992*.

1.3.6.5 *Effects of hydrocarbons on materials*

Hydrocarbon pollutants damage long chains of carbon atoms causing loss of tensile strength of polymers. Ozone forms new carbon chain links between parallel carbon chains so that the material becomes less elastic and more brittle.

1.3.7 Ammonia

Ammonia (NH_3), which is a pungent gas, is used as a raw material in large quantities by industries for the synthesis of ammonium nitrate, plastics, explosives, dyes and drugs. It is also used as a refrigerant.

1.3.7.1 *Sources of emission of ammonia*

Emission of NH_3 into the atmosphere from the biological degradation of proteins on soil surfaces occurs on a very large scale. This is known as 'NH_3 volatilisation' and compared to this, the industrial contribution is negligible. Atmospheric concentrations of NH_3 in temperate rural regions range from 5 to 10 ppm but are much higher near the equator. In urban regions, higher levels of NH_3 upto 280 ppm are recorded and it may be found in increasing levels close to industrial and intensive agricultural sources. In some developed countries, atmospheric levels of NH_3 are still rising with the accelerated use of artificial fertilizers and higher stocking rates of farm animals.

NH_3 readily forms cations or complexes of varying stability which affects the rate of volatilisation. Most important is the affinity of NH_3 for H_2O that results in ammonium ions – this is strongly enhanced by increased alkalinity. This means that there is an increased likelihood of NH_3 volatilisation at high pH.

$$NH_3 + H_2O \rightarrow NH_4^+ + OH^-$$

The variation of CO_2 levels or changes in temperature will have a marked influence on the NH_3–NH_4^+ relationship as both ionization of H_2O and the dissociation of $NH_3$are temperature dependent. Furthermore, the exchange of NH_3 between solution and the air above varies markedly with temperature.

In soils, NH_3 may be either adsorbed onto clay or organic particles and react with carbonyl and other acidic groups to form exchangeable salts, or it may combine with other organic components to form non-exchangeable products. In view of this, different soils have different rates of NH_3 volatilisation and these, in turn, are affected by their water contents, as OH^- ions are removed in the process of conversion of NH_4^+ to NH_3(the reaction $NH_3 + H_2O \rightarrow NH_4^+ + OH^-$ in reverse). As NH_3 is volatilised from the soil surface to the atmosphere, the soil solution becomes acidified at rates which depend on the soil buffer capacity. In view of this, volatilisation of NH_3 is more likely from soils where the acidity produced can be neutralized by high levels of carbonate or other forms of alkalinity. This explains why larger emissions of NH_3 occur from natural calcareous soils or after liming.

The anions present in applied fertilizers also affect the rate of NH_3 volatilisation from soils. Urea is frequently used as an alternative fertilizer because the enzyme urease, which is widely distributed in plants, microbes and soils, catalyses the hydrolysis of urea to bicarbonate and NH_4^+. As urease activity tends to be greater in soils with large organic contents and rather less in calcareous soils, the usual increase of NH_3 volatilisation from alkaline soils and decrease from acid soils is reversed when urea is used as a fertilizer.

NH_3 volatilisation also takes place at a higher rate from stored animal manures in stockyards or from sewage works. The factors involved are similar to those of applied slurries on fields. Periodic addition of fresh material to the tops of piles of manure or the agitation of sewage ponds, therefore, greatly accelerates rate of NH_3 volatillsation. Most NH_3 in the atmosphere arises from the direct hydrolysis of the urea in animal urine, other contributions being of less importance.

In Europe, as much as 10% of useful N is lost directly by NH_3 volatilisation and, in warmer climates, this can rise to as much as 30%. The amounts of N released into the atmosphere globally by NH_3 volatilisation are very large – between 115 and 245 million tons of nitrogen oxide. The background atmospheric levels of NH_3 over Belgium, Denmark, and the Netherlands, which are intensive arable and livestock-raising countries, are often around 25 ppm with peaks up to 75 ppm. The estimates of total N released into the atmosphere from these countries are especially high.

The large releases of NH_3 to the atmosphere over the last three decades are due to

(a) increased animal stocking levels,
(b) increased human population,
(c) increased use of artificial fertilizers, in the form of either NH_4^+ nitrate or urea, and
(d) decreased sinks for NH_3 or NH_4^+ uptake.

The first three go hand in hand. As material standards improve, humans move from plant-oriented to animal-based diets. This means more food has to be grown to feed animals and this can only be done by using more artificial fertilizers.

Much could be done to reduce N losses associated with the application of N fertilizers due to NH_3 volatilisation and run-off of excess nitrate into groundwater. Direct injection of anhydrous NH_3 or urea at the right depth into soil has not yet been extensively exploited. Even substituting urea for NH_4^+ in irrigation waters will reduce losses due to NH_3 volatilisation to below 2% in poorer regions where N is unduly expensive, and losses by this route tend to be greater because of higher temperatures.

1.3.7.2 *Removal mechanisms of ammonia*

Once in the atmosphere, NH_3 neutralizes sulphuric or nitric acids and, by decreasing acidity, promotes the oxidation of SO_2 to sulphate by O_3. Normally, atmospheric NH_3 has an average lifetime of 0.5 h before conversion to NH_4^+. At wind speeds of 10 m s^{-1}, therefore, a molecule of NH_3 travels about 18 km before it turns into NH_4^+.

Measurement of rates of NH_3 deposition are complicated because some intensively farmed lands give off more NH_3 than they receive. On the other hand, fluxes towards damp acidic ecosystems are considerable and cannot be accounted for by stomatal uptake alone as they form perfect sinks for NH_3. For example, the uptake rate of wet heathland in the Netherlands may be as high as 100 kg N ha^{-1} a^{-1}.

In most areas where atmospheric levels of NH_3 are high, elevated concentrations of SO_2 also occur. This leads to a phenomenon known as code position where fluxes are linked together. One explanation for this is that when NH_3 turns into NH_4^+, then oxidation of SO_2 into sulphate is enhanced. Removal of SO_2 from the atmosphere as ammonium sulphate at the surface then ensures that the high rates of code position of SO_2 and NH_3 are maintained. Similar processes occur in aerosol particles which form clouds. The hygroscopic nature of NH_4^+ and sulphate causes these particles to attract more water vapour so that they increase in size and eventually fall as rain, etc. Indeed, wet deposition of NH_4^+as ammonium sulphate in rainfall is the major removal pathway of NH_3 from the atmosphere.

1.3.7.3 *Effects of ammonia on plants*

Most plants are affected visibly by atmospheric NH_3 when there has been an industrial spillage and high local concentrations are achieved. Generally, no visible or invisible effects are detected on even the most sensitive species at levels of 70 ppb NH_3 or below. However, there are some claims that deposition of NH_3 and NH_4^+ on certain plants that prefer low levels of N induces leaching of potassium and magnesium which, over the long term, leads to mineral deficiencies.

At higher levels of NH_3, little effect on photosynthesis is detected but long-term exposures visibly injure sensitive conifers such as yew, spruces, and cypresses. However, this injury is usually associated with reduced frost hardiness and increased susceptibility to fungal attack. Vegetable and horticultural crops are generally more resistant. Cauliflower and Brussels sprouts, however, may show characteristic black spots at levels of 0.7 ppm NH_3 for 10 days as well as reduced frost hardiness.

The ecosystems most sensitive to nitrogen-based atmospheric compounds are those which contain species of plants which are specially adapted to low levels of N and nutrients (Table 1.24). There is increased competition from faster-growing plant species which were previously restricted in these ecosystems because of the low availability of N.

Ultimately, this process leads to the disappearance of the original plant species characteristic of such ecosystems –a loss of irreplaceable genetic resources.

Table 1.24 *Ecosystems thought to be especially sensitive to excessive inputs of nitrogen*

Type		Details
Wetlands	1.	Ombrotrophic mires i.e. bog-like areas only supplied with nutrients from rainfall, e.g. raised bogs
	2.	Mires in granitic areas or otherwise nitrogen-limited
Lakes	1.	Clear water lakes partially covered with certain species, e.g. water lobelia, quillwort, shore weed
	2.	Nutrient-lacking lakes with a high number of pond weeds
Others	1.	Heathlands with a high amount of lichen cover with limited mineral availability
	2.	Meadows with limited mineral availability used for extensive grazing and haymaking to which artificial fertilizers have never been added
	3.	High-altitude coniferous forests

Fears of too much N also reaching natural forest stands by similar processes have also been raised. Indeed, one possible explanation of recent forest decline suggests that too much N from the atmosphere encourages pathogens.

Vegetational changes due to excessive N input also exert changes upon animal populations. Fish are starved of O_2 because of competition with rapidly growing plants. Similarly, vigorous attacks of bark beetles occur in N-enriched forests.

1.3.7.4 Effects of ammonia on human health

Atmospheric NH_3 is not normally a hazard to health. Occasionally, industrial spillage releases large amounts of NH_3 which immediately reacts with the moist linings of the throat or the surface of the cornea to form ammonium hydroxide which then causes chemical burns. At much lower levels, the buffering of these layers is normally sufficient to absorb any NH_3 and very little penetrates into the lungs. Irritation of the eyes and throat occurs at levds of 350-700 ppm NH_3–well above ambient atmospheric levels. However, physiological changes have been detected at much lower levels. For example, at 16 ppm NH_3, levels of NH_4^+ and urea in the bloodstream are increased. The TLVs vary from country to country. Most are around 25 ppm NH_3 for an 8 h day.

1.3.8 Organic lead

1.3.8.1 Sources of emission of organic lead

Lead is discharged into the atmosphere in the organic form as tetraethyl lead ($(CH_3CH_2)_4Pb$) or trim ethyl lead ($(CH_3)_3Pb$) in unburnt or partially combusted fuel vapours. The amounts involved in developed countries were huge. In the late 1960s, for example, the amounts were 181 KT (Kilo Tons) of lead in the whole of the USA and 5 KT of lead in Los Angeles alone.

1.3.8.2 US-National trends in lead levels

Under the Clean Air Act, EPA sets and reviews national air quality standards for lead. Air quality monitors measure concentrations of lead throughout the country. EPA, state, tribal and local agencies use those data to ensure that lead is at levels that protect public health and the environment. EPA has tracked air quality trends for lead using data from this network of monitors. Nationally, average lead concentrations decreased dramatically after EPA's regulations reduced the lead content in on-road motor vehicle gasoline. The trends in lead emissions in USA in the period 1980-2015 are shown in Fig. 1.36.

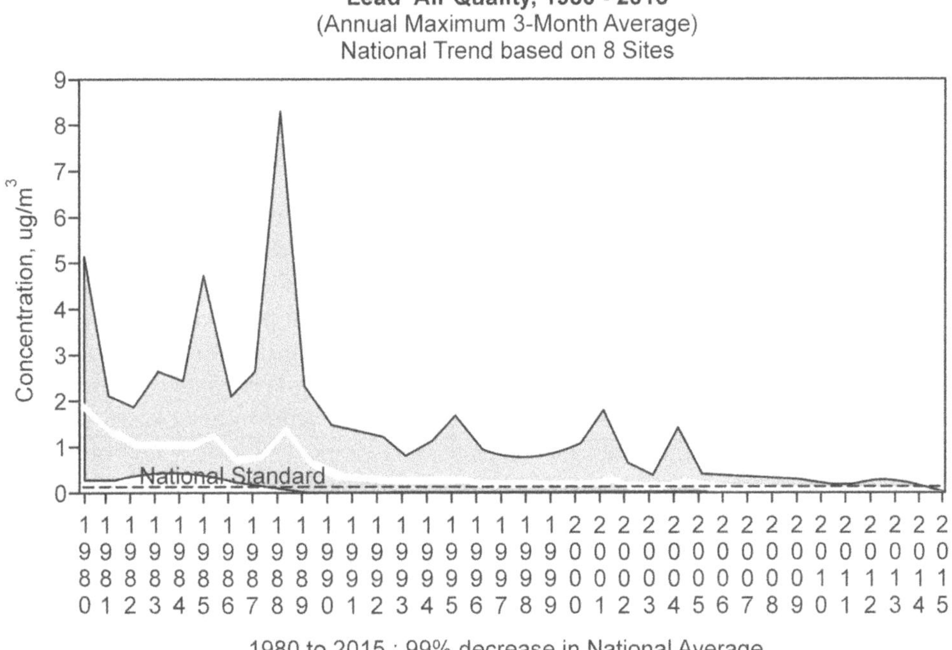

Fig. 1.36 *Trends in lead emissions in USA, 1980-2015 (USEPA,)*

In most developed countries, legislation to limit the amounts of tetraethyl and trimethyl lead, used as anti-knocking agents in vehicle fuels, has been progressively introduced and newer vehicles have been redesigned to use low lead or unleaded fuels. This has significantly reduced that fraction of lead entering the environment by airborne emissions in developed countries to less than one-tenth of the highest overall total in the past (Fig.1.36). However, the situation is unchanged or worsening in poorer developing countries that cannot afford the additional refining costs or newer vehicles.

1.3.8.3 Effects of lead on human health

Human intake of lead is mainly through food, especially from vegetables grown near busy roads. Plants take up very little lead from the soil because it is usually insoluble and tightly bound to soil particles. Most lead in plants enters through the stomata inorganic form. Once inside, some persists in organic form and the remainder is converted to inorganic lead. In certain circumstances, lead levels in lettuces, for example, approachl mg g-I which means that food intakes amounting to 300 mg of lead day^{-1} may be achieved. Of this, less than one-tenth (15-30 mg day^{-1}) is taken into the bloodstream from the digestive system.

A much higher proportion of lead (30-50%) is likely to directly enter the bloodstream by breathing for people living or working near busy roads, where airborne lead levels might range from 1 to 5 ppb. People who breathe 20 m^3 of air per day are likely to absorb 6-50 mg lead day^{-1} from their lungs into the blood stream--as much again as they might take up from their diet. Consequently, the current TLV for total airborne lead set at 0.15 ppb is frequently exceeded and the direct uptake of airborne lead often matches the dietary intake, although more usually it is 20-30% of the diet in the bulk of the population not living immediately near busy roads.

In contrast to the other major pollutants, lead is a cumulative poison. Another difference is that it is ingested through food and water, in addition to being inhaled. Of the portion ingested through food, approximately 5-10% is absorbed by the body. About 20-50% of the inspired portion is absorbed. The un absorbed portions are excreted in the faeces and urine. Lead is measured in the urine and blood for diagnostic evidence of lead poisoning. An early manifestation of acute lead poisoning is a mild anaemia. Fatigue, irritability, mild headache, and pallor in distinguishable from other causes of anaemia occur when the blood level of lead increases to 60-120 mg/100 g of whole blood. Blood levels in excess of 80 mg/100 g result in constipation and abdominal cramps. When an acute exposure results in blood levels of lead greater than 120 mg/100 g, acute brain damage (encephalopathy) may result. Such acute exposure results in convulsions, coma, cardiorespiratory arrest and death. Acute exposure may occur overa period of one to three weeks. Chronic exposure to lead may result in brain damage characterised by seizures, mental incompetence, and highly active, aggressive behavior. Weakness of extensor muscles of the hands and feet and eventual paralysis may also result. Atmospheric lead occurs as a particulate. The particle size range is 0.16-0.43 mm.

Symptoms due to organic lead differ from those of inorganic lead. Most effects of tetraethyl and trimethyl lead are upon the nervous system. They first show as sleeplessness and general irritability but, following heavy exposures, they progress to emotional instability and hallucinations accompanied by impaired vision and hearing. More often, at low persistent levels, they cause headaches, general fatigue and sometimes depression.

Inorganic lead poisoning also starts with fatigue and irritability but it is often accompanied by anaemia and abdominal pain. The anaemia is mainly due to the fact that one of the early enzymes of haem synthesis, d-amino-aevulinate dehydratase, is specifically inhibited by inorganic lead while RBCs show increased fragility and have shortened life spans. Chronic low level exposure of adults to inorganic lead may increase blood pressure and in children, disturb the metabolism of vitamin D which affects the growth of their long bones. Kidney disturbances are also associated with inorganic lead but the general slowing of nerve impulses occurs with both inorganic and organic forms of lead.

1.3.9 Hydrogen sulphide

The offensive smell of hydrogen sulphide (H_2S), well known as the odour of rotten eggs renders it a highly disturbing and objectionable pollutant.

1.3.9.1 Sources of emission of hydrogen sulphide

Hydrogen sulphide is emitted into the atmosphere by the degradation of industrial wastes in stagnant waters, swamps, and other areas where bacterial action reduces sulphur compounds to hydrogen sulphide, which is highly insoluble in water. It is estimated that decaying organic matter in the world emits 70 million tons of H_2S. Industrially, this gas comes from sewage treatment plants and the petroleum industry. The global emissions of sulphur by various activities are summarised in Table 1.25.

Table 1.25 *Total global emissions of sulphur[a] from both anthropogenic and natural sources*

Component	10^{12} mol y^{-1}	Percent
Anthropogenic[b]	2.40	73.9
Biomass burning[c]	0.07	2.1
Biogenic Marine[d]	0.48	14.8
Terrestrial[e]	0.01	0.3
Volcanic[f]	0.29	8.9

a--excludes sea-salt sulphate transfer, b-mainly SO_2 and carbonyl sulphide (OCS), c-not human-induced (the 95% that is initiated by humans is included in the component above), d-oceanic emissions are mainly CH_3SCH_3 (88%), the remainder being H_2S (<10%), and OCS plus CS_2 (<2%), e—emissions from land are mainly (>60%) from tropical regions (20°N-20°S) and consist primarily of OCS (47%),CH_3SCH_3 (27%) and H_2S (20%). f–mainly SO_2 and sulphates (>98%). (**Source:** *Bates, T. S., Lamb,B. K. Guenther, A., Dignon, J. and Stoiber, R. E. Sulphur emissions to the atmosphere from naturals ources, Journal of Atrn. Chemistry, 14,1992,315-337.*)

The standard levels of various countries are presented in Table 1.26.

Decay of organic matter in sewers and breakdown of animal and plant wastes by microbial activity are other major sources of H_2S and organic sulphides. Tanneries, glue and fur-dressing factories, abattoirs, waste treatment plants, and sugar-beet processing all produce significant quantities of H_2S which accounts for the fact that these are some of the industries least wanted in the neighbourhood. Tanning is very prone to accidental discharges because the first stage of hair removal uses a paste of sodium sulphide and the following chrome tanning process uses sulphuric acid. If care is not taken to prevent the two effluents from mixing, H_2S is liberated. The manufacture of paper, rayon and S dyes, and the vulcanising of rubber, are other industrial processes where H_2S may be generated. Special safety precautions are required in all these industries to protect workers. In areas around pulp mills, levels of H_2S may be as highas 11.5 ppm; even higher levels have been recorded inside them. However, in urban areas, levels of H_2S are normally lower than 5.4 ppb.

Table 1.26 *Standards for H_2S*

Country	State or town	Type of regulation	Max, ground level 30 min concentration, µg/m^3	
U.S.A	California New York State	Adverse level Standard		150 150
Germany		Standard		150
			(30 min in) 8 h up to 300)	
U.S.S.R		Standard	8 (20 min)	8
Czechoslovakia		Standard	15	5

The human threshold of odour detection for H_2S is between 0.15 and 1.5 ppb but these limits are lower for mercaptans. To a certain extent, sensitivity depends on the individual but the rotten egg smell is clearly apparent to almost everyone at about three times these levels. H_2S is almost always accompanied by other malodsrous substances such as methyl mercaptan, carbon disulphide, dimethyl monosulphide, and dimethyldisulphide. The 'quality' of the odour is changed accordingly – presumably from 'bad' to 'worse'.

Microorganisms of the soil, sewage treatment plants, as well as the plankton of the oceans, produce considerable quantities of CH_3SCH_3, CH_3SSCH_3, and CH_3SH. Estimates suggest that biogenic emissions of CH_3SCH_3, exceed those of H_2S and that a large proportion of these come from the oceans.

1.3.9.2 Effects of hydrogen sulphide on human health

Normally, the smell of H_2S is sufficient to warn humans of the presence of this gas at levels below 20 ppb and to cause them to retreat or to take remedial action. However, paralysis of the sense of smell occurs at higher levels. So victims may be unaware of the dangers (Table 1.27). High concentrations of H_2S are actually as toxic as hydrogen cyanide, HCN. In fact, both have the same TLV of 100 ppm and have similar effects. Respiratory failure occurs within seconds due to paralysis of the nervous control of breathing.

The toxicological problems associated with H_2S, like those of HCN, are caused by an inhibition of electron flow through cytochrome oxidase to O_2 in the mitochondria during respiration. The skin of victims of acute intoxication caused by H_2S is usually grey-green in colour as are the internal organs. Most of the metabolic consequences of sub lethal exposure to H_2S stem from this partial inhibition of cytochrome oxidase in mitochondria. This also increases the number and volume of RBCs. However, the main clinical effect is irritation of the lung passages and eyes which leaves the victim with pneumonia and conjunctivitis a few days after the exposure (Table 1.27).

Table 1.27 *Effects of hydrogen sulphide on humans*

Exposure range	Effects and symptoms
0-0.2 ppb	No discomfort
0.2-1.2 ppb	Odour detected
1.2-5.4 ppb	Consciousness of 'bad egg' smell
5.4 nl l^{-1} –10 ppm	Some discomfort and headache
10-20 ppm	Eye irritation threshold
20-50 ppm	Severe eye irritation and impairment
50-100 ppm	Characteristic eye damage called 'gas eye'
100-320 ppm	Loss of sense of smell, nausea and increased lung irritation
320-530 ppm	Lung damage and water accumulation
530-1000 ppm	Shortage of breath, stimulation of respiratory centre, convulsions, and chance of respiratory arrest
1000 ppm and above	Immediate collapse and respiratory failure, followed by death

Recovery from acute poisoning is usually without aftereffects but some cases show persistent clinical symptoms which result from the initial O_2 deprivation. TLVs for maximum exposure to H_2S set at 10 ppm H_2S for 8 h daily exposures would appear to be too high because these levels are also known to cause eye irritation.

Chronic problems due to exposure to H_2S are conjunctivitis, headaches, dizziness, diarrhoea, and loss of weight. Early French medical writers used the phrase *plomb desfosses* to describe the colic and diarrhoea of Paris sewermen probably from such a cause, which resembled that normally associated with lead poisoning.

1.3.10 Fluorides

Hydrogen fluoride, HF, is a highly corrosive and irritant gas. A typical fluoride concentration in the atmosphere is 0.05 mg/m^3. Because of its extreme toxicity, HF is a problem wherever processes involving fluorides take place, such as in the production of phosphate fertilizers, smelting of certain iron ores, and manufacturing of aluminium.

1.3.10.1 Sources of emission of fluorides

Fluorine is a gas so reactive that it does not occur naturally in elemental form. However, many fluoride-containing minerals such as fluorspar, cryolite, and certain apatites are used by industry. Some industries also produce HF either as a by-product or to form various useful fluoro-derivatives (Table 1.28).

Table 1.28 *Industrial and commercial processes involving fluorine compounds which may release fluoride and HF*

Emission Process	Processes using large amounts of fluorine-derivatives
Aluminium smelting	Clouding of electric bulbs
Steel production	Cut glass finishing
Phosphate fertilizers	Aviation fuel production
Enamel and pottery manufacture	Insecticides and rodenticides
Brick making	Separation of uranium isotopes
Missile propulsion	Synthesis of plastics
Beryllium, zirconium, tantalum and niobium purification	Aerosol, refrigerant and lubricant manufacture
Cleaning of castings	Wood preservation
Welding	Cement reinforcing
Sandstone and marble cleaning	Furniture cane bleaching
Cryolite, fluorspar and apatite mining	Water supplementation

Industrial emissions are superimposed upon significant natural background sources. Consequently, levels in both air and water supplies vary widely. The majority of rural and urban air monitoring sites record very low levels of atmospheric fluoride measured as total dissolved fluoride. Near phosphate fertilizer plants, aluminium smelters, or volcanoes, however, levels may rise above 200 ppm. Water supplies around these areas may also show elevated levels, well above the 1 ppm recommended as an optimum to provide an 'acceptable' incidence of dental caries and at the same time allow for the correct bone growth of children.

1.3.10.2 Effects of fluorides on plants

Fluoride deposition on plants not only causes them damage but may result in a second untoward effect. Grazing animals may accumulate an excess of fluoride, which mottles their teeth and ultimately causes them to fall out.

Problems associated with fluoride in plants are well known in relationship to fluorosis in farm animals. Animals grazing on pasture very close to brick works, smelters and phosphate fertilizer factories, or fed forage gathered from such areas, may show fluorosis, a condition also occasionally found in humans. The major recommendation has been to ensure that the yearly average fluoride content of herbage does not exceed 40 mg $l^{-1}y^{-1}$.

Application of lime to crops and herbage has long been known to be a practical means of reducing the effects of fluoride injury. Originally, it was thought that the lime caused the immobilization of the fluoride on the surfaces of the leaves as insoluble calcium fluoride. However, calcium chloride spraying has an alleviating effect similar to that of lime and recent studies have shown that the remedy actually relies upon additional calcium entering the leaves to interact with the fluoride inside and redress any calcium imbalances in the regulatory processes.

1.3.10.3 Accumulation by plants

Crop loss in the USA due to fluoride is ranked fourth in importance after O_3, SO_2 and nitrogen-based air pollutants. However, on a weight for weight basis, fluoride is the most phytotoxic of all atmospheric pollutants. Injuries to the most susceptible plants occur at concentrations between 10 and 1000 times lower than those of other air pollutants. Rates of uptake of fluoride into leaves are also faster than those of any other pollutant and go on to cause problems to animals feeding upon these plants.

Both gaseous and particulate fluorides are deposited on plant surfaces and some penetrate directly if the leaf is old or weathered. Nevertheless, the main access into a plant is through the stomata. An important feature of fluoride uptake and transport in plants is that it is later carried in the transpiration stream towards the leaf tips or margins where it accumulates and phytotoxic effects usually develop. Plant species show wide ranges of susceptibilities to fluoride but environment factors, such as light, temperature, humidity, water stress, etc., all influence plant response. Young conifers, gladioli, peaches and vines are especially sensitive while tea and cotton are very resistant.

There are several mechanisms, which reduce fluoride levels in plants. These include shedding of individual leaves or surface waxes, leaching by rain, or volatilisation. Fluoride levels are often lowest during summer months because of more favourable meteorological conditions for better dispersal of fluoride pollution and greater turnover of leaves in grass swards during summer.

There are many reports of changes in photosynthesis, respiration or metabolism of amino acids, proteins, fatty acids, lipids, and carbohydtes in plants due to fluoride. Certain enzymes are modulated by the presence or absence of fluoride but these do not explain the wide range of metabolic changes known to occur. These are due to interactions between fluoride and calcium or magnesium. Calcium and fluoride together, for example, stimulate phosphate uptake which means that calcium adsorption sites on cell membranes are involved in response to fluoride, Cytoplasmic calcium is a ubiquitous regulator of cell metabolism and many, but not all, of its effects are mediated by a calcium-binding protein calmodulin, which in turn stimulates a variety of enzymes. Moreover, calcium ions are known to affect the transport selectivity of membranes with respect to other substances. Because of this, fluoride exerts an effect on various regulatory activities (see Table 1.29) and this probably explains why it is so phytotoxic at such low concentrations.

Table 1.29 *Physiological effects of fluoride on plants*

Process	Disturbance	Likely cation interaction
Respiration & carbohydrate Metabolism	Glycolysis inhibited	Mg^{2+}
	Pentose phosphate pathway enhanced	Mg^{2+}
	Unusual mitochondrial swelling	Mg^{2+}
	Oxidative phosphorylation reduced	Mg^{2+}
Photosynthesis	Unusual chloroplast structure	Mg^{2+}
	inhibited pigment synthesis	Mg^{2+}
	Increased PEPC activity	Mg^{2+}
	Reduced electron flow	Ca^{2+}
Amino acid & protein metabolism	Increased in free amino acids and asparagine	Ca^{2+}/Mg^{2+}
	Decrease in ribosome sizes	Ca^{2+}/Mg^{2+}
Nucleic acid metabolism	Changes in transcription and translation	Ca^{2+}/Mg^{2+}
Fatty acid and lipid metabolism	Increased esterase activities	Ca^{2+}/Mg^{2+}
	Decreased unsaturated/saturated ratios	Ca^{2+}/Mg^{2+}
Other metabolic changes	Increase in peroxidase activities	Ca^{2+}/Mg^{2+}
	Decrease in acid phosphatase activity	Ca^{2+}/Mg^{2+}
Transport and translocation	Altered plasma membrane ATPases	Ca^{2+}
Fruit development	Poor fertilization and seed germination	Ca^{2+}
	Reduced pollen tube growth	Ca^{2+}
	Reduced seed number and fruit size	Ca^{2+}

Fluoride also forms magnesium-fluorophosphate complexes and, consequently, many enzyme pathways are adversely affected by fluoride. Most reactions involving ATP, for example, require additional magnesium complex to function correctly. If these natural complexes are also disturbed by the presence of additional fluoride then key reactions are inhibited.

Normally, soils contain 20-500 mg g^{-1} fluoride but, because it has limited solubility in soil water, uptake by roots is relatively low, and there is little relationship between soil fluoride and total plant fluoride content. Consequently, atmospheric sources of fluoride are more important than fluoride in groundwater in determining the amount of fluoride in or on a crop.

1.3.10.4 *Effects of fluorides on human health*

Food and drinks are the most important sources of human fluoride intake. Normally, these contain below 1 mg l^{-1} of fluoride. Tea, fish and other sea-foods are heavily laden exceptions. Other vegetables and cereals grown in areas subjected to high fluoride emissions may also be enriched in fluoride. The various physiological effects of fluoride on humans are shown in Table 1.30.

Table 1.30 *Physiological effects of fluoride on humans*

Process	Disturbance
Carbohydrate metabolism	Glycogen levels depleted, glycogen turnover depressed, phosphorylase activity reduced.
Lipid metabolism	Activation of acetate inhibited liver lipases, activated certain esterases inhibited.
Mineral metabolism	Interference in iron uptake; sulphite and phosphate counteract the inhibiting effect of Ca^{2+} upon intestinal absorption.
Hormonal balances	Effect on parathyroid function*.

*Calcium levels are influenced by the parathyroid hormone produced by the parathyroid and a hormonal derivative of vitamin D (called 1,2,5-dihydroxyl-cholecacalciferol) found in the liver and kidneys – both raise blood serum levels of calcium. Release of calcitonin from the thyroid, however, causes enhanced calcification of the bone tissues which then reduces blood calcium levels again.

1.3.11 Radon

1.3.11.1 Sources of emission of radon

Radon, which occurs naturally, is the heaviest known gas. It is colourless, odourless, almost inert, soluble in water (especially at low temperatures), and radioactive. Radon gas comes from a decay process of underground uranium ore. It is by far the most important source of ionizing radiation to affect humans. In most developed countries, radon accounts for 40-50% of the total ionizing radiation received by the population.

There are 27 isotopes of radon ($^{200-226}Rn$) but only three have half lives longer than an hour (^{210}Rn, 2.4 h; ^{211}Rn, 14.6 h and ^{222}Rn, 3.82 days). Of these, ^{222}Rn is the most important and arises from the decay of ^{238}U (Fig. 1.37)^{222}. Rn also decays into a series of radionuclides known as radon daughters or progeny. Principal among these are ^{214}Pb (half-life, 26.8 minutes), ^{210}pb (22.3 years), ^{210}Bi (5 days) and ^{210}Pb (138.4 days). The final product is ^{206}Pb which is non-radioactive. Most of the decays involved (Fig. 1.37) release α particles but some (e.g. ^{210}Pb or ^{210}Bi) emit β radiation.

Radon occurs naturally, although it may also build up in homes that were built on land with radioactive fill. Radon is a dangerous source of radiation and it has been estimated that radon gas is responsible for about 20,000 deaths per year from lung cancer in USA. The Environmental Protection Agency estimates that 8 million homes have critical levels of radon.

Sources of radon such as ^{238}U and ^{227}Ra are concentrated in certain acidic igneous rocks with a low melting point such as granite. It is, however, the distribution and accessibility of ^{238}U in such rocks that affect the emanation of radon. The number of faults and fissures in rocks affect the transport of radon from the solid rock into the intervening gas or liquid phases, and from there upwards – different overlying strata having different diffusion characteristics.

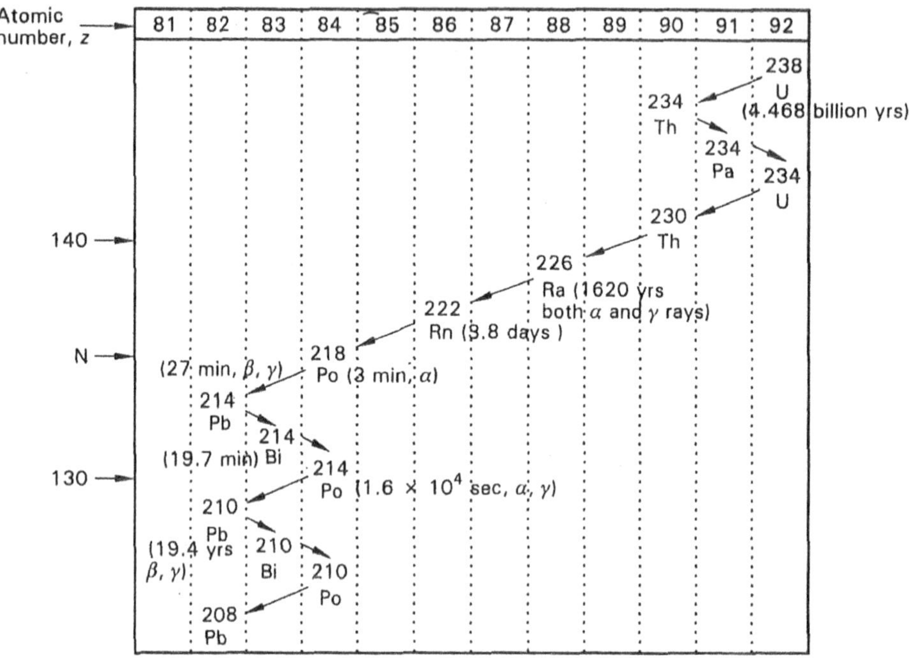

Fig. 1.37 *The major decay transitions of ^{238}U to radon and then into radon daughters, α particles are released when decays move to the left and lose two protons. β particles are formed when movement is to the right and the number of protons increases by one*

Radon emerges into the atmosphere by a variety of routes, including from the ground below and around the water supplies, and from natural gas or building materials; being denser than other gases, it tends to concentrate at low points. Consequently, lower storeys of buildings in radon-prone areas have higher levels of radon than upper floors. Further, most buildings have a slightly lower internal atmospheric pressure which forces radon inwards rather than outwards. This means that newer buildings which have fewer draughts are worse than older ones.

1.3.11.2 Effects of radon on human health

After seeping through the earth, the gas clings to particulate matter and is breathed by humans, causing damage in the lungs and eventually lung cancer.

The major hazard to humans of radon comes from breathing in radon daughters rather than radon itself. As a gas, almost as much radon comes out during expiration as went in during inhalation. Very little decays come out during expiration as went in during inhalation. Very little decays inside but being soluble, some does enter the bloodstream by diffusion. By contrast, non-gaseous radon daughters either adhere to dust particles or form clusters around water in aerosols. Both forms are inhaled but the radon daughters are deposited along the airways where they remain and decay. Consequently, the major deposition of radon daughter contaminated particles occurs in the pharynx and at the junction of the bronchi.

Their relatively short half lives (apart from ^{210}Pb) mean that most of the radiation dose is by the linings of air ways which then causes pulmonary fibrosis and cancer. The risk of lung cancer due to exposure of radon for people of different areas is shown in Fig. 1.38.

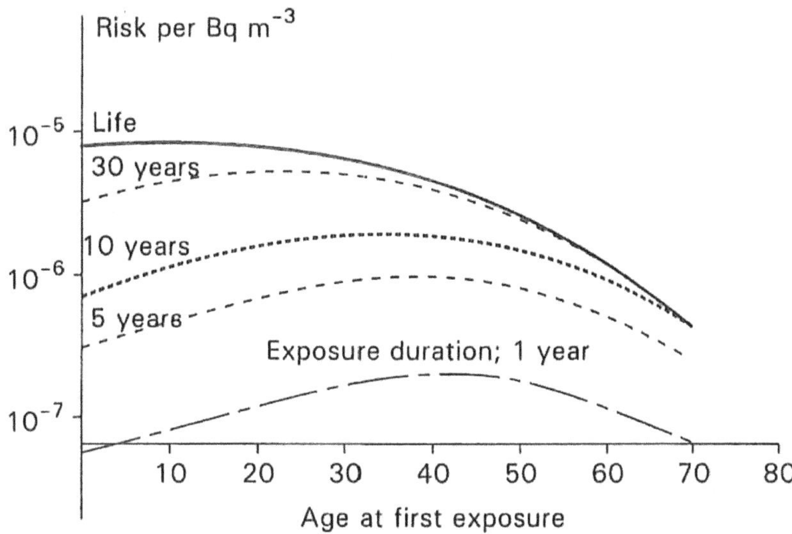

Fig. 1.38 *Lifetime risks of lung cancer due to exposure of radon (National Council on Radiation Protection and Measurement, 1984)*

A strong trend towards conserving energy at home has caused people to plug any leaks in order to prevent cold air from coming into their homes, and to keep warm air inside. By doing so, fresh air in these homes is limited, and those who were unknowingly exposed to the danger of radon contributed to the high concentration of radon gas by sealing it in their homes. An amount of 4 picocuries per litre of air, which is 4-trillionths of a curie, is suggested as a safety standard for home owners. One simple way of dissipating the gas would be to open doors and windows when feasible and bring in some air circulation; or by simply using a fan to dissipate this gas and get in some fresh air. It may be much more complicated, especially when some homes are built on landfills contaminated with radioactive waste. In some regions of the world, well waters contain radon gas.

Treatment, once the problem is recognized, is relatively simple. The ground below and around can be sealed off from the rest of the building but leaks may still occur. A much better solution is to create a small porous sump area at the base of a radon-prone building and then continuously pump out any gas that seeps into the building.

1.4 Secondary Air Pollutants-Photochemical Reactions – Ozone –PAN-Photochemical Smog

Secondary air pollutants are chemical products that are formed from the reactions of various primary air pollutants with one another; some of the important secondary pollutants are products of photochemical reactions.

When oxides of nitrogen, various hydrocarbons, and sunlight come together, they can initiate a complex set of reactions that produce a number of secondary pollutants known as photochemical oxidants.

1.4.1 Ozone and photochemical smog

Ozone is a major component of photochemical smog – the brownish haze that covers many urban areas during the hot, sunny days of summer. Ozone is produced by chemical

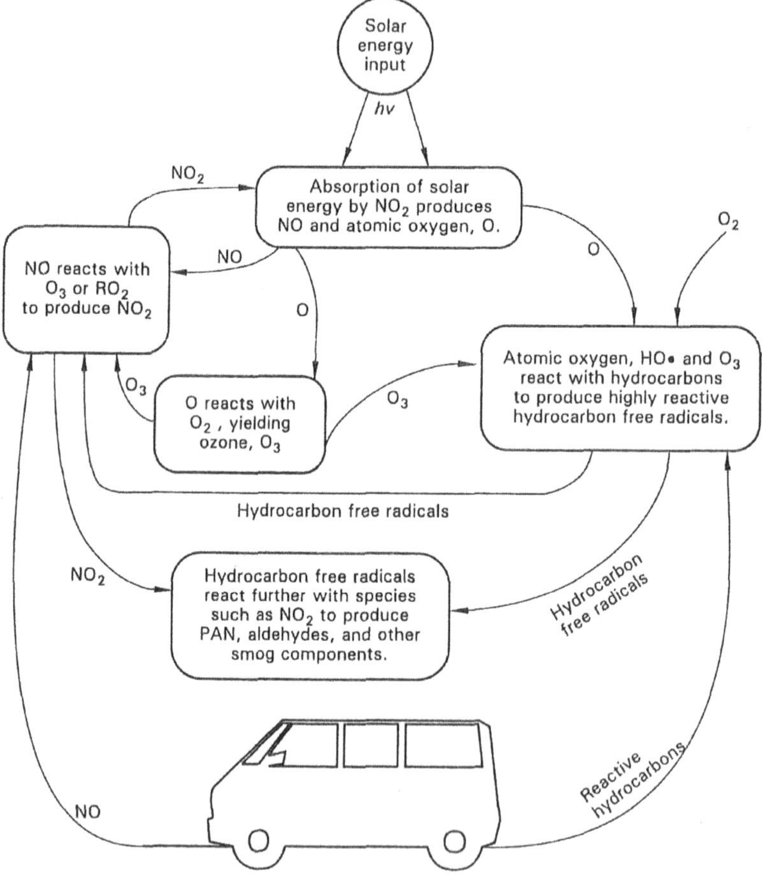

Fig. 1.39 *Generalized scheme for the formation of photochemical smog*

reactions between hydrocarbons and nitrogen oxides in the presence of sunlight – so the products of these reactions are called photochemical smog. The greater the intensity of sunlight and the warmer the day, the larger the amount of ozone produced. Since hydrocarbons and oxides of nitrogen are mainly generated by automobiles, the ozone levels gradually rise in the morning and peak between noon and 4 pm. On cloudy days, ozone production is reduced; after sunset, it stops altogether.

Photochemical smog has in addition to ozone, peroxyacetyl nitrates which are excellent oxidizing agents that readily react with many other compounds, including those found in living things, causing destructive changes. The general scheme for the formation of photochemical smog is shown in Fig. 1.39.

1.4.2 Conditions for photochemical smog

The conditions for the formation of photochemical smog are air stagnation, abundant sunlight, and high concentrations of hydrocarbons and nitrogen oxides in the atmosphere. Smog arises from photochemical reactions in the lower atmosphere by the interaction of hydrocarbons and nitrogen oxides released by exhausts of automobiles and some stationary sources. This interaction results in a series of complex reactions producing secondary pollutants such as ozone, aldehydes, ketones, and peroxyacetyl nitrates.

Photochemical smog is mainly an urban air pollution problem as the main contributing pollutants are from automobile emission sources. It was first observed in Los Angeles and has been found to be a severe problem in many cities. During the 1970s and 1980s it proved to be a daunting challenge to health officials in many cities like Denver, Salt Lake city, Milwaukee, Chicago, New York and Boston.

Photochemical smog is now a worldwide problem, especially in cities of lower latitudes which have high rates of population growth and rapid industrialisation. Currently, those most affected are Mexico city and Baghdad. Levels of O_3 in these and similar places are often over 100 ppb for considerable periods of time. Moreover, tropospheric O_3 is no longer just confined to summer. Significant wintertime levels are now being experienced in cities like Madrid and Athens during periods when stable, high pressure, sunny conditions occur. Tropospheric O_3 produced over populous areas spreads out in the manner of low-level clouds. In India Mumbai, and New Delhi are good candidates for photochemical smog.

Due to their climate and their geographic features, large metropolitan areas such as Log Angeles, Lake City, Phoenix, and Denver are troubled more with photochemical smog than the East Coast metropolitan areas. Most of these cities are located within a range of mountains. The prevailing winds are from the west. As cool air flows into a valley, it pushes the warm air upward. This warm air becomes sandwiched between two layers of cold air and acts like a lid on the valley, a condition known as a thermal inversion. The air is trapped in the valley. The warm air cannot rise further because it is covered by a layer of cooler air pushing down on it. It cannot move out of the area because of the ring of mountains. Without normal air circulation, smog accumulates.

Harmful chemicals continue to increase in concentration until a major weather change causes the warm air to move up and over the mountains.

Photochemical smog episodes have a distinctive daily rhythm. In the absence of wind around dawn, when urban activities are low, levels of CO, NO, and HCs increase as morning traffic builds up. Amounts of NO_2 then rise to a maximum about 1 ½ h after the peak of NO when the sun is reasonably high. The later appearance of aldehydes and then O_3 follow the disappearance of NO so that concentrations of O_3 are highest shortly after noon, just after the aldehyde peak has passed. This then allows other free radical chain reactions to form PANs and other irritants. As a rough approximation, one molecule of PAN is produced for every fifty O_3 molecules. In the late afternoon, returning traffic then generates more NO and NO_2 which immediately scavenges most of the O_3 and reactive aldehydes. This means that levels of nitrogen oxides do not follow their distinctive biphasic daily patterns normally found in windy and often variable, cloud covered conditions. Figure 1.40 presents the typical levels of the smog constituents at various times of the day observed in Los Angeles

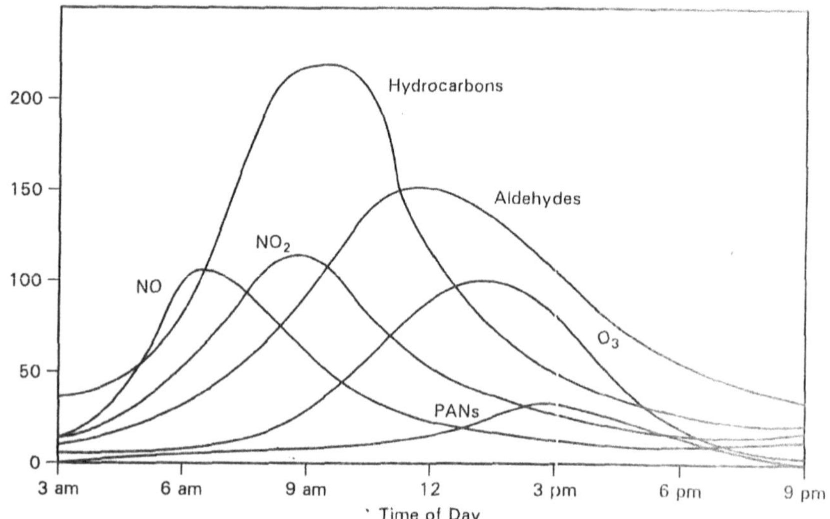

Fig. 1.40 *Development and decay of components in a typical smog over a large conurbation during a bright, warm working day with no wind. Typical values observed in Lar Angela on July 9, 1965*

The typical smog episode occurs in hot, sunny weather under low humidity conditions. The characteristic symptoms of the smog are the brown haze in the atmosphere, reduced visibility, eye irritation, respiratory distress and plant damage. The control of photochemical smog may require substantial reduction in NO_x produced in urban areas. At the same time it is also necessary to control the release of hydrocarbons from numerous mobile and stationary sources.

1.4.3 Sequence of photochemical reactions in photochemical smog

The NO – NO_2– NO_3 photochemical reaction sequence:

In the process of combustion NO will be formed.

$$N_2 + O_2 \rightarrow 2NO$$

The nitric oxide thus emitted can oxidize to NO_2.

$$2NO + O_2 \rightarrow 2NO_2$$

If sunlight is available, NO_2 can photolyse, and the freed atomic oxygen can then help to form ozone, as per the following pair of reactions:

$$NO_2 + h\,\nu\,(420) \rightarrow NO + O$$

$$O+O_2+M \rightarrow O_3+M$$

where $h\nu$ represents a photon ($\lambda< 0.38$ μm) and M represents a molecule (usually O_2 and N_2 since they are most abundant in air) whose presence is necessary to absorb excess energy from the reaction. Without M, the ozone would have too much energy to be stable, and it would dissociate back to O and O_2:

$$O_3 + NO \rightarrow NO_2 + O_2$$

Since this reaction is very rapid the concentration of O_3 will be low until the NO value decreases to a low value in the atmosphere. This set of reactions creates a cycle that is represented in Fig. 1.41.

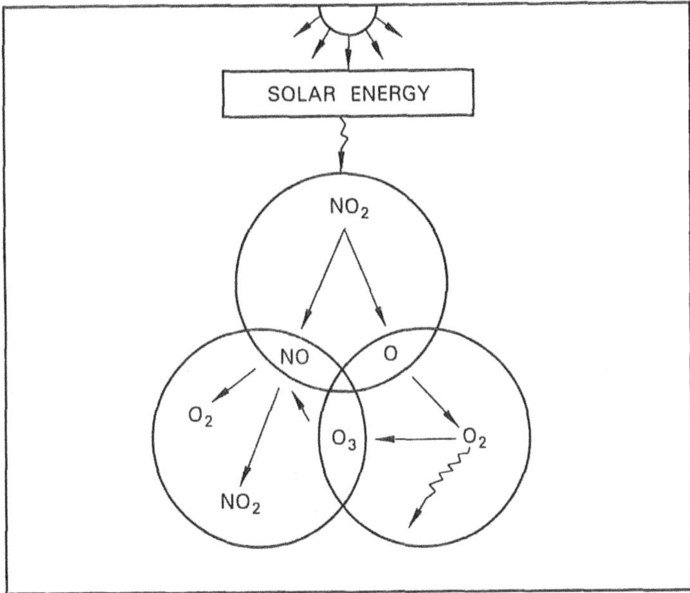

Fig. 1.41 *The cycle of chemical reactions initiated by the photochemical dissociation of* NO_2

If these were the only reactions involved, when they reach steady state, the rates of production and destruction of each of these important products of photochemical smog– NO, NO_2, and O_3– will be equal, and there would be no net change in their concentrations over time. Due to the rapid rate of reaction the NO concentration will not rise after smog sets in the day even though there are fresh emissions in the afternoon. However on many smoggy days, ozone levels are found to be much higher than those expected from the above reactions. This was found to be due to some other set of reactions involving hydrocarbons.

In the presence of certain hydrocarbons, the equilibrium set within the cycle of chemical reactions shown in Fig. 1.41, is destroyed and ozone begins to accumulate in the air.

For purely inorganic systems, kinetic calculations and experimental measurements cannot explain the rapid transformation of NO to NO_2 that occurs in an atmosphere under going photochemical smog formation and predict that the concentration of NO_2 should remain very low. However, in the presence of reactive hydrocarbons, NO_2 accumulates very rapidly by a reaction process beginning with its photodissociation, it may be concluded therefore, that the organic compounds form species which react with NO directly rather than with NO_2.

A number of chain reactions have been shown to result in the general type of species behaviour shown in Fig. 1.42.

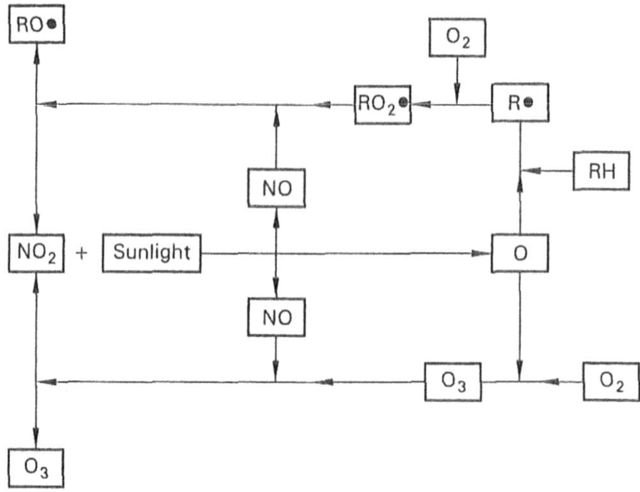

Fig. 1.42 *Interaction of hydrocarbons with the atmospheric nitrogen oxide photolytic cycle*

The greatest global problem with unsaturated hydrocarbons is their ability to promote the formation of photochemical smog in the presence of nitrogen oxides; strong sunlight and stable meteorological conditions.

$$RH + O \rightarrow R\bullet + \text{other products}$$

$$O_3 + RH \rightarrow RO_2\bullet$$

$$RH + OH\bullet \rightarrow R\bullet + H_2O$$

However, the reaction chains involved are long and complicated because one reaction involving a free radical generates another. This in turn, reacts to form a third, etc. Moreover, there are many starting points which finally rejuvenate OH• readily through chain reactions with a wide variety of hydrocarbons released into the atmosphere. Aldehydes and ketones also produce free radicals in strong light such as $RO_2\bullet$ radicals while O_3 itself may attack unsaturated HCs to produce similar free radicals including reactive aldehydes.

$$R\bullet + O_2 \rightarrow RO_2\bullet$$

$$RO\bullet + NO \rightarrow RO\bullet + NO_2$$

These reactions show how hydrocarbons help in the conversion of NO to NO_2 and the break up of O_2.

The mechanism of ozone build-up in the presence of reactive hydrocarbons is shown in Fig. 1.42. The ozone build-up is possible only when the ozone, formed by the reaction of oxygen atom (O) with O_2, does not get consumed again to restore NO_2 after reacting with NO as indicated in the figure.

The sequence of the reactions mentioned above is a chain reaction in which the products of one reaction are consumed in the next. Consequently, we may find that a large number of reactive free radicals are formed and accumulate in the air.

1.4.4 Role of hydrocarbons in photochemical smog

The evaporation of solvents and the incomplete combustion of fuels release a wide range of hydrocarbons into the atmosphere. The analysis of ambient air samples at various times has shown the presence of over 600 different atmospheric hydrocarbons (HCs) when smog has set in. These include acetylene, benzene, butanes, ethane, hexanes, pentanes, propane and toluene – all of which are characteristic of anthropogenic emissions.

1.4.5 Formation of free radicals in air

Free radicals are highly reactive chemical species that are formed by the photo dissociation of several inorganic and organic compounds. The main compounds forming free radicals of importance in air pollution are hydroxyl radicals, aldehydes, ketones, peroxyacetyl nitrates, hydrogen peroxide, organic peroxides and nitrous and nitric acids.

The hydroxyl radical plays a central role, via the peroxy radicals with which it is intimately related, in the production of ozone and hydrogen peroxide. It also contributes to the formation of sulphuric and nitric acids, in the atmosphere. Thus atmospheric processes leading to ozone formation will also tend to favour production of other secondary pollutants, including the strong acids, HNO_3 and H_2SO_4.

1.4.6 Hydroxyl and hydroperoxyl radicals in the atmosphere

The hydroxyl radical HO• is the single most important reactive intermediate species in atmospheric chemical processes. It is formed by several mechanisms (Fig. 1.43). At higher altitudes it is produced by photolysis of water.

$$H_2O + h\nu \rightarrow HO\bullet + H$$

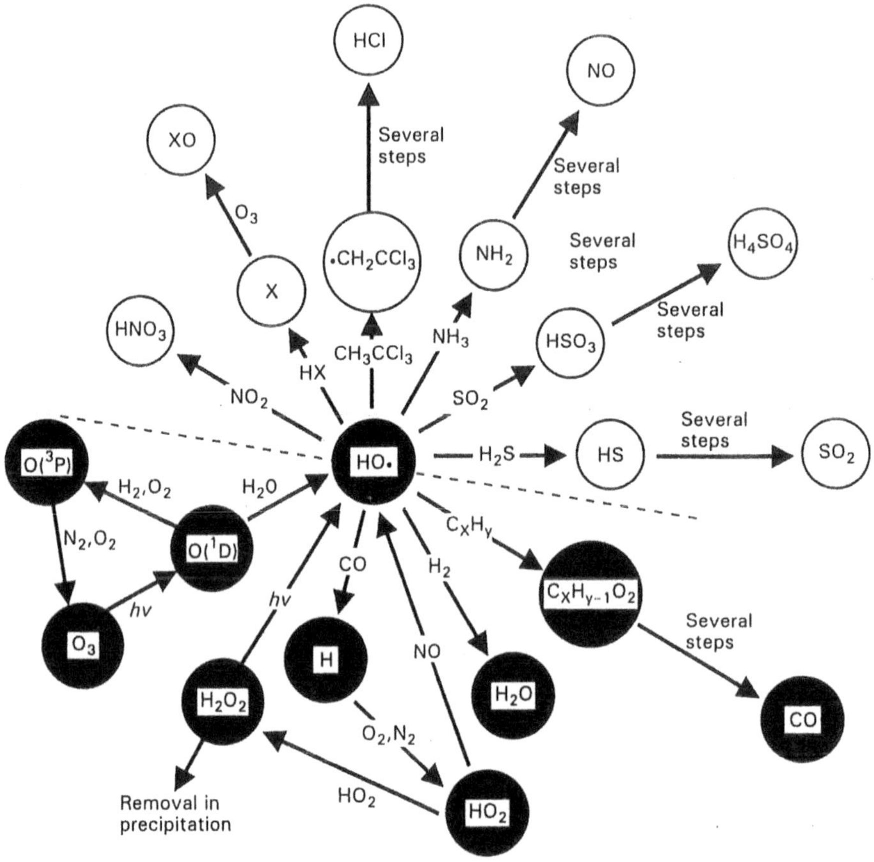

Fig. 1.43 *Control of trace gas concentrations by HO• radical in the troposphere*

In the presence of organic matter, the hydroxyl radical is produced in abundant quantities as an intermediate in the formation of photochemical smog. To a certain extent in the atmosphere, and for laboratory experimentation, HO• is made by the photolysis of nitrous acid vapour.

$$HONO + h\,V \rightarrow HO\bullet + NO$$

In the relatively unpolluted troposphere, hydroxyl radical is produced as the result of the photolysis of ozone.

$$O_3 + h\,V(\lambda < 315) \rightarrow O\bullet + O_2$$

The above is followed by the reaction of a fraction of the excited oxygen atoms with water molecules.

$$O\bullet + H_2O \rightarrow 2HO$$

The involvement of the hydroxyl radical in chemical transformations of a number of trace species in the atmosphere is summarized in Fig. 1.43 and some of the pathways illustrated are discussed below. Among the important atmospheric trace species that react with the hydroxyl radical are carbon monoxide, sulphur dioxide, hydrogen sulphide, lethane, and nitric oxide.

Hydroxyl radical is most frequently removed from the troposphere by reaction with methane or carbon monoxide.

$$CH_4 + HO\bullet \rightarrow H_3C\bullet + H_2O$$

$$CO + HO\bullet \rightarrow CO_2 + H$$

The highly reactive methyl radical, $H_3C\bullet$ reacts with O_2 to form the methylperoxy radical $H_3COO\bullet$.

$$H_3C\bullet + O_2 \rightarrow H_3COO\bullet$$

The hydrogen atom produced in the reaction, $CO + HO\bullet \rightarrow CO_2 + H$, reacts with O_2 to produce the hydroperoxyl radical.

$$H + O_2 \rightarrow HOO\bullet$$

The hydroperoxyl radical can undergo chain termination reactions such as:

$$HOO\bullet + HO\bullet \rightarrow H_2O + O_2$$

$$HOO\bullet + HOO\bullet \rightarrow H_2O_2 + O_2$$

or, reactions that regenerate the hydroxyl radical.

$$HOO\bullet + NO \rightarrow NO_2 + HO\bullet$$

$$HOO\bullet + O_3 \rightarrow 2O_2 + HO\bullet$$

It is estimated that the global concentration of hydroxyl radicals, averaged diurnally and seasonally, ranges from 2×10^5 to 1×10^6 radicals per cm^3 in the troposphere. The concentration of $OH\bullet$ radical is found to be higher in tropical regions due to their higher humidity and sunlight. Similarly the southern hemisphere has 20% higher levels of $HO\bullet$ than the northern hemisphere.

The hydroperoxyl radical $HOO\bullet$ is an intermediate in some important chemical reactions. In addition to its production by the reactions discussed above, in polluted atmospheres, hydroperoxyl radical is made by the following two reactions, starting with the photolytic dissociation of formaldehyde to produce a reactive formyl radical,

$$HCHO + h\nu \rightarrow H + HCO\bullet$$

$$HCO\bullet + O_2 \rightarrow HOO\bullet + CO$$

The hydroperoxyl radical reacts more slowly with other species than does the hydroxyl radical. The kinetics and mechanisms of hydroperoxyl radical reactions are difficult to study because it is hard to .retain these radicals free of hydroxyl radicals, which are summarized in Table 1.31.

Table 1.31 *Generalized chemical kinetic mechanism in photochemical box model*

1. $NO_2 \xrightarrow{hv} NO + O$	19. $OLEF + O \rightarrow RO_2 + ALD + HO_2$
2. $O + O + M \rightarrow O_3 + M$	20. $OLEF + O_3 \rightarrow RO_2 + ALD + HO_2$
3. $O_3 + NO \rightarrow NO_2 + O_2$	21. $OLEF + HO \rightarrow RO_2 + ALD$
4. $O_3 + NO_2 \rightarrow NO_3 + O_2$	22. $PARAF + HO \rightarrow RO_2$
5. $NO_3 + NO \rightarrow 2NO_2$	23. $ALD \xrightarrow{hv} 0.5RO_2 + 1.5HO_2 + 1.0CO$
6. $NO_3 + NO_2 + H_2O \rightarrow 2HONO_2$	24. ALD atmospheric layer deposition
7. $HONO \xrightarrow{hv} HO + NO$	25. $RO_2 + NO \rightarrow RO + NO_2$
8. $HO + NO \xrightarrow{O_3} HO_2 + CO_2$	26. $RO + O_2 \rightarrow ALD + HO_2$
9. $HO_2 + NO_2 \rightarrow HONO + O_2$	27. $RIO_2 + NO_2 \rightarrow PAN$
10. $HO_2 + NO \rightarrow HO + NO_2$	28. $RO + NO_2 \rightarrow RONO_2$
11. $HO_2 + NO_2 + M \rightarrow HOONO_2$	29. $RO_2 + O_3 \rightarrow RO + 2O_2$
12. $HOONO_2 \rightarrow HO_2 + NO_2$	30. $RIO_2 + NO \xrightarrow{O_3} RO_2 + NO_2$
13. $HO + HONO \rightarrow NO_2 + H_2O$	31. $PAN \rightarrow RIO_2 + NO_2$
14. $HO + NO_2 + M \rightarrow HONO_2 + M$	32. $AROM + HO \xrightarrow{O_3} RO_2 + R_2O_2 + 2ALD + CO$
15. $HO + NO + M \rightarrow HONO + M$	33. $R_2O_2 + NO \rightarrow R_2O + NO_2$
16. $HO_2 + O_3 \rightarrow HO + 2O_2$	34. $R_2O + O_2 \rightarrow ALD + HO_2 + 2CO$
17. $HO + O_3 \rightarrow HO_2 + O_2$	35. $R_2O_2 + O_3 \rightarrow R_2O + 2O_2$
18. $HO_2 + HO_2 \rightarrow H_2O_2 + O_2$	36. $RIO_2 + O_3 \rightarrow RO_2 + 2O_2$

Source: *K. L. Demerjian and K. L. Schere, 1979, Proceedings, Ozone/Oxidants: Interactions with Total Environment II (Pittsburgh: Air Pollution Association).*

Note: M stands for any available atom or molecular which by collision with the reaction product carries off the excess energy of the reaction and prevents the reaction product from flying a part as soon as it is formed.

Olefins (OLEF) with large number of double bonds also react photochemically to form free radicals (Table 1.31). On reacting with oxygen some free radicals form peroxyl compounds from which new peroxides or free radicals are produced (PARAF) that can cause polymerization of olefins.

Olefins are much more reactive, undergoing reactions with the hydroxyl radical,

$$R_2C{=}CR_2 + HO\bullet \xrightarrow[\text{rapid}]{\text{Very}} R{-}C(OH)(R){-}C\bullet(R)(R) \longrightarrow \text{Oxidation products}$$

Radical adduct

(where R may be one of a number of hydrocarbon moieties or an H atom) with oxygen atoms, or

$$R_2C{=}CR_2 + O \longrightarrow R{-}C(O\bullet)(R){-}C\bullet(R)(R) \longrightarrow \text{Oxidation products}$$

Biradical

with ozone,

$$R_2C{=}CR_2 + O_3 \longrightarrow R{-}C(R)(O{-}O{-}O){-}C(R){-}R \longrightarrow \text{Oxidation products}$$

Aromatic hydrocarbons, Ar-H, may also react with O and HO•. Addition reactions of aromatics with HO• are favoured. The product of the reaction of benzene with HO• is phenol, as shown by the following reactions sequence:

$$C_6H_6 + HO\bullet \longrightarrow C_6H_6(OH)\bullet$$

$$\text{C}_6\text{H}_6(\text{H})(\text{OH})\bullet + O_2 \longrightarrow C_6H_5OH + HOO\bullet$$

In the case of alkyl benzenes, such as toluene, the hydroxyl radical attack may occur on the alkyl group, leading to reaction sequences such as those of alkanes.

Aldehydes react with HO•

$$R-\overset{\overset{\displaystyle O}{\|}}{C}-H + HO\bullet + O_2 \longrightarrow R-\overset{\overset{\displaystyle O}{\|}}{C}-OO\bullet + H_2O$$

$$H_2C=O + HO\bullet + \frac{3}{2}O_2 \longrightarrow CO_2 + HOO\bullet + H_2O$$

Undergo photochemical reactions:

$$R-\overset{\overset{\displaystyle O}{\|}}{C}-H + h\nu + 2O_2 \longrightarrow ROO\bullet + HOO\bullet$$

$$H_2C=O + h\nu + 2O_2 \longrightarrow CO + 2HOO\bullet$$

The hydroxyl radical which reacts with some hydrocarbons at rates that are almost diffusion-controlled, is the predominant reactant in the early stages of smog formation.

Significant contributions are made by the hydroperoxyl radical (HOO•) and O_3 after smog formation is well under way.

One of the most important reaction sequences in the smog formation process begins with the abstraction by HO• of a hydrogen atom from a hydrocarbon and leads to the, oxidation of NO to NO_2 as follows:

$$RCH + O + OH\bullet \Leftrightarrow RCO\bullet \text{ (alyl alcohol)} + H_2O$$

This radical participates in a number of reactions leading to the formation of more reactive molecules like proxy alyl radical aldehydes and ketones which react with NO and hydrocarbons.

The alyl radical, RCO• reacts with O_2 to produce a peroxyl radical, RCOO•:

$$RCO\bullet + O_2 \rightarrow RCO_3, \text{ (per alyl alcohol)}$$

This strongly oxidizing species very effectively oxidizes NO to NO_2:

$$RCO_3 + NO \rightarrow RCOO\bullet + NO_2$$

$$RCO_3\bullet + NO_2 \rightarrow RCO_3\,NO_2 \text{ (PANs)}$$

This explains the once puzzling rapid conversion of NO to NO_2 in an atmosphere in which the latter is undergoing photo dissociation. The alkoxyl radical product, RO• is not as stable as ROO•. In cases where the oxygen atom is attached to a carbon atom that is also bonded to H, a carbonyl compound is likely to be formed.

Many free radicals like these that are generated in strong sunlight are responsible for the eye irritation of those exposed to photochemical smog.

Peroxyacetyl nitrate (PAN), one of the important constituent products is formed through a number of routes, most notably acetaldehyde oxidation as shown below.

$$CH_3\,CO_3\bullet + NO_2 \rightarrow CH_3\overset{\overset{\textstyle O}{||}}{C}\,OO\,N_2O$$

Where R is the methyl group, the product is peroxyacetyl nitrate. Although it is thermally unstable, peroxyacetyl nitrate does not undergo photolysis rapidly, reacts only slowly with the HO• radical, and has a low water solubility. Therefore, the major pathway by which it is lost from the atmosphere is thermal decomposition, the opposite of the above reaction.

Alkyl nitrates and alkyl nitrites may be formed by the reaction of alkoxyl radicals (RO•) with nitrogen dioxide and nitric oxide respectively:

$$RO\bullet + NO_2 \rightarrow RONO,$$

$$RO\bullet + NO \rightarrow RONO$$

Addition reactions with NO_2 such as these are important in terminating the reaction chains involved in smog formation. Since NO_2 is involved both in the chain initiation step and the chain terminating step, moderate reductions in NO_x emissions alone may not curtail smog formation and in some circumstances may even increase it.

1.4.7 Inorganic products from smog

Some of the important inorganic products from smog are sulphates and nitrates which are responsible for acidic precipitation, corrosion, reduced visibility, and adverse health effects. The oxidation of SO_2 to sulphate species takes place more rapidly (5-10% per hour) under smoggy conditions than in a clean atmosphere. A number of oxidants such as O_3, NO_3, and N_2O_5 as well as reactive radical species, particularly HO•, HOO•, O, RO• and ROO•, which are likely to be present in smog conditions oxidize SO_2 to sulphates.

$$SO_2 + O \text{ (from O, RO}\bullet\text{, ROO}\bullet\text{)} \rightarrow SO_3 \rightarrow H_2SO_4$$

$$HO\bullet + SO_2 \rightarrow HOSOO\bullet$$

The presence of HO• (typically at a level of 3×10^6 radicals/cm^3 but appreciably higher in a smoggy atmosphere) makes this a likely route. The addition of SO_2 to RO• or ROO• can yield organic sulphur compounds also.

Inorganic nitrates or nitric acid are also formed in smog reactions, particularly those involving N_2O_5 with water and NO_2 with hydroxyl radical. Nitric acid freed by these reactions reacts with ammonia in the atmosphere to form ammonium nitrate.

$$NH_3 + HNO_3 \rightarrow NH_4\,NO_3$$

Both nitric acid and nitrites are some of the damaging constituents of smog products which not only affect plant and animals but also corrode materials.

1.4.8 Adverse effects of photochemical smog

The major constituents of smog, with the exception of carbon dioxide, are powerful poisons. Many aspects of their toxic effects, both individually and in combination with one another, are not completely known. The danger, however, to people with susceptibility to respiratory and cardiac difficulties is widely recognised. Those with incipient or advanced bronchial asthma, chronic bronchitis, and pulmonary emphysema are apt to be hard hit. The various components of smog may affect people with different susceptibilities in different ways.

1.4.9 Short - and long-term effects

Photochemical smog causes irritation of the eyes, nose, throat, and chest. Eye irritation is not caused by O_3 but by PANs and trace free radical HCs. O_3 is a powerful oxidant that may injure the bronchiolar and alveolar walls of the lungs. Surface epithelial cells of the airways are damaged by O_3 and afterwards, these are replaced by thick cuboidal cells with few or small cilia (cell hairs). Apart from loss of cilia, additional changes to the epithelial (surface lining) cells include cytoplasmic vacuolation (formation of internal cellular spaces) and condensation of abnormal mitochondria.

Respiratory irritation and difficulty in breathing have been common, and the correlation of the oxidizing power of smog with its irritating action is having implications beyond the simple annoyance factor. The air passageways of the lungs do more than carry the air and waste gases to and from the air sacs. The passageways are lined with mucus-secreting epithelium and serve to condition the air before it reaches the alveoli. The air is changed in three important ways; it is warmed to body temperature, it is saturated with moisture, and it is filtered free from most of the dust particles and other foreign substances. The cells that form the epithelial lining are ciliated, that is, they have hair-like protoplasmic appendages that project into the passageways and, by continuous, coordinated, rhythmic action, work the mucus and entrapped foreign material upward to the throat where it is either swallowed or expelled. Both the mucous secreting ability and the ciliary action of the epithelium have been sensitive to adversities, and excessive exposure to foreign material such as smoke, dust, and toxic substances impair their functioning, causing an increase in susceptibility to various respiratory disorders.

The other constituent of photochemical smog that might be expected to affect health is NO_2.

Effects of O_3 are also made worse by exercise. It is now recommended that athletic activities should cease when levels of O_3 are high and that groups at risk (e.g. children who spend proportionately more time outdoors than any other group in the population) be moved indoors.

Epidemiological studies of enhanced respiratory infections and reduced lung functions in populations exposed to photochemical smogs are hampered by the fact that such surveys are dealing with complex and changing mixtures of O_3, nitrogen oxides, PANs, and other components of smog. Shorter-term dose response studies have indicated that, although nitrogen oxides and O_3 cause similar adverse effects on lungs, more than five times the level of NO_2 is required to elicit effects similar to that caused by a given level of O_3.

1.4.10 Effects on plants

Before the biological importance of air pollution was fully realized, it was found that plants grown in bad smog areas suffered injury from what often appeared to be obscure causes. It is now apparent that many plants have been more sensitive than humans to the acute effect of smog and in fact, can be used as indicators of the harmful level of air contaminants.

Two ingredients of photochemical smog have been recognized as being particularly injurious; ozone and PAN (peroxyacetyl nitrate). With PAN, injury has been appearing on some plants as a characteristic silvering on the bottoms of the leaves. Ozone, on the other hand, kills cells on the tops of the leaves and appears as flecking or stippling of the upper surfaces.

Still conditions during photochemical smog episodes provide some protection to vegetation because undisturbed boundary layers of still air around the leaves offer additional resistance to the diffusion of O_3. Later, as winds move tropospheric O_3 around, damage to vegetation may be experienced many kilometres away as these leaf boundary layers are stripped away.

Access of O_3 into the leaf takes place through the stomata although waxy leaf surfaces may be partially degraded by O_3. Most plants have open stomata during the day. Consequently, O_3 damage is initiated in the Light but cacti, etc., which only open their stomata at night to conserve water loss, have their period of maximum sensitivity to O_3 during periods of darkness.

Responses to O_3 are determined by a wide variety of environmental or genetic factors. Moist surfaces within the leaves allow O_3 to dissolve and diffuse down a concentration gradient similar to that of CO_2. Solubilities, rates of decomposition, and the pH values of the various media all influence the amount of O_3 that is taken up. O_3 is approximately one-third as soluble as CO_2 but 100-fold less soluble than SO_2.

Injury to plants from photochemical air pollution was reported in the mid-1940s when losses occurred to spinach and other sensitive plants in California. Since then, crop damage has taken place in many parts of the world. It is now impossible to raise orchids in metropolitan areas. Vegetable and citrus growing have been hit hard by injury and reduced yields.

Ornamental plants have not escaped the onslaught of air pollution, City parks have been becoming increasingly limited in the plants that can be successfully grown, and park superintendents find that they must be more restrictive in selecting types of plants that can tolerate smog. Home gardeners are usually puzzled by the failure of their favourite specimens to flourish. Injury is common to petunia, snapdragon, chrysanthemum, larkspur, carnation, orchid, pansy, rose and zinnia.

The principal effects of PAN have been the silvering, bronzing, and glazing of the lower leaf surface. Petunias are affected at very low concentrations (0.005 ppm for 8 h). Pinto beans are affected at 1.0 ppm for 30 minutes.

Not as much has been known about the effects of NO_2, but long-time exposure to 0.5 ppm has been said to damage tomato plants.

1.4.11 Effects on materials

Deterioration of materials by photochemical smog has been thought to be primarily the result of the ozone content of smog. Thus the most specific effect of smog on materials has been the cracking of rubber. In fact an early, specific analytical method for the determination of atmospheric ozone was the depth of cracks on stretched rubber bands. An important economic effect of smog in Los Angeles in the 1940s happened to be the deterioration of the side walls of automobile tires. Addition of an anti-ozonant has alleviated the problem. A variety of other rubber products (hoses, gaskets, wire insulation) are also still affected. Another important effect believed to be caused by ozone is the fading of dyes.

Natural polymers like rubber, cotton, cellulose, or leather, as well as paints, elastomers, plastics, nylon and fabric dyes, are all degraded by O_3. Only when the double bonds in their structures are protected by adjacent groups can attack be resisted. The electronegative chlorine atom adjacent to the double bond in neoprene is a good example of this type of protection. Similar protective mechanisms have been built into recent polymers but sacrificial autoxidants can also be incorporated. The annual global cost of this type of damage to materials is huge and constitutes just over 30% of that caused by all forms of atmospheric pollution to non-living materials.

Problems

1. The maximum one-hour carbon monoxide levels in Calcutta reach 35 ppm. Calculate the equivalent concentration in terms of mass fraction (w_p] and in milligrams per cubic meter at 25 °C and 1 atm.
2. The mean concentration of sulphur dioxide in Mumbai is 47 mg/m^3. What is the equivalent concentration in parts per million at 25 °C and 1 atm?
3. The exhaust gas from a motor vehicle shows a carbon monoxide concentration of 2% by volume. What is the concentration of CO in mg/m^3 at 0 °C and 1 atm?
4. A 1000 MW power plant of 25% thermal efficiency is proposed. The plant burns 3% sulphur coal with a heating value of 6000 kcal/kg and emits 64,000 m^3/min of flue gas. What is the concentration of SO_2 in the flue gas in ppm? Assume that the density of SO_2 is 1920 g/m^3.
5. A captive power plant of 1 MW with 40% thermal efficiency is to be designed which will use residual oil with 1% sulphur content. The emission factors for burning 1 m^3 of oil are: particulates = 1 kg, NO_2 = 10 kg and SO_2 = 16 S kg where S is the percent sulphur content of oil. What are the rates of emission of particulates, NO_2 and SO_2 to the atmosphere if no control devices are provided? The heating value of the oil is 8000 kcal/kg, and its density may be taken as 950 kg/m^3.
6. Consider a new 38% efficient, 600 MW power plant burning 5.79 kW/kg coal containing 1 % sulphur. What would be the maximum allowable SO_2 emission rate (kg/day) and how efficient must the scrubber be?
7. In the above power plant, how many kg per day of sulphur would be released from the fuel? If the scrubber were 90% efficient, how many kg per day of sulphur and SO_2 would be released?
8. A new coal fired power plant has been built using a sulphur emission control system that is 70% efficient. What maximum percent sulphur content can the fuel have if 9.66 kW/kg coal is burned? If 5.79 kW/kg coal is burned, what maximum sulphur content can the fuel have?

1.4.12 US-National trends in ozone levels

Using a nationwide network of monitoring sites, US-EPA has developed ambient air quality trends for ozone. Trends are shown here for the 8-hour ozone standards. Under the Clean Air Act, EPA sets and reviews national air quality standards for ozone. Air quality monitors measure concentrations of ozone throughout the country. EPA, state, tribal and local agencies use that data to ensure that ozone is at levels that protect public health and the environment. Nationally, average ozone levels declined in the 1980's, leveled off in the 1990's, and showed a notable decline after 2002 (Figure 1.44).

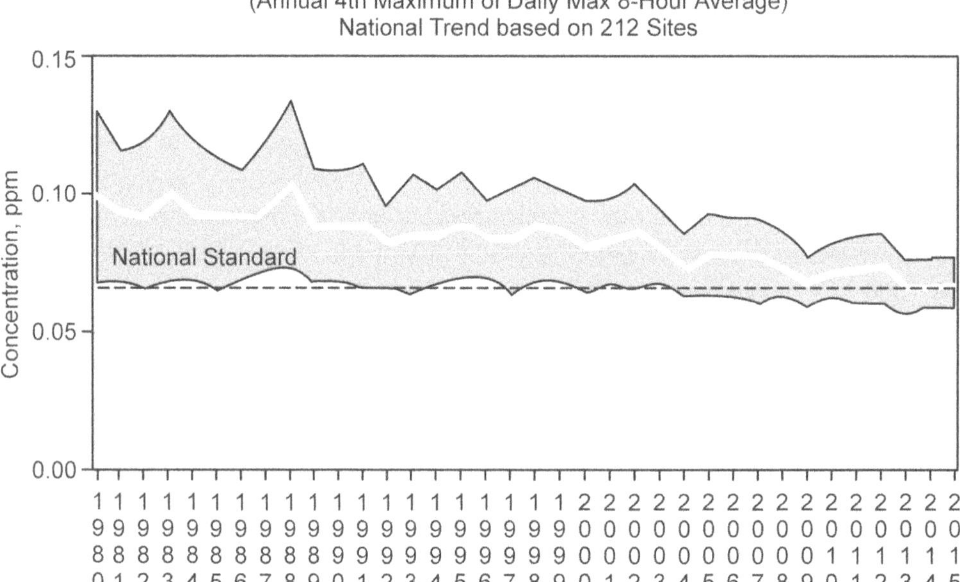

Fig. 1.44 *Trends on Ozone Levels from 1980-2015*

Questions

Short Answer Questions

1. List out the emission sources of Air Pollutants.
2. How does the food and agricultural sector contribute to air pollution sources? Name a few pollutants.
3. How are air pollutants classified based on their source of pollution?
4. Outline the atmospheric cycle of a pollutants.
5. Classify different types of particulates based on the particle size.
6. Brief how human health is affected by particulate pollutants.
7. Show the formation mechanism of sulfur oxides in the atmosphere.
8. Write a short note on emission sources of hydrocarbons.

9. Brief about Carcinogenic air pollutants and their significance.
10. Explain about radon emission and its sources.

Essay Questions

1. Explain in detail about the gases produced by Anthropogenic activities
2. Explicate the different Pathological effects caused due to pollutant gases.
3. Describe the removal mechanisms employed for nitrogen oxides.
4. What is Blood biochemistry and pollution? Explain in detail.
5. List out the Physiological effects of fluoride on plants in detail.
6. Schematize the formation of Photochemical smog.
7. Analyze the sequence of photochemical reactions in photochemical smog.
8. Explain in detail about Hydroxyl and hydroperoxyl radicals in the atmosphere.
9. Analyze the inorganic products produced in smog. Give the chemical reactions.

References

1. USEPA 1977. Air quality criteria for lead. Environmental protection agency, Washington DC.
2. Archives of Environmental Health 1973.27:134-137.
3. J. B. Upham, 1967. Atmospheric concentrations in two metropolitan areas. J. of *air pollution control association.* 17:398.
4. R. A. Wadden and P. A. Scheff, 1983. *Indoor air pollution characterization, prediction and control.* Wiley Interscience, New York.
5. Vallero D. Fundamentals of Air Pollution. 4th ed. California, USA: Academic Press; 2007.
6. WHO. Database: Outdoor Air Pollution in Cities; 2003-2010. Available from: http://www.who.int/phe/health_topics/outdoorair/databases/cities-2011/en/
7. Thom GC, Ott WR. Air Pollution Indices: A Compendium and Assessment of Indices Used in the United States and Canada. 6th ed. Michigan, USA: Ann Arbor Science; 1976.
8. Chen TM, Gokhale J, Shofer S, Kuschner WG. Outdoor air pollution: Nitrogen dioxide, sulfur dioxide, and carbon monoxide health effects. Am J Med Sci. 2007;333:249–56.
9. U. S. Environmental Protection Agency. Metropolitan area trends. Washington, DC: US Environmental Protection Agency; 1996
10. Harrison RM. Sources of air pollution. Air Quality Guidelines Global Update 2005. Copenhagen, Denmark: World Health Organization Regional Office for Europe; 2006. pp. 9–30.

11. Hewitt C, Jackson AV. Sources of air pollution. In: Jackson AV, editor. Handbook of Atmospheric Science: Principles and Applications. 2nd ed. Malden, MA, USA: Blackwell Science Ltd; 2007. pp. 124–55.
12. Nakano T, Otsuki T. Environmental air pollutants and the risk of cancer. Gan To KagakuRyoho. 2013;40:1441–5.
13. Kampa M, Castanas E. Human health effects of air pollution. Environ Pollut. 2008;151: 362–7.
14. Makri A, Stilianakis NI. Vulnerability to air pollution health effects. Int J Hyg Environ Health. 2008;211:326–36.
15. Weisel CP. Assessing exposure to air toxics relative to asthma. Environ Health Perspect. 2002;110(Suppl 4):527–37.
16. Brunekreef B, Beelen R, Hoek G, Schouten L, Bausch-Goldbohm S, Fischer P, et al. Effects of long-term exposure to traffic-related air pollution on respiratory and cardiovascular mortality in the Netherlands: The NLCS-AIR study. Res Rep Health Eff Inst. 2009;139:5–71.
17. Cairncross EK, John J, Zunckel M. A novel air pollution index based on the relative risk of daily mortality associated with short-term exposure to common air pollutants. Atmos Environ. 2007;41:8442–54.
18. Johnson D, Ambrose S, Bassett T, Bowen M, Crummey D, Isaacson J, et al. Meanings of environmental terms. J Environ Qual. 1997;26:581–9.
19. Lovett GM, Tear TH, Evers DC, Findlay SE, Cosby BJ, Dunscomb JK, et al. Effects of air pollution on ecosystems and biological diversity in the eastern United States. Ann N Y Acad Sci. 2009;1162:99–135
20. Mellouki A, George C, Chai F, Mu Y, Chen J, Li H. Sources, chemistry, impacts and regulations of complex air pollution: Preface. J Environ Sci (China) 2016;40:1–2.
21. Catcott EJ. Effects of air pollution on animals. Monogr Ser World Health Organ. 1961;46:221–31.
22. Veras MM, Caldini EG, Dolhnikoff M, Saldiva PH. Air pollution and effects on reproductive-system functions globally with particular emphasis on the Brazilian population. J Toxicol Environ Health B Crit Rev. 2010;13:1–15
23. Schneider SH. The greenhouse effect: Science and policy. Science. 1989;243:771–81.
24. Loomis D, Huang W, Chen G. The International Agency for Research on Cancer (IARC) evaluation of the carcinogenicity of outdoor air pollution: Focus on China. Chin J Cancer. 2014;33:189–96.
25. Kjellstrom T, Lodh M, McMichael T, Ranmuthugala G, Shrestha R, Kingsland S. Air and water pollution: Burden and strategies for control. - Jamison DT, Breman JG, Measham AR, Alleyne G, Claeson M, Evans DB, et al., editors. Disease Control Priorities in Developing Countries. 2nd ed. Washington, DC: World Bank; 2006. pp. 817–32.
26. Zhou M, Liu Y, Wang L, Kuang X, Xu X, Kan H. Particulate air pollution and mortality in a cohort of Chinese men. Environ Pollut. 2014;186:1–6.
27. Pelucchi C, Negri E, Gallus S, Boffetta P, Tramacere I, La Vecchia C. Long-term particulate matter exposure and mortality: A review of European epidemiological studies. BMC Public Health. 2009;9:453.

28. Jerrett M, Finkelstein MM, Brook JR, Arain MA, Kanaroglou P, Stieb DM, et al. A cohort study of traffic-related air pollution and mortality in Toronto, Ontario, Canada. Environ Health Perspect. 2009;117:772–7]
29. Gorai AK, Tuluri F, Tchounwou PB. A GIS based approach for assessing the association between air pollution and asthma in New York state, USA. Int J Environ Res Public Health. 2014;11:4845–69
30. McCarthy JT, Pelle E, Dong K, Brahmbhatt K, Yarosh D, Pernodet N. Effects of ozone in normal human epidermal keratinocytes. ExpDermatol. 2013;22:360–1.
31. Lippmann M. Health effects of ozone. A critical review. JAPCA. 1989;39:672–9544.
32. Hatch GE, Slade R, Harris LP, McDonnell WF, Devlin RB, Koren HS, et al. Ozone dose and effect in humans and rats. A comparison using oxygen-18 labeling and bronchoalveolar lavage. Am J RespirCrit Care Med. 1994;150:676–83.
33. American Academy of Pediatrics Committee on Environmental Health. Lead exposure in children: Prevention, detection, and management. Pediatrics. 2005;116:1036–46.
34. Ko FW, Hui DS. Air pollution and chronic obstructive pulmonary disease. Respirology. 2012;17:395–401.
35. Nogueira JB. Air pollution and cardiovascular disease. Rev Port Cardiol. 2009;28:715–33.
36. Steenhof M, Janssen NA, Strak M, Hoek G, Gosens I, Mudway IS, et al. Air pollution exposure affects circulating white blood cell counts in healthy subjects: The role of particle composition, oxidative potential and gaseous pollutants – The RAPTES project. InhalToxicol. 2014;26:141–65.
37. Sun Q, Yue P, Ying Z, Cardounel AJ, Brook RD, Devlin R, et al. Air pollution exposure potentiates hypertension through reactive oxygen species-mediated activation of Rho/ROCK. ArteriosclerThrombVasc Biol. 2008;28:1760–6.
38. Leary PJ, Kaufman JD, Barr RG, Bluemke DA, Curl CL, Hough CL, et al. Traffic-related air pollution and the right ventricle. The multi-ethnic study of atherosclerosis. Am J RespirCrit Care Med. 2014;189:1093–100.
39. Van Hee VC, Adar SD, Szpiro AA, Barr RG, Bluemke DA, Diez Roux AV, et al. Exposure to traffic and left ventricular mass and function: The multi-ethnic study of atherosclerosis. Am J RespirCrit Care Med. 2009;179:827–34. [PMC free article] [PubMed]
40. Newman NC, Ryan P, Lemasters G, Levin L, Bernstein D, Hershey GK, et al. Traffic-related air pollution exposure in the first year of life and behavioral scores at 7 years of age. Environ Health Perspect. 2013;121:731–6
41. Singh B, Maibach H. Climate and skin function: An overview. Skin Res Technol. 2013;19:207–1
42. Vierkötter A, Schikowski T, Ranft U, Sugiri D, Matsui M, Krämer U, et al. Airborne particle exposure and extrinsic skin aging. J Invest Dermatol. 2010;130:2719–26.
43. Drakaki E, Dessinioti C, Antoniou CV. Air pollution and the skin. Front Environ Sci. 2014;2:11.
44. Potera C. More human, more humane: A new approach for testing airborne pollutants. Environ Health Perspect. 2007;115:A148–51.
45. Jacobson MZ. Review of solutions to global warming, air pollution, and energy security. Energy Environ Sci. 2009;2:148–73.

46. Faiz A, Weaver CS, Walsh MP. Air Pollution from Motor Vehicles: Standards and Technologies for Controlling Emissions. 1st ed. Washington, DC: World Bank Publications; 1996.
47. Wong EY, Gohlke J, Griffith WC, Farrow S, Faustman EM. Assessing the health benefits of air pollution reduction for children. Environ Health Perspect. 2004;112:226–32
48. Friedlander, S.K. 1973. Chemical element balances and identification of air pollution sources. Environmental Science and Technology 7:235-240
49. Murphy, J.J., M.A. Deluki, D.R. McCubbin, and H.J. Kim. 1999. The cost of crop damage caused byozone air pollution from motor vehicles. Journal of Environmental Management 55:273-289.NAPAP (National Acid Precipitation Assessment Program). 2005. National Acid Precipitation Assessment, Program Report to Congress: An Integrated Assessment.NRC (National Research Council). 2002. Estimating Public Health Benefits of Air Pollution Regulations. Washington, D.C.: National Academy Press.

CHAPTER 2

Indoor Air Pollution

2.1 Introduction

Indoor air pollution is by far the biggest environmental problem of the world. Every year, 4.3 million people die due to the exposure to household air pollution caused by indoor open fire. It is predominantly women and young children who are killed by indoor air pollution. The World Health Organization estimates that 4.3 million people a year die prematurely from illness attributable to the household air pollution caused by the inefficient use of solid fuels like wood, crop residues, dung, charcoal, and coal. Around 50% of people, almost all in developing countries, rely on solid fuels like coal and biomass in the form of wood, dung and crop residues for domestic energy. These materials are typically burnt in simple stoves with very incomplete combustion. Consequently, women and young children are exposed to high levels of indoor air pollution every day.

There is consistent evidence that indoor air pollution increases the risk of chronic obstructive pulmonary disease and of acute respiratory infections in childhood, the most important cause of death among children under 5 years of age in developing countries. Evidence also exists of associations with low birth weight, increased infant and perinatal mortality, pulmonary tuberculosis, nasopharyngeal and laryngeal cancer, cataract, and, specifically in respect of the use of coal, with lung cancer.

Thus indoor air pollution is a major global public health threat requiring greatly increased efforts in the areas of research and policy-making. Research on its health effects should be strengthened, particularly in relation to tuberculosis and acute lower respiratory infections. A more systematic approach to the development and evaluation of interventions is desirable, with clearer recognition of the interrelationships between poverty and dependence on polluting fuels.

2.2 Sources of Indoor Air Pollution

.The sources of indoor air pollution can be broadly categorized as combustion appliances, tobacco smoking, building materials, consumer products, metabolic activity and cooking. Some of them are described in this chapter. Typical sources of indoor air pollutants grouped by origin are presented in Table 2.1.

2.2.1 Combustion appliances

In cold countries, residential heating is very important and it is done by burning combustion materials like wood, coal, oil and gas in various types of furnaces. In Western countries, the fuel is burnt either outside the building or inside with flues that vent outside, to heat the water that is circulated in the rooms through pipes and/or heat exchangers. In developing and underdeveloped countries, the fuel is burnt directly in the rooms resulting in increased pollutant levels indoors in winter. In India, the rich and the upper middle class normally use electrical appliances for heating their rooms, whereas, the lower middle class and the poor use coal, wood or oil. In southern India, except in hill stations, residential heating is normally not in practice.

Pollutant emissions from residential wood combustion sources are more significant when compared to emissions from other fuel combustion sources like oil and gas. Wood combustion produces CO, NO_x, hydrocarbons, and respirable particles, and some emissions that are suspected carcinogens, such as benzo-α-pyrene. While most wood smoke is released to the outdoors, where it has become a particularly significant cause of ambient air pollution in some areas, some fraction remains inside.

Table 2.1 *Typical sources of indoor air pollutants, their sources and exposure guidelines*

Pollutant	Sources of indoor air pollutants	Exposure guidelines for average concentrations
SPM	Fuel combustion, cigarette smoking	55-110 $\mu g/m^3$ annual 150-350 $\mu g/m^3$ for 24 h
SO_x	Fuel combustion, kerosene, coal, kitchen furnaces	80.8 $\mu g/m^3$ annual 365 mg/m^3 24 h
NO_x	Fuel combustion, tobacco smoke, kerosene, coal, LPG	100 mg/m^3 annual
CO	Fuel combustion, tobacco smoke, kerosene, LPG	10 mg/m^3 for 8 h 40 mg/m^3 for 1 h
CO_2	Combustion, metabolic activity	-
Volatile organics	Combustion, volatilization, metabolic activity solvent, evaporation paints, pesticides, insecticides, fungicides, fumigants, adhesives, cosmetics, solvents, refrigerants, cooking vapours, cleaning sprays, paints, various furnaces	-
PAH	Combustion, tobacco smoke, polishers, cleaning agents, cooking vapours	-
Ozone	Photocopying, Laser printers	235 $\mu g/m^3$ 1 h once a year

Table 2.1 ***Contd...***

Pollutant	Sources of indoor air pollutants	Exposure guidelines for average concentrations
Ca, Cl_2, Si, Cd, F_2	-	
Rn and its daughters	Building materials, stone etc.	0.01 μg/m^3 working levels0.02 annual
HCHO	Tobacco smoke particle board, insulation, furnishings, carpets, urea, formaldehyde, other construction material	120 mg/m^3
Asbestos, mineral & synthetic fibres	Fire retardants, acoustic, thermal or electric insulation products	0.2 fibres longer than 5 μm
NH_3	Metabolic activity, cleaning products	-
Aerosols	Consumer products like perfume sprays	-
Hg	Fungicides in paints, spills in dental care facilities, laboratories, thermometer breakage	-
Viable organisms	Infections	-
Allergens	House dust, animal dander	-

(**Source:** *N. L. Nagda, H. E. Rector and M. D. Koontz, 1987. Guidelines for monitoring indoor air quality. Hemisphere. Washington DC.)*

Table 2.2 indicates the air pollutant emissions from various fuels used for residential heating. Emission rates of various indoor air pollutants by various types of kitchen burners and house heaters are given in Table 2.3 while their decay constants or re activities are represented in Table 2.4.

Table 2.2 *Air pollutant emissions from various fuels used for residential heating (kg/645 kW)*

Source	CO	NO_x (as NO_2)	SO_x (as SO_2)	Particulates	Bap (mg/645 kW)
Gas	0.02	0.08	0.006	0.02	0.08
Oil	0.04	0.09	0.3-7	0.07	0.9
Coal	3.5	0.1	0.5-7	0.6	2500
Wood	3.0	0.25	-	1.3	45

Table 2.3 *Some measured polluted emission rates from various sources*

Source	Pollutant emission rate, (mg/h)			
	CO	NO_x[a]	SO_2	HCHO
Gas range	1900	52	0.9	23
Oven	1840	83	1.5	16
One top burner				
Kerosene heater[b]				
Convective	71	122	-	1.1
Radiant	590	15	-	4.0
One cigarette (side stream smoke)[c] (mg)	86	0.05	-	1.44

(a) NO_x reported as N (b) New portable heaters, warm-up emissions not included, SO_2 not measured; (c)
Source: *National Research Council 1981. Indoor air pollutants. National Academy Press, Washington DC.*

Table 2.4 *Decay constants or reactivities, K*

Pollutant	K (l/h)
CO	0.0
NO	0.0
NO_x (as N)	0.15
HCHO	0.4
SO_2	0.23
Particles (<0.5 mm)	0.48
Radon	7.6×10^{-3}

Source: *G. W. Traynor, J. R. Allen, M.G. Apte, J.F. Dillworth, J. R. Girman, C. D. Hellowell and J. F. Koontz Jr., 1982. Indoor air pollution from portable kerosene-fire stoves, and wood-burning furnaces. Lawrence Berkeley Lab (LBL), 14027.*

2.2.2 Consumer products

Many consumer products produce different types of pollutants, especially Volatile Organic Compounds (VOC) in significant concentrations. Ozone from photocopiers, methanol from wet-copy machines, pesticides from fumigation, solvent evaporation from paints, polishes, paintings and cleaning agents, formaldehyde from various materials and draperies, aerosols from perfume sprays, air fresheners, etc. are some of the indoor pollutants generated from consumer products.

In an estimate, more than 250 volatile organic compounds were identified in quantities more than 1ppb in indoor environments. Volatile organic compounds are complex mixtures of many substances and their concentrations vary with time, room ventilation and proximity of the source. Table 2.5 shows the types of pollutants released by various classes of consumer products. Polycyclic aromatic hydrocarbons most of which are known to be carcinogenic are found in different cosmetic products like cold creams, cleaning creams, suntan oils, body lotions, lipstick, etc. Cosmetic raw materials such as mineral oils, refined petroleum products, vegetable oils, etc. also contain carcinogenic PAH.

Table 2.5 *Pollutants generated indoors from consumer products*

Consumer product	Pollutant/class
Paints	
Protective coatings	Polyurethane
Stains	Hydrocarbons, others
Metallic coatings	Hydrocarbons, resins
High temperature coatings	Hydrocarbons, resins
Primer	Toluol
Adhesives	Petroleum distillates
Varnishes	Petroleum distillates
Thinners	Methylene chlorides, naphtha

Table 2.5 ***Contd…***

Consumer product	Pollutant/class
Furniture polishes	Dinitrobenzene, 1,1,1-trichloroethane petroleum distillates, silicone, wax, morpholine
Paintings	Turpentine
Personal use	
Deodorant sprays	Hydrated aluminium chloride, isopropyl myristate talc, trigylcerides
Hair spary	Polyvinyl-pyrolide resins, ethanol, lanolin
Shaving foams	Stearic acid, triethanol amine, menthol, glycerine
Insecticides	
Wasp and hornet	Carbamate derivatives, petroleum distillates
Ant and roach	Pyrethrins, petroleum distillates
Mosquito and fly repellents	Dicarboxamine, petroleum distillates
Moth	Dicarboxamine, petroleum distillates
House and garden	Cyclopropane, petroleum distillates
General insects	Toluamide
Pet sprays	
Dog flea repellent	Para dichlorobenzene
Cat trainer	Para dichlorobenzene
Cat flea spray	Carboxy methyl carbonate
Cleaning agents	
Oven cleaners	Sodium hydroxide, hydroxyl ethyl cellulose, poly oxyethylene, fatty ethers
Drain cleaners 1,1,1	Trichloroethane, petroleum distillates
Tile cleaners	Tetra sodium methylene diamine
Window cleaners	Sodium nitrite, isopropyl alcohol, ethylene glycol, ammonium hydroxide
Spot remover	Perchloroethylene
Others	
Disinfectant sprays	Triisorporpanol amine, morpholine
Air fresheners	Propylene glycol, morpholine, ethanol

2.2.3 Cigarette (or tobacco) Smoke

Tobacco smoke contains numerous known or suspected carcinogens, including benzene, hydrazine, benzo-α-anthracene, benzo-α-pyrene (BaP), and nickel. Smoke particles are small, averaging about 0.2 mm, so they are easily carried into the deepest regions of the lungs. Smokers, of course, voluntarily expose themselves to the greatest concentrations (as when a smoker takes a puff – mainstream smoke), although non-smokers can be exposed to significant amounts as well from a cigarette simply smouldering in the air (side stream smoke). Other indoor air pollutants arising from tobacco smoke include carbon monoxide, nicotine, nitrosamines, a crolein and other aldehydes.

2.2.4 Building materials

The choice of building materials has a substantial influence on indoor air quality as the materials directly interact with air or define the temperature and humidity of buildings. Building materials release certain pollutants like formaldehyde and radon indoors. These may also be generated outdoors but their impact outdoors is not much due to their low concentrations. On the other hand, the impact of these pollutants on indoor air quality especially in tight and energy-efficient structures may be significant. Formaldehyde is a highly water soluble gas that can absorb on wood and other hygroscopic surfaces, from which it can evaporate after a dormant period. Formaldehyde is a common component of several types of condensation type synthetic resins. School buildings are especially problematic and have led to well-documented complaints in Canada, Germany and Russia. The high occupant density in schools and the high metabolic rate of children combine to provide a hot and humid atmosphere that is conducive to formaldehyde release. However, in India, there is limited utilization of formaldehyde synthetic resins indoors, especially in normal residential dwellings.

Radon can be emitted from some earth derived building materials such as brick, concrete, and tile. It can also exist in groundwater, to be released when that water is aerated, for example, during a shower, (the radon risk in water is from inhalation of the released gas, not from drinking the water itself). Radon has also been detected in some natural gas supplies, so modest amounts may be released in cooking. By far the most important source of radon, however, is soils and rocks that contain radium. The radon gas formed as radium decays can work its way to the surface. From there it can enter a house through the floor, especially if there is stack driven infiltration. Radon rich, high permeability soils are considered to be causing the highest levels of indoor radon concentrations. The emanation rate from soil seems to range from about 0.1 pCi/m^2s to over 100 pCi/m^2s, with an average value of 1 pCi/m^2s.

Indoor concentrations of radon tend to rise in structures in which the entrapped radon is sealed from the outdoors as a part of energy conservation measures. For example, land over uranium or phosphate rock deposits tends to have high radon emission rates. The rate of radon release from the earth varies from place to place as a function of the composition and the geological structure of the earth crust, composition of soil, source strength, infiltration and atmospheric pressure. Its release increases during low atmospheric pressure. It is reported that indoor concentrations of ^{222}Rn in energy efficient houses are in the range of 0.6-22 millicuries/m^3 and for other constructions the range is 0.005-3.1 millicuries/m^3.

2.3 Some Factors Influencing Indoor Air Quality

The indoor air quality will respond to changes in outdoor air quality and is influenced by various factors such as permeability of structures, ventilation of the building, generation of pollutants in indoor environments, nature of pollutants, pollutant depletion mechanisms, meteorological factors, etc.

2.3.1 Permeability of structures

In cold countries, measures to conserve energy in heating and cooling the buildings invariably call for measures to reduce the permeability of the structures; these measures in turn tend to isolate the indoor environments from the outdoors. The relative permeability of structures determines the ability of outdoor pollutants to penetrate the interior and the rate at which pollutants generated indoors are diluted with outdoor air.

2.3.2 Ventilation

The windows, doors and other openings which determine the ventilation of a building control the air exchange between the indoors and outdoors. If the ventilation is poor, the oxygen content of the indoor environments is limited and the quality of indoor air is depleted. People living or working in a poorly ventilated building may exhibit symptoms like drowsiness, nausea, sneezing, headache and dizziness. The set of symptoms is called the "sick building syndrome". It is recommended by the American Society of Heating, Refrigeration and Air-Conditioning Engineers (ASHRAE) that the minimum volume of outdoor air entering a building for ventilation purposes must be15-20 m^3 per occupant.

2.3.3 Meteorological factors

Meteorological factors tend to have secondary effects on indoor air quality. The indoor-outdoor temperature relationship influences the penetration of outdoor pollutants and dilution of pollutants generated indoors. Depending on the location of windows and doors the wind can have a profound effect on the indoor-outdoor air exchange of a building. Moisture, whether originating indoors or outdoors, may affect the relative permeability of the structure (causing swelling of window and door frames, reducing leakages) and can enhance pollutant deposition rates for gases such as SO_2 and O_3.

2.4 An indoor Air Quality Model

Indoor air quality in a building can be modelled using a box model. In this simple model the building will be treated as a single, well-mixed box with sources and sinks for the pollutants in question. This box model can be expanded to include several boxes, each characterized by uniform pollutant concentrations. A two-box model, for example, is sometimes used for radon estimates, where one box is used to model radon concentrations within the living space of a dwelling and the other box models the air space beneath the house.

Far quantifying the air pollutants in a building consider the simple, one-box model (Fig. 2.1) where the sources of pollution within the building can be characterized by various emission rates. Further, ambient air entering the building can also bring new sources of pollution, which adds to whatever may be generated inside. These pollutants may be removed from the building by infiltration or ventilation, or they may be non-conservative and decay with time. In addition, if there is a mechanical air cleaning system, some pollutants may be removed as indoor air is passed through the cleaning system and returned. To help keep the model simple, we will ignore such mechanical filtration.

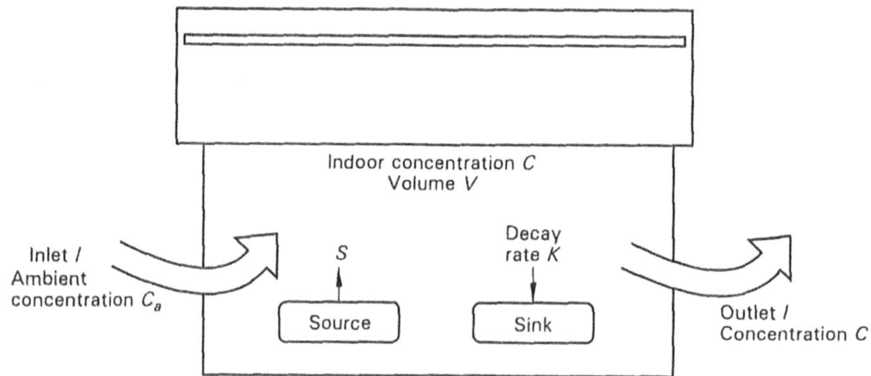

Fig. 2.1 *Box model for indoor air pollution*

For a well-mixed condition, a mass balance equation for the pollutant levels in the building can be written as:

(Rate of increase of the pollutant concerned in the box) = (Rate of pollutant entering the box) – (Rate of pollutant leaving the box) – (Rate of decay in the box)

$$V\frac{dC}{dt} = S + C_a IV - CIV - KCV \qquad .....(2.1)$$

where

V = volume of space in building (m^3/air change)
I = air exchange rate (ach)
S = pollutant source strength (mg/h)
C = indoor concentration (mg/m^3)
C_a = ambient concentration (mg/m^3)
K = pollutant decay rate or reactivity (l/h)

By setting $dC/dt = 0$, the steady state solution can be written as

$$C(\infty) = \frac{S/V + C_a I}{I + K} \qquad .....(2.2)$$

$$C(t) = \frac{S/V + C_a + I}{\left(1 - e^{-(1+kt)}\right)} + C(0)e^{-(I-K)t} \qquad .....(2.3)$$

where C(0) = the initial concentration in the building.

Pollutants like CO and NO_x, can be treated as if they were conservative, that is they do not decay with time or have significant reactivities; thus $K = 0$. A special case for conservative pollutants, corresponding to negligible ambient concentration ($C_a = 0$), and zero initial indoor concentration, is

$$C(t) = \frac{S}{IV}\left(1 - e^{-It}\right) \quad \text{.....(2.4)}$$

Problem 2.1

Estimation of pollution concentration in a room: When a kerosene heater is burnt for 2 h in a test chamber with a 54.0 m^3 volume and an infiltration rate of 0.25 ach, the concentration of CO reaches 50 mg/m^3. Initial CO in the best carbon is zero, and the ambient CO level is negligible throughout the heating period. Considering CO as a conservative pollutant, estimate the rate at which the heater emits CO. Estimate the steady state concentration if the heater was placed in a small home to heat 180 m^3 of space with 0.4ach infiltration rate.

Solution:

Rearranging (2.4) to solve for the source rate *S* gives

$$S = \frac{IVC(t)}{1-e^{-It}} = \frac{0.25\ \text{ach} \times 54.0\ \text{m}^3/\text{ac} \times 50\ \text{mg}/\text{m}^3}{1-e^{-0.25/\text{h} \times 2\text{h}}} = \frac{675}{0.40} = 1687.5\ \text{mg}/\text{h}$$

In the small home, the steady state concentration would be

$$C(\infty) = \frac{S}{IV} = \frac{1687.5\ \text{mg}/\text{h}}{0.4\ \text{ach} \times 180\ \text{m}^3/\text{ac}} = 23.43\ \text{mg}/\text{m}^3$$

This is in excess of the ambient standard for an 8 h average of 10 mg/m^3.

Problem 2.2

In a single level house the soil under the house emits 2.0 pCi/m^2s of radon gas. Assuming that all of this gas finds its way through the floor and into the house of 500 m^2 of floor space, with an average ceiling height of 2.6 m and an air change rate of 0.9 ach, calculate the radon concentration with steady state condition. The ambient concentration is negligible and the decay rate of radon is given as $K = 7.6 \times 10^{-3}$/h.

Solution:

We can find the steady state concentration using Equation 2.2. The decay rate of radon is given as $K = 7.6 \times 10^{-3}$/h.

$$C(\infty) = \frac{S/V + C_a I}{1+K}$$

$$\frac{\left(2\text{pCi}/\text{m}^2\text{s} \times 3600\,\text{s}/\text{h} \times 500\,\text{m}^2\right) / \left(500\,\text{m}^2 \times 2.6\,\text{m}\right)}{0.9/\text{h} + 7.6 \times 10^{-3}/\text{h}} = 2.769 \times 10^3\ \text{pCi}/\text{m}^3$$

2.5 Indoor Air Pollution in Underdeveloped and Developing Countries

In underdeveloped and developing countries like India, especially in the rural areas and small towns and cities, housewives or cooks spend a few hours per day in the kitchen.

The study of indoor air pollution in kitchens is very important because kitchen environments are especially susceptible to a high degree of air pollution from two angles, i.e. fuel combustion and processing of food items. Even today the indoor air quality in the kitchens of many rural homes in developing countries like India, is more or less the same as the smoky conditions that must have occurred in unventilated caves thousands of years ago.

2.5.1 Fuel combustion

Traditional fuels like wood, dried cattle dung, crop residues, etc., are generally used as fuel in rural areas and also in slums of urban areas for cooking purposes. In semi-urban areas like developed villages, small towns, and slums of cities, coal and kerosene are used for cooking. In the urban areas, with the improvement of living habitats, developments in technology and modification of the kitchen designs, traditional fuels have been substituted by modern energy sources like gas and electricity. It is estimated that about one half of the world's households use traditional fuels for cooking.

In a study carried out by the Planning Commission, Government of India, it was estimated that firewood is the major source of fuel (68.5%) followed by oil products (LP gas) (16.9%), cattle dung (8.3%), coal (2.3%), electricity (0.6%) and others (3.4%). The source-wise domestic energy consumption in rural India is presented in Table 2.6.

Table 2.6 *The source-wise rural domestic energy consumption in India*

Source	Energy usage	Per capita energy consumption in India		
		Purchased	Collected	Home-grown
Non-commercial fuels (80%)				
Fire wood	68.5	12.7	64.7	23.1
Cattle dung	8.3	5.1	26.2	68.7
Commercial fuels (20%)				
Oil products	16.9	100	0.0	0.0
Coal	2.3	65.1	34.9	0.0
Electricity	0.6	100	0.0	0.0
Others	3.4	8.9	61.0	30.1

Cooking dominates the energy consumption in the rural sector (64%) followed by agriculture needs (22%). For cooking, the maximum demand is met by traditional and non-commercial energy sources (98%) like firewood, cattle dung, and crop residues and the rest from commercial energy sources like coal, kerosene, oil and electricity. The rural energy use pattern in India is presented in Table 2.7 of the above mentioned energy sources, wood and cattle dung constitute the primary domestic energy sources for cooking in rural India. Data regarding per capita traditional fuel consumption in the different regions in rural parts of India are available.

Generally cooking in rural areas is performed using a stove called 'chula', consisting of just a horseshoe shaped block made by bricks and mud; the chula is designed to fit the cooking pots. The chula types range from a simple three-side brick design to relatively complicated stoves with a chimney, dampers and two or three pot holes. Although a large fraction of households cook indoors, only a small fraction have enclosed stoves with flues that vent combustion products outside.

Table 2.7 *Rural energy use pattern in India*

Use	Commercial		Non-commercial energy		Energy Kcal × 10^4	Total %
	Fuel	%	Fuel	%		
Cooking	Coal and others	2	Wood, cattle dung crop ressidues	98	7.31	64.0
Agriculture	Fertilizer	14	Human energy	77	2.52	22.0
	Diesel	3	Animal energy			
	Electricity	6	Manure			
Lighting	Kerosene	87	Negligible		0.48	4.0
	Electricity	13	Human energy	100	0.35	3.0
Transportation	Negligible		Human energy			
			Animal energy			
			Bicycles			
Others	Negligible		Fire wood	100	0.76	7.0
			Charcoal			
			Human energy			

The air pollutant emissions from fuel combustions depend on the fuel type and combustion conditions. In fact, the wood combustion emission factors for particulates, carbon monoxide and other pollutants exceed even that of coal, the dirtiest of the fossil fuels. Only few actual measurements have been made of indoor air quality, especially in kitchen environments in rural areas of developing countries.

2.5.2 Food processing

Frying of edible oils during food preparations is an important activity in the indoor (kitchen) environments of all classes of people in rural as well as urban areas in India. This is a general practice in almost all Asian countries. Such frying may contribute different toxic and carcinogenic substances in significant concentrations.

Problems

1. A single story home with infiltration rate of 0.5 ach has 200 m^2 of floor space and a total volume of 500 m^3. If 0.6 pCi/m^2s of radon is emitted from the soil and enters the house, estimate the steady state indoor radon concentration.
2. If the house in Problem 1 had been built as a two-story house with 100 m^2 on each floor and the same total volume, what would be the estimated radon concentration?
3. If an individual lived in the house given in Problem 1, estimate the life time risk of cancer from the resulting radon exposure. Use Nero's estimate that a house concentration of 1.5 pCi/L leads to a lifetime risk of 0.3 percent.
4. Consider a "tight" 300 m^3 home with 0.2 ach infiltration rate. The only source of CO in the home is the gas range and the ambient concentration of CO is always zero. Suppose there is no CO in the home at 6 pm, and then the oven and two burners are on for 1 h. Assume that the air is well mixed in the house and estimate the CO concentration in the home at 7 pm and again at 10 pm.
5. A convective kerosene heater is tested in a well mixed 27 m^3chamber having an air exchange rate of 0.39 ach. After 1 h of operation, the NO concentration reached 4.7 ppm. Treating NO as a conservative pollutant:
   (a) Estimate the NO source strength of the heater (mg/h).
   (b) Estimate the NO concentration that would be expected in the lab 1 h after turning off the heater.
   (c) If this heater were to be used in the home described in Problem 4, what steady state concentration of NO would you expect to be caused by the heater?
6. Sidestream smoke from a cigarette contains roughly 100 μg of benzo-α-pyrene. How many cigarettes per day would yield enough sidestream smoke to increase the cancer risk for a non-smoker by 1×10^{-6} if that individual spends all of his time indoors in a 150 m^3 home having 0.5 ach? Assume that benzo-α-pyrene has a half-life of 3 days and assume that its ambient concentration is negligible.

2.6 Indoor Air Pollution-Health Effects

The majority of households in developing countries burn biomass fuels in open fireplaces, consisting of such simple arrangements as three rocks, a U-shaped hole in a block of clay, or a pit in the ground, or in poorly functioning earth or metal stoves. Combustion is very incomplete in most of these stoves, resulting in substantial emissions which, in the presence of poor ventilation, produce very high levels of indoor pollution (1).

Many of the substances in biomass smoke can damage human health. The most important are particles, carbon monoxide, nitrous oxides, sulphur oxides (principally from coal), formaldehyde, and polycyclic organic matter, including carcinogens such as benzo [a] pyrene (2). Particles with diameters below 10 microns (PM10), and particularly those less than 2.5 microns in diameter (PM2.5), can penetrate deeply into the lungs and appear to have the greatest potential for damaging health (3).

Indoor concentrations of particles usually exceed guideline levels by a large margin: 24-hour mean PM10 levels are typically in the range 300–3000 mg/m3 and may reach 30 000 mg/m3 or more during periods of cooking (4–16).The mean 24-hour levels of carbon monoxide in homes using biomass fuels in developing countries are in the range 2–50 ppm; during cooking, values of 10–500 ppm have been reported

A health effect is determined not just by the pollution level but also, and more importantly, by the time people spend breathing polluted air, i.e. the exposure level. Exposure refers to the concentration of pollution in the immediate breathing environment during a specified period of time. This can be measured either directly through personal monitoring or indirectly by combining information on pollutant concentrations in each microenvironment where people spend time with information on activity patterns (17). Information on such patterns is very important for understanding the dynamic relationship between levels of pollution and behior. The Mechanisms of action of some key pollutants in smoke from domestic sources which cause health risks like respiratory infections asthma lung cancer etc. are presented in Table 2.8.

Table 2.8 *Mechanism of health risks of pollutants in smoke*

Pollutant	Mechanism	Potential health effects
Particules (small particles less than 10 microns, and particularly less than 2.5 microns aerodynamic diameter)	• Acute: bronchial irritation, inflammation and increased reactivity • Reduced mucociliary clearance • Reduced macrophage response and (?) reduced local immunity (?) Fibrotic reaction	• Wheezing, exacerbation of asthma • Respiratory infections • Chronic bronchitis and chronic obstructive pulmonary disease • Exacerbation of chronic obstructive pulmonary disease
Carbon monoxide	• Binding with heamoglobin to produce carboxy heamoglobin, which reduces oxygen delivery to key organs and the developing fetus.	• Low birth weight (fetal carboxy-heamoglobin 2 – 10% or higher) • Increase in perinatal deaths

Table 2.8 ***Contd...***

Pollutant	Mechanism	Potential health effects
Polycyclic aromatic hydrocarbons, eg. Benzo [a] pyrene	• Carcinogenic	• Lung cancer • Cancer of mouth, nasophgyix and larynx
Nitrogen dioxide	• Acute exposure increases bronchial reactivity • Longer term exposure increases susceptibility to bacterial and viral lung infections	• Wheezing and exacerbation of asthma • Respiratory infections • Reduced lung function in children
Sulphur dioxide	• Acute exposure increases bronchial reactivity • Longer term: difficult to dissociate from effects of particles	• Wheezing and exacerbation of asthma • Exacerbation of chronic obstructive pulmonary disease, cardiovascular disease
Biomass smoke condensates including polycyclic aromatics and metal ions	• Absorption of toxins into lens, leading to oxidative changes	• Cataract

2.7 Strategies to be adopted to Reduce Serious Health and Environmental Problems caused by Indoor Air Pollution

1. The immediate and widespread introduction of improved cook stoves, household biogas plans retained heat cookers, solar cookers the production of briquettes from biomass waste, and other local clean solutions. The introduction of these efficient cooking technologies should include the intensive training of trainers, local volunteers and end users in their use and production using locally available materials, The development of these programs should include input from local community leaders in order to determine the availability of commonly used building materials and fuels, and to respect local cooking traditions and cultural preference.
2. Funding for coordinated international research programs should be provided for the development of even more efficient low cost, clean cooking technologies.
3. Funding should be provided for public information campaigns to increase the acceptance of clean cooking technologies in the developing world. National leaders and celebrities should be recruited to help popularize these technologies by emphasizing their health benefits and their convenience.
4. Subsidies to be provided for clean fuels like LPG and clean cooking

Questions

1. What is Indoor Air Pollution? Explain in detail.
2. What are the major sources of indoor air pollution? Explain.
3. Explain how the combustion appliances influences indoor air pollution?
4. List various indoor air pollutants and their sources.
5. list different consumer products and air pollutants released by them.
6. Explain the influence of building materials on indoor air pollution.
7. Does food processing increases indoor air pollution. Explain.
8. Discuss the factors influencing the Indoor Air Quality.
9. Write a short note on Indoor Air Quality in restaurants.
10. Explain the modeling procedure for Indoor Air Quality.
11. How indoor air pollution influences human health.
12. Explain preventive methods to reduce indoor air pollution.
13. When a kerosene heater is burnt for 3h in a test chamber with a volume of 70.0 m^3 and an infiltration rate of 0.35 ach, the concentration of CO reaches 45 mg/m^3. Initial CO in the best carbon is zero, and the ambient CO level is negligible throughout the heating period. Considering CO as a conservative pollutant, estimate the rate at which the heater emits CO. Estimate the steady state concentration if the heater was placed in a small home to heat 200 m^3 of space with 0.35ach infiltration rate.
14. Ina single story house the soil under the house emits 1.5 pCi/m^2s of radon gas. Assuming that all of this gas finds its way through the floor and into the house of 650 m^2 of floor space, with an average ceiling height of 1.9 m and an air change rate of 0.8 ach, calculate the radon concentration with steady state condition. The ambient concentration is negligible and the decay rate of radon is given as $K = 7.1 \times 10^{-3}/h$.

References

1. Air quality guidelines for Europe. Copenhagen, World Health Organization Regional Office for Europe, 2000
2. De Koning HW, Smith KR, Last JM. Biomass fuel combustion and health. Bulletin of the World Health Organization, 1985, 63: 11–26.
3. United States Environmental Protection Agency. Revisions to the National Ambient Air Quality Standards for Particles Matter. Federal Register, July 18 1997, 62: 38651–38701.
4. Smith KR et al. Air pollution and the energy ladder in Asian cities. Energy, 1994, 19: 587–600.

5. McCracken JP, Smith KR. Emissions and efficiency of improved wood burning cook stoves in highland Guatemala. Environment International, 1998, 24: 739–747.
6. Air quality guidelines for Europe. Copenhagen, World Health Organization Regional Office for Europe, 2000 (in press).
7. Anderson HR. Respiratory abnormalities in Papua New Guinea children: the effects of locality and domestic wood smoke pollution. International Journal of Epidemiology, 1978, 7: 63–72.
8. Collings DA, Sithole SD, Martin KS. Indoor wood smoke Pollution causing lower respiratory disease in children. Tropical Doctor, 1990, 20: 151–155.
9. Martin KS. Indoor air pollution in developing countries. Lancet, 1991, 337: 358.
10. Nor boo T et al. Domestic pollution and respiratory illness in a Himalayan village. International Journal of Epidemiology, 1991, 20: 749–757.
11. Saksena S et al. Patterns of daily exposure to TSP and CO in the Garhwal Himalaya. Atmospheric Environment, 1992, 26A: 2125–2134.
12. Ellegard A. Cooking fuel smoke and respiratory symptoms among women in low-income areas of Maputo. Environmental Health Perspectives, 1996, 104: 980–985.
13. Robin LF et al. Wood-burning stoves and lower respiratory illnesses in Navajo children. Pediatric Infectious Diseases Journal, 1996, 15: 859–865.
14. Ellega°rd A. Tears while cooking: an indicator of indoor air pollution and related health effects in developing countries. Environmental Research, 1997, 75: 12–22.
15. Albalak R, Frisancho AR, Keeler GJ. Domestic biomass fuel combustion and chronic bronchitis in two rural Bolivian villages. Thorax, 1999, 54 (11): 1004–1008.
16. Albalak R et al. Assessment of PM10 concentrations from domestic biomass fuel combustion in two rural Bolivian highland villages. Environmental Science and Technology, 1999, 33: 2505–2509.
17. Zhang J et al. Carbon monoxide from cook stoves in developing countries: 2. Potential chronic exposures. Chemosphere-Global Change Science, 1999, 1: 367–375

CHAPTER 3

Air Pollution-Episodes-Preventive Measures-Air Quality Management Approaches, Emission Inventories, Present Global and National Scenarios

3.1 Air Pollution Episodes

A critical examination of air pollution episode (1) will provide important lessons for the management of air quality. Table 3.1 lists some prominent air pollution episodes that have occurred in different places of the world resulting in great damage to people, animal life, vegetation and materials.

Table 3.1 *Air pollution episodes and mortality*

Period	City and country	Deaths reported
1873, Dec. 9-11	London, England	250
1880, Jan. 26-29	London, England	1000
1892, Dec. 28-30	London, England	NA
1930, December	Meuse Valley, Belgium	63
1948, October	Dohora, Pennsylvania, USA	70
1948, Nov. 26-Dec. 1	London, England	800
1950, November	Poza Rica, Mexico	22
1952, Dec. 5-9	London, England	4000
1953, November	New York, USA	250
1956, Jan. 3-6	London, England	1000
1957, Dec. 2-5	London, England	800
1958	New York, USA	

Table 3.1 ***Contd...***

Period	City and country	Deaths reported
1959, Jan. 26-31	London, England	250
1962, Dec. 5-10	London, England	700
1963, Jan. 7-22	London, England	700
1963, Jan. 9-Feb. 12	New York, USA	400
1966, Nov. 23-25	New York, USA	170
1976, July	Sevesco, Italy	
1984, December 2	Bhopal, India	2500
1986, April	Chernobyl, Ukraine	2500

If we examine air pollution episodes, it will be clear that they occur mostly in un favourable meteorological conditions which do not facilitate dispersion of air pollutants, such as fog and lower height inversions and stagnating air masses in steep valleys. In nuclear disasters the effects of the air pollutants are on a continental scale requiring international monitoring authorities for monitoring the air quality.

3.1.1 Meuse valley episode of Belgium in 1930

Meuse valley of Belgium extends over an area of 24 km with 100 m high hills on either side above the river. The intense industrial activity in the valley in the form of steel mills, sulphuric acid plants, fertilizer plants, blast furnaces, and glass factories makes it a typical candidate for an air pollution disaster.

In December 1930, when anti-cyclonic conditions with still air set in over the whole of Belgium, a haze hung over Meuse valley for a period of 3-4 days. This resulted a large number of people taking ill, and the death of about 60 old people. The main symptoms of the illness were throat irritation, cough, breathlessness and constriction of the chest. The death rate was about 10-15 times the normal rate. A combination of high levels of SO_2, SO_3, and SPM was responsible for the illness and deaths.

3.1.2 Donora episode of Pennsylvania, USA in 1948

Donora which is covered with fog due to anticyclonic and temperature inversion conditions experienced serious air pollution conditions in October 1948. This town is located in a river valley, 48 km south of Pittsburgh, has a range of hills of 120 m height (340 m above sea level) on three sides and is part of a highly industrialized area with industries like chemical plants, a large zinc reduction plant and a sulphuric acid plant. On October 27, 1948, a high concentration of air pollutants stagnated under foggy conditions; this prevailed upto October 31. The pollutants were then washed out due to precipitation by rain.

During these four days nearly 47% of the population became ill due to cough and cold, 21.5% had constriction of chest, 17% suffered from headache and 12.3% had eye irritation. 70 people aged 48-52 years died from the third day. The death rate from1948 to 1958 rose to one person every third day.

Autopsies on the patients who died indicated acute irritative changes in the lungs characterised by capillary dilations, and haemorrhage resulting in death. Patients with severe cardio-respiratory problems suffered the most. The combination of air pollutants, SO_2 (0.5-2.0 ppm), SO_3 and SPM was responsible for this disaster.

3.1.3 London episodes

London, situated in the Thames valley of southern England, experienced a series of air pollution episodes because of its frequent anticyclonic atmospheric conditions, temperature inversions and clusters of industrial complexes. Particularly in winter, a thick fog covers many parts of the British Isles; air pollutants concentrate under these adverse atmospheric conditions resulting in frequent air pollution episodes. About ten such episodes have occurred in the period, 1873-1963, affecting several lakhs of people.

Notable among these episodes is the 1952 episode from December 4-9 when there was a dense fog over London with severe anticyclonic conditions and temperature inversion. Under these conditions the fog did not blow away but continued with a sunken foot and changed the atmosphere from white to brown and then a gloomy black haze. Due to the drastic reduction in visibility buses could ply at a speed of only 2 km/h; 12 h after the fog set in, the concentration of air pollutants in the atmosphere began affecting people exposed to it and caused serious health problems: respiratory problems, throat and nasal diseases, cough, sneezing, etc. It is estimated that nearly 4000 deaths occurred in this period and in the next two months 8000 more deaths were recorded with a good correlation between the death rate and concentration of smoke and SO_2 (0.7-1.3 ppm). The combined synergic action of sulphur oxides and particulate matter was responsible for the disaster.

In the winter of 1956, another episode of a similar nature occurred resulting in 1000 more deaths. The reason for the high concentrations of sulphur dioxide and SPM is the cluster of industries around London and domestic heating by coal and coke containing an average of 1.5% sulphur. To control these emissions, British authorities passed the Clean Air Act in 1956 which restricted the use of high sulphur fuels in critical areas of the city. The forcible shift to alternative energy sources like electricity, oil or low sulphur coal though expensive, helped in reducing the pollutants in urban air drastically.

3.1.4 Poza Rica disaster of Mexico in 1950

On November 24, 1950, in Poza Rica, a town of 15,000 inhabitants, in the Gulf of Mexico, an accident at one of the local factories which recovers sulphur from natural gas released hydrogen sulphide into the ambient air. The release which lasted for only 30 minutes was concentrated under fog and still atmospheric conditions with temperature inversion. 22 people were killed and 320 people hospitalized.

3.1.5 TCDD accident at Sevesco, Italy in 1978

In July1978, at Sevesco, Italy, an explosion took place in a chemical plant manufacturing herbicides. A white cloud of poisonous gas 2,3,7,8 tetrachloridebenzodioxin (TCDD) came out and engulfed the nearby town of Sevesco. TCDD is very poisonous to man. Buildings, ground and soil were contaminated. After three weeks of the accident, the Italian government evacuated about 800 people from the worst-affected areas. However, the dioxin pollution continued to spread and particulates were deposited over a widespread area. About 200 people including children suffered from skin diseases and some of them, from liver troubles. About 1 % of the babies born after the accident were deformed and also premature. According to experts, dioxin will continue to contaminate soil and water and affect water resources and natural biological cycles.

3.1.6 Bhopal gas tragedy of India in 1984

Bhopal, the capital city of the state of Madhya Pradesh in India, with a population of 7 lakhs experienced a serious air pollution disaster on December 2, 1984 when a major accident occurred in a Union Carbide operated pesticide plant. This plant was producing methyl isocyanide, CH_3NCO, a basic ingredient for making pesticides. It is very toxic, highly inflammable and fatal at high concentrations (>5 ppm); it can cause blindness and respiratory problems at lower concentrations. MIC with a boiling point of 39.1°C, density of 0.96 g/m^3 and vapour pressure of 348 at 20 °C will be generally in the liquid state under pressure, Aldicarb, carbofuran, carbonyl and methyl are the widely used pesticide derivatives of MIC. It is also an intermediate for serin and termic herbicides. [2, 3]

At 11 pm, December 2, 1984, a pressure build-up in 15000 gallon tubes containing 45 tons of MIC (methyl isocyanate) was observed by a worker at the Union Carbide plant in Bhopal. MIC leaked from the valve into the Bhopal air and being heavier than air drifted into the city covering an area of 25 sq. km. Inhabitants near the plant were the first victims. In a few hours, the whole city experienced the toxic effects of MIC.

The still air with a heavy mist at a temperature of 15 °C in the night did not help in the dilution of MIC by dispersion; people were thus exposed to the toxic gas for a longer period. Within a few hours, hundreds of people were found dead and thousands were suffering from breathlessness, irritation of eyes and throat and desperately seeking medical help.

The first symptom is choking. The lungs then eventually fill up with fluid and asphyxiation ensues. In most cases of moderate exposure to MIC, the effects are treatable. For a number of days after the accident, people kept streaming into the hospitals, seeking help, while coughing and gasping for breath, or convulsed with violent spasms. Within a few weeks, hospitals filled up, and patients gathered in the corridors and on the grounds outside. Tents were erected.

About 2500 people died in a week and about two lakh people exposed to the deadly gas in Bhopal and neighbouring towns fell ill. A large number of animals were also killed by the MIC gas. People who survived are likely to be suffering from asthma, emphysema, or bronchitis. The poisoning could have resulted in permanent damage to the liver and kidneys and the central nervous system.

The US government conducted some tests relating to the Bhopal gas leak disaster which revealed that a sudden high level of exposure to MIC causes persistent lung damage in animals. Many Bhopal victims are permanently disabled and vulnerable to a number of respiratory diseases.

The immediate cause of the tragic gas leak was a rise in pressure in an MIC storage tank. The storage tank was partially buried in the ground, and the outside surface was being cleaned when temperatures started to rise. Although the tank had a capacity of 35 tons, it contained 15 tons of MIC at the time. The rise in pressure ruptured a safety valve, leading to MIC emission into the atmosphere. At that point, the scrubber, which uses caustic soda to neutralize MIC, should have been triggered into operation. But apparently it failed, allowing the lethal gas to move through the flaring tower. At this tower, the poisonous gas should have been incinerated and transformed into harmless gases prior to spewing into the atmosphere. This also failed, probably because of the great surge in pressure and the overwhelming quantities of the escaping gas. The refrigeration unit that controlled the temperature was not functioning and had been out of operation for several months. After nearly 40 minutes of the emission of the deadly gas into the atmosphere, the leak was stopped by hosing down the tank with cold water, causing the safety valve to close after the pressure had decreased.

An exothermic reaction must have taken place within the tank where MIC was usually stored under a blanket of dry nitrogen. Such reactions occur when MIC combines with moisture to form methylamine and carbon monoxide. The important question was, how did the water get into the tank to react with the deadly MIC? A gauge indicating the pressure of the tank was removed. A water hose line could have been connected accidentally or on purpose to the tank, and thus water was mixed with the deadly chemical to cause an exothermic reaction. A number of experts mentioned that a disgruntled employee could have mixed the water with MIC in the tank as an act of sabotage.

A human error, or mechanical failure, or sabotage could have caused this tragedy. Safety measures that were supposed to protect against such a disaster did not work. The plant was also temporarily closed for maintenance two weeks prior to the accident. The storage tanks and the pipes as well as the scrubber connecting them were also under repair. The leak did not show up on monitors, simply because a critical panel in the control room had been removed. [4, 5]

Studies of the residue taken from the storage tank indicated that a higher than average percentage of chloroform existed. Chloroform is used as a solvent in the production of methyl isocyanate. The excessive amount of chloroform could have resulted from the reaction occurring at higher than recommended temperatures.

Human error and negligence are paramount factors in this accident. It is evident from the fact that a number of accidents at the Bhopal plant had taken place since it first went into operation in 1977; the facility endured six accidents in six years before the large-scale tragedy.

3.1.7 The Chernobyl nuclear disaster in 1986

In the early hours of April 26, 1986 the Chernobyl nuclear reactor, 18 km from the town of Chernobyl, located 128 km north of the city of Kiev in the Ukraine, exploded resulting in the worst accident in the history of nuclear energy. [6]

This was the result of a series of blunders made by plant operators and engineers the previous day. They had deliberately disabled a number of automatic safety and warning systems in a disastrous sequence of events for the purpose of conducting an experiment to determine how long the steam-driven turbines could run and generate electricity if power was shut down unexpectedly. This experiment was carried out in spite of the fact that a number of diesel generators were provided as a back-up system to be used at Chernobyl in case of an emergency to supply the needed energy to operate the pumps and other critical control mechanisms.

The control rods were lowered into the core of the reactor by the time plant operators– one of the terrible, calculated, deadly steps responsible for the accident. Soon the energy output level dropped from 3,200 megawatts (MW) to 1,600 megawatts. Then the emergency cooling system was shut off because it would draw power and affect the test results and in this process, all the safety regulations were violated.

They wanted to continue the supply of electrical energy to the region, so the reactor was kept running at 50% of the normal power level for nine hours, while the emergency cooling system was still out of operation. By 11:10 p.m. on April 25, the monitoring systems were adjusted to low power levels, but the computer was not programmed to maintain power at 700 MW to 1,000 MW. The power then dropped to the dangerously low level of 30 MW, making the reactor more difficult to control and thus extremely unstable. To increase the power output, most of the control rods were withdrawn. After the power dropped so low for the test, the operators took steps to increase it.

In principle, 30 rods must be inserted into the reactor's nuclear fuel assembly at all times for the purpose of regulating the nuclear chain reaction. However, operators took out all but about 8 of these rods. In addition to this, they closed another safety system which would have automatically shut down the nuclear reactor as soon as the turbines stopped operating–a violation of the most sacred rule.

At this point, radioactive xenon gas had built up to higher levels in the reactor. Xenon absorbs neutrons and "poisons" the nuclear reaction, thus dangerously lowering the power level and making the reactor more difficult to control. The operators had difficulty in stabilizing the power output, but they were eventually able to bring it back up to 200 MW.

By 1 a.m. on April 26, the operators activated all eight pumps to ensure adequate cooling after the test. Since they found out that the water level was too low, and since they added main coolant pumps, they created a critical new problem. Excessive flow rates of cold water into the reactor caused condensation of the steam and further destabilized the nuclear reactor. The combination of low power and high flow rates necessitated several manual adjustments. Due to the wide fluctuations in water and pressure levels, the operators were afraid that the reactor would shut itself down for safety reasons; this would have prevented their experiment from being carried out. To avoid this, they blocked the remaining emergency warning signals for water and pressure levels. By doing so, they brought the plant to the maximum point of danger, and then started their dangerous experiment.

To prevent steam from reaching the turbine-generating unit which was being tested, the operators shut off the steam valves leading to it, in the early hours on April 26. The moment they started the tests by shutting off the steam valves a series of dangerous events took place in a split second. Water pumps slowed down; lower cooling water flow rates resulted. Steam was building up at dangerous rates thus triggering a runaway reaction. Only 40 seconds after starting their test, they realized that a critical situation was at hand. Futile attempts were made to stop the reaction by inserting control rods which have 5 metres of graphite at their ends, back into the nuclear reactor, but it was too late. The additional moderator and the water it displaced accelerated the reaction. In the next 4 seconds, power zoomed to 100 times the operating maximum. The uranium fuel disintegrated and burst through its zirconium alloy cladding, bringing itself into direct contact with the cooling water.

Two enormous explosions occurred. The first was probably due to the excessive steam pressure causing tubes in the reactor core to burst and the reactor vault to disintegrate, Steam and graphite reacted, producing free hydrogen and carbon monoxide, These gases combined with oxygen to cause another big explosion, which blew off the top of the building and ignited the graphite. Sixteen hundred water pipes were burst. The radioactive discharge was equivalent to a kiloton-strong nuclear explosion.

It has been estimated that 10% of the more than 150 ton pure core of the Chernobyl reactor escaped into the atmosphere. Radioactive dust and gases formed a dense cloud surging nearly 4.8 km into the atmosphere.

This prevailing winds carried the radioactive isotopes thousands of kilometers away from the site of the tragedy and spread them over a wide area. The explosions started more than 30 fires in the plant area. The reactor core and the graphite surrounding it burned at temperatures in excess of 1535 °C. The searing heat became an inferno, propelling a deadly plume far into the upper atmosphere.

To prevent the flames from spreading to an adjoining reactor, thousands of civilians risked their lives and futures to engage in one of the greatest peacetime mobilisations in history. They dumped more than 5,000 tons of boron, lead, and sand over the fires in a futile attempt to extinguish them. Many lost their lives and many more will suffer from the radiation effects. Defying the deadly radioactive emissions, they halted the spread of fire by 5 a.m. on Saturday, April 26. However, the core of the reactor continued to rage in flames for days.

The explosion and destruction came from nuclear reactor No. 4, which knocked off a lead cap weighing around 1,000 tons. It also ripped wide open the building's side roof. Several tons of the uranium oxide fuels and fission products were spewed into the atmosphere. These included cesium-137 and iodine-131.

The two republics most affected by the tragic accident were Ukraine and Byelorussia. To prevent accumulation of excessive exposure to radiation, lead was used to shield the vault, boron carbide was used to absorb the neutrons, and dolomite was used to generate carbon dioxide so that the fires would be smothered and the graphite would be extinguished. Sand and clay were dumped to help cut off oxygen supply and filter out radionuclide's. Five thousand tons of materials were deposited. Fear was genuine about the possibility of the reactor sinking and thus contaminating groundwater. A tunnel was dug beneath the building so that it could be reinforced with concrete. Robots and remote control bulldozers were used to detect and remove radioactive matter. Where radiation was high, robots ceased to be useful as their electronics stopped functioning. By mid-summer, many tons of concrete were being dumped on the site and in November1986, the No. 4 nuclear reactor was encased in a concrete tomb which can safely contain the dangerous material of the reactor for many, many hundreds of years before nuclear decay quiets down in its radioactive core.

After about six months of hard labour in 1987, the ruined reactor could be encased in concrete and decontamination of the surrounding area was almost completely accomplished.

Over 100,000 people from the vicinity of Chernobyl will have to be monitored for the rest of their lives for possible damage from radiation. They were all resettled outside the nuclear disaster zone. The centre of this tragedy is expected to remain unfit for habitation for many years to come.

3.1.8 Impact all over the world

Radiation alarms were sounded in Sweden, 1280 km away from the accident on April 28, 1986. The Swedes exhausted all possible avenues for leaks or mishaps, and decided that the dangerous radiation in the atmosphere must have emanated from elsewhere. Attention was then directed toward their powerful neighbour, the former Soviet Union. Winds coming from that direction carried the deadly radioactive emissions over the Black Sea, across the Ukraine, over the Baltic, and into the Scandinavian countries. [7]

When the Soviets were asked about the mysterious radiation, they were assertive in declaring "there is nothing wrong in the Soviet realm". Officials in the Scandinavian countries insisted that something was terribly wrong, and that it had to be coming from the direction of the former Soviet Union! The denials went on for six more critical days after the tragic accident. Then the former Soviet Union acknowledged in a four-sentence statement: "An accident has taken place at the Chernobyl power station and one of the reactors was damaged. Measures are being taken to eliminate the consequences of the accident. Those affected by it are being given assistance. A government commission has been set up".

3.2 Air Pollution–Preventive Measures

3.2.1 WHO air quality guidelines

Air pollution is the world's largest single environmental health risk, as recognised by the World Health Organisation (WHO) based on a significant number of epidemiological studies which have reported the adverse impacts of air pollution exposure on human health in recent years. The WHO estimates that outdoor air pollution exposure leads to approximately 3.7 million premature deaths annually worldwide, with South-East Asia and the Western Pacific Regions carrying the largest air pollution-related health burden. In order to tackle global air quality, the WHO recently passed a landmark resolution focusing on it's technical capacity to support Member States in taking action on air pollution, building their capacity to implement the WHO air quality guidelines and advance research into air pollution's health effects

The 2005 WHO Air quality guidelines (AQG's) are designed to offer global guidance on reducing the health impacts of air pollution. The guidelines first produced in 1987 (8) and updated in 1997 (9) had a European scope. The new (2005) guidelines apply worldwide and are based on expert evaluation of current scientific evidence. They recommend revised limits for the concentration of selected air pollutants: particulate matter (PM), ozone (O_3), nitrogen dioxide (NO_2) and sulfur dioxide (SO_2).

Table 3.2 *WHO Guidelines (Tolerance values) for major Air Pollutants (Source WHO)*

Air Pollutant	Long-term mean value	Short-term mean value
$PM_{2.5}$ PM_{10}	10 μg/m_3 annual 20 μg/m_3 annual	25 μg/m_3 24-hour 50 μg/m_3 24-hour
O_3	-	100 μg/m_3 8-hour
NO_2	40 μg/m_3 annual	200 μg/m_3 1-hour
SO_2	20 μg/m_3 24-hour	500 μg/m_3 10-minute

Preventive Measures

Air pollution prevention measures aim at the preservation of environmental quality by prescribing the tolerated degree of pollution, leaving it to the local authorities and polluters to devise and implement actions to ensure that this degree of pollution will not be exceeded. An example of legislation within this approach is the adoption of ambient air quality standards, very often, on air quality guidelines (8) (WHO 1987) for different pollutants; these are accepted maximum levels of pollutants (or indicators) in the target area (e.g., at ground level at a specified point in a community) and can be either primary or secondary standards.

Primary standards (10) (WHO 1980) are the maximum levels consistent with an adequate safety margin and with the preservation of public health, and must be complied with within a specific time limit;

Secondary standards are those judged to be necessary for protection against known or anticipated adverse effects other than health hazards (mainly on vegetation) and must be complied "within a reasonable time".

Air quality standards are short - medium - or long-term values valid for 24 hours per day, 7 days per week, and for monthly, seasonal or annual exposure of all living subjects (including sensitive subgroups such as children, the elderly and the sick) as well as non-living objects; this is in contrast to maximum permissible levels for occupational exposure, which are for a partial weekly exposure (e.g., 8 hours per day, 5 days per week) of adult and supposedly healthy workers.

3.2.2 International agencies role in global air pollution preventive measures

International agencies such as the World Health Organization (WHO), the World Metrological Organization (WMO) and the United Nations Environment Programme (UNEP) have instituted monitoring and research projects in–order to clarify the issues involved in air pollution and to promote measures to prevent further deterioration of public health and environmental and climatic conditions. The Global Environmental Monitoring System GEMS/Air (WHO/UNEP 1993 (11) is organized and sponsored by WHO and UNEP and has developed a comprehensive programme for providing the instruments of rational air pollution management (Figure 3.1).

The importance of this programme is development of a global database of urban air pollutant concentrations of sulphur dioxides, suspended particulate matter, lead, nitrogen oxides, carbon monoxide and ozone. As important as this database, is the provision of management tools such as guides for rapid emission inventories, programmes for dispersion modeling, population exposure estimates, control measures, and cost-benefit analysis. In this respect, GEMS/Air provides methodology review handbooks (12) (WHO/UNEP 1994, 1995), conducts global assessments of air quality, facilitates review and validation of assessments, acts as a data/information broker, produces technical documents in support of all aspects of air quality management, facilitates the establishment of monitoring, conducts and widely distributes annual reviews, and establishes or identifies regional collaboration centres and/or experts to coordinate and support activities according to the needs of the regions. (WHO/UNEP 1992, 1993, 1995). The Global Atmospheric Watch (GAW) programme (13) (Miller and Soudine 1994) provides data and other information on the chemical composition and related physical characteristics of the atmosphere, and their trends, with the objective of understanding the relationship between changing atmospheric composition and changes of global and regional climate, the long-range atmospheric transport and deposition of potentially harmful substances over terrestrial, freshwater and marine ecosystems, and the natural cycling of chemical elements in the global atmosphere/ocean/biosphere system, and anthropogenic impacts thereon.

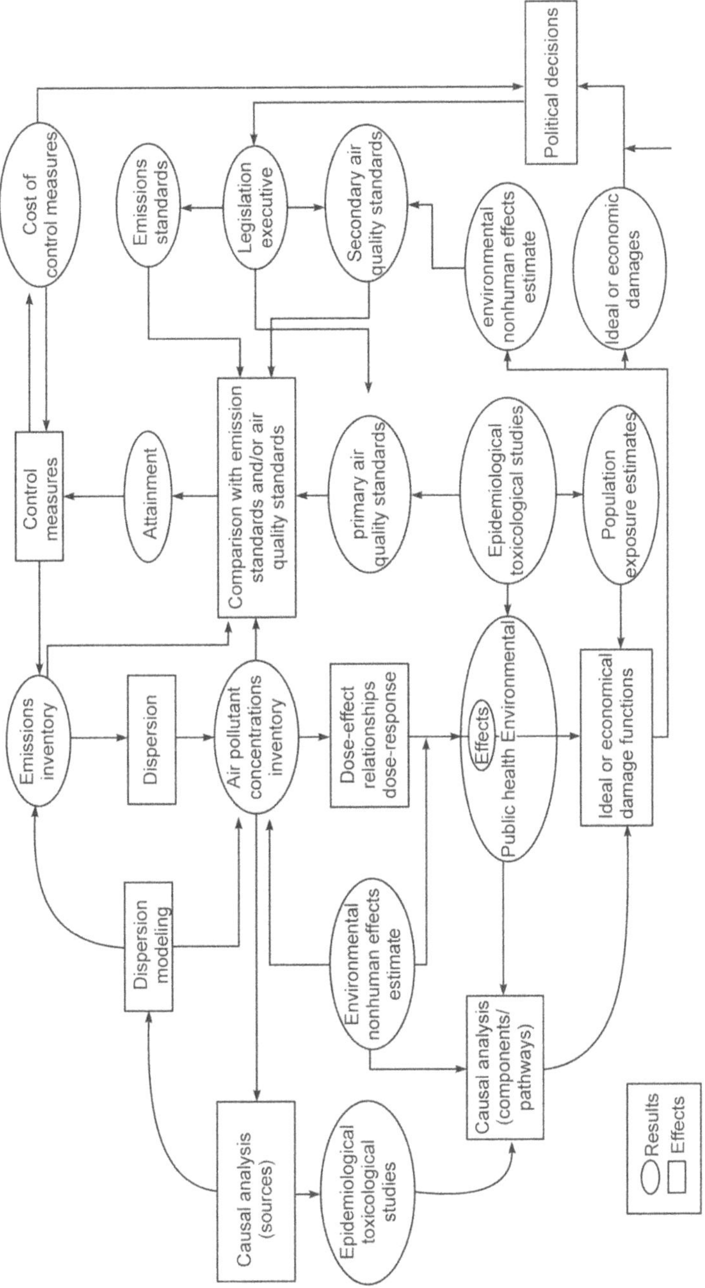

Fig. 3.1 *Global environmental monitoring system/air pollution prevention programs*

The GAW programme consists of four activity areas:

(i) the Global ozone Observing System (GO_3OS), (ii) global monitoring of background atmospheric composition, including the Background Air Pollution Monitoring Network (BAPMoN); (iii) dispersion, transport, chemical transformation and deposition of atmospheric pollutants over land and sea on different time and space scales and (iv) exchange of pollutants between the atmosphere and other environmental compartments; and integrated monitoring. One of the most important aspects of the GAW is the establishment of Quality Assurance Science Activity Centres to oversee the quality of the data produced under GAW.

3.3 Air Quality Management Approaches

Air Quality management aims at the elimination, or reduction to acceptable levels, of airborne gaseous pollutants, suspended particulate matter and physical and, to a certain extent, biological agents whose presence in the atmosphere can cause adverse effects on human health (e.g., irritation, increase of incidence or prevalence of respiratory diseases, morbidity, cancer, excess mortality) or welfare (e.g., sensory effects, reduction of visibility), deleterious effects on animal or plant life, damage to materials of economic value to society and damage to the environment (e.g., climatic modifications). The serious hazards associated with radioactive pollutants, as well as the special procedures required for their control and disposal, also deserve careful attention.

Unless there is adequate control, the multiplication of pollution sources in the modern world may lead to irreparable damage to the environment and mankind. Air pollution management, requires a multidisciplinary approach as well as a joint effort by private and governmental entities.

3.3.1 Approaches with control measures at source

Typical approaches in air quality management are control measures at the source, for example,

- enforcement of the use of catalytic converters in vehicles or of emission standards in incinerators,
- land-use planning and shut-down of factories or reduction of traffic during un favourable weather conditions.

The best air quality management approach stresses that the air pollutant emissions should be kept to a minimum; this is basically defined through emission standards for single sources of air pollution and could be achieved for industrial sources, for example, through closed systems and high-efficiency collectors.

An emission standard is a limit on the amount or concentration of a pollutant emitted from a source. This type of legislation requires a decision, for each industry, on the best means of controlling its emissions (i.e., fixing emission standards).

3.3.2 Clean air implementation plan approach

The air quality management approach proposed by Schwela and Koth-Jahr 1994 (14) is to derive a clean air implementation plan (or air pollution abatement plan) which consists of the following elements:

- description of area with respect to topography, meteorology and socioeconomy · emissions inventory
- comparison with emission standards
- air pollutant concentrations inventory
- simulated air pollutant concentrations
- comparison with air quality standards
- inventory of effects on public health and the environment
- causal analysis
- control measures
- cost of control measures
- cost of public health and environmental effects
- cost-benefit analysis (costs of control vs. costs of efforts)
- transportation and land-use planning
- enforcement plan; resource commitment
- projections for the future on population, traffic, industries and fuel consumption
- strategies for follow-up

3.3.3 Emissions inventory; comparison with emission standards

The emissions inventory is a most complete listing of sources in a given area and of their individual emissions, estimated as accurately as possible from all emitting point, line and area (diffuse) sources. When these emissions are compared with emission standards set for a particular source, first hints on possible control measures are given if emission standards are not complied with. The emissions inventory also serves to assess a priority list of important sources according to the amount of pollutants emitted, and indicates the relative influence of different sources—for example, traffic as compared to industrial or residential sources. The emissions inventory also allows an estimate of air pollutant concentrations for those pollutants for which ambient concentration measurements are difficult or too expensive to performa clean air implementation plana clean air implementation plan

3.3.4 Assessment of emission flows and emission factors

The next step of clean air implementation plan is to determine the types and quantities of pollutants that will be released from the project sources, i.e., the emission rates of all pollutants. Using the emission rates of pollutants and the collected meteorological data, one can easily calculate the incremental ground concentration of the pollutant under consideration, for which an emission inventory has to be taken, that is, the air quality, emission levels, emission factors and pollution impacts have to be estimated which require data on the following aspects.

(a) Actual and allowable emission rates (g/sec) and operating rates.
(b) Maximum design load short-term emission rate (g/sec).
(c) Associated emission characteristics as a function of load for maximum, average and normal operating conditions.
(d) Associated stack characteristics such as location, stack height, stack diameter, exit velocity, and stack gas temperature.
(e) Area source emissions (rates, size of area, height of area source).
(f) Location, dimensions of building to determine potential building downwash.
(g) Associated variables such as boiler size, boiler parameters, operating conditions and pollution control equipment.
(h) Anticipated growth changes.

In addition to concentration of the pollutants one must also estimate mass flow for assessing air quality. Mass flows can be estimated by multiplying concentration with volume flows, and can be expressed either as emission flow (which is the mass of the pollutant emitted per unit time) or pollutant flux (which is the mass of the pollutant received by the receptor like any plant/animal human being/matter per unit time). The product of concentration and time is also called as pollutant dose which is the integrated value of pollutant flux received by the receptor in a particular time range.

The typical flow sheet for assessing the emission flow and factors for exhaust gases is shown in Fig. 3.2, which can be used for exhaust gases from industrial furnaces, industrial process operations, etc. However for exhaust gas pollutants from automobiles, individual vehicles in traffic flow driving patterns also have to be accounted for assessing the pollutant emissions like CO, hydrocarbons; root combustion conditions are also to be examined as they are the products of incomplete combustion.

3.3.5 Estimation of gaseous pollutant emissions from fuel composition

Sulphur dioxide: If the sulphur content of the fuel is known, the SO_2 emissions can be calculated from fuel flows based on the following mass balance equations taking place in combustion and fuel.

$$1 \text{ mol S} + 1 \text{ mol } O_2 \rightarrow 1 \text{ mol } SO_2$$

$$32 \text{ g S} + 2 \times 16 = 32\text{g } O_2 \rightarrow 32 + 32 = 64 \text{ g } SO_2$$

So 1 g of S + 1 g of $O_2 \rightarrow$ 2 g SO_2; which can be used for calculating SO_2 emissions when a fuel is burnt.

Example: Estimate the SO_2 released when 10 kg of diesel oil containing 0.15% sulphur is burnt in a furnace.

10 kg of diesel oil contains = 15 g of sulphur

1 g of sulphur in diesel oil gives 2 g of SO_2

15 g sulphur = 2 × 15 = 30 g of SO_2

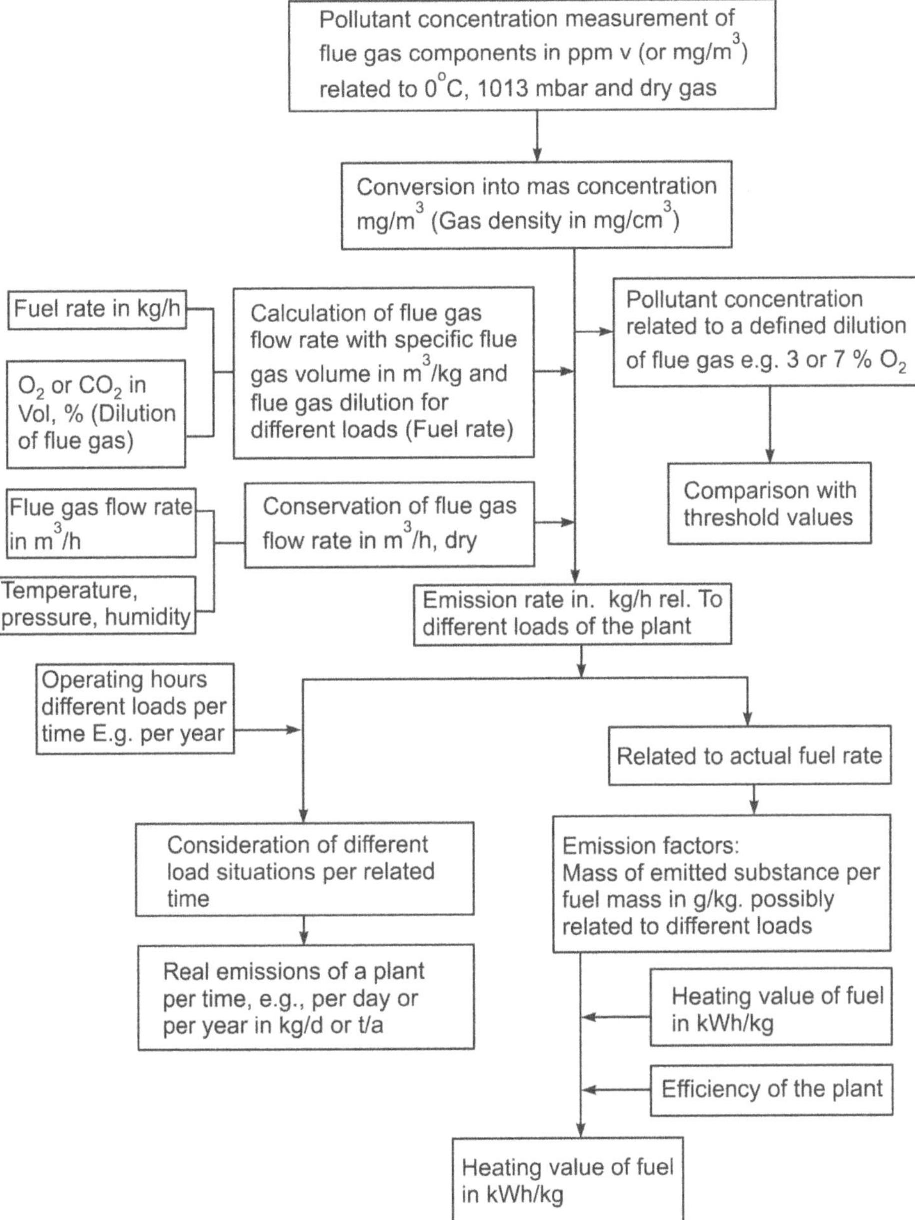

Fig. 3.2 *Determination of emission rates and emission factors from concentration measurements of flue gases–diagrammatic example for furnaces*

Thus based on the inventory of the individual fuels and their sulphur content burnt in various activities in a given area the SO_2 emission can be estimated. However any inaccuracy in the sulphur content data and flue gas desulphurization process installed in any fuel burning activity may result in inaccuracy in SO_2 emissions mass flow calculation based on the above equality.

3.3.6 Estimation of NO_x, HC and heavy metals emission from fuel composition

Since NO_x emissions will be generated not only from fuel nitrogen but also from oxidation of atmospheric nitrogen (which depends on furnace temperature) it is not possible to estimate NO_x emissions from fuel composition. Similarly the emission of incomplete combustion products like hydrocarbons cannot be computed from fuel consumption as they depend on combustion process conditions. However the heavy metal emissions can be roughly estimated (as metal oxides) particularly in heavy oils; light oils contain very low levels of heavy metals.

3.3.7 Air pollutant concentrations Inventory; Comparison with air quality standards

The air pollutant concentrations inventory summarizes the results of the monitoring of ambient air pollutants in terms of annual means, percentiles and trends of these quantities. Compounds measured for such an inventory include the following:

- sulphur dioxide
- nitrogen oxides
- suspended particulate matter
- carbon monoxide
- ozone
- heavy metals (Pb, Cd, Ni, Cu, Fe, As, Be)
- polycyclic aromatic hydrocarbons: benzo (a) pyrene, benzo (e) pyrene, benzo (a) anthracene, dibenzo (a,h) anthracene, benzoghi) perylene, coronen
- volatile organic compounds: n-hexane, benzene, 3-methyl-hexane, n-heptane, toluene, octane, ethyl-benzene xylene (o-,m-,p-), n-nonane, is opropyl benzene, propyl benezene, n-2-/3-/4-ethyltoluene, 1, 2, 4-/1, 3, 5-trimethyl benzene, trichloromethane, 1,1,1 trichloroethane, tetrachloromethane, tri-/tetrachloro-ethene.

Comparison of air pollutant concentrations with air quality standards or guidelines, if they exist, indicates problem areas for which a causal analysis has to be performed in-order to find out which sources are responsible for the non-compliance. Dispersion modelling has to be used in performing this causal analysis.

3.3.8 Air quality assessment

Air Quality Assessment (AQA) begins with a knowledge of existing air quality conditions that depend on ambient pollutant concentrations, pollutant sources and their locations, meteorology and climatology and local topographical and physical conditions affecting pollutant dispersion other words, one must know what air pollutants are present in the area under consideration and in what quantities, where the pollutants come from, how they will be dispersed and where they will go. Such information must be readily available and if necessary, may have to be collected using monitoring systems.

3.3.9 Pollution indices in air monitoring

In most natural situations the effects of air pollution are a result of the combined effects of various different pollutants rather than only one single pollutant. The magnitude of air pollution in such cases is difficult to assess on the basis of the concentration of individual pollutants alone, but the goal can be achieved by using certain indices. The kind and number of pollutants for calculating the indices can be selected depending upon their predominance in the ambient air.

Air pollution index (API) should indicate the gross level of pollution with reference to the standard limits of the individual pollutants. It should be easy to understand and should include the major air pollutants. If the use of indices is to be made on a national level, API need to include always the same air pollutants to maintain comparability.

3.3.9.1 Calculation of air pollution indices

There are a number of methods which are used to calculate the air pollution indices.

I. In one of the methods, the important individual pollutants are compared with their ambient air quality standards in per cent terms and the Air Pollution Index (API) is calculated by taking their average. If there are three pollutants then

$$API = 1/3\sum_{1=1}^{3} A_1$$

$$A_1 = C_1 / S_1 \times 100$$

where C is the concentration of the pollutant I and S1 is the standard value of the pollutant in ambient air.

II. In another method, ratios of three important air pollutants to their ambient air quality standards are first obtained and then the average of their sum is multiplied by 100.

$$API = \frac{1}{3}\frac{I_{PM}}{S_{SPM}} + \frac{I_{SO}}{S_{SO}} + \frac{I_{CO}}{S_{CO}} \times 100 \qquad .....(3.1)$$

where, I_{PM}, I_{SO} and I_{CO} represent individual values of particulate matter, sulphur dioxide and carbon monoxide, and S_{PM}, S_{SO}, and S_{CO} their ambient air quality standards.

III. In another method, sub-indices (A_1) are first obtained for five air pollutants by giving them arbitrary numbers according to their range of ambient concentrations as indicated in Table 3.3 the air pollution index is then calculated as the sum of these sub-indices.

$$API = \sum_{i=1}^{5} A_j \qquad .....(3.2)$$

Table 3.3 *Sub-index value (A_1) for five important air pollution parameters to calculate air pollution index*

Air pollution parameters	Sub-Index values (A_1)					
	2	**4**	**8**	**12**	**16**	**20**
CO (ppm)	0–1	1–2	2–4	4–6	6–8	8–35
NO_2	0–0.005					
SO_2		0.01	0.02	0.06	0.10	0.20
Oxidants (ppm) coefficient of haze	0–0.5	0.5–1.0	1–2	2–3	3–4	4–5
Visibility	12–24	8–12	6–8	4–6	2–4	0–2

IV. The air pollution index can also be calculated on the basis of a single parameter taking into consideration only the dominant pollutant. For example, index for sulphur dioxide can be represented as follows.

$$API = \frac{1_{SO}}{S_{SO}} \times 100 \qquad .....(3.3)$$

where I_{SO} and S_{SO} are ambient SO_2 concentration and its air quality standard respectively. Similarly, the ozone index can be calculated in highly congested traffic areas. For industrial areas the index can also be calculated by taking two parameters into account, i.e., sulphur dioxide and particulate matter.

Table 3.4 *Air quality rating scale*

Index range	Quality of air
0–25	Clean air
26–50	Light air pollution
51–75	Moderate air pollution
76–100	Heavy air pollution
Above 100	Severe air pollution

As all the above indices have been calculated on the per cent basis, a common rating scale given in Table 3.4 can be used to indicate the air quality.

3.3.9.2 The pollution standards index (PSI)

In 1976, as a result of the coordinated efforts of a number of U.S. agencies, a Pollution Standards Index (PSI) was developed (Table 3.5). This index makes it possible to compare the air quality of different urban areas and its potential threat to human health. The USEPA has established acceptable standards of air pollution–National Ambient Air Quality Standards (NAAQS). If a pollutant in a city is at the standard, it is given an index rating of 100. Any index value below 100 suggests relatively clean air with minimal health effects. However, as the index value rises above 100, the air is progressively more polluted and the health effects correspondingly severe. Any index value of 400 or above suggests air pollution. It may cause premature death of the sick and elderly. Under those conditions, all persons are advised to remain indoors. PSI values for various atmospheric contaminants are frequently reported in daily local newspapers in major urban areas. Based on these values the pollution levels in different cities can be evaluated Figure 3.3

Table 3.5 *Pollution Standards Index (PSI) breakpoints*

Index	1-h O_3 μg/m^3	8-h CO μg/m^3	24-h TSP μg/m^3	24-h SO_2 μg/m^3	TSP × SO_2 10^3 μg/m^3	1-h NO_2 μg/m^3
0	0	0	0	0	0	–
50	118	5	75	80	–	–
100	235	10	260	366	–	–
200	400	17	375	800	65	1130
300	800	37	625	1600	261	2260

The actual calculation of a day's PSI value is done based on the values reported in Table 3.5 which gives individual PSI numbers corresponding to various pollutant concentrations. Individual PSI sub-indices are computed for each of the pollutants in the table using linear interpretation between index break points. The highest PSI sub-index determines the overall PSI. In addition to TSP (Total Suspended Particulates), the product of TSP × SO_2 is also given as the two pollutants when present together act synergistically producing effects greater than their individual effects. The general health effects likely to occur and precautions to be taken under various PSI values with ambient air are summarized in Table 3.6

Problem 3.1

Calculation of PSI of a city.

In New Delhi on one day the following concentrations of pollutants were observed. 1-h O_3 275 μg/m^3; 8-h CO 10 μg/m^3; 24-h TSP 200 μg/m^3; 24-h SO_2 300 μg/m^3 Estimate the PSI and give its nature.

Solution: Examining the corresponding PSI values for various pollutants in Table 7.12, we have PSI

For 1-h O_3 275 ug/m^3> 100

8-h CO 10 ug/m^3 = 100
24-h TSP 200 ug/m^3< 100
24-h SO_2 300 ug/m^3< 100

The product of SO_2 and TSP = 200 × 300 = 60 × 10^3 $(ug/m^3)^2$ and this value is lower than the value given for TSP × SO_2 in Table 7.12; there is no sub-index for the product of SO_2 × TSP. So the PSI corresponds to CO and O_3 only which can be calculated through the values given in Table 7.12.

$$\text{PSI for CO} + O_3 = 100 + \frac{275-235}{400-235}(200-100) = 124.24$$

So the air quality is unhealthy.

Table 3.6 *Atmospheric pollution standards index*

	Pollutant levels (in micrograms per litre)		General health effects	Cautionary statements
	Index value	**Air quality level**		
Hazardous	500	Significant harm	Premature death of ill and elderly. Healthy people will experience adverse symptoms that affect their normal activity.	All persons should remain indoors. All persons should minimize physical exertion and avoid traffic.
	400	Emergency	Premature onset of certain diseases, in addition to significant aggravation of symptoms and decreased exercise tolerance in healthy persons.	Elderly persons and persons with diseases should stay indoors and avoid physical exertion. General population should avoid outdoor activity.
	300	Warning	Significant aggravation of symptoms and decreased exercise tolerance in persons with heart or lung disease, with widespread symptoms in the healthy population.	Elderly persons and persons with heart or lung diseases should stay indoors and reduce physical activity.
Unhealthy	200	Alert	Mild aggravation of symptoms in susceptible persons, with irritation symptoms in the healthy population.	Persons with heart or respiratory ailments should reduce physical exertion and outdoor activity.
Unhealthy	100	National ambient air quality standard		
Moderate	50	50 per cent of National ambient air quality standard		
Good	0			

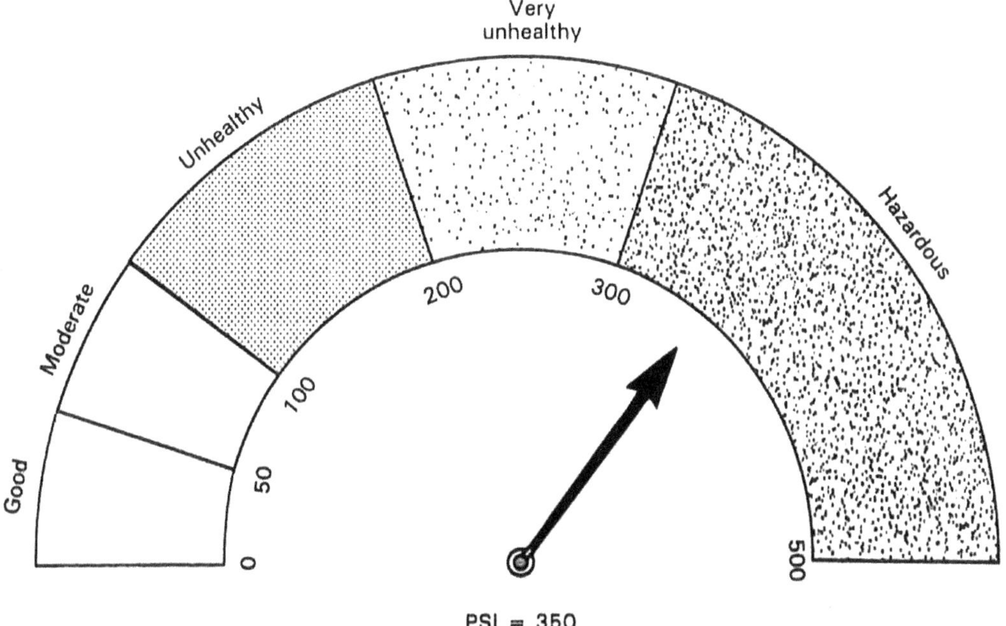

Fig. 3.3 *Pollution standards index*

3.3.9.3 Simulated air pollutant concentrations; comparison with air quality standards

Starting from the emissions inventory, with its thousands of compounds which cannot all be monitored in the ambient air for economy reasons, use of dispersion modeling can help to estimate the concentrations of more "exotic" compounds. Using appropriate meteorology parameters in a suitable dispersion model, annual averages and percentiles can be estimated and compared to air quality standards or guidelines, if they exist.

3.3.10 Inventory of effects on public health and the environment; Causal analysis

Another important source of information is the effects inventory (15) (Ministerium for Umwelt 1993), which consists of results of epidemiological studies in the given area and of effects of air pollution observed in biological and material receptors such as, for example, plants, animals and construction metals and building stones. Observed effects attributed to air pollution have to be causally analysed with respect to the component responsible for a particular effect—for example, increased prevalence of chronic bronchitis in a polluted area. If the compound or compounds have been fixed in a causal analysis (compound-causal analysis), a second analysis has to be performed to find out the responsible sources (source-causal analysis).

3.3.11 Control measures; cost of control measures

Control measures for industrial facilities include adequate, well designed, well installed, efficiently operated and maintained air cleaning devices, also called separators or collectors. A separator or collector can be defined as an "apparatus for separating any one or more of the following from a gaseous medium in which they are suspended or mixed: solid particles (filter and dust separators), liquid particles (filter and droplet separator) and gases (gas purifier)". The basic types of air pollution control equipment (discussed further in "Air pollution control") are the following:

- *for particulate matter:* inertial separators (e.g., cyclones); fabric filters (bag houses); Electrostatic precipitators; wet collectors (scrubbers)
- *for gaseous pollutants:* wet collectors (scrubbers); adsorption units (e.g., adsorption beds); after burners, which can be direct fired (thermal incineration) or catalytic (catalytic combustion).

Wet collectors (scrubbers) can be used to collect, at the same time, gaseous pollutants and particulate matter. Also, certain types of combustion devices can burn combustible gases and vapours as well as certain combustible aerosols. Depending on the type of effluent, one or a combination of more than one collector can be used.

The control of odours that are chemically identifiable relies on the control of the chemical agent(s) from which they emanate (e.g., by absorption, by incineration). However, when an odour is not defined chemically or the producing agent is found at extremely low levels, other techniques may be used, such as masking (by a stronger, more agreeable and harmless agent) or counteraction (by an additive which counteracts or partially neutralizes the offensive odour).

It should be kept in mind that adequate operation and maintenance are indispensable to ensure the expected efficiency from a collector. This should be ensured at the planning stage, both from the know-how and financial points of view. Energy requirements must not be overlooked. Whenever selecting an air cleaning device, not only the initial cost but also operational and maintenance costs should be considered. Whenever dealing with high-toxicity pollutants, high efficiency should be ensured, as well as special procedures for maintenance and disposal of waste materials

3.3.12 Fundamental control measures for industries

Substitution of materials: substitution of less toxic solvents for highly toxic ones used in certain industrial processes; use of fuels with lower sulphur content (e.g., washed coal), therefore giving rise to less sulphur compounds and so on.

3.3.13 Modification or change of the industrial process or equipment

In the steel industry, a change from raw ore to polluted sintered ore (to reduce the dust released during ore handling); use of closed systems instead of open ones; change of fuel heating systems to steam, hot water or electrical systems; use of catalysers at the exhaust air outlets (combustion processes) and so on.

3.3.14 Modifications in processes, as well as in plant layout

This may also facilitate and/or improve the conditions for dispersion and collection of pollutants. For example, a different plant layout may facilitate the installation of a local exhaust system; the performance of a process at a lower rate may allow the use of a certain collector (with volume limitations but otherwise adequate). Process modifications that concentrate different effluent sources are closely related to the volume of effluent handled, and the efficiency of some air-cleaning equipment increases with the concentration of pollutants in the effluent. Both the substitution of materials and the modification of processes may have technical and/or economic limitations, and these should be considered.

3.3.15 Adequate housekeeping and storage

strict sanitation in food and animal product processing; avoidance of open storage of chemicals (e.g., sulphur piles) or dusty materials (e.g., sand), or, failing this, spraying of the piles of loose particulate with water (if possible) or application of surface coatings (e.g., wetting agents, plastic) to piles of materials likely to give off pollutants.

3.3.16 Adequate disposal of wastes

Avoidance of simply piling up chemical wastes (such as scraps from polymerization reactors), as well as of dumping pollutant materials (solid or liquid) in water streams. The latter practice not only causes water pollution but can also create a secondary source of air pollution, as in the case of liquid wastes from sulphite process pulp mills, which release offensive odorous gaseous pollutants.

3.3.17 Maintenance

Well maintained and well-tuned internal combustion engines produce less carbon monoxide and hydrocarbons.

3.3.18 Work practices

Taking into account meteorological conditions, particularly winds, when spraying pesticides.

By analogy with adequate practices at the workplace, good practices at the community level can contribute to air pollution control - for example, changes in the use of motor vehicles (more collective transportation, small cars and so on) and control of heating facilities (better insulation of buildings in order to require less heating, better fuels and so on).

3.3.19 Control measures in vehicle emissions

Adequate and efficient mandatory inspection and maintenance programmes enforced for the existing car fleet, and programmes of enforcement of the use of catalytic converters in new cars, aggressive substitution of solar/battery-powered cars for fuel powered ones, regulation of road traffic, and transportation and land use planning concepts.

Motor vehicle emissions are controlled by controlling emissions per vehicle mile travelled (VMT) and by controlling VMT itself. Emissions per VMT can be reduced by controlling vehicle performance - hardware, maintenance - for both new and in use cars. Fuel composition of leaded gasoline may be controlled by reducing lead or sulphur content, which also has a beneficial effect on decreasing HC emissions from vehicles. Lowering the levels of sulphur in diesel fuel as a means to lower diesel particulate emission has the additional beneficial effect of increasing the potential for catalytic control of diesel particulate and organic HC emissions.

Another important management tool for reducing vehicle evaporative and refueling emissions is the control of gasoline volatility. Control of fuel volatility can greatly lower vehicle evaporative HC emissions. Use of oxygenated additives in gasoline lowers HC and CO exhaust as long as fuel volatility is not increased.

Reduction of VMT is an additional means of controlling vehicle emissions by control strategies such as

- use of more efficient transportation modes
- increasing the average number of passengers per car
- spreading congested peak traffic loads reducing travel demand.

While such approaches promote fuel conservation, they are not yet accepted by the general population, and governments have not seriously tried to implement them.

All these technological and political solutions to the motor vehicle problem except substitution of electrical cars are increasingly offset by growth in the vehicle population. The vehicle problem can be solved only if the growth problem is addressed in an appropriate way.

3.3.20 Cost of public health and environmental effects; cost-benefit analysis

The estimation of the costs of public health and environmental effects is the most difficult part of a clean air implementation plan, as it is very difficult to estimate the value of lifetime reduction of disabling illnesses, hospital admission rates and hours of work lost. However, this estimation and a Transportation and Land-Use Planning

The pollution problem is intimately connected to land-use and transportation, including issues such as community planning, road design, traffic control and mass transportation; to concerns of demography, topography and economy; and to social concerns. In general, the rapidly growing urban aggregations have severe pollution problems due to poor land-use and transportation practices. Transportation planning for air pollution control includes transportation controls, transportation policies, mass transit and highway congestion costs. Transportation controls have an important impact on the general public in terms of equity, repressiveness and social and economic disruption - in particular, direct transportation controls such as motor vehicle constraints, gasoline limitations and motor vehicle emission reductions. Emission reductions due to direct controls can be reliably estimated and verified. Indirect transportation controls such as reduction of vehicle miles travelled by improvement of mass transit systems, traffic flow improvement regulations, regulations on parking lots, road and gasoline taxes, car-use permissions and incentives for voluntary approaches are mostly based on past trial and error experience, and include many uncertainties when trying to develop a viable transportation plan.

National action plans incurring indirect transportation controls can affect transportation and land-use planning with regard to highways, parking lots and shopping centres. Long-term planning for the transportation system and the area influenced by it will prevent significant deterioration of air quality and provide for compliance with air quality standards. Mass transit is consistently considered as a potential solution for urban air pollution problems. Selection of a mass transit system to serve an area and different modal splits between highway use and bus or rail service will ultimately alter land-use patterns. There is an optimum split that will minimize air pollution; however, this may not be acceptable when non-environmental factors are considered.

Land-use planning for air pollution control includes zoning codes and performance standards, land-use controls, housing and land development, and land use planning policies. Land-use zoning was the initial attempt to accomplish protection of the people, their property and their economic opportunity. However, the ubiquitous nature of air pollutants required more than physical separation of industries and residential areas to protect the individual. For this reason, performance standards based initially on aesthetics or qualitative decisions were introduced into some zoning codes in an attempt to quantify criteria for identifying potential problems.

The limitations of the assimilative capacity of the environment must be identified for long-term land-use planning. Then, land-use controls can be developed that will prorate the capacity equitably among desired local activities. Land-use controls include permit systems for review of new stationary sources, zoning regulation between industrial and residential areas, restriction by easement or purchase of land, receptor location control, emission density zoning and emission allocation regulations.

Housing policies aimed at making home ownership available to many who could otherwise not afford it (such as tax incentives and mortgage policies) stimulate urban sprawl and indirectly discourage higher-density residential development. These policies have now proven to be environmentally disastrous, as no consideration was given to the simultaneous development of efficient transportation systems to serve the needs of the multitude of new communities being developed. The lesson learnt from this development is that programmes impacting on the environment should be coordinated, and comprehensive planning undertaken at the level where the problem occurs and on a scale large enough to include the entire system.

Land-use planning must be examined at national, provincial or state, regional and local levels to adequately ensure long-term protection of the environment. Governmental programmes usually start with power plants ting, mineral extraction sites, coastal zoning and desert, mountain or other recreational development. As the multiplicity of local governments in a given region cannot adequately deal with regional environmental problems, regional governments or agencies should coordinate land development and density patterns by supervising the spatial arrangement and location of new construction and use, and transportation facilities. Land-use and transportation planning must be interrelated with enforcement of regulations to maintain the desired air quality. Ideally, air pollution control should be planned for by the same regional agency that does land-use planning because of the overlapping externalities associated with both issues.

3.3.21 Enforcement plan, resource commitment

The clean air implementation plan should always contain an enforcement plan which indicates how the control measures can be enforced. This implies also a resource commitment which, according to a polluter pays principle, will state what the polluter has to implement and how the government will help the polluter in fulfilling the commitment.

3.3.22 Projections for the future

In the sense of a precautionary plan, the clean air implementation plan should also include estimates of the trends in population, traffic, industries and fuel consumption in order to assess responses to future problems. This will avoid future stresses by enforcing measures well in advance of imagined problems.

3.3.23 Strategies for follow-up

A strategy for follow-up of air quality management consists of plans and policies on how to implement future clean air implementation plans.

3.3.24 Role of environmental impact assessment

Environmental impact assessment (EIA) is the process of providing a detailed statement by the responsible agency on the environmental impact of a proposed action significantly affecting the quality of the human environment. EIA is an instrument of prevention aiming at consideration of the human environment at an early stage of the development of a programme or project.

EIA is particularly important for countries which develop projects in the framework of economic reorientation and restructuring. EIA has become legislation in many developed countries and is now increasingly applied in developing countries and economies in transition.

EIA integrative in the sense of comprehensive environmental planning and management considering the interactions between different environmental media. On the other hand, EIA integrates the estimation of environmental consequences into the planning process and thereby becomes an instrument of sustainable development. EIA also combines technical and participative properties as it collects, analyses and applies scientific and technical data with consideration of quality control and quality assurance, and stresses the importance of consultations prior to licensing procedures between environmental agencies and the public which could be affected by particular projects. A clean air implementation plan can be considered as a part of the EIA procedure with reference to the air.

In the US there are federal regulations termed "new source permitting" that require a new major source (or an existing major facility undertaking major modification) to obtain preapproval under the New Source Review (NSR) program before commencing construction of an air pollution source. The federal requirements vary for different areas of the country depending on whether the air quality in the area complies with the National Ambient Air Quality Standards (NAAQS). Areas designated, as being in attainment will be required to adhere to the requirements under the Prevention of Significant Deterioration (PSD) program. Areas that are designated as being non-attainment will be required to adhere to stricter requirements under the Non-attainment Area (NAA) program.

3.3.25 Prevention of significant deterioration

Prevention of Significant Deterioration (PSD) program permitting requires that an applicant analyze all technically feasible emission control alternatives and demonstrate that the emission rate(s) proposed is reflective of the best available control technology (BACT). A BACT analysis evaluates each emission control alternative relative to energy, environmental and economic impacts. The procedure is called a top-down BACT, meaning that the lowest possible emission rate must be considered first. The control technique with the lowest emission rate may only be dropped from consideration if there are legitimate energy, environmental or economic reasons. If so, the emission control technique with the next lowest emission rate may then be considered. The progression continues until an emission rate is identified as BACT. Another PSD requirement is to evaluate the projected impact of the new emissions on the existing ambient air levels in order to show that the attainment area will not fall into non-attainment.

3.4 Air Pollution Preventive Measures Implemented by Different Countries

3.4.1 Air pollution preventive measures in USA

Air pollution is a leading environmental issue and a key public health threat in urban centers in USA. In the 1960s, 1970s, and 1990s, the United States Congress enacted a series of Clean Air Acts which significantly strengthened regulation of air pollution. The Clean Air Act sets numerical limits on the concentrations of a basic group of air pollutants and provide reporting and enforcement mechanisms.

Air pollution control acts and regulations in UK and USA, which are adopted by many other countries, are summarized in Table 3.7 In the United States, the Clean Air Act, 1963 was the first modern environmental law enacted by the United States Congress. The Clean Air Act, 1970 reviewed and amended by Congress in 1975, 1977, and 1990 forms the basis of the federal air pollution control program of the United States Air pollution stipulating the National ambient air quality standards for public health protection. Standards have to be maintained by application of control technologies to reduce emissions with the aim of improved air quality, and public health protection has to be given priority over cost and technological capability.

Figure 3.4 shows the general structure and contents of the Clean Air Act while Fig. 3.5 presents information on the implementation of the regulations. The USEPA is responsible for implementing the Clean Air Act and for carrying out the program. There are, for instance, nationwide regulations directly concerning wide spread products such as motor vehicles, gasoline (petrol) and some major industries.

Table 3.7 *Air pollution control acts and regulations–global scenario*

Country	Name of act	Year of act	Remarks
UK	First Alkali Act	1863	The acts was later extended to other processes not covering alkali maintenance only
	Public Health Act	1936	Laid down regulations, in relation to the type of food to be used
	Clean Air Act	1956	Specification regulations, emission standards. The act also regulates emission from industry
	Motor Vehicles (construction area) Regulations Act	1973	Regulates construction
USA	Air Pollution Control Act	1955	First federal legislation

Table 3.7 ***Contd...***

Country	Name of act	Year of act	Remarks
	Clean Air Act	1963	Federal financial assistance but problem left to the states
	Amended Clean Air Act	1965	First federal emission standards for cars
	Air Quality Act	1967	First requirement for ambient standards, but these are to be set by the states
	Amended Air Quality Act	1970	First national ambient standards–to be set up by federal action; first national emission standards for certain pollutants; continued state implementations.

The EPA set up New Source Performance Standards (NSPS) that placed emission limits on new industrial plants and on existing ones that were substantially modified. Other regulations contained in the State Implementation Plans (SIPs) are addressed to the executive authorities of the individual states. They are to observe them when, for example, granting operating permits for industrial plants. SIPs are subject to EPA approval and if a plan is not acceptable, the EPA is required to provide an implementation plan which the state is then required to enforce.

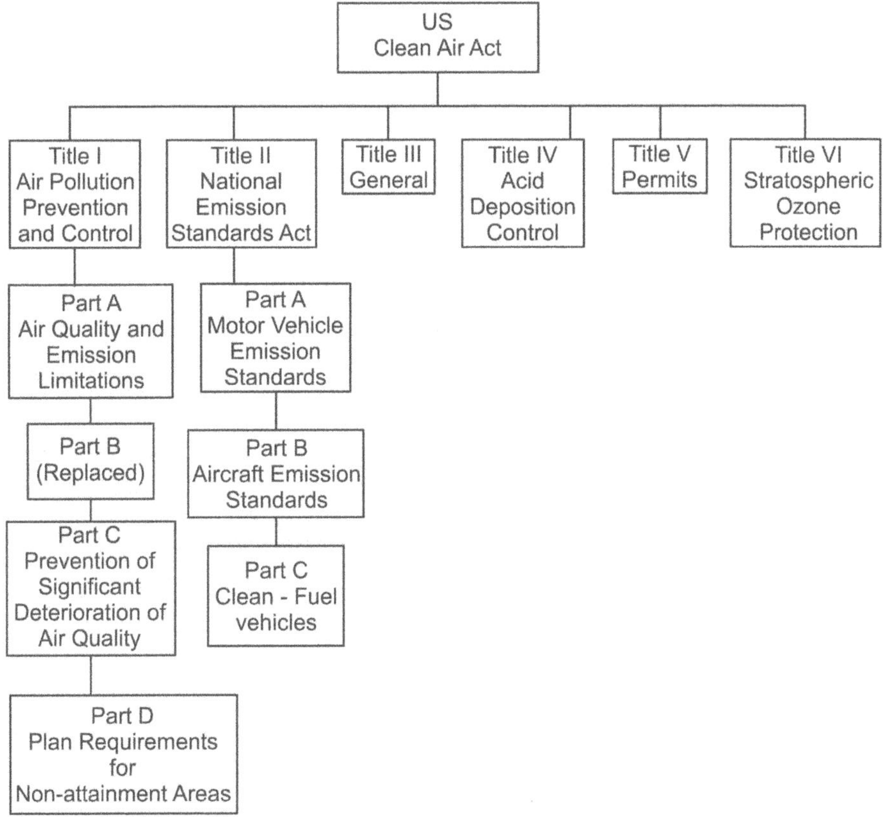

Fig. 3.4 *Structure and subjects of the US Clean Air Act*

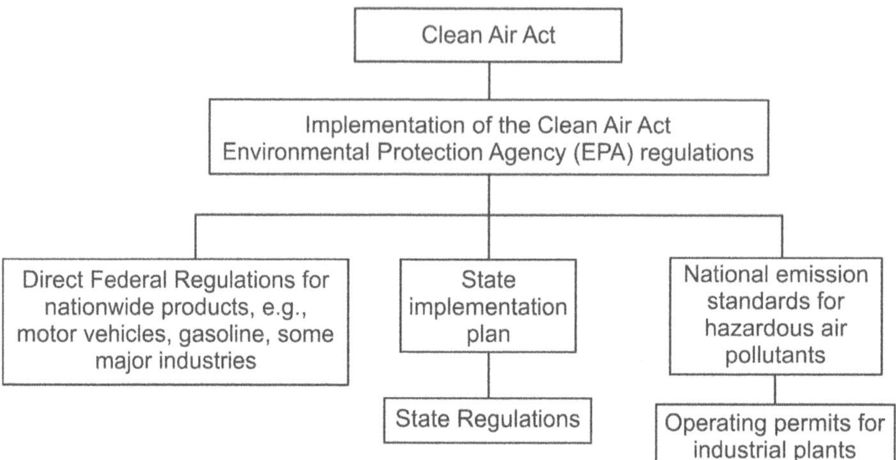

Fig. 3.5 *Procedures for implementing the US Clean Air Act (from legislation to operating permits)*

US Environmental Protection Agency (EPA) has implemented a series of rules that have dramatically reduced emissions from the transportation, off-road, and stationary source sectors have driven continued national improvements in air quality. As required by the federal Clean Air Act, the U.S. Environmental Protection Agency (U.S. EPA) sets standards for particulate matter (PM2.5), nitrogen dioxide (NO_2) and sulfur dioxide (SO_2) emissions from large fossil-fuel combustion sources and also designates areas that fail to meet health-based standards for air quality established by the agency (16). Most recently, between 2006 and 2010, the U.S. EPA phased-in stringent emissions standards for heavy duty diesel vehicles such as trucks and for gasoline passenger vehicles while also reducing the sulfur content of diesel fuel and gasoline (17, 18). These new emissions standards, once fully implemented, are expected to significantly reduce air pollution from key transportation sectors and reduce their associated public health impacts.

The broadest analysis of the implementation of control strategies for air pollution was conducted by the U.S. Environmental Protection Agency in the late 1990s (19) (Krupnick and Morgenstern 2002). The analysis developed a hypothetical scenario for 1970 to 1990, assuming that the real costs for pollution control during this period could be compared with the benefits of reduced mortality and morbidity and avoided damage to agricultural crops brought about by the reduction of major air pollutant levels across the country during this period. The study estimated reduced mortality from dose-response relationships for the major air pollutants, assigning the cost of each death at the value of statistical life and the cost of morbidity in relation to estimated health service utilization.

The analysis showed a dramatically high cost-benefit ratio and inspired debate about the methodologies used and the results. One major criticism was of the use of the value of statistical life for each death potentially avoided by the reduced air pollution. A recalculation using the life-years-lost method reduced the benefits for deaths caused by PM from US$16,632 billion to US$9,100 billion (19) (Krupnick and Morgenstern 2002). The recalculated figure is still well above the fifth percentile estimate of benefits and does not undermine the positive cost-benefit ratio reported. Thus, if a developing country were to implement an appropriate control strategy for urban air pollution, it might derive significant economic benefits over the subsequent decades. The country's level of economic development, local costs, and local benefit valuations will be important for any cost-benefit assessment. (9) WHO's (2000) air quality guidelines are among the documents that provide advice on analytical approaches.

3.4.2 Air pollution preventive measures in UK

Air pollution in the United Kingdom is monitored and regulated by Department for Environment, Food and Rural Affairs (DEFRA), (20) which sets air quality targets for particulates, nitrogen dioxide and ozone to be implemented by local government representatives responsible for the management of air quality in cities, where air quality management is the most urgent.

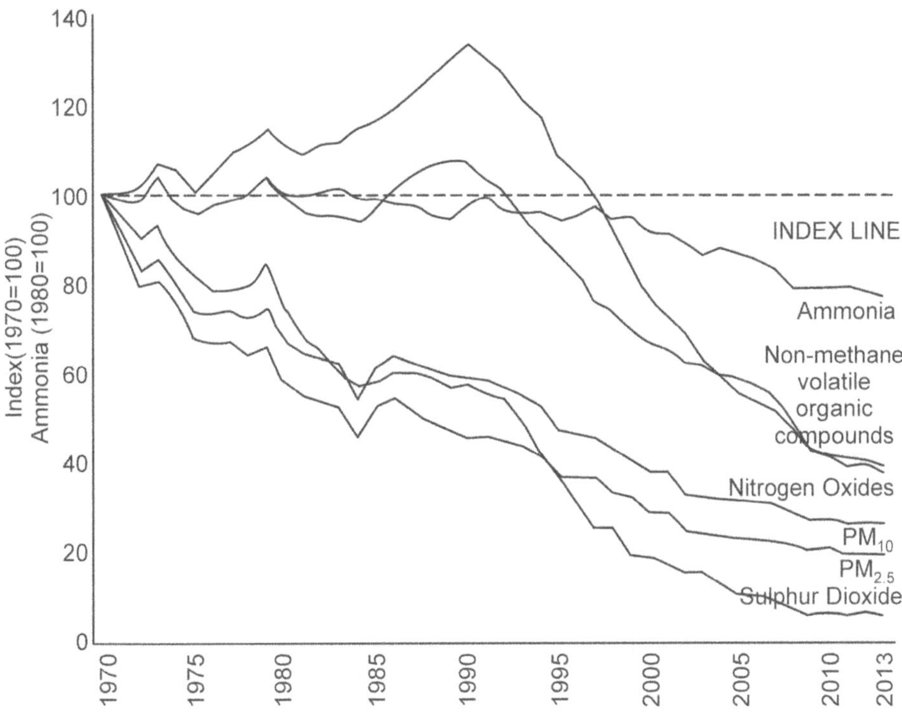

Fig. 3.6 *Trends in UK sulphur dioxide, nitrogen oxides, non-methane volatile organic compounds, ammonia and particulate matter (PM_{10}, $PM_{2.5}$) emissions 1970–2013*

3.4.2.1 England air management

If a local authority finds an area where the targets are not likely to be met, it must declare it an Air Quality Management Area (AQMA) (21) and produce a Local Air Quality Action Plan to improve the air quality. DEFRA has published a list of local authorities with AQMAs (22) the action plan may include measures for idle reduction of vehicle engines. An example is the Metropolitan Borough of Dudley. The implementation of rigorous air quality management plans by DEFRA resulted in reduction in critical air pollutant levels in UK (Figure 3.6)

UK has established an air quality network where levels of the key air pollutants (23) are published by monitoring centers (24). Air quality in Oxford, Bath and London (25, 26) is particularly poor. One controversial study (27) performed by the Color Gas Company and published in the Guardian newspaper compared walking in Oxford on an average day to smoking over sixty light cigarettes.

More precise comparisons can be collected from the UK Air Quality Archive which allows the user to compare a cities management of pollutants against the national air quality objectives set by DEFRA in 2000. (23)

Localized peak values are often cited, but average values are also important to human health. The UK National Air Quality Information Archive offers almost real-time monitoring of "current maximum" air pollution measurements for many UK towns and cities. This source offers a wide range of constantly updated data, including:

Hourly Mean Ozone (μg/m^3)

Hourly Mean Nitrogen dioxide (μg/m^3)

Maximum 15-Minute Mean Sulphur dioxide (μg/m^3)

8-Hour Mean Carbon monoxide (mg/m^3)

24-Hour Mean PM10 (μg/m^3 Grav Equiv)

DEFRA acknowledges that air pollution has a significant effect on health and has produced a simple banding index system that is used to create a daily warning system that is issued by the BBC Weather Service to indicate air pollution levels DEFRA has published guidelines for people suffering from respiratory and heart diseases.

3.4.3 Air pollution preventive measures in China

Air pollution in China is primarily regulated by the Law on the Prevention and Control of Atmospheric Pollution 1987 ('the LPCAP') (30). The aim of the LPCAP is to prevent and control air pollution. The LPCAP was last amended in 2000. Some affected people are concerned that LPCAP provisions are based on out-of-date data. For example, the LPCAP focuses on regulating activities that release large particulate matter into the atmosphere, such as coal burning and certain waste disposal (31) the LPCAP is less concerned with regulating activities that release fine particulate matter into the air, such as metal processing and smelting. This raises concerns for some affected people who feel that fine particulate matter may pose a greater risk to human health than large particulate matter. The Ministry of Environmental Protection began a review of the LPCAP in 2006. The Ministry submitted a revised version of the law to the State Council in 2010. On 31 October 2013, the 12th National People's Congress Standing Committee announced that the revised version was scheduled for deliberation within the Committee's term of office (32).

In addition to the LPCAP, there are numerous policies dealing with air pollution in China. On 29 October 2012, the State Council published its 12th Five-Year Plan on the Prevention and Control of Atmospheric Pollution in Key Areas. The Plan addresses fine particulate matter and introduces a new policy focus on improved air quality. This indicates a shift away from a focus on control of pollution causing activities (33). On 10 September 2013, the State Council issued the Action Plan on the Prevention and Control of Atmospheric Pollution ('the Action Plan'). The broad aim of the Action Plan is to improve China's overall air quality within five years. Other aims include a significant reduction in smog days and a dramatic improvement in regional air quality. Targeted regions include Bejing-Tianjin-Hebei, Yangzi River Delta and Pearl River Delta. The Action Plan sets 2017 as the achievement date for two major goals: a 10 percent reduction in concentrations of large particulate matter in major cities; up to a 25 percent decrease in concentrations of fine particulate matter in targeted regions.

The Action Plan contains 10 major sections. Each section sets out measures to address specific problems or further certain goals. The following provides a brief overview of each section.

Section 1: Increase the effort of comprehensive control, reduce multi-pollutants emissions The aim of Section 1 is to reduce the discharge of pollutants by:

- regulating the use of small coal-fired boilers
- accelerating desulfurization, de-nitrification and dust removal
- retrofit projects in key industries
- improving the quality of fuels
- controlling urban dusts
- eliminating high polluting yellow-sticker vehicles and old vehicles

Section 2: Adjust and optimize the industrial structure, promote industrial upgrading and restructure

This section introduces measures to:

- regulate new capacities in high energy consuming and high polluting industries
- accelerate the elimination of backward productivity in 21 key industries, such as iron and steel, cement, electrolytic aluminium, glass, et cetera
- reduce excess capacity
- halt illegal production projects in industry with over capacity

Section 3: Accelerate industrial technology transformation; improve the innovation capability Section 3 promotes:

- the strengthening of scientific and technological development and promotion
- clean production
- a 30 percent reduction of emission intensity in key industries by 2017
- the development of a circular economy
- the fostering of energy saving and environmental protection industries

Section 4: Accelerate the adjustment of energy structure and increase clean energy supply the plan here is to:

- limit the overall consumption of coal
- decrease the share of coal in total energy consumption to less than 65 percent by 2017
- accelerate clean energy utilization by increasing the supply of alternative energies like natural gas and coal-based methane
- promote the clean use of coal
- improve energy efficiency

Section 5: Strengthen energy-saving environmental thresholds and optimize industrial layout Section 5 aims to strengthen the use of energy-saving environmental protection indicators by regulating industry actions. For example, projects which have not passed energy audits and environmental impact assessments will not be approved for construction, provided with land or credit support or supplied electricity or water.

Section 6: Elaborate the role of market mechanism and improve environmental economic policies

Tasks here include:

- The active promotion of new energy-saving and emission reduction mechanisms that integrate constraints with incentives
- increasing pollution charges
- encouraging banking and financial institutions to increase credit support for projects concerning the prevention and control of air pollution

Section 7: Integrate law and regulation systems, supervise and manage strictly based on the law

Section 7 is designed to achieve the Section 2 aim of 'promoting industrial up grading and restructure' through the improvement of law and regulations. The focus here is on accelerating revision of the LPCAP and:

- considering possible criminal penalties for those who cause malicious pollution
- accelerating the establishment and revision of emission standards in key industries

- strictly implement environmental disclosure requirements
- publishing urban air quality ranking at provincial level

Section 8: Establish a regional corporation mechanism and integrate regional environmental management

This section establishes a 'Beijing-Tianjin-Hebei-Yangzi River Delta-Pearl River Delta Regional Corporation Mechanism'. The Regional Mechanism allows provincial governments and State Council to:

- Collaboratively resolve significant regional environmental problems
- better implement air pollution prevention and control measures
- report on the progress of prevention and control measures
- determine periodical work requirements, priorities and major tasks

Section 9: Establish monitoring and warning system, and properly cope with heavy pollution weather

Section 9 requires local governments to include a Heavy Pollution Weather Emergency Response in their emergency management system. Responses may include:

- production limits
- temporary closure of heavy polluting enterprises
- temporary closure of primary and secondary schools
- control over motor vehicle use
- relevant meteorological interventions

Section 10: Clarify the responsibilities of the government and society, encourage the public to participate in environmental protection

Section 10 vests local governments with responsibility for air quality within their jurisdictions. It requires relevant governmental sectors to closely corporate with one another and to take unified action against air pollution. It acknowledges the role of enterprises and the public in tackling air pollution.

Joint Efforts to Combat Air Pollution

The Action Plan is the Chinese government's toughest plan so far to combat air pollution. It is likely to serve as the political benchmark for the nation's efforts to improve air quality. Several challenges may arise in the implementation of the Plan. Firstly, implementation requires coordination and cooperation at the national, provincial, municipal and public level. In addition, implementation may require significant economic investment and industry reform. Consequently, it is uncertain to what extent the Action Plan will be implemented and to what extent it will deliver practical value to the people of China. On the positive side, the Ministry of Environmental Protection argues that the

Plan is likely to benefit some industries, such as manufacturing and clean energy, and increase Gross Domestic Product by 1942.2 billion RMB ($319 billion). (34)

Just after the Action Plan was published, the Ministry of Finance allocated 5 billion RMB ($820 million) to the elimination of air pollution in six cities and provinces including Beijing, Tianjin, Hebei, Shanxi, Shangdong, Neimenggu and Hebei.13. Several national Ministries and administrative bodies cooperated to develop and issue Detailed Rules on the Implementation of the Action Plan in Beijing-Tianjin-Hebei and Adjacent Regions. The Detailed Rules specifically address actions to reduce fine particulate matter. On 5 September 2013, Hebei province issued a document outlining 10 specific measures to eliminate air pollution.14. One of these measures is the establishment of an Environmental Police team comprised of public security and environmental protection personnel. (35) Regional governments have also worked to review or create laws that complement the Action Plan. For example, the Shanghai Municipal Council has passed the Measures for the Implementation of LPCAP. The draft document is deliberated by the Standing Committee of Shanghai Municipal People's Congress. On 20 November 2013, the Standing Committee of Beijing Municipal People's Congress considered the Regulations on the Prevention and Control of Atmospheric Pollution in Beijing for the third time. The Regulations were deliberated by Beijing Municipal People's Congress in early 2014 (36).

As recognized by Section 10 of the Action Plan, air pollution cannot be eliminated without public participation. Effective public participation requires proper access to relevant information. To facilitate this access, the Chinese government has set up an official air quality monitoring programme and made relevant data available to the public.

The proposed Action Plan looks at the prevention and control of air pollution at the national level. Its implementation depends upon different groups at different levels working successfully together. These groups include:

- central, provincial and municipal governments
- industry bodies
- commercial entities
- the public

Action Plan can play an important role in the elimination of air pollution. The next step is to consider translating this policy into law. The stabilization, nominalization and institutionalization of air pollution control and prevention also depend upon the strengthening of environmental law. China is currently in the process of revising its Environmental Protection Law. These revisions may help address China's air pollution problem. Similarly, the revised LPCAP has a significant impact on China's air quality (Figure.3.7)

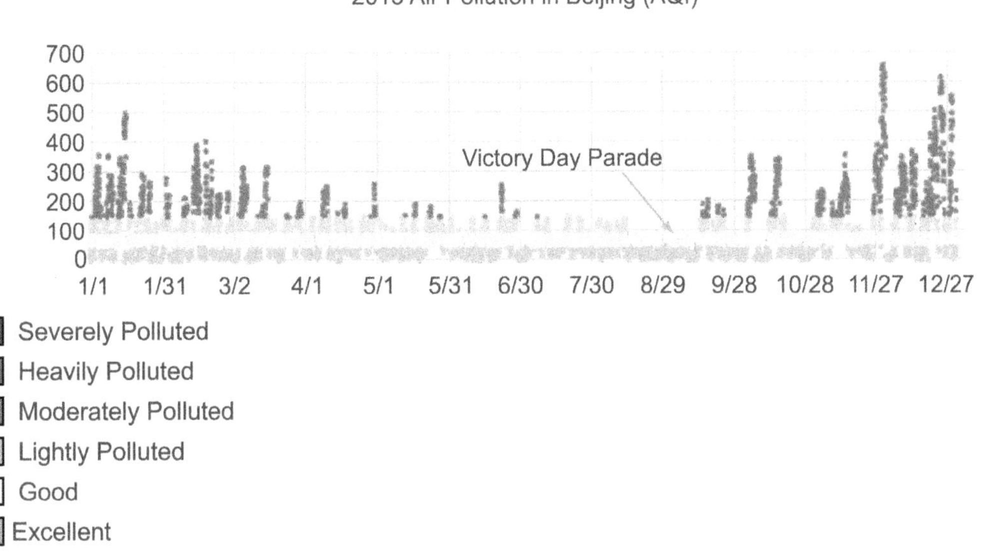

Fig. 3.7 *2015 Air pollution in Beijing as measured by Air Quality Index (AQI)*

3.4.4 Air pollution preventive measures in India

Air pollution in India is quite a serious issue with the major sources being fuel wood and biomass burning, fuel adulteration, vehicle emission and traffic congestion.

3.4.4.1 Air pollution control strategies in India

India has a relatively extensive set of regulations designed to improve both air and water quality. The environmental policies have their roots in the Water Act of 1974 and Air Act of 1981. These acts created the Central Pollution Control Board (CPCB) and the State Pollution Control Boards (SPCBs), which are responsible for data collection and policy enforcement, and also developed detailed procedures for environmental compliance. The CPCB and SPCBs quickly advanced a national environmental monitoring program (responsible for the rich data underlying our analysis (37, 8). The Ministry of Environment and Forests (MoEF), created in its initial form in 1980, was established largely to set the overall policies that the CPCB and SPCBs were to enforce. Starting in 1987, India's Central Pollution Control Board (CPCB) began compiling readings of Nitrogen Dioxide (NO_2), Sulfur Dioxide (SO_2), and particulate matter with diameter less than 100µm (PM). The data were collected as a part of the National Air Quality Monitoring Program (NAMP), a program established by the CPCB to help identify, assess, and prioritize the pollution controls needs in different areas, as well as to help in identifying and regulating potential hazards and pollution sources. The CPCB reports that there are currently 365 functional air quality monitoring stations in 127 Indian cities, there has been much movement and reclassification of these monitors over the years. (39)

The Air (Prevention and Control of Pollution) Act was passed in 1981 to regulate air pollution and there have been some measurable improvements.

Thus the Government has formulated several rules, established many agencies and developed an administrative mechanism for implementation of Air Pollution (Prevention and Control of Pollution) Acts.

3.4.4.2 Administrative structure for environmental protection (39)

The administrative structure in India has a Central Government at the federal level and State Governments at the State or provincial levels. The mandates and the administrative controls at Central and State levels on various subjects have been well defined in the constitution of India. There are strong linkages between the Central and State Governments for smooth functioning of the Governments and implementation of various laws & Acts. With regard to the environmental pollution control, the linkage between Central and State Governments are complementary. The administrative structure for environmental protection in India is presented in Fig. 3.8

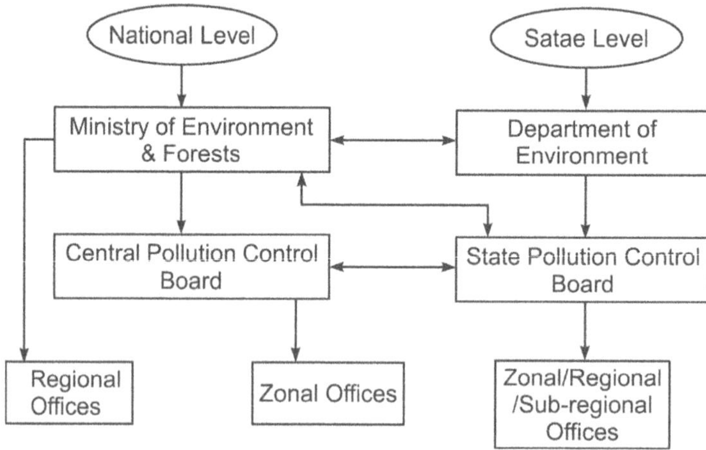

Fig 3.8 *Administrative structure for environment protection*

The Ministry of Environment & Forests (MoEF) is the nodal agency, in the administrative structure of the Central Government, for planning, promotion and co-ordination of environmental & forestry programmes. Various tasks for protection of environment are being fulfilled through environmental impact assessment; eco-regeneration; assistance to organisations implementing programmes on environmental protection; promotion of research, training & education; international co-operation; dissemination of environmental information; and creation of environmental awareness among all the sectors of the country's population. The departments of environment, at

State levels, have similar functions in respective States. The Central Pollution Control Board (CPCB) was created in the year 1974 at the federal level under the provisions of the Water (Prevention & Control of Pollution) Act, 1974. Subsequently, the CPCB was entrusted with the implementation of various provisions of the Air (Prevention & Control of Pollution) Act, 1981 and Environment (Protection) Act, 1986. The CPCB, which is an autonomous agency and fully funded by MoEF, is the nodal agency at the Central level for pollution control planning & execution and formulating ambient environment & source-specific emission as well effluent discharge standards at the national level.

The Central Pollution Control Board has the following statutory functions:

1. To improve the quality of air and to prevent, control or abate air pollution in the country.
2. To advise the central government on any matter concerning the improvement of the quality of air and the prevention, control or abatement of air pollution.
3. To plan and execute a nationwide programme for the prevention, control or abatement of air pollution and arrange training programmes.
4. To coordinate the activities of the State Boards, provide technical assistance, carryout and sponsor investigations and research relating to the prevention and control of air pollution.
5. To organize through mass media, a comprehensive programme regarding prevention and control of air pollution.
6. To collect, compile, publish technical and statistical data on air pollution and the measures devised for its prevention and control and prepare manuals, codes or guides.
7. To lay down standards for the quality of air.
8. To establish or recognize laboratories.

The State Pollution Control Boards are assigned the following duties by the Air Act:

1. To plan and implement a comprehensive programme for prevention and control of air pollution in the states.
2. To advise the state government on any matter concerning air pollution including sitting of any industry which is likely to cause air pollution?
3. To collect and disseminate information and collaborate with the Central Board in organizing training programmes and mass education programmes.
4. To inspect any control equipment, industrial plants and give by order such directions to such persons as may be necessary to take steps for prevention and control of air pollution.

5. To inspect air pollution control areas, assess the quality of air therein and take appropriate steps for prevention and control of air pollution in such areas.
6. To lay down, in consultation, with the Central Board, standards for emission of air pollutants into the atmosphere from industrial plants and automobiles.
7. To establish or recognize laboratories.

3.4.4.3 *Strategy for pollution control*

The strategy for pollution control in India involves three different approaches:

(i) Classification of Pollution sources
(ii) tackling of the pollutants;
(iii) Environmental auditing
(iv) controlling the pollution at source; and
(v) Tackling of the polluted areas.

A combination of these approaches is an effective way to protect the environment.

3.4.4 4 *Classification of pollution sources*

As per sections 19–26, the operative sections of the Air Act, air pollution sources to be classified into three classes: (1) Major air polluting industries (2G industries listed in the Schedule attached to the Act) (2) Automobiles excluding ships or aircraft (3) Other sources including domestic sources. It should be noted that the approach to pollution control, problems in implementation and the success of enforcement will vary significantly for these three sources. So different approaches have been proposed for these classes of sources as detailed below:

Under Section 21 of the Air Act, major air polluting industries require comprehensive and specific pollution control and measuring systems. The pollution control boards have to identify and implement these measures to be regularized and monitored by grant of consent after verifying the implementation of requirements and conditions laid down in the consent. The consent should clearly specify the equipment to be taken up by the industry within a stipulated time.

3.4.4.5 *Tackling of the pollutants*

This is aimed to minimise the quantity of the pollutants itself. The waste minimisation techniques include process modifications, use of clean/low waste technologies, conservation of energy & natural resources, recycling of wastes, recovery of valuable products out of wastes, etc. The reduction of lead content in motor spirit, reduction of sulphur content in diesel, coal beneficiation, use of cleaner fuel in the industrial processes, etc. are typical examples of the direct control of the pollutants. This approach is well reflected in the policies & programmes of Government of India, particularly on environmental auditing and promotion of cleaner technologies. (40-43)

3.4.4 6 Environmental auditing

The Government of India has made environmental auditing mandatory and the industries are required to submit environmental statements every year starting from 1992-93. Besides a regulatory requirement, the industrial concerns should feel that they have a responsibility for abatement of pollution and should use this tool as a self-regulatory mechanism. This not only assists in regulatory compliance with laws & regulations but also helps in evaluating performance of process and pollution control systems and serve as an eye-opener for waste minimisation. This would result in reduction in losses and pollution load and, in turn, potential cost-savings. An annual statement, constantly, helps in identifying and focusing attention on the areas of concern, practices that need to be changed and plans to deal with the adverse effects.

3.4.4.7 Controlling pollution at source

After tackling the pollutants for minimising their generation, the next strategic approach is to control the pollution at source. It is always better to control the pollution at source before it goes into the environment. Under this strategy, many schemes & programmes including fixing of emission standards, action plan for controlling the pollution from highly polluting industrial sectors, etc. have been formulated and implemented.

3.4.4.8 Cleaner production technologies

Small-scale industries are a special feature of Indian economy. However, at times, it becomes difficult for the small-scale industries to afford the costs towards pollution control. As such, it is essential for the industrial sectors to explore the possibilities of adopting better & cleaner process technologies giving no/low pollution. Government provides financial assistance and technical support for identification and demonstration of such technologies. The development and adoption of cleaner technologies including environment-friendly biotechnology are also promoted.

3.4.4.9 Tackling the pollutants generated from vehicles

A vehicular emission, which is a cause of major concern in urban areas, is a deadly cocktail of poisonous gases and particulate, which affect the human beings, vegetation and buildings. Realizing the gravity of the problem, CPCB has paid special attention towards the curtailment of emission coming out of the vehicular exhaust. The strategy, for tackling the pollutants emitted from automobiles, involves reduction of pollutants at two stages viz. (i) pre-combustion stage, where the quality of fuel can be upgraded; and (ii) combustion stage, where engine modifications are required.

Under Section 20 of the Air Act the State Pollution control Boards have the statutory obligation to control, pollution due to automobiles. The state government can take up measures for control of air pollution from automobiles by insisting on some design changes or installation of suitable appliances even at the time of manufacturing of vehicles and give necessary instructions to the regional transport authorities for monitoring the implementation of these changes to keep the vehicles which do not meet these requirements out of the roads. Further as per Section 19.3 the state government in consultation with the pollution control boards can also take necessary steps like restricting the use of certain types of fuels like unleaded petrol, liquefied natural gas, etc., to particularly high traffic intensity regions.

3.4.4.10 Improvement in fuel quality

In-order to improve the fuel quality, based on the recommendations of CPCB, specifications for motor gasoline and diesel have been notified under the Environment (Protection) Act, 1986. These specifications came into effect from April 1996. The specifications, prescribed in view of the adverse effect of benzene, lead, sulphur and aromatic content on human health are provided in Tables 3.8 and 3.9 CPCB also issued directions to various oil refineries for implementing these specifications (43, 44). The schedules for implementation of specifications for improved fuel quality with regard to sulphur in diesel and lead content in gasoline are provided in Tables 3.10 and 3.11

Table 3.8 *Specifications for motor-gasoline for emission related parameters*

S. No.	Characteristics	Requirement	Method of Test (Ref. parts of IS-1448)
1.	Reid vapour pressure at 38°C, kpa	35 – 70	P:39
2.	Benzene, 0% by voume max.	5.0*	P:104
3.	Lead content (as Pb) g/I max.	0.15** (low leaded) 0.023 (unleaded)	P.38
4.	Sulphur, % by mass, max.	0.10 (unleaded) 0.20 (leaded)	P:34
5.	Potential gum, g/m^3, max.	50	ASTM 873.8
6.	Gum (solvent washed), max.	40	P:29
7.	Oxygenates Content : Ether (MTBE, ETBE), Alcohol, % by volume, max.	15	-
8.	Phosphorus	Absent	ASTMD 3231

* 3.0% Benzene by volume, maximum, in metro cities by 2000AD

** 0.15 g/l of lead by December, 1996 for entire country; 0.013 g/l of lead by April, 1995 in 4 metro cities, by December, 1998 in all State capitals, Union Territories and major metro-cities and by December, 2000 for entire country.

Table 3.9 *Specification of diesel fuel for emission related parameters*

S. No.	Characteristics	Requirement	Method of Test (Ref. parts of IS-1448)
1.	Density at 15O c, Kg/m3	820 to 880*	P:32
2.	Cetane Number, min.	45.0**	P:9
3.	Distillation 85% by volume recovery at C max 95% by volume recovery at C max	 350 370	P.18

Table 3.9 ***Contd...***

S. No.	Characteristics	Requirement	Method of Test (Ref. parts of IS-1448)
4.	Sulphur, % by mass, max.	0.50***	P:33

* 820-850 by 2000 AD , ** 48 by December 1998 (except in the refineries-Digboi, Guwahati and BRPL) , *** (i) 0.50% by mass by April 1996 in 4 metro-cities and Taj Trapezium; (ii) 0.25% by mass by October 1996 in Taj Trapezium; and (iii) 0.25% by mass by April 1999 throughout the country.

Note: Above specifications apply to HSD only; and for new refineries coming during or after 1997 the specifications applicable by 2000 AD for existing refineries, will be applicable by 1997.

Table 3.10 *Diesel sulphur phase out programme*

Phase	Target		
	Date	Levels	Areas
I	April 96	Low Sulphur (0.5%)	Four metros and Taj Trapezium
II	August 97	Low Sulphur (0.25%)	Delhi and Taj Trapezium
III	April 98	Low Sulphur (0.25%)	Metro Cities
IV	April 99	Low Sulphur (0.25%)	Entire Country

On the recommendation of CPCB, the Ministry of Environment & forests notified the specifications for 2T oil, which became effective from April 01, 1999. The specifications required use of low smoke 2T oil, which is one of the causes of pollution from 2stroke engines. The specifications for 2T oil are given in table 3. 12 The excessive use of 2T oil in petrol used for two-stroke engines leads to emission of excessive smoke, particularly fine particulate matter from exhaust which are harmful to human health and environment.

In-order to prevent excessive use of 2T oil, pre-mixed 2T oil dispensers have been installed in all the petrol filling stations of Delhi and sale of loose 2T oil has been banned from December 1998.

Table 3.11 *Gasoline lead phase out programme*

Phase	Target		
	Date	Levels	Areas
I	June 1994	Low Sulphur (0.15 g/l)	Cities of Delhi, Mumbai, Chennai and Calcutta
II	April 01, 1997	Unleaded (1.013 g/l) (+ low leaded)	Cities of Delhi, Mumbai, Chennai and Calcutta
III	January 01, 1997	Low leaded	Entire country
IV	September 01, 1998	Unleaded	NCT Delhi
V	December 31, 1998 (advanced to September 01, 1998)	Unleaded	All other capitals of states/UTs and other major cities.
VI	January 01, 1999	Unleaded only (0.013)	NCR
VII	April 01, 2000	Unleaded (0.013 + low leaded)	Entire country

Table 3.12 *Specifications for 2T oil for emission related parameters*

Performance parameter	Standard index* (min.)	Test procedures
Exhaust smoke	85	JASO M342-92

$$*Smoke\ index = \frac{Reference\ Oil}{Candidate\ Oil}$$

3.4.4.11 Modifications in engine design

During 1990-91, for the first time, mass emission norms for vehicles at manufacturing stage have been notified. The manufactures, for meeting these norms, did not require any major modifications. The emission norms along with fuel quality specifications, laid down in 1996, required modifications in the engine design particularly in regard to crankcase emissions and evaporative emissions control.

From April 1995, new passenger cars were allowed to register only if these were fitted with catalytic converters. Emission norms for such cars were tightened by 50 percent as compared to 1996 norms. The testing method for passenger cars norms was changed to cold start from hot start from April 1998, which is a stricter procedure than the previous one. The norms for the year, 2000 notified in August 1997 under the Motor Vehicle Rules, require major modifications in the engine design specially in regard to fuel injection system in passenger cars and fitting of catalytic converters in two- stroke engines. These standards are akin to EURO-I norms adopted in the European countries in 1992. With the recent directions of the Humble Supreme Court of India, passenger cars (both petrol and diesel) are required to meet at least EURO-I norms from June 1999 and from April 2000 only such vehicles meeting EURO-II norms are being registered in the National Capital Region. The mass emission standards for diesel and petrol-driven vehicles are provided in tables3.13 and 3.14 respectively.

Table 3.13 *Mass emission standards for diesel driven vehicles*

Category of vehicle	**CO (gm/kwh)**			**HC (gm/kwh)**			**NOx (gm/kwh)**			**PM (gm/kwh)**
	1991	1996	2000	1991	1996	2000	1991	1996	2000	2000
Heavy vehicles (3.5 tonnes GVM & above)	14.0	11.2	4.5	3.5	2.4	1.1	18.0	14.4	8.0	0.36
Light vehicles (less than 3.5 tonnes)	14.0	11.2 or 5.0-9.0*	4.5 or 2.75-6.9*	3.5	2.4 or 2.0-4.0*+	1.1 or 0.97-1.7*+	18.0	14.4 or 2.0-4.0*+	8.0 or 0.97-1.7*+	0.61 or 1.14-0.25*

in g/km + HC+NO_x emissions

Table 3.14 *Mass emission standards for petrol driven vehicles*

Category of vehicle	CO (gm/kwh)			HC (gm/kwh)	PM (gm/kwh)	
	1991	1996	2000	1991	1996	2000
Passenger cars	14.3-27.1	8.68-12.4	2.72	2.0-2.9	3.0-4.36	0.97
Two-wheelers	15.0-35.0	4.5	2.0	10.0-12.0	3.6	2.0
Three-wheelers	40.0	6.8	4.0	15.0	5.4	2.0

3.4.4.12 Use of cleaner fuel

The Government of India has also taken up use of CNG, an eco-friendly fuel, for use of petrol driven vehicles and an action plan is being prepared for conversion of all the three wheelers plying in Delhi to use CNG or other alternate cleaner fuels. In Delhi, most of the Government vehicles have already been converted to CNG dual fuel mode. In addition, replacement of a part of existing fleet of city busses with new busses having dedicated CNG engines has already been initiated and about 10,000 CNG-driven buses are expected to operate in Delhi by March 2001.

3.4.4.13 Clean coal initiatives

Among the fossil fuels, coal has acquired the dubious distinction of being the dirtiest one. Such an attribute is on account of environmental damage and pollution problems caused during mining, processing, end use and wastes of coal. Besides other problems, emissions of fly ash during combustion of coal and huge quantities of ash generated from boilers of coal-based power plants and industrial houses are of major concern. The problems are attributed, mainly, to the poor quality of coal and use of such coal in power sectors. Due to drift origin of Indian coal, inorganic impurities are intimately mixed in the coal matrix, resulting in difficult beneficiation characteristics. Though the sulphur content in Indian coal is generally less than 0.6%, the ash content is very high. The consumption of coal during 1996-97 was about 290 million tones, which is expected to increase to 416 tones during 2001-2002. Over 200 million tones of coal reach the consumers with ash content averaging 40%.

About 70% of total coal production is used in the power generation sector and 70% of the total installed capacity of power generation in the country is from coal based thermal power plants. Only 20% of total coals, transported to the power plants, are of superior grade with ash content of 24% or less and remaining 80% are of inferior grade with ash ranging from 24 to 45%. Increased dependence of power sector on the inferior quality of coal has resulted into higher emissions of particulate matter and generation fly ash. The fly ash poses land as well as air pollution. On the recommendation of a committee, the Government of India has promulgated a Gazette Notification on use of beneficiated/blended coal containing ash not more than 34% with effect from June 2001 in the power plants located beyond 1000 kms. From pit head and located in critically polluted areas, urban areas and ecologically sensitive areas. The power plants using FBC (CFBC, PFBC & AFBC) and IGCC combustion technologies are exempted to use beneficiated coal irrespective of their location.

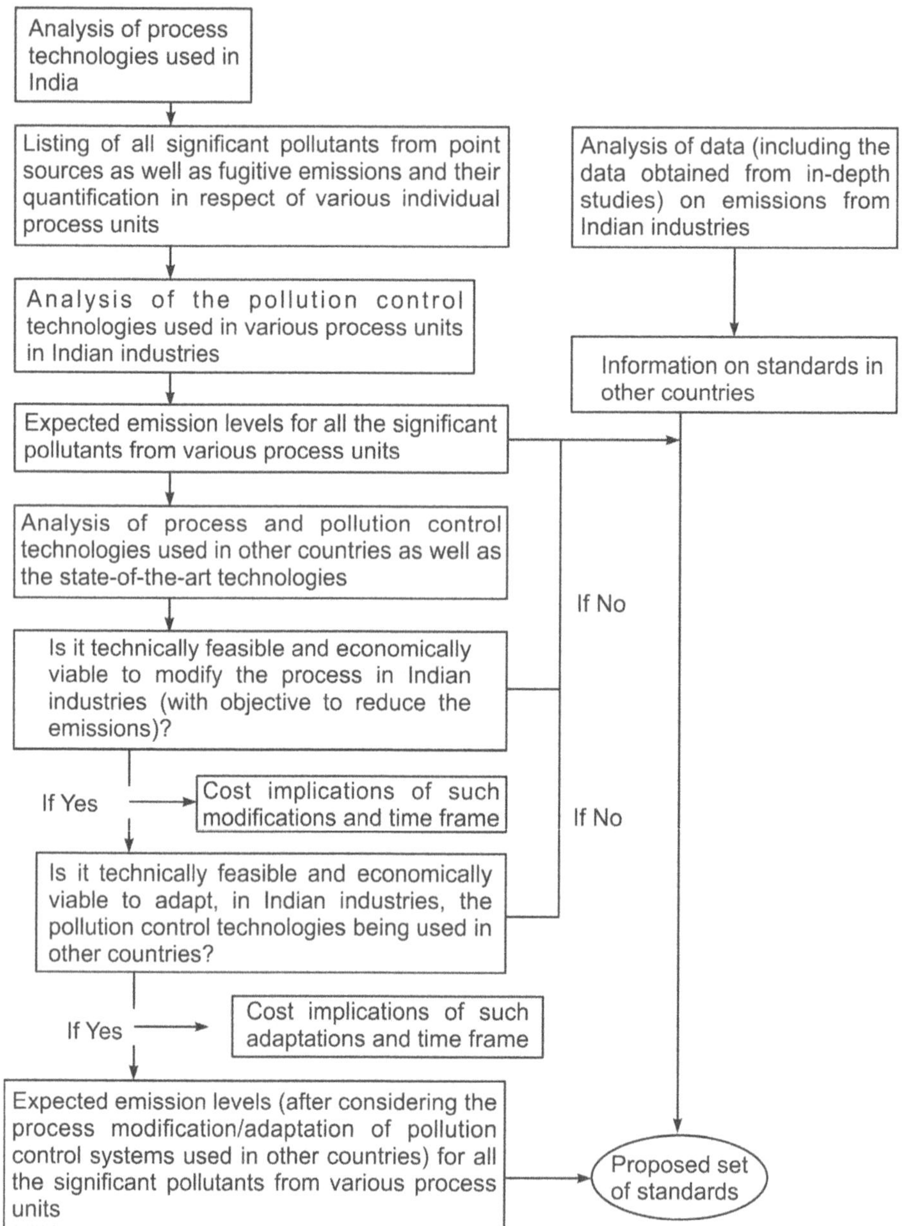

Fig. 3.9 *Flow-scheme of the methodology for development of emission standards*

3.4.4 14 Formulation of emission standards

Emission standards

As per Section 17(1) of the Air Act, the Central Board also has to lay down standards for emission of air pollutants into the atmosphere from industries, plants and automobiles and for the discharge of any air pollutant into the atmosphere from any other source, what so ever not being a ship or an aircraft. This section also provides for laying down of different standards for emission for different industrial plants considering the quality and composition of emissions of air pollutants into the atmosphere from them.

Thus, there are two types of limits on the emissions from major polluting industries:

(i) Standards laid down under Sec. 17 (1) (g), applicable over the entire state, which the industry concerned will have to comply with, wherever it may be located.

(ii) Specific emission limits that may be prescribed by the pollution control boards in the consent conditions under Sec. 21(5). These limits should generally be more stringent than the standards set under Sec. 17 (l) (g).

The central and state boards are evolving industry specific emission standards duly considering the prevailing ISI standards, and the guidelines given by the Policy Planning Committee set up by the Central Board. Emission standards should not only be easily enforceable but also be achievable with the technology available in the country.

The standard methodology adopted for developing industry-specific standards for discharge of pollutants including emissions is shown in Figure 3.9

3.4.4.15 Emission standards for air polluting industries in India

Generally, two main aspects are taken into consideration for development of standards. One relates to the adverse effects on human health & environment and other to the achievability of limits of pollutants by incorporation of appropriate pollution control measures. The emission standards for about 60 industrial sectors have been notified so far. The standards are based on the advisability and availability of indigenous air pollution control technology. For some of the pollutants which are not very critical and for which no indigenous control technology is available, for example. SO_2 from thermal power plants, the height of the stack and dispersion are considered.

3.4.4.16 Status of air pollution control in major air polluting industries

Concerned with the increasing menace of environmental pollution, the Government of India has evolved a 15-point Action Plan with particular emphasis on industrial pollution control within a time targeted schedule. The present air pollution control status of five major polluting industries in the country is given in Table 3.15

Table 3.15 Major polluting industries in India

Industry	No. of industrial units	Complying with standards	Not complying with standards	No. of units closed
Thermal power plants	69	28	40	01
Cement	97	81	11	05
Iron and steel	07	01	06	–
Non-ferrous metallurgica industries (Cu, Zn)	06	02	04	–
Aluminium	07	05	02	–

3.4.4.17 Air pollution emissions–industrial sources and limits

The stationary sources of air pollutants include a large number of industries, smaller metal smelting units, and municipal incineration devices. The same types of pollutants like suspended particulate matter, various oxides of nitrogen, sulphur and carbon are included in industrial air pollutants also. Table 3.16 presents the major industry-wise air pollutants with their sources.

Tall chimneys or stacks themselves work as a preliminary control device. Stack heights throughout the world are being increased to provide better dispersion. Air pollution control equipment include electrostatic precipitators, fabric filters, centrifugal separators and wet scrubbers (Table 3.17).

Table 3.16 *The majority industry-wise air pollutants with their sources*

Industry	Sources	Pollutant
Thermal power	Boiler	Fly ash, smoke, CO, CO_2& SO_2
Iron and steel	Blast furnace	Smoke, fumes, dust, CO, CO_2
	Coke-oven	Smoke, dust, cyanides, H_2S, CO_2, H_2
Cement	Rotary kiln	Ash, fumes, SO_2& dust
	Dryer & Storage	Dust
	Boiler	Ash, CO_2, CO
Pulp and paper	Digester & multiple effect evaporator	Methyl mercaptan, methyl sulphide, methyl disulphide, terpenes, H_2S
	Lime kiln	H_2S, $CaCO_3$, dust
	Recovery furnace	Sodium sulphide, Na_2CO_3, SO_2, CO
	Smelter	CO_2, NaS
	Bleaching	Cl_2, calcium hypochlorite
	Caustizing unit	$CaCO_3$
Sugar	Lime kiln	CO_2, dust
	Boiler	CO, CO_2, ash, smoke
	Chemical	Smoke & particulates
	Neutralizer	H_2PO_4, SO_2

Table 3.16 ***Contd...***

Industry	Sources	Pollutant
Petroleum	Cracking unit	Oxides of Sulphur
	Regenerator	Oxides of sulphur, hydrocarbons, oxides of nitrogen, NH_3, CO & fine catalyst
	H_2S flare	Oxides of sulphur, oxides of nitrogen. HCl, soot, Cl_2, cyanides NH_3
	Incinerator	CO, smoke, ash, particulate matter
	Storage tank	Hydrocarbons, odour
	Pump house, cooling tower & compressor	Hydrocarbons, & oxides of nitrogen
	Heater	Oxides of nitrogen
Super phosphate	Reactor	CO_2, SO_2, steel and silica
	Acidulator & curing operation	Tetra fluoride, fluorides
Phosphate	Cooler & dryer	HF, dust, fumes & SiF_4
	Granular & mixing equipment	NH_4Cl, NH_3
H_2SO_4	Absorption system & burners	SO_2, SO_3, H_2SO_4, fumes, aerosol, mist, oxides of nitrogen, acid mist
HNO_3	Absorption tower	Oxides of nitrogen, acid mist
Copper	Furnace	CO, oxides of sulphur, dust
	Convertors & roaster	Oxides of sulphur, zinc oxide, mercaptans & organic sulphides
Lead	Blast furnace	SO_2, mercaptans & organic sulphides
	Concentrator	Fumes of volatile meteal
	Mining & milling	Dust
Sodium carbonate	Absorber	NH_3
	Furnace	CO_2, ash, smoke
	Rotary dryer	Dust

Table 3.17 *Important characteristics of some cleaning devices*

Type	Collection efficiency (Percentage wt. basis)	Pressure drop (mm of water)	Power consumed BHP per 100 m^3/min
Electrostatic precipitator	80 to 99.5	2 to 12	0.2 to 0.5
Fabric filter	95 to 99.5	50 to 150	0.8 to 2.6
Mechanical collector			
(dry centrifugal type)	50 to 95	25 to 100	1 to 2.5
Scrubber (wet collection)	75 to 99	50 to 400	2.0 to 8.0

As per the Air Prevention & Control of Pollution Act, 1981 in India, all industries are being asked to control emission levels within specified limits (Table 3.18).

Table 3.18 *Limits for emissions into air from certain industries*

Industry	SO_2	SO_3	Fluorine based gases	Particulates	NO_x as NO_2	Hydro-carbons (as CH_4)	Smoke
Phosphoric acid							
Existing plants			1.5 kg/t of P_2O_5				
New plants			0.6 kg/t of P_2O_5				
Single super phosphate							
Existing plants			0.5 kg/t of product	500 mg/Nm^3			
New plants			0.1 kg/t of product	500 mg/Nm^3			
Triple super phosphate							
Existing plants			0.3 kg/t of product	4 kg/t			
New plants			0.07 kg/t of product	4 kg/t			
Sulphuric acid (100 %)							
Existing plants							
<200 t/day	16 kg/t	5 kg/t					
>200 t/day	12 kg/t	5 kg/t					
New plants							
<200 t/day	12 kg/t	5 kg/t					
>200 t/day	4 kg/t	0.5 kg/t					
Nitric acid and nitrogenous fertilizers							
Existing plants				6 kg/t product	24 kg/t HNO_3	5 kg/t of product	
New plants				0.5 kg/t product	3 kg/t HNO_3	5 kg/t of product	

Table 3.18 ***Contd...***

Industry	SO_2	SO_3	Fluorine based gases	Particulates	NO_x as NO_2	Hydro-carbons (as CH_4)	Smoke
Petroleum refineries							
Catalytiwc regeneration units	5.5 kg/t						Other than while smoke not darker than No.2 of Ringlemann's chart for >5 min in any consecutive 60 min &>6h in any 10 day period
Other units	3 kg/t			125 mg/Nm^3			-do-
Cement plants							
<200 t/d in the protected areas				250 mg/Nm^3			
In other areas				400 mg/Nm^3			
>200 t/d in protected areas				150 mg/Nm^3			
In other areas				250 mg/Nm^3			
Thermal power plants							
<200 MW in protected areas				150 mg/Nm^3			
In other areas				350 mg/Nm^3			
>200 MW in protected areas				150 mg/Nm^3			
In other areas				150 mg/Nm^3			
Iron and steel							
Sintering plant				150 mg/Nm^3			
During oxygen lancing				400 mg/Nm^3			
Normal operations (during steel making)				150 mg/Nm^3			

Table 3.18 ***Contd...***

Industry	SO_2	SO_3	Fluorine based gases	Particulates	NO_x as NO_2	Hydro-carbons (as CH_4)	Smoke
Calcium carbide							
Kiln				250 mg/Nm^3			
Arc furnace				150 mg/Nm^3			
Smelting							
Concentrator				150 mg/Nm^3			
Smelter & converter		No release (all SO_2 and SO_3 should go for H_2SO_4 manufacture)					
Aluminium plants							
Calcination				250 mg/Nm^3			
Al smelting			1 kg/t of Al	150 mg/Nm^3			

Standards are based on:

1. Sulphuric acid and phosphatic fertilizers (IS:8635-1977)
2. Nitric acid and nitrogenous fertilizers (IS:9005-1978)
3. Petroleum refineries (IS:8636-1977)
4. All remaining industries (Central Pollution Control Board, Emission regulations Part I 1984, and Part II 1985)

In addition, the Government has made it mandatory to fit catalytic converters in passenger cars from April 01, 1995 in four metros and notified the emission standards for such vehicles (Table 3.19). Similarly, emission standards for exhaust from in-use vehicles are also prescribed, which are presented in Table 3.20).

As yet another step to tackle the pollution at source, 15-year old commercial vehicles, which are grossly fuel inefficient & high polluters, are being phased out since December 31, 1998 in the capital city of Delhi.

Table 3.19 *Emission standards for catalytic converter fitted vehicles*

Vehicles	CO % by volume	Absorption coefficient	Bosch unit	Haritridge unit
Petrol driven				
* 2/3 Wheelers	4.5	-	-	-
* 4 Wheelers	3.0	-	-	-
Diesel driven				
* Full load	-	3.25	5.2	75
* Free acceleration	-	2.5	-	65

Table 3.20 *Standards for emissions from in-use vehicles*

Cubic capacity of engine (cm^3)	CO (gm/km)	HC + NO_x (gm/km)
< 1400	4.34	1.50
1400 to 2000	5.60	1.92
> 2000	6.20	2.16

3.4.4.18 Action plan for highly polluting industrial sectors

With the aim of intensifying pollution control programmes in highly polluting industrial sectors, an action plan was formulated in the year 1991. CPCB & MoEF identified 17 highly polluting industrial sectors and a total of 1551 large industries belonging to these sectors for priority actions. CPCB, in association with SPCBs, rigorously followed up implementation of pollution control programmes in these industries and results have been quite encouraging. Out of 1551 industries, in the year 1991, only about 150 had the requisite pollution control systems but now in a period of 9 years time the number of industries having requisite pollution control systems has increased to 1320. While only 67 have yet to comply with the regulatory requirements, 164 units are closed either due to penal action or due to some other reasons. The year-wise progress of pollution control status in these industries is presented in Fig. 3.10.

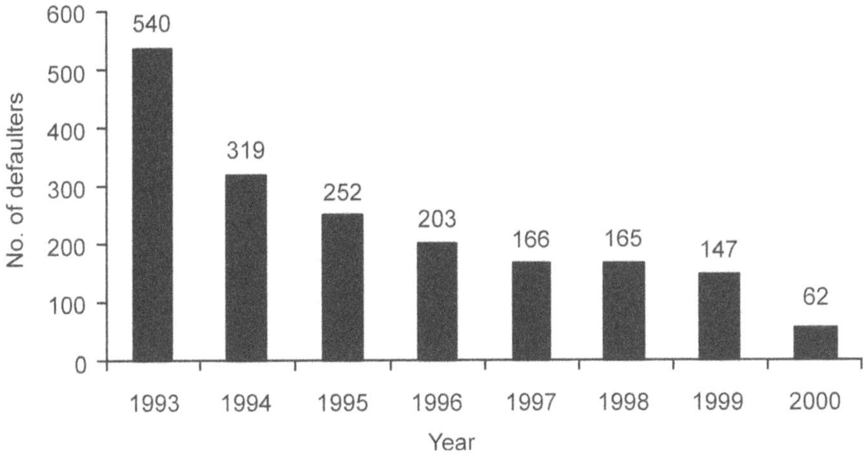

Fig. 3.10 *Year-wise progress of pollution control status in 17 categories of industries*

3.4.4.19 Tackling of the polluted areas

The third strategy involves continuous monitoring of air quality in various parts of the country, taking steps for improving air quality in critically polluted areas/hot spots and preventing other areas from getting polluted using environmentally sound developmental planning.

3.4.4.19.1 Monitoring air quality (45-53)

CPCB, as in integral part of the air pollution control programme, initiated a nationwide network of National Ambient Air Quality Monitoring in the year 1984 with seven monitoring stations. Since then the number of monitoring stations have increased, steadily, over the period and now 290 stations covering 92 cities/towns are in operation. The programme is being carried out with the help of SPCBs and some other agencies. The three criteria pollutants viz. sulphur dioxide (SO_2), oxides of nitrogen (NO_x) and suspended particulate matter (SPM) are being monitored regularly at all the monitoring locations. Besides this, in 10 metro-cities of the country, additional parameters such as respire able particulate matter (RPM), respire able lead & other toxic trace metals, hydrogen sulphide (H_2S), ammonia (NH_3) and polycyclic aromatic hydrocarbons (PAHs) are also being monitored.

3.4.4.19.2 Preparation of area emission inventories: spatial and temporal variations of pollutant emissions

For the "Preparation of Area Emission Inventories" in an urban air shed one has to first categorize the various emission sources or activities which fall into different groups like a) Automobiles, b) Industrial processes, industrial furnaces, power plants, etc.,

c) Domestic furnaces, small scale industrial operations, etc.

The area under study has to be divided into several grids of 1 sq.km and emission inventories for each grid per annum have to be prepared and mapped by overlaying on corresponding topographic maps with different colour codes for different ranges of pollutant concentrations. This will give a clear idea about which portions of a study area have high emissions and which have low emissions. For preparing emission inventories for general common emission air pollutants like particular matter (PM-10), SO_2, NO_x, CO. Total Suspended Particulate Matter (TSP) are generally monitored while for environmental pollutants like O_3, Pb, hydrocarbons, chlorine and fluorine organic compounds are monitored. For actually estimating the emissions from different constituent sources of any grid area, the following guidelines will be useful.

3.4.4.19.3 "Preparation of area emission inventories" from vehicular traffic

For assessing the air pollution emissions from vehicular traffic in a given traffic scenario the flow sheet given in Fig. 3.11 can be used. By adopting standard emission factors for each type of vehicle one can calculate the emission loads of each pollutant. Traffic survey inventories giving the traffic loads for various important roads in the grid area under study could include estimation of traffic volumes and their composition with reference to different categories of vehicles and their temporal variations. Using appropriate emissions factors for various fuels, emission flows can be easily estimated.

Based on traffic load data, driving pattern and appropriate emission factors, the annual loads of air pollutants like SO_2, NO_x, and CO can be estimated in tons per km for each street. After measuring each street and the number of streets in a grid area under study, the total air pollution loads coming from vehicular traffic can be estimated by multiplying the specific emission rate of each street by its length. Summing of the annual emission of each street for all the streets in the grid will give the annual emission of pollutant in to n/km^2 due to vehicular traffic in the grid area under study.

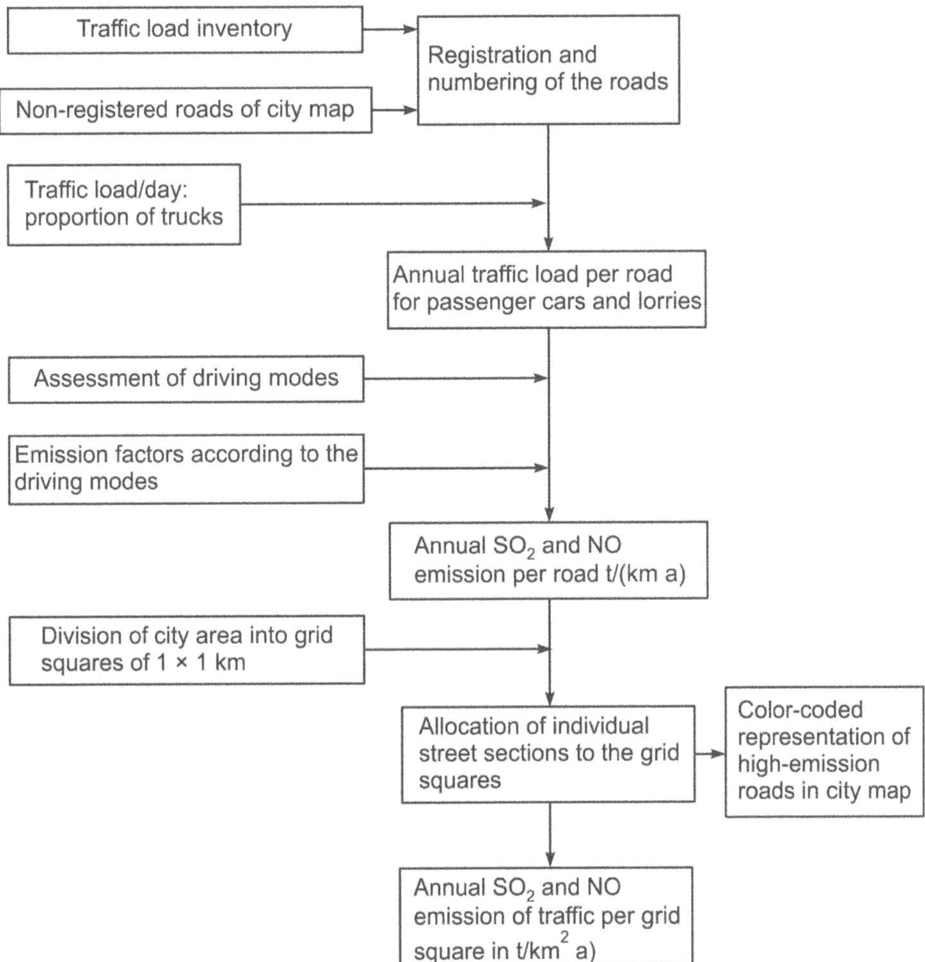

Fig. 3.11 *Flow chart for determining the SO_2 and NO_x emissions from traffic per road and per square kilometer in a city (also valid for other compounds)*

3.4.4.19.4 Estimation of air pollution emissions from industrial sources in the grid area under study

The emission of air pollutants from various industrial sources involving a number of plant operations can be estimated either by direct measurement (wherever possible) of emission flows and multiplying them with operation hours or in directly by inventor sing the amount and nature of the fuel consumed in each industrial operation like industrial furnace operation, process operation, etc. Multiplying with specific emission factors of each fuel with fuel consumed in each operation, the annual emissions of each idew1ndustrial operation can be estimated.

Based on the geographical location of the industry, the pollutants emissions from various industrial sources can be allocated to the appropriate grid area and corresponding maps in the grid with appropriate colour codes for different emission concentrations can be prepared, which will help in rapid qualitative assessment of areas under high emission levels and low emission levels from industrial sources.

3.4.4.19.5 Preparation of area emission inventories from domestic and small scale industrial activity in a grid area

An emission inventory from domestic sources can be prepared by surveying nearly 70% of the households in a grid area by individually interviewing and collecting data on types of fuel and annual fuel consumption. However this kind of survey will be tedious and costly particularly for large towns. An examination of different types of fuel supplied in a given grid area from different supply sources gives a rough estimate of different types of fuel consumed in a unit time and by multiplying by a suitable factor for each fuel, the emissions from domestic activity can be estimated.

The contribution of air pollutant emissions from small scale industries can be estimated by examining the type of material produced, type of fuel used in the furnaces, etc., and using appropriate emission factors and operating hours.

3.4.4.19.6 Assessment of combined annual emissions in a grid area and preparation of maps with spatial distribution of pollutants sources with different colour codes

The pollutant emissions from different categories of sources should be combined for each sq.km grid area and their spatial dispersion can be mapped with different colour codes based on the spatial variation in the emission level.

It is very important to note that though the quantity of emissions of a pollutant from a particular source may be high, its relative effectiveness as a pollutant depends on the height of the source from which it is discharged into the atmosphere. For example, the quantity of NO_x, emissions from a power plant in a grid may be very high but its relative effectiveness as a pollutant in the ambient air may be less when compared to NO_x released from vehicular traffic as in the former case, the NO_x will be released from a stack height of 250 m while in the latter case the NO_x will be directly released in the respiratory space. So one has to evaluate the relative effectiveness of each air pollutant released into the ambient air from different source heights in different weather conditions. Thus based on dispersion calculations and pollution measurements, the relative effective pollution load of each pollutant coming from different sources in a grid area can be assessed.

3.4.4.20 Temporal distribution of air pollutants in a grid area

Temporal variations in air pollution levels occur due to changes in meteorological conditions and variations in strength of source emissions with operating hours. A critical examination of the temporal resolution of air pollution in a grid area will be very useful in understanding the relation between pollution emissions and deposition mechanisms.

The diurnal variation study also greatly helps in preparing traffic management plans. On weekdays diurnal variations in vehicular traffic will have peaks of rush in the morning and in the evening, while on holidays the diurnal variation will be governed by shopping trips, picnic trips, holiday trips, etc.

The monitoring of pollutants is carried out for 24 hours averaging period (4-hourly sampling for gaseous pollutants & 8-hourly sampling for SPM) with a frequency of twice a week. The data, so generated, are scrutinized, statistically analyzed to get the information on air quality in various parts of the country. The air quality trends for the period 1994-98 in the industrial areas of four metropolitan cities in respect of three criteria pollutants viz. SO_2, NO_x and SPM are provided in Figs. 3.12 through 3.14 this clearly indicates positive results of various initiatives and programmes taken up by pollution control agencies for improvement in air quality.

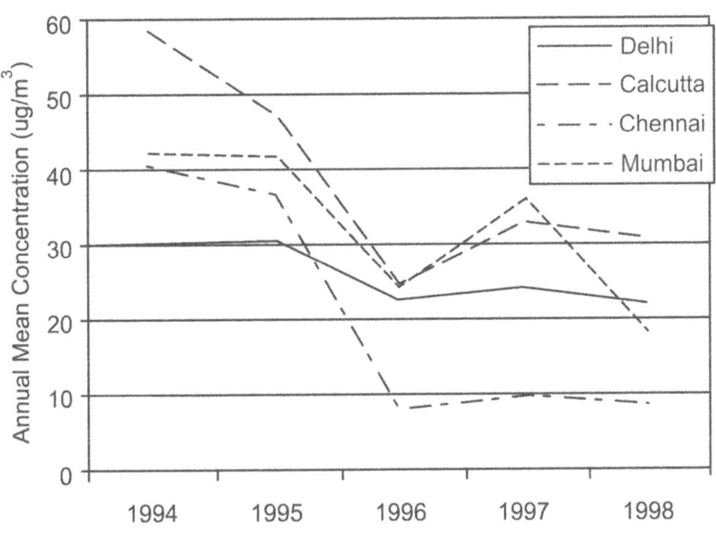

Fig. 3.12 *Ambient air quality trend for sulphur dioxide in four metro cities*

3.4.4.21 Ambient air quality standards

Empowered by Section 16(h) of the Air Act to lay down standards for the quality of ambient air for the entire country, the Central Pollution Control Board first notified the ambient air quality standards in its notification dated 16-12-1982. The entire country is divided into three main areas: (a) Industrial and mixed uses areas (b) Residential and rural areas (c) Sensitive areas.

CPCB laid down standards for ambient air quality with regard to various pollutants in respect of industrial, residential and sensitive areas are given in table 3.21 which provide the basis for protecting the public health, vegetation, animals and national heritage (monuments) from the adverse effects of air pollution within adequate margins of safety. This helps in continuous evaluation of air quality and deciding & taking appropriate actions for controlling the air pollution.

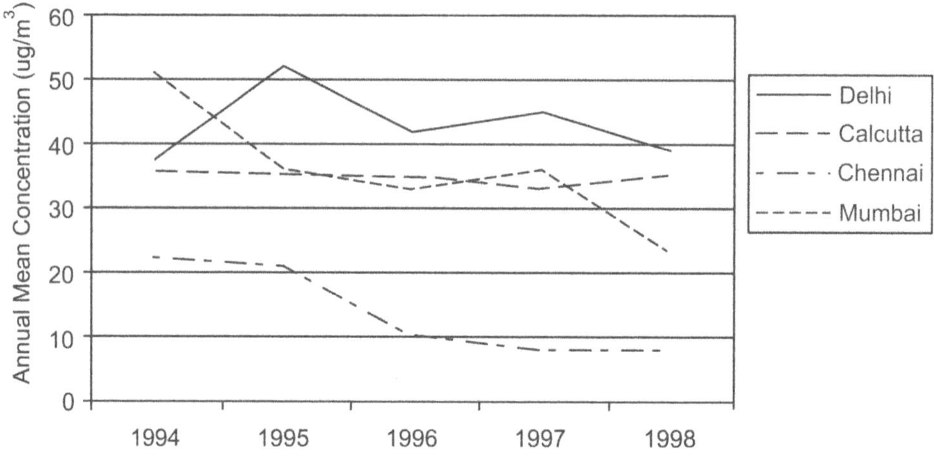

Fig. 3.13 *Ambient air quality trend for nitrogen dioxide in four metro cities*

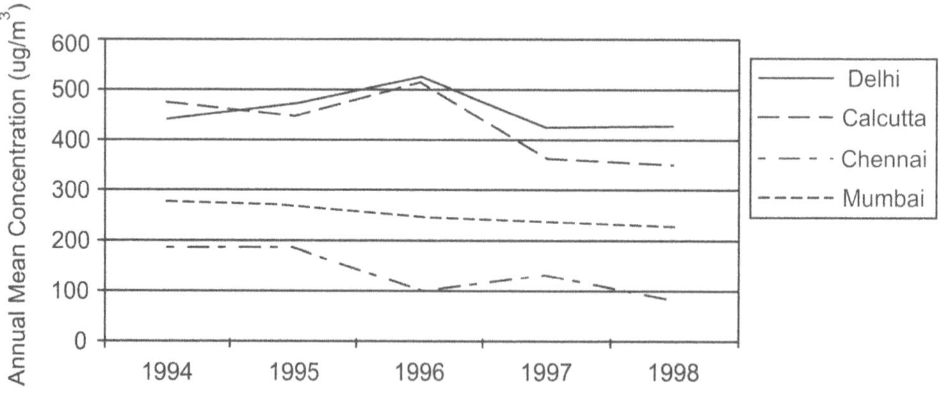

Fig. 3.14 *Ambient air quality trend for suspend for suspended particulate matter in four metro cities*

Table 3.21 *National ambient air quality standards*

Concentration in Ambient air					
Pollutant	**Time weighted average**	**Sensitive area**	**Industrial area**	**Residential, rural and other areas**	**Method of measurement**
Sulphur dioxide (μg/m^3)	Annual* 24-hours**	15 30	80 120	60 80	*Improved West & Gacke method *Ultraviolet Fluorescence
Oxides of nitrogen as NO_2 (μu/m^3)	Annual* 24-hours**	15 30	80 120	60 80	*Jacob & Hochheiser modified (Na-arsenite) method *Gas phase Chemiluminiscence
Suspended particulate matter (size less than to μm) μg/m^3)	Annual* 24-hours**	70 100	360 500	140 200	*High volume Sampling (average flow rate not less than 1.1 m/minute)
Respirable particulate matter (size less tan 10 mm) (mg/m^3)	Annual* 24-hours**	50 75	120 150	60 100	*Resprable Particulate Matter Sampler
Lead (mg/m3)	Annual* 24-hours**	0.50 0.75	1.00 1.50	0.75 1.00	*AAS Method after sampling using EPM 2000 or equivalent filter paper
Corbon monoxide (μg/m^3)	8-hours** 1-hour	1.0 2.0	5.0 10.5	2.0 4.0	*Non-dispersive Infrared Spectroscopy

* Annual arithmetic mean of minimum 104 measurements in a year taken twice a week 24-hourly at uniform interval.

** 24/8 hourly values should be met 98% of the time in a year. However, 2% of the time, it may exceed but not on two consecutive days.

3.4.4.22 Action plan for critically polluted areas

The air pollution control in critically polluted areas is an integrated approach towards environmental management through control at source. This involves concerted efforts in identified polluted areas/hot spots through the various agencies concerned. This approach was initiated by CPCB in early 1989 with the identification of 10 critically polluted areas in the country. Subsequently, the number of such areas was increased to 24. These areas have been extensively surveyed, the extent of pollution assessed and sources contributing to the problem are identified. Subsequently, action plans, clearly indicating the actions and the concerned agencies required to take these actions along with time targets, have been formulated. These action plans are being implemented and rigorously followed up. Besides this, new industrial units in the problem areas are required to comply with location-specific standards for stringent environmental quality objectives.

3.4.4.23 Zoning atlas for siting of industries

Much of the problems of pollution and costs can be avoided by preventing indiscriminate siting of industrial units. An important prerequisite for judicious siting of industries is to consider the environmental profiles of the proposed sites. In-order to delineate the areas suitable for siting of industries of different categories based on the existing environmental features, a project for preparation of district-wise Zoning Atlas has been taken up by CPCB in collaboration with SPCBs and the German Technical cooperation (GTZ). Under the programme, which was initiated in 1994, zoning atlases have been prepared for about 60 districts and it is proposed to cover the entire country at the earliest. This would, surely, prevent the deterioration of environmental quality caused by unplanned and haphazard siting of industries.

3.5 Air Pollution Trends - Global Scenario

3.5.1 Global emissions of air pollutants

To provide a rough estimate of global emissions of many air pollutants in 1987 as base line data are given in Table 3.22 which tremendously increased at present due to increased industrialization.

3.5.2 Global urban air pollution trends

More than 80% of people living in urban areas around the world that monitor air pollution (now covering 3000 cities in 103 countries) are exposed to air quality levels that exceed WHO limits. While all regions of the world are affected, populations in low-income cities are the most impacted. According to the latest urban air quality database, 98% of cities in low and middle income countries with more than 100,000 in habitants do not meet WHO air quality guidelines. However, in high-income countries, that percentage decreases to 56%.

Table 3.22 *Estimated global emissions of major air pollutants*

Species	Major sources	Approximate annual flux (millions of tons)	Average half-life in atmosphere (days)	Global Emission (tons) Man-made	Natural	Atmospheric concentration (ppm)
CO_2	Respiration	100,000	2,500	1.4×10^{10}	1012	305 (daytime) 320 (average)
	Burning of biomass, fossil fuels	10,000	2,500			400 (nights)
CO	Burning of biomass, fossil fuels	1,000	75	400×10^6	3364×10^6	0.1
VOC	Man-made isoprene, terpenes from plants	100 800	1–1,000 < 1	88.0×10^6	170×10^6	
NO_x	Soils, burning of biomass, fossil fuels	100	4	60×10^6	430×10^6	0.1 to 2 ppb
NH_3	Industrial and biological nitrogen-fixation	100	9	4×10^6	5300×10^6	6 to 20 ppb
SO_2& SO_4	Sea spray, fossil fuels, smelting	90	1 – 4	132×10^6	130×10^6	0.2 ppb
H_2S & organic sulphur	Biogenic, anthropogenic	90	1 – 900	3×10^6	1.1×10^6	0.2 ppb
SPM	Wind erosion, fires volcanoes, human sources	10,000	1 – 1,000	100×10^6	1500×10^6	–
Hydrocarbons						
CH_4	Wetlands, rice, paddy textiles, termites, ruminant animals	400	3600	–	1450×10^6	–

Source: *USEPA council of Environmental Quality Report, 1987a.*

Using new, high-resolution global satellite maps of air quality indicators, NASA scientists tracked air pollution trends over the last decade in various regions and 195 cities around the globe. According to recent NASA research findings, the United States, Europe and Japan have improved air quality due to rigorous emission control regulations, while China, India and the Middle East, with their fast-growing economies and expanding industry, have seen more air pollution.

WHO was able to compare a total of 795 cities in 67 countries for levels of small and fine particulate matter (PM10 and PM 2.5) during the five-year period, 2008-2013. PM10 and PM 2.5 include pollutants such as sulfate, nitrates and black carbon, which penetrate deep into the lungs and into the cardiovascular system, posing the greatest risks to human health. Data was then analysed to develop regional trends. (8)

3.5.3 Key trends from 2008-2013

- Global urban air pollution levels increased by 8%, despite improvements in some regions.
- In general, urban air pollution levels were lowest in high-income countries, with lower levels most prevalent in Europe, the Americas, and the Western Pacific Region.
- The highest urban air pollution levels were experienced in low-and middle-income countries in WHO's Eastern Mediterranean and South East Asia Regions, with annual mean levels often exceeding 5-10 times WHO limits, followed by low-income cities in the Western Pacific Region.
- In the Eastern Mediterranean and South –East Asia Regions and low-income countries in the Western
- In the African Region urban air pollution data remains very sparse, however available data revealed particulate matter (PM) levels above the median

According to the new WHO database, levels of ultra-fine particles of less than 2.5 microns (PM2.5s) are highest in India, which has 16 of the world's 30 most polluted cities.

China, which has been plagued by air pollution, has improved its air quality since 2011 and now has only five cities in the top 30.

Nine other countries, including Pakistan and Iran, have one city each in the worst 30. For the larger, but slightly less dangerous PM10 particles, India has eight cities in the world's top 30.

Nigeria, Saudi Arabia and Pakistan each have two cities in the top 10. The true figure for the growth in global air pollution is likely to be worse because only a handful of African cities monitor their levels.

The most polluted city in the world, according to the WHO data, is Onitsha, a fast-growing port and transit city in south-eastern Nigeria that recorded levels of nearly 600 micrograms per cubic meter of PM10- around 30 times the WHO recommended level of 20 micrograms per cubic meter.

Air pollution levels were generally much lower for cities in developed countries with Sydney, New York and London registering 17, 16 and 22 micrograms per cubic meter for PM10 respectively. However, the data only includes measurements for particulates and does not include forms of air pollution such as NO_2 and ozone.

The new data, drawn from city and academic records, shows a rapid deterioration in air quality as low-income cities grow unchecked and populations become unable to escape clouds of smog and soot from transport, industry, construction sites, farming and wood-burning in homes.

Outdoor air pollution causes more than 3 million deaths a year - more than malaria and HIV/Aids - and is now the biggest single killer in the world. The toll is expected to double as urban populations increase and car numbers approach 2bn by 2050.

Air pollutants such as sulphates, nitrates and black carbon penetrate deep into the lungs and into the cardiovascular system, posing the greatest risks to human health, says the UN.

As urban air quality declines, the risk of stroke, heart disease, lung cancer and chronic and acute respiratory diseases, including asthma increases for the people who live in them. "When dirty air blankets our cities the most vulnerable urban populations - the youngest, oldest and poorest - are the most impacted," said Flavia Bustreo, WHO assistant director general.

Encouragingly, there is evidence from the WHO data that many cities are addressing air pollution. More than half of the monitored cities in high-income countries and more than one-third of those in low- and middle-income countries reduced their air pollution levels by more than 5% in five years. Delhi, one of the most polluted cities in the world, has banned large diesel cars from going into the city center.

Measures taken by cities include reducing industrial smokestack emissions, increasing the use of renewable power sources like solar and wind, and prioritizing rapid transit, walking and cycling networks in cities. Many cities are also committed to reducing car traffic and diesel vehicles in particular.

The UN's third outdoor air pollution database suggests the cleanest cities in the world are generally small, wealthy and situated far from industrial centers. Muonio in Finland, a town above the Arctic circle, has the world's purest recorded urban air, recording just 2 micrograms per cubic meter of PM 2.5 pollution and 4 micrograms per cubic meter of PM10. It is closely followed by Norman Wells in Canada, Campisabalos in Spain and Converse County, Wyoming in the US. Of 52 UK towns and cities included in the UN database, Port Talbot in south Wales, a hub for the UK steel industry, is the most polluted, a head of London, Glasgow, Southampton and Leeds. The cleanest UK city in the WHO list is Inverness, followed by Bournemouth, Newcastle and Sunderland.

The most polluted city in Australia, according to the data, is Gerald ton, a major seaport on the west coast, north of Perth. The most polluted city in the United States is the inland city of Visalia Porterville in California.

3.5.4 Air quality trends in USA

The "State of the Air 2016" (54) found continued improvement in air quality in 2012–2014, showing lower levels of year-round particle pollution and ozone. Still, more than half of all Americans are exposed to unhealthful levels of these pollutants. The "State of the Air 2016" report shows that cleaning up pollution continues successfully in much of the nation. In the 25 cities with the worst pollution, the majority saw improvements from last year. Many saw their lowest levels ever of year-round particle pollution or ozone pollution.

State of the Air 2016 shows that more than one in two people had unhealthy air quality in their communities. Yet, even as most cities experienced strong improvement, too many cities suffered worse episodes of unhealthy air. While most of the nation has much cleaner air quality than even a decade ago, a few cities reported their worst number of unhealthy days since the report began, including some that experienced extreme weather events. The "State of the Air 2016" report provides evidence that a changing climate will make it harder to protect human health.

The "State of the Air 2016" report shows that, even with continued improvement, too many people in the United States live where the air is unhealthy for them to breathe. Despite that continued need and the nation's progress, some people seek to weaken the Clean Air Act, the public health law that has driven the cuts in pollution since 1970, and to undermine the ability of the nation to fight for healthy air.

The "State of the Air 2016" report looks at levels of ozone and particle pollution found in official monitoring sites across the United States in 2012, 2013 and 2014. The report uses the most current quality-assured nationwide data available for these analyses. The report examines particle pollution (PM 2.5) in two different ways: averaged year round (annual average) and over short-term levels (24-hour). For both ozone and short-term particle pollution, the analysis uses a weighted average number of days that allows recognition of places with higher levels of pollution. For the year-round particle pollution rankings, the report uses averages calculated and reported by the U.S. Environmental Protection Agency (EPA). For comparison, the "State of the Air 2015" report covered data from 2011, 2012, and 2013.

Due to stronger standards for pollutants and for the sources of pollution, the United States has seen continued reduction in ozone and particle pollution as well as other pollutants for decades.

Figure 3.15 (from the data of US EPA) shows that since 1970, the air has gotten cleaner while the population, the economy, energy use and miles driven increased greatly. As the economy continues to grow, overall air emissions that create the six most-wide spread pollutants continue to drop. Overall, the best progress came in the continued reduction of ozone and year-round particle pollution, thanks to cleaner power plants and increased use of cleaner vehicles and engines. Continued progress to cleaner air remains crucial to reduce the risk of premature death, asthma attacks and lung cancer. Many cities reduced their ozone pollution in 2012–2014 below that reported in 2011–2013.

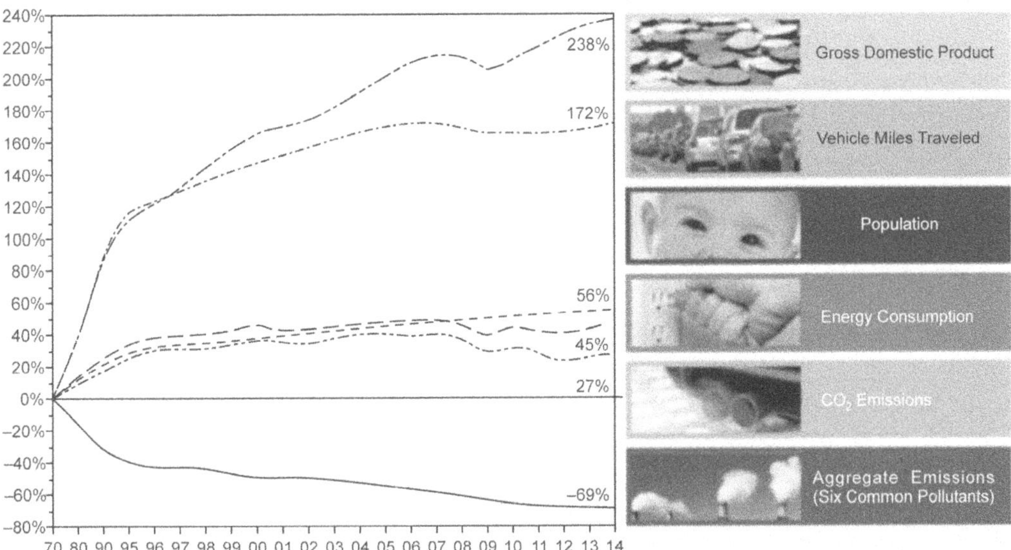

Fig. 3.15 *Air pollution emissions have dropped steadily since 1970 due to implementation of the clean air act.*

As the economy continues to grow, emissions that contribute to the most wide spread pollutants continue to drop. (Source: U.S. EPA, Air Quality Trends, 2016).

In 2015, EPA updated and strengthened the national ozone standard, officially recognizing that ozone is unhealthy to breathe at lower levels than previously thought. In preparing "State of the Air 2016," the Lung Association re examined all the ozone data for all prior years, back to 1996 – 1998 covered by the first report in 2000, using the new standard. Even using that more protective standard, six of the 25 most ozone-polluted cities reported their fewest unhealthy ozone days ever.

3.5.5 Air pollution scenario in India

Increase in technological, industrial and agricultural advancement, coupled with increases in population growth, has triggered the deterioration of environmental quality throughout the world. Rapid growth in India has brought though many benefits the environment has suffered, exposing the population to serious air pollution 55 [10]. The consequences of pollution have led to poor urban air quality in many economic Indian cities. The air pollution and the resultant air quality can be attributed to emissions from transportation, industrial and domestic activities. The air quality has been, therefore, an issue of social concern in the backdrop of various developmental activities.

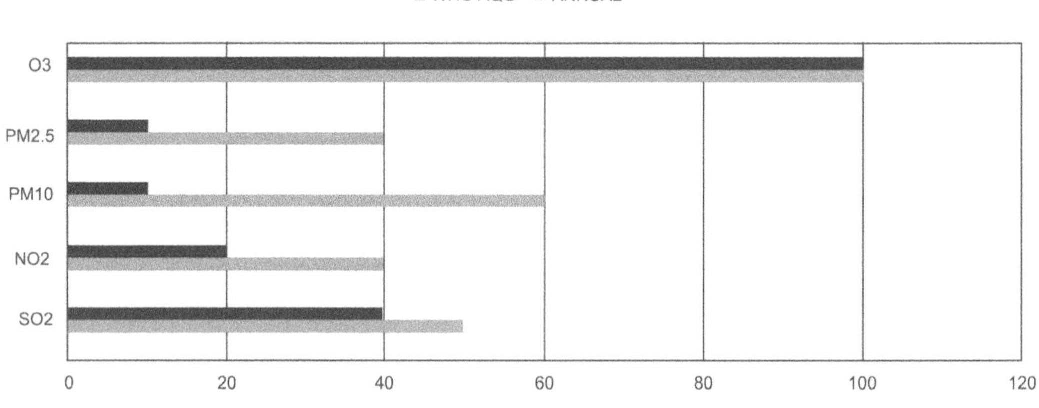

Fig. 3.16 *AQM in India (CIA-ASIA 2011)*

3.5.5.1 Air pollution trends in India

India is now ahead of China and everyone else in having the dubious distinction of possessing the world's most polluted air (Fig. 3.16). Earlier this year an Environment Performance Index (EPI) report placed India at the bottom of the 132 countries assessed (56). Air in some Indian cities has particulate matter (PM) levels 5 times above safety limits (57). The PM_{10} levels are the biggest cause of health concern. In 2007, nearly 52% (58) cities were at critical PM_{10} levels (≥1.5 times limit). Higher PM_{10} levels were found in northern Indian cities with continuous increases in Mumbai, Faridabad, Lucknow, Bangalore and Delhi. (According to WHO report 2010) in recent times, SO_2 levels have dropped in most cities except for some high level areas such as Khurja, Nashik, Jamshedpur and Chandrapur. However, NO_2 levels are on the rise all over India because of the rise in the number of vehicles (59-65). The ambient PM concentrations fell quite steadily over the sample timeframe, from 252.1μg/m^3 in 1987-1990 to 209.5μg/m^3 in 2004-2010. This represents about a 15 percent reduction in PM. The SO_2 trend line is flat until the late 1990, and then declines sharply. Comparing the 1987-1990 to 2004-2010 time periods, mean SO_2 decreased from 19.4 to 12.2μg/m^3. NO_2 appears more volatile over the start of the sample period, but then falls after its peak in 1997 [66].

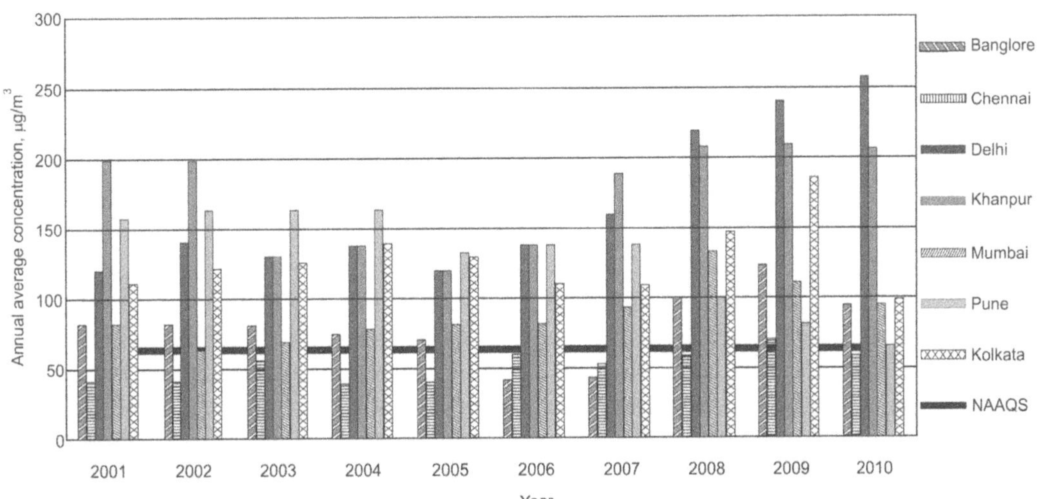

Fig. 3.17 *Concentration of RSPM in residential and industrial areas in major cities in India*

The 10th percentiles of PM and SO_2 pollution both declined by about 13 percent from 1987-1990 to 2004-2010. However, the distribution of NO_2 across cities does not appear to change much over the sample time frame. In fact, if anything, the distribution of pollution appears to have worsened, with the 10th percentile rising from 8.5 to 10.4µg/m^3 and the 90th percentile rising from 42.6 to 47.0µg/m^3 [66].

3.5.5.2 Air pollution status in Indian cities (66-71)

The air quality data generated by the Central Pollution Control Board (CPCB) for 2007 under the National Air Quality Monitoring Programme (NAMP) presents deadly facts about air pollution levels in Indian cities. Centre for Science and Environment has analysed the official data to assess the state of air quality and trend in Indian cities. The most widely monitored pollutants in India are particulate matter (PM), nitrogen dioxide (NO_2), sulphur dioxide (SO_2), and on a limited scale carbon monoxide. Almost half of the total cities monitored under NAMP have critical levels of PM_{10}. CPCB classifies cities as critically polluted if the levels of criteria pollutants are more than 1.5 times the standards. Levels up to 1.5 times the standards are labeled high. Levels that reach up to 50 per cent of the standards are moderate. Indian cities are reeling under heavy particulate pollution with 52 percent of cities (63 cities) hitting critical levels (exceeding 1.5 times the standard), 36 cities with high levels (1–1.5 times the annual standard) and merely 19 cities are at moderate levels, which is 50 per cent below the standard.

The air pollution trends in seven cities were analyzed for three major pollutant levels in RSPM, NO_2, SO_2. The seven cities selected for the present study are Bangalore (Karnataka); Chennai (Tamil Nadu), Delhi (Delhi), Khanpur (Uttar Pradesh), Mumbai(Maharashtra), Pune (Maharashtra), Kolkata (West Bengal).

3.5.5.3 Status of RSPM

The annual average concentration of major cities in India shows a fluctuating trend in RSPM levels, (Figure.3.17 indicates increasing trend in Delhi, Mumbai and Kolkata but for Chennai it shows decreasing trend. The reason for high particulate matter levels is attributed by vehicular traffic, Diesel/Kerosene generator sets, small-scale industries, biomass incineration, suspension of traffic dust, commercial and domestic use of fuels.

3.5.5.4 Status of NO_2

The annual average concentration of major cities in India shows a fluctuating trends in RSPM levels but indicates decreasing trend in NO_2 (Figure 3.18) but the concentration is above the NAAQS is observed in Banglore, Delhi, Khanpur, Mumbai and pune. In many cities the levels declined up to 2004 however there is an increasing trend observed again in past two years. Southern Indian cities show a rising trend especially in Bangalore, cities in western India are relatively better off with almost constant to declining NO_2 levels, though the levels indicate an increasing trend in Delhi and Khanpur.

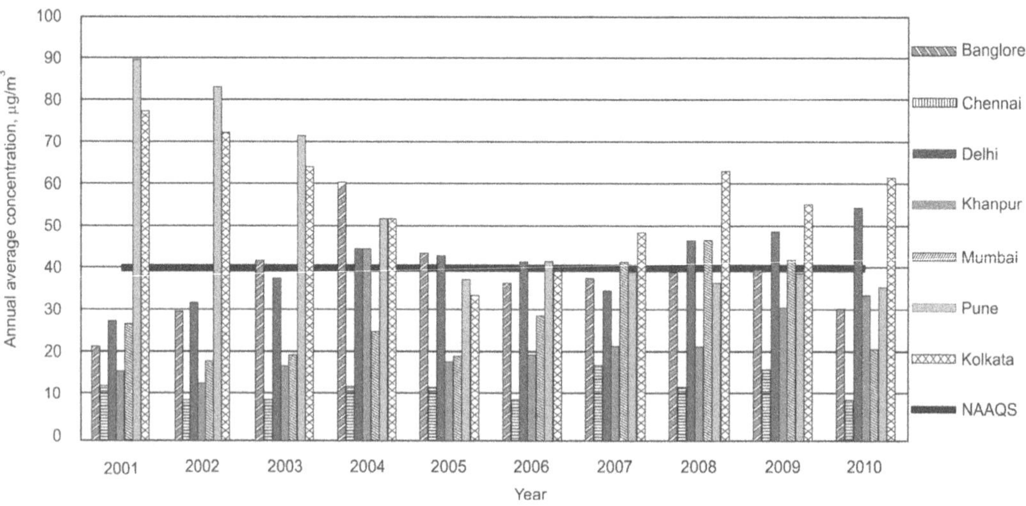

Fig. 3.18 *Concentration of NO_2 in residential and industrial areas in major cities in India*

3.5.5.5 Status of SO_2

It has been observed that, trend of Sulfur dioxide is not considered a problem in India. Its levels in most cities are already very low and declining (Figure 3.19). However, Pune has moderate levels. As per the literature, during 2010 National Ambient Air Quality Standard (NAAQS annual average) was not exceeded at any monitoring station in residential and industrial areas. SO_2 levels at 75 per cent of the monitoring stations in industrial areas and 90 per cent of the monitoring stations in residential areas were less than 25 μg/m^3, although SO_2 levels at none of the monitoring stations exceeded the annual average standard. (Figure .3.20)

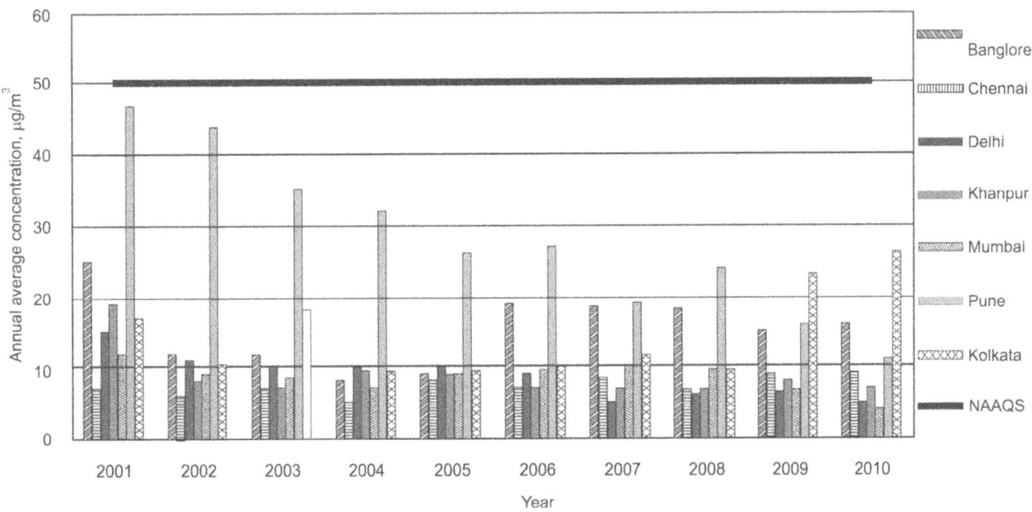

Fig. 3.19 *Concentration of SO_2 in residential and industrial areas in major cities in India*

Air quality monitoring data clearly showed lower concentrations of gaseous pollutants (SO_2 and NO_x) and higher concentrations of SPM and RPM in the ambient air. The incidence of higher rate of respiratory diseases among the people of selected cities could be attributed to higher concentration of SPM and RPM in the ambient air. Under long term exposure, there is correlation between particle concentrations and mortality form lung disease. The literature review shows that that Suspended Particulate Matter (SPM) is the main problem of urban air pollution in India. The main cause of air pollution is mainly due to fast growing population of automobiles and poor traffic control, congested roads. To overcome this development and planning of city and public awareness play very important role to reduce the ambient air pollution. Various measures such as implementation of Bharat Stage-III norms have been strictly implemented to mitigate ambient RSPM which have been shown exponential increase in number of vehicles. Central air pollution board is reported that, NO_2 levels at 35 % of the monitoring stations in industrial and residential areas were low. High levels of NO_2 were observed at 48% and Critical levels of NO_2 were observed at 13% of the locations. The decreasing trend in sulphur dioxide is been observed in India these levels may be due to various measures taken such as reduction of sulphur in diesel, use of LPG instead of coal as domestic fuel conversion of diesel vehicles to CNG.

But according to CPCB almost half of the total cities monitored under National Air Quality Monitoring Programme (NAMP) have critical levels of PM10. Air quality monitoring data clearly showed comparatively lower concentration of gaseous pollutants (SO_2, NO_2) and higher concentration of SPM and RPM in the ambient air. (Figure 3.20)

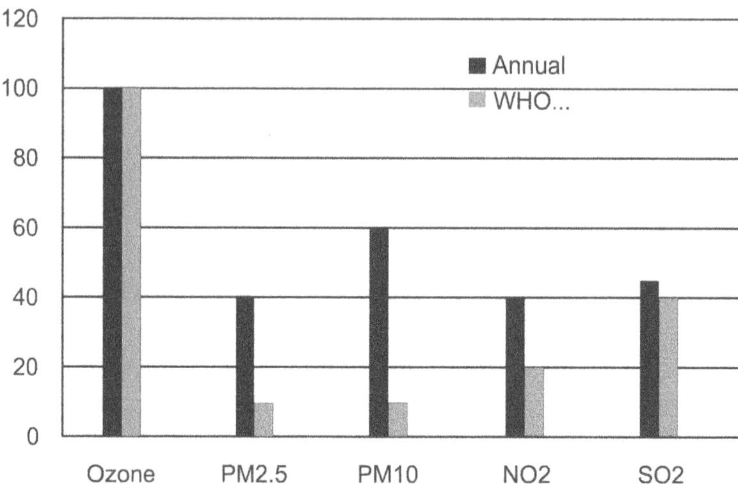

Fig. 3.20 *Comparison between revised NAAQs with WHO industrial residential, rural& other areas and WHO AQG*

However, the 2016 Environmental Performance Index ranked India 141 out of 180 countries. In 2015, Ministry of Environment, Forest and Climate Change, Government of India launched National Air Quality Index to the public involving IIT Kanpur. The same can be tracked at http://envfor.nic.in/content/national-air-quality-index.

Questions

1. Write a short notes on present Global & National scenarios of air pollution.
2. Write a notes on Bhopal gas tragedy of India in 1984
3. What are the guidelines of WHO on air quality.
4. Explain Air quality measurements approach.
5. Explain the procedure for assessment of emission flows and its factors.
6. How to estimate the gaseous pollutant emissions from fuel composition.
7. What are the fundamental control measures for industrial air pollution.
8. Discuss the strategies of air pollution preventive measures followed USA,UK and China.
9. What are the strategies INDIA is following to control air pollution.
10. Improvement of fuel quality decreases pollution, Explain.
11. How to plan to assess extent of air pollution in apolluted area?
12. Estimate the SO_2 released when 9 kg of diesel oil containing 0.2% sulphur is burnt in a furnace.

13. How much quantity of NO_2 will release when 12 kg of diesel oil containing 0.15% Nitrogen is burnt in a furnace.
14. Estimate the CO released when 10 kg of diesel oil containing 76 % sulphur is burnt in a furnace

References

1. Draggon, S. J.J. Cohressen and R. E. Morrison (Eds), 1987. Environmental impacts on human health, the agenda for long-term research and development. Praeger, New York.
2. N. Basta, 1985. US CPI Start to feel effects of Bhopal Tragedy. *Chemical Engineering.* March, 27-33.
3. M. Heylin, 1984. Indian's chemical tragedy. Death toll at Bhopal still rising. *Chemical Engineering.* December, 5-7.
4. 3-M. "Union Carbide mobilizes resources to control damage from gas leak." *Wall Street Journal.* December 1984.
5. 3-M. "India's disaster. The night of death." *Time.*17 December 1984 20-38.
6. A loss of coolant started the chain of events resulting in the Chernobyl accident. *Chemical Engineering.* May 1986.
7. T. R. Detwyler, 1971. *Man's impact on environment.* McGraw Hill Book Co., New York.
8. Air quality guidelines for Europe. Copenhagen, World Health Organization Regional Office for Europe, 1987 (WHO Regional Publications, European Series, No. 23).
9. Air quality guidelines for Europe, 2nd ed. Copenhagen, World Health Organization Regional Office for Europe, 2000 (WHO Regional Publications, European Series, No. 91).
10. World Health Organization (WHO). 1980. Glossary on Air Pollution. European Series, No. 9. Copenhagen: WHO Regional Publications
11. Global Environmental Monitoring System (GEMS/Air). 1993. A Global Programme for Urban Air Quality Monitoring and Assessment. Geneva: UNEP
12. World Health Organization (WHO) and United Nations Environmental Programme (UNEP). 1994. GEMS/AIR Methodology Reviews Handbook Series. Vol. 1-4. Quality Insurance in Urban Air Quality Monitoring, Geneva: WHO.1995a. City Air Quality Trends. Vol. 1-3. Geneva: WHO.1995b. GEMS/AIR Methodology Reviews Handbook Series. Vol. 5. Guidelines for GEMS/AIR Collaborative Reviews. Geneva: WHO
13. Miller, JM and A Soudine. 1994. The WMO global atmospheric watch system. Hvratski meteorolski casops is 29:81-84
14. Air Quality Management Source Book for Policy Makers in Developing Cities ; Schwela and Koth-Jahr 1994

15. Ministerium für Umwelt, R und L Nordrhein-W. 1992. Krebsrisiko Durch Luftverunreinigungen, Studie Des Länderausschusses Fur Immissions schutz [Cancer Risk from Air Pollutants, Study of the State Committee for Environment Protection]. Dussel dorf: North Rhine-Westphalian Ministry for Environment, Regional Planning and Agriculture.
16. U.S. Environmental Protection Agency. Integrated Science Assessment of Ozone and Related Photochemical Oxidants (Final Report).
17. U.S. Environmental Protection Agency, Washington, DC, EPA/600/R-10/076F, 2013. Mar TF, Koenig JQ. Relationship between visits to emergency departments for asthma and ozone exposure in greater Seattle, Washington. Ann Allergy Asthma Immunol. 2009; 103: 474-479. Villeneuve PJ, Chen L, Rowe BH, Coates F. Outdoor air pollution and emergency department visits for asthma among children and adults: A case-crossover study in northern Alberta, Canada. Environ Health Global Access Sci Source. 2007; 6: 40.
18. Medina-Ramón M, Schwartz J. Who is more vulnerable to die from ozone air pollution? Epidemiology. 2008; 19: 672-679.
19. The future of benefit-cost analyses of the Clean Air Act. Krupnick A1, Morgenstern R.Annu Rev Public Health. 2002; 23:427-48. Epub 2001 Oct 25
20. Public Health England, Estimating local mortality burdens associated with particulate air pollution, DEFRA-April 2014,
21. "Air pollution is now a public health emergency" The Independent, 19 January 2016.
22. Directive 2008/50/EC on Ambient Air Quality and Directive 2001/81/EC on National Emissions Ceilings for certain pollutants. The 1999 Gothenburg Protocol to Abate Acidification, Eutrophication and Ground-level Ozone also sets national emission ceilings for 2010–2020 on four pollutants (sulphur dioxide (SO_2), nitrogen oxides
23. Defra, Draft plans to improve air quality in the UK: Tackling nitrogen dioxide in our towns and cities, UK overview document, September 2015
24. UN health agency re-classifies diesel engine exhaust as 'carcinogenic to humans', UN news centre press release, 12 June 2012
25. The Scottish Government, Cleaner Air for Scotland, the road to a healthier future, November 2015.
26. Defra, Defra plans to improve air quality in the UK: tackling nitrogen dioxide in our towns and cities, UK overview document, September 2015, para 8
27. Royal College of Physicians and Royal College of Paediatrics and Child Health, Every breath we take, February 2016 14 Joint Nature Conservation Committee, (AQU12), paras 3.3 -3.6
28. Local Government Association (AQU 27) para 2
29. Amber Rudd declines to reveal details of secretive 'clean growth' committee Business Green, 5 February 2016
30. Found at: http://society. people. com.cn/n/2013/1104/c86800-23427944.html. As amended by the Standing Committee of the National People's Congress respectively in 1995 and 2000.
31. Found at: http://www.gov.cn/ziliao/flfg/2005-08/05/content_20945. htm.

32. Found at: http://www.npc.gov.cn/npc/xinwen/syxw/2013-10/31/content_1812101.htm. 10
33. Found at: http://www.mep.gov.cn/gkml/hbb/bwj/201212/t20121205_243271.htm. 11 Found at: http://www.gov.cn/zwgk/2013-09/12/content_2486773.htm.
34. Found at: http://news.xinhuanet.com/fortune/2013-11/27/c_125768942.htm. 13 Found at: http://www.gov.cn/jrzg/2013-10/14/content_2506559.htm. 14 Special Projects and Measures, found at: http://env.people.com.cn/n/2013/0905/c101022811817.html.
35. Ibid. The recent progress of the projects can be found at http://www.gov.cn/gzdt/201310/13/content_2505855.htm.
36. Found at: http://fzb.sh.gov.cn/fzbChinese/page/legalinfo/locallegalinfo24449
37. Pollution Control Acts, Rules & Notifications issued there under; CPCB Publication; PCLS/2/1992; CPCB, Delhi; 1992.
38. Hadden, Susan G. (1987), "Statutes and Standards for Pollution Control in India". Economic and Political Weekly, 22(16)
39. Sankar, U. (2004). "Law and Institutions Relating to Environmental Protection in India." The India: Environmental Management Capacity Building Technical Assistance Project, the World Bank
40. Ambient Air Quality Monitoring Status - 1997; CPCB Publication; NAAQMS/11/1999-2000; CPCB, Delhi; 1999. Annual report of Central Pollution Control Board, 1998-99; CPCB, Delhi; 1999.
41. Clean Coal Initiatives; CPCB Publication; CPCB, Delhi; 2000.
42. Parivesh News letter on Air Pollution & its Control; Vol. 2 (i); CPCB, Delhi; 1995.
43. Parivesh News letter on Vehicular Pollution; Vol. 4 (iii); CPCB, Delhi; 1997.
44. Standards for Liquid Effluents, Gaseous Emissions, Automobile Exhaust, Noise and Ambient Air Quality; CPCB Publication; PCLS/44/1994-95; CPCB, Delhi; 1995.
45. Parivesh News letter on Zoning Atlas for Siting of Industries; Vol. 3 (i); CPCB, Delhi; 1996.
46. Sharma, Anuja and Roychowdhury, Anuja. (1996), "Slow Murder: The deadly story of vehicular pollution in India". Centre for Science and Environment, New Delhi.
47. Sharma H.C. (1994), "Environmental Pollution Compliance." CBS, Publisher, New Delhi, India.
48. CPCB, (2000). "Air Quality Status and trends in India, National Ambient Air Monitoring Series: NAAQMS". Central Pollution Control Board, New Delhi.
49. World Bank, (199s6). "India's Environment: taking stock of plans, programs and priorities. South Asia Regional Office, The World Bank, pp 319.
50. CPCB, (2010). "National Ambient Air Quality Status and Trends in India". Central Pollution Control Board, Ministry of Environment and Forest, New Delhi.
51. Hosamane, Sateesh N. and Dr. Desai. (2013). "Urban Air Pollution Trend in India- Present Scenario". International Journal of Innovative Research in Science, Engineering and Technology. Vol.2 Issue 8. pp 3738 -3747.

52. CPCB, Air quality trends, http://www.cpcb.nic.in/Trends.php, 2007.
53. The World Bank report, "India: green growth- overcoming India's environment challenges to promote development" (2013). http://www.world bank.org/en/news/pressrelease /2013/07/17/india-green-growth-neccessary-and-afford able-for-india-says-newworld. Lal, D.S. (2007). "Climatology." Sharda Pustak Bhawan Publishers & Distributors, Allahabad.
54. State of the Air 2016-American Lung Association Report
55. The world bank report, "India: Green growth – overcoming India's environment challenges to promote development", 13th July-2013. http://www.worldbank.org/en/news /feature/2013/07/17/green-growth-overcoming-india-environment-challenges-promote-development.
56. Yanrui Wu., "Understanding economic growth in China and India: A comparative study of selected issue", world scientific publishing Co. Ltd.
57. Agrawal, M., "Effects of air pollution on agriculture: An issue of national concern", National Academy Science Letter 28: 93-106, 2005.
58. Brulfert, G.T., Cholleta, J.-P., Jouveb, B., Villard, H., "Atmospheric emission inventory of the Maurienne valley for an atmospheric numerical model", Science of the Total Environment 349, 232–248, 2005.
59. Parra, R., Jimenez, P., Baldasano, J.M., "Development of the high spatial resolution EMICAT2000 emission model for air pollutants from the north-eastern Iberian Peninsula (Catalonia, Spain)", Environmental Pollution 140 (2), 200–219, 2006.
60. HaidongKan , Bingheng Chen, Changhong Chen, QingyanFu, Minghua Chen, "An evaluation of public health impact of ambient air pollution under various energy scenarios in Shanghai, China", Atmospheric Environment 38 (2004), 95–102.
61. Agrawal, M., "Effects of air pollution on agriculture: An issue of national concern", National Academy Science Letter 28, 93- 106, 2005.
62. Ta-Chen Su, Szu-Ying Chen, Chang-Chuan Chan, "Progress of ambient air pollution and cardiovascular disease research in Asia", Progress in Cardiovascular Diseases 53 (2011), 369–378.
63. Discussion Paper Air Quality Standards, July 2010, Prepared for the National Environment Protection Council.
64. World Health Organization, "World Health Report 2002: Reducing risk, Promoting healthy life", World Health Organization. Geneva, Switzerland. Available from www.who.int/whr/2002/en/.
65. Cohen AJ, Anderson HR, Ostro B, Pandey KD, Krzyzanowski M, Künzli N, Gutschmidt K, Pope CA III, Romieu I, Samet JM, Smith KR., Urban air pollution. Comparative quantification of health risks: global and regional burden of disease attributable to selected major risk factors" (Ezzati M, Lopez AD, Rodgers A, Murray CJL, eds.). pp. 1153–1433. World Health Organization, Geneva, Switzerland, 2004.
66. Urban air pollution, Current Science, vol. 77, NO. 3, 10 AUGUST 1999. [13] Air quality management: Air Quality Guidelines, WHO, Geneva, 1999.

67. Ambient Air Quality Monitoring Status - 1997; CPCB Publication; NAAQMS/11/1999-2000; CPCB, Delhi; 1999.

68. Annual report of Central Pollution Control Board, 1998-99; CPCB, Delhi; 1999.

69. Clean Coal Initiatives; CPCB Publication; CPCB, Delhi; 2000. 4. Parivesh News letter on Air Pollution & its Control; Vol. 2 (i); CPCB, Delhi; 1995. 5. Parivesh News letter on Vehicular Pollution; Vol. 4 (iii); CPCB, Delhi; 1997.

70. Parivesh News letter on Zoning Atlas for Siting of Industries; Vol. 3 (i); CPCB, Delhi; 1996. 7. Pollution Control Acts, Rules & Notifications issued there under; CPCB Publication; PCLS/2/1992; CPCB, Delhi; 1992.

71. Standards for Liquid Effluents, Gaseous Emissions, Automobile Exhaust, Noise and Ambient Air Quality; CPCB Publication; PCLS/44/1994-95; CPCB, Delhi; 1995.

CHAPTER 4

Air Pollution and Global Climatic Changes

4.1 Introduction

The high levels of air pollutants seem to be causing global climatic changes. Global warming, ozone depletion and acid rain are three important climatic problems that most scientists agree are the result of high levels of air pollution, but there is much less agreement among them about the magnitude of the problems and/or the need for prompt (and expensive) action. The evidence that global warming has been occurring due to air pollution is not strong enough to be distinguished from the annual random variation of climate with complete certainty. The evidence for destruction of the ozone layer over Antarctica seems strong, suggesting that the stratospheric ozone problem is severe and may cause significant effects on a time scale of years rather than decades or centuries. The effects of acid rain are seen visibly in many districts and urban areas.

4.2 Global Warming

4.2.1 The greenhouse effect

The earth receives a very large input of energy daily from the sun and maintains a steady state by giving off the bulk of this energy at the same rate.

When solar radiation (mainly visible and long-wave UV) enters the atmosphere, some is immediately reflected back into space by clouds, etc., but when about half reaches the planetary surface, most of this is reflected back as lower energy infrared (IR) radiation. A small proportion of this (10%) passes directly out into space but the main component is absorbed by certain gaseous molecules in the atmosphere. These molecules then re-radiate a proportion of this absorbed IR energy in all directions, some out to space and some back to the earth's surface. Also, the warming of the earth's surface, directly by solar radiation and indirectly by re-radiated IR, causes both evaporation of water and upward convection of air, both of which transfer energy from the earth's surface back into the atmosphere. The overall effect is a warming, blanket-like process around the earth's surface known as the greenhouse effect, illustrated in as Fig. 4.1.

4.2.2 The radiation balance of the earth

The solar spectrum incident on the earth's surface has a peak of 483 nm (visible region), indicating that the surface temperature of the sun is approximately 6000 K. The earth's spectrum shows a peak of about 10,000 nm (infrared region), with the surface temperature of the earth about 290 K (17°C). Wien's displacement rule, gives the wavelength at which a blackbody spectrum peaks, as a function of absolute temperature:

$$\lambda \max(\mu m) = \frac{2898}{T(K)} \qquad \text{..... (4.1)}$$

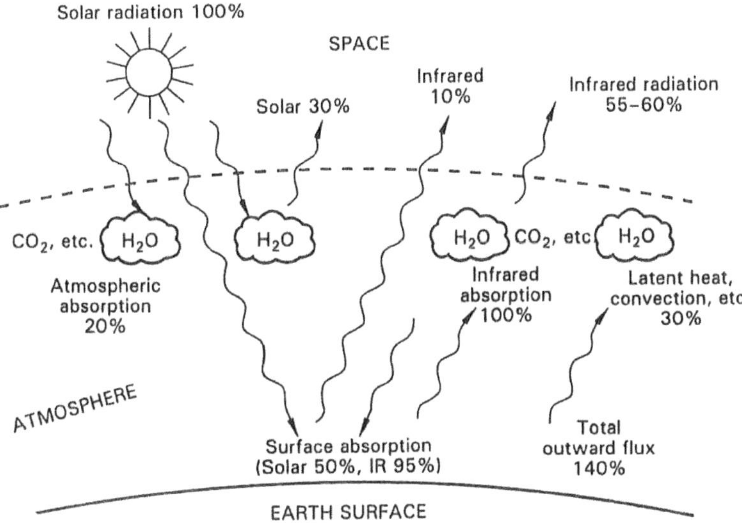

Fig. 4.1 *Fluxes of radiation in and out of the earth's atmosphere which collectively give rise to the greenhouse effect*

With a temperature around 6000 K, the sun can be considered as a blackbody with a peak radiation of 0.5 μm wavelength in the visible region. The earth's emission has a peak wavelength of about 10.1 μm. Figure 4.2(a) gives the two spectra of the sun and the earth normalized to have the same areas. It can be seen that nearly all the incoming solar energy is in wavelengths less than 4 μm, while the outgoing energy radiated by the earth is essentially in wavelengths greater than 4 μm. With such a small overlap, the solar energy can be termed as being short wavelength radiation, while the energy radiated from the earth's surface is long wavelength or thermal radiation. In other words, the earth absorbs radiation mainly in the visible region, but emits radiation at the same rate in the infrared region (2-40 μm, with a maximum at 10 μm).

Fig. 4.2(b) shows the contribution to the absorptive properties of the atmosphere by CH_4, N_2O, O_2 and O_3, CO_2, and H_2O. We can see that O_2 and O_3 block all light with wavelengths less than 0.28 μm and that CO_2 blocks all light with wavelengths more than 15 μm.

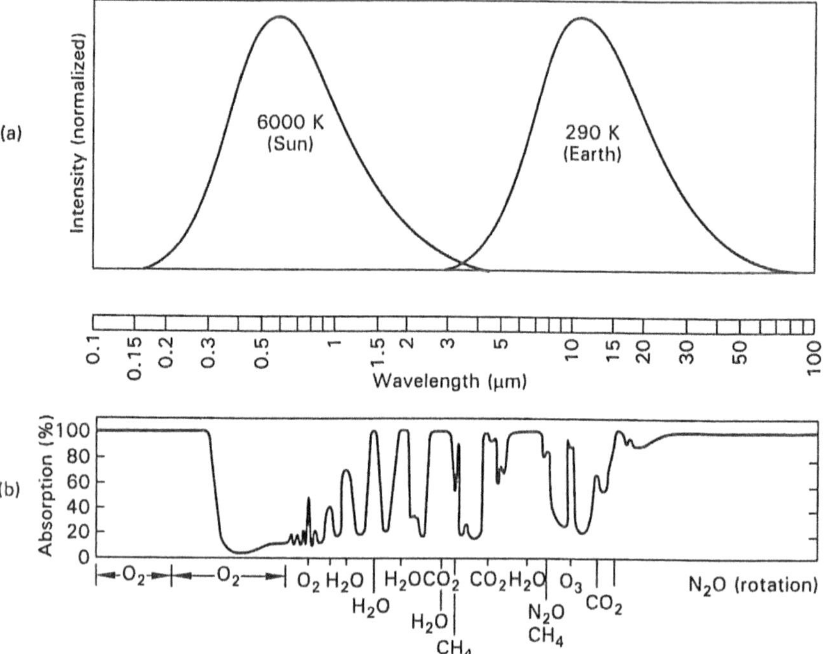

Fig. 4.2(a) *Normalized blackbody radiation curves for the sun and earth;* **(b)** *Atmospheric absorption on a clear day [1]*

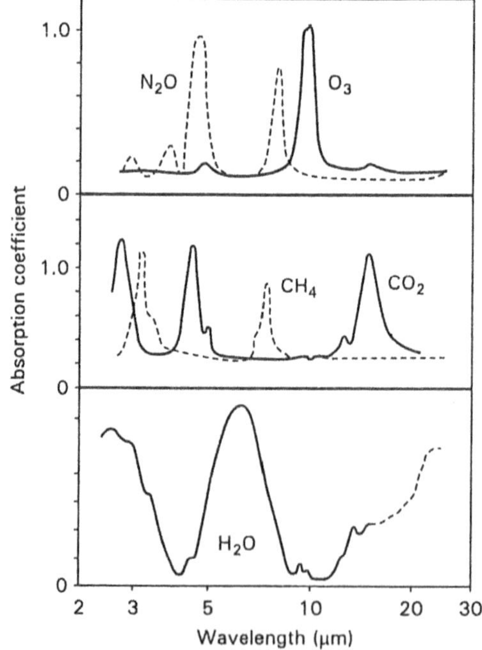

Fig. 4.3 *Individual absorption spectra of the different greenhouse gases*

The combined absorption spectrum for the whole atmosphere is largely transparent in the wavelength range 0.3- 0.7 μm, and has various "windows" at other wavelengths, of which the most important is the two-part window between about 8 μm and 12 μm (Fig. 4.3).

Comparing the lower and upper parts of the figure, we see that sunlight comes to the surface practically unimpeded except in cloudy areas, whereas the peak radiation from the earth is close to the 8-12 μm window, which is not as wide nor as completely open as the window for solar energy. The atmosphere is thus less transparent for outgoing infrared energy than it is for incoming solar energy.

4.2.3 Earth's temperature without the greenhouse effect

Assuming the earth as a blackbody and that all the incoming solar radiation is absorbed, we have

$$E_a = S\pi R^2 \qquad \text{..... (4.2)}$$

Where E_a is the rate of absorption, R is the radius of the earth, and S is the solar constant which is equal to 1372 W/m^2. Considering the earth as an isothermal body with temperature T(K) everywhere, the energy radiated E_r from this blackbody, earth, with surface area $4\pi R^2$ is given by the Stefan-Boltzmann equation:

$$E_r = \sigma 4\pi R^2 T^4 \qquad \text{..... (4.3)}$$

where σ, the Stefan-Boltzmann constant = 5.67×10^{-8} W/m^{2}K^4

As the earth's temperature is constant and not changing with time the rate at which energy is absorbed from the sun can be equated to the rate at which energy is radiated to outer space.

So $$S\pi R^2 = \sigma 4\pi R^2 T^4 \qquad \text{..... (4.4)}$$

Since the albedo (the reflected portion of the incoming solar radiation) is only 30% of the incoming solar radiation is reflected and setting the absorbent energy as equal to reradiated energy under equilibrium conditions and a uniform temperature for the earth.

Equation 4.4 can be written as

$$S(1-\alpha)\pi R^2 = \sigma 4\pi R^2 T_e^4 \qquad \text{..... (4.5)}$$

where α is the albedo and T_e is the effective temperature.

$$T_e = \left(\frac{S(1-\alpha)}{4\sigma}\right)^{1/4} \qquad \text{..... (4.6)}$$

Substituting the appropriate values for S, α, and σ, we get

$$T_e = \left(\frac{1372 \text{ W/m}^2 (1-0.3)}{4 \times 5.67 \times 10^{-8} \text{ W/m}^2\text{K}^4}\right)^{1/4} = 255\,\text{K}$$

If there is no greenhouse effect on earth, then its temperature would be 255 K, i.e. an average temperature of –18 °C, or about 0°F. The greenhouse effect can be quantified for the effective temperature T_e, with the actual surface temperature T_s as per Equation 4.6:

$$\text{Greenhouse effect (°C)} = T_s - T_e = 280 - 255 = 33°\text{C} \qquad .....(4.7)$$

So the greenhouse effect adds 33 °C to the temperature of the surface of the earth.

By applying the same concepts to Venus and Mars, their equivalent temperatures can be calculated as shown in Table 4.1. Even though the atmosphere of Mars is mainly composed of carbon dioxide, there is no greenhouse effect. The atmospheric pressure on Venus, on the other hand, is nearly 100 times that of earth, and its atmosphere is 97% CO_2; a very large greenhouse effect is observed on Venus.

Table 4.1 *Application of the simple blackbody model to compute equivalent temperatures, compared with actual surface temperatures. Mars shows almost no greenhouse effect while on Venus it is quite pronounced*

Planet	Insolation S (W/m^2)	Albedo	Equivalent temperature T_e(K)	Actual temperatures T_s (K)
Earth	1372	0.30	255	288
Venus	2613	0.75	232	700
Mars	589	0.15	2.17	220

4.2.4 Energy flows between the earth's surface, its atmosphere, and outer space

Let us consider the earth, its atmosphere, and outer space as three separate regions and normalize energy flows between these regions by expressing them in terms of rates per unit of surface area of the earth. Considering the total amount of solar radiation striking the earth as $S\pi R^2$ distributed over the entire surface of the earth, the incoming solar radiation is equal to

$$\frac{\text{Incoming solar radiation}}{\text{surface area of earth}} = \frac{S\pi R^2}{4\pi R^2} = \frac{S}{4} = \frac{1372\text{W}/\text{m}^2}{4} = 343\ \text{W}/\text{m}^2 \qquad .....(4.8)$$

As the albedo is 30%, the amount of incoming radiation reflected back into space is

$$\text{Incoming radiation reflected} = 0.30 \times 343\ \text{W/m}^2 = 103\ \text{W/m}^2$$

Of the above amount, it is estimated that 89W/m² is reflected off the atmosphere itself while the remaining 14 W/m² is reflected off the earth's surface.

The incoming radiation absorbed by the atmosphere of the earth is = 0.70 × 343 W/m² = 240 W/m², of which 86 W/m² is absorbed by the atmosphere and the remaining 154 W/m² is absorbed by the surface of the earth.

$$= 0.70 \times 343\ \text{W/m}^2 = 240\ \text{W/m}^2$$

If global temperatures are unchanging with time, then the rate at which the earth and its atmosphere absorb energy from space must equal the rate at which energy is being returned to space. So, the earth and its atmosphere must radiate 240 W/m^2 back into space. If the earth's surface were at 255 K, it would radiate 240 W/m^2, which would be just enough to balance the incoming energy. However, as the greenhouse gases absorb most of that outgoing 240 W/m^2, the required energy balance cannot be achieved. Therefore to force enough energy through the atmosphere to create the necessary balance, the temperature of the earth's surface must be higher than 255 K.

If we treat the earth as a blackbody, we can use Equation 4.3 to estimate the rate at which energy is radiated from the earth's surface towards the atmosphere. The temperature of 288 K is usually considered to be the global average, which is the temperature of the air just above the surface of the earth. The actual temperature of the solid and liquid surface of the earth itself is considered to be about 2 °C higher.

Substituting in Equation 4.3, T= 290 K

$$\frac{\text{Energy radiated by surface}}{\text{Suface area}} = \frac{\sigma 4\pi R^2 Ts^4}{4\pi R^2} = \sigma Ts^4$$

$$= 5.67 \times 10^{-8}\ \text{W/m}^2\ \text{K}^4 \times (290\text{K})^4$$

$$= 401\ \text{W/m}^2 \qquad \ldots..(4.9)$$

Out of 401W/m^2, only 20W/m^2 passes directly though the atmosphere, mostly through the atmospheric window while the remaining 381 W/m^2 is absorbed by greenhouse gases in the atmosphere and 344 W/m^2 will be radiated back to the surface by the atmosphere.

Fig.4.4 shows the energy flow. The rate of energy gain should equal the rate of energy loss in each of the three regions: the earth's surface, its atmosphere, and outer space.

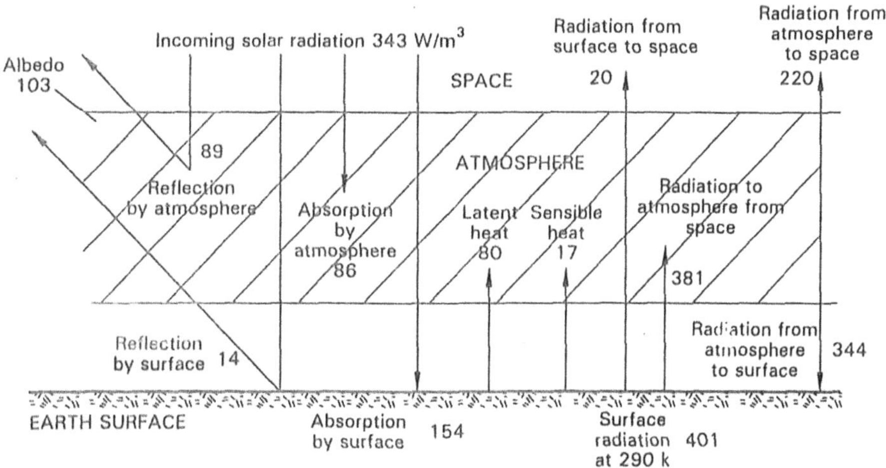

Fig. 4.4 *Global average energy flows between the earth's surface, its atmosphere and outer space (Units are watts per square metre of surface area)*

Rate of energy gain = Rate of energy loss

Earth's surface: 154 + 344 = 80 + 17 + 20 + 381 (in W/m^2)
Atmosphere: 86 + 80 + 17 + 381 = 220 + 344 (in W/m^2)
Space: 103 + 20 + 200 = 343 (in W/m^2)

The above accounts for the entire energy flow.

From Fig, 4.2 it is evident that the gases in the atmosphere, CO_2, CH_4, N_2O, and H_2O, all absorb some radiation in the 8-12 μm window. If the concentration of any or all of these gases in the atmosphere increases, the 8-12 μm window would become less transparent thus, making the atmosphere's ratio of outgoing to incoming radiation decline. In turn, the average temperature of the earth would rise, depending on the radiative forcing of each greenhouse gas which varies with its concentration.

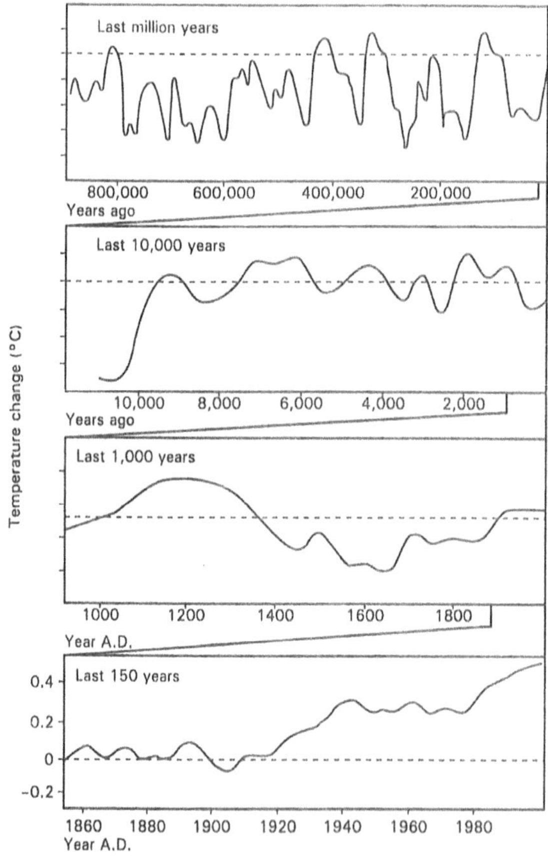

Fig. 4.5 *Changes in mean global temperatures in the past over different time scales until the present time (All values of temperature change are expressed relative to that in year 1990 shown as a horizontal dashed line)*
(Source: *Hansen and Lebediff. Geophysical Research Letters. 15:323-326. © American Geophysical Union.)*

Water vapour and the warming effect of carbon dioxide combine together to have a long range impact on the global climate. As the surface temperature increases with increase in the levels of carbon dioxide, the evaporation of surface water increases, thereby raising the temperature further.

4.2.5 Past temperatures and carbondioxide/methane levels

Probable temperatures in the past over different time scales are shown in Fig. 4.5. By themselves, they cannot indicate future trends, but past records of greenhouse gases such as CO_2 and CH_4 established by measurements taken from ice cores agree with local temperatures which have existed at the poles (Fig. 4.5). These indicate that the large swings ($\pm$ 3.5 °C) in temperature across the various ice ages and interglacial periods correlate closely with the then existing CO_2 and CH_4 concentrations. However, they still tend to reflect rather than to explain why the earth moves into an ice age and out again.

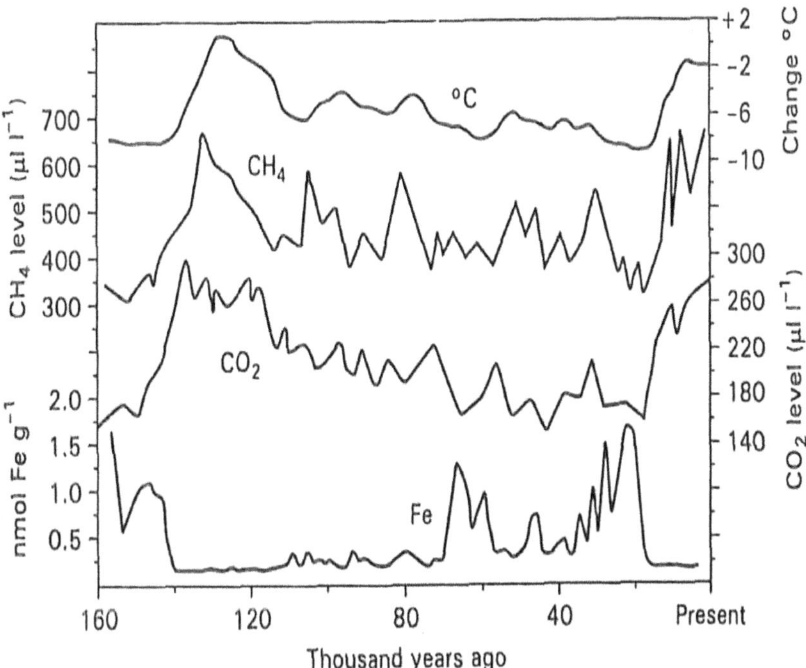

Fig. 4.6 *Correlation of local temperatures over the last 1,60,000 years with analyses of ice core samples from Antarctica (Vostok) for trapped CO_2 and CH as well as with iron deposition*

(**Source:** *Billi et al., 1986. The greenhouse effect, climatic change and ecosystems. ©John Wiley & Sons, Inc.*)

The climatic record, based on the studies of ice cores taken from Greenland and Antarctica, suggests that CO_2 may serve as a kind of global thermos which resets the earth's temperature as the CO_2 level rises and falls. Since the last great ice age ended about 10,000 years ago, we have enjoyed a warm interglacial period. Fig. 4.6 shows the remarkable correlation between atmospheric CO_2 levels and Antarctic surface temperatures. During glacial periods the CO_2 level is low while in the warmer interglacial periods, the CO_2 level is very high [4]. Fig. 4.7 shows how changes in the tropospheric levels of carbon dioxide correlate closely with changes in the earth's mean surface temperature and thus its climate, although other factors also influence global climate.

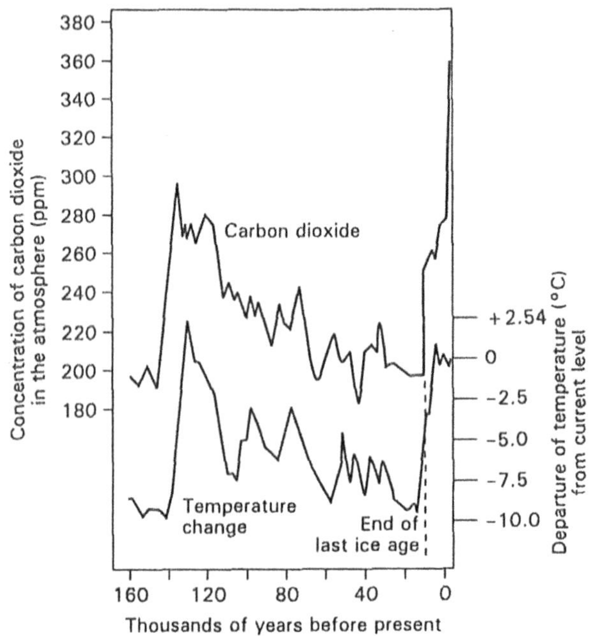

Fig. 4.7 *Estimated long-term variations in mean global surface temperature and average tropospheric carbon dioxide levels over the past 1,60,000 years*

(**Source:** *J. T. Houghton, G.J. Jenkins, J. J. Ephramus (eds), 1990. Climatic change: the IPCC scientific assessment. Cambridge University Press, Cambridge.)*

4.2.6 Current trends in the levels of atmospheric secondary gases; their relative contributions to global warming

An international committee of scientists commissioned by different world governments provided a definitive verdict on what is happening to the weather around the globe in the past decade. The report states that the world has been warming up since the 1980s. The world has already warmed up by about 0.5 °C. It is expected to get 4 °C warmer by the year 2030.

4.2.6.1 Carbon dioxide

The first assessments of the carbon dioxide content of the atmosphere at the beginning of the Industrial era indicate [5, 6] that the levels of CO_2 in the lower atmosphere (troposphere) are rising year by year (Fig. 4.8). During the last ice age (Fig. 4.5), CO_2 levels were below 300 ppm, and were still around 280 ppm before the industrial revolution; but they have risen to 360 ppm by 1990. Over the period 1860-1960, the rise in CO_2 levels was 9% a^{-1} but by the year 2000, the increase over the century as a whole will have been 25% a^{-1}.

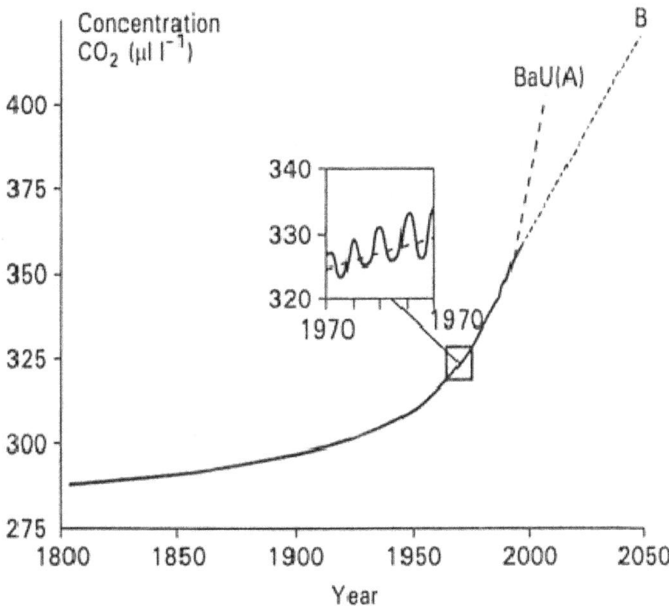

Fig. 4.8 *Global changes in atmospheric CO_2 levels with time (The solid part of the graph represents actual measured values and the dashed regions on the right are those most likely to occur in the future either with 'business as usual' [BaU (A)] or if the IPCC predictions of Scenario B (B) are brought into reality by international implementation. Typical seasonal fluctuations are shown in the inset figure. Values in the northern summer are lowest because of the higher photosynthetic activity of northern forests)*

(**Source:** *Neftal et al., 1985. Nature Vol. 315 © Macmillan Magazines.)*

According to present information, the overall average temperature rises by 0.5°C (0.9°F) for each increase in the carbon dioxide content by 18% with a net increase of 1 °C by the end of 2000.

The current global trend in deforestation and increased combustion of fossil fuels have a cumulative effect on the net increase in carbon dioxide content. Forests are areas where a great deal of photosynthesis occurs. Moreover, they maintain vast reservoirs of fixed but readily oxidizable carbon in the form of wood and humus. Overall they serve to maintain a balance in the level of atmospheric CO_2.

The major sink for CO_2 is the ocean which contains the bulk of dissolved carbon dioxide as bicarbonate Table 4.2. Covering some 70% of the planet, the oceans act as a gigantic flywheel in the earth's machinery. As a huge reservoir of carbon, the oceans regulate the atmospheric CO_2 levels and hence the global climate. Another important sink is the biomass, namely, the living green plants, in which the photosynthesis rate is known to accelerate with increase in carbon dioxide levels.

Table 4.2 *Estimated sources and sinks of CO_2*

Sources/Sinks	Range (Tg C/year)
Sources	
Ocean	102700-106500
Land	8700-120000
Fossil fuel	4500-5500
Land use conversion	0-2600
Sinks	
Oceans	106000-108000
Land	100000-140000

Source: *A primer on energy. Dept of Energy, USA*

Most scientists now agree that the global temperature increases are due to various human activities (Fig. 4.9). By burning large quantities of fossil coal or oil [Figs. 4.10(a) and (b)], and destroying tropical rain forests at the rate of 0.5 ha s^{-1} and replacing them with grasslands and waste scrub, human activity has tipped the net balance of the natural carbon cycle in favour of more CO_2 being returned to the atmosphere than at any time since the end of the last ice age (Fig. 4.5). When expressed in global quantities of carbon exchanged over one year the full extent of this imbalance can be appreciated [8, 9]. The 5 Tt of fossil fuels burned in 1976, for example, contributed an extra 2.3 ppm CO_2 to the atmosphere. Of this total, 1.6 ppm was either absorbed by the oceans or added to the biomass, leaving a net increase of 0.7 ppm CO_2 in the atmosphere by the end of 1976.

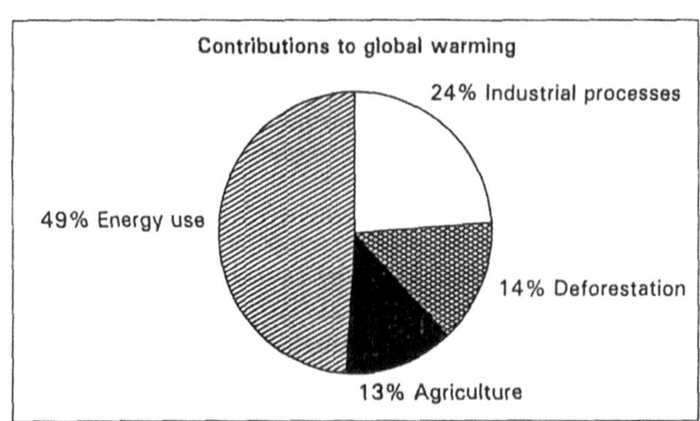

Fig. 4.9 *Contributions to global warming by different types of human activities in 1990*

(**Source:** *Data from World Resources Institute.)*

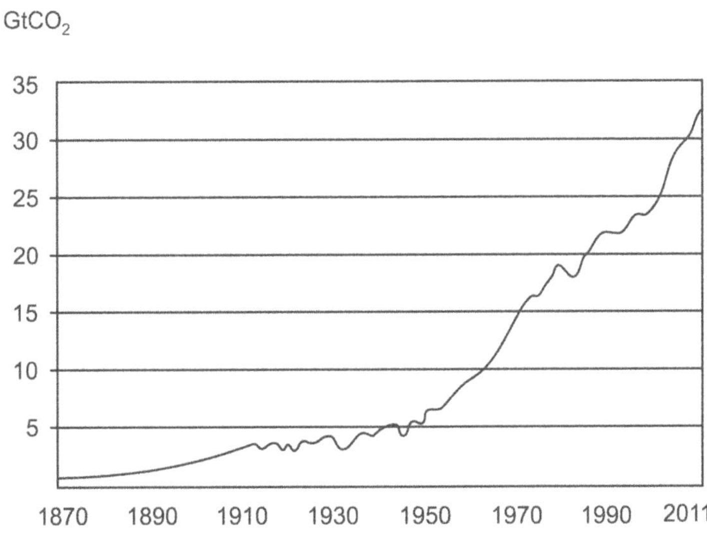

Fig. 4.10(a) *The global emission of CO_2 as a result of the combustion of fossil fuels*

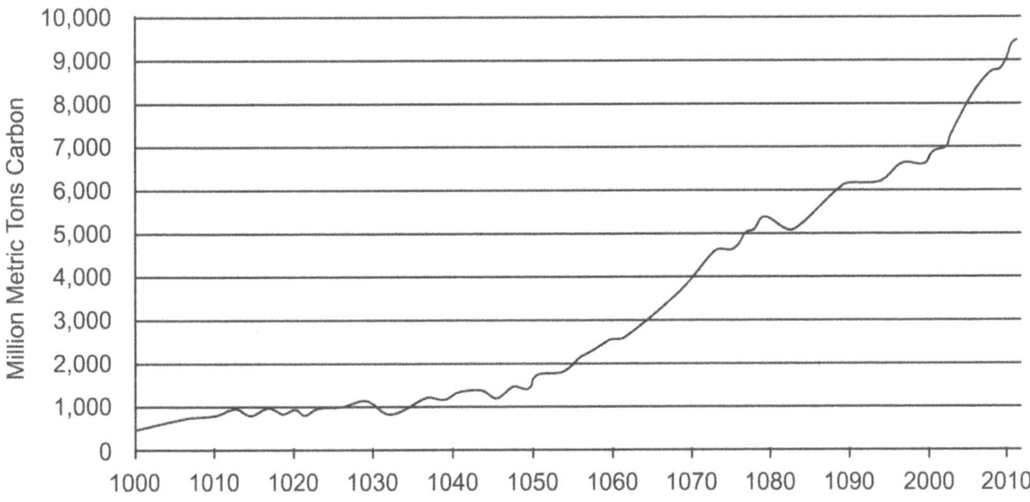

Fig. 4.10(b) *The global emission of CO_2 from deforestation and the burning of fossil fuels.*

If current trends continue, the pre-industrial CO_2 concentrations will have doubled by the year 2075. Computer models predict that doubling the atmospheric CO_2 could cause global temperatures to rise by 1.5-4.5 °C. This may not sound like much change, but the difference between the temperature now and the last ice age about 10,000 years ago when glaciers covered much of North America is only about 5 °C.

Problem 4.1

If the total mass of air equals 5.1×10^{18} kg and the density of air at standard temperature and pressure (STP:0 °C and 1 atm) is 1.29 kg/m^3, calculate the total carbon content of the atmosphere corresponding to 370 ppm of CO_2.

Solution: One g mole of CO_2 contains 44 g (12 + 2 × 16) and each mole at STP occupies a volume of 22.4×10^{-3} m^3. At 370 ppm, the concentration of CO_2 (by weight at STP) is given by

$$CO_2 = \frac{370 \text{ m}^3 \text{ CO}_2}{1 \times 10^6 \text{m}^3 \text{ air}} \times \frac{\text{mol}}{22.4 \times 10^{-3} \text{m}^3 \text{ CO}_2} \times \frac{44 \text{ g}}{\text{mol}} = 0.726 \text{ g/m}^3$$

Since 44 g of CO_2 contains 12 g of C, the total amount of carbon in the atmosphere =

$$C = \frac{0.726 \text{ g CO}_2}{\text{m}^3 \text{air}} \times \frac{12 \text{ g of C}}{44 \text{ g of CO}_2} \times \frac{5.1 \times 10^{18} \text{ kg air}}{1.29 \text{ kg/m}^3 \text{ air}} = 7.83 \times 10^{17} \text{ g}$$

$= 7.83 \times 10^9$ tons or 783 gigatons (Gt)

Problem 4.1 provides a very convenient conversion factor between ppm CO_2 concentration and tons of atmospheric carbon:

$$1 \text{ ppm CO}_2 = 1 \text{ ppm CO}_2 \times \frac{783 \text{ Gt C}}{370 \text{ ppm CO}_2} = 2.12 \text{ Gt C} \quad .....(4.10)$$

For understanding the relationship between future carbon emissions and expected changes in atmospheric CO_2 concentrations, Equation 4.10 will be very useful. However, as some carbon added to the atmosphere will be absorbed by the oceans or taken up by plants during photosynthesis, not all emissions will result in increased carbon dioxide concentrations in the atmosphere. So we have to represent the fraction of emissions that remain in the atmosphere by a quantity called the airborne fraction.

$$\text{Airborne fraction} = \frac{\Delta C \text{atmosphere}}{\Delta C \text{emissions}} \quad .....(4.11)$$

where ΔC atmosphere is the change in carbon content of the atmosphere, and ΔC emissions is the total amount of carbon added to the atmosphere.

It has been observed that fossil fuel combustion is mainly responsible for atmospheric carbon and it has been reported that as per the 1987 emission rates, fossil fuel combustion was adding 5.6 billion tons of carbon to the atmosphere per year. Further, carbon locked in the earth's biomass also contributes to atmospheric carbon when it is reduced. When forests are cleared and soil degrades, carbon that had been locked into vegetation or humus is released, adding to the CO_2 emissions from fossil fuels. The contribution from biomass is estimated to be 0.9 Gt C/year while that from fossil fuels, is 5.6 Gt/ year; when they are added together, the total emission rate is 6.5 Gt/ year. If all this emission is into the atmosphere, the increase in CO_2 would be

$$6.5\ \text{GtC/yr} \times \frac{1\ \text{ppm}}{2.12\,\text{GtC}} = 3.1\,\text{ppm}\,CO_2\ /\ \text{year}$$

As it has been observed in the last few decades that carbon dioxide is actually increasing at about 1.5 ppm/year, the airborne fraction can be estimated.

$$\text{Airborne fraction} = \frac{1.5\ \text{ppm/yr}}{3.1\ \text{ppm/yr}} 0.48 = 0.5$$

The airborne fraction of carbon is found to vary in the range 0.4-0.7, with 0.5 being a commonly used ratio.

Problem 4.2

The concentration of CO_2 in 1990 was 355 ppm, while in 2000 it was 370 ppm. Expressing the growth of CO_2 levels with a simple exponential function, calculate the exponential rate of growth. Calculate the year in which the atmospheric CO_2 will reach 560 ppm, i.e., double the pre-industrialization value.

Solution: Expressing the concentration in exponential form

$$C = C_0 e^{rt}$$

We have

$$370 = 335\ e^{t(2000-1990)} = 320\ e^{10r}$$

or

$$r = \frac{1}{10} = \ln\frac{370}{355} 0.0041 = 0.41\ \text{percent/yr}$$

If growth continues at 0.41% year, the time required to raise the concentration from the 2000 level of 370 to 560 ppm (double the pre-industrial value):

$$2 \times 280 = 370\ e^{0.0041t}$$

$$t = 1/0.0041\ \text{In}\ 560/370 = 101\ \text{yr.}$$

That is, it would be reached in the year 2000 + 101 = 2101

4.2.6.2 Emissions of carbon from fossil fuels

88% of the (non-biomass) world energy needs are supplied by fossil fuels while the balance comes from hydroelectric generation and nuclear power plants. The estimated fuel based, total and per – capita CO_2 emissions of certain developed and developing countries based on their industrial and urban activities are shown in Figure 4.11

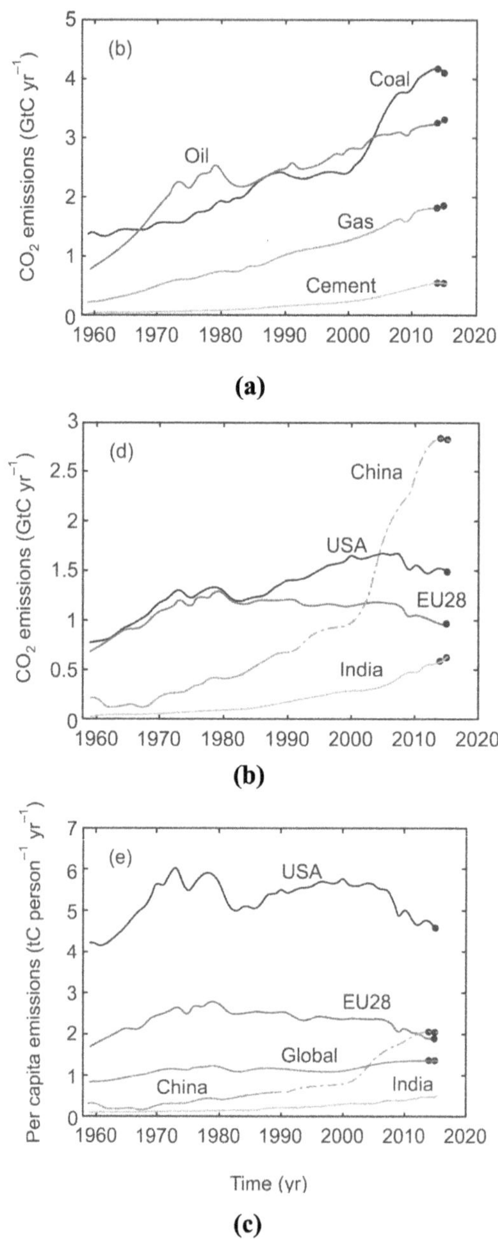

Fig. 4.11 *Estimated CO_2 emissions based on fossil fuel usage, total and per-capita emissions of certain developed and developing countries between 1960–2020*

The calculated carbon emissions from use of fossil fuels by certain countries are given in Table 4.3. As can be seen, the United States emits 22% of the total carbon though it has less than 5% of the world's population; on a per capita basis, it emits nearly five times as much carbon per person as the world averages, and nearly double the amount emitted by other developed countries.

Table 4.3 *Carbon emissions from fossil fuels by different countries*

KYOTO PARITIES WITH TARGET (1)	1990 $MtCO_2$	2013 $MtCO_2$	% change 90-13	Kyoto Target	OTHER COUNTRIES	1990 $MtCO_2$	2013 $MtCO_2$	% change 90-13	Kyoto Target
	8,269.8	**6,873.6**	**–16.9%**	**–4.6%** (2)		**99.8**	**58.2**	**–41.6%**	
Europe	***3,107.0***	***2,7705***	***–10.0%***		***Non-participating***				
Austria	56.2	65.1	16.0%	–13%	***Annex 1 Parties***	*5,454.6*	6,006.1	*10.1%*	
Belgium	106.2	89.1	–16.1%	–7.5%	Belarus	99.8	58.2	–41.6%	–8%
Denmark	51.0	38.8	–23.8%	–21%	Canada (2)	419.0	536.3	28.0%	–6%
Finland	53.5	49.2	–8.1%	0%	Cyprus	3.9	5.6	44.6%	none
France [4]	345.5	315.6	–8.7%	0%	Malta	2.3	2.3	0.6%	none
Germany	940.3	759.6	–19.2%	21%	Turkey	127.1	283.8	123.3%	none
Greece	69.9	68.9	–1.4%	+25%	United States	4,802.5	5,119.7	6.6%	–7%
Iceland	1.9	2.0	7.0%	+10%					
Ireland	30.1	34.4	14.1%	+13%	*Other Regions*	*6,158.9*	18,020.8	*197.7%*	none
Italy	389.3	338.2	–13.1%	6.5%	Africa	529.0	1,074.7	103.2%	none
Luxembourg	10.7	9.8	–9.0%	28%	Middle East	534.9	1,647.5	208.0%	none
Netherlands	144.9	156.2	7.9%	6%	N-CECD Eur. & Eurasia (5)	621.0	526.3	–15.3%	none
Norway	27.5	35.3	28.5%	+1%	Latin America (5)	812.8	1,579.3	94.3%	none
Portugal	37.9	44.9	18.6%	+27%	Asia (excl. China) (5)	1,444.4	4,178.9	189.3%	none
Spain	202.6	235.7	16.3%	+15%	China (Incl. Hong Kong)	2,216.9	9,023.1	307.0%	none
Sweden	52.1	37.5	–28.0%	+4%					
Switzerland	40.7	41.5	2.0%	–8%	**INTL MARINE BUNKERS**	371.5	608.8	63.9%	
United Kingdom	547.7	448.7	–18.1%	–12.5%	**INTL AVIATION BUNKERS**	258.8	490.4	89.5%	
European Union - 15	3,037.9	2,691.7	–11.4%	–8%	**WORLD**	20,623.0	32,189.7	56.1%	
Asia Oceania	***1,330.0***	***1,654.5***	***24.3%***						
Australia	259.6	388.7	49.7%	+8%					

Table 4.3 ***Contd…***

KYOTO PARITIES WITH TARGET [1]	1990 $MtCO_2$	2013 $MtCO_2$	% change 90-13	Kyoto Target	OTHER COUNTRIES	1990 $MtCO_2$	2013 $MtCO_2$	% change 90-13	Kyoto Target
	8,269.8	**6,873.6**	**–16.9%**	**–4.6%** [2]		**99.8**	**58.2**	**–41.6%**	
Japan	1,049.3	1,235.1	17.7%	–6%					
New Zealand	21.7	30.7	41.3%	–0%					
Economies in Transition	***3,831.3***	***2,448.6***	***36.1%***						
Bulgaria	74.6	39.3	–47.3%	–8%					
Croatia	20.7	16.0	–22.5%	–5%					
Czech Republic	150.3	101.1	–32.7%	–8%					
Estonia	36.0	18.9	–47.6%	–8%					
Hungary	65.7	39.5	–39.9%	–6%					
LaMa	18.8	6.9	–63.1%	–8%					
Lithuania	32.2	10.7	–66.7%	–8%					
Poland	344.8	292.4	–15.2%	–6%					
Romania	168.3	68.8	–59.1%	–8%					
Russian Federation	2,163.2	1,543.1	–28.7%	0%					
Slovak Republic	54.8	32.4	–40.9%	–8%					
Slovenia	13.5	14.3	5.9%	–8%					
Ukraine	688.4	265.0	–61.5%	0%					

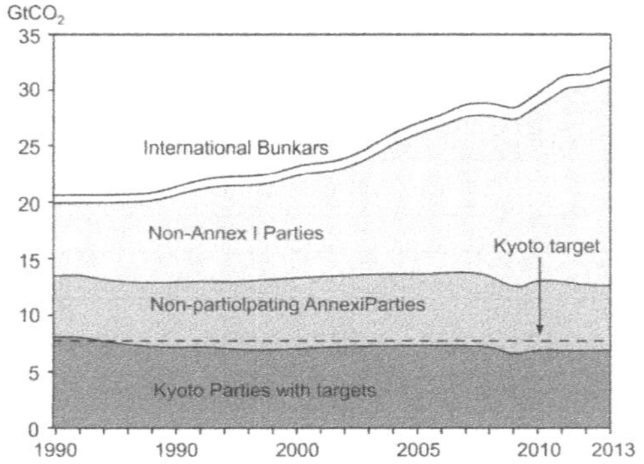

$GtCO_2$
International Bunkars
Non-Annex I Parties
Kyoto target
Non-partiolpating AnnexiParties
Kyoto Parties with targets
0 5 10 15 20 25 30 35
1990 1990 2000 2005 2010 2013

1. The actual country targets apply to a basket of six greenhouse gases and allow sinks and international credits to be used for compliance. The overall "Kyoto target" is estimated for this publication by applying the country targets to IEA data for CO_2 emissions from fuel combustion, and is only shown as an indication. The overall target for the combined EU-15 under the protocol is-8%, but the member countries have agreed on a burden-sharing arrangement as listed. The country composition and specific reduction targets shown refer to those agreed to number the first commitment period of the Kyoto Protocol (2008-2012). Reduction targets and the composition of Parties that have agreed to targets differ under the second commitment period of the Kyoto Protocol (2013-2020).
2. On 15 December 2011, Canada withdrew from the Kyoto Protocol. The is action became effective for Canada on 15 December 2012.
3. Emissions from Monaco are included with France.
4. Composition of regions differs from elsewhere in this publication to take into account countries that are not Kyoto Parties.
5. The Kyoto target is calculated as percentage of the 1990 CO_2 emissions from fuel combustion only, therefore it does not represent the total target for the six-gas basket. This assumes that the reduction targets are spread equally across all gases.

Key Point: The existing targets under the Kyoto Protocol are not sufficient to achieve global emissions reductions.

Significant differences in carbon emission factors exist, for different fossil fuels (Table 4.4). Nearly 75% more carbon is released per unit of energy when coal is burned instead of natural gas while synthetic oil or gas made from coal or shale oil produces about 80% more carbon than coal.

From the data presented in Table 4.4 one can infer that one way to reduce carbon emissions is to switch from coal to oil or natural gas. A power plant based on gas emits 42% less carbon than a power plant based on coal. Coal burning is the main source for the increase of CO_2 in the atmosphere and unless we change to solar energy or nuclear power the CO_2 levels will continue to increase with increase in the use of coal.

Table 4.4 *Carbon emission factors for various fuels, and 1985 world fuel and emission data*

Factor	Natural gas	Conventio nal oil	Coal	Synthetic fules			Non – fossil fuel
				Oil from shale	Oil from coal	Gas from coal	
10^6ton C/quad	14.5	20.8	25.2	50.2	40.7	42.9	0
Qiads/yr	60	120	36	0	0	0	36
Carbon emissions (Gt C/yr)	0.9	2.5	2.2	0	0	0	0

The world's reserves of fossil fuels expressed both by energy content and carbon content are presented in Table 4.5.

Table 4.5 *Energy and carbon content of the world's fossil fuel reserves and resources*

Fossil fuels	**Gas**	**Oil**	**Coal**
Reserves			
Quads	3600	4100	22000
Gt C	52	85	550
Resources			
Quads	10000	11000	280 000
Gt C	145	230	7100

Problem 4.3

Estimate the increase in atmospheric carbon dioxide that would result from complete combustion of the entire resource for the fossil fuels individuals at an airborne fraction 50%.

Solution: The total global resources of coal (Table 4.5) is 7100 GEC.

Using Equation 4.10 we can estimate the corresponding contribution to atmospheric CO_2 in ppm assuming 50% as airborne.

Increase in atmospheric CO_2 by burning all the coal:

$$\text{Increase} = 7100 \times 0.50 \ \frac{1\ \text{ppm of CO}_2}{2.12\,\text{GtC}} = 1675\,\text{ppm}\,\text{CO}_2$$

Increase in atmospheric CO_2 by burning all the natural gas:

$$\text{Increase} = 145 \times 0.50 \ \frac{1\text{ppm of CO}_2}{2.12\,\text{GtC}} = 34\,\text{ppm CO}_2$$

Increase in atmospheric CO_2 by burning all the oil

$$\text{Increase} = 230 \times 0.50 \ \frac{1\text{ppm of CO}_2}{2.12\,\text{GtC}} = 54\,\text{ppm}\,\text{CO}_2$$

The above data indicates that the threat for increasing global CO_2 levels lies with coal burning.

4.2.7 Other greenhouse gases – methane, CFCs, water vapour, nitrous oxide

Carbon dioxide is not the only gas that could cause climate warming. Methane, chlorofluorocarbons (CFCs), nitrous oxide, and other trace gases also absorb infrared radiation and warm the atmosphere (Fig. 4.12). Although rarer than CO_2, some of these gases trap heat much more effectively. Methane, for instance, absorbs twenty to thirty times as much, molecule for molecule, as CO_2 and CFCs absorb approximately 20,000 times as much [7, 8].

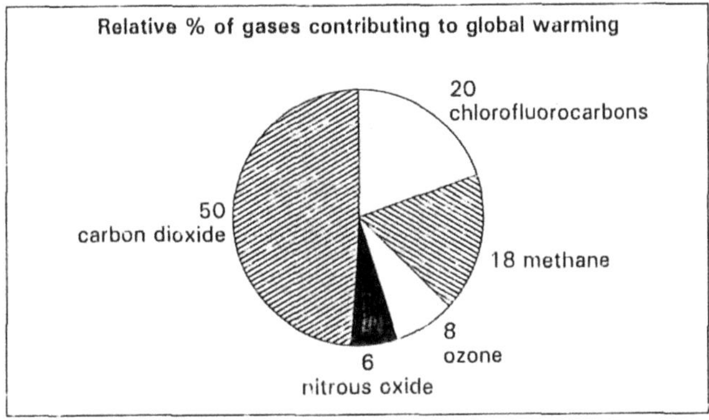

Fig. 4.12 *Relative contribution to global warming (% of expected climate change) by anthropogenic releases of gases into the atmosphere (Notice that while far less methane and chlorofluorocarbons are released than CO_2, they still are very powerful "greenhouse" gases.)*

4.2.7.1 Methane

Methane levels in the atmosphere are also rising as a result of various human activities. Currently, atmospheric concentrations are around 1.75 ppm. Table 4.6 lists the various sources and sinks of CH_4 in terms of their global significance. Most emissions of CH_4 arise from methanogenic bacteria in anaerobic environments but a small proportion of it is also of abiogenic origin, from deeper inside the earth's crust.

Table 4.6 *Estimates of global annual turnover of CH_4 [9]*

Natural sources	Mt a^{-1}	Anthropogenic sources	Mt a^{-1}	Sinks	Mt a^{-1}
Wetlands	120-200	Landfill	30-70	Tropospheric OH oxidation	375-475
Termites	5-150	Biomass burning	30-100	Stratospheric decomposition	35-50
Oceans	7-13	Paddy fields	70-170	Soil absorption	10-30
Ruminants (wild)	2-6	Gas venting, pipeline leaks and flaring	25-50		
Tundra	1-5	Coal mining	10-35		
Others.e.g. volcanoes	0-80	Sewage operations	5-70		

Methane is produced by intestinal bacteria in animals, anaerobic decomposition in wet rice paddies, pipe leaks, decaying wastes in landfills, and releases from coal mining. Atmospheric methane is increasing by 1% per year.

CFCs: Fig. 4.13 presents the global warming potentials of a number of CFCs, HCFCs, and HFCs relative to CFC-11 which was given an arbitrary value of 1 on each axis. The areas of the circles are proportional to the atmospheric lifetimes of each substance.

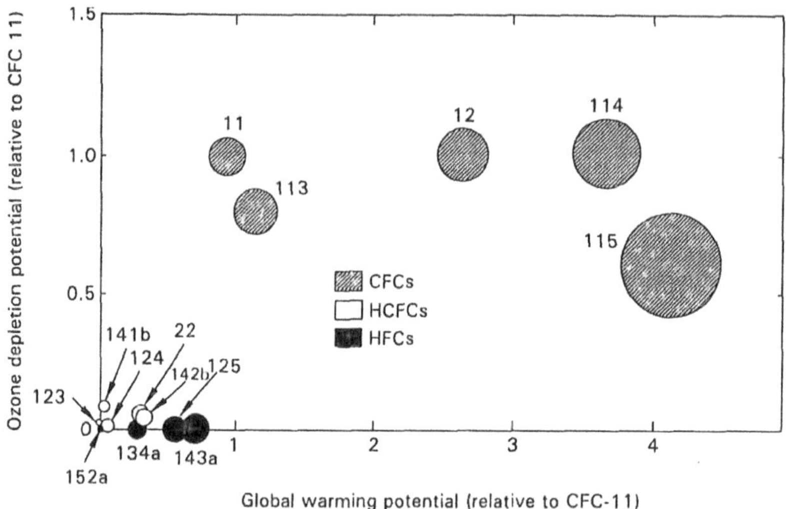

Fig. 4.13 *Global warming potential of CFCs, HCFCs, HFCs (relative to CFC-11)*

Chlorofluoro carbons, used as spray propellants, degreasing agents and refrigerants, have been accumulating at 5% each year. CFCs contain relatively large amounts of chlorine and if completely halogenated have long atmospheric lifetimes. Further as they absorb strongly radiation in the range 7-13 μm, they have great potential to affect both global warming and stratospheric ozone depletion, For example, CFC-11 is 77% chlorine, has strong absorption bands at 9.22 and 11.82 μm, and its atmospheric residence time is estimated in the range 60-110 years. CFC-12 is 59% chlorine, absorbs at 7.68, 9.13 and 19.93 μm and its atmospheric residence time is in the range 55-400 years. As can be seen by the smaller size of the circles in Fig. 4.13, the HCFCs have short atmospheric lifetimes. They break down relatively quickly and thus have only modest potential to affect either the ozone layer or cause global warming. However, HFCs with no chlorine atom can deplete the ozone layer, and have some potential to affect global warming.

4.2.7.2 Water Vapour

Water vapour is another important greenhouse component. It is an IR absorber/radiator both inside and outside clouds. Levels of water vapour, in turn, are highly dependent upon temperature, especially with respect to the capability of air to absorb water vapour. In fact, the contribution of water vapour to the greenhouse effect is a function of the square of the vapour pressure. Also, the flux of latent heat into the atmosphere increases during precipitation. The net effect is to magnify the greenhouse effect at the tropics.

4.2.7.3 Nitrous oxide

Nitrous oxide is another naturally occurring greenhouse gas that has been increasing (by 0.2%) annually due to human activities. It is released into the atmosphere mostly during the nitrification portion of the nitrogen cycle.

$$NH_4^+ \longrightarrow N_2 \longrightarrow N_2O \longrightarrow NO_2^- \longrightarrow NO_3^-$$

The reverse reaction of denitrification is no longer considered a significant source of atmospheric N_2O. Combustion of fossil fuels and consumption of nitrogen fertilizers are thought to be the two important human activities leading to increase in nitrous oxide levels. It apparently has no significant tropospheric sinks.

The trends in the rate of increase in important greenhouse gases and their relative contribution to the greenhouse effect in the atmosphere are presented in Table 4.7. Some of these exist for long periods in the atmosphere, e.g., CO_2, N_2O and CFCs.

Table 4.7 *Past, present, and future details of important greenhouse gases*

Parameter	CO_2	CH_4	N_2O	CFCs
Concentration 1850 1982	280 ppm 345 ppm	1150 ppb 1790 ppb	285 ppb 305 ppb	0 ppb 0.4 ppb
Expected 2075	526 ppm	4402 ppb	478 ppb	3.8 ppb
Annual concentration increases, %	0.4	1	0.2	5
Relative greenhouse efficiency, CO_2^{-1}	1	25	230	1500
Current greenhouse contribution, %	57	12	6	25
Anticipated rate between 1985 and 2075, %	0.57	1.0	0.5	2.5
Equilibrium temperature change 1850-2075, °C	3.12	0.62	0.28	0.93 = tot 4.95 °C
Total temperature change, %	63	12	06	19 = tot 100%
Life-time, years	100	10	150	100

Thus CO_2 predominantly contributes about 57% to overall warming though it has less warming potential than other greenhouse gases (a molecule of O_3, for example, absorbs 2000 times as much energy as a CO_2 molecule, and a CFC molecule absorbs 15500 more). CO_2 dominates as it occurs in much larger quantities in nature than one other gases.

The equilibrium temperature increase ΔT (°C) contributed by a mixture of carbon dioxide, nitrous oxide, methane, CFC-11 and CFC-12 is given by

$$\Delta T = \Delta T_d \times \ln\left(\frac{(CO_2)}{(CO_2)_0}\right) + 0.057\left[(NO_2)^{0.5} - (NO_2)_0^{0.5}\right] + 0.19[CH_4]^{0.5} - (CH_4)_0^{0.5}$$

$$+ 0.14\left[(CFC-11) - (CFC-11)_0\right] + 0.16\left[(CFC-12) - (CFC-12)_0\right] \quad \ldots\ldots (4.12)$$

where concentrations are expressed in ppb. The subscript (0) denotes the reference concentration for each compound. Using this relation and the data given in Table 4.7, the expected equilibrium temperature increase from 1850 to 2075 contributed by each greenhouse gas was calculated (Table 4.8).

Table 4.8 *Temperature changes caused by individual greenhouse gases*

Gas	Concentration in 2075	Equilibrium temperature change °C (1850-2075)	% of total temperature change
CO_2	576 ppm	3.12	63
CH_4	4402 ppb	0.62	12
N_2O	478 ppb	0.28	6
CFC-11	2.28 ppb	0.32	7
CFC-12	3.80 ppb	0.61	12
Total		4.95	100%

The data in Table 4.8 shows that approximately two-thirds of the 5°C increase in equilibrium temperature in 2075 (over the pre-industrial times) will be contributed by CO_2. Figure 4.14 depicts the relative contribution of each greenhouse gas to the increase in the equilibrium temperature. Other greenhouse gases actually contribute more to each decade's temperature increment than carbon dioxide in the later decades. Thus controlling future increases in temperature requires stringent controls on carbon dioxide and the other greenhouse gases as well – in particular, the chlorofluorocarbons.

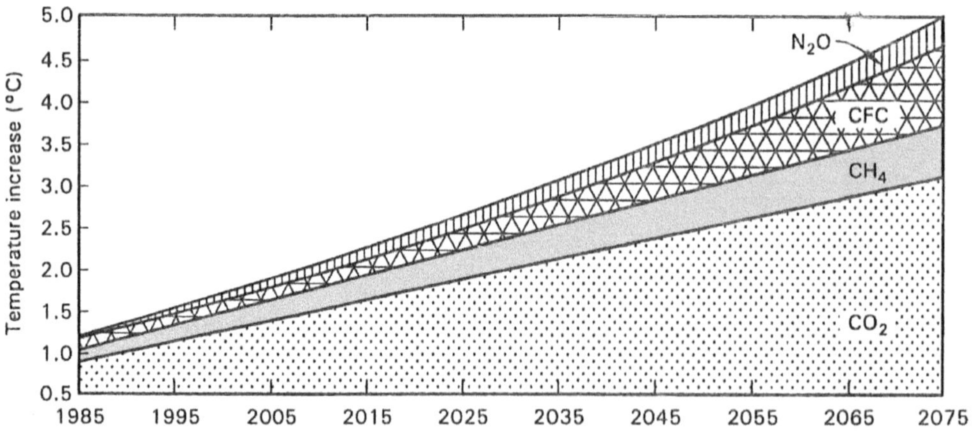

Fig. 4.14 *Increase in the equilibrium temperature (over pre-industrial values)*

The largest quantities of greenhouse gases are produced by the developed countries of the world Table (4.9). Together, the United States, the former Soviet Union, Europe and Japan are responsible for about two-thirds of all potential global warming.

4.2.8 Predicted effects of global warming

We don't yet know enough about how the earth works to accurately predict the possible effects of our emissions of greenhouse gases on global and regional climates or on the biosphere.

4.2.8.1 A warmer global climate

According to greenhouse theory, any warming on earth should be accompanied by cooling in the stratosphere (the second layer of the earth's atmosphere) and the mesosphere (the layer of the atmosphere above the stratosphere). Some cooling of the stratosphere has been observed, but most of this is probably the result of ozone depletion; recent measurements indicate a 2 °C drop in the temperature of the lower mesosphere.

To date, any temperature changes possibly caused by an enhanced greenhouse effect have been too small to exceed normal short-term variations in mean atmospheric temperature caused by volcanic eruptions, air pollution, and other climatic factors.

Mean global temperature never varies more than a degree, either side of 12 °C. Should this change, even by a degree, then the consequences are likely to be considerable. For example, 6000-8000 years ago, a temperature rise of 2 °C was responsible for melting the ice sheets spread across both hemispheres and brought about the end of the last ice age.

Warming or cooling by more than 1 °C over a few decades would be disastrous for the earth's ecosystems and for human economic and social systems. Such rapid climatic changes would alter conditions so quickly that some species, especially plants, would not adapt to the altered conditions or migrate from the areas. These changes might also shift areas in which it is possible for people to grow food. Many areas might become uninhabitable because of drought, or because of floods following a rise in average sea levels.

A warmer troposphere would have different consequences for different peoples and species. Some places would get drier, some wetter. Some would get hotter, others cooler.

4.2.8.2 Some possible effects of a warmer global climate

Rise in sea levels: Currently, global sea levels are rising a few centimetres per decade. Scientists project that two-thirds of this rise is the result of global warming. Recent studies indicate the remaining third may be due to cutting down of forests, draining of wetlands, and large-scale pumping of ground water. Mean global temperatures are likely to rise by 1.4 °C (± 0. 7°) by year 2030, and by 2.1 °C (± 0.8 °C) by year 2050 (Fig 4.15).The global mean sea level which has risen 10-15 cm in the last century may rise by another 10-20cm by 2025 and 50-200cm by 2100 [10]. These predictions of sea level rise are less than those of a few years ago but the rise will still cause extensive flooding of low-lying areas in the world, especially during storms. Most of this increase in sea level will come about from an increase in sea volume due to thermal expansion and less from the melting of polar ice. If warming at the poles caused ice sheets and glaciers to melt even partially, global sea levels would rise even more.

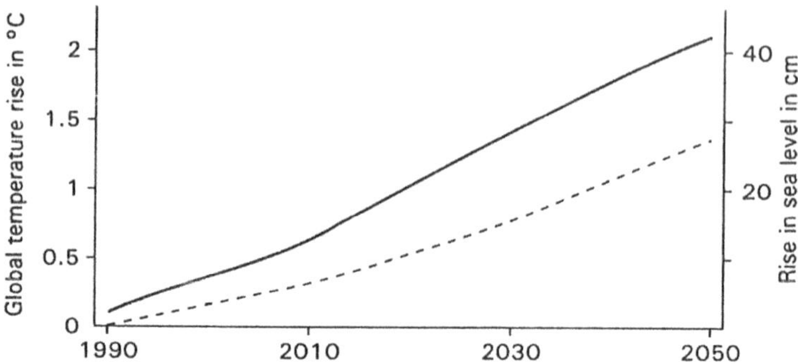

Fig. 4.15 *Likely changes in mean global temperatures and sea levels (IPCC predictions)*

About one-third of the world's population and more than a third of the world's economic infrastructure are concentrated in coastal regions. Thus, even a moderate rise in sea level would flood low-lying areas (many of which contain major cities), as well as lowlands and deltas where crops are grown. It would also destroy wetlands and coral reefs and accelerate coastal erosion. The regions most at risk are the river delta areas of Bangladesh, India, Egypt, Thailand, and China (most vulnerable first) but many small islands will also suffer. The immediate effect of inundation will be crop destruction but more importantly, the surrounding groundwater will become increasingly saline and large areas of productive land will be rendered permanently useless.

Some possible effects of a 1-metre (3.3 foot) rise in sea levels by 2035

1. The seas will move 30 metres (100 feet) farther inland along many portions of India's coasts. Crores of rupees of coastal properties would be lost in USA. The city of Charleston, South Carolina, alone would suffer $650 million in damage by the year 2035. Similar flood damage would probably be experienced in Boston, Philadelphia, New York city, Baltimore, Washington, Norfolk, Miami, St. Petersburg, and New Orleans.
2. 20% of the most populated areas of India and Bangladesh would be inundated.
3. Large tracts of agricultural land would be destroyed because of salinisation. Changing weather patterns could make huge areas infertile or uninhabitable. The result could well be more massive movements of displaced people than any in the history of this planet.
4. Many of the world's most productive rice-growing areas would be destroyed by flood waters, including the fertile, highly populated deltas of the Indus (Pakistan), Ganges (Bangladesh), and Yang Tze (China) rivers.
5. Many aquifers along the Gulf and Atlantic coasts of the United States, which now serve as a source of drinking water for many American cities, would become contaminated by salt water from the ocean.
6. The increased salinity of many American estuaries, such as the Chesapeake Bay and Delaware River estuaries, would destroy their value as breeding and nursery grounds for many valuable species of game and commercial fish.

7. Many plants and animals could become extinct because of the rapid change in temperature.

Some possible effects of a 2-metre (6.6 foot) rise in sea levels

The coastal areas around the globe where about a quarter of the world's population now lives, and especially in poor countries lying in the tropical belt of the earth, would be severely affected. Some of the best beaches in the world would be lost for ever, thus affecting tourism.

Maldives, off the west coast of India, Marshall Islands in the Pacific and some islands in West Indies, would disappear from the globe for ever. The future of the Maldives, comprising over 2000 inhabited and 1,000 uninhabited islands, about 600 km south west from the Indian coastline, is hanging delicately in balance. The sea waters around the capital, Male, have already reached optimum levels. One island in the Maldives has already gone under the Arabian Sea. A 2-metre rise could wipe out these atolls from the face of the earth, while a 1-metre rise could cause heavy tidal wave activity during storms and cyclones. The 2.45 lakh people there, when the sea devours the islands, are destined to emerge as the world's first environmental refuges.

Bangladesh, dominated by the Ganges-Brahmaputra-Meghna delta, is massively populated and is painfully poor. It is already plagued by floods; with the sea level rising by 2 metres, it would lose 16% of its land area. Egypt will be in similar trouble. It could lose 15% of the land under farming.

Fig. 4.16, generated by a computer, shows the present Indian coastline and the future coastal areas that will be submerged by a 2 m rise in sea level.

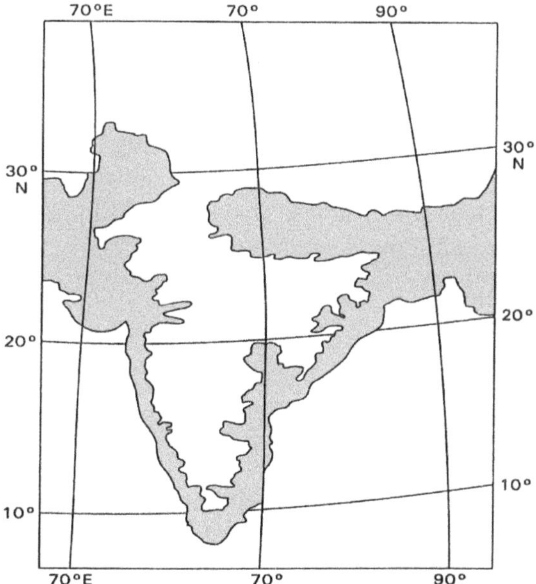

Fig. 4.16 *The present Indian coastline and the future one of the sea level rises by 2 metres. Shadowed region is likely to be submerged*

4.2.8.3 Changes in food production

Food productivity could increase in some areas and drop in others. Past archaeological evidence and computer models indicate that climate belts and thus tolerance ranges of plant species (including crops) would shift northward by 100-150 kilometres (60-90 miles) or 150 metres vertically for each 1 °C rise in global temperature. Computer models have projected drops in the global yield of key food crops ranging from 30% to 70%. With current knowledge, we can't predict where changes in crop-growing capacity might occur or how long such changes might last.

4.2.8.4 Reduction in water supplies

Lakes, streams, and aquifers in some areas that have watered ecosystems, croplands, and cities for centuries could shrink or dry up altogether, forcing entire populations to migrate to areas with adequate water supplies, if they could. We can't say with much certainty where this may happen.

4.2.8.5 Changes in forests

Forests in temperate and sub-arctic regions would be forced to move toward the poles or to higher altitudes leaving more grasslands and shrublands in their wake. However, tree species can move only through the slow growth of new trees along forest edges, typically about 0.9 kilometre per year or 9 kilometres per decade. If climate belts moved faster than this or if migration were blocked by cities, roads, or other barriers built by people, entire forests of oak, beech and other trees could die and release CO_2 into the atmosphere as they decompose. According to Oregon State University scientists, projected drying from global warming could cause massive fires in up to 90% of habitats, injecting huge amounts of CO_2 into the atmosphere.

4.2.8.6 Reduction in biodiversity

Large-scale forest diebacks would also cause extinction of species that couldn't migrate to new areas. Fish would die as temperatures soared in streams and lakes, and as lowered water levels concentrated pesticides. Any shifts in regional climate would threaten many parks, wildlife reserves, wilderness areas, wetlands, and coral reefs, wiping out many current efforts to stem the loss of biodiversity.

4.2.8.7 Weather extremes

In a warmer world, prolonged heat waves and droughts would become the norm in many areas. As the upper layers of sea water become warmer, hurricanes and typhoons would occur more frequently and blow more fiercely.

4.2.8.8 Threats to human health

A warmer world would disrupt supplies of food and fresh water, displacing millions of people and altering disease patterns in unpredictable ways. The spread of tropical climates from the equator would bring malaria, encephalitis, yellow fever, dengue fever, and other insect-borne diseases to formerly temperate zones. Sea-level rise could spread infectious disease by flooding sewage and sanitation systems in coastal cities.

4.2.8.9 *Some Observed Global Features that can be Related to Global Warming*

The average global temperature in July 1998 was 16 °C, 0.63 °C higher than the long-term average for July and 0.22 °C higher than the previous average up to July in 1997. July 1998 was the hottest July in the history of the world since meteorologists started keeping records more than 118 years ago. It was also the hottest month on record,

Despite the repeated warnings of scientists that the greenhouse effect could eventually have highly adverse effects on human society, the general public, as well as policy makers, were generally unconcerned about the threat until 1988. In 1988, Americans experienced unusual drought, searing heat, floods, and hurricanes – events that scientists predicted would occur with increasing frequency and severity as the levels of green house gases rise. The climatic disasters of 1998 may or may not have been valid signals that the green house effect was finally in evidence.

4.2.8.10 *Global warming effects in India*

According to Dr. G. B. Pant, working at the Indian Institute of Tropical Meteorology, Pune, India has also warmed up somewhat in the last three decades. While the average temperature had risen by 0.4 °C for the better part of the 20th century, the rate has climbed to 0.6 °C in the last 30 years. Days are hotter and the nights are the same. This has increased the difference between the maximum and minimum temperatures. So the weather is changing.

In 1988, the rainfall in India was above average. India has been enjoying good monsoons for the last 10 years. Even in 1998, the rainfall has been above the average. Is the greenhouse effect (GHE) the reason behind this phenomenon? No one is sure.

One global climate model studying changes in summer soil moisture suggests a wetter India. Some scientists, therefore, blame global warming and its consequences, for the heavy rainfall in July and August 1998 in the Himalayas, and the landslides that occurred there. The loss of human lives due to landslides in 1998 was rather heavy. One such landslide in July near Rudraprayag in Uttar Pradesh took the lives of 69 tourists.

4.2.9 Global warming: control measures being taken

Many steps are being taken around the world to combat the threat of global warming. In the United States, $1 billion was earmarked for global warming research for the year 1991. According to recent studies by the National Academy of Sciences and the Congressional Office of Technology Assessment, the United States could reduce its greenhouse gas emissions by 25-35% of 1990 levels using existing technology at low or no net cost, and by 48% at low to moderate cost. A 1993 study estimated that a doubling of current atmospheric levels of greenhouse gases by the middle of the next century could cost the U.S. economy alone about $60 billion per year and the world, upto $335 billion per year, assuming temperatures continued to rise.

The United States and many other countries have committed to phasing out the production of CFCs, the potent greenhouse gases. In addition, billions of trees will be planted around the world to help combat the excessive production of CO_2, as trees absorb CO_2 in the air. The Montreal protocol signed by 40 countries in September 1987 embodies the strategy to eventually phase out chlorofluorocarbons (CFCs) by the year 2000. In the Indian context, the Montreal protocol seeks to phase out production of CFCs in the developing countries by the year 2010 (a grace period of 10 years). The cost of switching over to CFC substitutes is estimated to be between Rs. 35,000 and 60,000 million depending on the lagtime for the changeover. India cannot have access to the technology to produce the replacement refrigerants as it is patented by Du Pont and a few other multinational corporations.

The agreement adopted at the UN climate conference (Berlin, April 1995) includes a tool to give poor countries their best access yet to rich nations' technologies. It is a mechanism set up so that clean energy technologies can be transferred to the developing world, with benefits for both parties and the global environment.

Nearly 120 nations adopted a mandate at the Berlin meet on strengthening the 1992 Rio Climate Treaty. It obliges developed countries to set 'objectives' for further emission cuts by 1997.

On the scientific side, tremendous research is going on with remarkable achievements in alternative energy sources – less polluting, more efficient systems. To actually reduce atmospheric CO_2 levels nothing much is going on. Use of solar generated building material from sea water as a sink for carbon is of course one such technology. Simple prudence and being more efficient can alter matters to an extent, but the rising trend of CO_2 emissions is something about which not much can be done now. Global warming responses should be linked to existing programs for energy, conservation, soil and water conservation, sustainable agriculture, and sustainable urban development.

4.3 Stratospheric Ozone Depletion

4.3.1 The ozone layer

The second global environmental problem of great concern is the possible destruction of the stratospheric ozone layer. At ground level, ozone, O_3, is a strong eye and respiratory irritant and a major component of photochemical smog. It may also act as a greenhouse gas. In the stratosphere, 10 to 20 km above the earth's surface, is a layer of low-density air containing 300-500 ppb of ozone. Fig. 4.17 shows the various layers of the atmosphere surrounding the earth including the stratospheric ozone layer. Fig. 4.18 depicts how the ozone layer filters out the lethal ultraviolet radiation in the solar radiation penetrating the atmosphere.

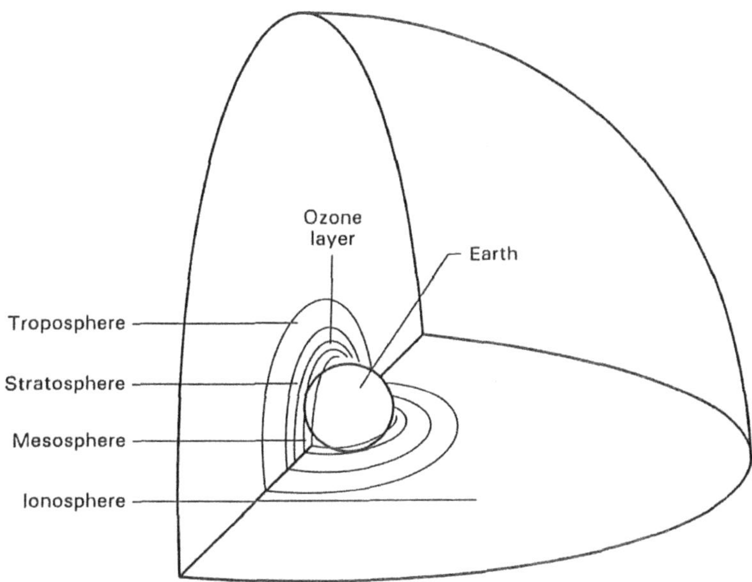

Fig. 4.17 *Ozone layer and other layers of the earth's atmosphere*

As shown in Fig. 4.17 the earth's atmosphere can be divided into a series of spherical layers. The layers have a distinctive vertical temperature gradient. The troposphere extending from 0-20 km above the earth's surface is heated from below by solar radiation reflected back by the earth's surface while the stratosphere (with the lower limit at an altitude of 8-16 km depending on latitude and extending 45-50 km from the earth's surface) is heated from above by the absorption of ultraviolet (UV) light by its ozone layer. (Figure 4.18)

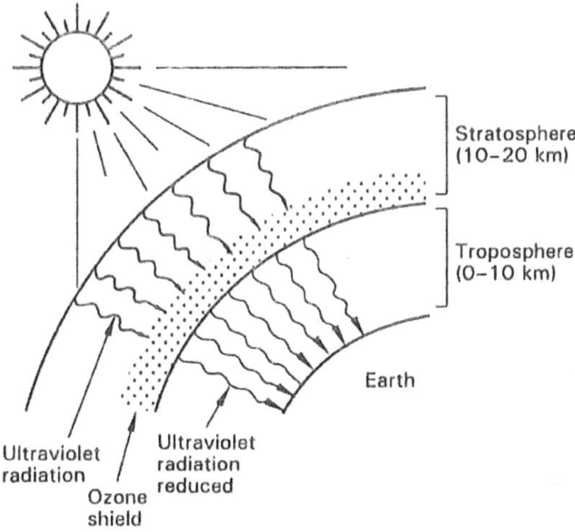

Fig. 4.18 *The ozone layer in the stratosphere shields the earth from potentially lethal ultraviolet radiation*

4.3.2 Relative penetration of different types of UV radiation from sunlight

Only certain frequencies of solar radiation penetrate the entire atmosphere and reach the surface of the earth (Fig. 4.19). Very long radio waves do not get much further down than 50 km above sea level.

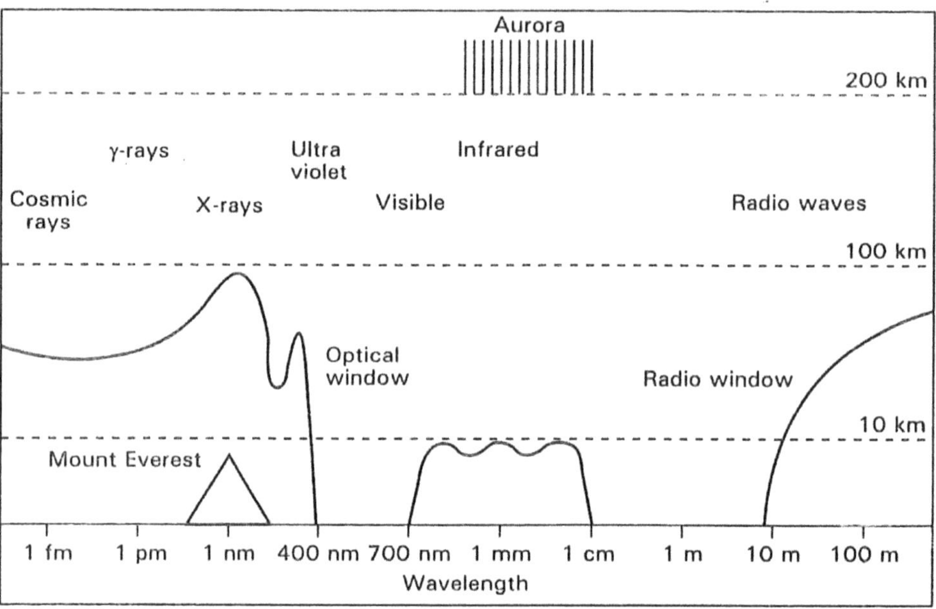

Fig. 4.19 *Relative penetration of different types of solar radiation into the atmosphere (Only certain radio frequencies and radiation around that of visible light penetrate as far as sea level.)*

Most incoming infrared radiation is completely absorbed above 10 km from the earth's surface (i.e. just above the height of Mount Everest) but small amounts of short wavelength infrared radiation, all the visible spectrum, and longer wavelength ultraviolet radiation reach sea level. Shorter wavelengths than these (x-rays, y-rays and most cosmic rays) however, are screened out between 10 to 100 km above the earth's surface (Fig. 4.19). Conventionally, UV light is subdivided into three categories (Table 4.9) and it is only UV-A and sometimes UV-B that penetrate as far as the surface of the earth. Fortunately, the biologically more harmful UV-C is filtered out completely.

4.3.3 Atmospheric UV penetration shields

A variety of factors influence the atmospheric penetration of UV-B, the solar radiation known to cause most of the adverse effects on biological systems. The major atmospheric barriers to UV-B penetration are the O_3 molecules in the stratosphere but other environmental factors also affect UV-B fluxes. These include tropospheric pollutants (other than O_3), sunspot activity on the sun (which may increase stratospheric O_3 levels by nearly 2%) and reflections of UV-B from surfaces, clouds and aerosols. All are highly variable and account for the fact that long-term records of UV-B fluxes vary and often conflict. These large variations have prevented the establishment of a reliable global model to predict trends in UV-B flux. Indeed, networks of UV-B monitors, usually placed in urban areas for convenience, often record decreases of UV-B due to increases in tropospheric pollutants and cloudiness. The only consistent trends of enhanced fluxes of UV-B due to stratospheric O_3 depletion have been recorded by instruments in polar regions.

Table 4.9 *Types of radiation in or near the visible window of the earth's atmosphere*

Wavelength	Type	Remarks
> 13 μm	Long wave infrared	Molecules like CO_2 and CFCs in the atmosphere absorb infrared (IR) radiation between 13 and 17 μm, reflected back from the earth to provide the blanket or greenhouse effect of global warming.
700 nm-3 μm	Shortwave infrared	13 μm is the upper limit for IR radiation to penetrate to the earth's surface from the sun. This IR band heats up surfaces on the planet and also permits reflection of IR of < 3 μm back into space.
400-700 nm	Visible spectrum	Blue (400 nm) and red (700 nm) components are most important for photosynthesis while vision is most sensitive in green-yellow (500 nm) light.
315-400 nm	UV-A	Long wave ultraviolet, not normally associated with injury to biological systems. Certain photoreceptors of animals and plants operate in this region.
280-315 nm	UV-B	The wavelengths of sunlight likely to cause biological effects. There is some dispute over the upper limit: 315 nm is in the UNEP Report (1991); others quote 320 nm.
<280 nm	UV-C	These wavelengths are completely absorbed by the atmosphere before sunlight reaches the surface of the earth. They are more disruptive to biological systems than UV-B and are commonly used for sterilization procedures.

4.3.4 How the ozone is produced

According to Chapman the following reactions are responsible for the dynamic equilibrium of ozone formation [11].

1. $O_2 + h\nu \rightarrow O + O$
2. $O + O_2 + M \rightarrow O_3 + M$
3. $O_3 + h\nu \rightarrow O + O_2$
4. $O + O_3 \rightarrow O_2 + O_2$

Thus ozone is produced by the absorption of solar ultraviolet radiation by oxygen molecules which then photo-disassociates to form oxygen atoms (O). These O atoms in turn react with O_2 in a three-body reaction to form O_3. The O_3 produced is subsequently photo-disassociated by both UV and visible light. Finally the chemical loss for odd oxygen is the reaction between O and O_3 to re-form two O_2 molecules. Since reaction (2) becomes slower with increasing altitude, while reaction (3) becomes faster, atomic oxygen predominates at higher altitudes and O_3 is favoured at lower ones. Reactions (2) and (3), in which the odd oxygen is interchanged are much faster than reactions (1) and (4) in which the odd oxygen is created or destroyed. If n_1, n_2, n_3 and n_m are the concentrations of O, O_2, O_3 and all molecules respectively and k_2 and k_4 are the reaction rates of reactions (2) and (3) and T_i and T_3 are the photolysis rates for reactions (1) and (3), then the equilibrium ozone concentration is given by

$$n_3 = n_2\left[(T_1 k_2 n_m)/(T_3 k_4)\right]^{1/2} \quad .....(4.13)$$

Until a few years ago, the ozone layer was in a state of dynamic equilibrium. In other words, ozone-forming and ozone-depleting reactions proceeded at similar rates, and therefore the amount of stratospheric ozone remained fairly constant. It is true, however, that there might be variations in abundance as high as 10% from year to year. Even in a given year, the amount of ozone in the stratosphere might be two to three times greater in one place than another.

4.3.5 Ozone distribution and temperature profile in the atmosphere

The distribution of ozone in the stratosphere has a profound effect on the outer atmospheric temperature; the strong temperature inversion in the stratosphere makes the air dynamically stable and resistant against vertical mixing. So once the pollutants enter the stratosphere they will have long lifetimes and can build up to globally unacceptable levels. Fasry and Dobson are the first scientists who measured atmospheric ozone by spectrometric techniques and pointed out that ozone occurs in trace quantities throughout the atmosphere with peak concentration in the stratosphere at an altitude of 20-25 km (Fig. 4.20). Observations of the total amount of ozone in a vertical column have established the average column abundance and the general pattern of latitudinal, longitudinal, seasonal, and meteorological variations. Total ozone is usually stated in Dobson units (DU) or matm cm, related to the thickness of an equivalent layer of pure ozone at standard pressure and temperature. If the ozone is in a column at atmospheric pressure, it would be about 3 mm thick, i.e., the average total amount of ozone is 300 DU. Seasonal and latitude variations at sub-polar latitudes are about $\pm$ 20% of this value. The annual average total ozone is a minimum of approximately 260 DU at equatorial latitudes and increases poleward in both hemispheres to a maximum of about 400 DU at sub-polar latitudes. The high latitude maximum results from the transport of ozone from the equatorial middle and upper stratospheres, the region of primary production, to the lower stratosphere in polar regions. About 90% of all the ozone in the atmosphere resides in the stratosphere.

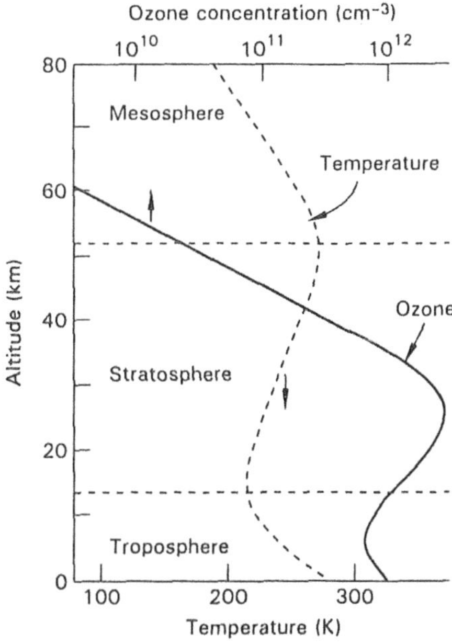

Fig. 4.20 *Ozone distribution and temperature profile in the atmosphere*

4.3.6 Destruction of ozone By CFCs and other gases

In 1974, F. S. Rowland, of the University of California, reported that the chlorofluorocarbon gases (CFCs) used to propel deodorants, hair sprays, shaving creams, insecticides from spray cans, and used in refrigerators and air conditioners were accumulating in the stratosphere and depleting the ozone there. The total amount of CFCs released into the global atmosphere is considerable [12]. In 1986, the U.S. used 320 million kilograms (700 million pounds) of CFCs.

The destruction of the ozone layer is a complex process. It is mostly caused by elemental chlorine atoms; the mechanism apparently involves two reactions (Fig. 4.21).

$$Cl + O_3 \rightarrow ClO + O_2$$

$$ClO + O_3 \rightarrow Cl + 2O_2$$

$$2O_3 \rightarrow 3O_2$$

Other reactions are going on in the stratosphere that modify and compete with these two, but if we ignore other reactions, add these two reactions, and cancel like terms, we see that the overall reaction is with no net consumption of Cl atoms. Thus one Cl atom can convert many ozone molecules to ordinary oxygen molecules. One sees estimates of 10^4–10^6 O_3 molecules destroyed by one Cl atom. This mechanism is often referred to as catalytic destruction of ozone, because the chlorine atom acts as a non-consumed catalyst for the reaction.

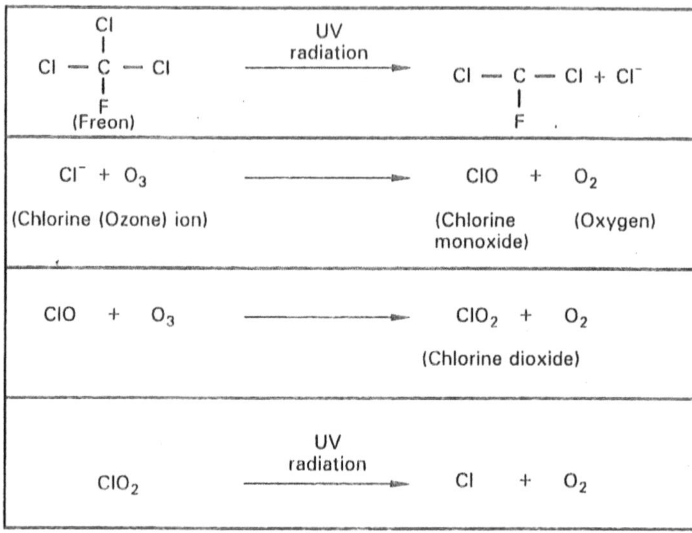

Fig. 4.21 *Photo-dissociation of Freon 11 (When struck by UV radiation, chlorofluorocarbons release chlorine ions. These chlorine ions combine with ozone to make chlorine monoxide plus oxygen. The chlorine monoxide reacts with another ozone molecule to make chlorine dioxide, which dissociates into oxygen and a reactive chlorine ion that starts the cycle all over again.)*

4.3.7 Anthropogenic sources of ozone-destroying chemicals

Man's capability to produce nitrogen and chlorine compounds at levels equalling or exceeding natural sources has led to numerous investigations aimed at assessing the impact of human activities on the ozone layer. Starting about 1900, man began releasing into the atmosphere synthetic chlorine-containing compounds in significant amounts. Those like methyl chloride that have hydrogen atoms can be attacked in the atmosphere by the OH radical; for this reason most of them do not survive to reach the stratosphere. Carbon tetrachloride, CCl_4, has no hydrogen; most of it is believed to reach the stratosphere and to participate in the destruction of the ozone layer. Its world production has been more or less constant over the past 60 years.

CFCs (compounds containing chlorine, fluorine, and carbon, commonly called Freons) were first developed by General Motors for use in household refrigerators. One of their virtues is their chemical inertness; they are non-toxic, non-flammable, invisible, tasteless, odourless, non-almost everything else. They replaced toxic sulphur dioxide and ammonia in household refrigerators. This replacement saved many lives.

The CFCs which have attracted most attention in ozone depletion in are CFC-11 ($CFCl_3$) and CFC-12 (CF_2Cl_2). The lifetime of CFC-11 is about 77 years, while that of CFC-12 is 139 years [13]. Hence any reduction in the release of CFCs will have little effect in the immediate future. Model predictions of ozone column changes due to increases in CFCs have varied widely during the last ten years. If CFC emissions continue at the 1980 rates long enough to obtain a steady-state, 8 ppb of chlorine would result.

Calculated decreases in column ozone, when chlorine is increased from 1 ppb to 8 ppb, are in the range 3-9%, depending on the model used. In all models, the major decrease occurs in the upper stratosphere, at an altitude of about 40 kms. As chlorine is increased, it exceeds a certain threshold where each additional chlorine atom destroys progressively more and more ozone. If chlorine increases to 15 ppb, the models then predict ozone column changes of 9-22%. This non-linearity for large chlorine perturbations can have significant implications for the interpretation of effects of changing CFC emissions. The most abundant organ chlorine compounds in the atmosphere are CCl_2F_2, CCl_3F, CH_3Cl, CCl_4 and CH_3CCl_3. The total atmospheric content of organ chlorine compounds has grown from about 0.8 ppb in 1950 to 1.0 ppb in 1970 and 3.5 ppb in 1987 (WMO-NASA, 1988). The current rate of increase exceeds 1 ppb per decade.

CFCs are not the only ozone-eaters, a few other chemicals can release highly reactive chlorine and bromine atoms if they end up in the stratosphere and are exposed to intense UV radiation. For example: (1) bromine-containing compounds called halons and HBFCs (both used in fire extinguishers) and methyl bromide and (2) carbon tetrachloride and methyl chloroform, or 1,1,1, trichloroethane (used as a cleaning solvent for clothes and metals and in more than 160 consumer products, such as correction fluid, dry cleaning sprays, adhesives, and other aerosols).

Wofsy et al. (1975) suggested that trace amounts of bromine might efficiently catalyse ozone destruction [14]. The predominant anthropogenic sources are CH_3Br and CF_3Br. Some sources, such as CF_2BrCl and $C_2H_4Br_2$, are entirely man-made. CH_3Br is used as a soil fumigant and other compounds of bromine are used as fire retardants and fuel additives. An increase in CH_3Br from 20 to 100 ppt would result in a decrease in total ozone of 30%. The largest relative change in ozone occurs at an altitude of about 15 km, with a secondary peak at 40 km.

Some other gases can attack the ozone layer, e.g., NO from stratospheric airplanes and relatively inert N_2O, if we release much of it at ground level. NO, released by high flying aircraft, can contribute to ozone depletion by the reaction

$$NO + O_3 \quad \rightarrow \quad NO_2 + O_2$$

which is swift and practically irreversible. But it is not a catalytic reaction like the chlorine reaction; one NO molecule only destroys one O_3 molecule, so much more NO is needed per unit of damage than Cl.

Nuclear explosions produce copious quantities of NO_x, which may be injected into the stratosphere. Early work predicted O_3 reductions of up to 70% in the northern hemisphere. There are two significant sources of N_2O which result from human activity: combustion and the use of nitrogenous fertilizers that enhance biogenic emissions of N_2O. A doubling in N_2O concentrations could give a 10-16% global ozone depletion. The depletions vary with altitude, being greatest at higher altitudes. For a more realistic increase of 20% in N_2O, total ozone decreases are in the range 1.1-2.6%. Small increases in O_3 are calculated in the upper troposphere and lower stratosphere due to the increased efficiency of the CH_4-NO_x-smog reactions from the added NO_x.

4.3.8 Gases that increase ozone levels

Carbon dioxide concentrations are increasing largely as a result of burning fossil fuels. Increased CO_2 leads to lower stratospheric temperatures, and hence a slowing down of temperature dependent ($O + O_3$, $NO + O_3$) ozone destruction reactions. This results in a net increase in stratospheric ozone concentrations. Doubling CO_2 concentrations leads to a temperature decrease of 7-10 K at an altitude of about 40 km, and a corresponding increase in ozone of 9-19%. Increase in the total ozone column will be in the range 1.2-3.5%.

Increases in total ozone of 0.9-3.0% are also predicted for a doubling of CH_4 concentrations. For such an increase in CH_4, stratospheric OH is predicted to increase by up to 50% and HO_2 by 100%. The resulting increase in HNO_3 and HNO_4 production reduces NO_2 concentrations and a subsequent reduction in the NO_x catalytic cycle. Methane is currently increasing due to the growth in rice farming and cattle farming.

4.3.9 Antarctic ozone depletion

The protective ozone layer showed a hole over 50% of the area of the Antarctica continent, according to a picture taken by the Nimbus-7 satellite on October 5, 1987. This hole covers approximately 7 million square kilometres (Fig. 4.22). It shows the lowest ozone value ever observed according to the officials of the National Science Foundation, USA.

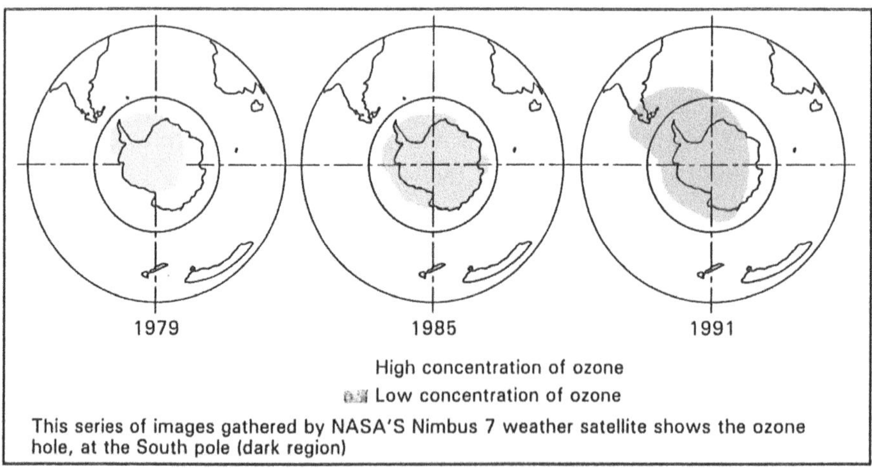

Fig. 4.22 *Seasonal thinning or loss of ozone*

Measurements from instruments aboard the Nimbus-7 satellite showed that the region of ozone depletion was larger than the Antarctic continent; balloon-borne instruments indicated that the greatest depletion occurred in the altitude range 12-24 km.

The variation of ozone with altitude above Halley Bay is shown in Fig. 4.24. In mid-August the profile was near normal but in less than two months, 97% of the ozone between the altitudes of 14 and 18 km had been destroyed.

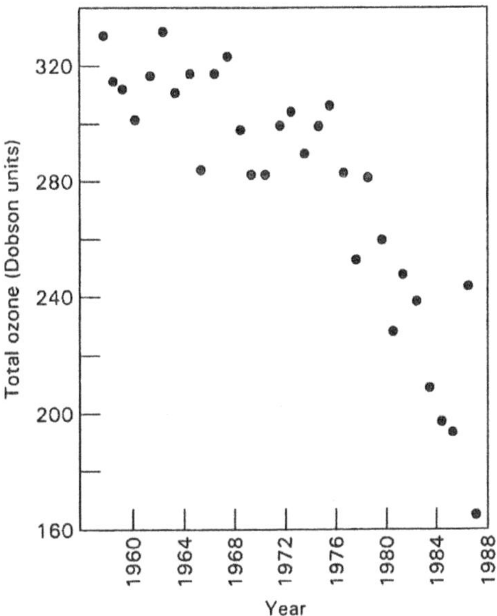

Fig. 4.23 *Mean October value of column ozone over Halley Bay (76°S) since 1957 (adapted from SORG, 1988) [13]*

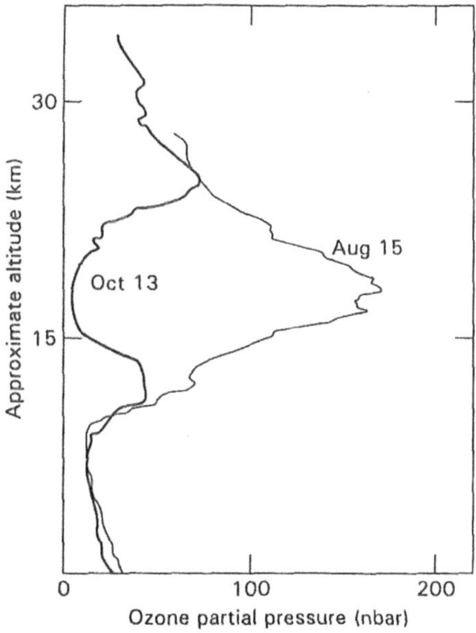

Fig. 4.24 *Vertical ozone profiles above Halley Bay for 15th August, 1987 and 13th October, 1987 (adapted from SORG, 1988) [13]*

Fig.4.23 shows the monthly mean values of the ozone column for October over Halley Bay. Prior to the mid-1970s the concentration of ozone was approximately 300 DU. By 1985 the total ozone had fallen to below 200 DU and yet greater depletions were observed in 1987. The Antarctic ozone hole of 1989 was deep and long lasting.

The Antarctic ozone hole of 1990 was deep and lasted right through December.

The Antarctic ozone hole of 1991 was deep and long lasting.

The data from the worldwide network of ozone measuring stations are reported to the World Ozone Data Centre in Toronto and are published monthly.

The following factors are believed to be responsible for the preferential depletion of the ozone layer over the Antarctic rather than the Arctic:

1. The Antarctic stratosphere is much colder. The decrease in temperature enables the formation of Polar Stratospheric Clouds (PSCs), below 20 km.
2. Ozone absorbs sunlight, causing the characteristic increase of temperature with increase in altitude, in the stratosphere. If Ozone is being depleted, the air is cooler, hence this further adds to the favourable conditions for the formation of PSCs and stabilising the vortex.
3. The longevity of the Antarctic vortex is another factor, enhancing favourable condition for the depletion of ozone. The vortex is a ring of rapidly circulating air that confines the ozone depletion, it remains in fact throughout the polar winter, well into midspring whereas the vortex in the Arctic disintegrates by the time the polar spring (March-April) arrives.

4.3.10 Typical happenings in the winter months leading to the ozone hole over the antarctic

June: Antarctic winter begins, the vortex develops and the temperature falls enough for PSCs to form.

July and August: PSCs denitrify and dehydrate the stratosphere through precipitation; hydrochloric acid and chlorine nitrate react on cloud surfaces to free chlorine and winter temperatures drop to their lowest point.

September: Sunlight returns to the centre of the vortex as the austral spring begins and PSCs disappear because of increasing temperature; ClO-ClO and ClO-BrO catalytic cycles destroy ozone.

October: Lowest levels of ozone are reached.

November: Polar vortex breaks down, ozone rich air from mid latitudes replenishes the Antarctic stratosphere and ozone poor air spreads over the Southern hemisphere.

4.3.11 Arctic ozone hole

Of late, the ozone hole has been increasingly evident over the Arctic as well. The Arctic ozone hole which swept across Britain in March '96 is the deepest ever seen in the Northern Hemisphere. Scientists claim that it has been caused in part by a dramatic cooling of the upper atmosphere in the northern latitudes over the past two years.

The ozone depletion over the northern hemisphere has been increasing steadily from the winter of 1992. Apart from the build-up of ozone depleting chemicals, the main cause is the increasingly cold temperature in the Arctic stratosphere which encourages the formation of Polar Stratospheric Clouds. Data on temperature in the upper atmosphere collected by weather forecasters and analysed by Steve Pausen at the Free University in Berlin show that, at an altitude of 20 km, the winter air has been colder during the 1990s than at any time since recording.

Various theories have been postulated to explain this ozone depletion including dynamical air motions near the pole, a connection with the 11-year cycle in solar activity, and a chemical mechanism. It is now recognised that the depletion is caused by manmade chlorine pollution of the atmosphere in combination with unique winter meteorological conditions.

The circulation of the winter stratosphere over Antarctica and the surrounding oceans is dominated by the polar vortex. This is the region of very cold air surrounded by westerly winds which may reach speeds of $\sim$100 ms^{-1} or more by the spring. Air within the vortex is sealed off from that at lower latitudes and the vortex acts as a containment vessel in which chemical reactions may occur in near isolation. Temperatures within the vortex fall below 190 ^{o}K, cold enough in the lower half of the stratosphere for Polar Stratospheric Clouds (PSC) to form. These clouds remove active nitrogen from the stratosphere that would otherwise remove active chlorine and also provide surfaces for heterogeneous chemical reactions.

Some scientists have maintained that even if the protective ozone blanket was partially destroyed, the increase of UV radiation would be absorbed by clouds and atmospheric pollution. However, two Canadian scientists recently reported that the amount of UV radiation in the air over Toronto, Ontario, increased 5% each year from 1989 to 1993, coincident with the annual thinning of the ozone shield.

In 1988 the Ozone Trends Panel of the National Aeronautics and Space Administration concluded: "The weight of evidence strongly indicates that man-made chlorine compounds are primarily responsible for the hole". Recent studies indicate that the ozone breakdown is accelerated in the presence of ice crystals and sunlight. Although the greatest amount of ozone thinning has been caused by man-made CFCs, sometimes a natural event, such as a volcanic eruption, will contribute to the depletion of ozone. In 1992, the volcano Mt. Pinatubo, in the Philippine Islands erupted and injected trillions of tiny sulphate droplets into the atmosphere. It is no coincidence that in the fall of 1993, the ozone hole over Antarctica was the largest ever recorded, according to David J. Hofmann of the National Oceanic and Atmospheric Administration. More than 70% of the ozone blanket was destroyed in a 23 million square kilometre area, almost the size of the entire North American continent. Scientists theorised that the trillions of the tiny sulphate droplets injected by the eruption provided the surface on which the ozone depleting reactions could occur. The implications are disturbing. As Rowland stated: "There's a distinct worry that what's happening in Antarctica could happen here".

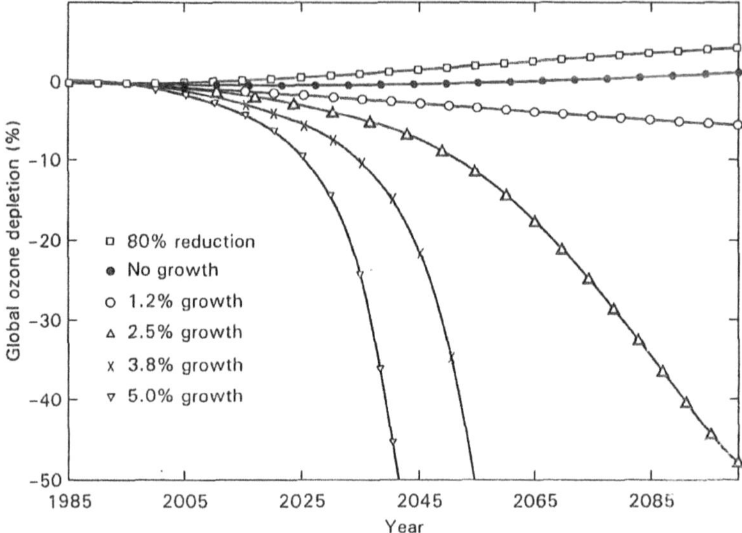

Fig. 4.25 *Average global depletion for six scenarios (Ozone depleting gases (CFCs) are assumed to grow at the indicated rates until 2050, and are constant thereafter. Greenhouse gases that counter ozone depletion are assumed to grow at historically extrapolated rates.)*

Rowland proved to be prophetic. In 1994 scientists reported a dramatic worsening of the ozone depletion problem, not only in the polar regions but over much of the Northern hemisphere. Rumen Bujkov of the U.N. International Ozone Commission, reported the ominous news at a 1994 meeting of the American Association for the Advancement of Science. According to Bujkov, levels of ozone had dropped 10% globally over the past 25 years. More serious for Americans, however is that the ozone shield over the United States was thinned by 15% during the winters of 1992 and 1993. On some days the depletion was as high as 45%. Similarly high levels of ozone depletion have been recorded over Europe and Siberia.

The importance and the gravity of the situation led to an international conference in Montreal, Canada, sponsored by the United Nations Environment Program, where 24 countries signed a milestone accord to cut the production and use of ozone-destroying chemicals by 50% around the year 1999. The chairman of the conference declared, "There has never been an agreement like this on a global scale". The administrator of the United States Environmental Protection Agency said, "The signing of the accord shows an unprecedented degree of cooperation among nations of the world in balancing economic development and environmental protection". Fig.4.25 shows the average global depletion for six scenarios of levels of ozone depleting gases.

4.3.12 Effects of ozone depletion

The ozone layer reduces the amount of UV-B reaching the earth's surface; and thinning of this layer increases the amount of UV-B reaching the earth. Increases in UV-B can cause adverse effects on human health, terrestrial and aquatic plants, and a degradation of materials.

The biological response to increased ultraviolet radiation is often represented by an action spectrum. The relative ability of various wavelengths to cause biological harm can be understood from the action spectrum of the DNA molecule (Fig. 4.26). DNA is highly sensitive to shorter wavelengths of UV, and this part of sunlight is effectively blocked by ozone. For longer wavelengths in the UV-B portion of the spectrum, the ozone shield is less effective, but DNA is correspondingly less sensitive. However, the intermediate range of wavelengths, (300-315 nm) which are blocked less by ozone cause greatest damage to DNA.

UV-induced changes in DNA, are found to be responsible for skin cancer which may be either the more common non-melanoma (basal cell or squamous cell carcinomas) type, or the much more life-threatening malignant melanoma. Non-melanoma skin cancer is more prevalent among people with fair complexion, is most likely to occur on areas of the skin habitually receiving greatest exposure to the sun, and is associated with areas of the world having the greatest amount of sunshine. Though it is rarely fatal, it often causes disfigurement. In the United States, there are about 400,000 new cases of non-melanoma skin cancer each year, with a fatality rate of about 1%.

It was observed that at higher latitudes there is less surface level UV-B, and non-melanoma skin cancer rates decrease by roughly a factor of 2 for each 10° increase in latitude. Based on these observations, it was predicted that for every 1%decrease in ozone, the added UV-B exposure would cause the incidence rate of non-melanoma skin cancer to increase by approximately 2-3% [15]. By simply moving 160 kilometres (100 miles) closer to the equator, UV-B exposure is expected to increase by an amount comparable to what would be expected by a 5% thinning of the ozone layer.

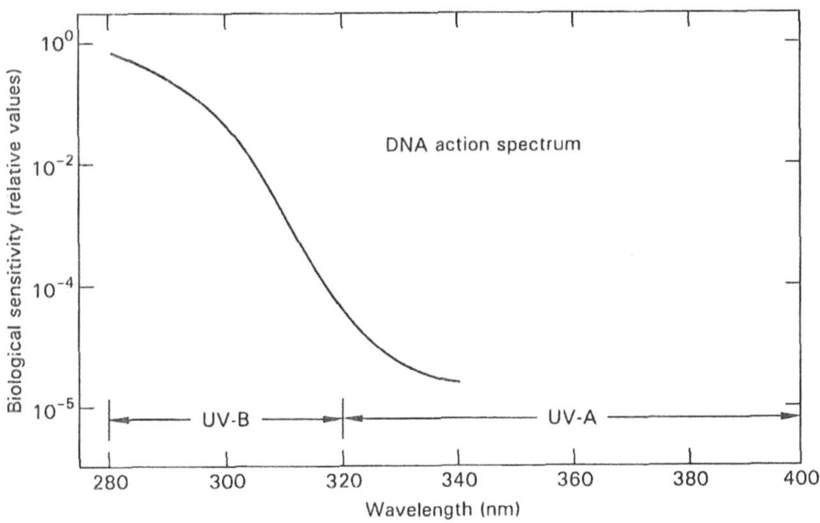

Fig. 4.26 *Action spectrum for damage to the DNA molecule (The vertical scale is a relative measure of biological sensitivity to radiation at each wavelength)*

The relation between malignant melanoma and sunlight is more well-established from facts than the link between non-melanoma skin cancer and sunlight. Malignant melanoma death rates among whites in the United States have been increasing by about 2-3% annually for the past two decades in spite of improved survival rates among those who receive treatment. Ocular damage (cataracts and retinal degeneration) and immune system suppression are some of the other health problems linked with exposure to UV light. It is also observed that crops and terrestrial ecosystems would be adversely affected by increasing UV exposure. UV-B radiation deleteriously affected crop yield and quality [15]. Aquatic ecosystems would be affected as well. Various experiments, for example, have shown that UV-B radiation damages fish, shrimps, and crab larvae, as well as copepods, and plants essential to the marine food web [15].

Scientists have estimated that if the release of CFCs into the atmosphere continues at the present rate, by the year 2025 there will be 1.4 million additional cases of skin cancer worldwide. There will also be an increased frequency of cataract, a clouding of the lens of the eye that results in blurred vision. Exposure of laboratory animals to increased UV radiation reduces their ability to fight bacterial infections. It is probable that humans suffer similar effects. Scientists estimate that the increased UV levels now experienced in the United States will reduce cotton production by 3% and will decrease corn and wheat yields as well. With the world's highest rate of skin cancer, in Australia, television stations now broadcast daily UV levels and warnings for fair-skinned Australians to stay inside during bad spells and to protect themselves from the sun's rays with hats, clothing, and sunscreens when they go out during the daytime. Such protection is required by law for school children.

Chile, the only populous country located under the Antarctic ozone hole, has experienced a fourfold increase in malignant melanoma since 1980. Levels of skin cancer and cataracts are also increasing rapidly in the United States, presumably from increased exposure to the sun.

Table 4.10 shows the medical conditions known to be induced by enhanced UV-B and their likelihood of increase with further stratospheric O_3 depletion.

Table 4.10 *Medical conditions known to be induced by enhanced UV-B and their likelihood of increase with further stratospheric* O_3 *depletion*

Area affected	Condition or agent	Individual types	Radiation amplification factor*
Skin	Erythema sunbum		1.7
	Non-melanomatous cancer	Basal cell carcinoma, Squamous cell carcinoma, Lip cancer, Salivary gland cancer, Superficial spreading, malignant lentigo, (Hutchinson's freckles)	 1.4 1.6 1.2
	Melanoma (cutaneous, malignant)	Nodular Unclassified	
	Xeroderma pigmentosa Skin ageing	(condition worsended) Elastosis	

Table 4.10 ***Contd...***

Area affected	Condition or agent	Individual types	Radiation amplification factor*
Eye	Corneal damage	Acute photokeratitis (snow-blindness)	1.1 0.7
	Lens Abnormalities	Nuclear cataracts Posterior subcapsular Cataracts Cortical cataracts Presbyopia (reading glasses required earlier)	
	Other effects	Intraocular melanoma	
Infections	Viruses	Measles Chicken pox Herpes simplex HIV-I activation	0.8
	Protozoa	Leishmaniasis Malaria	
	Bacteria	Tuberculosis Leprosy	
	Fungi	Candidiasis	

***Source:** *1991 UNEP report.*

4.3.13 Effects On Air Quality

Reduction of stratospheric ozone and the concomitant increases of UV-B radiation penetrating to the lower atmosphere result in higher photodissociation rates of key trace gases that control the chemical reactivity of the troposphere. This can increase both production and destruction of ozone (O_3) and related oxidants such as hydrogen peroxide (H_2O_2), which are known to have adverse effects on human health, terrestrial plants, and outdoor materials. Changes in the atmospheric concentrations of the hydroxyl radical (OH) may change the atmospheric lifetimes of climatically important gases such as methane (CH_4) and the CFC substitutes.

Trends in the photodissociation rate coefficient of tropospheric O_3, about + 0.36 ± 0.04% per year in the Northern Hemisphere and + 0.40 ± 0.05% per year in the Southern Hemisphere, have been estimated from satellite measurements of the ozone column between 1979 and 1992. The corresponding model calculated changes in tropospheric chemical composition are nonlinear and sensitive to the prevailing levels of nitrogen oxides (NO_x). In polluted regions (high NO_x) tropospheric O_3 is expected to increase, reaching potentially harmful concentrations sometimes more important than damaging effects of UV-B. These changes can have important implications for plant competitive balance, herbivory, plant pathogens and biogeochemical cycles. These ecosystem level effects can be anticipated, but not easily predicted or evaluated. Research at the ecosystem level for solar UV-B is barely beginning. Other factors including those involved in climate change such as increasing CO_2 also interact with UV-B. Such reactions are not easily predicted, but are of obvious importance in both agriculture and in nonagricultural ecosystems.

4.3.14 Effects on aquatic ecosystems

More than 30%of the world's animal protein for human consumption comes from the sea, and in many countries, particularly the developing countries, this percentage is significantly higher. As a result, it is important to know how increased levels of exposure to solar UV-B, radiation might affect the productivity of aquatic systems.

In addition, the oceans play a key role with respect to global warming. Marine phytoplankton are a major sink for atmospheric carbon dioxide, and they have a decisive role in the development of future trends of CO_2 concentrations in the atmosphere.

Phytoplankton form the foundation of aquatic food webs. Marine phytoplankton are not uniformly distributed throughout the oceans of the world. The highest concentrations are found at high latitudes while, with the exception of upwelling areas, the continental shelves, the tropics and subtropics have 10 to 100 times lower concentrations. In addition to nutrients, temperature, salinity and light availability, the high levels of exposure to solar UV-B radiation that normally occur within the tropics and subtropics may play a role in phytoplankton distributions.

Phytoplankton productivity is limited to the euphotic zone, the upper layer of the water column in which there is sufficient sunlight to support net productivity. The position of the organisms in the euphotic zone is influenced by the action of wind and waves. In addition, many phytoplankton are capable of active movements that enhance their productivity and. therefore, their survival. Exposure to solar UV-B radiation has been shown to affect both orientation mechanisms and motility in phytoplankton, resulting in reduced survival rates for these organisms.

Solar UV-B radiation has been found to cause damage to early developmental stages of fish, shrimp, crab, amphibians and other animals. The most severe effects are decreased reproductive capacity and impaired larval development. Even at current levels, solar UV-B radiation is a limiting factor, and small increase in UV-B exposure could result in significant reduction in the size of the population of consumer organisms.

4.3.15 Effects on biogeochemical cycles

Increases in solar UV radiation could affect terrestrial and aquatic biogeochemical cycles thus altering both sources and sinks of greenhouse and chemically important trace gases, e.g., carbon dioxide (CO_2), carbon monoxide (CO), carbonyl sulfide (COS) and possibly other gases, including ozone. These potential changes would contribute to biosphere-atmosphere feedbacks that attenuate or reinforce the atmospheric buildup of these gases.

In terrestrial ecosystems increased UV-B could modify both the production and decomposition of plant matter with concomitant changes in the uptake and release of atmospherically important trace gases. Decomposition processes can be accelerated when UV-B photodegrades surface litter or, retarded when the dominant effect is on the chemical composition of living tissues resulting in reduced biodegradability of buried litter. Primary production can be reduced by enhanced UV-B but the effect is variable between species and even cultivars of some crops. Likewise, photoproduction of CO from plant matter is species dependent and occurs more efficiently from dead than living matter.

4.3.16 Effects on materials

Synthetic polymers, naturally occurring biopolymers, as well as some other materials of commercial interest are adversely affected by solar UV radiation. Application of these materials, particularly plastics, in situations which demand routine exposure to sunlight is only possible through the use of light-stabilizers and/or surface treatments to protect them from sunlight. Any increase in solar UV-B content due to partial ozone depletion will therefore accelerate the photogradation rates of these materials, limiting their service lifetimes outdoors.

4.3.17 Solutions to protect the ozone layer

If the model and measurements of ozone depletion and its possible effects are correct, it is highly essential to stop producing any ozone depleting chemicals now. Even after stopping the production of ozone depletion chemicals, the models indicate that we still will have to wait 50-100 years for the ozone layer to return to 1985 levels, and another 100-200 years for full recovery.

Substitutes are already available for most uses of CFCs, and others are being developed (Table 4.11). CFCs in existing air conditioners, refrigerators, and other products must be recovered and in some cases re-used until the substitutes are phased in. This will be expensive, but if ozone depletion should become more severe because of our activities, the ecological, health, and financial costs will be much higher.

Table 4.11 *CFC substitutes*

Types	Pros	Cons
HCFCs	Breakdown faster (2-20 years). Pose about 90% less danger to ozone layer Can be used in aerosol sprays, refrigeration, air conditioning, foam. and cleaning agents.	Are greenhouse gases. Will still deplete ozone, especially if used in large quantities. Health effects largely unknown. HCFC-123 causes benign tumours in the pancreas and testes of male rats and may be banned for use in aerosol sprays, foam, and cleaning agents. May lower energy efficiency of appliances.
HFCs	Breakdown faster (2-20) years. Do not contain ozone-destroying chlorine can be used in aerosol sprays, refrigeration. air conditioning, and foam.	Are greenhouse gases. Safety questions about flammability and toxicity still unresolved. May lower energy efficiency of appliances. Production of HFC-134a, a refrigerant substitute. vields an equal amount of methyl chloroform, a serious ozone depleter.
Hydro-carbons	Cheap and readily available. can be used in aerosol sprays, refrigeration, foam and cleaning agents	Can be flammable and poisonous. Some increase ground-level pollution.
Ammonia	Simple alternative for refrigerators; widely used before CFCs.	Toxic if inhaled. Must be handled carefully.
Water and steam	Effective for some cleaning operations and for sterilizing medical instruments.	Creates polluted water that must be treated. Wastes water unless the used water is cleaned up and reused.
Terpenes	Effective for cleaning electronic parts	None
Helium	Effective coolant for refrigerants, freezers and air conditioners.	This rare gas may become scarce if use is widespread, but very little coolant is needed per appliance.

UNEP took up the issue of ozone depletion in 1976 and adopted a simple but effective approach. It first concentrated on assessment of the problem by convening a meeting of experts on the ozone layer in 1977. On the recommendations of this meeting, UNEP set up a Coordinating Committee of the Ozone Layer in cooperation with the World Meteorological Organisation. CCOL consisted of the leading experts of the world on the ozone layer. Over the next ten years, this committee provided regular assessments of the state of the ozone layer and coordinated further research. These reports attracted international interest among people and governments and UNEP provided the framework for harnessing this interest to promote international action. It marshalled the information, options and legal expertise needed for decision-making. This process led to the Vienna Convention for the Protection of the Ozone Layer in 1985 and the Montreal Protocol on Substances that deplete the Ozone Layer in 1987.

The Montreal Protocol was signed by 40 countries. An agreement was reached between the manufacturers and consumers on the issue of CFCs. The CFCs were classified under two groups-I and II. The consumption and production of group-II CFCs were to be stopped at once. For group-I CFCs, it was agreed upon to implement:

(a) a freeze on consumption at 1986 levels by 1990
(b) a cut to 80% by 1994
(c) a further cut to 50% by 1999
(d) production could be raised by 10% by 1990 but production had to be reduced to 90% by 1994 and to 65% by 1999.

Although reluctant initially, India and China, the two potentially important CFC consumers finally agreed to sign the Montreal Treaty. In the Indian context, the treaty seeks to phase out the production of CFCs in the developing countries by 2010. The cost of switching over to CFC substitutes is estimated to be between Rs. 35,000 and Rs. 60,000 million depending on the lagtime for the changeover.

4.4 Acid Rain

4.4.1 What is acid rain?

Acid rain refers to atmospheric precipitation that has dissolved in it strong mineral acids like sulphuric acid, nitric acid, and in some locations even hydrochloric and hydrofluoric acids bringing down the pH to less than 5.6. Acid deposition is the accumulation of potential acid-forming particles on surfaces.

Anthropogenic sources like power plants, industries and vehicles and agriculture, and natural sources like vegetation, volcanoes, geysers and lightning produce the basic ingredients of acid rain, namely, sulphur dioxide (SO_2) and nitrogen oxides (NO_x) (Fig. 4.27). Amongst natural sources of acid rain, other than volcanoes and geysers which contribute SO_x and NO_x, the most significant is formic acid. Biomass burning due to forest fires causes emissions of formic acid (HCOOH) and formaldehyde gets oxidized to formic acid in the atmosphere. Oxidizing agents such as ozone, hydroxyl ions or hydrogen peroxide, along with water, are necessary to convert the sulphur dioxide or nitrogen oxides to sulphuric acid and nitric acid. Various reactive hydrocarbons (HCs) encourage the production of oxidizing agents. In time, these oxides combine with water vapour in the atmosphere and return to the earth's surface in the form of acid rain. The acid forming agents are classified as wet or dry. Wet reactions occur in the atmosphere and come down to the earth as some form of precipitation: acid rain.

4.4.2 Dry or gas phase conversion

This predominates in the vicinity of the emission sources and involves SO_2, NO_x, sulphates and nitrates returning to the earth in the form of gases and particles, usually in the vicinity of the sources of pollution. The gases and particles are dry deposited, and can be converted to acid through adsorption (where SO_2 and NO_2 enter the stomata-the leaf pores that control the passage of gases into and out of the plant-and are oxidised to acid inside the leaf). Acids can also be formed directly on the surface of vegetation or on the ground when dry-deposited material is dissolved in rain or snow.

Millions of metric tons of sulphuric acid and nitric acid fall in rain on the earth each year, principally in the United States, Canada and Europe. The concentrations of these acids are too weak to cause burns, but they do produce other severe effects. Depending on a variety of biological, chemical, and meteorological factors, soils, forests, lakes, rivers, animals, plants and buildings are susceptible in varying degrees to the effects of these pollutants, acting alone or in conjunction with other factors. Acid rain causes serious health problems and brings death to lakes and rivers, along with destruction to structures and the ecosystem as a whole.

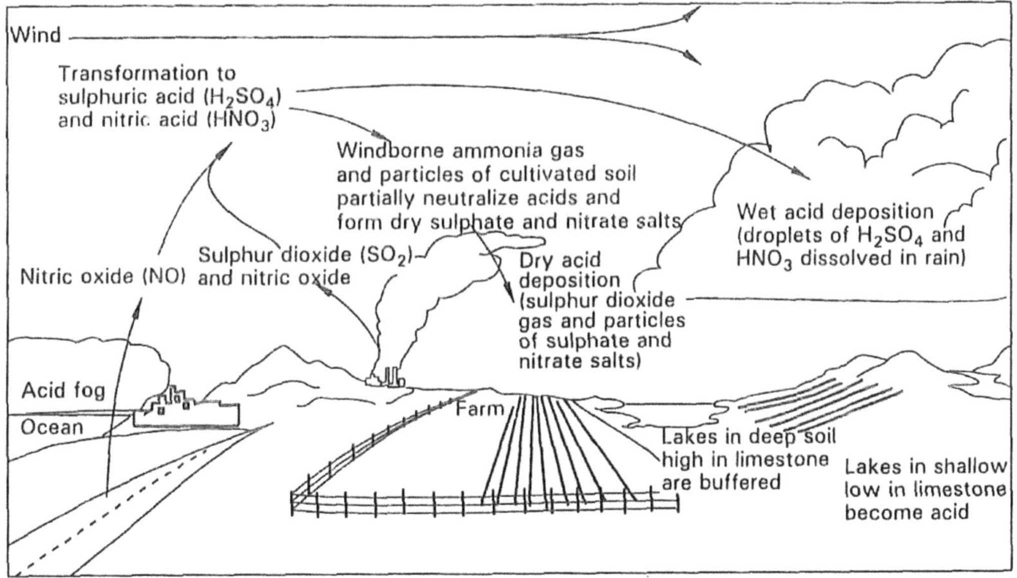

Fig. 4.27 *Acid rain*

4.4.3 Chemicals involved in acid rain

4.4.3.1 Bicarbonate Acidity

If one excludes the possible contributions of SO_2 and nitrogen oxides from volcanoes, swamps, lightning strikes, etc., then the major natural gas in the atmosphere contributing to acidity is carbon dioxide.

1. $CO_2 + H_2O \leftrightarrow H_2CO_3$
2. $H_2CO_3 \leftrightarrow HCO_3^- + H^+$
3. $HCO_3^- \leftrightarrow CO_3^{2-} + H^+$

Carbon dioxide reacts with H_2O to form both bicarbonate and carbonate, as seen in reactions (1) and (2). But as the pK_a of reaction (1) forming carbonate is as high as 10.3 (i.e. the concentrations of carbonate and bicarbonate match), then reaction (2) forming bicarbonate from CO_2 has by far the greater influence upon the acidity of the natural atmosphere. The amount of gas in solution is a function of the partial pressure which in turn is proportional to the mole fraction (i.e. the number of moles of a component divided by the total number of moles in the mixture). In the case of CO_2, this means that if the partial pressure of atmospheric CO_2, is 0.00035 atmospheres, then Henry's Law constant *(K_H)* is as follows:

$$K_H = \frac{[H_2CO_3]}{[CO_2gas]} = 3.79 \times 10^{-2}\, mol\, 1^{-1}\, atm^{-1}$$

and the equilibrium constant of reaction (2) is given by

$$K_{4.9} = \frac{[H^+][HCO_3]}{[H_2CO_3]} = 4.5 \times 10^{-7}\, mol\, 1^{-1}$$

By combining and rearranging these two expressions, we get the following equation

$$[HCO_{3-}] = \frac{[CO_2\, gas] \times K_H \times K_{4.5}}{[H^+]}$$

Thus, if the concentration of bicarbonate in pure water is equal to the H^+ concentration, then by substitution the following is obtained:

$$[H^+]^2 = [CO_2\, gas] \times K_H \times K_{4.9}$$
$$= 0.00035 \times 3.79 \times 10^{-2} \times 4.5 \times 10^{-7}$$
$$= 5.97 \times 10^{-12}\, mol^2\, 1^{-2}$$

Therefore

$$[H^+] = 2.44 \times 10^{-6}\, mol\, 1^{-1}$$

and hence

$$pH = -log[H^+] = 5.61$$

This means that however clean the atmosphere and disregarding any wind-borne alkaline materials, natural cloud water is always slightly acidic.

4.4.3.2 Sulphur dioxide

Fossil fuels were formed when dead animals and plants decomposed and were broken down into liquids (oil), gases (natural gas) and solids (coal and peat). Hence they contain sulphur absorbed by those animals and plants from their environment. The sulphur content varies by location; crude oil can have a sulphur content of 0.1-3%, while coal can have a sulphur content of 0.5-10% by weight. 'Hard' coals such as anthracite are normally low in sulphur, while 'soft' coals (bituminous, sub-bituminous and brown coal, or lignite) can be much higher. When these fossil fuels are burned, or sulphide ores are smelted (roasted) to extract metal, the long-stored sulphur is released as sulphur dioxide, an invisible gas with pungent odour.

Much of the SO_2 comes from power generation using coal energy. In the United States electric power plants are responsible for nearly 70% of SO_2.

4.4.3.3 Oxidation of sulphur dioxide

In the atmosphere sulphur dioxide gets oxidized to sulphur trioxide which reacts with water or alkalies to give sulphuric acid or sulphates.

The acid or sulphates, in particular ammonium sulphate and bisulphate, occur as aerosols. The process is illustrated by the following equations:

1. $SO_2 + H_2O \rightarrow H_2SO_3$
2. $H_2SO_3 \rightarrow H^+ + HSO_3^-$
3. $HSO_3^- \rightarrow H^+ + SO_3^{2-}$

If bisulphite is oxidized to bisulphate within water droplets then the whole relationship between the uptake of SO_2 and the ionization of sulphurous acid (reactions (1) and (2) is disturbed and the balance swings in favour of more SO_2 equilibrating with cloud water and resulting in more acidity. Moreover, there are other oxidations within cloud droplets which are important acid forming events in addition to the gas-phase reactions involving OH radicals that produce acidity.

Many oxidants in the atmosphere, such as O_3, hydrogen peroxide (H_2O_2), peroxyacetyl nitrates or other free radicals (e.g., $CH_3O_2.H$ and $CH_3COO_2.H$) do not react with gaseous SO_2 to any great extent. However, when they are dissolved in cloud water, they readily oxidize dissolved bisulphite ions which then forms sulphate, resulting in more acidity.

The oxidation step proceeds in one or more of the three ways, namely, catalytic oxidation, photochemical oxidation and oxidation by radicals.

4.4.3.4 Catalytic oxidation

The reaction

$$2SO_2 + 2H_2O + O_2 \rightarrow 2H_2SO_4, \; \Delta H = 600 \text{ KJ}$$

is slow in clean air but is catalysed by aerosols containing metal ions like Mn^{2+} and Fe^{3+}. Surfaces such as buildings may also act as catalytic centres. The reaction is swifter in places where the relative humidity is more than 32%. The solubility of SO_2 in water is a function of pH and can be described by the equilibria:

$$SO_2\ (g) + H_2O \leftrightarrow SO_2\ (aq)$$

$$SO_2\ (aq) + H_2O \leftrightarrow H_2SO_3\ (aq)$$

$$H_2SO_3 + H_2O \leftrightarrow H_3O^+ + HSO_3^-$$

$$HSO_3^- + H_2O \leftrightarrow H_3O^+ + SO_3^{2-}$$

$$2HSO_3 + H_2O \leftrightarrow S_2O_5^{2-} + H_2O$$

The sulphur species in water is either covalent or ionic, the latter being more soluble due to the high dielectric constant of water.

4.4.3.5 Photochemical oxidation

SO_2 absorbs solar radiation producing excited states, which then react by a variety of routes to give SO_3.

4.4.3.6 Oxidation by free radicals

Hydrocarbons and nitrogen dioxide, both increase the rate of oxidation of SO_2 to free radicals. These are mostly components of automobile emissions or secondary products.

$$SO_2\bullet\ + O_3 \rightarrow SO_3 + O_2$$

This reaction is slow in the gaseous phase and rapid in solution:

$$SO_2 + NO_2 \rightarrow SO_3 + NO$$

Other feasible reactions are:

$$SO_2 + HO_2\bullet \rightarrow SO_3 + HO\bullet$$

$$SO_2 + RO_2\bullet \rightarrow SO_3 + RO \quad \text{(where R indicates } CH_3C = 0\text{)}$$

$$HO\bullet + SO_2 \rightarrow HOSO_2\bullet$$

$$HOSO_2\bullet\ + O_2 \rightarrow HOSO_2O_2$$

$$HOSO_2O_2 + NO \rightarrow HOSO_3 + NO_2 + H_2SO_4\bullet$$

4.4.3.7 Nitrogen oxides

Fossil fuel combustion is also the main source of non-natural nitric oxide and nitrogen dioxide, known collectively as nitrogen oxides. Combustion oxidizes both the nitrogen in fuel and some of the nitrogen naturally present in the air (which is 80% nitrogen). The quality and composition of the nitrogen oxides depends on how combustion is performed and at what temperatures. Burning will normally oxidize 5-40% of the nitrogen in coal, 40-50% of that in heavy oil, and nearly 100% of that in light oil and gas. The higher the temperature, the greater the amount of nitrogen oxides formed. Once again, the proportion of emissions from different sources varies regionally. Road vehicles account for 50% of emissions in the European Union and 38% of emissions in the United States and Canada, while power stations account for 20% of emissions in the EU, 37% of emissions in the United States, and 14% of emissions in Canada. NO_x is a critical factor in many Asian and Latin American cities, where a combination of heavy traffic and local wind patterns can often recirculate nitrogen oxides and other pollutants, creating severe photochemical smog. The still and stable air over many Southeast Asian cities creates ideal conditions for the build-up of pollutants from road traffic and industry, the density of which is increasing rapidly as these countries industrialize.

Nitrogen oxides and carbon particles produced by automobiles and other motor vehicles are so plentiful in some large cities that the air is coloured by them. Nitrogen oxides also enter the air from the breakdown of agricultural fertilizers. These eventually form nitric acid in moist air. Like sulphuric acid, nitric acid corrodes many metals, and strong concentrations of nitric acid can cause burns on skin.

4.4.3.8 Ammonia

The third primary pollutant causing acid rain is ammonia to which more attention has been paid only since the late 1980s. A by-product of animal manure and fertilizer use, ammonia is mainly a problem in regions using intensive agricultural techniques, notably Western Europe. It contributes to acidification by causing cloud water to become more alkaline, thereby increasing the rate of conversion of SO_2 and NO_x to acids. It can also be oxidized by microbial activity in soil to form nitric acid.

The acids in the atmosphere react with ammonia and other cations forming sulphates and nitrates which fall to the surface of the earth in the form of particulates. It is estimated that about 40% of the particles present in the atmosphere over the north polar regions are in the form of sulphates.

Calculating emissions is difficult, since they depend on a variety of different factors, such as the nitrogen content of animal feed and the conditions under which wastes are spread. Nevertheless, UN figures suggest that Europe produces about 800,000 tons of NH_3 emissions every year. Hence it is now being taken more seriously by western European policy makers as an element in the broader problem of acidification. This contrasts with the situation in the United States, where ammonia has so far drawn little attention.

4.4.3.9 Volatile organic compounds

This is a collective term for a wide range of compounds containing carbon, but is usually taken to include hydrocarbons, oxygenates such as alcohol, and methane, although there is some debate about how to define them. The UN defines VOCs as "all organic compounds of anthropogenic nature other than methane that are capable of producing photochemical oxidants by reaction with nitrogen oxides in sunlight", but a definition as broad as this makes it difficult to measure emissions and to identify trends in the production of VOCs. They come mainly from road vehicles, the chemical and petroleum industries, paints and glues and the use of solvents in industrial processes.

4.4.3.10 Other chemicals

In some situations hydrochloric acid vapour and mists of other acids such as phosphoric acid may also be present responsible for acid rain. These gases dissolve in falling rain making it more acidic than natural rain. Acid fog is formed when chemicals pollutants are dissolved in very moist air. This causes changes in the pH of the air in the same way that acid rain changes the pH of the soil or the water in lakes and rivers.

4.4.4 Acid rain-global scenario

As the fossil fuel consumption is highest in the industrial centres of Europe and North America, the emission of air pollutants is also greatest in these areas, and acid rain has, at least until recently, been essentially a Northern hemisphere problem.

In industrialized countries, the major sources of SO_2 (Table 4.12) are coal- or oil fired power stations, industries that use fossil fuels, metal smelters, oil refineries and heating installations, although the relative contributions of different sources vary from place to place.

Table 4.12 *The ten biggest point sources of Sulphur in Europe*

Name	Facility	Country	Annual emissions
Maritsa east	Power Station	Bulgaria	3500000
Afsin-Elbistan	Power Station	Turkey	288000
Puenies	Power Station	Spain	271000
Montsegorsk	Smelter	Russia	212000
Nikel	Smelter	Russia	211000
Teruel	Power Station	Spain	183000
Belchatow	Power Station	Poland	168000
Janschewalde	Power Station	Germany	157000
Boxberg	Power Station	Germany	149000
Pruneroy	Power Station	Czech Republic	137000
Total			2126000

Source: *Mark Barrett and Rodri Protheroe, 1994. Sulphur emissions from large point sources in Europe.* Colchester: Pollen Consultancy

While power stations account for two-thirds of SO_2 emissions in the United States and 58% of emissions in the European Union, they account for only 22% of emissions in Canada. where smelters play a much bigger role. A recent study in Europe revealed that the worst sulphur pollution comes from a relatively small number of point sources, mainly power stations (Table 4.12). The 100 biggest emitters (one third of them in Western Europe) between them accounted for 42% of European SO_2 emissions, with the ten biggest alone accounting for more than two million tons of SO_2 emissions each year.

However, the balance is changing as industrialized states reduce their consumption and as fossil fuel consumption increases in newly industrializing countries (NICs) such as China, India, Brazil, and South Africa. China, for example, relies on fossil fuels for nearly 98% of its energy needs, and saw its coal consumption grow by 178% between 1973 and 1993. In the same period, coal consumption grew by 101% in South Africa, by 193% in India, by 322% in Brazil and by nearly 650% in Taiwan.

In Asia the situation is about to reach alarming proportions (Fig. 4.28). Given the fact that the two most populous countries of the world, China and India, have substantial reserves of coal, the continent remains a region where coal use is expected to grow substantially in the coming decades. Fortunately, for India, the sulphur content of coal found in the country is consistently low. But this is not the case with China. The Rains-

Asia model, indicates the possibility of a 315% increase in SO_2 emissions over present levels in India by the year 2020.

Sulphur dioxide emissions are expected to increase almost three-fold in China and more than four-fold in India between 1990 and 2020

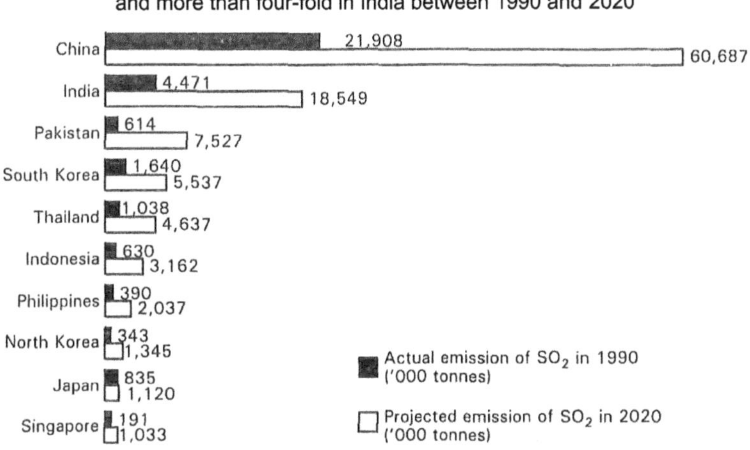

Fig. 4.28 *Acid rain in Asian countries*

Global emissions of acid causing air pollutants like CO, NO_x, SO_2 in the period 1960-2010 are shown in Figure 4.29.

CO and NOx global emissions

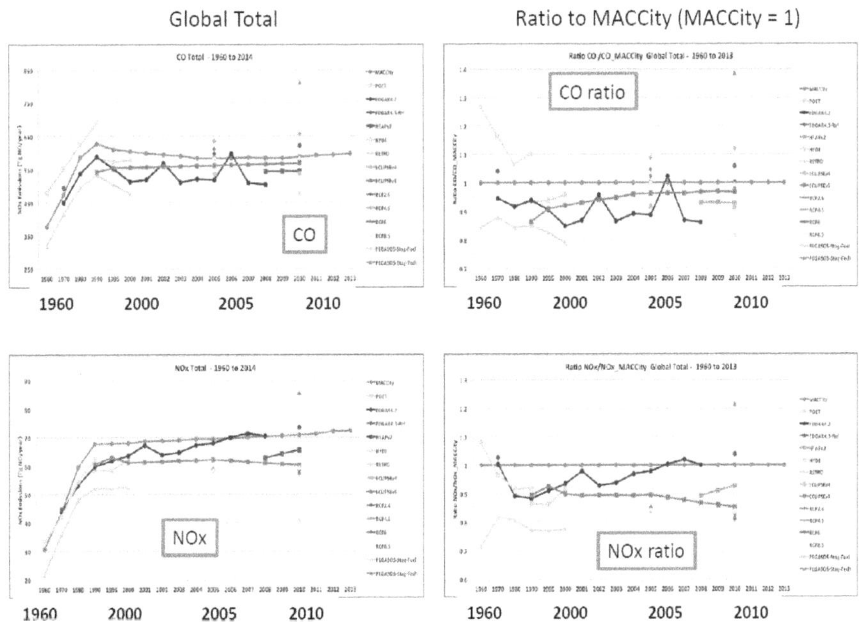

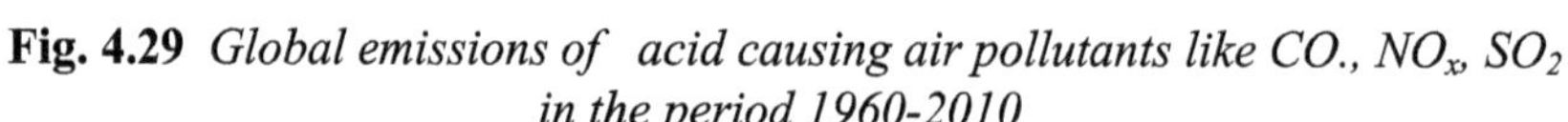

Fig. 4.29 *Global emissions of acid causing air pollutants like CO., NO_x, SO_2 in the period 1960-2010*

4.4.4.1 Acid rain pollution in India

India enjoys the dubious distinction of releasing the maximum pollutants in the atmosphere after China. Total sulphur emissions are expected to rise from 4,400 kilotons (kt) in 1990 to 6,500 kt in 2000, 10,900 kt in 2010 and 18,500 in 2020. The areas at risk in India exposed to acid rain in 1995 and the predicted increase in area by 2010 are given in Fig. 4.30.

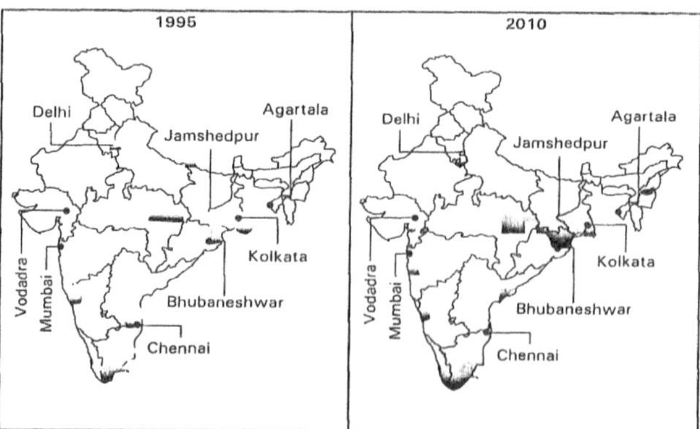

Fig. 4.30 *Areas at risk in India, exposed to acid rain (incidence of acid rain in 1995 and the predicted increase of area affected in India by 2010)*

In India, the first report of acid rain came from Mumbai in 1974. Instances of acid rain are being reported from metropolitan cities. A typical study conducted at the heavily industrialised Udyogamandal area in Kerala State revealed that it often receives acid rain; the pH going as low as 4.3. A study conducted at Trombay, Mumbai revealed high incidence of acidic and sulphate components in air. In India annual SO_2 emission has almost doubled in the last decade due to increased fossil fuel consumption and lowering of soil pH is reported from North-eastern India, coastal Karnataka and Kerala, and parts of Orissa, West Bengal and Bihar. Levels of noxious fumes and dust are more than twice the permissible levels in Agra; they are corroding the marble of the well-known monument Taj Mahal, causing it to lose its grace and glitter. This phenomena has been described as 'marble cancer'.

4.4.4.1.1 *Safe Areas*

There are two types of 'safe areas' where acid rain is not a problem. One comprises areas that simply do not receive acid rain or the gaseous oxides of sulphur and nitrogen due to their location. Almost all of the Southern hemisphere is thus protected, as also most of the tropics and parts of the Northern hemisphere. The other type is areas that receive acid precipitation but can withstand it. Many areas provide a natural resistance (buffering) to acidification with counter action offered by alkaline soils or limestone beds which neutralise acid inputs. Other forms of buffering are offered in the mid-western United States, where alkaline dust blown from the west neutralizes acid rain before it reaches the ground. Several regions in India also have good buffering capacity but if such regions are persistently showered with acid rain the soil may gradually lose its neutralising ability.

4.4.5 Impact of acid rain

4.4.5.1 Damage to ecosystems-forests

The effects of acid rain on ecosystems are often difficult to quantify. Intense sulphur dioxide pollution around smelters is known to cause the death of many kinds of trees and other vegetation. But this is an extreme case and may not be directly comparable to less intense acid rain. However, in many parts of the world, acid rain is suspected of causing the death of many forests and reducing the vigour and rate of growth of others. In central Europe, many forests have declined significantly, resulting in the death of about 6 million hectares of trees. Northeastern North America has also seen significant tree death and reduction in vigour, particularly at higher elevations. Some areas have had 50% mortality of red spruce trees.

A clear link between the decline of the forests and acid rain is difficult to establish, but several hypotheses have been formulated. Sulphur dioxide and ozone are known air pollutants and cause direct damage to plants. Acid rain affects trees and undergrowth in forests in several ways, causing reduced growth or abnormal growth. The typical growth-decreasing symptoms are:

(a) Discoloration and loss of foliar biomass (yellowing and browning of needles and leaves).

(b) Loss of feeder-root biomass, especially in conifers.
(c) Decreased annual increment (width of growth rings).
(d) Premature senescence (aging) of older needles in conifers.
(e) Increased susceptibility to secondary root and foliar pathogens.
(f) Death of herbaceous vegetation beneath affected trees (all within the drip line from the canopy).
(g) Prodigious production of lichens on affected trees (all within the drip line from the canopy).
(h) Death of affected trees.

The abnormal growth symptoms are:

(a) The active shedding of needles and leaves while still green, with no indication of disease.
(b) Shedding of whole green shoots, especially in spruce.
(c) Formation of stork's nest crown in young white fir.
(d) Altered branching habit and greater than normal production of adventitious (out-of-place) shoots.
(e) Excessive seed and cone production year after year.
(f) Diseased trees show growth at top, not near bottom with very brittle branches.

In addition water stress symptoms may also be caused:

(a) Altered water balance.
(b) Increased incidence of wet wood disease.

4.4.5.2 Aluminium toxicity

This theory argues that the natural acidification of forest soils is accelerated by acid deposition, which mobilizes heavy metals in the soil, notably aluminium. The fine roots of trees are attacked, preventing their roots from taking up enough nutrients or water to survive. Once the roots are damaged the rest of the tree suffers. Support for this theory was offered by the study in the 1980s of sugar maple decline in Ontario, Canada; soils in the region are acidic, and declining maple trees were found to have fine-root systems with higher concentrations than healthy trees.

Evidence that forest decline follows dry summers (when soils become more acidic and nutrients more scarce) has also lent credence to this theory,

Natural forms of aluminium in rocks and soils are highly variable, existing as silicates, oxides and a wide variety of other forms. Normally, cations like calcium and magnesium are rapidly exchanged from rock and soil by H^+ions in a reversible and rapid reaction which serves to buffer soil pH (reaction (3)).

1. $CaCO_3 + H_2O + CO_2 \rightarrow Ca^{2+} + 2HCO_3^-$
2. $CaA1_2Si_2O_8 + 2H_2CO_3 + H_2O \rightarrow Ca^{2+} + 2HCO_3^- + Al_2Si_2O_5(OH)_4$
3. $[Soil] = Ca + 2H^+ \rightarrow H\text{–}[Soil]\text{–}H + Ca^{2+}$

When, however, soils continue to receive large inputs of acidity, this exchange process is limited by the availability of calcium and magnesium in the soil particles. Once these are exhausted, then the release of Al^{3+} occurs in exchange for H^+ions. Consequently, acid rain-accelerated weathering releases aluminium hydroxide from the surfaces of soil and rock particles. Aluminium hydroxide has both basic and acidic properties in the presence of strong bases or acids and is called an ampholyte. Actually, the release of aluminium hydroxide from soil surfaces is more accurately described as the dissolution followed by a proton transfer. Further transformations and dissolutions then follow. Consequently, acidity does not disappear when aluminium is mobilized but it is transferred to a progressively weaker cationic acid.

$$Al^{3+} + 3\ OH^- \rightarrow Al(OH)_3 \rightarrow H^+ + H_2AlO_3^-$$

$$Al\ (OH)_3 + 3H_2O \rightarrow Al(OH)_3(H_2O)_3^0$$

$$Al\ (OH)_2(H_2O)_3^0 + H^+ \rightarrow Al(OH)_2(H_2O)_4^+$$

$$Al\ (OH)_2(H_2O)_4^+ + H^+ \rightarrow Al(OH)(H_2O)_5^{2+}$$

$$Al\ (OH)\ (H_2O)_5^{2+} + H^+ \rightarrow Al(H_2O)_6^{3+}$$

If excess acidity still remains then the mobilization of iron oxides may also occur as hydrated forms of Al^{3+} $Al(OH)^{2+}$ and $Al(OH)_2^+$ are lost to run-off waters draining into streams, rivers and lakes.

4.4.5.3 General stress

Though the concentrations of SO_2, NO_x, oxidants, hydrocarbons and heavy metals are too small individually to cause damage, when combined, however, they may decrease the production of carbohydrates in plant leaves. thereby decreasing the vitality of roots and leaves. This lowers root production, reducing the ability of trees to withstand attacks from fungi, bacteria, viruses and insects, and making them more susceptible to the stresses of drought, frost and wind.

4.4.5.4 Excess nutrients or nitrogen

This argues that forest ecosystems have been receiving excessive doses of 16 elements considered essential for plant growth. The increased deposition of nitrogen in particular has encouraged increased growth and excessive demand for nutrients. Nutrient shortages in turn lead to an increased susceptibility to frost and root-disease fungi, changes in root-shoot ratios, and possible nitrogen fixation.

4.4.5.5 The Ammonium hypothesis

Nitrogen in the form of ammonium or nitrate is an important plant nutrient, but most plants have adapted through evolution to exist with very low levels of nitrogen. However, acid deposition may build up nitrogen in soils and plants, leading to saturation. If tree growth is accelerated, trees may form large cells that are easily attacked by wind and drought.

4.4.5.6 Damage to Aquatic Ecosystems

Freshwater Aquatic ecosystem: The effect of acid rain on the chemistry of surface water is determined by the geochemistry, geomorphology and hydrodynamics of the watershed and the water body. These factors determine the capacity of water to neutralize acids and absorb or release metals.

Table 4.13 compares the characteristics of naturally acidic lakes and anthropogenically acidified natural lakes.

Table 4.13 *Characteristics of naturally acidic lakes and anthropogenically acidified natural lakes*

Naturally acidic lakes	Anthropogenically acidified lakes
Brown to yellow colour caused by humic substances (peat, tannings, etc.).,	Very clear water caused by reduced primary productivity (i.e., lack of phytoplankton), precipitation of organic matter by aluminium ,and dissolution of iron and manganese colloids. Inorganic acids (particularly sulphuric and nitric acids) predominate in these lakes.
Concentration of dissolved organic carbon is high while transparency is low.	Dissolved organic carbon concentrations are low (seldom exceeding a few milligrams per liter) whereas the transparency is high.
Low pH (below 5) but well buffered; these lakes generally have a greater acid neutralizing capacity (by organic and inorganic buffers and actions of microorganisms). This suggests that brown-water lakes would be less sensitive to the acidifying effects of precipitation.	Poorly buffered: once the bicarbonate alkalinity is exhausted, the pH decreases rapidly, giving little time for the species to adapt.
Abound with aquatic life Presence of humic, fulvic, and other organic acids keeps availability of toxic metals in check.	Some of the more sensitive taxa, such as blue-green algae, some bacteria, snails, mussels, crustaceans, mayflies, and fish are either reduced or eliminated. No such cushion available.

In natural water bodies, metals such as zinc, cadmium, nickel, manganese, chromium. copper and mercury form complexes with naturally occurring ligands such as fulvic acids and phosphates. In the complexed form the bioavailability, and consequently the toxicity, of these metals is greatly reduced. pH also affects the mobility of the metals, present in sediments, towards the overlying water through direct dissolution; the solubility of most metal compounds in water gets lower as the pH is increased. When a lake gets acidified the metal-fulvic acid complexes are destabilized releasing metal ions. Reduction in pH also increases mobility of the metal compounds from the sediments to the overlying water. The combined effect of these two phenomena is increase in the concentration of the metals in water.

Acidified lakes are most likely to be found in regions where bed rock and soil are low in naturally occurring acid-neutralizing materials. In regions where limestone is abundant, acid precipitation gets neutralized easily. Acidification causes adverse effects on all trophic levels of aquatic biota in surface water, from decomposers to fish and amphibians. Table 4.14 shows the effects of decreasing pH on aquatic organisms. As lakes become acidified the diversity of phytoplankton species declines due to stresses on acid-sensitive organisms. Molluscs, crustaceans, ephemeroptera and plecoptera appear to be highly sensitive to acidification. Invertebrates in acidic lakes may accumulate toxic trace elements that could then be passed on to vertebrates in the food web. Acidification of aquatic habitat also has some effect on amphibians (Fig. 4.31).

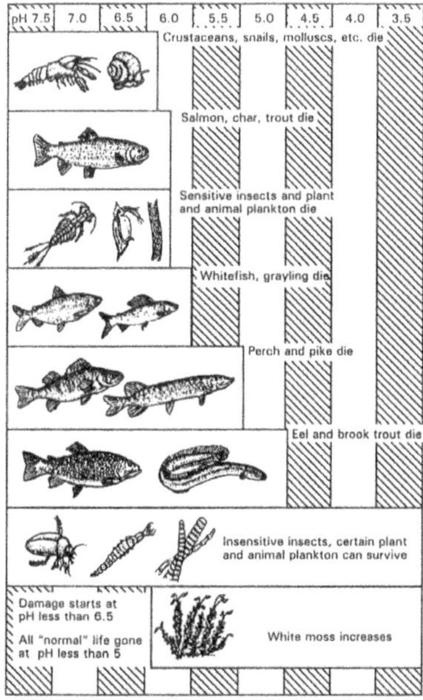

Fig. 4.31 *Effects of water acidity on aquatic organism*

Declining yields of Atlantic salmon due to acidification was observed in acidified (pH 5.1) rivers in Norway [Fig. 4.32(a) and Nova Scotia [Fig. 4.32(b)].

Fig. 4.32(a) *The declining yield of Atlantic salmon from seven acidified rivers in southernmost Norway with an average pH of 5.1, compared with 68 rivers in the rest of the country with an average pH of 6.6 (upper curve, left vertical axis), Modified from Velvestad et al.(1976)*

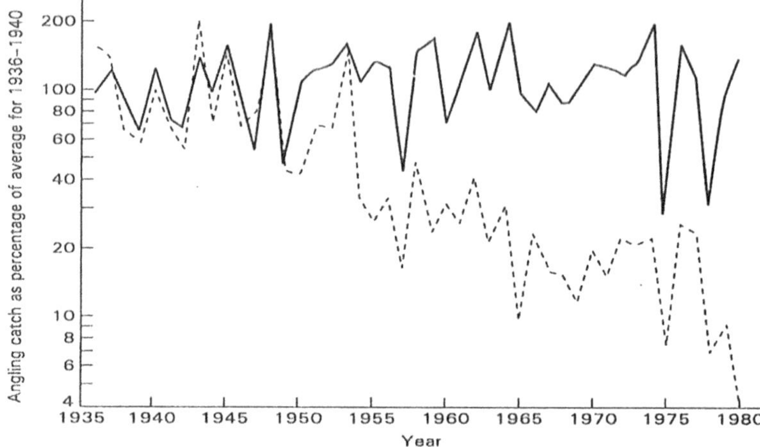

Fig. 4.32(b) *The catch of Atlantic salmon by sport fishing in Nova Scotia rivers. The data were standardized to facilitate the comparison of acidic and less acidic rivers. (–) Mean for 12 rivers with pH>5.0 (1980). (–) Mean for 10 rivers with pH <5.0 (1980). From Watt et al. (1983)*

Table 4.14 *Effects of decreasing pH on aquatic organisms*

pH	Effect
8.0-6.0	Decreases of less than one-half of a pH unit in the range of 8.0 to 6.0 are likely to alter the biotic composition of lakes and streams to some degree in the long run. However, the significance of these slight changes is not great. Decreases of one-half to one pH unit (a three-fold to ten-fold increase in acidity) may detectably alter community composition. Some species may be eliminated.
6.0-5.5	Decreasing pH from 6.0 to 5.5will reduce the number of species in lakes and streams. Among remaining species, significant alterations in the ability to withstand stress may occur. Reproduction of some salamander species is impaired.
5.5-5.0	Below pH 5.5, number and diversity of species will be reduced, reproduction is impaired and many species will be eliminated. Crustaceans, zooplankton, phytoplankton, molluscs, amphiploid, most mayfly species, and many stone fly species will begin to be eliminated. In contrast, several invertebrate species tolerant to low pH will become abundant. Overall, invertebrate biomass will be greatly reduced. Certain higher aquatic plants will be eliminated.
5.0-4.5	Below pH 5.0,decomposition of organic detritus will be impaired severely, Moat fish species will be eliminated.
4,5 and below	In addition to exacerbation of the above changes, many forms of algae will not survive at a pH of less than 4.5

4.4.5.7 *Leaching of soil nutrients*

Under natural conditions, sulphate is often a major anion in soil water. Faster leaching of sulphate increases the loss of cations like potassium, magnesium and calcium. Similarly, increased input of acidity increases the exchange between protons and bound potassium, magnesium or calcium ions. This causes the latter to become more available for leaching. Consequently those soils most sensitive to acid rain are non-calcareous sandy soils with a natural pH of around 6 and low in colloidal material through which water percolates easily. However, acid soils with low cation content show the largest loss of nutrient ions even though the pH change is minimal.

4.4.5.8 *Magnesium deficiency*

Acid deposition may add nitrogen to soil while leaching out magnesium and calcium from needles and soils. The loss of magnesium may account for the yellowing of spruce needles commonly seen in Germany, since magnesium plays a central role in the structure of the chlorophyll molecule. The loss of magnesium and calcium from the leaves may cause the roots to work overtime to replace these crucial nutrients, and the resulting transfer of these nutrients from the soil increases the hydrogen/calcium ratio in the soil, and hence the acidity.

4.4.5.9 *Micro-organisms*

pH determines whether any microbial species can proliferate in a particular environment and the rate at which it can reproduce. Most bacteria and protozoa have pH optima near neutrality; most fungi prefer an acid environment, most blue-green bacteria prefer an alkaline environment.

4.4.5.10 *Wildlife*

It is postulated that acid rain can indirectly affect wildlife by allowing metals bound in soils and sediments to be released into the aquatic environment, from where toxic substances may be ingested by wildlife, like birds, that feed in such environments. Other indirect effects of acid rain on wildlife are loss or alteration of food and habitat resources. The elimination of fish from a lake, for example may mean that loons or mergansers will lack food during the breeding season.

4.4.5.11 *Human health*

Fish are not the only organisms that are adversely affected by acid deposition. Humans are vulnerable as well. One of the greatest air pollution disasters the world has known occurred in London in 1952. The killer fog that shrouded the city for five days contributed to the death of more than 4,000 people from bronchitis, pneumonia, and heart disease. Although pH measurements of the fog were not recorded at the time, scientists today estimate that the fog was a highly diluted sulphuric acid mist with a pH of about 1.4-1.9! But that was almost five decades ago. According to the Office of Technology Assessment, more than 50,000 premature deaths are caused annually by sulphate-laden air. Acid precursors are the second largest cause of lung disease (after smoking) in the opinion of Dr. Philip J. Landrigan of the Mount Sinai School of Medicine in New York City.

Today more than 40 million people in the United States, or about one in six, are drinking water with lead levels above 20 ppb, the level considered safe by the EPA. Where did the lead come from? It apparently was leached from galvanized water pipes by acidic drinking water. Such intake of lead by millions of Americans is cause for considerable concern because medical researchers have shown lead to be one of the factors responsible for high blood pressure and heart attacks in adults and brain damage in children.

Acid rain has a wide range of effects on human health. The obvious ones are exposure to bad odours, reduced visibility, irritation of the skin, eyes and the respiratory tract. Some direct effects include chronic bronchitis, pulmonary emphysema and cancer. Some indirect effects include food poisoning via drinking water and food. Increase in levels of toxic heavy metals like manganese, copper, cadmium, and aluminium also affect human health.

Some health experts believe that acid rain may also be indirectly responsible for many cases of Alzheimer's disease. This is a disease of the elderly characterized by degradation of the brain and severe loss of memory. Chemical analysis of the brains of people who died from the disease have revealed relatively high levels of aluminium. Some researchers believe that aluminium may have caused the disease. The involvement of acid deposition is strongly suggested.

4.4.5.12 Buildings and materials

Historical monuments such as Taj Mahal in India have been affected by what has been termed 'stone cancer'. Stone cancer is yet another effect of acid rain, caused by corrosive action on building materials. Numerous other materials are harmed by acid rain or the gases responsible for acid rain (Table 4.15).

The rate of material corrosion is affected by three main factors.

4.4.5.13 The relative reactivity of the material to acidity:

Stone and masonry that contain calcium carbonate or calcium sulphate are particularly susceptible. The pigments in paint can make paint-work susceptible, and humidity encourages metal corrosion.

4.4.5.14 Meteorological conditions and physical force of the delivery of pollutants

Humidity, air temperature, wind speed, and the intensity, duration and amount of rainfall are all important.

Experiment Human health's have found that the deposition of SO_2 to marble and limestone increases exponentially with humidity, giving cause for particular concern for buildings in humid tropical and sub-tropical regions.

4.4.5.15 The physical location and shape of the material

The erosion of urban buildings may be worse at street level because of the greater concentration of pollutants, and stone carved into complex shapes tends to erode more quickly than flat, vertical stone because it retains pollutants longer.

Table 4.15 *Acid rain damage to materials*

Materials	Type of impact	Principal air pollutants	Other environmental factors	Mitigation measures
Metals	Corrosion, tarnishing	Sulphur oxides and other acidic gases	Moisture, air, salt, particulate matter	Surface plating or coating, replacement of material, removal to controlled environment
Building stone	Surface erosion soiling, black crust formation	Sulphur oxides and other acid gases	Mechanical erosion, particulate moisture, temperature fluctuations, salts, vibration, CO_2, microorganisms	Cleaning, impregnation with resins, removal to controlled environment
Ceramics and glass	Surface erosion, soiling, black crust formation	Acid gases, especially those containing fluoride	Moisture	Protective coating, replacement with more resistant material, removal to controlled atmosphere
Paints and organic coatings	Surface erosion, discoloration, soiling	Sulphur dioxides, hydrogen sulphides	Moisture, sunlight, ozone, particulate matter, mechanical erosion, microorganisms	Repairing, replacement with more resistant material
Paper	Embrittlement, discoloration	Sulphur oxides	Moisture, physical wear, acidic materials introduced in manufacture	Synthetic coatings, storing in controlled atmosphere, deacidification, encapsulation, impregnation with organic polymers
Photographic materials	Microblemishes	Sulphur oxides	Particulate matter, moisture	Removal to controlled atmosphere
Textile dyes	Fading, colour change	Nitrogen oxides, ozone	Light, temperature	Replacement, use of substitute materials, impregnation with polymers
Leather	Weakening, powdered surface	Sulphur oxides	Physical wear, residual acids introduced in manufacture	Removal to a controlled, atmosphere, consolidated with polymers, or replacements
Rubber	Cracking	Ozone	Sunlight, physical wear	Add antioxidants to formulation, replace with more resistant materials

The costs of acid pollution to the world's cultural heritage we causing increasing concern. The end of the cold war and political changes in eastern Europe have drawn new attention to the need to protect the spectacular architectural heritage of cities such as

Crakow in Poland, Prague in the Czech Republic, Bratislava in Slovakia, and Sofia in Bulgaria. There are more than 100,000 stained glass windows in Europe, some of them as old as 1000 years. There is evidence that much of this glass survived in relatively good condition until the turn of the century, and even survived the second World War, when many stained-glass windows were taken down and stored.

The green India 2047 project of TERI has estimated that India is already losing between 11% and 26% of agricultural output on account of soil degradation. Acid rain would increase this figure significantly. The prospect of increasing consumption of coal in Asia will add insult to injury.

4.4.6 Acid rain: control measures

Most effects of acid rain on fish and wildlife cannot be directly countered without reducing or eliminating the sources of pollution. However some measures may be taken locally to counter the acidification of streams, soil, and rest of the environment. Some of the measures are:

4.4.6.1 Buffering

The practice of adding a neutralizing agent to the acidified water to increase the pH is one of the important control measures. Usually lime in the form of calcium oxide and calcium carbonate is used.

4.4.6.2 Breeding of hatchery fish In acidified waters

This practice replenishes fish population that would otherwise decline, thus helping to maintain an ecological balance within acidified lakes.

4.4.6.3 Prevention of the emission of excess SOx and NOx from industries:

Laws should be formulated by the government to regulate the emissions. Some measures are

(a) Decreasing emissions of SO_2 from power stations by burning less fossil fuel, using alternative energy sources like tidal waves, wind, hydro-power, etc.
(b) Using low sulphur fuel
(c) Desulphurization
(d) Decrease in emission of NO_x from power stations
(e) Modification of engines

Problems

1. The solar flux S striking a planet will be inversely proportional to the square of the distance from the planet to the sun, That is, $S = k/d^2$, where k is some constant, and d is the distance.
   (a) Use the data given in Table 4.7 to estimate k, given the distance from the sun to the earth is 150×10^9 m.
   (b) Use your result from (a) to estimate the distance from the sun to Venus.

(c) Mercury is 58×10^9 m from the sun, and has an albedo equal to 0.06. Estimate its equivalent temperature.

2. The average rate at which energy is used to evaporate water is given as 80 W/m^2. Knowing that the latent heat of vaporization of water is about 2.61×10^6J/kg and that the surface area of the earth is about 5.1×10^{14} m^2, estimate the total world annual precipitation in m^3/year.
3. The concentration of CO_2 in 1974 was about 330 ppm and in 1984 it was 345 ppm. Using a simple exponential growth model, if the rate of growth that prevailed during that 10-year period remains constant, in what year would CO_2 concentration reach 560 ppm, double its pre-industrialization value?
4. The United States derives about 25% of its energy from coal, 45% from oil, 20% from gas, and 10% from non carbon-emitting sources.
   (a) Estimate the ratio of tons of carbon emitted per quad of energy.
   (b) Suppose all of the coal consumption is replaced with nuclear and renewable energy systems, but everything else remains constant (total energy and distribution of sources). Estimate the new ratio of carbon to energy.
   (c) Suppose that transition takes 100 years. Modelled as an exponential change, what rate of growth in the ratio of emissions to energy, $r_{em/en}$, would have prevailed?
5. Suppose in 100 years the world is virtually out of natural deposits of oil and gas and energy demand is double the current rate of 300 quads/year. At that time, assume that 28% of world energy comes from coal, 20% from synthetic gas from coal, 40% from synthetic oil from shale, and the remainder is derived from non-carbon emitting sources (that is, our demands for different fuels, is percentagewise, the same as now).
   (a) What would be the carbon emission rate then (Gt/year)?
   (b) If carbon emissions grow exponentially from the current year of 5.5 Gt/year to the rate found in (a) what rate of growth r would have prevailed during that 100 years?
   (c) If the airborne fraction is 50%, how much of the carbon emitted in the next 100 years would remain in the atmosphere?
   (d) What would be the atmospheric concentration of CO_2 in 100 years if there is no net contribution from biomass (and there are 730 Gt of carbon in the atmosphere now)?
   (e) If the equilibrium temperature increase for a doubling of CO_2 is 4 °C, what would be the equilibrium temperature increase in 100 years if the initial concentration is 345 ppm?
6. Based on a 1980 world carbon emission rate of 5.2 Gt/year, a maximum emission rate equal to 22 Gt/year, and total fossil fuel carbon ever emitted equivalent to the consumption of 200,000 quads of coal, draw a graph of carbon emissions versus time using a Gaussian emission function. In what year would maximum emissions occur?
7. For total carbon emissions of fossil fuels equivalent to the consumption of 200,000 quads of coal, and using an airborne fraction of 73% (this may seem high, but it is the result of the diminishing effectiveness of the oceans to absorb excess carbon),

estimate the ultimate concentration of CO_2 added to the "natural" 280 ppm, except for our assumption that the 73% of the carbon emitted that stays in the atmosphere is there forever (we neglect the approximately 1000-year time constant associated with the gradual assimilation of carbon in the oceans).

8. For the two scenarios in the table below:

Pop. (%)	**GNP/person R_{GNP} (%)**	**Energy/GNP E_{en}/GNP(%)**	**C/energy $R_{em/en}$(%)**	**Airborne fraction (%)**	**Equilibrium ΔT_d**
1.0	0.3	−2.0	−0.7	40	3°C
1.5	1.5	−0.2	0.4	50	3°C

   (a) Predict the global equilibrium temperature increase in 70 years, if the initial CO_2 concentration is 345 ppm, the initial emission rate is 5.5 Gt/year, and the initial atmospheric carbon content is 730 Gt.
   (b) If those growth rates continue, in what year would the CO_2 concentration be equal to double the initial 345 ppm?

9. Assume the following data for CFC-11: atmospheric residence time = 75 years, atmospheric concentration in 1985 = 0.24 ppb; 1985 emission rate = 3,42,000 t/yr.
   (a) If emissions rates remain constant, what would be the ultimate, steady-state concentration?
   (b) What drop in emission rate would be required to maintain existing concentrations?
   (c) If emissions immediately drop to 50% of their initial (1985) level, sketch the concentration versus time and indicate the ultimate, steady state level that would be achieved.

10. Solve problem 9 for CFC-13, which has an atmospheric residence time of about 90 years, a 1985 concentration of 0.032 ppb, and a 1985 emission rate of about 165,000 t/year.

11. If CFC-11 and CFC-12 emissions remain at their current rate, their ultimate concentrations in the atmosphere would be about 1 ppb for CFC- 11 and 3.1 ppb for CFC-12. What equilibrium global temperature increase would be expected due to each of these greenhouse gases?

Questions

1. What is green house effect? Discuss the Relative Contributions of various green house gases to Global Warming
2. Discuss the predicted Green House Effect on (i) a warmer global climate, (ii) Changes in food production, (iii) Reduction in water supplies, (iv) Changes in forests, and (v) Reduction in biodiversity,
3. Explain Stratospheric ozone depletion .Discuss how CFCs water vapour, nitrous oxide, Methane, Nitrous oxide contribute to ozone depletion

4. Discuss Effects of Ozone Depletion, on (i) air quality, (ii) aquatic ecosystems, (iii) biogeochemical cycles, (iv) materials and solutions to Protect the Ozone Layer,
5. What is Acid Rain? Discuss Global- Indian Scenario and Impact of Acid Rain, damage to ecosystems &forests

References

1. J. M. Wallace and P.V.H obbs, 1977. Atmospheric science. Academic Press, Orlando.
2. T. R. Oke, 1987. Boundary layer climates. 2nd edition. Metheun. London.
3. W. C. Castle (Ed.), 1982. Carbondioxide review. Oxford University Press, New York.
4. J. M. Barnola, D. Raynaud, Y.S. Korotkenica and C. Lorius, 1987. Vostok ice core provide 160,000 year record of atmospheric CO_2. Nature.329. 1st October.
5. A. Neftal, E. Moor, H. Oeshger and B. Stuffer, 1985. Evidence from polar ice cones further increase in atmospheric CO_2 in the last two centuries. Nature.315.
6. C. D. Keeling and R. B. Bacataw, 1977. Impact of industrial gases on climate. National Research Centre, Washington DC.
7. A. Lacis, J. Hansen, P.Lee, T.M. Tchell and S. Lebedef, 1981. Greenhouse effect of trace gases 1970-1980. Geophysical Research Letters.
8. V. Ramanathan, R. J. Cicerone, H. B. Singh and J. T. Keehl, 1985. Trace gas trends and their potential role in climatic change. Journal of Geophysical Research. 90(D3):5547-5565.
9. R. Allan and N. Welban, 1994. Air pollution and climatic change. Addison Wesley Longman Ltd.
10. J. G. Tutus, 1986. The cause and effect of sea level rise. Vol. I. USEPA and UNEP, Washington DC.
11. S. Chapman, 1930. A theory of upper atmospheric ozone, Mens.R. Meterol. Sc.3, 103-125.
12. M. J. Molina and F. S. Rowland. 1974. Stratospheric sink for chlorofluoromethanes-chlorinated catalyzed destruction of ozone. Nature.249. 810-812.
13. SORG. United Kingdom Stratospheric ozone review group-second report, HMSO, London.
14. S. C. Wofsy et al., 1975. The chemistry of atmospheric bromine. Geophys. Res. Lett. 215-218.
15. J.S. Hofmann, 1987. Assessing the risks of trace gases that can modify the stratosphere. Vol.1. US EPA, Washington DC.

CHAPTER 5

Meteorological aspects of Dispersion of Air Pollutants

5.1 Atmosphere and Changes in the Global Climate

5.1.1 The structure of the atmosphere

The atmosphere may be broadly divided into four regions, as shown in Table 5.1. It extends up to 500 km with the temperature varying from a minimum of –2 °C to a maximum of 1200 °C.

Table 5.1 *Major regions of the atmosphere*

Region	Altitude range, km	Temperature range, °C	Important chemical species
Troposphere	0-11	15 to –56	N_2, O_2, CO_2, H_2O
Stratosphere	11-50	–56 to –2	O_2
Mesosphere	50-85	–2 to –92	O_2, NO^+
Thermosphere	85-500	–92 to 1200	O_2^+, O^+, NO^+

The troposphere contains 70% of the mass of the atmosphere. The upper limit of the troposphere may vary by a kilometre or more, depending on factors such as temperature, nature of terrestrial surface, etc. Air is far from uniform, with respect to density and temperature. Density decreases exponentially with increasing altitude. The composition of the troposphere is more or less homogeneous in the absence of air pollution, mainly due to the constant circulation of air masses in this region. The water content, however, varies due to the global energy flow arising from imbalances in the heating and cooling rates between the equator and the poles.

The temperature in the troposphere decreases uniformly with increasing altitude. The air near the ground level is heated by radiation from the earth. The cold layer (–56 °C) at the top of the troposphere is called the tropopause, and marks a region of temperature inversion, i.e. a transition from negative to positive lapse rates (slope of temperature altitude curve, Fig. 5.1).

The stratosphere is a calm layer having a positive lapse rate. The temperature increases with increase in altitude, with a maximum of –2 °C at the upper limit of the stratosphere. As discussed earlier, ozone in this region absorbs ultraviolet radiation and raises the temperature causing a positive lapse rate. The residence times of molecules or particles in the stratosphere are quite long because of the slow mixing in this region. If the pollutants can somehow reach or are injected into the stratosphere, they pose long-term global hazards compared to their impact in the much denser troposphere as discussed in detail in Chapter 4.

The mesosphere shows a negative lapse rate, i.e. temperature falls with increasing altitude. This is due to low levels of ultraviolet absorbing species, particularly ozone.

In the thermosphere immediately above the mesosphere, the temperature rises once again, giving a positive lapse rate (maximum temperature:1200°C). Here, the atmospheric gases, particularly oxygen and nitric oxide, split into atoms and also undergo ionization after absorption of solar radiation in the far ultraviolet region.

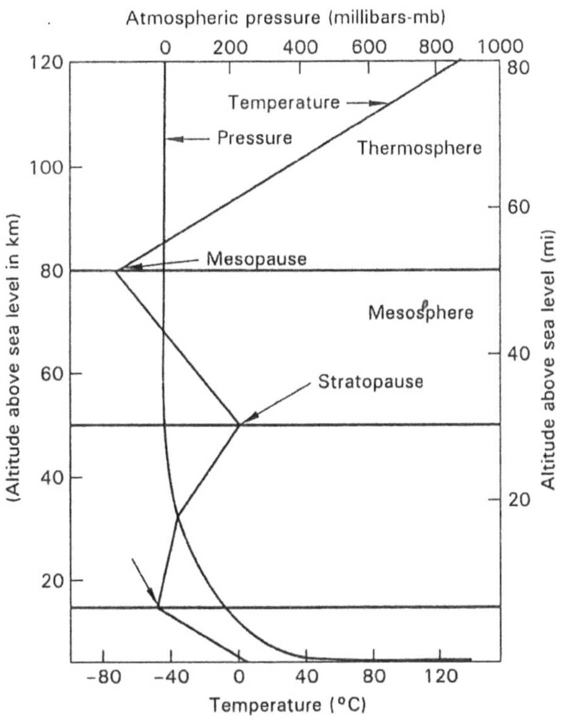

Fig. 5.1 *Temperature-altitude profile of the atmosphere*

Climate and global air circulation are also affected by the properties of air and water. When heated by the sun, ocean water evaporates and removes heat from the oceans to the atmosphere, especially near the hot equator. This moist, hot air expands, and becomes fairly narrow vortices that spiral upward, creating an area of low pressure at the earth's surface. As this moisture-laden air rises, it cools and releases moisture as condensation (because cold air can hold less water vapour than warm air). The heat released when water vapour condenses, radiates into space. The resulting cooler, drier air becomes denser, sinks (subsides), and creates an area of high pressure. As this air mass flows across earth's surface, it picks up heat and moisture and begins to rise again. The resulting small and giant convection cells circulate air, heat, and moisture both vertically and from place to place in the troposphere, leading to different climates and patterns of vegetation (Fig. 5.4).

Air has a specific weight of about 1.2 kg/m^3 at NTP conditions and hence the whole atmosphere presses down upon the earth beneath it; atmosphere at a latitude of 45° and at a temperature of 0 °C exerts a pressure of 1.03 kg/cm^2 which is equivalent to 1 bar or one standard atmosphere. 1 atmospheric pressure = 1.03.kg/cm^2 = 10.3 m of water =760 mm Hg = 1030 millibars = 12.88 m of kerosene). The lines connecting points of equal atmospheric pressure are known as isobars.

5.1.2 Global air circulation and biomes

The two most important factors determining the climate of an area are temperature and precipitation. The temperature and precipitation patterns that lead to different climates are caused mostly by the way air circulates over the earth's surface. The following factors determine the patterns of global air circulation:

(a) Even slight changes in solar output, slight planetary shifts in which the earth's axis wobbles (a 22,000-year cycle) and tilts (a 44,000-year cycle) as the earth revolves around the sun, and minute changes in the shape of its orbit around the sun (a 100,000-year cycle) result in long-term variations in the amount of solar radiation received by the earth.

(b) Air is heated more strongly at the equator than at the poles where sunlight strike's at an angle. The uneven heating of the earth's surface explains the existence of hot tropical regions near the equator, cold polar regions, and temperate regions with intermediate average temperatures.
As long as the earth's surface is heated unequally, air will move in an attempt to balance the inequalities. Hadley suggested that on a nonrotating earth the air movement would take the form of one large convection cell in each hemisphere as shown in Fig. 5.2. The more intensely heated equatorial air would rise and move pole ward. Eventually this upper level flow would reach the poles, where it would sink and spread out at the surface and return to the equator. As the cold polar air approached the equator, it would be reheated and rise again. Thus the proposed Hadley circulation for a nonrotating earth has upper-level air flowing pole ward and surface air flowing equatorward. When we add the effect of the earth's rotation, the Coriolis effect causes the surface flow to become more or less easterly (toward the west) and the flow aloft to be westerly.

(c) The earth's axis is tilted. As the earth makes its annual revolution, various regions are tipped toward or away from the sun creating opposite seasons in the northern and southern hemispheres.

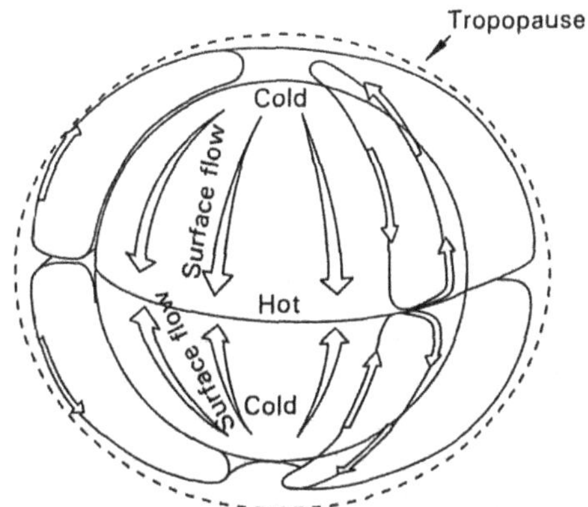

Fig. 5.2 *Global circulation on a nonrotating earth*

(d) Due to the Coriolis effect, while the earth rotates on its axis it prevents air currents from moving from the north and the south; forces in the atmosphere created by this rotation deflect winds to the right in the northern hemisphere.

5.1.2.1 *Driving forces for the horizontal movement of winds*

The principal driving forces for the horizontal movement of wind in the atmosphere, are uneven heating of the earth's surface modified by the effect of the earth's rotation (Coriolis force) and the frictional resistance offered by the ground and the sea.

Averaged over the year, the solar heat flow to the earth's surface at the equator is 2.4 times that at the poles. The atmosphere moves in response to this difference in heating, and in so doing transports heat from the tropics to the poles, resulting in a temperature gradient from the equator to the poles.

As the equator is directly under the sun, the air there will be heated, become buoyant, and rise. As that air approaches the tropopause, it begins to turn, some heading north and some south. The 18th century meteorologist, George Hadley, postulated that the air would continue to the poles before descending. In actuality, it descends at a latitude of about 30 ^{o}C and then returns to the equator, forming what is known as a Hadley cell. In similar fashion, though not as distinct, there are Hadley cells, linked as in a chain, between the 30° and 60° latitudes, and between the 60° latitude and the poles. The descending air at 30° latitude creates a persistent high pressure zone with a corresponding lack of clouds or rainfall that is a contributing factor to the creation of the world's great deserts. Any odd number of such cells could exist in each hemisphere, on the earth. There are normally three cells in each hemisphere, as shown in Fig. 5.3.

In the zone between the equator and roughly 30° latitude the circulation closely resembles the convective model used by Hadley for the whole earth; so the name Hadley cell is generally applied to it. Here the surface flow is equator ward, whereas the flow aloft is poleward. Near the equator the warm rising air that releases latent heat during the formation of cumulus towers is believed to provide the energy to drive this cell. These clouds also provide the rainfall that maintains the lush vegetation of the equatorial rain forests. As the upper flow in this cell moves pole ward, it begins to subside in a zone between the 20° and 35° latitudes. Two factors are thought to contribute to the general subsidence found here. First, as this flow moves away from the stormy equatorial region where the release of latent heat of condensation keeps the air warm and buoyant, radiation cooling would result in increased density of the air aloft. This factor could account for the subsidence. Satellites that monitor the radiation emitted in the upper troposphere have supplied data to support this mechanism for subsidence. Second, because the Coriolis effect becomes stronger with increasing distance from the equator, the once poleward directed winds are deflected into a nearby west-to-east flow by the time they reach 25° latitude resulting in a restricted pole ward flow of air. Thus the Coriolis effect causes a general pile-up of air (convergence) aloft. In either case, general subsidence occurs in the zone between 20° and 35° latitudes. This subsiding air is relatively dry, for it has released its moisture near the equator. In addition, the effect of adiabatic heating during descent further reduces the relative humidity of the air.

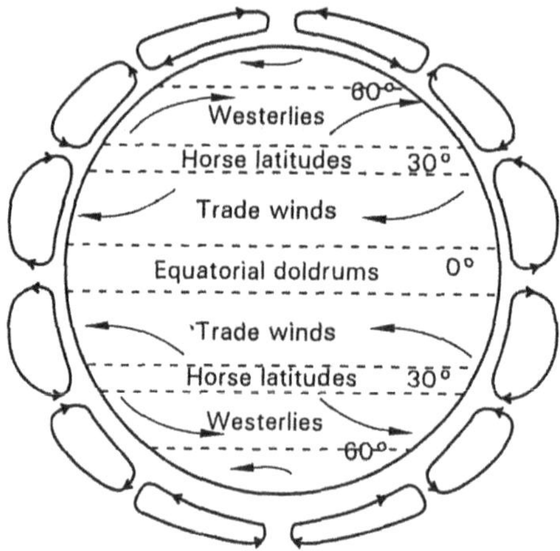

Fig. 5.3 *Schematic representation of the general circulation of the atmosphere-Three cell model (***Source:** *Frederick K. Lutgens and Edward J. Tarbuck,* 1992. The Atmosphere*) [1]*

Consequently, this subsidence zone is the site of the world's subtropical deserts. Winds are generally weak and variable near the centre of this zone of descending air.

In the northern hemisphere, the circulation cells sketched at the edges of the figure indicate a south-to-north flow at high altitudes and a north-to-south flow at the surface in the tropical and polar cells with oppositely directed flows in the temperate cell.

There are seven boundaries between the cells on the globe, one at the equator and two in each hemisphere and two at the poles (Fig. 5.3). At the boundary at the equator and the two between the temperate and polar cells the air is rising; at the boundaries between tropical and temperate cells and at the poles (which are "hole in the donut" boundaries), the air is sinking. The rising air will be cooled due to adiabatic expansion and produces rain, while sinking air is heated and becomes relatively dry.

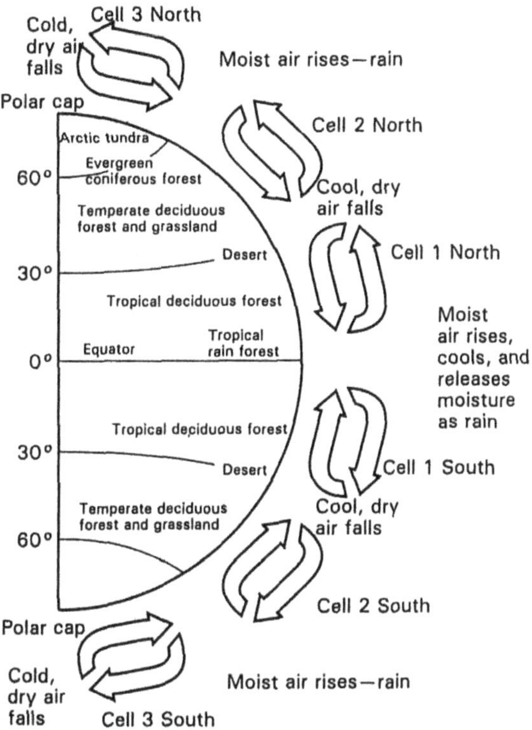

Fig. 5.4 *Global air circulation and biomes (Source: G. Tyler,1995. Miller Junior Environmental Science: Working with Earth. Wadsworth Publishing Company, Behnuti, California.) [2]. (Heat and moisture are distributed over the earth's surface by vertical convection currents that form into six large convection cells (called Hadley cells) at different latitudes. The direction of the airflow and the ascent and descent of air masses in these convection cells determine the earth's general climatic zones. The uneven distribution of heat and moisture over the planet's surface leads to the forests, grasslands and deserts that make up the earth's biomes.)*

The rising boundaries (equator and temperate-polar boundaries) are regions of higher than average rainfall; most of the world's rain forests are located near the equatorial rising zone, and the great temperate forests are near the temperate-polar rising zones.

The sinking boundaries (the poles and the tropical temperate boundaries) are regions of lower than average rainfall; most of the world's great deserts are located near the temperate-tropical sinking zones (Fig. 5.4).

Winds directly under regions of subsiding air often are light and variable. Rising air at the equator creates doldrums where the winds at about 60° north latitude, called the polar front, tend to bless the southward flow of cold polar air. The southern hemisphere has more stable wind patterns because it has more ocean and less landmass than the northern hemisphere.

5.1.2.2 Pressure gradient

The pressure gradient force is the primary driving force of the wind. As an unbalanced force, it causes air to accelerate from regions of higher pressure to regions of lower pressure. However some other force, or forces like friction oppose the pressure gradient force and moderate the airflow.

Although friction significantly influences air flow near the earth's surface, its effect is negligible above a height of a few kilometres. Above a few kilometres the Coriolis force is responsible for balancing the pressure gradient force and thereby directing airflow. Figure 5.5(a) shows how a balance is reached between these opposing forces. When the Coriolis force is exactly opposite to the pressure gradient force the air flow is said to be geostrophic winds.

Thus wind above a few kilometres can be considered geostrophic; that is, they flow in a straight path parallel to the isobars at speeds that can be calculated from the pressure gradient. The major deviation from true geostrophic winds involves the flow along highly curved paths, a topic considered next.

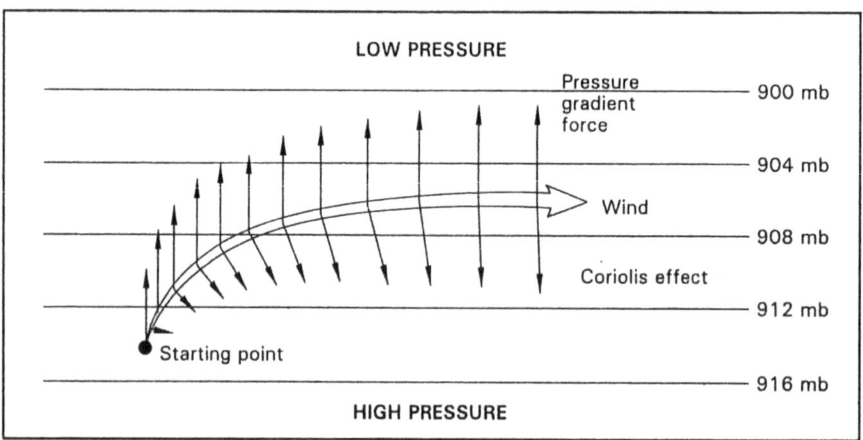

Fig. 5.5(a) *The geostrophic wind. Upper-level winds are deflected by the Coriolis force until the Coriolis force just balances the pressure gradient force. Above 600 meters, where friction is negligible, these winds will flow nearly parallel to the isobars and are called geostrophic winds*

It can be seen in weather maps that isobars are not generally straight; instead they make broad sweeping curves. Occasionally the isobars connect to form roughly circular cells of either high or low pressure. Thus, unlike geostrophic winds that flow in a straight path, winds around cells of high or low pressure follow curved paths in order to parallel the isobars. Winds of this nature, which blow at a constant speed parallel to curved isobars, are caused gradient winds.

It can be seen in Fig, 5.5(b) how the pressure gradient force and Coriolis effect combine to produce gradient winds. The pressure gradient force is directed toward the centre of low pressure. As soon as the flow begins, the Coriolis force causes the air to be deflected. In the northern hemisphere, where the Coriolis force deflects the flow to the right, the resulting wind blows counter-clockwise about a low pressure cell [Fig. 5.5(b)]. Conversely, around a high-pressure cell the outward-directed pressure gradient force is opposed by the inward-directed Coriolis force and a clockwise flow results [Fig. 5.5(b)]. Because the Coriolis force deflects the winds to the left in the southern hemisphere, the flow is reversed there---clockwise around low-pressure centers and counter-clockwise around high-pressure centers.

It is common practice to call all centers of low pressure as cyclones and the flow around them cyclonic. Cyclonic flow has the same direction of rotation as the earth: counter-clockwise in the northern hemisphere and clockwise in the southern hemisphere. Centers of high pressure are frequently called anticyclones and exhibit anticyclonic flow (opposite that of the earth's rotation). Whenever isobars curve to form elongated regions of low and high pressure, these areas are called troughs and ridges, respectively.

The flow about a trough is cyclonic; the flow around a ridge is anticyclonic.

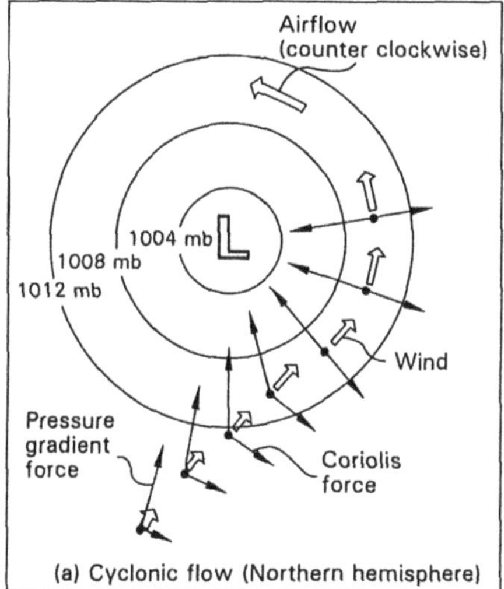

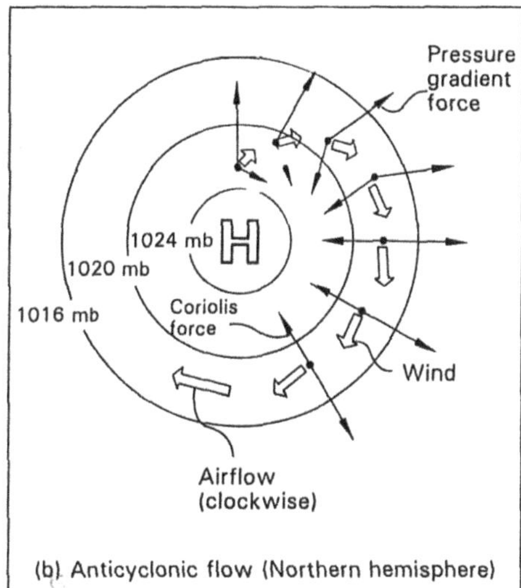

Fig. 5.5(b) *Idealized illustration showing expected airflow aloft around low- and high pressure centers*

In order for a surface low to exist for any reasonable time, compensation must occur at some layer aloft. Surface convergence could be maintained, for example, if divergence (spreading out) aloft occurred at a rate equal to the inflow below. Figure 5.5(c) diagrammatically shows the relationship between surface convergence (inflow) and the divergence (outflow) aloft that is needed to maintain a low-pressure center. Surface convergence about a cyclone causes a net upward movement. The rate of this vertical movement is slow, generally less than 1 kilometre per day. Nevertheless, because rising air often results in cloud formation and precipitation, the passage of a low is generally related to unstable conditions and "stormy" weather. Divergence aloft may occasionally even exceed surface convergence, thereby resulting in intensified surface inflow and increased vertical motion. Thus divergence aloft can intensify these storm centers as well as maintain them. On the other hand, inadequate divergence aloft permits surface low to "fill" and weaken the accompanying cyclone. As often as not, it is divergence aloft that first creates the surface low. Spreading out (divergence) aloft initiates up flow in the layer directly below, and it eventually works its way to the surface where inflow is encouraged.

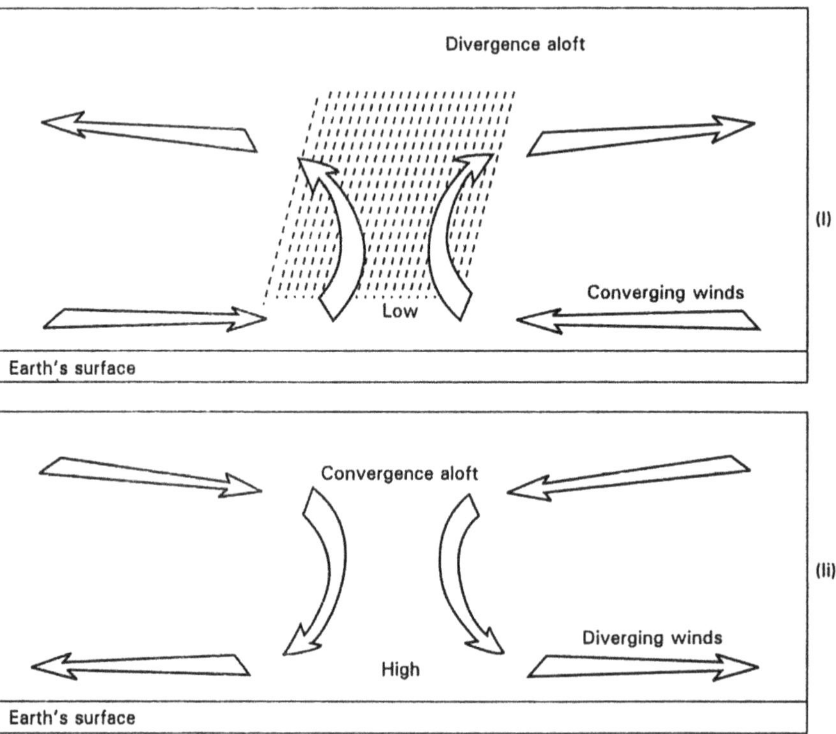

Fig. 5.5(c) *Schematic of the airflow associated with cyclones and anticyclones. (i) Converging winds and rising air are associated with a low, or cyclone (ii) Highs or anticyclones are associated with descending air and diverging winds*

Like their cyclonic counterparts, anticyclones must also be maintained from above. The mass outflow near the surface is accompanied by convergence aloft and general subsidence of the air column Fig. 5.5(c) (i)]. Because descending air is compressed and warmed, cloud formation and precipitation will not occur in an anticyclone resulting in fair weather. The relationship between high pressure zones (anticyclones) and low pressure zones (cyclones) is given in Table 5.2.

Table 5.2 *Behavior of high and low-pressure areas*

	High	Low
Pressure, compared with average atmospheric pressure	High, typical 1020 to 1030 mb	Low, typically 980 to 990 mb
Average vertical air motion near the ground	Sinking	Rising
Behavior of moisture in the air they contain	Evaporates, causing clear skies	Condenses, causing clouds and precipitation
Winds they generate	Out from the center, clockwise in the northern hemisphere, weak	In toward the center, counter-clockwise in the northern hemisphere, strong

5.1.3 Frontal weather

The boundary between two air masses of different temperatures and densities is called a front. A front may be moving or stationary.

5.1.3.1 Cold front

When cooler air displaces warmer air, we call the moving boundary a cold front. Since cold air tends to be denser than warm air, a cold front will hug the ground and push under warmer air as it advances (Fig. 5.6). As warm air is forced upward, it cools adiabatically and its cargo of water vapour condenses and precipitates. Upper layers of a moving cold air mass move faster than those in contact with the ground because of surface friction or drag, so the boundary profile assumes a curving, "bull nose" appearance. The region of cloud formation and precipitation will be relatively narrow. Cold fronts generate strong convective currents and often are accompanied by violent surface winds and destructive storms. An approaching cold front generates towering clouds called thunderheads that reach into the stratosphere where the jet stream pushes the cloud tops into a characteristic anvil shape. The weather after the cold front passes is usually clear, dry, and invigorating.

5.1.3.2 Warm front

If the advancing air mass is warmer than local air, a warm front results. Since warm air is less dense than cool air, an advancing warm front will slide up over cool, neighbouring air parcels, creating a long, wedge-shaped profile with a broad band of clouds and precipitation. Gradual uplifting and cooling of air in the warm front avoid the violent updrafts and strong convection currents that accompany a cold front. A warm front will have many layers of clouds at different levels, The highest layers are often wispy cirrus (mare's tail) clouds that are composed mainly of ice crystal. They may extend 1,000 km ahead of the contained zone with the ground and appear as much as forty-eight hours before any precipitation. A moist warm front can bring days of drizzle and cloudy skies.

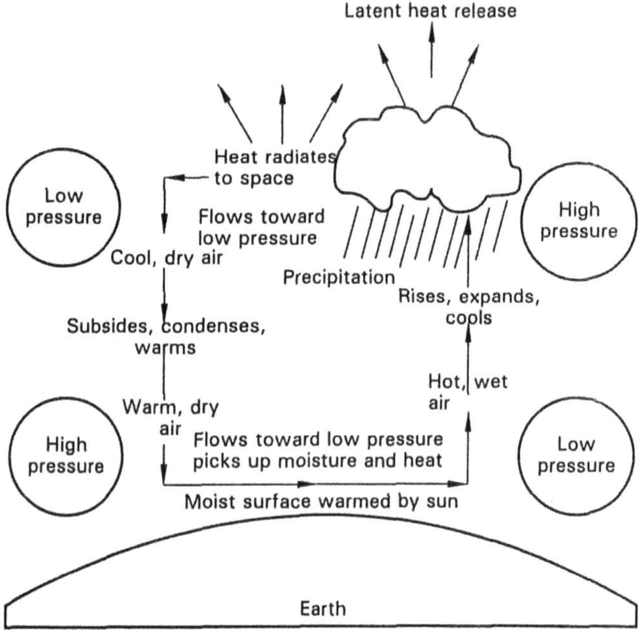

Fig. 5.6 *Convection currents and latent energy cause atmospheric circulation and redistribute heat and water around the globe*

When the transition zone between warm and cold air forces does not move one way or the other, it is called a *stationary factor;*

5.1.3.3 The influence of the ground and the sea

The surface of the ground heats and cools rapidly from day to night and from summer to winter because solid ground is a poor conductor of heat. The surfaces of oceans and lakes heat and cool slowly mainly because their surface layers are stirred by winds and by natural convection currents, thus conducting heat up and down. Solid ground is not stirred by wind or convection currents, so heat cannot be conducted up and down; its surface temperature changes more rapidly than that of bodies of water. Thus the heating (or cooling) of the air layer adjacent to solid ground is much faster than that of air over bodies of water. Probably the most spectacular example of this phenomenon is the monsoon weather of India and parts of East Africa. The summer sun warms the air above India more than the air over the surrounding oceans, which causes strong upward motion of the air over India. Moist air from over the surrounding warm oceans flows inward to fill the low-pressure region caused by this rising air (Fig. 5.7). This moist air rises, cools, and forms the monsoon rains on which Indian agriculture depends. As the summer heats the semiarid regions of India, a strong low-pressure system develops. Flow of cool air from the north is blocked by the Karakoram and Himalayan mountains.

A continuous flow of moisture-laden air from the subtropical high pressure area over the Arabian Sea sweeps around the tip of India and up into the Gangetic Plains and the Bay of Bengal. As the onshore winds are driven against the mountains and rise, the air cools and drops its load of water. During the four months of the monsoon, which usually lasts from June until September, 1,000 cm of rain may fall on Nepal, Bangladesh, and western India. The same phenomenon leads to summer thunderstorms over the southwestern United States, but they are not as strong as those of the Indian monsoon.

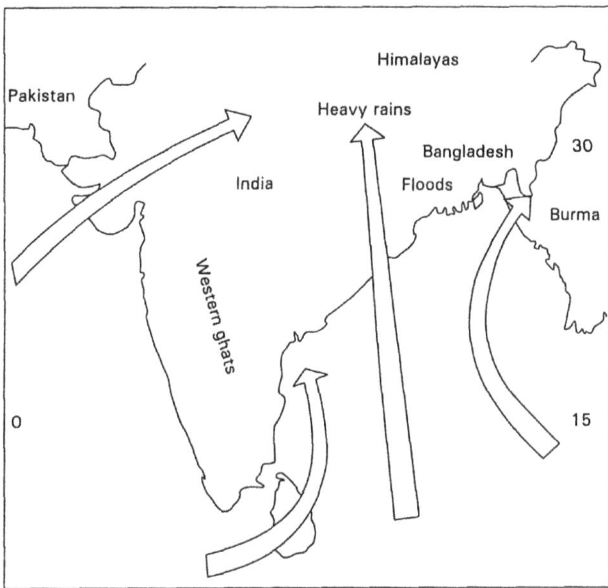

Fig. 5.7 *Summer monsoon air flows over the Indian subcontinent*

5.1.3.4 Ground level wind patterns and velocities

A number of local variations both in space and time are likely to influence the ground level wind patterns. The highest ground-level wind velocities are those in tornadoes, up to 320 km/h (89.54 m/s). The average ground-level wind velocity in most of North America (day and night, summer and winter) is about 16 km/h (4.5 m/s). Wind speed less than about 3.2 k m/h is known as calm. If one cannot feel a 3.2 km/h wind, the only way one can tell which way the wind is blowing is to observe the behavior of leaves, flags, smoke, or steam plumes, which will show that there is a wind even if one cannot feel it. Most weather services report any wind less than about 3.2 km/h as "calm" because the wind-measuring instruments (anemometers) become unreliable far such low velocities.

Wind speed increases with elevation, most of the time, in most of the troposphere. At ground levels ground friction slows down wind. Typically wind will reach its frictionless velocity (called the geostrophic or gradient velocity) at about 500 m above the ground. The region below this elevation, where ground friction plays a significant role, is the planetary boundary layer. The ground-level wind velocity is largely determined by how well this layer is coupled to the fast-moving geostrophic layer above it. When the atmosphere is stable or has an inversion, there is little vertical movement; and the coupling between the planetary boundary layer and the geostrophic wind is weak. Thus, inversions and stable atmospheres are normally associated with low ground-level wind velocities. Different types of winds and their values on the Beaufort scale are summarized in Table 5.3.

Table 5.3 *Beaufort scale for various wind types on the basis of velocity and effects*

Beaufort's scale	Wind type	Wind velocity (km/h)	Specifications
0	Calm	< 1	
1	Light air	1-5	Smoke rises vertically
		(3)	Direction of wind shown by smoke drift but not by wind vanes
2	Light breeze	6-11 (9)	Wind felt on face; leaves rustle; ordinary vane moved by wind
3	Gentle breeze	12-19 (16)	Leaves and small twigs in constant motion; wind extends light flag
4	Fresh breeze	20-28 (24)	Raises dust and loose paper; small branches are moved
5	Moderate breeze	29-38 (34)	Small trees in leaf begin to sway; crested wavelets form on inland waters
	Strong breeze	39-49 (44)	Large branches in motion; whistling heard in telegraph wires; umbrellas used with difficulty
6	Near gale	50-61 (55)	Whole trees in motion; inconvenience felt in walking against the wind. Breaks twigs of trees; generally impedes progress
7	Gale	62-74 (68)	Slight structural damages occurs (chimney pots and slate removed)
	Strong gale	75-88 (82)	Trees uprooted; considerable structural damage occurs
8	Storm	89-102 (96)	Rarely experienced; accompanied by widespread damage
	Violent storm	103-117 (110)	
9	Hurricane	> 117	

Figures given in parentheses represent mean values.

5.2 Air Pollutants and Influence of Meteorological Factors on their Removal/Transportation

5.2.1 Influence of air pollution on climate

As discussed in chapter 4, Air pollutants generated by natural and anthropogenic sources have a tremendous effect in moderating the climatic conditions.

Air pollution can reduce sunlight penetration, decrease air temperature, affect cloud formation, and alter precipitation levels. Such dramatic climatic changes may disrupt terrestrial and aquatic ecosystems.

5.2.2 Particulate matter and sunlight

Particulate matter can reduce the amount of sunlight that reaches the earth because particles scatter and absorb solar radiation. Washington DC and Los Angeles, for example, receive 10% less sunlight than they did at the beginning of this century, when the air was relatively unpolluted.

This phenomenon is not limited to urban areas because wind currents disperse particulates widely. The pollution from a city, may cover an area 50 times the size of the urban source. Such a reduction in solar energy-the power base of all ecological systems–could conceivably have profound ecological effects by reducing plant growth.

The darkened skies of cities force urban residents to turn on lights in homes and shops some what earlier in the day resulting in an increase in a nation's yearly light bills (and incidentally causing more pollution as well).

5.2.3 Moisture, humidity and air pollutants

In the hydrological cycle the evaporation of water from surface water bodies transfers water vapour into the atmosphere, which finally condenses at higher altitudes forming clouds. These clouds precipitate as rain or snow. In this process some moisture is maintained in the atmosphere.

The extent of moisture content which air can hold is a measure of humidity and depends on temperature. The temperature at which condensation begins is known as *dew point.* Absolute humidity is a measure of the amount or weight of water vapour in a unit volume of air (grams per kg of air or grams per m^3 of air). Relative humidity (RH) is the ratio of the amount of water in the air at a given temperature to the maximum amount it could hold at that temperature; expressed as a percentage. If RH exceeds 65%, air inside the room feels sticky and uncomfortable. RH below 30% results in unpleasant dry conditions. Exposure to low humidities can cause drying of the nasal mucosa which may predispose one to infection, exemplified by sore throat and cough.

Pollutants emitted from industrial smokestacks and motor vehicles form clouds above cities, especially in winter. To make matters worse, particulates (soot and dust) in urban air serve as condensation nuclei that absorb tiny moisture droplets. Particulates, therefore, can promote cloud formation and increase rainfall. Now, if human generated air pollution did indeed induce rainfall, we would expect more rainy days from Monday to Friday, when factories are operating and pollutants are generated, than on weekends, when plants are closed. This is precisely the case. For example, in Paris, France the average daily rainfall on weekdays is 31 percent higher than on weekends. Industrial contaminants, such as particulates, generated in Chicago are blown to the general region of LaPorte, Indiana, 48 km to the east, where they trigger considerable amounts of rainfall.

The discharge of particulate matter to the atmosphere through human and natural sources is tabulated for comparison in Table 5.4. The table shows the tremendous loading of particulate matter into the atmosphere each year. An approximate calculation shows that if the particle loading in the atmosphere increases by 50%, the average temperature of the earth will decrease by 0.5-1 °C. The same order of temperature drop will result from particle-induced cloud formation.

Table 5.4 *Worldwide addition of particulate matter to the atmosphere from human and natural sources (Annual production in million tons)* [3]

Materials	Natural	Man-made
Total particles	773-2200	184-415
Dust and smoke	–	10-90
Salt, forest fires	438-1100	–
Sulphate particles	130-200	130-200
Nitrates	30-25	140-700
Hydrocarbons	15-20	75-200

5.2.4 Transport and removal of air pollutants

The ambient levels of air pollutants which are discharged annually into the atmosphere from various sources throughout the world are fairly constant. This indicates that there are certain exchange mechanisms working between the atmosphere and the earth, through which the pollutants are continually removed. These removal mechanisms for particles and gases can be classified into the following categories.

(a) Wet removal by rainfall or precipitation
(b) Dry removal by sedimentation, impaction. Chemical reaction in the atmosphere to produce aerosols and/or absorption on aerosols with subsequent removal.
(c) Interaction at land and ocean surfaces.
(d) Transportation/dispersion based on meteorological conditions.

5.2.4.1 Wet removal by rainfall

Precipitation has a two-fold cleansing action on the pollutants discharged into the atmosphere. It helps to remove the gaseous pollutants that are soluble in water. It also acts as a scrubbing fluid for the removal of gaseous pollutants, airborne radioactive wastes and particulate matter from the atmosphere. Thus it accelerates the deposition of pollutants helping to disperse them easily and brings the pollutants downwards forming a thick blanket that finally sinks to the ground, leading to very high ground- level concentrations.

Inside the clouds, the contaminants may serve as condensation nuclei on which droplets condense causing precipitation and removing the pollutant, which is known as rain-out. The removal of pollutants below the cloud level by falling rain is known as wash-out, which is the most effective scavenging process for both particulate and gaseous pollutants in a global sense.

Aitken particles whose size is less than 0.1 mm will be captured by cloud droplets by Brownian diffusion, and the cloud droplets (typically 30 mm) in turn grow in size by coalescence and precipitate removing the pollutant by rain-out mechanisms. For particles larger than 2 mm wash-out is effective. The removal efficiency is influenced by the cross-sectional area of the raindrop and the intensity of rainfall. Particles of the size of 500 mm are not collected by raindrops as they will be brushed aside by the diverging air ahead of the drop. For removing acid droplets and sulphate particles formed after chemical reactions in the atmosphere, wet precipitation is an effective natural operative mechanism.

5.2.4.2 *Dry deposition*

Gravity settling is one of the removal mechanisms in which particulate matter smaller than 0.1 mm often coagulate through mutual collisions to form larger aggregates which are effectively removed. Though atmospheric turbulence is the driving force for the diffusive motion of particles, Brownian motion is the major mechanism of coagulation. For particles whose Brownian motion is less pronounced, atmospheric turbulence plays a significant role. According to Stokes' law [4]:

$$v_t = \frac{g d_p^2 (\rho_p - \rho_a)}{18\mu_a}\left[1 + \frac{2C}{d_p P}\right] \qquad \text{.....(5.1)}$$

where

v_t = terminal settling velocity
d_p = particle diameter
ρ_p and ρ_a = density of particle and air, respectively
μ_a = viscosity of air
P = air pressure
C = constant [when p is given in millibars and d_p in centimeters, $C = 0.0084$]

From Equation 5.1 it is evident that particle size is one of the controlling factors which determines the rate of sedimentation; particles greater than 10 mm have high settling rates and so a short residence time in the atmosphere. Further it is also dependent on particle's density, to a certain extent. Irregular particle shapes, turbulence in the course of large particles; and atmospheric vertical velocities cause deviation from Stoke's law. So for effective removal from the atmosphere, small particles necessarily should be aggregated.

Inertial impaction, dispersion and diffusion also contribute to the removal of particulate matter in addition to sedimentation; overall, dry deposition accounts for about 20% of the total particulate removal from the atmosphere.

5.2.4.3 Removal by interaction with land and ocean surfaces, chemical reactions

Gaseous pollutants like SO_x, NO_x, CO, etc. come into contact with the earth's surface by atmospheric turbulence where they interact with the ocean surface, vegetation and upper layers of the soil, and are removed by absorption or chemical reactions.

For example, as per certain estimates [5], nearly 4×10^7 tons of SO_2 will be removed yearly by oceanic absorption while the absorption on solid surfaces removes nearly 5.6×10^{17} tons per year; it is estimated that global emissions of SO_2 are 14×10^7 tons/year. Similarly CO is mainly removed by biological action in soil while oxides of nitrogen are removed by the ocean.

5.2.4.4 Transportation/dispersion of air pollutants based on meteorological conditions

Air pollutants naturally get transported/dispersed in the atmosphere due to the natural phenomenon of convection. In the diluted atmosphere a number of physical and photochemical transformations takes place.

5.2.4.5 Factors affecting transportation/dispersion of air pollutants

The factors that affect the transportation, dilution, and dispersion of air pollutants can be categorized in terms of (i) meteorological/atmospheric conditions (ii) characteristics of emission (plume) point sources (iii) wind speed and such effects of natural terrain and anthropogenic land-use structures and (iv) nature of pollutant materials.

5.3 Meteorological aspects of Air Pollution Dispersions

Convection is one of the main natural processes through which the air pollutants generated by man-made sources get dispersed in the atmosphere. A number of physical, photochemical transformation takes place in the atmosphere. The stability conditions of the atmosphere determine the extent of dispersion; higher instability of the atmosphere facilitates greater dispersion.

5.3.1 Adiabatic lapse rate

The rate of change of air temperature with altitude will reveal the ease with which the pollutants can disperse vertically into the atmosphere. Air at a given altitude may have physical forces acting on it such that it remains at that elevation; then the air is termed as stable and this mainly depends on the temperature profile. Stable air will not allow the dispersion and dilution of air pollutants. The air may be unstable allowing rapid vertical mixing, permitting vertical pollutant dispersal and diffusion. So vertical stability of the atmosphere is an important factor which determines the ability of the atmosphere to dilute emissions; hence it is crucial to air quality.

To understand atmospheric stability, we must examine the relationship between pressure and temperature with respect to altitude.

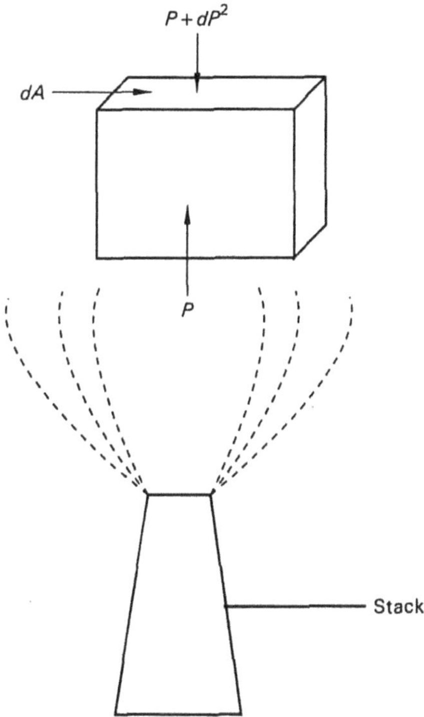

Fig. 5.8 *Schematic of plume coming from a stack as a parcel or packet of air moving vertically in the atmosphere*

5.3.2 Pressure versus altitude

Consider a plume coming from a stack as a parcel or packet of air moving vertically in the atmosphere (Fig. 5.8) with imaginary boundaries (area *d*A and height *dz*). For static equilibrium the net force acting on this parcel will be zero. If '*w*' is the specific weight of air (1.2 kg/m^3) at NTP and if *dp* is the net elemental pressure acting in the downward direction, then

$$\Sigma F = 0 = w.\ dA.dz + dP.dA, \text{ so } dP = -wdz \quad .....(5.2)$$

Since *w* is always positive, pressure decreases with increasing altitude.

$$dP = -wdz \text{ or } dP/dz = -w \quad .....(5.3)$$

However, the specific weight *w* is a product of density and acceleration due to gravity.

So we have

$$\frac{dP}{dz} = -\rho g \qquad \text{.....(5.4)}$$

where ρ is the atmospheric density and g is acceleration due to gravity. As w is always positive and pressure decreases with increasing altitude, if we assume air is an ideal gas, then we have

$$PV = \text{n } RT \text{ and } \rho = M/V \qquad \text{.....(5.5)}$$

or $$V = nRT/P$$

where M = gram molecular weight of the gas.

$$\rho = \frac{MP}{nRT} \qquad \text{.....(5.6)}$$

Substituting Equation 5.6 for ρ in Equation 5.4, we have

$$\frac{dP}{P} = -\frac{Mg}{nRT} dz \qquad \text{.....(5.7)}$$

For isothermal atmosphere by integrating Equation 5.7 we get

$$P = P_0 \exp\left(-\frac{gM}{RT} z\right) \qquad \text{.....(5.8)}$$

where P_0 is the pressure at ground level (1.013 bars). Equation 5.8 gives the relation for decreasing pressure with altitude which is a typical characteristic of compressible fluids such as air.

5.3.3 Temperature versus altitude

Air pollution meteorology is in part a consequence of the thermodynamic processes of the atmosphere. One such process is adiabatic expansion and contraction. An adiabatic process is one that takes place with no addition or removal of heat and with sufficient slowness so that the gas can be in equilibrium at all times.

In the vertical movement of gaseous air pollutants, the adiabatic decrease in temperature with altitude is important. It can be explained by assuming the concept of movement of a small sphere of air as a parcel. In the vertical movement of the air parcel, we can imagine that its temperature/pressure and volume will be changing. The air parcel expands as it rises in the atmosphere and goes through a region of decreasing pressure and thus does work on the surroundings. There will be no heat transfer between the air parcel and the surroundings as the process occurs rapidly and so the internal energy decreases resulting in decrease in temperature.

As per the first law of thermodynamics, for gaseous substances where the kinetic energy and potential energy do not undergo any change, we can write

$$dQ = dU + dW \qquad \text{.....(5.9)}$$

where dQ is the heat added to the parcel per unit mass, dU is the change in the internal energy of the system which is equal to $C_v\ dT$ (where C_p is the specific heat at constant volume), and W is the work done on the system which is equal to PdV. Substituting these terms in Equation 5.10 we have

$$PdV + C_v dT = dQ \qquad .....(5.10)$$

For an ideal gas we have the following relation

$$PV = nRT \qquad .....(5.11)$$

where n = number of moles and R is the gas constant. The derivative of a product can be written as

$$d(PV) = PdV + VdP \qquad .....(5.12)$$

Combining Equations 5.11 and 5.12, we have

$$PdV + VdP = nRdT \qquad .....(5.13)$$

$$PdV = nRdT – VdP \qquad .....(5.14)$$

from thermodynamics, the relation between C_v = heat capacity at constant volume and C_p = the heat capacity at constant pressure is given by

$$C_v = C_p - nR \qquad .....(5.15)$$

$$C_v\, dT = (C_p – nR)\, dT \qquad .....(5.16)$$

Substituting Equation 5.14 and Equation 5.16 in 5.10 we have

$$dQ = (C_p - nR)\, dT + nRdT - VdP \qquad .....(5.17)$$

which can be reduced to

$$dQ = C_p dT - VdP \qquad .....(5.18)$$

where dQ = heat added to the parcel per unit mass (J/kg)

C_p = specific heat at constant pressure, that is the amount of heat required to raise the temperature of 1 kg of air by 1 °C while holding its pressure constant = 1005 J/kg °C.

dT = incremental temperature change (K)

V = volume per unit mass (m^3/kg)

dP = incremental pressure change in the parcel (Pa)

If one assumes that as the parcel moves there is no heat transferred across its boundary, which is true for any reversible adiabatic process, then dQ = 0 and by re arranging Equation 5.18 we have

$$\frac{dT}{dP} = \frac{V}{C_p} \qquad .....(5.19)$$

Equation 5.19 describes how atmospheric temperature would change with air pressure, while Equation 5.8 gives the relation between atmospheric pressure and altitude.

Expressing the rate of change in temperature with altitude as a product, and substituting in Equations 5.4 and 5.18 gives

$$\frac{dT}{dz} = \frac{dT}{dP}\frac{dP}{dz} = \frac{V}{C_p}(-\rho g) \qquad \text{.....(5.20)}$$

however, since V is the volume per unit mass and ρ is mass per unit volume, the product $V \times \rho = 1$, and the expression simplifies to

$$\frac{dT}{dz} = \frac{-g}{C_p} \qquad \text{.....(5.21)}$$

The negative sign indicates that temperature decreases with increasing altitude. Substituting the constant g = 9.806 m/s^2, and the constant volume specific heat of dry air at room temperature, C_p= 245 cal/kg °C, or 1005 J/kg °C in Equation 5.21 yields

$$\frac{dT}{dz} = \frac{-9.806\,\text{m/s}^2}{1005\,\text{J/kg°C}} \times \frac{1\,\text{J}}{\text{kg}-\text{m}^2/\text{s}^2} = -0.00976\,°\text{C/m} \qquad \text{.....(5.22)}$$

To keep matters simple, dT/dz is given a special name: the dry adiabatic lapse rate Γ.

$$\Gamma = -\frac{dT}{dz} = 9.76°\text{C/km} = 5.4°\text{F/km} = 5.4°\text{F/1000m} \qquad \text{.....(5.23)}$$

Equation 5.23 indicates that if a parcel of dry air moves up or down, it will cause its temperature to change by 9.76 °C/km or essentially 1 °C/100m. This can be used as a reference against which actual ambient air temperature gradients can be compared [Fig. 5.9(a)]. If the actual air temperature decreases faster than the adiabatic lapse rate with increasing elevation, the air will be unstable and rapid mixing and dilution of pollutants will occur. If the actual air temperature drops more slowly than the adiabatic lapse rate, the air will be stable and pollutants will concentrate [Fig. 5.9(b)].

For a polytrophic atmosphere, Equation 5.24 will be a better model for describing the variation of temperature with pressure and altitude.

$$T = T_0\,(P/P_0)^{(n-1)/n} \qquad \text{.....(5.24)}$$

Substituting in Equation 5.24 in Equation 5.7 and differentiating we get

$$dT/dz = -(n-1)/n\ g/R \qquad \text{.....(5.25)}$$

which is the equation that describes the variation of temperature with altitude in a polytrophic model.

Actually in the atmosphere upto 10 km of the troposphere, the temperature decreases with increasing altitude at a rate of 6.4 °C/km. Substituting this value in Equation 5.24:

$$dT/dz = 6.4\ °\text{C}/1000\ \text{m} = -(n-g)/R \qquad \text{.....(5.26)}$$

from which $n = 1.23$.

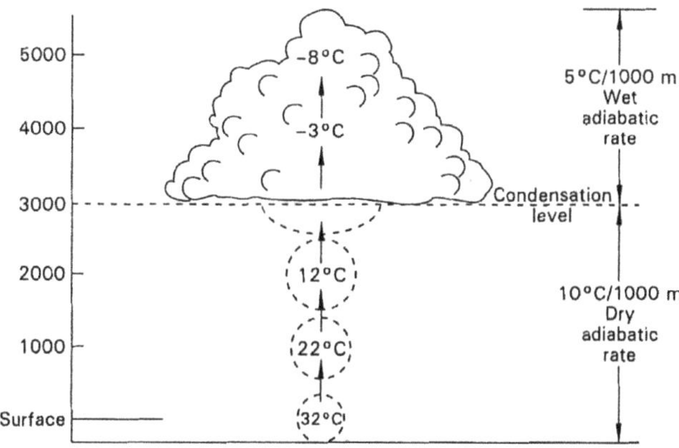

Fig. 5.9 (a) *Rising air cools at the dry adiabatic rate of* 10°C *per* 1000 *meters until the air reaches the dew point and condensation (cloud formation) begins. As air continues to rise, the latent heat released by condensation lowers the rate of cooling. The wet adiabatic rate is therefore always less than the dry adiabatic rate*

However in the stratosphere the temperature is constant upto 25 km, for which the isothermal model based on $n = 1$ is applicable. (Fig. 5.10)

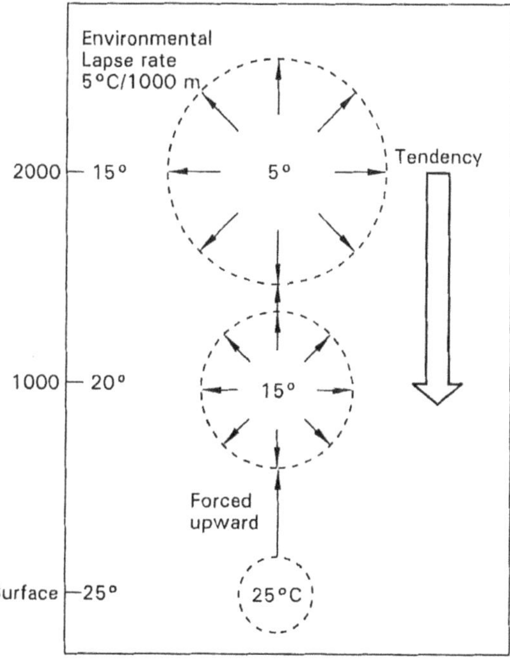

Fig. 5.9(b) *Schematic representation of a stable atmosphere. Note that the air near the surface is potentially cooler than the air at elevation and therefore resists upward motion*

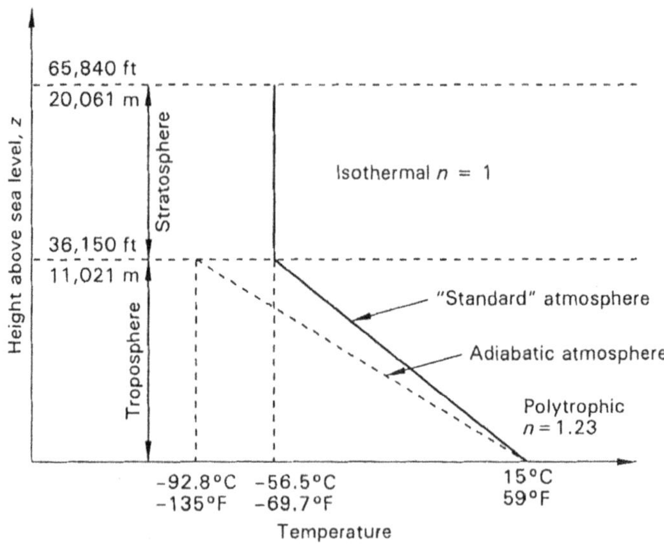

Fig. 5.10 *Comparison of the temperature altitude profiles in the adiabatic atmosphere and the standard atmosphere*

5.3.4 Effect of humidity

Equation 5.21 was derived by assuming that the. parcel of air could be treated as an ideal gas that could be moved without heat transfer taking place between it and its surroundings, which is true. However it was also assumed that the air was dry, but this may or may not be a true assumption. If air has some water vapour in it, C_p changes slightly from the value assumed.

A rising parcel of air in the presence of humidity, may cool until the partial pressure of water vapour equals the saturated vapour pressure of water. Then some of the water vapour will condense if sufficient nucleation sites are present and release the latent heat. If λ is the latent heat of water per gm then the release of this heat is given by

$$dQ = -\lambda\, m\, dw \qquad .....(5.27)$$

where m is mass of the dry air in the parcel and dw is the ratio of water vapour which is condensing to the true mass of dry air and moving further. In the presence of humidity, Equation 5.21 can be written as

$$\frac{dT}{dz} = -\left[\frac{g}{C_p} - \frac{\lambda}{C_p}\frac{dw}{dz}\right] - \frac{g}{\left[(1-w)C_p + wC_{p_w}\right]} \qquad .....(5.28)$$

where Cp_w is the specific heat of water vapour at constant pressure which between -63 to 27°C has a value approximate 440 cal/kg while the saturated absolute humidity at this temperature is 0.00079 μg of water vapour per 0.45 kg of dry air so w = 0.00079/0.00079 + 1.

$$\text{So } dT/dz = \tau = \frac{9.81}{\sqrt{(-0.00079)245 + 0.00079(440)(4.184)}} = 0.0009564\ ^\circ\text{C/m}$$

which is considered as the wet adiabatic lapse rate, though the entire process is not completely adiabatic. A typical conditional instability with an environmental lapse rate in between adiabatic lapse rate and wet adiabatic lapse rate is shown in Fig. 5.11.

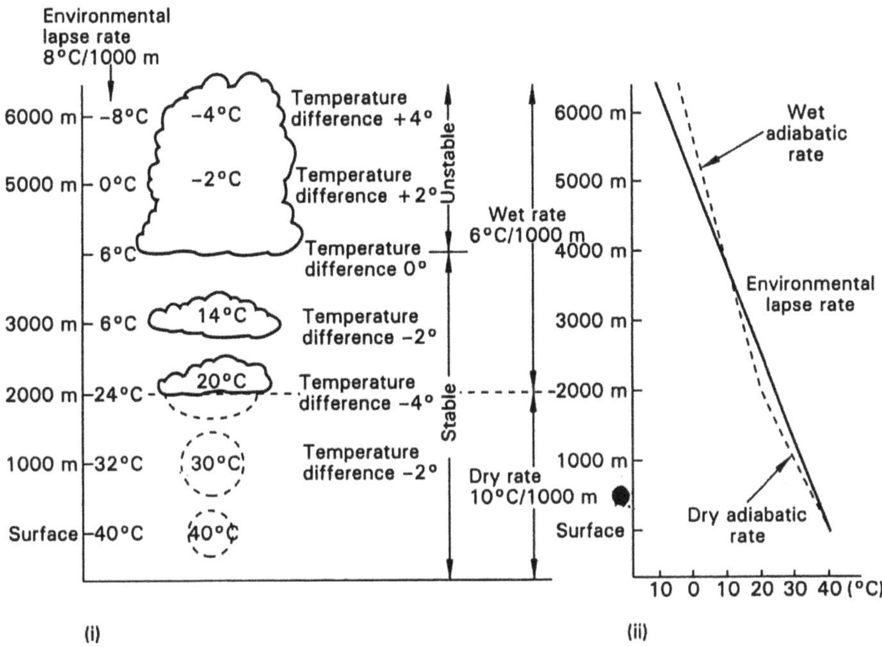

Fig. 5.11 *Conditional instability illustrated by using an environmental lapse rate of 8 °C per 1000 meters that lies between the dry adiabatic rate and the wet adiabatic rate. (i) The rising parcel of air is cooler than the surrounding air below 4000 meters and warmer above 4000meters (ii) Graphic representation of the conditions shown in part*

Since the condensing water has negligible effect on temperature, *dw/dz* is negative for a rising parcel in which condensation is taking place and so the last term in Equation5.28 becomes positive in the presence of humidity (the wet adiabatic lapse rate). The temperature decrease with increasing altitude will be less than that for dry air. As the saturated vapour pressure of water increases rapidly with temperature, the value of *dw/dz* changes with temperature and so the wet adiabatic lapse rate is not a constant and will be independent of altitudes (Fig. 5.12). In warm tropical air the wet adiabatic lapse rate is approximately 1/3 of the dry adiabatic lapse rate.

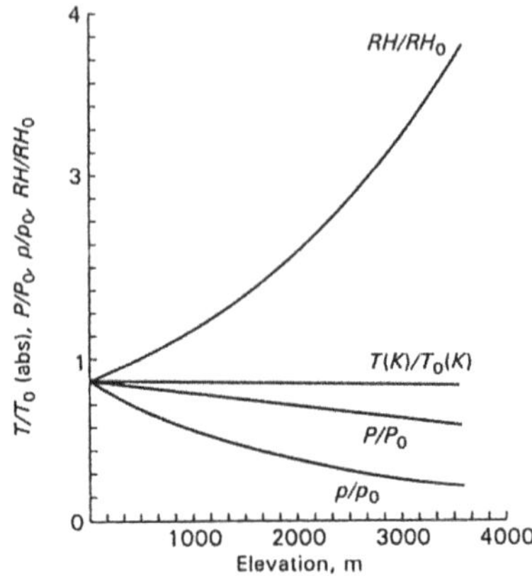

Fig. 5.12 *Changes with elevation of temperature, total pressure of pure water and relative humidity, for the standard atmosphere. The zero subscript indicates conditions at z =0*

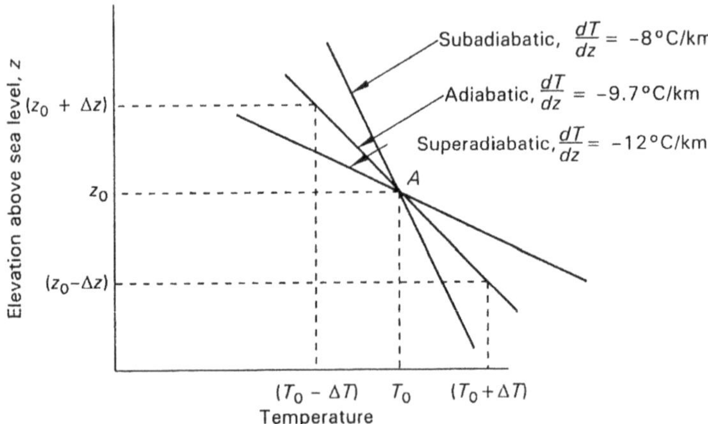

Fig. 5.13 *Behaviour of an air parcel displaced up or down in an atmosphere with an adiabatic, a subadiabatic and a superadiabatic lapse rate*

If the surrounding air has the adiabatic lapse rate, then whether the parcel is moved up or down, it will reach the same temperature as the surrounding air at its new location (Fig. 5.13). In any such displacement the pressures of the displaced air and surrounding air will be practically equal; if they were not, the higher-pressure parcel would expand rapidly against the lower-pressure one.

5.3.5 Atmospheric stability

The nature of atmospheric stability can be understood from the typical temperature-altitude profiles depicted in Fig. 5.14 which shows various types of lapse rate curves.

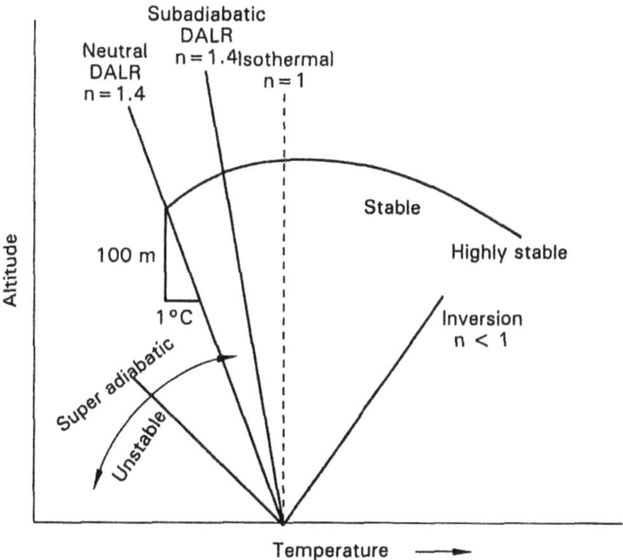

Fig. 5.14 *Atmospheric temperature-altitude profiles and atmospheric stability*

SALR: Super-Adiabatic Lapse Rate; DALR: Dry Adiabatic Lapse Rate; WALR: Wet Adiabatic Lapse Rate; CS: Conditional Stability

When the environmental lapse rate and the dry adiabatic lapse rate are under exactly equal conditions, a rising parcel of air will have the same pressure, temperature and density of the surroundings and would experience no buoyant force. The atmosphere is termed as neutrally stable where a displaced mass of air neither tends to return to its original position nor tends to continue its displacement.

When the environmental lapse rate is greater than the dry adiabatic lapse rate, the atmosphere is considered as super-adiabatic. The rising parcel of air, cooling at the adiabatic rate, will be warmer and less dense than the surrounding environment. Under these conditions the parcel of air becomes more buoyant and tends to continue its upward motion. The vertical motion is enhanced by buoyancy. Such an atmosphere is called unstable and in such a case the air from different altitudes mixes thoroughly and dilutes the pollutants in the air. This is a favourable condition for pollution control.

If the surrounding air has a super-adiabatic lapse rate, then if the parcel of air is displaced upward it will follow the adiabatic curve; so it will be warmer and less dense than the surrounding air. Thus, gravity will drive it further upward. If it is displaced downward it will be colder and denser than the surrounding air, so that gravity will drive it further downward. Whatever disturbs it from its original location, gravity will continue to move it in the direction of that displacement, and with an increasing velocity. Thus if the lapse rate is super adiabatic, the condition is unstable and the vertical air motion is spontaneous if any minute disturbance triggers it.

If the surrounding air has an environmental lapse rate less than the dry adiabatic lapse rate, it is said to be sub-adiabatic and the displaced parcel of air will have a different fate. If it is displaced upward, it will continue to follow the adiabatic curve, because its motion will be adiabatic, so that it will be colder, and denser than, the surrounding air. Gravity will drive it back down. If it is displaced downward, it will be warmer and less dense than the surrounding air, so gravity will drive it back up. Whatever disturbs its location, gravity will move it back toward its original location (Fig. 5.15).

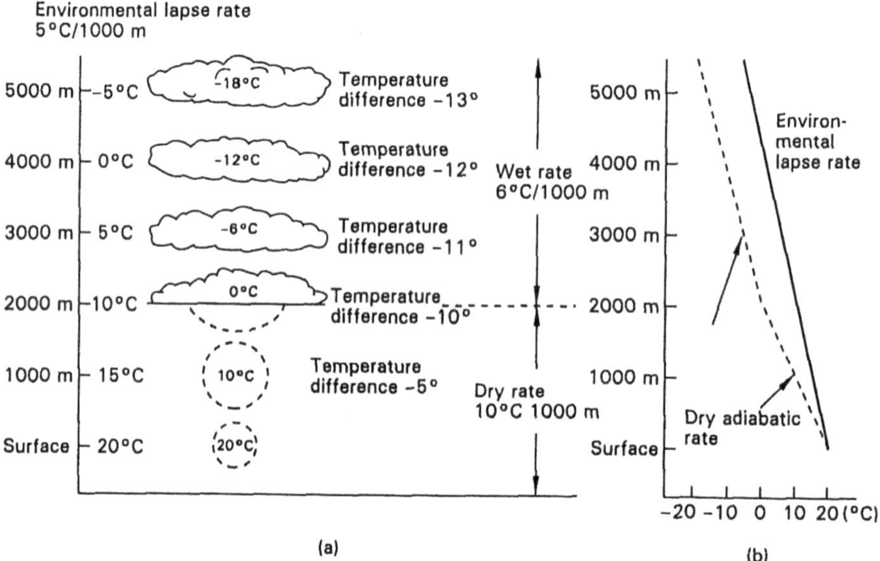

Fig. 5.15 *Environmental Lapse Rate*

Such an atmospheric condition is called as stable and the lapse rate is said to be sub-adiabatic. Under stable conditions there is very little vertical mixing and pollutants can only disperse very slowly. As a result, their levels can build up very rapidly in the environment.

5.3.6 Temperature inversions

Inversion occurs when temperature increases with altitude and the lapse rate in such a condition is known as the negative lapse rate. This is the extreme case of stable atmosphere. Practically no mixing of the pollutants takes place under stable atmospheric conditions.

Temperature inversions play a significant role in air pollution meteorology. Within an inversion the air is stable against buoyant vertical motion. That stability also lessens the exchange of wind energy between the air layer near the ground and high altitude winds, so that both horizontal and vertical dispersion of pollutants are hindered (Fig.5.16).

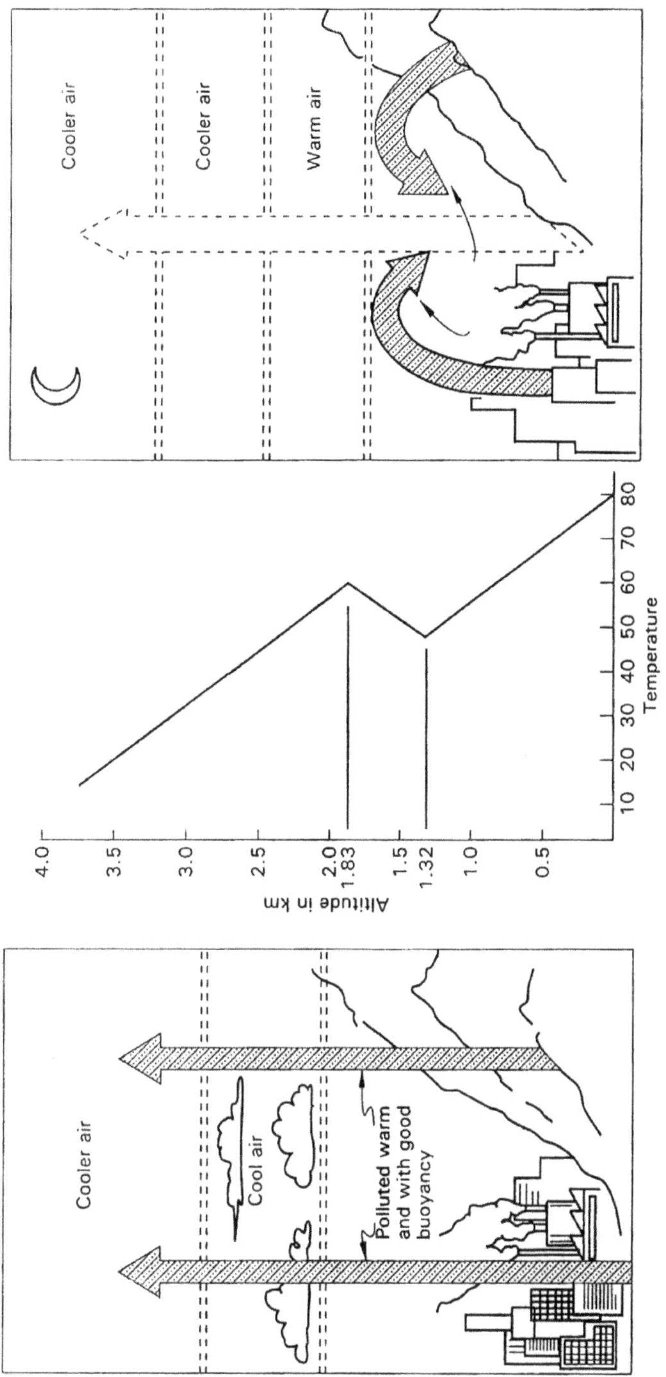

Fig. 5.16 *Temperature inversion upsets the natural tendency for upward air movement*

Inside the inversion the situation is extremely stable; vertical disturbances are strongly damped out. Above the inversion, in the region with the standard lapse rate, the situation is mildly stable; vertical disturbances are damped, but not nearly as strongly as inside the inversion.

Inversion layers based on their nature of formation which can be termed as subsidence inversions, radiation inversions and advection inversions, *Subsidence inversion* is usually associated with subtropical anticyclones where the air is warmed by compression as it descends in a high pressure system and achieves a temperature higher than that of the air underneath. If the temperature increase is sufficient, an inversion will result.

Anticyclones are semi-permanent in nature and can cause subsidence inversions to last for months at a time (Fig. 5.17). These semi-permanent highs are the result of the general atmospheric circulation patterns.

Air in the middle of a high pressure zone and descending air on the edges, will be rising. Air near the ground moves outward from the centre while air aloft moves to wards the centre from the edges. The result is a massive vertical circulation system as air in the centre of the system falls, experiences greater pressure and is compressed and heated. If at elevation $z_2$it is heated adiabatically to $T_2 = T_1 + T\ (z_1 - z_2)$ as is often the case, this compressive heating warms the descending air to a higher temperature than the air below, whose temperature is dictated primarily by conditions on the ground. The result is an inversion, located anywhere from several hundred metres above the surface to several thousand metres, that lasts as long as the high pressure weather system persists.

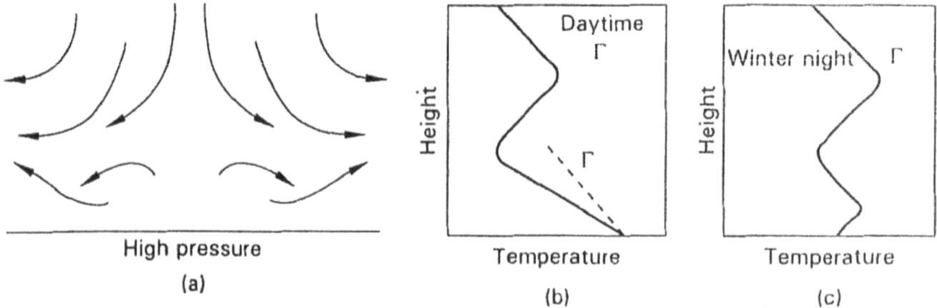

Fig. 5.17 *Subsidence inversions (a) As air descends it is warmed adiabatically creating clear skies and an upper level inversion (b) During the day, air under the inversion may be unstable due to solar warming of the surface (c) Radiation inversions may form at night under subsidence inversion, especially in winter*

Since subsiding air is getting warmer, it is more and more able to hold water vapour as it descends. Without sources of new moisture, its relative humidity drops; thus there is little chance for clouds to form. The result is that high pressure zones create clear, dry weather with lots of sunshine during the day and clear skies at night. Clear skies allow solar warming of the earth's surface. This helps create super-adiabatic conditions under the inversion during the daytime and hence, good mixing. At night, the surface can cool quickly by radiation, which may result in a radiation inversion located under the subsidence inversion.

Radiation inversion occurs due to the normal diurnal cooling cycle. The ground cools quickly by radiational heat transfer after sunset due to which the lowest layer of air in contact with the surface loses sensible heat through conduction and small scale mixing; this will result In temperature inversion between the cool low level air and the warmer air above, in the first few hundred metres above the land surface on a clear or slightly cloudy day with low of average winds [Fig. 5.18(a)]. The subsidence inversion is potentially more serious than radiation inversion because the latter usually dissipates quite rapidly after sunrise.

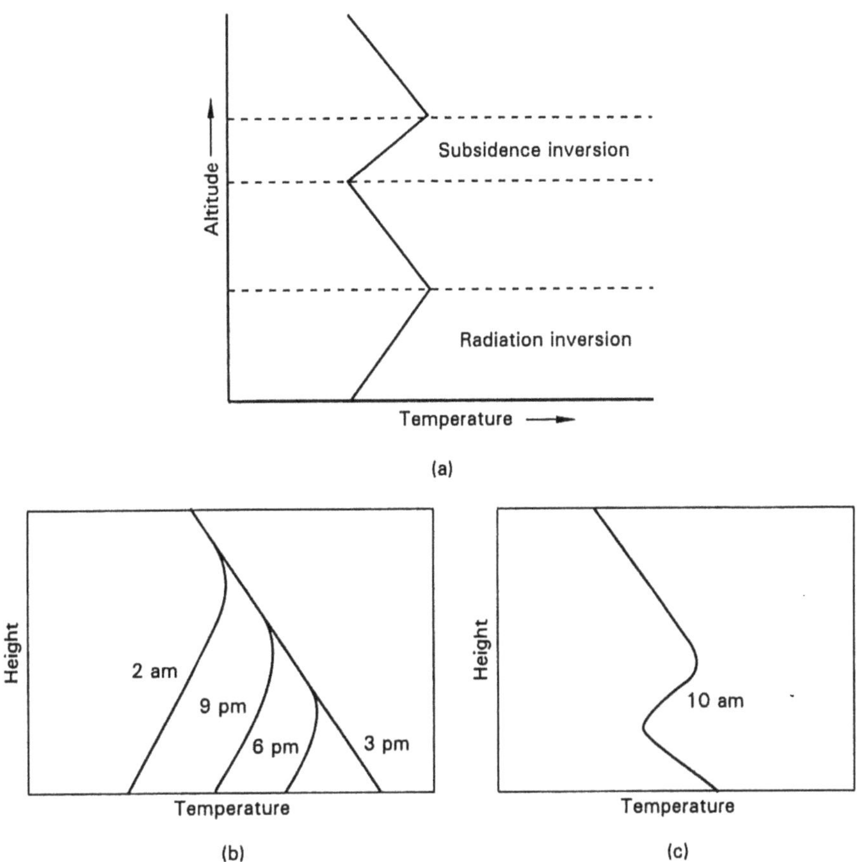

Fig. 5.18(a) *Temperature-altitude profiles for radiation and subsidence inversion;* **(b)** *Development of a radiation inversion;* **(c)** *The subsequent erosion of the inversion*

Figure 5.18(d) shows a strong radiation inversion at dawn, as would be caused by radiation to the night sky in a dry region. In more humid areas, atmospheric moisture partly blocks the loss of heat from the ground to outer space, and a thin or partial cloud also reduces this heat loss. The resulting weaker inversions can serve as quasi mixing layers within which there may be substantial mixing, but whose tops retard further upward mixing. Shortly before sunset the ground surface begins to cool by radiation, and to cool the air layer nearest it. By sunset there will be weak inversion close to the ground. All night this inversion will grow in strength and size, until by dawn of the next day the temperature profile will be practically the same as that shown for dawn in Fig. 5.18(a).

When the sun comes up, it heats the ground surface, which heats the layer of air above it, by conduction, convection, and radiation, that layer heats the next layer above it, and so on. There will be a layer of warmed air near the ground, in which the lapse rate is practically the adiabatic lapse rate. At its top, this layer encounters the remainder of the previous night's inversion. Rising air from below cannot penetrate that inversion, for stability reasons. But at the boundary it mixes with the inversion, slowly destroying it, so that by four hours after sunrise the warmed air layer will have grown and almost eliminated the inversion [Fig.5.18(c)]. By mid-afternoon, enough heat will be transfer red from the warmed ground surface to the adjacent air so that the inversion is gone. The heated air, which now has an adiabatic lapse rate, extends to perhaps 6000 ft, where it encounters the more stable air above, with a lapse rate at or near that of the standard atmosphere. In the few hundred feet closest to the ground the lapse rate is even greater than the adiabatic lapse rate (i.e. super-adiabatic), and the air is unstable.

Advective inversion is formed when warm air moves over a cold surface or cold air [Fig. 5.18(e)]. The inversion can be ground-based in the former case, or elevated in the latter case. An example of an elevated advective inversion occurs when a hill range forces a warm land breeze to flow at high levels and a cool sea breeze flows at low levels in the opposite direction.

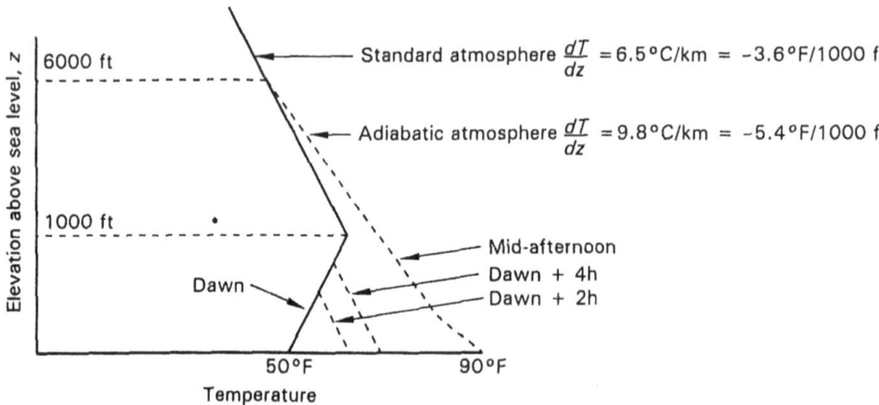

Fig. 5.18(d) *Vertical temperature distribution at various times on a cloudless day with low or average winds in a dry climate*

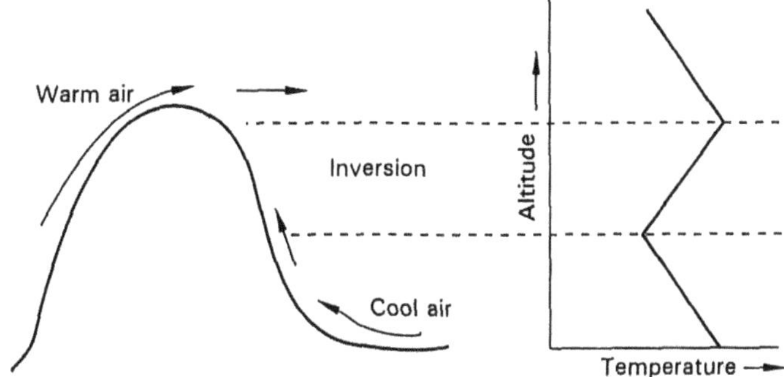

Fig. 5.18(e) *Elevated advective inversion*

Valleys and low-lying areas are particularly affected by this type of nocturnal inversions because denser, colder air tends to sink down beneath the warmer air. The next day sunlight destroys the inversions as the earth is warmed and the air previously stratified by inversions is overturned by convective currents.

All inversions either at ground level or at higher elevations, inhibit atmospheric mixing and thus lead to the accumulation of pollutants. In summer, with clear skies, heating of the ground by the sun will normally eliminate an inversion in daytime. However, local effects like cool onshore breezes, may be powerful enough to maintain inversions. In winter the sun is often not strong enough to eliminate such inversions, and they may persist until a major storm brings winds strong enough to overcome the local topographic effects and sweep them away. If the inversion is strong enough to form a fog in a valley, it will reflect away sunlight, making the inversion persist longer than it would without the fog.

5.3.7 Mixing height

Once air pollutants are released into the atmosphere the amount of air available for dilution and dispersion will depend on wind speed and the extent to which emissions start to rise into the atmosphere. As the atmospheric temperature profile determines the stability of the atmosphere, it also effects the vertical dispersion emission coefficient. So the vertical height of the atmosphere upto which the mixing takes place varies diurnally, from season to season, and is also affected by topographical features. The greater the height of vertical discharge, the larger the volume of the atmosphere available to dilute the pollutant concentration.

If we can have some measure of the thickness of the air layer near the surface of the earth in which we could expect to have turbulence and effective dilution taking place it will be very useful. The depth of the convection mixing layer is known Maximum Mixing Depth (MMD) which is measured as average values for a period of one month; these values are termed as Mean Maximum Mixing Depth (MMMD).

Consider, on a day with good sunshine, a parcel of air at ground level being warmed by convection; due to its buoyancy, air will rise. The parcel will cool at a rate of 10 °C/ 1000 m as it moves upward and expands adiabatically. As long as its temperature is warmer than that of the surrounding air, it will continue to rise; and when the parcel air cools down to a temperature less than the surroundings it will come down. At the point where the temperature of the air parcel becomes equal to that of the surrounding air, it will stop moving upward. If the temperature profile of the atmosphere is known, then one can determine the elevation to which the parcel will ascend by simply projecting the ground temperature upward, at the adiabatic lapse rate, until it crosses the ambient temperature profile. The point at which the dry adiabatic lapse rate crosses the temperature profile of the atmosphere, is called the mixing depth (sometimes called the mixing height) of the atmosphere.

In practice the daily temperature profile, usually at night or in the early morning, will be obtained using a balloon equipped to transmit the necessary data by radiational measurement and the dry adiabatic temperature line is then drawn from the surface, with a starting point equal to the average maximum surface temperature for the month. The elevation where these lines cross is known as the mean maximum mixing depth (USEPA Report, 1972) [Fig. 5.19(a)]. The maximum mixing depth for different environments is shown in Figs. 5.19(b), (c), (d). In Fig. 5.19, we can see that in the morning the mixing height must be much lower and it grows during the day. Similarly, we would expect that the mixing height would be larger in summer than in winter (Table 5.5).

Table 5.5 *Typical values of mixing heights [6]*

	Mixing height, m	
	Range	**Average**
Summer morning	200-1100	450
Summer afternoon	600-4000	2100
Winter morning	200-900	470
Winter afternoon	600-1400	970

In general the mixing depths are low at night and increase with increasing surface temperature in the day. If the inversion sets at night, the MMD values up to 2000 and 5000 m are common, and may even become zero in the day. The minimum mixing depth, before sunrise can be obtained by adding 5 °C to the minimum surface temperature as shown in Fig. 5.20.

The concept of mixing height can be understood by examining Fig. 5.19 when in the mid-afternoon condition, there will be vigorous vertical mixing from the ground to about 6000 ft, and then negligible vertical mixing above that height. The rising air columns that provide good vertical mixing induce large-scale turbulence in the atmosphere. This turbulence is three-dimensional, so it also provides good horizontal mixing. Pollutants released at ground level will be mixed almost uniformly up to the mixing height, but not above it. Thus the mixing height sets the upper limit to the dispersion of atmospheric pollutants.

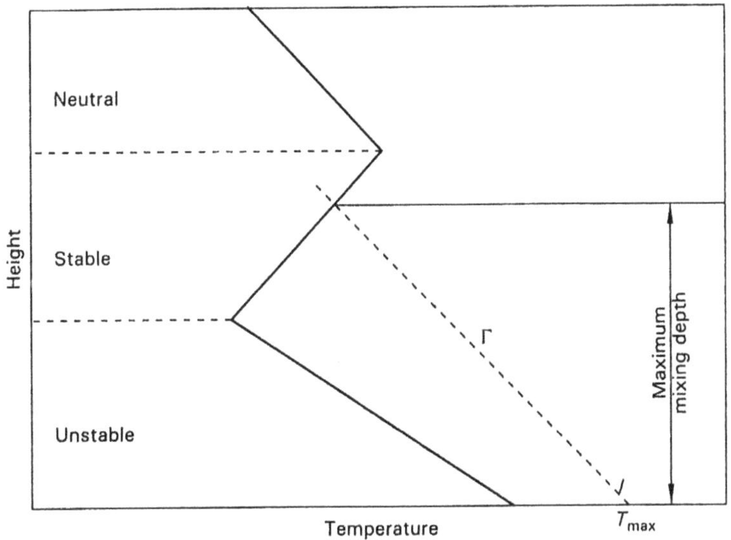

Fig. 5.19(a) *The maximum mixing depth is obtained by projecting the dry adiabatic lapse rate line to the point of intersection with the atmospheric temperature profile. Notice that the mixing depth does not necessarily correspond to the layer of unstable air*

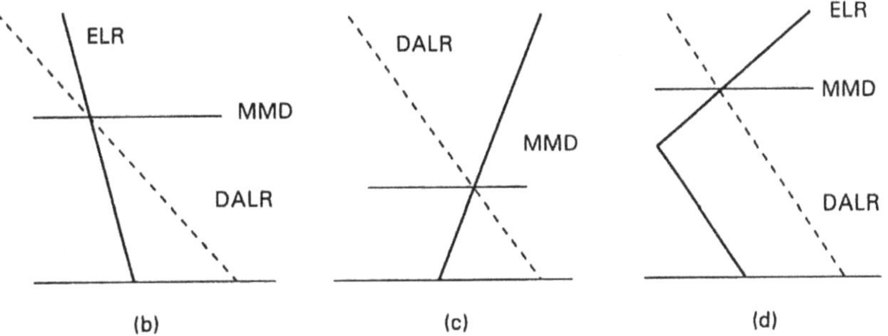

Fig. 5.19(b),(c),(d) *Maximum mixing depth in different environments*

5.3.8 Ventilation coefficient

Ventilation coefficient expressed in m^2/s is a measure of atmospheric dispersion capability which can be obtained by multiplying maximum mixing depth with the average wind speed within the mixing depth. Values of ventilation coefficient less than about 6000 m^2/s are considered indicative of high air pollution potential [7].

As wind speeds increase with increasing height, it may be necessary to estimate wind at higher altitudes than the standard 10 m weather station anemometer. For altitudes less than a few hundred metres above the ground, the following equation can be used for calculating wind speed

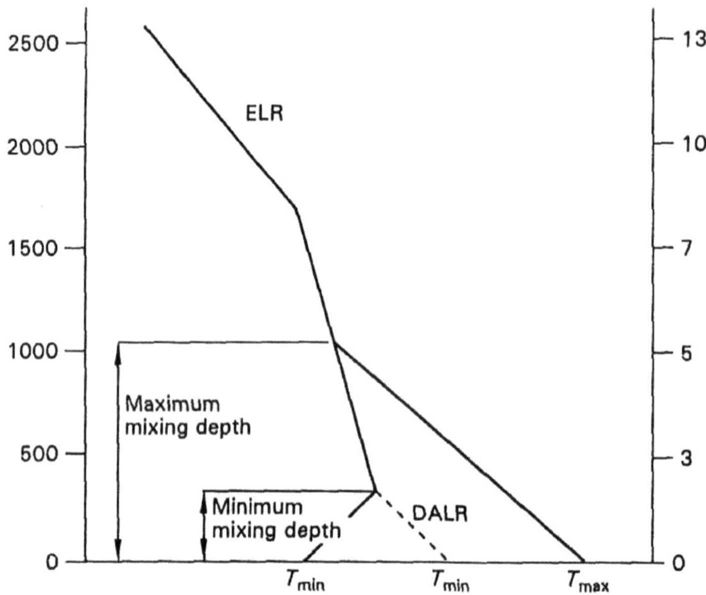

Fig. 5.20 *Determination of mixing height at different surface temperatures of the day*

$$\frac{u_1}{u_2} = \left(\frac{z_1}{z_2}\right)^p \quad .....(5.29)$$

where u_1 and u_2= wind speed at the higher and lower elevation, respectively,

z_1 and z_2= higher and lower elevation, respectively

p = a dimensionless parameter that varies with atmospheric stability.

p values will be larger with increasing stability in the atmosphere, and increase in wind speed with increase in elevation. Further the topography also affects the value of p. The rougher the terrain, the more the wind speed near the surface will differ from the wind speed higher up.

Table 5.6 gives values for p that are recommended by the EPA for rough surfaces in the vicinity of the measurement anemometer [8]. For smooth terrain, flat fields or over bodies of water, the values of p given in Table 5.6 should be multiplied by a factor of 0.6.

Table 5.6 *Wind profile exponent p for rough terrain*

Stability class	Description	Exponent, p
A	Very unstable	0.15
B	Moderately unstable	0.15
C	Slightly unstable	0.20
D	Neutral	0.25
E	Slightly stable	0.40
F	Stable	0.60

Source: *W.B. Peterson 1978. User guide for PAL; A Gaussian'splume algorithm for point area and line sources. USEPA, Triangle Park N.C. [8]*

For smooth terrain p should be multiplied by 0.6.

Estimation of ventilation coefficient

Problem 5.1

A city with a rough terrain has an isothermal atmospheric profile at 27 °C and its maximum surface temperature is found to be 37 °C while the average wind speed is 4.0m/s at a height of 10 m. Determine its mixing height and ventilation coefficient.

Solution: The dry adiabatic lapse rate is 1 °C/100 m, so projecting that lapse rate from 37 °C at the surface until it reaches the 27 °C isothermal means the mixing depth will be 1000m as shown in Fig. 5.21. Since the temperature profile is isothermal, "slightly stable" stability class, with p = 0.4 can be taken. To estimate the average wind speed at 1000 m mixing depths, it is necessary to calculate the wind speed at half way point, 500 m, for which

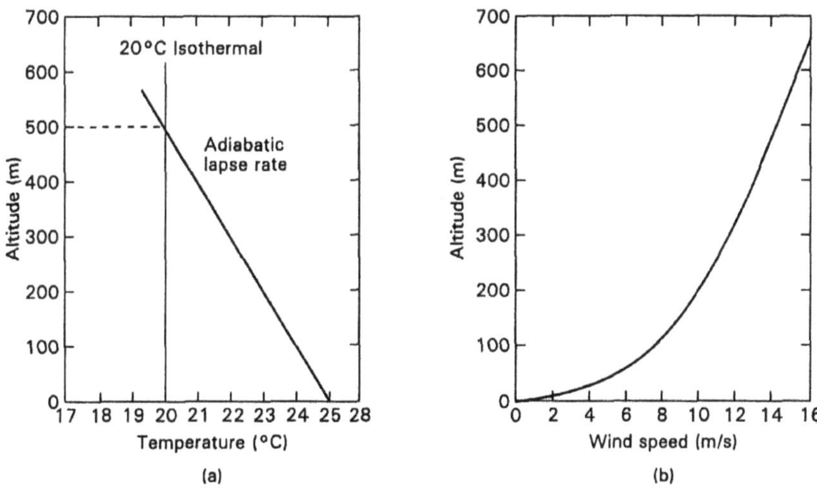

Fig. 5.21 *Temperature and wind speed for above example*

$$\frac{u_1}{3.0} = \left(\frac{500}{10}\right)^{0.4} = 14.34\,\text{m/s}$$

Ventilation coefficient = mixing depth × average wind speed at half of mixing depth = 500 × 14.34 = 7170 m^2/s

5.3.9 Effect of wind speed, direction, and topographic effects on dispersion of pollutants

Although the atmospheric condition (stability or instability) is the most important overall indicator of dispersion of pollutants coming from the stack, wind movement and associated turbulence also have significant effects. For example, eddies arising from the turbulent motion of the wind could cause changes in plume dispersion [Figs. 5.22(a) and (b)].

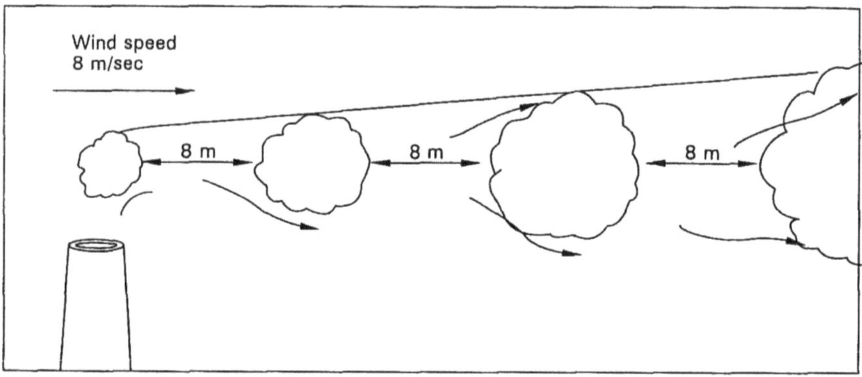

(a)

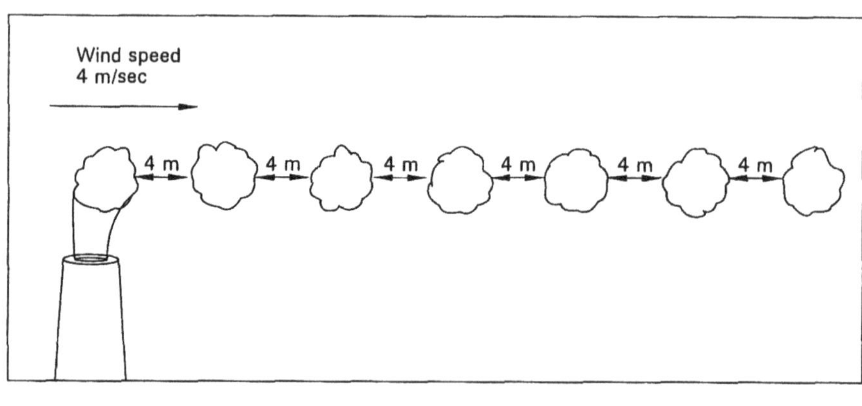

(b)

Fig, 5.22(a),(b) *The effect of wind speed on the dispersion of pollutants*

The wind direction determines the direction in which the contaminated gas stream will move across local terrain. Wind speed affects the plume rise and the rate of mixing or dilution of the contaminated gases as they leave the point of discharge. An increase in wind speed will decrease the plume rise by bending the plume over more rapidly.

The decrease in plume rise tends to increase the pollutant's ground level concentration. On the other hand, an increase in wind speed will increase the rate of dilution of the effluent plume, tending to lower the downwind concentrations. Under different conditions, one or the other of the two wind speed effects becomes the predominant effect. These effects, in turn, affect the distance downwind of the source at which the maximum ground level concentration will occur.

The horizontal transport of air pollutants has been a function of the wind speed direction, air turbulence and topography.

5.3.10 Topographic effects

An obstruction can also play an important role in determining plume rise and dispersion because it can disturb or deflect the airflow. The obstructions could be in the form of natural irregularities like hills and depressions and unusual ground roughness a well as tall structures in large urban areas. Generally, the effects of these obstructions are adverse due to downwash and increased ground concentrations.

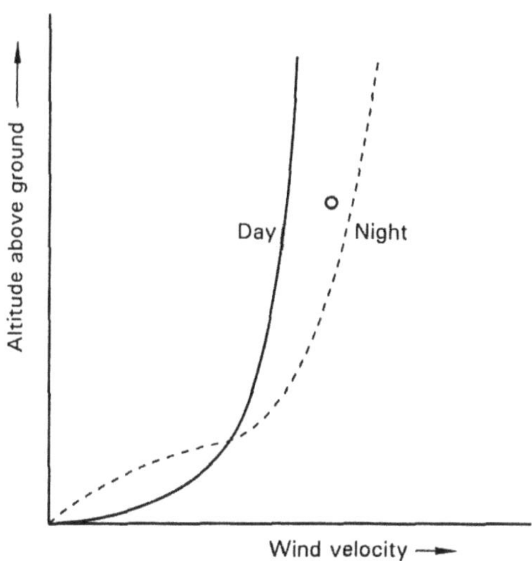

Fig. 5.22(c) *Wind velocity profiles during day and night*

Air movement near the surface of the earth is hindered by friction, which depends on surface roughness and this layer extends to about 1 km above the earth's surface. In addition to surface roughness, the time of the day influences the wind velocity profiles of the day. Solar heating produces thermal turbulence or eddies which set up convective currents so that turbulent mixing is increased resulting in a more flat velocity profile in the day than in the night [Fig. 5.22(c)]. The thermal stability of the atmosphere also influences the thermal turbulences which will be maximum on a clear sunny day in the afternoon and minimum at night or in the early morning.

Mechanical turbulence is another type of turbulence produced by shearing stress generated by air movement over the earth's surface, which increases with increasing surface roughness. The topographic effects on the wind velocity profiles are shown in Fig. 5.23. For smooth surfaces, the air velocity profile becomes very steep near the ground. For rougher surfaces such as those in urban areas there is more mechanical turbulence; the velocity profile becomes less steep, reaching deeper into the atmosphere.

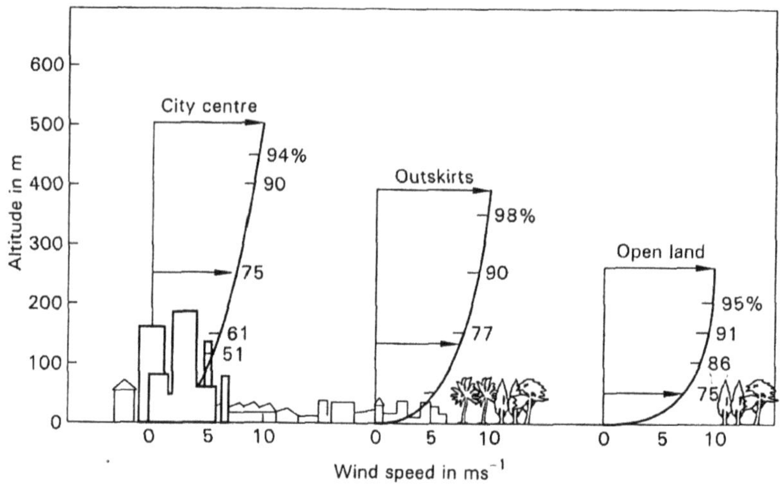

Fig. 5.23 *Profile of wind velocity in different terrains*

Mountains and valleys influence wind direction and magnitude as well as other meteorological parameters. On a clear night the ground is cooled by radiation to outer space, and a layer of air adjacent to it will be colder and hence denser than the air above it. If the ground were perfectly flat, this layer would be perfectly flat, and gravity would not tend to move it. But if the ground is not flat, then this denser layer will tend to flow downhill. The steeper the hill, the faster it flows. In any valley cold air flows down to the bottom, and then the collected cold air flows down the valley in the same direction that the stream or river in the valley flows. The mountain valley wind system is illustrated in Fig. 5.24.

Thus plume received from a stack in a valley will be confined within the valley due to the regularly changing wind patterns, resulting in pollutant concentrations building up to dangerous levels. During the day the opposite occurs. The sun-heated ground heats the air adjacent to it, which then rises by buoyancy. Normally one side of the valley will be more strongly heated by the sun than the other, so the air will begin to rise on that side, causing a rotating flow with its axis along the axis of the valley. This is superimposed on a net upward motion of the heated air, causing an up slope and up valley daytime flow that is generally not as strong as the night flow.

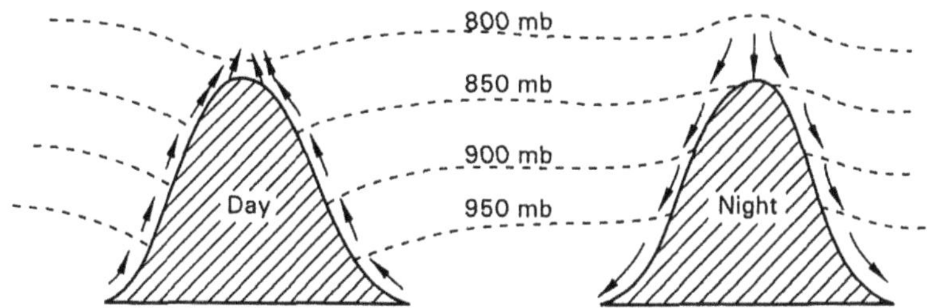

Fig. 5.24 *Mountain-valley wind system*

Land and sea breeze is a typical example of the effect of surface topography on the wind speed and direction (Fig. 5.25). No pressure gradients exists in the early morning hours as the conditions are calm, while by afternoon the air over the surface of the coast becomes warmer than at over the adjacent ocean resulting, in upper-level air movement from land to ocean and a low-level movement from the sea. However, the pressure gradient and wind reverses due to a more rapid cooling of the air layer at over the land which in turn sets up the land breeze. The pollutant emissions released with in the inflow night layer move inland during the day and return to the high up layers at night due to diurnal cycles which contribute significantly to the pollution problem of some large coastal cities where the concentrations inland are much higher than expected.

Any irregular terrain surrounding a stack can also influence markedly the plume nature. For example, in a deep valley the plume discharged from the stack will be carried to the ground when the wind blows over a cliff as shown in Figs. 5.26(a-d).This is known as "downwash" and occurs if the stack does not extend up to the rim of the valley.

5.3.11 Dust domes and heat islands

Dust dome is caused by a unique atmospheric circulation pattern that depends on the marked temperature differences between the city proper and outlying countryside

(Fig. 5.27). Although the average annual temperature of a city might be only 1.0 °C higher than that of the surrounding rural areas on a given day, occasionally a city may actually be 2.0 °C warmer. Meteorologists call this a heat island.

The air is warmed up in the city because of large numbers of industrial furnaces, cars, human bodies, heat-absorbing surfaces, etc. This warm air moves upward and eventually cools. It may then sink, because of its greater density, move back into the city along with other cool air from the countryside, and be warmed again. As a result, a circular flow of air occurs. Any dust in the air will remain suspended above the city in a mushroom-like formation known as a dust dome.

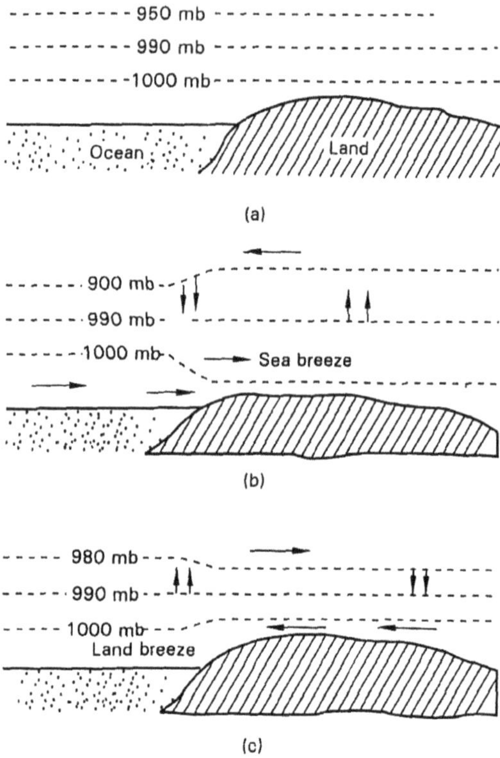

Fig. 5.25 *Sea and land breezes-the dashed lines represent isobars*

5.3.12 Weather maps

The condition of the atmosphere at any one time, important in weather forecasting, is usually shown on weather maps-maps based on measurements and observations at many stations throughout the country. Measurements of atmospheric pressure, temperature, humidity, wind speed and direction; amount, type and height of clouds; precipitation; visibility; and other special conditions in the atmosphere are plotted on weather maps. Plotted data are analysed by drawing isobars, i.e. lines of constant pressure, at 3-millibar intervals. The boundary lines (fronts) between different air masses are drawn and the type of front indicated. Shaded areas reveal incidence and type of precipitation; arrows indicate direction of the wind. Isotherms are usually drawn for 0 °F and 32 °F.

Even a cursory examination of these maps will reveal whether atmospheric conditions in a given area will be favourable for dispersion of pollution. For example, when a cold front moves into a warmer area the air next to the ground will become heated and results in an unstable atmospheric condition (pollutants will disperse readily). But an air mass approaching a colder area will get cooled near the ground, and a stable atmosphere will develop in the surface layers.

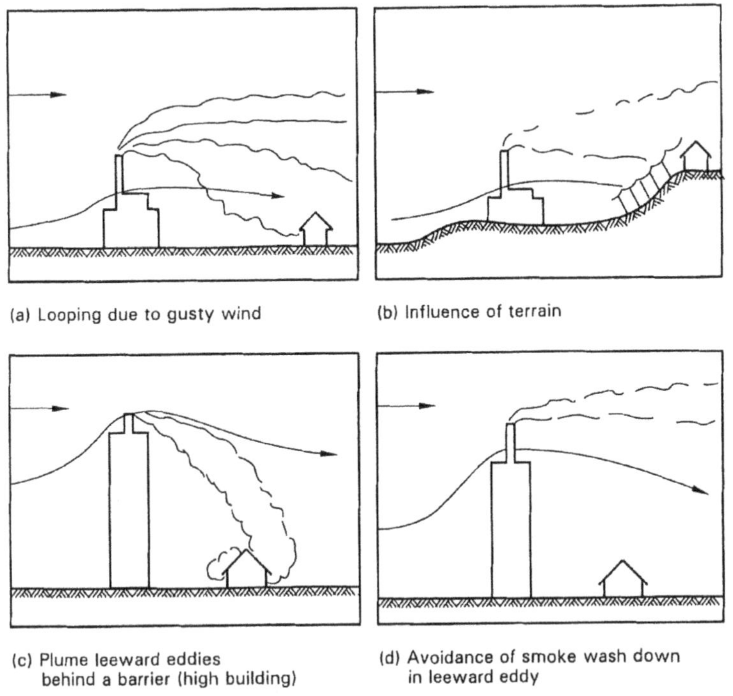

Fig. 5.26(a–d) *Specific meteorologic and topographic influences on the dispersion of smoke plumes, wind direction*

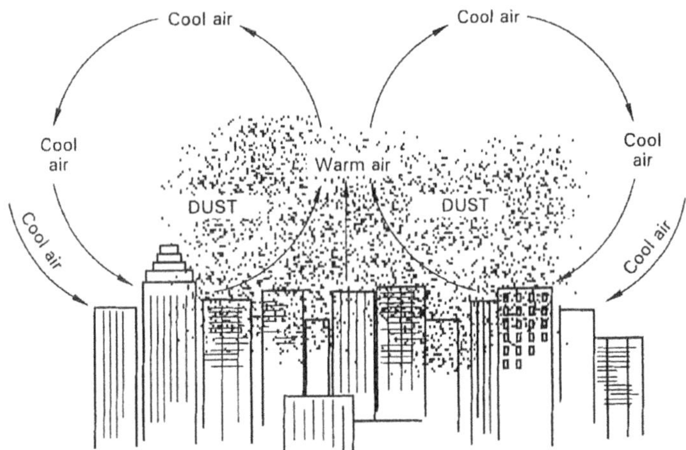

Fig. 5.27 *Heat island effect of a city*

The wind velocities shown on the map will be those of the gradient wind. These winds will be generally along the isobars and at a height (2000 ft and above) sufficiently high to be unaffected by surface friction. Flow will be clockwise around high-pressure areas and counter-clockwise around low-pressure areas. Near the ground, friction affects the speed and direction of the gradient wind so that surface winds generally blow between 20° and 30° across the isobars toward the low-pressure centre and at half the speed of the gradient winds.

5.3.12.1 Wind roses

Wind rose is the diagram designed to indicate the distribution of wind direction prevalent in a given place for a large period. In addition to daily weather maps, monthly maps of average wind data such as direction, duration and intensity are also prepared in maps which are known as wind roses. The wind roses are very useful for the quick estimation of pollutants in different directions and will vary from season to season and city to city. Normally wind roses will be prepared from data for a period of at least 10 years for getting realistic average data. A circle with data prepared by lines in eight main and eight secondary directions of the compass will be incorporated in a wind rose. The frequency of wind from a particular direction is indicated by the length of the line; the frequency of calm conditions is entered in the centre. Wind speeds from each direction are indicated by certain shades in the wind rose which exists over a period of time or season. Appropriate scale is used to represent percentage frequencies of wind directions and appropriate index shades, lines, etc. Meteorological services regularly prepare wind roses like those shown in Figs. 5.28(a-c). These summarize the frequency of winds of varying velocities and directions at one location. Normally one speaks of and plots a wind in terms of the direction from which it comes. A west wind blows from west to east. The wind rose in Fig. 5.28(b) is for Cincinnati, USA; that in Fig. 5.28(c) is for Hyderabad. The most common winds, governed by local topographic effects are the up and down valley winds, north and south. The strongest winds come from the northwest and are associated with the passage of winter storms.

Wind roses are particularly useful in determining areas and periods of stable high pressure atmospheric conditions conducive to serious air pollution problems. The high pressure area (anticyclone) will be the condition always responsible for extreme cases of local air pollution. When a high-pressure area becomes stagnant in one locality for several days, conditions for pollution buildup will be extremely favorable. In the central portion of the "high" winds will be very light and the atmosphere will be quite stable. It was exactly the situation that existed at Donora and the disaster potential could have been foreseen, at least at the end of the second day of these conditions.

In addition to large-scale effects of air movement, those of a micro-meteorological nature must not be overlooked. Local circulation and temperature variations in valleys and on the slopes of hills and mountains are extremely important from an air pollution viewpoint; similarly, the presence of buildings (with resultant eddies) and type of groundcover affect air movement. In estimating the wind direction at any time and any location, one can use the following guidelines.

1. Major, rapidly moving storms and fronts overwhelm local influences; local ground-level winds blow the way that the major storms dictate.

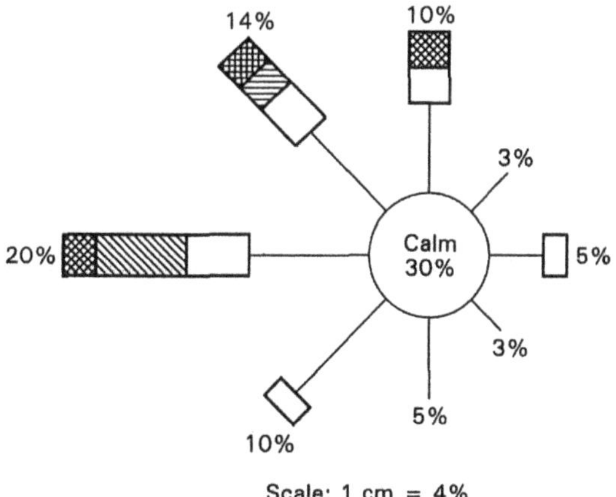

Fig. 5.28(a) *General wind rose parameters*

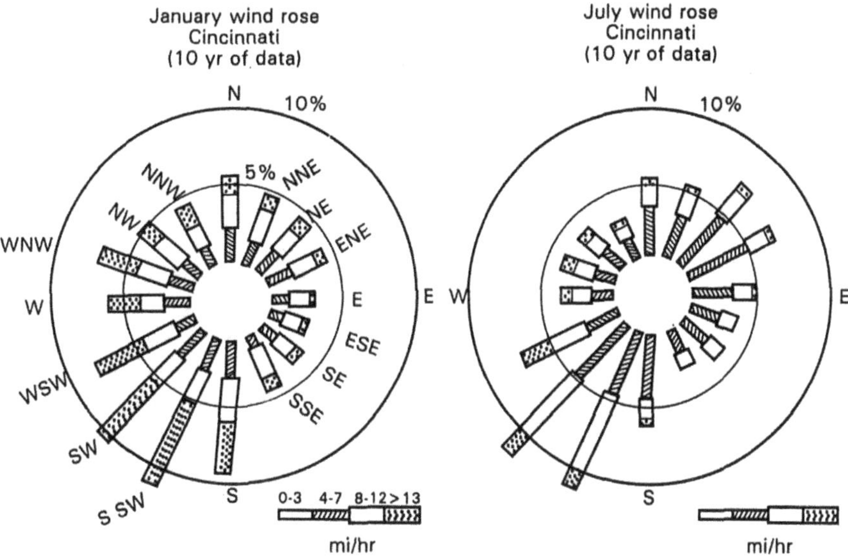

Fig. 5.28(b) *January and July wind roses, Cincinnati. The monthly distributions of wind direction and wind speed are summarised on polar diagrams. The positions of the species show the direction from which the wind was blowing; the length of the segments indicates the percentage of the speeds in various groups*

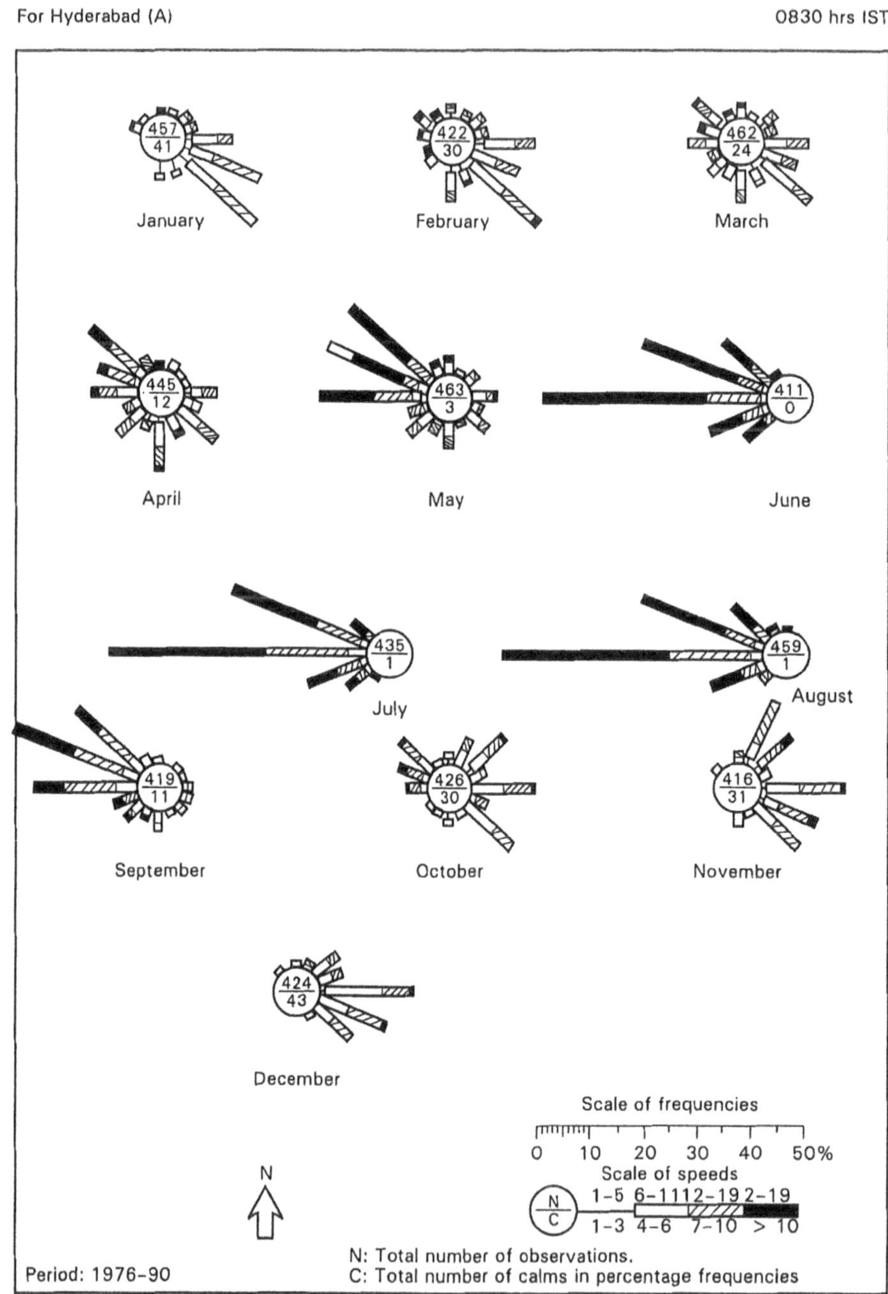

Fig. 5.28(c) *Wind roses for Hyderabad*

2. In deep valleys the daily alternation-wind up the valley in the daytime, down at night-overcomes most other influences and determines most of the local flow when no major storm or frontal passage dominates. The valley effect is greater in deep valleys than in shallow ones, in steep valleys than in gentle ones, at night than in the daytime, and under conditions of light wind and clear sky than of strong wind or cloudiness.
3. Onshore and offshore breezes dominate when there is no major storm. They are more likely to control the wind direction in light wind, clear sky conditions than in the opposite conditions, and more likely to control in the daytime than at night.

5.3.13 Pollutant wind roses: Pollutant-wind correlation

In air quality measurements wind direction and wind speed should always be measured to find out from which direction atmospheric pollutants are transported to the measuring site. How representative of a larger or smaller area the measured wind directions are depends on the local topography, and must be checked in each individual case. In the diagram the measured wind directions (the most frequent values per each half hour respectively) can, for example be divided into 12 or 16 sectors and frequency (in %) of the measuring valuables occurring in the individual sectors can be plotted in the circular coordinates. In this way the wind rose of the measuring period is obtained which can be divided according to wind speed section.

If every measured pollutant value is matched according to the wind direction (in 12 or 16 sectors), if the concentration's mean value is formed for each sector and if then these mean values are entered in the circular coordinates over the wind direction, then the so-called pollutant wind rose is obtained, Fig. 5.29. If there are bulges in the concentration curve in certain sectors of the pollutant wind rose, then increased atmospheric pollutant concentrations are carried along with winds coming from these directions.

The following can be recognized from the pollutant wind rose shown in Fig. 5.29, the highest NO and NO_2 concentrations occur during winds from northerly directions. The lowest concentrations occur during southwesterly winds.

A slight increase in the nitrogen oxide wind rose occurs during southeasterly winds, implying sources in this direction.

Pollutant wind roses can also be drawn up for different wind speed classes. Figure 5.29 shows a wind compass diagram divided into three wind speed classes. It can be seen, that during weak winds (0.5-2.0 m/s) higher nitrogen oxide concentrations occur than during stronger winds (2.0–5.0m/s).During strong winds (> 5 m/s) which solely come from southwesterly directions in this area, only very low nitrogen oxide concentrations occur.

The pollutants wind roses shown here do not provide information on the frequency of the concentrations determined in the individual sectors. It could well be possible that the wind directions during which high concentrations were measured occurred only very rarely.

If the concentrations of the pollutant wind roses are multiplied by the corresponding frequency of the wind direction (affecting duration) and then divided by the total concentration mean value for all sectors, a pollutant dose wind rose is obtained. The values thus calculated provide information on which percentage contribution is made by a wind direction sector to the mean pollutant load at the measuring site. Detailed information to he representation of wind direction-related pollutant loads including examples of such pollutant dose wind roses is given in by Baumuller and Reuter.

One can also determine the total load by adding up all concentrations measured as half hourly mean values in the measuring period; also the concentrations of each wind direction sector can be added up and their contribution to the total load be determined. In this way, too, a pollutant dose wind rose is obtained representing the contribution of each wind direction to the total load. There is no difference between this pollutant dose wind rose and the one according to Baumuller and Reuter relating to the mean value. The pollutant dose wind rose of Fig. 5.29 is based on the same measurements as those represented in Fig. 5.29. In this pollutant wind rose low concentrations, too, if they occur often enough contribute to the bulging of the dosage curve (e.g., direction southwest) owing to the high frequency of this wind direction. By comparison, the elevated concentrations measured during southeasterly winds Fig. 5.29 occur so rarely that they practically do not show up in the dose wind rose.

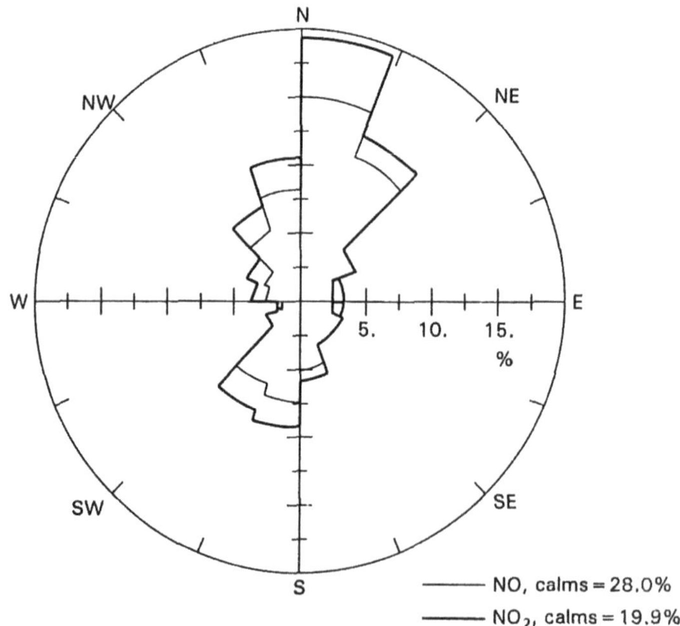

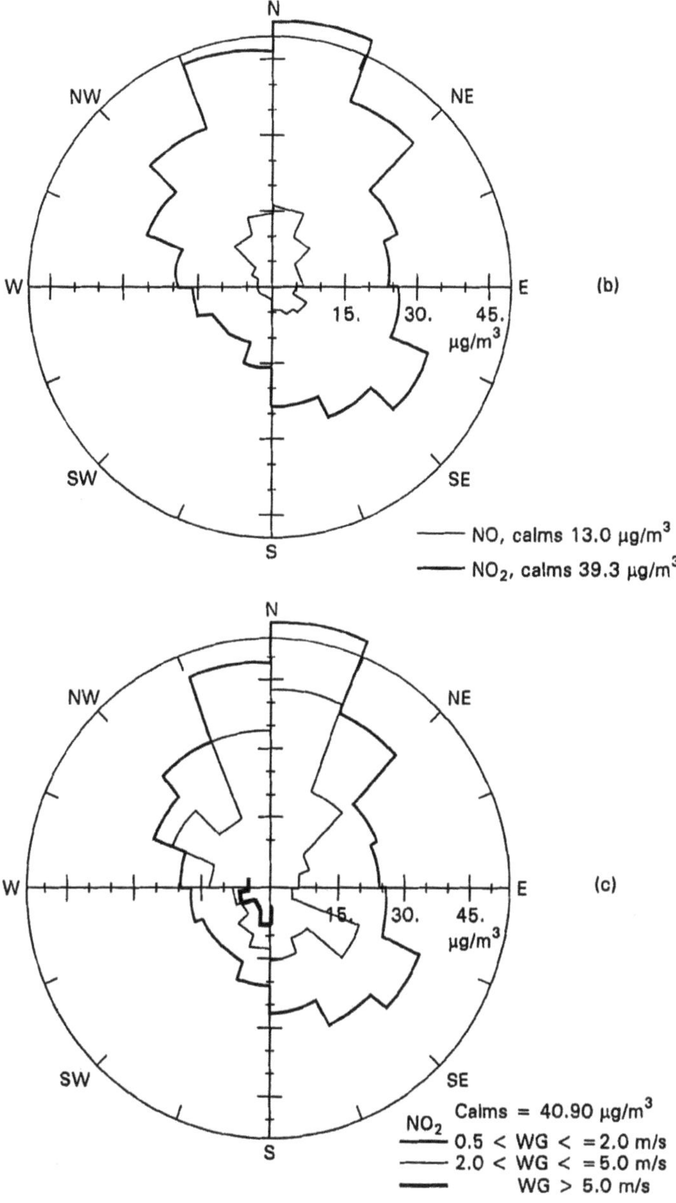

Fig. 5.29(a), (b) & (c) *Typical pollutant wind rose for* NO_x

5.3.14 Plume characteristics and plume behaviour

All industrial emissions are released through stacks (chimneys) of varying heights. These represent the stationary sources of air pollution with the exhaust gases dispersed over a region. Normally the plume or parcel of gaseous effluents released from a stack will be at a higher temperature and have drifted to the surrounding environment due to which it will be moving upward with buoyancy action.

Atmospheric stability and the wind turbulence conditions existing will determine the behaviour of the plume emitted from a stack of any industry.

5.3.14.1 Various types of plumes

The geometrical configuration of plumes [Fig. 5.30(a)] diffusing into the atmosphere was found to be a function of the air vertical stability [9,10] and their characteristics are discussed in the following pages.

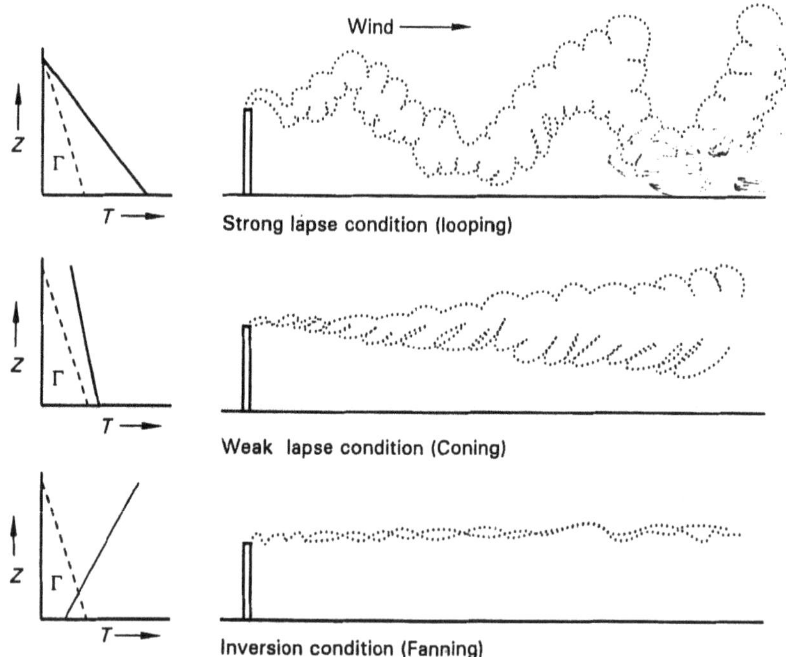

Fig. 5.30 (a) *Plume behaviour under various conditions of stability (dry adiabatic lapse rate, ambient rate)*

5.3.14.2 Looping

Looping is characteristic of unstable and turbulent conditions. The plume alternately ascends and descends or descends and ascends after leaving the stack. The high degree of turbulence rapidly disperses the plume but high concentrations of the exhaust gases may occur close to the chimney if the plume reaches the ground.

Looping plumes occur when there has been a super-adiabatic lapse rate and solar heating. The large thermal eddies in the unstable air may bring the plume to the ground level periodically making it necessary to use higher stacks. In general, however, the dilution of the plume with the surrounding air occurs arher rapidly. Looping occurs during warm seasons with clear skies when the solar insolation is high.

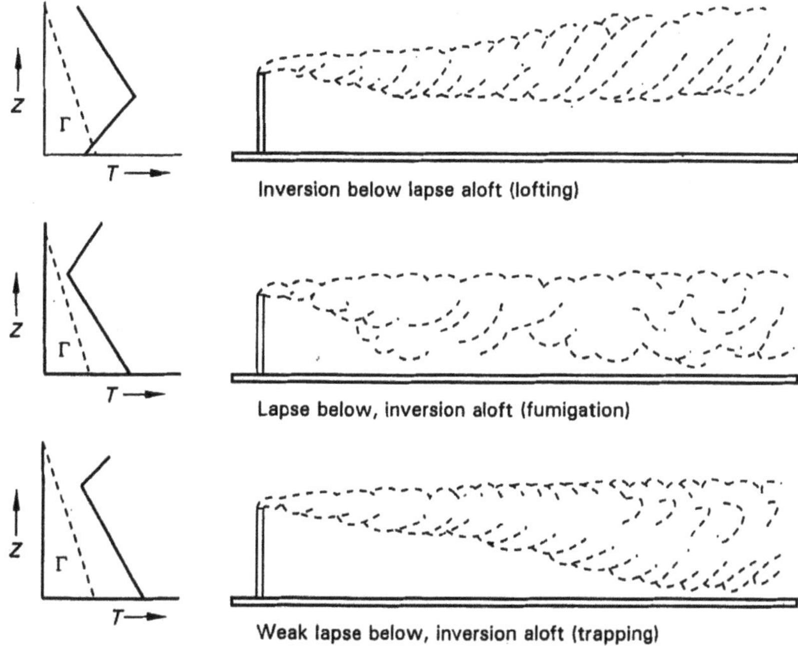

Fig. 5.30 (b) *Plume behaviour under various conditions of stability (dry adiabatic lapse rate, ambient rate)*

5.3.14.3 Coning

Coning is characteristic of neutral or near neutral stability. The plume is shaped like a cone and dispersion is slower than for looping, and reaches the ground only at greater distances. It occurs with overcast or night skies and moderate wind speeds.

A coning plume results when the vertical air temperature gradient has been between dry adiabatic and isothermal, the air being slightly unstable with some horizontal and vertical mixing occurring. Coning is most likely to occur during cloudy or windy periods. Under coning conditions high pollutant concentrations occur near ground level at larger distances and diffusion models work well for estimating dispersion of pollutants.

5.3.14.4 Fanning

Fanning is characteristic of stable conditions. The temperature gradient is positive, which represents a continuous state of inversion both above and below the plume. Under these conditions the vertical plume spread is minimized since the air is very stable. The plume meanders for long distances horizontally. Although it does not lead to high ground concentrations, fumigation conditions invariably set in if the plume hits an elevated terrain, and high ground level concentrations could result.

Fanning plumes spread out horizontally but do not mix vertically. Fanning plumes take place when the air temperature increases with altitude (inversion). The plume rarely reaches the ground level unless the inversion is broken by surface heating or the plume encounters a hill. At night, with light winds and clear skies, fanning plumes are most probable.

Fanning occurs when the plume is dispersed in the presence of very light winds as a result of strong atmospheric inversions. The stable lapse rate suppresses the vertical mixing, but not the horizontal mixing entirely. If the density of the plume is not significantly different from that of the surrounding atmosphere, the plume travels parallel to the ground in a flat, straight ribbon downwind, and on occasions plumes in a stable layer may be observed for 10 to 20 km downwind. For high stacks, fanning is considered a favourable meteorological condition because the plume does not contribute to ground pollution. A fanning plume is often observed at a height and in the early morning in all seasons.

5.3.14.5 Fumigation

The preceding three types of plume behaviour are observed under conditions of uniform lapse rate. When the lapse rate changes from stable to unstable a situation usually arising when an inversion is breaking up in the early morning when the sun comes up---a condition known as **fumigation** takes place.

If a pollutant source is located in a region that has a strong, ground-based inversion, then its plume of pollutants will be trapped in the inversion and will travel with the local wind with very little dilution, as sketched on the right of Fig. 5.31. In the left side of that figure we see the variation of atmospheric temperature as a function of time. At 6 a.m. there is a strong ground-based radiation inversion, caused by nocturnal cooling of the ground. As soon as the sun rays hit the ground, the ground's temperature rises, and an unstable layer is formed that eats away at the bottom of the inversion. By 9 a.m. the inversion is gone. Returning to the right side of this figure, we see that when the unstable layer reaches the plume, at perhaps 8.30 a.m., the plume will mix down to the ground. In this instance the plume will not have been diluted much from its initial concentration, so that the ground level concentration at that point and that time will be surprisingly high. The high concentration will not last long, but such short, intense exposures can damage crops, etc, This kind of event is called a fumigation, which most likely occurs with clear skies and light winds.

Fumigation can also occur if the plume from a shoreline source is carried inland by a stable onshore breeze. As the breeze passes inland it encounters warmed air from the solar-heated soil, which mixes in from below; if the stability of the shore breeze is strong enough, it will prevent the plume, that mixing from below may pull the plume to the ground, producing a fumigation.

Fumigation causes the high pollutant concentration plume reach the ground level along the full length of the plume and is caused by a super-adiabatic lapse rate beneath an inversion. The super-adiabatic lapse rate at the ground level occurs due to solar heating. This conditions is favoured by clear skies and light winds.

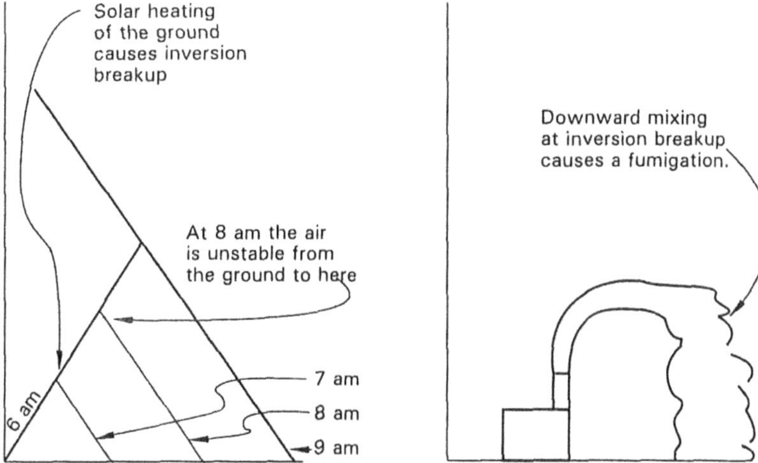

Fig. 5.31 *The development of a fumigation*

5.3.14.6 Lofting

Lofting is characteristic of inversion below the stack level and somewhat similar to fanning, although it occurs at a lower wind velocity. Lofting plumes diffuse upwards but not downwards and occur when there is a super-adiabatic layer above a surface inversion. A lofting plume will generally not reach the ground surface.

The conditions for the lofting plume are the inverse of those for fumigation; the lapse rate in the upper portion of the plume is unstable and that in the lower portion stable. When the pollutants are emitted above the inversion layer, they are dispersed vigorously in the upward direction since the top of the inversion layer acts as a barrier to the movement of pollutants towards the ground. Lofting is the most favourable plume type as far as ground level concentrations are concerned and is one of the major goals of tall-stack operation.

5.3.14.7 Trapping

Trapping refers to conditions where the plume is caught between inversions and can only diffuse between a limited vertical height. Trapping is similar to fumigation in that the plume is below an inversion; however, the plume does not reach the ground level along the plume length.

The trapped plume occurs when the plume effluent is caught between two inversion layers. The diffusion of effluent is severely restricted to the unstable layer between the two stable regions. Trapping may also be associated with subsidence inversions lasting for several days, where almost all emissions are trapped below the inversion layer creating one of the worst pollution situations.

5.4 Air Pollution Dispersion Models

5.4.1 Air pollution dispersion modelling

Air pollution dispersion modelling is a mathematical technique for predicting ambient air pollutant concentrations resulting from specific emissions. The impacts of emission sources from a particular source on the ambient air can be predicted by air quality models if information on meteorology, terrain, and present air quality are known.

The basic problem of air pollution dispersion modelling is to estimate concentrations of one or more pollutants and their variation in the space and time in relation to the meteorological conditions, removal and transformation processes and natural implications (Fig. 5.32).

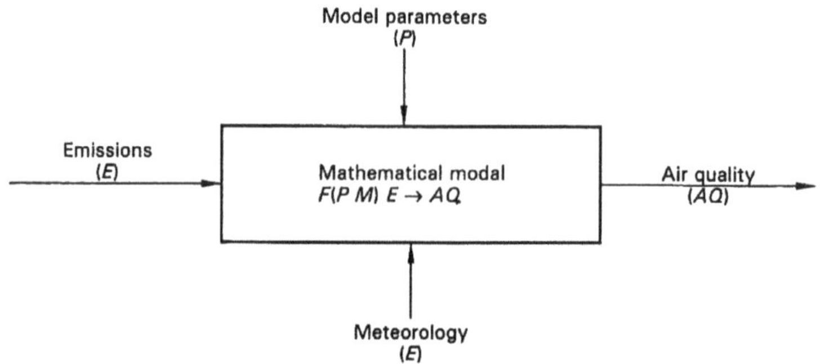

Fig. 5.32 *Schematic structure of an air quality model*

An overview of the various air pollution models is presented in Fig. 5.33.

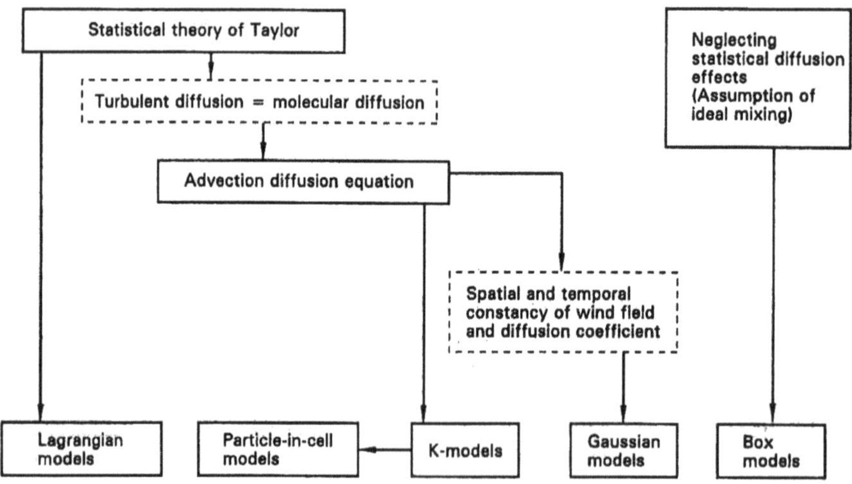

Fig. 5.33 *Overview of various air pollution models*

The dispersion of air-borne substances in the atmosphere depends on two different mechanisms:

(1) Carrying along with existing wind flows (advection), (2) distributed by diffusion movements due to atmospheric turbulence, the movements being superimposed on the average wind field. The course of pollutant dispersion can be calculated by also taking in to account turbulence effects determined by terrain features (orography), the structure of the ground (type and extent of building, vegetation, roughness), weather conditions, wind velocity and diurnal variations in radiation intensity (atmospheric layers).

5.4.2 Flow and turbulence models

Flow models can further classified into diagnostic and prognostic simulation models which will be used depending on the task and objective concerned.

5.4.2.1 Diagnostic models

One of the main applications of diagnostic models is for simulating the actual situation of a wind field based on a statistically proven series of measurements and a known experimentally investigated framework of conditions. This type of model is used mainly for long-term investigations and case studies during which a wide variety of dispersal situations is simulated.

5.4.2.2 Prognostic models

In this model the flows will be embedded in a larger scale floating flow field (geostrophic wind). Further, they will also be able to take into account smaller scale phenomena (topographically or thermically induced turbulences).

They can predict the development of a flow field in a spatial scale of 2.5to 2500km and time-wise from just a few hours to several days.

The mathematical treatment of flow processes with prognostic models is based on a set of partial non-linear differential equations. A three dimensional grid, for which this equation system must be solved with numerical methods, is superimposed on the model area where a flow field is to be determined. By determining the suitable coordinate system and choosing the grid points carefully, the model's resolving power can be adapted to the requirements given by the orography of the area under investigation.

5.4.3 Area source deterministic models

This model uses mainly mathematical meteorological calculation methods originating from the conservation theorems of mass energy and impulse, while the mathematical description of turbulence effects are described by the statistical theories of Taylor. Based on these theories, the physical interrelationships of turbulent wind velocity variations which overlay the medium altitude flow will be evaluated.

5.4.4 Lagrange models

The Lagrange models (Monte Carlo-simulation models, particle simulation, trajectory models) are based on the static diffusion theory which assumes that air pollutants are always carried along with the average wind flow, while diffusion is statistically random.

The path of an air parcel containing pollutants (particles) is calculated in a flow field for a longer period of time, and the temporal and spatial movements of the air are evaluated in the model. In the course of this movement an air element absorbs pollutants on its path across sources and releases them over sinks. If this process is simulated for particles which are released by certain emission sources, pollutant concentration can be determined from the spatial distribution of these particles according to a random given dispersion time at a certain location.

5.4.5 K-models

K-models are based on the fact that turbulent diffusion can be treated analogous to molecular (Brownian) diffusion and the turbulence-dependent substance transport is treated as being proportional to the concentration gradient of a pollutant gas between inlet and outlet of a controlled volume. The proportionality factor is, in this case, the turbulent exchange coefficient *K,* for which this type of model is named. As turbulence-induced movements of air parcels differ in vertical and horizontal directions, they must be assigned different diffusion coefficients.

The central equation for K-models is the advection-diffusion equation which can be derived from the continuity equation in general formulation:

$$\frac{\partial c(\overline{x},t)}{\partial t} = -\overline{u}\,(\overline{x},t).\nabla\left(K(\overline{x},t).\nabla c(\overline{x},t)\right) + Q(\overline{x},t) + S(\overline{x},t) \qquad \text{.....(5.30)}$$

where

$\overline{x}$ location vector,

t time,

$c(\overline{x},t)$ pollutant concentration at location x at time t,

$\overline{u}(\overline{x},t)$ time and location dependent (medium altitude) wind field,

∇ Nabla operator,

$K(\overline{x},t)$ diffusion tensor containing as diagonal elements diffusion coefficients in all three coordinate directions,

$Q(\overline{x},t)$ mass flow of time and location dependent pollutant sources,

$S(\overline{x},t)$ mass flow of time and location dependent pollutant sinks.

The differential equation for the description of time-related change in pollutant concentrations at a defined location at a certain time is composed of a transport term(advection term) caused by the driving average wind field, a diffusion term taking into account turbulence-induced movements and spatially and temporally varying source and sink terms.

Generally formulated, the advection-diffusion equation cannot be solved completely (mathematically definitely). For this reason numerical processes must be employed to solve this equation system. The differential quotients of the advection-diffusion equation are then replaced by differential quotients between spatial and temporal points, where by a comprehensive system of equations is available for the calculation of concentrations closely linked in space and time. By following a certain framework of conditions, this system can be solved step by step.

Depending on the type of concept used to solve the advection-diffusion equation one differentiates between different types of K-models:

When working with Euler's grid models a suitable, fixed spatial grid is imposed on the distribution area and the space to be investigated is partitioned into a multitude of individual volume elements. At the individual grid points the advection-diffusion equation is served with the help of finite difference equations for discrete sections of time. As these computing operations are highly time-intensive and require large memory capacities, the use of Euler's models is largely restricted to scientific questions.

In particle-in-cell models diffusion velocity is determined via the concentration gradient and imposed on the advection movement determined by the wind field. With the thus derived pseudo transport velocity' the continuity equation follows on from the advection equation via a formal simplification. This continuity equation can then be solved in a fixed spatial grid with numerical processes.

5.4.5.1 Fixed box models

The simplest one of the above class is known as the box model, which equates ambient concentrations to the mass of pollutants within a specified volume of the atmosphere divided by the volume. More complex dispersion models attempt to simulate the mass balance within individual parcels of air. These models employ equations which simulate the various processes that influence this mass balance. Such processes includes transport. turbulent diffusion, and reaction of pollutant species. The complexity of dispersion models depends upon the assumptions one is willing to make. Models may be formulated to describe the behaviour of non-reactive, inert pollutants. A steady-state assumption is commonly employed in many dispersion models. The assumption says that the processes affecting the mass balance do not change with time.

Consider an urban airshed which can be represented by a rectangular box, such as is shown in Fig. 5.34 with base dimensions length L and width W, and height H and oriented such that wind, with speed u, is normal to one side of the box. The height of the box will be fixed based on atmospheric conditions, which will be normally the mixing height and rate of emission per unit area, of q_s (g/m^2– s).

The air is blowing into the box on the upwind side with pollutant concentrations C_m and for simplicity, we can assume:

(a) That no pollution is lost from the box along the sides parallel to the wind or from c the top.
(b) That the pollutants are rapidly and completely mixed in the box, creating a uniform average concentration.
(c) That the pollutants do not react, decay, or fall out of the airstream. For more advanced versions of the box model these restrictions can be modified.

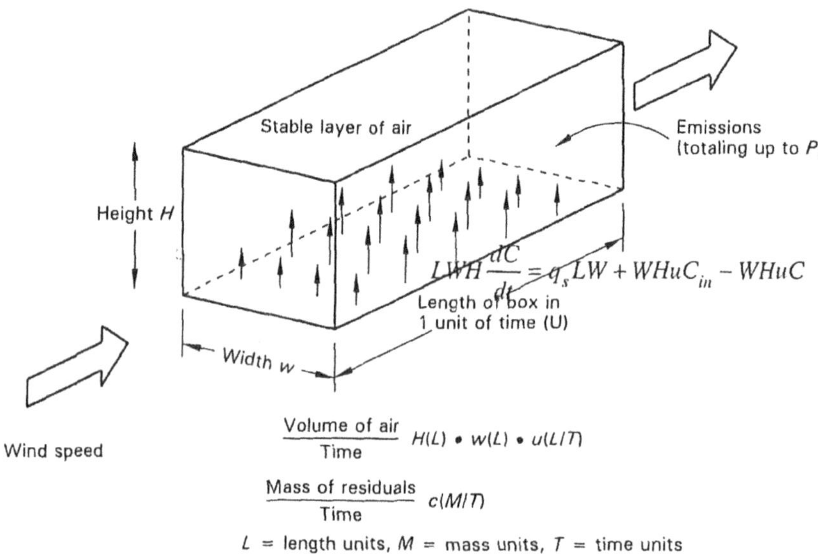

Fig. 5.34 *Rectangular city box, showing meaning of symbols used in the fixed box model*

The amount of pollution in the box will be equal to the volume of the box times the concentration, *LWHC* while the rate at which air will be entering and leaving the box will be equal to the product of wind speed and area of the box *WHu*. The rate at which pollutant will be entering the box is $WHuC_{in}$ while the rate at which it leaves the box is given by *WHuC*. If we assume the pollutant is conservative, then we can write the following mass balance for the box.

(rate of change of pollution in the box)

= (Rate of pollution entering the box) – (rate of pollution leaving the box)

where C = pollution concentration in the airshed

C = concentration in the incoming air

q_s = emission rate per unit area

H = mixing height

L = length of airshed

W = width of airshed

u = average wind speed against one edge of the box

The steady state solution to Equation 5.30 can be obtained by simply setting $dC/dt=0$ so that

$$C(\propto)=\frac{q_s L}{uH}+C_{in} \qquad .....(5.31)$$

The steady state concentration of the pollutant in the box is proportional to the emission rate and inversely proportional to the ventilation coefficient (the product of mixing depth and wind speed).

To compute the air pollutant concentration using this model the following major simplifying assumptions are to be made:

1. The air shed is a rectangle with dimensions W and L and with one side parallel to the wind direction.
2. Atmospheric turbulence produces complete and total mixing of pollutants up to the mixing height H and no mixing above this height.
3. This turbulence is strong enough in the upwind direction so that the pollutant concentration is uniform in the whole volume of air over the city and not higher at the downwind side than the upward side. This assumption is quite contrary to what we observe in nature but permits a great simplification of the mathematics.
4. The wind blows in the x direction with velocity u. This velocity is constant and is independent of time, location, or elevation above the ground. This is also contrary to observation; wind speeds increase with elevation. Here we use the average u between that at ground and that at H.
5. The concentration of pollutant in the air entering the city (at $x = 0$) is constant and is equal to C_{in} for background concentration. Concentrations in this model are usually in units of g/m^3 or micrograms/m^3.
6. No pollutant leaves or enters through the top of the box, nor through the sides that are parallel to the wind direction.
7. The pollutant in question is sufficiently long-lived in the atmosphere so that the destruction rate is zero.

With these assumptions, one can now evaluate all the terms in Equation 5.30 for a city or air shed with a volume of WLH. Because all of the assumptions indicate that flows and emission rates are independent of time, we see that this is a steady state situation in which nothing is changing with time. For any steady state situation in any application of the general balance equation, the accumulation rate is zero, so the term to the left of the equation is zero.

Equation 5.30 is a great simplification of what really occurs in nature. However, all of the important variables enter, with the correct signs and powers. It correctly indicates that the upwind concentration for a long-lived pollutant is additive to the concentration produced by the city, and that the latter increases with increase in q and L and decreases with increase in u and H. If the air entering the box is not clean we have to add the effect of incoming concentration also.

By far the worst of the foregoing assumptions is the third-that the concentrations at the upwind and downwind edges of the city are the same and independent of time which may not be true. Holzworth [6] developed a somewhat more complex form of the equation by replacing this assumption with a more realistic one.

Equation 5.30 can be solved to obtain the time-dependent increase in pollution above the city. Letting *C*(0) be the concentration in the air shed above the city (the box) at time *t* = 0 the solution becomes

$$C(t)=\left(\frac{q_s L+}{uH}C_{in}\right)1-e^{-ut/L}+C(0)e^{-ut/L} \qquad .....(5.32)$$

If it is assumed that no pollution is brought by the incoming wind and if the initial concentration in the box is zero then Equation 5.32 simplifies to

$$C(t)=\frac{q_s L}{uH}\left(1-e^{-ut/L}\right) \qquad .....(5.33)$$

where t = *L/u* the exponential function becomes $e^{-1}$and the concentration reaches 63% of its final value and *t* is termed as time constant or ventilator *t* or on residence time.

The second worst assumption is that the emissions are uniformly distributed over the area of the city (i.e. q is constant over the whole city). If indeed we have perfect mixing within the box, this assumption makes no difference. But if we drop the perfect mixing assumption, this uniform emission assumption becomes the worst. Frequently we have information on the variation of *q* from place to place in the city. For example, we would assume that for most pollutants *q* is low in the suburbs and much higher in industrial areas. Hanna [11] has presented a modification of the simple box model that incorporates the same idea presented by Holzworth and also allows one to divide the city into subareas and apply a different value of *q* to each.

The simple fixed-box model of Equation 5.30 and its modifications, as well as most of the others presented, predicts concentrations for only one specific meteorological condition. To find the annual average concentration of some pollutant, we would have to use the frequency distribution of various values of wind direction, *u* and H, compute the concentration from Equation 5.30 for each value, and then multiply by the frequency and sum to find the annual average; that is,

$$\left(\begin{array}{l}\text{Annual average}\\ \text{concentration}\end{array}\right) \sum_{\text{overall all meteorologies}} \left(\frac{\text{concentration for that meterology}}{\text{Frequency of occurrence of that meteorology}}\right)$$

Estimation of CO in city due to motor vehicles

In an urban air shed which is a 20 km square, 20000 cars with length 20 km are on the road, each being driven 40 km between 4 pm and 6 pm, and each emits CO for 5 g/km. On a clear winter evening a radiation inversion is restricting the mixing height to 30.0m. The wind is bringing in clean air at a steady rate of 2.0 m/s along the edge of the city. Estimate the CO concentration at 7 pm using the above model if there was no CO in the air at 4 pm, and the only source of CO is cars. Further also assume that CO is conservative and that there is complete and instantaneous mixing in the box.

Solution: q_s emission per m^2 which is equal to

$$q_s = \frac{20,000 \times 40 \text{ km/car} \times 5\text{g/km}}{\left(20 \times 10^3 \text{m}\right)^2 3600\text{s/hr} \times 3\text{hr}} = 0.1389 \times 10^{-6} \text{g/m}^2\text{s}$$

The concentration after 3 h (10800 s) will be .....(5.34)

$$C(t) = \frac{q_s L}{\mu H}\left(1 - e^{-ut/2}\right)$$

$$C_{(3h)} = \frac{0.1389 \times 10^{-6} \text{g/m}^2\text{s} \times 20 \times 10^3 \text{m}}{2.0 \text{ m/s} \times 30.0 \text{ m}}\left[1 - \exp\left(\frac{2.0 \text{ m/s} \times 10800\text{s}}{20000 \text{ m}}\right)\right]$$

$$= 4.6 \times 10^{-3} \text{g/m}^3 \times 0.923 = 4.245 \times 10^{-3} = 4.245 \text{ mg/m}^3$$

This value is lower than the 1 h standard 4 mg/m^3 or 8 h standard 10 mg/m^3 which indicates that the build up of pollution loads are acceptable.

5.4.6 Gaussian plume model

For the steady-state concentration downwind from a continuous point source the Gaussian plume model was presented by Sutton [3] and further developed by Pasquill [12] and Gifford [13]. In this model the concentration distribution perpendicular to the plume axis is assumed to be a Gaussian distribution in both horizontal and vertical planes and the extent of plume growth in these planes is measured by the standard deviation.

In the Gaussian plume approach the expanding plume has a Gaussian, or Normal, distribution of concentration in the vertical (z) and lateral (y) directions.

Figure 5.35 depicts the nature of the plume coming out. The coordinate system has been set up to show a cross-section of the plume, with z representing the vertical direction and x being the distance directly downwind from the source.

As stack emissions will be at elevated temperatures they will have some initial upward velocity and. buoyancy and the plume before it takes up a symmetrical path about the centre line, will travel some distance downwind and the centre line will be above the actual stack height. The highest concentration of pollutants will be spread along this centre line which will decrease in the concentration values as we go further and further away. This model assumes that the pollutant concentration follows a normal distribution in both the vertical planes, as is shown in the figure, and in the horizontal plane and treats the emissions as if they came from a virtual point source along the plume centre line, at an effective stack height H.

The average steady state pollutant concentrations are related to the source strength, wind speed, effective stack height, and atmospheric conditions. These can be derived from the nature of various factors involving gaseous diffusion in three-dimensional space.

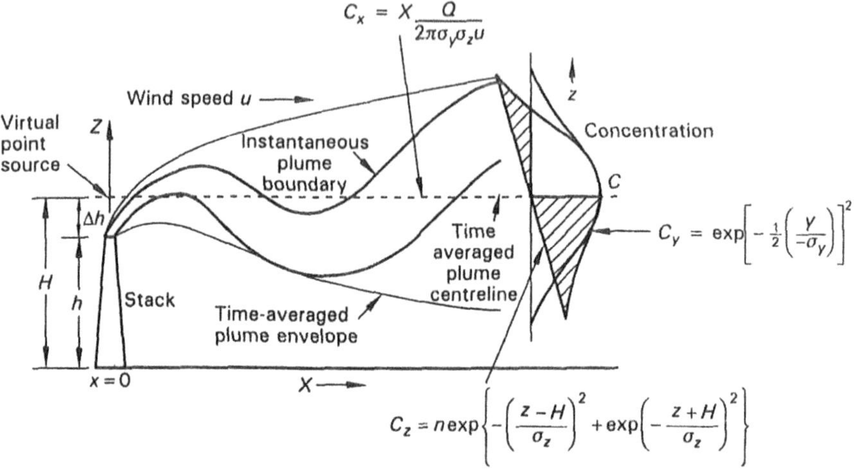

Fig. 5.35 *The instantaneous plume boundary and a time-averaged plume envelope*

The Gaussian plume equation for the concentration C (in units of μgm^{-3} for example) at any point (x, y, z) in the three-dimensional coordinate system of the plume is then given by Turner [14]

$$C(x,y,z) = \underbrace{\frac{Q}{2\pi u \sigma_y \sigma_z}}_{\text{central line}} \exp\left[\underbrace{\left(-y^2/2\sigma_y^2\right)}_{\text{horizontal}}\right] \quad .....(5.35)$$

$$\underbrace{\left(\exp-(z-H)^2/2\sigma_z^2 + \exp-(z+H)^2/2\sigma_z^2\right)}_{\text{vertical}}$$

$C(x, y, z)$ = centre line concentration of pollutant A at any point (x, y, z) in space, $\mu g/m^3$

Q = source strength i.e. pollutant emission rate, ug/sec

u = average wind speed at the source level (at stack height), m/sec

H = Effective stack height, m

H = $h + \Delta h$ where h = physical height of stack, in metres

Δh = plume rise, in metres

$\sigma_{□}$ and σ_z are the standard deviations of the concentration of the pollutant at A in the horizontal crosswind and vertical directions respectively. These dispersion coefficients vary with the time of sampling, the basic time of sampling being one hour. σ_y and σ_z increase with downwind distance, x signifying that the dilution increases with distance. From the Gaussian plume Equation 5.35 one can calculate the downwind Ground Level Concentration (GLC) of pollution emitted from a point source which is normally required for pollution measurement plans and implementation of control technologies. As per Equation 5.35, GLC is dependent on source Q expressed in ug/m^3, decreases with increasing stack height and is inversely proportional to the wind speed. For a given situation the downwind GLC can be calculated and appropriate reduction in the source or increase in the stack length can be suggested for achieving acceptable down wind GLC values for the pollutant.

Equation 5.35 has three terms corresponding to the (1) the centre line concentration as a function of downwind distance, (2) the lateral spread relative to the centre line value, (3) the vertical spread relative to the centre line value, which are shown pictorially in Fig. 5.35.

$$\text{Top} = C_x = \frac{Q}{2\Pi\sigma_y\sigma_z u}$$

$$\text{Middle} = C_y = \exp\left[-\frac{1}{2}\left(\frac{y}{\sigma_y}\right)^2\right]$$

$$\text{Bottom} = C_z = \exp\left[-\left(\frac{z-H}{\sigma_z}\right)^2\right] + \exp\left[-\left(\frac{z+H}{\sigma_z}\right)^2\right] \qquad .....(5.36)$$

5.4.6.1 Estimation of dispersion coefficients σ_y, σ_z

The dispersion coefficients σ_y, σ_z are standard deviations of plume concentration in crosswind σ_y and vertical σ_z directions. They reflect the growth dimension of a Gaussian plume as a function of atmospheric stability, distance travelled, travel time from the source and are influenced by source height and surface roughness.

In addition to other physical parameters like coordinate values *x, y, z*, emission rate *Q*, effective height *H*, travel speed etc., the values of dispersion coefficients are necessary for precise estimation of the concentration of the pollutant emitted from a continuous plume.

Turbulent motions or eddies in the atmosphere vary in size and intensity; the greater their size and/or intensity the more rapid is the plume growth and hence the dilution of the pollutants. Small-scale turbulent motions tend to dominate the plume growth close to the point of emission where the plume is still relatively small and the larger scale eddies dominate at greater distances. Furthermore, the small and the larger eddies are associated respectively with short and larger time scales. Consequently, σ_y and σ_z increase in value with distance from the source; also they increase with the time or sampling period over which they have been measured. This latter point means that it is essential to state the sampling period to which σ_y and σ_z apply, especially if comparisons are being made between calculated and measured concentrations; ideally the two periods should be identical. Most values of σ_y and σ_z to be found in the literature are for sampling periods in the range 3-6 minutes. It is also evident that σ_y and σ_z are dependent on atmospheric stability, being smallest when the atmosphere is most stable (category G) i.e. when atmospheric turbulence is least, increasing to their greatest values in highly turbulent very unstable conditions (category A). The underlying surface roughness elements also play a part, σ_y and σ_z increasing with increasing surface roughness so that for a given distance downwind of a chimney σ_y and σ_z will be larger in, for example, an urban area than in an area of open relatively flat agricultural area.

σ_y and σ_z values for different atmospheric conditions representing different stability classes viz. A to F calculated by Turner are given in (Table 5.7). Stability classes *A* to *C* represent unstable atmospheric conditions; while E and F represent stable conditions class *D* represents neutral condition. Regardless of the wind speed, class D is assumed for overcast conditions, day or night.

Table 5.7 *Atmospheric stability classifications*

Surface wind speed	Strong	Day solar insolation Moderate	Slight	Night cloudiness	
				Cloudy	Clear
<2	A	A-B	B	E	F
2-3	A-B	B	C	E	F
3-5	B	B-C	C	D	E
5-6	C	C-D	D	D	D
>6	C	D	D	D	D

A-very unstable; B-moderately unstable; C-slightly unstable; D-neutral; E-slightly stable; F-stable

While using these tables the following restrictions should be kept in mind clearly.

1. The sampling time should be kept nearly 10 min while estimating the concentrations from these charts.
2. σ_y and σ_z values correspond to open terrain and flat grassland and necessary corrections should be incorporated for other topographic changes.

By consulting these tables one can estimate the pollutant dispersion from a given source for a typical stability class of the region and identify which type of atmospheric conditions will result in greater level concentrations. Table 5.8 lists σ_y and σ_z values for the six stability classes for some arbitrarily chosen downwind distances.

Table 5.8 *Approximate values of σ_y and σ_z as a function of downwind distance for various stability classes in metres.*

Distance (km)	Stability Classes and σ_y values						Stability Classes and σ_z values					
	A	B	C	D	E	F	A	B	C	D	E	F
0.2	51	37	25	16	12	8	29	20	14	9	6	4
0.4	94	69	46	30	22	15	84	40	26	15	11	7
0.6	135	99	66	43	32	22	173	63	38	21	15	9
0.8	174	128	85	56	41	28	295	86	50	27	18	12

Table 5.8 ***Contd...***

Distance (km)	Stability Classes and σ_y values						Stability Classes and σ_z values					
	A	B	C	D	E	F	A	B	C	D	E	F
1.0	213	156	104	68	50	34	450	110	61	31	22	14
2.0	396	290	193	126	94	63	1953	234	115	51	34	22
4.0	736	539	359	235	174	117		498	216	78	51	32
8.0	1367	1001	667	436	324	218		1063	406	117	70	42
16	2540	1860	1240	811	602	405		2274	763	173	95	55
20	3101	2271	1514	990	735	495		2904	934	196	104	59

Source: *D. B. Turner. 1970. Workbook of atmospheric dispersion estimates. Washington D.C. [14]*

5.4.6.2 Advantages and limitations of Gaussian model

This model is easy to understand, efficient in computer running time, appealing conceptually and results agree well with experimental data.

The Gaussian plume model is based on the following assumptions:

(a) The pollutant plume coming from a stack will have Gaussian distribution in both horizontal and vertical planes.

(b) σ_y and σ_z represent the standard deviations of the concentrations of the plume in horizontal crosswind and vertical directions respectively.

(c) The plume will be affected by the wind speed existing at the source level.

(d) The rate of emission of the pollutant from the source (Qg/sec) is continuous and uniform.

(e) Steady-state conditions prevail through the dispersion and factors affecting the dispersion of the pollutant do not change with time and space.

(f) The pollutants emitted from the stack are inert and do not undergo any chemical reaction in the atmosphere.

(g) Perfect reflection of the plume at the underlying surface, i.e. no ground adsorption.

(h) The turbulent diffusion in the *x*–direction is neglected relative to advection in the transport (*x*) direction, which implies that the model should be applied for average wind speeds of more than 1 m/s (u > 1 m/s).

(i) The coordinate system has it *x*-axis in the direction of the flow, and the *v* (lateral) and *w* (vertical) components of the time averaged wind vector are set to zero.

(j) The Gaussian plume equation is derived based on the assumption that the terrain in which plume is dispersed is flat.

(k) All variables are averaged over a period of about 10 min, which implies that for different averaging times, corrections to the equation have to be made.

5.4.6.3 Limitations

1. The Gaussian plume model does not take into consideration the formation of different stability layers at different heights in the atmosphere and at different times of the day.
2. For varying terrain characteristics such as terrain roughness, mountains, valleys, land increases and water bodies, the model needs corrections.
3. Free convection regions and strong wind shears like change of wind directions and change of wind speed with heights are not taken into consideration.
4. Downwind, concentrations of the pollutant can be calculated only for shorter distances and for shorter travel times.

Though this model gives good values for downwind distances between approximately100 m and 2000 m, for distances below 100 m, the value has to be corrected by considering the details of the wind flow pattern around the source structure. Similarly, for distances greater than 1000 m, the local terrain features and meteorological variability have to be taken into consideration.

The dispersion coefficients σ_y and σ_z used in the equation are not very accurate. Also, the model neglects the turning of the wind due to frictional effects which leads to spread of the plume in the cross-wind direction. Also, any absorption or deposition of pollutants when the plume reaches the ground is neglected. If the plume is reactive, the model gives inaccurate results. Hence, care is required in using the equations.

The steady state assumption implies that the Gaussian plume equation can be applied only for shorter distances (order of 10 km) and for shorter travel time (order of 2h).

Despite their disadvantages, the Gaussian plume models are still the most commonly applied method for calculating dispersion of pollution from point sources.

As the Gaussian plume equation is applicable only for shorter sampling time of few minutes, for estimating the concentrations from a point source for longer sampling times, the following equation can be used

$$C_A = C_B \left(t_B / t_A \right) k$$

where C_A = concentration of pollutant at the sampling time, t_A; C_B= concentration of the pollutant at the shorter sampling time, t_B and k is a constant having a value between 0.17 and 0.2. The short-term models can be used to calculate concentration for time periods of 30 min to one day.

Pasquill [12] and Gifford [13] also have derived analytical expressions (Equations 5.37 and 5.38) for calculating the dispersion σ_y and σ_z coefficients for any downward distance at different stability categories from the weather conditions.

$$\sigma_y = \alpha x^b \quad .....(5.37)$$

$$\sigma_z = cx^d + f \quad .....(5.38)$$

where x is the downwind distance expressed in km and b has a value of 0.894. The values of a, c, d and f for different categories are functions of downwind distance x and are given in Table 5.9.

The following equations can be derived from the basic Gaussian plume equation for estimating ground level concentrations of the pollutants at different situations.

1. Equation for estimation of ground level concentrations (GLC) with plume having no vertical dispersion. In this case $z = 0$ and Equation 5.35 becomes

$$C(x, y, 0, H) = \frac{Q}{\pi\sigma\ y\sigma\ zu} \exp\left\{-\left(\frac{y^2}{2\sigma^2 y}\right)\right\} \exp\left\{-\left(\frac{H^2}{2\sigma^2 z}\right)\right\} \quad .....(5.39)$$

2. Equation for the estimation of ground level concentration of the pollutant along the central line of the plume emitted from the elevated source (stack).

   In this case z = 0 and y = 0 and Equation 5.35 becomes

$$C(x,0,0,H) = \frac{Q}{\pi\sigma\ y\sigma\ zu} \exp - \left\{\left(\frac{H^2}{2\sigma^2 z}\right)\right\} \quad .....(5.40)$$

   However if the emission source (stack) is also at ground level Equation 5.40 simplifies to

$$C(x,0,0,0) = \frac{Q}{\pi\sigma y\sigma zu} \quad .....(5.41)$$

3. Calculation of maximum ground level concentrations (MGLC) from the stack of height H,

Table 5.9 *Values of constants to be used in equation as a function of downwind distance and stability condition*

Stability	$x \le 1$ km			$x \ge 1$ km			
	a	c	*d*	*f*	c	*d*	*F*
A	213	440.8	1.941	9.27	459.7	2.094	–9.6
B	156	106.6	1.149	3.3	108.2	1.098	2.0
C	104	61.0	0.911	0	61.0	0.911	0
D	68	33.2	0.725	–1.7	44.5	0.516	–130
E	50.5	22.8	0.678	–1.3	55.4	0.305	–34.0
F	34	14.35	0.740	–0.35	62.6	0.180	–48.6

Source: *D. O. Martin. 1976. J. Air Pollution Control Assoc. 26. No. 145. [15]*

The MGLC for any given stack height H can be calculated by using Equation 5.40 and substituting the following relation between the spreading coefficients σ_y and σ_z, Equations 5.37 and 5.38 for various stabilities given in Table 5.9 and at downwind distance of maximum concentration assuming the relation $\sigma_y = \sigma_z$ we will have

$$\sigma_z = H\sqrt{2} \qquad .....(5.42)$$

Substituting for $\sigma_z = H/\sqrt{2}$ in Equation 5.40 we get the equation for MGLC

$$C_{\max} = \frac{Q}{u}\left(\frac{Cu}{Q}\right)_{\max} \qquad .....(5.43)$$

Using the stability classification and effective stack height as parameters Turner has prepared a number of curves [Fig. 5.36(a) and (b)] to find the distance downwind to get the maximum concentrations. A normalized concentration $(Cu/Q)_{max}$ was calculated from which the maximum concentration can be estimated using the following expression [Fig. 5.36(c)]

$$C_{\max} = \frac{Q}{u}\left(\frac{Cu}{Q}\right)_{\max} \qquad .....(5.44)$$

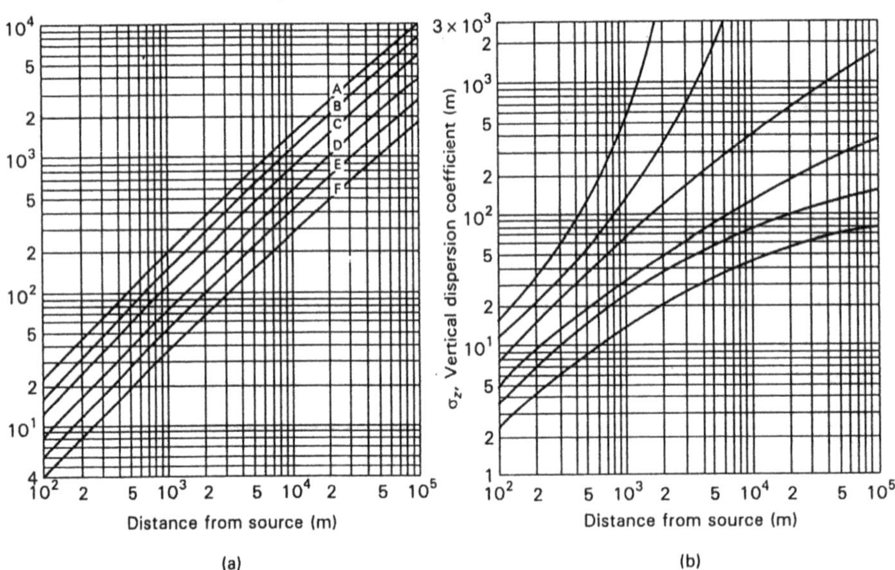

Fig. 5.36 *Gaussian dispersion coefficients as a function of distance downwind*

(a) *Horizontal dispersion coefficient* σ_y ; **(b)** *Vertical coefficient* σ_z

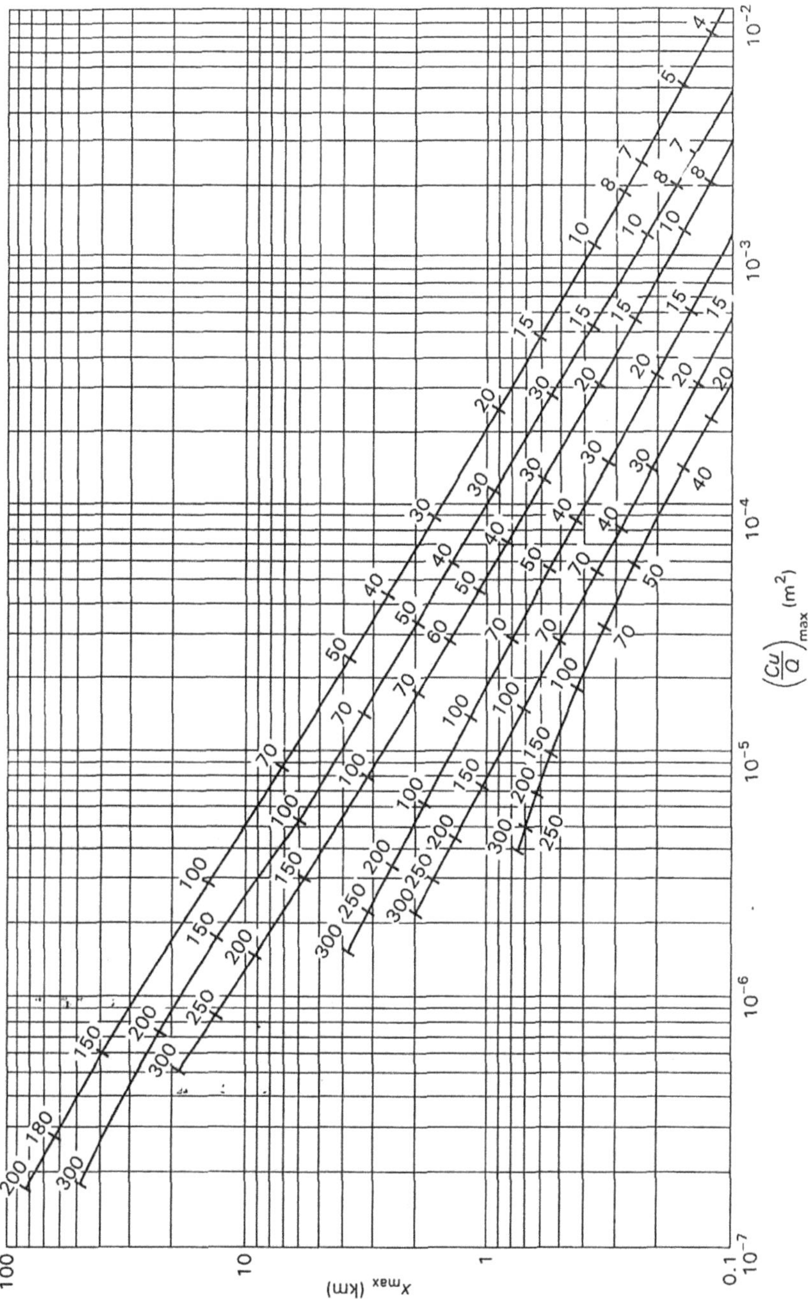

Fig. 5.36(c) *To determine the peak downwind plume concentration, enter the graph at the appropriate stability classification and effective stack height (number above the lines, in metres) and then move across to find the distance to the peak, and down, to find a parameter from which the peak concentration can be found*

Problem 5.2

It is proposed to establish a new coal fired 1000 MW power plant which works with 35% efficiency. If coal with 4 percent sulphur with a heat content of 14 000 Btu/lbemits SO_2 at the rate of 143000 Ib/day from a stack with an effective height equal to 200 m, estimate the ground level concentration of SO_2 at a downwind distance of 4 km.

The wind speed at 10 m height is 3.5 m/s and it is a cloudy summer day. Also estimate at what downwind distance maximum concentration of SO_2 can be observed.

Solution: The power plant emits SO_2 emissions at the rate of 143000 Ib/day. Converting that to μg/s gives

$$Q = 143000\,\text{lb}/\text{day} \times \frac{1000\,\text{g}}{2.2\,\text{lb}} \times \frac{10^6\,\mu\text{g}}{\text{g}} \times \frac{1\,\text{day}}{86400\text{s}} = 7.52\times10^8\,\mu\text{g}/\text{sec}$$

From Table 5.7 and the given wind speed and solar conditions, we find that the stability classification is C. Note that Table 5.7 uses the wind speed as measured at the standard 10 m anemometer height. The wind speed in the Gaussian plume equation requires that we estimate the wind at the effective stack height. To estimate wind speed at 200 m we can use Equation 5.29 which requires an estimate of the wind parameter p.

From Table 5.6 for stability class C the approximate p is found to be 0.20. Winds at 400m then can be estimated to be

$$u_1 = u_2\left(\frac{z_1}{z_2}\right)^p = 3.5\left(\frac{200}{10}\right)^{0.20} = 6.371\,\text{m}/\text{s}$$

At 4 km downwind the dispersion coefficients are σ_y = 359 m and σ_z = 216 m. substituting these values in Equation 5.40 with y = 0 (since we are interested only in the downwind) concentration gives

$$C(x,y) = \frac{Q}{\pi u \sigma_y \sigma_z}\exp\left(\frac{-H^2}{2\sigma_z^2}\right)$$

$$C(4,0) = \frac{7.52\times10^8\,\mu\text{g}/\text{s}}{\pi\times6.37\text{m}/s\times359\text{m}\times216\text{m}}\left(\exp\frac{-(200)^2}{2\times(216)^2}\right) = 311.35\,\mu\text{g}/\text{m}^3$$

So, the power plant would add 311.5 μg/m^3 to whatever SO_2 level is already there from other sources. The ambient air quality standards for SO_2 concentrations must be less than 80 μg/m^3. So under the given conditions the power plant would contribute greater than permissible levels of SO_2.

5.4.6.4 Estimation of stack height

Usually gaseous discharges will be made from chimney stacks which will be of sufficient height to disperse the pollutant in the atmosphere. These gaseous discharges are emitted into the atmosphere at high temperatures which gives rise to the plume due to the buoyancy giving the vertical momentum under windless conditions. However, in windy conditions very frequently, it is bent as a result of the wind that is usually present. The plume rise due to buoyancy of high temperature gaseous discharges increases the height of the stack by a factor Δh and the effective stack height increase in such conditions is given by $H = h, + \Delta h$ which is used in Gaussian plume calculations (Fig. 5.35). Many times the increased stack height Δh due to plume rise contributes significantly in reducing the ground level concentrations in the downwind direction.

5.4.6.5 Estimation of plume rise

A large number of techniques have been reported in the literature for estimating plume rise, which give different results. The EPA model based on work by Briggs assumes that the plume rise depends on both momentum and buoyancy, and as the exhaust gases will have molecular weight close enough to that of air, the buoyancy due to density differences can be neglected.

Plume rise Δh is a function of exit gas velocity, temperatures of ambient air and stack gas, emission rate, wind speed, atmospheric stability and shear of the wind speed, *dul dz*. Plume rise typically increases the effective stack height by a factor of 2 to 10 times and can reduce the maximum ground level concentration of pollutants by a factor of as much as 100.

The effect of varying effective stack height for a given atmospheric stability can be understood by considering the following example.

Example

A power plant is emitting SO_2 at the rate of $Q = 6.47 \times 10^8$ µg/s into atmosphere at wind speed of 4.9 m/s. $\sigma_y = 359$ m and $\sigma_z = 216$ m. The effect of variation of effective stack height and atmospheric stability on the minimum ground level concentration (MGLC) are plotted in Figs. 5.37(a) and (b) respectively. It can be seen from the figure that the least stable atmosphere produces higher ground level concentration down wind near the stack while the turbulence in the unstable atmosphere brings the plume to earth quickly. This high peak value near the stack reduces very quickly as the down wind distance increases. The stable atmosphere will have a lower peak but beyond a few kilometres the concentration will increase than for an unstable atmosphere.

The relationship between the buoyancy force on a parcel of exhaust gas and its temperature, exit velocity, and stack diameter is given by the expression proposed by Perkins [16]

$$F = gr^2 v_s \left(1 - \frac{T_a}{T_s}\right) \qquad .....(5.45)$$

where F = buoyancy flux parameter

g = acceleration due to gravity

r = inside radius of the stack, m

v_s = stack gas exit velocity, m/s

T_s = stack gas temperature, K

T_a = ambient temperature, K

For neutral or unstable conditions in the atmosphere (stability categories A–D) the following equation should be used to estimate plume rise as per Briggs' model [17]

$$\Delta h = \frac{1.6 F^{1/3} x_f^{2/3}}{u} \quad .....(5.46)$$

where Δh = plume rise, m

u = wind speed at stack height, m/s

x_f = distance downwind to point of final plume rise, m.

Use $x_f = 120\ F^{0.4}$ if $F \geq 55\ \text{m}^4/\text{s}^3$

$x_f = 50 F^{5/8}$ if $F < 55\ \text{m}^4/\text{s}^3$

For stable windy conditions (stability categories E and F), use the following

$$\Delta h = 2.4\left(\frac{F}{uS}\right)^{1/3} \quad .....(5.47)$$

where S is a stability parameter with units of s^{-2} given by

$$S = \frac{g}{T_a}\left(\frac{dT_a}{dz} + \Gamma\right) \quad .....(5.48)$$

where Γ is the adiabatic lapse rate, 0.01 °C/m. The derivative dT_a/dz represents the actual rate of change of ambient temperature with altitude.

5.4.6.6 Holland's equation for buoyant plumes

Holland [I8] proposed an equation of plume rise, given below, for gaseous discharges at a temperature (50 °C) higher than ambient temperature, which was widely used in many plume studies because of its simplicity. It is assumed that for hot gaseous discharges the buoyancy factor will be made more greater than vertical momentum for plume rise.

$$\Delta h = \frac{\left(1.5\ V_s D_s + 4\times10^{-5} Q_H\right)}{\mu} \quad .....(5.49)$$

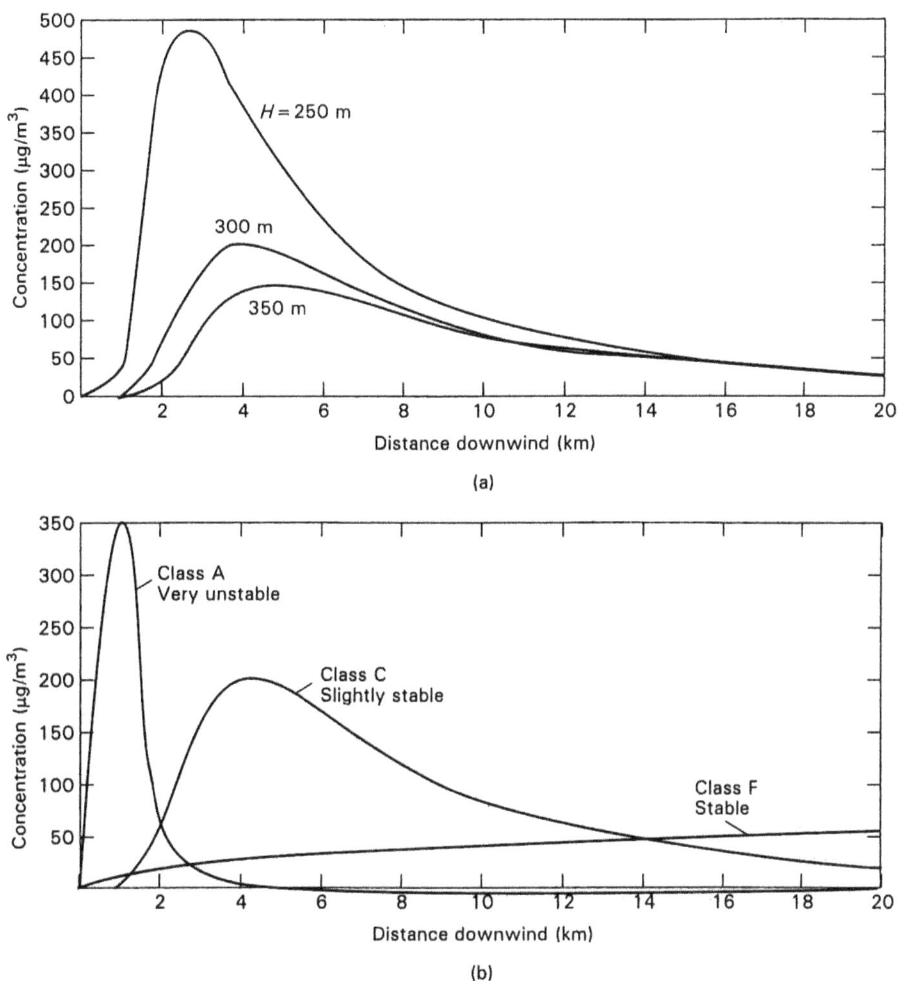

Fig. 5.37 *The effect of variations in key parameters on SO_2 plume for the 1000 MW coal plant* **(a)** *Effect of effective stack height for constant stability classification;* **(b)** *Effect of stability classification given a constant effective stack height*

where

V_s = Stack gas exit velocity, m/s
D_s = diameter of stack outlet, m
μ = wind velocity, m/s
Q_H = heat emission rate, cal/s and
Δh = plume rise above the stack, m

The equation was further modified by Turner as

$$\Delta h = \frac{V_s D_s}{\mu}\left(1.5 + 2.68 \times 10^{-3}\, p_a \frac{T_s - T_a}{T_s} D_s\right) \qquad .....(5.50)$$

where

p_a= atmospheric pressure, millibars;

T_a and T_s are the absolute temperatures of' the ambient air and stack gas respectively.

Holland further recommended correction factors for use of his equation in different stability conditions. The correction factors are 1.15, 1.10, 1.00 and 0.85 for stability classes A–B, C, D and E–F, respectively

5.4.6.7 Plume rise under stable and calm conditions

Under windless calm condition the plume rises to some height where the buoyancy force is completely dissipated and for such condition the recommended equation is

$$\Delta h = 5.0\, F^{1/4} S^{-3/8} \qquad .....(5.51)$$

In actual practice Δh may be calculated and the smaller value can be used. Generally the plume Δh is due to two fractions u. z buoyancy and vertical momentum as shown in equation

$$\Delta h = \Delta h \text{ buoyancy } \Delta h \text{ momentum} \qquad .....(5.52)$$

Moses and Carson [19] recommended the following equation for different atmospheric conditions.

Unstable: (dT/dz<–12.2)

$$\Delta h = 3.47\frac{V_s D_s}{u} + 10.53\frac{Q_H^{1/2}}{u} \qquad .....(5.53)$$

Neutral: ($-12.2 < dT/dz < -1.5$)

$$\Delta h = 3.35\frac{V_s D_s}{u} + 5.541\frac{Q_H^{1/2}}{u} \qquad .....(5.54)$$

Stable: ($dT/dz >= -1.5$)

$$\Delta h = 1.04\frac{V_s D_s}{u} + 4.58\frac{Q_H^{1/2}}{u} \qquad .....(5.55)$$

where dT/dz is the environmental lapse rate in °C/km and Q_H is in kcal/s.

In all the above equations discussed the plume rise is inversely proportional to wind speed and directly proportional to emission rate and square root of the heat of combustion of the gas. These equations also constitute at least one constant which has to be evaluated experimentally.

5.4.6.8 Non-buoyant plumes

The American Society of Mechanical Engineers [20] has proposed the following equation for plume rise for non-buoyant plumes for which lapse temperature close to ambient temperature is less than 50 °C with exit speed of 10 m/s.

$$\Delta h = \left(\frac{V_s}{\mu}\right)^{1.4} \times D_s \qquad .....(5.56)$$

The Bureau of Indian Standards has suggested that the following Brigg's formula be used for the computation of plume rise (IS: 8829).

(a) For effluents with heat release of $>10^6$cal/sec.

(b) For effluents with lesser heat contents i.e. for effluents with momentum as the dominant contributor, $\Delta h = 3\ W_o d/u$

where,

Q_H = Heat release rate, calories/sec
W_o = Effluent velocity, m/sec and
d = Inner diameter of stack at exit, m

It should be stated that these plume rise formulae may not be recommended for tall chimneys of height in excess of 300 m.

5.4.6.9 Estimation of minimum stack height

From earlier discussions it is understood that the maximum ground level concentration of any pollutant at a given downwind direction decreases with increasing stack height and is inversely proportional to the square of its effective stack height. The minimum stack height can be calculated using the following equation prescribed by CPCB.

(a) $H_{min} = 14\ Q^{0.3}$ where Q is SO_2 emission rate in kg/hour

(b) For PM, $H_{min} = 74\ Q^{0.27}$ where Q is particulate emission rate in tons per hour

For any given maximum ground level concentration (A expressed in mg/m^3) of a pollutant, the minimum effective stack height, H can be calculated using the equation

$H = (2.71\ Q/u\ A)^{1/2}$ where Q = pollutant emission rate in kg/day.

However the minimum stack height prescribed by CPCB for any stack is 30 m.

5.4.6.10 Downwash

Large structures distort the atmosphere and interfere with wind flow to some extent. These atmospheric distortions usually take the form of a wake, which consists of a pocket of slower, more turbulent air. If a plume is emitted near a wake, it is usually pulled down because of the lower pressure in the wake region. This effect is termed downwash. When downwash occurs, the plume is brought down to the ground near the emission source more quickly.

A wake that causes downwash usually occurs as the result of one of three physical conditions: 1) the stack, referred to as stack-tip downwash, 2) local topography, or 3) nearby large structures or building downwash, Figure 5.38 shows examples of each of these conditions.

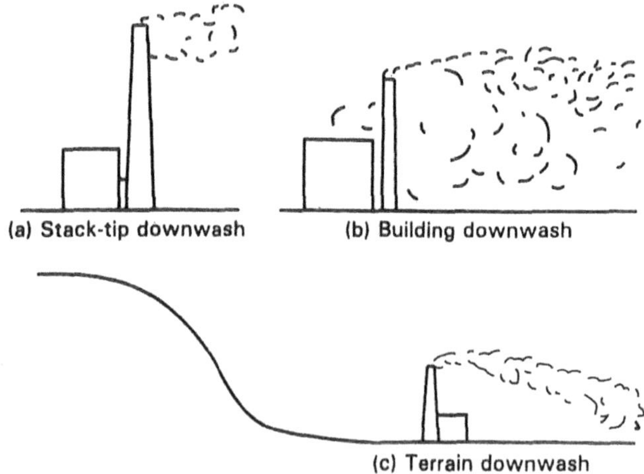

Fig. 5.38 *Physical conditions that cause downwash. Reprinted from G. A. Briggs, 1969.* Plume rise. *U.S. Atomic Energy Commission Critical Review Series T10-25075, Clearinghouse for Federal Scientific and Technical Information [9]*

To avoid the stack-tip downwash, the ratio of emission velocity (v_s) to wind speed velocity at the stack height (u_s) has to be considered in the stack design. If $v_s < 1.5\ u_s$, then the physical stack height should be adjusted by the following equation:

$$h_{ad} = 2d_s\ [v_s/u_s - 1.5]$$

if $v_s/u_s > 1.5$, then stack-tip downwash is avoided since the exhaust gas is emitted from the stack at sufficient velocity to clear the downwash area on the downwind side of the stack. If $v_s/u_s < 1.0$, downwash will probably occur, possibly seriously. For inter mediate values of v_s/u_s, downwash may occur depending on the ambient conditions at the time.

5.4.6.11 Downwash caused by topography

Large hills or mountains can change the normal wind patterns of an area. If the stack is located closely downwind of a hill above stack height, the air flowing off the hill can cause the plume to impact closer to the stack than normal as shown in Fig. 5.38(c). To model these situations physical models in wind tunnels are often used. A recently developed model, the complex terrain dispersion model plus algorithms for unstable situations(CTDM-PLUS), employs a critical hill height calculation that determines if the plume impacts the hill or follows the uninterrupted laminar flow around it (USEPA,1989).

5.4.6.12 Building downwash

Large structures surrounding the stack also affect ambient wind conditions. The boundaries of the wake region resulting from surrounding structures are nor sharply defined. They depend on the three-dimensional characteristics of the structure and are time dependent. The extent of distortion depends extensively upon building structure geometry and wind direction. Generally, a single cylindrical structure (e.g. a free standing silo) has little influence on the wind flow compared to a rectangular structure.

Figure 5.38(b) shows the building downwash that occurs when the plume is drawn in to a wake from nearby structure. Two zones exist within the downwash area of the structure. The first zone, which extends approximately three building heights downwind, is the cavity region where plume entrapment can occur. The second zone, which extends from the cavity region to about ten times the lesser dimensions of the height or projected width, is the wake region where turbulent eddies exists as a result shows an example of these zones. Generally, the cavity region concentration is higher than the wake region concentration due to plume entrapment.

To avoid building downwash, the EPA has developed a general method for designing the minimum stack height needed to prevent emissions from being entrained into any wake created by the surrounding buildings. The equation below gives the stack height.

$$H_{GEP} = H = 1.5(L)$$

where H_{GEP}= the GEP (Good Engineering Practice)stack height

H = the maximum height of an adjacent or nearby structure

L= lesser dimension of the height or projected width of an adjacent or nearby structure.

The projected width of a structure is the exposed area perpendicular to the wind as shown in Fig. 5.39. It is necessary to check all structures within $5L$ of the stack for their possible effect on the plume. The range of influence for a given structure is defined by Tickvart [21] as $2L$ upwind of the structure, $5L$ downwind, and $0.5L$ on the sides parallel to the wind flow.

5.4.6.13 Effect of inversion on GLC

However all these formulas which are empirical in nature and are applicable for estimation of Δh based on statistical averages are not useful for estimation of Δh where inversions occur.

The Gaussian plume equation is derived based on the assumption that the vertical temperature profile is a simple straight line which is not the case if there is any inversion set in at the effective stack height which may differ many times.

Turner [14] suggests the following modified Gaussian equation to estimate concentrations downwind under an inversion which is derived based on the assumption that the downwind distance from the source is at least twice the distance to where the plume first interacts with the inversion layer.

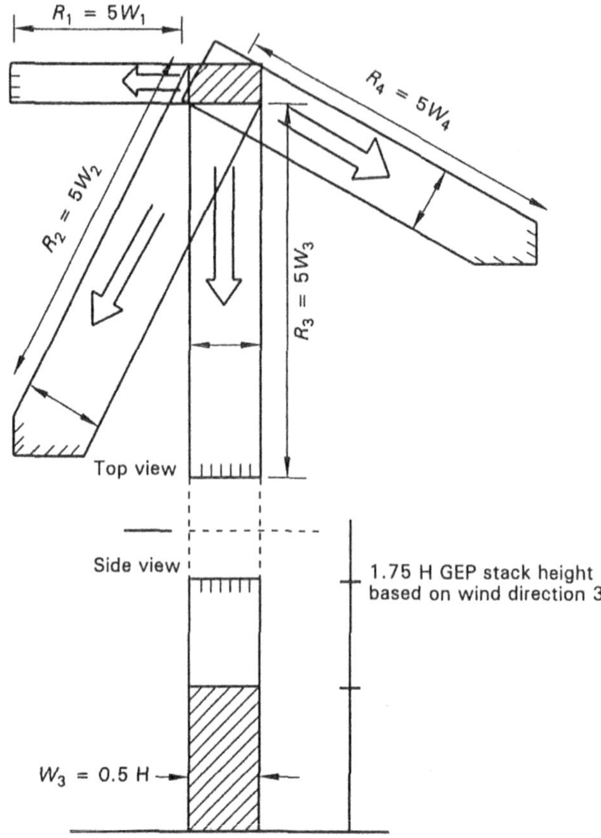

Fig. 5.39 *GEP determination of projected structure width and associated region of adverse influence for a structure in four different wind directions. [28] (Reprinted from U.S.Environmental Protection Agency (EPA)), 1985, Guideline for determination of good engineering practice stack height [Technical support document for the stack height regulations] [Revised], EPA-450/4-80-023R,Research Triangle Park, N.C.: USEPA, Office of Air Quality Planning and Standards[June]*

$$C(x, y) = \frac{Q}{(2\pi)^{1/2} u \sigma_y L} \exp\left(\frac{y^2}{2\sigma_y^2}\right) \quad \text{for } x\, 2 \geq X_L \qquad .....(5.57)$$

where L = elevation of the bottom of the inversion layer (m)

X_L = the distance downwind where the plume first encounters the inversion layer Equation 5.57 is applicable only for distances $x \geq 2X_L$. That distance, X_L occurs at the point where the vertical dispersion coefficient, σ_z is equal to

$$\sigma_z = 0.47\,(L - H)\ (\text{at } x = X_L) \qquad .....(5.58)$$

Once σ_z is found from Equation 5.58 the distance X_L can be estimated using Fig.5.36 or using Equation 5.56 and Table 5.9. For distances $x \leq X_L$, the standard Gaussian plume Equation 5.35 can be used to estimate downwind concentrations. For distances between X_L and $2X_L$, concentrations can be estimated by interpolating between the values computed for $x = X_L$ and $x = 2X_L$ (Fig. 5.40).

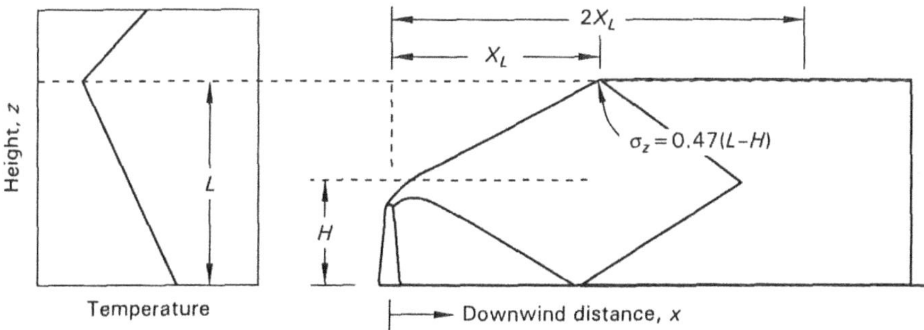

Fig. 5.40 *Plume dispersion under an elevated inversion. Equation 5.57 applies for distances greater than $2X_L$ downwind, where X_L occurs at the point where $\sigma_z = 0.47\ (L - H)$*

5.4.6.14 Estimation of concentration of the pollutant dispersed through a stack under an inversion loft

Problem 5.3

In a coal-fired boiler a stack with effective height 150 m, emits SO_2 at the rate of 400 g/s. The wind speed at 10 m above the surface is at 6.5 m/s while at the stack height it is 20 m/s. On a clear summer day at 12 noon an inversion layer sets in at 400 m. Calculate the ground level SO_2 concentration at a distance downwind twice that where reflection begins to occur from the inversion.

Solution: From Equation 5.54 we get σ_z at the distance X_L

$\sigma_z = 0.47\ (L - H) = 0.47 \times (400–150) = 141$ m.

For estimates of X_L Fig. 5.36 can be used if the stability classification is below the inversion. From Table 5.8, a clear summer day with 6.5 m/s wind speed corresponds to stability class *C*. Entering Fig. 5.36 on the vertical axis at $\sigma_z = 141$ m, then going across to the class *C* line and dropping down to the horizontal axis or using Equation 3.5 with Table 5.8, we will get a value of X_L.

To find the concentration at a distance $x = 2X_L = 2 \times 1750$ m $= 3500$ m, we can use Equation 5.57 but first we need to estimate σ_y at that point. Using Fig. 5.35 at $x = 3500$m, and class *C* we can estimate σ_y to be about 320 m. The concentration directly downwind corresponds to y = 0, so

$$C(2X_L, 0) = \frac{Q}{(2\pi)^{1/2} u\sigma_y L} = \frac{400 \text{ g/s}}{(2\pi)^{1/2} 20 \text{ m/s} \times 320 \text{ m} \times 400 \text{ m}}$$

$$= 62.3 \times 10^{-6} \text{g/m}^3 = 62.3 \text{ µg/m}^3$$

Infinite line source

For an area source or an infinite line source like a group of industries located along a highway or river along which the pollutants are emitted continuously, the ground-level downwind concentration is given by

$$C(x, y, 0) = \frac{2q}{\sqrt{2\pi\sigma_z}\ u \sin\beta} \exp\left[-H^2 / 2\sigma_z^{\ 2}\right] \quad .....(5.59)$$

where q = Q/L = source strength per unit distance and β= angle between wind direction and line source. When β= 90°, the above equation can be simplified (sin β= 1) to

$$C_{xy} = \frac{2Q}{\sqrt{\pi - 2\mu}} \exp\left[-H^2 / z^{-2}\right] \quad .....(5.60)$$

However if β is less than 45° this equation cannot be used.

Problem 5.4

In a plain terrain at 10 m above the ground, the wind speed was found to be 6 m/s. Determine the wind speed at height of 200 m, when the change of temperature with altitude is (a) –20 °C/km (b) + 10.0 °C/km.

The wind speed at any elevation z_1 is given by the Equation 5.29.

$u_1/u_2 = (z_1/z_2)^p$

where u_1 is the wind speed at altitude z_1 and u_2 is the wind speed at altitude z_2.

Solution:

(a) As the lapse rate is –20 °C/km it comes to 2 °C per 100 m and the environmental lapse rate corresponds to super adiabatic lapse rate, which indicates the atmosphere is unstable [category-A] for which the p value from Table 5.6 is 0.15.
So $u_1 = u_2 \times (z_1/z_2)^p = 6 \times (200/10)^{0.15} = 9.40$ m/s.

(b) When the lapse rate is +10°C/km which comes to 1°C per 100 m and the atmosphere is considered as highly stable [category-F] for which p value is 0.6.
So $u_1 = u_2 \times (z_1/z_2)^p = 6 \times (200/10)^{0.6} = 36.2$ m/s.

Problem 5.5

A thermal power plant uses coal that contains 5% sulphur and burns at the rate of 5tons per hour and discharges effluents through a stack 200 m high and 1 m inside diameter.

The other parameters are: wind speed at 10 m above the ground is 4.0m/s and at the effective stack gas velocity = 85 m/s, air temperature = 35 °C, barometric pressure =1000 mbar, stack gas velocity = 10 m/s, and stack gas temperature is 150 °C. The atmospheric conditions are moderately to slightly stable. (a) Determine the ground level concentration of SO_2 at a distance 2 km downwind from the stack along the plume axis (b) Assuming an effective stack height of 85 m but with the same wind velocity, determine the ground-level concentration at this distance. (c) What is the average concentration at the equivalent stack section at this point?

Solution:

(a) $$C_{xy} = \frac{Q}{\pi\mu\sigma y\sigma z}\exp\left[\frac{-y^2}{2\sigma^2 y}\right]\exp\left[\frac{-H^2}{2\sigma^2 z}\right]$$

$S + O_2 \rightarrow SO_2$

Then,

$$Q = SO_2 \text{ produced} = 5\ (1000)\ (0.5)\ (SO_2/S) = 5\ (1000)(0.5)\left[\frac{32+32}{32}\right]$$

500 kg/h = 0.1388 kg/sec.

Effective stack velocity = $H = h + \Delta h = 200 + \Delta h$

$$\Delta h = \frac{v_s d}{u}\left(1.5 + 2.68(10^{-3})p\frac{\Delta T d}{T_s}\right)$$

$v_s = 10$ m/s; $d = 1.0$ m; $\Delta T = 150 - 35 = 115$ K, $h = 8.5$ m/s

$T_s = 150 + 273 = 423$ K; $p = 1000$ mbar

$$\Delta h = \frac{10\text{m/s}(1\text{ m})}{8.5\text{ m/s}}\left\{1.5 + 2.68(10^{-3})(1000)\text{ m}\frac{115\text{ K}(1\text{ m})}{423\text{ K}}\right\} = 2.68\text{ m}$$

$H = h + \Delta h = 200\text{ m} + 2.68\text{ m} = 202.68$

$\sigma_y = ax^{0.894}$

$\sigma_z = cx^d + f$

From Table 5.9,

$$a = \frac{50.5+34}{2} = 42.45\ c = \frac{55.4+62.6}{2} = 59$$

$$d = \frac{0.305+0.18}{2} = 0.24\ f = \frac{-34.0-48.6}{2} = -41.3$$

$$\sigma_y = 42.25(2.0)^{0.894} = 78.51\text{ m}$$

$$\sigma_z = 59.0(2.0)^{0.24} - 41.3 = 28.38\text{ m}$$

when $H = 202.68$

(a) $$C = \frac{0.1358 \text{ kg / sec}}{3.17 \times 8.5 \text{ m/sec} \times 78.51 \text{ m} \times 28.38 \text{ m}} \times \exp\left[-\frac{0^2}{2(78.51)^2}\right] e\left[\frac{202.68}{2(28.38)}\right]$$

$$= 2.26 \times 10^{-3} \text{kg / m}^3 \times 8.68 \times 10^{-12} = 19.6 \times 10^{-6} \mu\text{g / m}^3$$

(b) When $H = 85$, C = 2.26 × $10^3 \mu g/m^3$ × 0.01 12 = 25.89 $\mu g/m^3$

(c) When $H = 0$ and $y = 0$ i.e., at the centre section of the stack, $C = 2260$ $\mu g/m^3$

Problem 5.6

A coal fired power plant releases from its stack SPM at a rate of 2.3 g/s. The stack height is 60 m while the temperature of the stack gases is 160 °C and the ambient air temperature is 30 °C. The wind velocity at the stack height is 2.5 m/s, while the stack gas velocity is 5.0 m/s. The stack diameter is 3.5 m. The atmospheric pressure is 1.005 bar. The wind speed at 10 m height from the ground is 1.95 m/s. Estimate the ground level concentration of the pollutants at 1, 2, and 4 km downwind distance.

Solution: The Turner's formula for plume rise is

$$\Delta h = v_s \times (d_s/u)\ 1.5 + 2.68(T_s - T_a/T_s) \times p_a d_s$$

where v_s = stack gas velocity

d_s = diameter of the stack

u = wind speed at 10 km/h

T_s = stack gas temperature

T_a = ambient air temperature

So

$$\Delta h = 5 \text{ m/s} \times (3.5 \text{ m}/1.95 \text{ m}) \times 1.5 + 2.68[(160\ °\text{C} - 130\ °\text{C})/160\ °\text{C}] \times 1.005 \times 3.5 \text{ m}$$

$$= [5(1.79) \text{ m} \times 1.5 + 2.68(0.187)] \times 3.575$$

$$= (13.42 + 0.501) \times 3.575 = 41.763$$

Effective stack height, $H = 60 + 41.763 = 101.763$ m

So, $C(x, 0, 0) = (Q/\pi u \sigma_y \sigma_z) \exp\text{-}1/2\ [H/\sigma_z)^2]$

$\sigma_y\sigma_z$ values for category–F, stability from Table (5.8) are

1 km: $\sigma_y = 34$, $\sigma_z = 14$, 2km: $\sigma_y = 63$, $\sigma_z = 22$,3km: $\sigma_y = 117$, $\sigma_z = 32$ respectively.

$$C(1, 0, 0) = 2.3/3.14 \times 1.95\ (34 \times 14) \exp[-0.5(101.763/14)^2] = 4.08\ \mu g/m^3$$

$$C(2,\ 0,\ 0) = \left(\frac{2.3 \times 7.89 \times 10^{-4} \times 5177.55}{3.15 \times 1.95 \times 63 \times 22}\right) \exp\left[-0.5 \times \frac{101.763}{14}\right]^2$$

$$= 2.73 \times 10^{-4}(5177.88) = 1.4135 \times 10^{-4} \mu\text{g / m}^3 = 1.4135\ \mu\text{g / m}^3$$

$$C(4,\ 0,\ 0) = \left(\frac{2.3}{3.15 \times 1.95 \times 117 \times 32}\right) \exp\left[-0.5 \times \frac{101.763}{14}\right]^2 = 0.5177 \mu\text{g / m}^3$$

Problem 5.7

A boiler with stack of 200 m height and 0.4 m diameter is releasing flue gas at a velocity of 15.5 m/s at a temperature of 170°C. The wind speed at the stack height is 6m/s and ambient temperature is 35 °C. Estimate the plume rise (a) when the environmental lapse rate is –2.5 °C/100 m (b) +2.0 °C/100 m.

Solution: Equation 5.48 for plume rise is

$\Delta h = 1.6\ F^{1/3} X_f^{2/3}/u$ where F is buoyancy flux parameter $= g\ r^2 v_s (1 - T_a/T_s)$

(Equation 5.45)

$F = 9.81 \times 2^2 \times 15.5\ [1-(308/443)] = 188.54.$

X_f is the distance downwind to the final plume rise.

$X_f = 120 \times F^{0.4} = 120 \times (188.54)^{0.4} = 109.22$

(a) For unstable condition,

$\Delta h = (1.6 \times 188.54^{1/3} \times 109.22^{2/3})/6 = (1.6 \times 0.75 \times 1.22)/6 = 0.244$ m

(b) For stable condition,

$\Delta h = 2.4 \times F/(uS)^{1/3}$ where $S = g/T_{a},(dT_a/d_z) + \tau$

$S = 9.81/308(0.01) + 0.01 = 3.18 \times 10^{-3}$

$\Delta h = (2.4\ (185.5)/6 \times 3.18 \times 10^{-3})^{1/3} = 144.27$ m

Problem 5.8

A coal fired boiler burns 8 tons of coal (containing 5% sulphur)/h and discharges through stack, the effluent gases containing sulphur dioxide. The effective stack height is 175 m while the wind speed at the top of the stack is 5 m/s. When the atmosphere is unstable [category-C], estimate (a) maximum ground level concentrations at the cross-section of the stack (b) at 4 km downwind distance (c) GLC at 0.5 km crosswind distance.

Solution: Emission of sulphur from the coal = $S + O_2 \rightarrow SO_2$

32g + 32g = 64g

Rate of emission of sulphur for 8 tons of coal per hour coal containing 5% sulphur

$= 8 \times 10^3 \times 5/100 = 400$ kg/h.

So, 32 g of sulphur releases 64 g of SO_2, the emission rate of SO_2

$= 400 \times 64/32 = 800$ kg/h

$= 800 \times 1000/3600 = 222.2$ g/s

At category C of atmospheric stability $\sigma_y = 359$ m, $\sigma_z = 216$ m

So we have

(a) C at the cross-section of the stack

$$= \frac{Q}{\pi\mu\sigma_y\sigma_z} = \frac{222.2 \text{ g/s}}{3.14 \times 5 \text{ m/s} \times 359 \text{ gm} \times 216 \text{ m}}$$

$$= 1.825 \times 10^{-4} \ \mu\text{g/m}^3 = 182.5 \ \mu\text{g/m}^3$$

(b) At 4 km downward distance,

C_x (4, 0, 0) = (222.2/3.14 × 359 X 216 × 5) exp [–(175)^2/2 × (216)^2]

= 182.5 μg/m^3 × 0.7202 = 131.4 μg/m^3

(c) GLC at 4 km distance and 0.5 km cross wind distance

C_x (4000, 500, 0) = (222.2/3.14 ×359 × 216 × 5) exp [-0.5 (500/359)^2 exp[-0.5 (175/216)^2 = 182.5 μg/m^3× 0.38 × 0.72 = 49.932 μg/m^3

Problem 5.9

In a coal fired boiler 250 tons of coal (4.2% S) was burned per day which emits flue gases with a velocity of 15 m/s, with a particulate concentration of 6500 mg/m^3. Estimate a maximum stack height required. The total flue gas flow was 20 m^3/s.

Solution: The minimum stack height required as per CPCB formula is

For SO_2, $H = 14 \times q^{0.3}$; F or particulates $H = 74 \times q^{0.27}$

SO_2 release per day = 250 × 10^3 kg/day × 4.2/100 × 64/32 = 875 kg/h

Minimum stack height required for SO_2 = 14 × 875$^{0.3}$ = 106.8 m

Particulates release per day = 250 × 1000 × 6.5 × 20 × 3600 = 0.468 ton/h

Minimum stack height required for particulates = 74 × 0.468$^{0.27}$ = 60.3 m

So minimum stack height will be the maximum of the above i.e. 106.8 m

Problem 5.10

A 750 MW coal fired power plant has a 200 m stack with inside radius 3 m. The exit velocity of the stack gases is estimated at 15 m/s, at a temperature of 160 °C (433 K). Ambient temperature is 30°C and winds at stack height are estimated to be 4 m/s. Estimate the effective height of the stack if (a) the atmosphere is stable (E) with temperature increasing at the rate of 2 °C/km (b) the atmosphere is slightly unstable, class C.

Solution: First find the buoyancy parameter F from Equation 5.45

$$F = gr^2 v \left(1 - \frac{T_a}{T_s}\right)$$

$$F = 9.8 \text{ m/s}^2 \times 3\text{m}^2 \times 20 \text{ m/s} \times \left(1 - \frac{303}{433}\right) = 182.28 \text{ m}^4/\text{s}^3$$

(a) When the atmosphere is stable *(E)*, we need to use Equations 5.47 and 5.48.

$$S = \frac{g}{T_a}\left(\frac{dT_a}{dT_s} + T\right)$$

$$S = \frac{98\ \text{m/s}^2}{303\ \text{K}}(0.002 + 0.01)\ \text{k/m} = 0.000388/\text{s}^2$$

$$\Delta h = 2.4\left(\frac{F}{uS}\right)^{1/3}$$

$$= 2.4\left(\frac{176\ \text{m}^4/\text{s}^3}{4\ \text{m/s} \times 0.00039/\text{s}^2}\right)^{1/3} = 270.77$$

So the effective stack height is $H = h + \Delta\text{h} = 200 + 270.77 = 470.77$

(b) With an unstable atmosphere class C, we need to use Equation 5.46, since $F > 55\text{m}^4/\text{s}^3$, the distance downwind to the point of final plume rise that should be used is

$X_f = 120\ \text{F}^{0.4} = 120 \times 176.4^{0.4} = 949.26$ m

$\Delta h = 1.6\ \text{F}^{1/3} x_f^{2/3}/\text{u} = 1.6(176.4)^{1/3}\ (949.26)^{2/3}/5 = 0.466$ m

and the effective stack height is $H = 200 + 0.466 = 200.466$ m

The Briggs model, then predicts a very substantial plume rise relative to the actual stack height that is quite sensitive to atmospheric conditions.

5.4.7 Gaussian puff model

The Gaussian plume model is applicable only for sources that are emitting continuously or for time periods to or greater than the travel times from the source to the receptor. However, the solution also is needed for the cases of instantaneous release of pollutant. Such an alternative approach has developed and is known as a "puff" model (for the presentation of the theory).

The puff model applies to instantaneous sources, where the release time is short compared with the travel time. The centre of the puff is moving along the wind trajectory. The concentration distribution inside the puff is assumed to be Gaussian. The dimensions of the puff are described by dispersion coefficients as a function of the travel time of the puff. The Gaussian puff formula for the ground level concentration downwind from a release from height H is given as:

$$C(x\ y\ 0\ t\ H) = \frac{Q\Delta t}{8(\pi t)3/2(k_x k_y k_z)^{1/2}} \exp\left\{-\frac{1}{4}\left(\frac{x^2}{k_x} + \frac{y^2}{k_y} + \frac{(z-H)^2}{k_z}\right)\right\} \quad \text{.....(5.61)}$$

where t = time elapsed since the release, Δt = time duration of release which is very small, k_x, k_y, k_z, are eddy diffusion coefficients in the x,y, z directions.

The puff formulation overcomes some problems connected with the plume models, as it enables calculations of pollutant concentration in near calm conditions using both small-scale and large diffusion processes.

5.5 Air Pollution Computer Dispersion Models

The US-EPA recommends the following two computer packages for simulation of non-reactive chemicals (e.g., SO_2):

- AERMOD: http://www.epa.gov/scram001/dispersion_prefrec.htm#aermod
  AERMOD is a steady-state Gaussian plume model. It uses a single wind field to transport emitted species. The wind field is derived from surface, upper-air, and onsite meteorological observations. AERMOD also combines geophysical data such as terrain elevations and land use with the meteorological data to derive boundary layer parameters such as Monin-Obukhov length, mixing height, stability class, turbulence, etc.

  AERMOD is today replacing the ISC models for most regulatory applications in the US.
- CALPUFF:ttp://www.epa.gov/scram001/dispersion_prefrec.htm#calpuffCALPUFF is a non-steady state Lagrangian puff dispersion model. The advantage of this model over a Gaussian-based model is that is can realistically simulate the transport of substances in calm, stagnant conditions, complex terrain, and coastal regions with sea/land breezes.
- CALPUFF is particularly recommended for long-range simulations (e.g., more than 50 miles) and studies involving the assessment of the visual impact of plumes.

With the development of the VISTAS Version 6 model, CALPUFF can use sub-hourly meteorological data and run with sub-hourly time steps. This version of CALPUFF is appropriate for both long-range and short-range simulations.

5.5.1 Photochemical modeling

Photochemical air quality models have become widely recognized and routinely utilized tools for regulatory analysis and attainment demonstrations by assessing the effectiveness of control strategies. These photochemical models are large-scale air quality models that simulate the changes of pollutant concentrations in the atmosphere using a set of mathematical equations characterizing the chemical and physical processes in the atmosphere. These models are applied at multiple spatial scales from local, regional, national, and global.

Some examples of photochemical models are the following:

- CMAQ: http://www.epa.gov/asmdnerl/CMAQ/index.html
  The primary goals for the Models-3/Community Multiscale Air Quality (CMAQ) modeling system are to improve

1. the environmental management community's ability to evaluate the impact of air quality management practices for multiple pollutants at multiple scales and
2. the scientist's ability to better probe, understand, and simulate chemical and physical interactions in the atmosphere. The newest Models-3/CMAQ version 4.5 is now available for download from the CMAS Website:

http://www.cmascenter.org/download/release_calendar.cfm?temp_id=99999

- CAMX: http://www.camx.com/

  The Comprehensive Air quality Model with extensions is a publicly available open-source computer modeling system for the integrated assessment of gaseous and particulate air pollution. Built on today's understanding that air quality issues are complex, interrelated, and reach beyond the urban scale, CAMx is designed to

  - Simulate air quality over many geographic scales

    Treat a wide variety of inert and chemically active pollutants:
    - Ozone
    - Inorganic and organic PM2.5/PM10
    - Mercury and toxics

    Provide source-receptor, sensitivity, and process analyses
  - Be computationally efficient and easy to use

  The U.S. EPA has approved the use of CAMx for numerous ozone and PM State Implementation Plans throughout the U.S, and has used this model to evaluate regional mitigation strategies.

- UAM: http://uamv.saintl.com/

  The Urban Airshed Model® (UAM®) modeling system, developed and maintained by Systems Applications International (SAI), is the most widely used photochemical air quality model in the world today. Since SAI's pioneering efforts in photochemical air quality modeling in the early 1970s, the model has undergone nearly continuous cycles of application, performance evaluation, update, extension, and improvement. Other photochemical models have been developed during this long period, but no model today is more reliable or technically superior.

- CALGRID: http://www.arb.ca.gov/eos/soft.htm#calgrid

5.5.2 Other models

Many additional models are available either for regulatory applications or for R&D studies

Meteorological Models

- CALMET:

  http://www.src.com/html/calpuff/calpuff1.htmhttp://www.breeze-software.com/prod/brzSoftware.asp?P=CALMET

CALMET is a meteorological diagnostic model that combines data from surface stations, upper-air stations, over-water stations, precipitation stations, with geophysical data like land use, terrain elevations, albedo, etc., to produce a fully 3-dimensional diagnostic gridded wind field for the duration of the CALPUFF simulation. This wind field is then passed into CALPUFF and is used to transport the emitted substances.

CALMET can link to prognostic meteorological models (i.e., MM5, ETA, RUC2, RAMS) and use their data to produce the gridded wind field.

- MM5: http://www.mmm.ucar.edu/mm5/

  The PSU/NCAR mesoscale model (known as MM5) is a limited-area, non hydrostatic, terrain-following sigma-coordinate model designed to simulate or predict mesoscale atmospheric circulation. The model is supported by several pre- and post-processing programs, which are referred to collectively as the MM5 modeling system. The MM5 modeling system software is mostly written in Fortran, and has been developed at Penn State and NCAR as a community mesoscale model with contributions from users worldwide.

  The MM5 modeling system software is freely provided and supported by the Mesoscale Prediction Group in the Mesoscale and Microscale Meteorology Division, NCAR.

- RAMS: http://rams.atmos.colostate.edu/rams-description.html

  RAMS, the Regional Atmospheric Modeling System, is a highly versatile numerical code developed by scientists at Colorado State University for simulating and forecasting meteorological phenomena, and for depicting the results. Its major components are:

  1. An atmospheric model which performs the actual simulations
  2. A data analysis package which prepares initial data for the atmospheric model from observed meteorological data
  3. A post-processing model visualization and analysis package, which interfaces atmospheric, model output with a variety of visualization software utilities.

5.5.2.1 Plume rise modules

Most air pollution models include a computational module for computing plume rise, i.e., the initial behavior of a hot plume injected vertically into a horizontal wind flow. In particular, AERMOD includes PRIME (Plume Rise Model Enhancements): http://www.epa.gov/scram001/7thconf/iscprime/tekpapr1.pdf

PRIME is an algorithm for simulating plume rise effects, including downwash as the plume travels over buildings.

5.5.2.2 Particle models

Particle models are based on Lagrangian methods for simulating atmospheric diffusion. In these models, plumes are represented by thousands (even hundreds of thousands) of "fictitious" particles, which often move with semi-random trajectories in order to recreate the random components of atmospheric turbulence. These high-resolution models are particularly useful for simulating short-term releases from sources with highly variable emission rates in complex dispersion scenarios. Particle models are capable of simulating very short-term concentrations (e.g., 1-minute averages). Examples are listed below:

- Kinematic Simulation Particle (KSP) Model in CALPUFF http://www.src.com/html/verio/download/CALPUFF_UsersGuide.pdf
- MONTECARLO (Zannetti and Sire, 1999)

5.5.2.3 Deposition modules

Many air pollution models include a computational module for computing the fraction of the plume deposited at the ground as a consequence of dry and wet deposition phenomena.

5.5.2.4 Odor modeling

The mechanisms of dispersion of odorous chemicals (e.g., mercaptans) in the atmosphere are the same as the dispersion of other pollutants. However, when multiple pollutants are emitted, masking and enhancing effects may occur. In this case, the relationship between concentrations of individual chemicals and odor is not well defined and odor must be characterized in terms of an odor detection threshold value for the entire mixture of odorous chemicals in the air. This is why, in odor modeling applications, it is often preferred to express the emission in "odor units".

If the pollutant of interest is chemically uniform (e.g. CO, O_3, SO_2), then there is no way to distinguish between sources. But if the pollutant is particulate matter (either TSP or PM_{10}) that consists of a wide variety of chemical species, then by analysing the chemical composition one can make some inferences about the sources. The result of such analysis is called a *source apportionment or chemical mass balance* [30].

One of the most common receptor models is the Chemical Mass Balance (CMB) Model EPA-CMBv8.2. This modelis one of several receptor models that have been applied to air quality problems over the last two decades. Based on an effective-variance least squares method (EVLS), EPA has supported CMB as a regulatory planning tool through its approval of numerous State Implementation Plans (SIPs), which have a source apportionment component. CMB requires speciated profiles of potentially contributing sources and the corresponding ambient data from analyzed samples collected at a single receptor site. CMB is ideal for localized nonattainment problems and has proven to be a useful tool in applications where steady-state Gaussian plume models are inappropriate, as well as for confirming or adjusting emissions inventories.

5.5.2.5 *Modeling of adverse effects*

Special models or mathematical techniques are available to calculate the adverse effects of air pollution. These models include:

- Health effects (e.g., cancer risk) http://cfpub.epa.gov/ncea/cfm/recordisplay.cfm?deid=116283
- Visibility impairment http://vista.cira.colostate.edu/IMPROVE/Publications/Principle/EPA_Report/chapter5.pdf
- Global effects, such as climate change http://www.epa.gov/appcdwww/apb/globalchange/
- Damage to materials
- Ecological damages

Suitability of Models

The extent to which a specific air quality model is suitable for the evaluation of source impact depends upon several factors:

1. Meteorological and topographic complexities of the area
2. Source configuration
3. Level of detail and accuracy needed for the analysis
4. Technical competence of those undertaking such simulation modeling
5. Resources available
6. Detail and accuracy of the database emissions inventory, meteorological data, air quality data

Models are highly specialized tools. A model applied improperly, or with inappropriately chosen data, can lead to serious misjudgments regarding the source impact or the effectiveness of a control strategy. In general, the more parameters a model includes, the more accurately the result will represent the real situation.

If the parameters necessary for a particular model are unknown, however, that model should not be used. As modeling efforts become more complex, it is increasingly important that they be directed by highly competent individuals with a broad range of experience and knowledge in air quality meteorology.

The procedures and techniques for determining the acceptability of a model for an individual case is contained in the document entitled Interim Procedures for Evaluating Air Quality Models, EPA 450/4-84-023 or NTIS document PB84-106060, and is also discussed in Michigan Administrative Rule 240 entitled "Required Air Quality Models."

5.5.3 Screening and refined computer models for dispersion of air pollutants

The methods for air quality dispersion modeling should be considered in two levels: (1) screening techniques, and (2) refined dispersion models.

Screening techniques are relatively simple methods that provide conservative estimates of the air quality impact of a specific source or source category. The purpose of screening is to eliminate the need for further detailed modeling for sources that clearly will not cause or contribute to ambient concentrations in excess of specific air quality criteria. If the predicted maximum impact from the screening model exceeds the specified criteria, more sophisticated models need to be applied.

Refined models consist of analytical techniques that provide more detailed treatment of physical and chemical atmospheric processes, require more detailed and precise input data, and provide more specialized concentration estimates. Theoretically, refined models give a more accurate estimate of source impact and the effectiveness of control strategies. These models can also be used to evaluate engineering changes (e.g., stack height and location) that may bring the source into compliance

Table 5.10 contains the refined models preferred by the EPA in simple terrain. These models are for use with criteria pollutants from stationary sources although other applications are possible.

A refined model provides a more detailed analysis of the parameters and thus give a more accurate estimate of fine model demands more specific input data. The specific data can include topography, better receptor grid resolution, downwash or other plume adjustments and pollutant decay or decomposition.

Table 5.10 *Preferred models of USEPA for selected application in simple terrain*

	Land use	**Model***
Short-term (1-24 h)		
Single source	Rural	CESTER
	Urban	RAM
Multiple source	Rural	MPTER
	Urban	RAM
Complicated sources**	Rural/Urban	ISCST
Buoyant industrial line sources	Rural	BLP
Long-term (monthly, seasonal or annual)		
Single source	Rural	CRSTER
	Urban	RAM
Multiple source	Rural	MPTER
	Urban	CDM 2.0 or RAM***
Complicated sources***	Rural/Urban	ISCLT
Buoyant industrial line sources	Rural	BLP

*Several of these contain options which allow them to be interchanged; **, ***

5.5.4 Multiple cell models or urban airshed models for the application of Gaussian plume model for area sources

Multiple cell model is one of the presently widely used models involving complex simultaneous reaction expressions for Gaussian plume model e.g. the Urban Airshed Model, or UAM, [25,26] in which the airspace over a city or region is divided into a number of multiple cells. In this model each cell is treated separately from the others and mostly used for ozone, or other secondary pollutants which have very small residence times in ambient air.

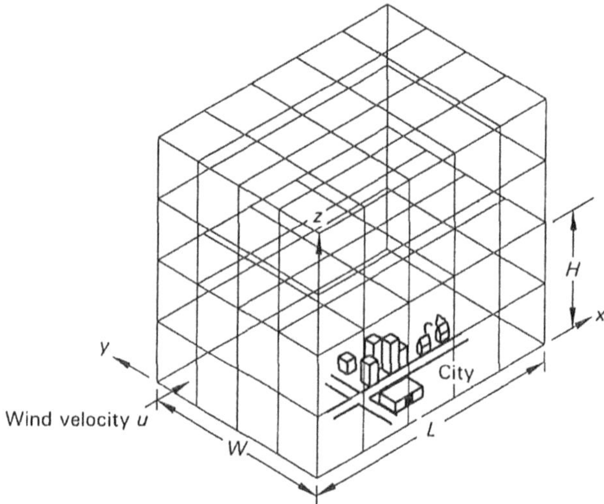

Fig. 5.41(a) *Division of the airspace over a city into cells for the Urban Air shed Model (UAM)*

In the UAM model each cell will have uniform grid sizes in the x and y directions, about 2 to 5 km each way for the entire city. There will be normally four or six layers, half below the mixing height and half above in the vertical direction and with the variation of the mixing height over the day, and from location to location within the city their boundaries will also change. For a large city with 2 km grid spacing there can be thousands of cells assumed in Fig. 5.41(a). For a city for each cell all the times in Equation 5.32 will be included with an assumed initial distribution of pollutants and pollutant precursors in all of the cells. The concentration will be calculated after sometime. After a typical time of 3 to 6 minutes, the program calculates the change in concentration of the pollutant of interest and its precursors in each of the cells by numerically integrating Equation 5.32. This computation requires data or an estimating procedure for the wind velocity and direction at the centre of each cell (to calculate the flows in and out across the boundaries) plus emission estimates for each of the ground-level cells, plus a subprogram to compute the chemical transformations during the time step in any cells, plus a subprogram for deposition of the pollutant from the ground-level cells. Instead of solving all the terms in Equation 5.32 simultaneously, in the UAM the changes in concentration due to flows with the winds across the cell boundaries will be computed and using the concentrations from the end of the previous time step, then the changes due to chemical reactions in the cell will be calculated. For estimating the concentration in each cell at the end of the time period the results obtained in the first two operations will be added.

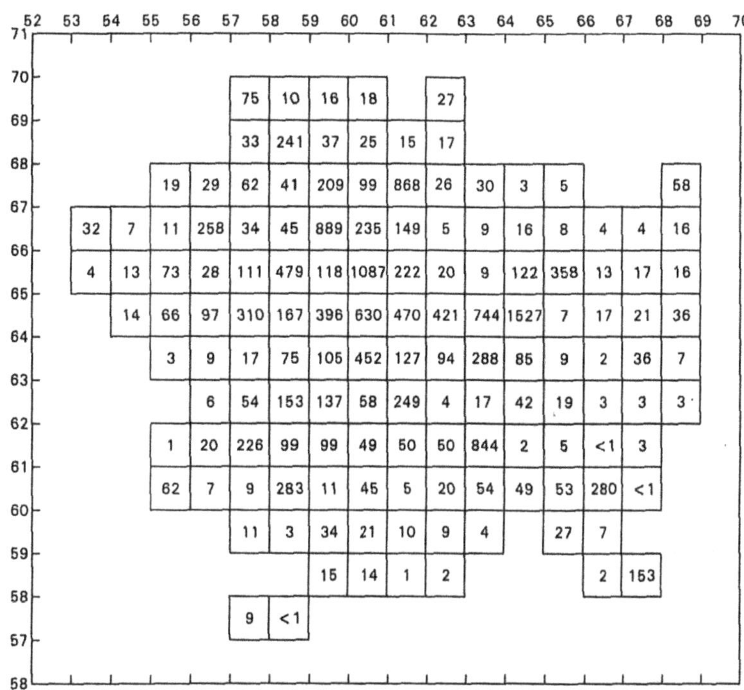

Fig. 5.41(b) *Total SO_2 emissions for a typical city area (tons/yr) in 1 × 1 km grid squares*

For this model to simulate the data for a day or few days in an urban area a complete history of the wind pattern, solar inputs, and emissions is required. If these data are not available, the program has ways of estimating them. A day on which the measured pollutant (usually ozone) concentration was the maximum for the past year or past few years is noted. Then by using the past record of the wind speeds and directions, solar inputs, and estimated emissions for that day the model will be run. Then the adjustable parameters will be modified until the calculated concentrations match well with the observed ambient concentrations for that day. These model simulations will be repeated; the model is rerun with different emission rates or distributions, corresponding to proposed or anticipated future situations, and the meteorology for that day. A prediction of the worst day situation under the proposed future emission pattern can be evaluated by this model.

Multiple cell models are found to be the best models that show promise for being able to give useful guidance on the photochemical oxidant problems like ozone. For pollutants that do not change rapidly in the atmosphere, (e.g. SO_2) this model is not in use since large data are required for this model.

5.5.5 Mobile and line source modelling

Mobile sources are difficult to model because the source is moving and may not be continuously emitting pollutants at a constant emission rate. Nonetheless, environmental engineers often model these sources using line source techniques that assign emissions at fixed points along a line. The USEPA has developed a program called MOBILE 5bto calculate emissions from mobile sources based on EPA tests of emissions from a variety of vehicles classes and types. In certain cases, environmental engineers can use the ISC3 models to estimate concentrations at receptors, but the EPA has identified the California line (CALINE3), CAL3HCR and the buoyant line and point source (BLP) models as the most appropriate for calculating emission concentrations from line sources.

CALINE3 model

The CALINE3 model [27] was designed to estimate highway traffic concentrations of non-reactive pollutants for the state of California. The model is a steady state, Gaussian-based mode1 capable of predicting 1 h to 24 h average concentrations from urban or rural roadway emission sources.

CAL3QHC model

While the CAL3QHC model [27] is not officially part of the Appendix A models, it is a recommended model for CO emissions in the Guidelines on Air Quality Models (USEPA 1993). The model predicts carbon monoxide concentrations at signalized intersections. In effect the model determines the increase in emissions and the resultant concentrations during queuing periods at stop lights. A version of the model, referred to as CAL3QHC has recently been created to process up to a year of meteorology, vehicle emission, traffic volume, and signalization data.

BPL model

The BPL model [28] is a Gaussian based, plume dispersion model that deals with plume rise and downwash effects from stationary line sources. It was originally designed to simulate emissions from an aluminium reduction plant. The model can predict short-term concentrations in rural areas for buoyant, elevated line sources. However, it does not treat deposition or settling of particles.

Accuracy of models

For assessing air quality impacts or for elucidating chemical and transport mechanisms, it is important to assess the accuracy the model can reproduce for observed concentrations. Detailed model validation exercises can be expensive, involving considerable resources in measurements of pollutants at many locations and time-scales, and the appropriate meteorological variables over the domain of the model. In general the more complex the model, the more complex is the validation required. However, the confidence one has in the output of a model depends very much on the questions one is attempting to answer. The prediction of a peak concentration at a specific location will place different demands on a model such as answering a question whether or not controlling a particular category of source in a region would have a beneficial or an adverse effect. In general, long-term (e.g. annual) average concentrations can be predicted with greater confidence than short-term (hourly or less) averages.

5.5.6 Models frequently used in air permit applications-screening techniques

The simplest model used by AQD is a screening technique called the Ambient Impact Ratio (AIR) Matrix, which is contained in the AQD's Administrative Rule 227. The matrix is a simplified, conservative estimate of the emission rate in lbs/hr divided by the maximum ambient impact that emission rate would cause in micrograms per cubic meter. It is used to demonstrate compliance of the discharge of toxic air contaminants with Rule 225. Since the matrix portrays a conservative estimate, there is justification for using refined models for a more accurate and lower estimate.

The MDEQ AQD typically uses two dispersion models for more in depth analysis: **SCREEN3 and ISC3. SCREEN3** are the simplest to use but may produce conservative results. However, it is the U.S. Environmental Protection Agency's (USEPA's) only guideline model to consider cavities and flares.

Industrial Source Complex 3 (ISC3) is appropriate for sources located in both simple terrain, where the terrain features are all lower in elevation than the top of the stack of the source, and in complex terrain, where terrain exists above the stack top. It is the most commonly used model by the AQD.

5.5.6.1 Ambient Impact Ratio (AIR) matrix

Description

The ambient impact ratio (AIR) matrix allows the user to determine the emission rate of a toxic air contaminant (TAC) that would cause a maximum predicted ambient air impact equal to a screening level. This emission rate is derived by multiplying the screening level by the appropriate AIR value. Emission rates which do not exceed that rate are determined to be in compliance with the health-based screening level under R 225.

Use of the AIR matrix requires information pertinent to the dispersion characteristics of emission source, namely, the distance to the nearest secured property line, the height of the stack, and the height of the influential building. The AIR matrix can not be used if any of the following provisions apply:

(a) the stack height is less than 10 feet.

(b) if the influential building height is more than 100 feet.

(c) if there are terrain elevations that are more than 25% of the discharging stack height within a distance of 500 feet from the stack.

(d) if the analysis is being conducted for elevated receptors, (i.e., hospital air intakes.) Instructions for the use of the AIR matrix are, as follows:

1. Determine the height of the discharging stack from ground level in feet (Hs).
2. Determine the height of the influential building in feet (Hb). This is done by first identifying all buildings, including buildings on-site and off-site, located within a distance of five times their height from the discharging stack. Then, determine which building is the highest. This is the influential building, with height (Hb) in feet. If there are no buildings within 5 Hb, then a building 2.5 times lower than the stack height must be assumed.
3. Determine the ratio of the stack height to the influential building height by dividing the stack height, in feet, by the influential building height, in feet, for example, Hs/Hb.
4. Determine the minimum distance, in feet, from the discharging stack to the secured property line. If there is no secured property line, than a distance of 25 feet is used.
5. Determine the appropriate value from the AIR matrix. This is done by selecting the column with the appropriate influential building height and Hs/Hb ratio, and selecting the row with the appropriate minimum distance to the secured property line. If the influential building height is between values in the column headings, then use the lower value or interpolate between values in the column headings. If Hs is less than Hb, then set the influential building height equal to the stack height and use the 1.25 Hs/Hb column. If Hs/Hb is between 1 and 1.25, then select the 1.25 column. If Hs/Hb is between 1.25 and 1.75, then use the 1.25 column or interpolate between the 1.25 and 1.75 columns. If Hs/Hb is between 1.75 and 2.5, then use the 1.75 column or interpolate between the 1.75 and 2.5 columns. If Hs/Hb is greater than or equal to 2.5, then use the 2.5 column. If the minimum distance to the secured minimum distance to the secured property line is between 2 distances in the row headings, then use the lower value, for example, if the distance is 250 feet, then use the 200-foot distance row in the matrix.

   The value thus derived from the body of the matrix is the ratio of the annual averaged hourly emission rate divided by the maximum annual ambient impact, in units of (lbs/hr)/(ug/m3). This value is referred to as the annual AIR.
6. The annual averaged hourly emission rate ratio (annual AIR) is adjusted as necessary for shorter averaging times, consistent with the averaging times for the screening levels. This adjustment is done, as follows:

   24-hr. AIR (lbs/hr)(ug/m3) = annual AIR × 0.091

   8-hr. AIR (lbs/hr)(ug/m3) = annual AIR × 0.046 1-hr. AIR (lbs/hr)(ug/m3)10 = annual AIR × 0.02

7. Determine the maximum emission rate that would comply with the health-based screening level and averaging time. This is done by multiplying the screening level, in ug/m3, by the AIR value for the appropriate averaging time. The result is the highest emission rate, averaged over the averaging time period that would be in compliance with the screening level. If a source's maximum hourly emission rate does not exceed this, then the screening level would not be exceeded. If the emission is intermittent, then the emission rate can be averaged over the applicable averaging time as long as the averaged emission rate is not less than 10% of the maximum hourly emission rate, as specified in Rule 227(2).
8. In the special case of TAC emissions from multiple stacks, determine the AIR value for each stack and select the lowest value among them. Then proceed as in step (6).
9. Atmospheric Chemistry, Concentration dispersion and trajectory Models
    1. Weather Research Forecast (Chemistry) (WRF – Chem) Model
    2. Hybrid Single Particle Lagrangian Trajectory Model (HYSPLIT)

WRF-Chem: The Weather Research and Forecast (WRF) model coupled with Chemistry (104, 105) provides the capability to simulate chemistry and aerosols from cloud scales to regional scales. WRF-Chem is a community model, led by NOAA/ESRL with contributions from National Center for Atmospheric Research (NCAR), Pacific Northwest National Laboratory (PNNL), EPA, and university scientists (http://www.wrf-model.org/WG11). WRFChem is an online model which includes the treatment of the aerosol direct and indirect effect. Standard gas phase chemistry, additional chemistry options are available with a preprocessing tool based on Kinetic Pre-Processor (KPP).For aerosols, the model offers the choice between bulk, modal, and sectional schemes. The Volatile Basis Set (VBS) approach is also available for the modal and sectional aerosol approaches to treat Secondary Organic Aerosol formation. Among other options MEGAN may be used for biogenic emissions, two pre-processors are available for wildfires (injection heights are being calculated online).WRF-Chem is used for research applications or for forecasting of air quality, volcanic ash dispersion, and weather.

The Weather Research and Forecasting (WRF) model is a next generation mesoscale meteorological model. WRF is a non hydrostatic model with two dynamic cores (Advanced Research WRF (ARW) and Non-hydrostatic Mesoscale Model (NMM)) and physical parameterizations so that it can be used to simulate atmospheric processes over a wide range of spatial and temporal scales (115). WRF-chem (105) simulates trace gases and particulates simultaneously with the meteorological fields using the mass and scalar conserving flux form of the governing equations and a terrain-following mass vertical coordinate system. The meteorological and air quality components of the model employ the same time transport schemes, vertical mixing parameterizations, and time step for transport and vertical mixing. The WRF-chem includes the RADM2 photochemical mechanism (117) and the MADE/SORGA Maerosol mechanism

(100, 112). The modular framework of WRF, however, also permits it to be used for research purposes so that different gas-phase and aerosol modules can be evaluated and compare dusing the same emissions and meteorology. The flow chart or WRF-Chem is shown in Figure 5.42.

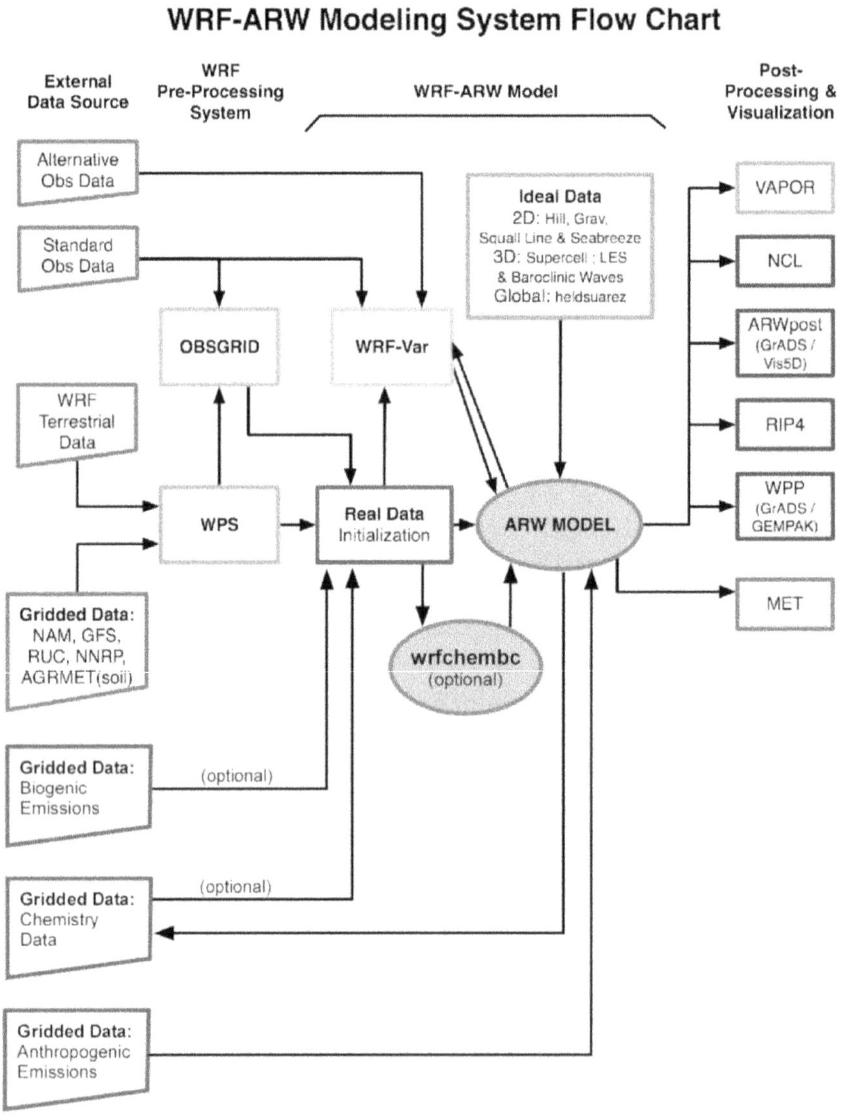

WRF-ARW Modeling System Flow Chart

Fig. 5.42 *WRF-Chem flow chart.*

The main meteorology and chemistry/aerosol interacting processes and effects, which could be considered in online. The meteorological parameters affecting chemistry, worth mentioning effects of altered meteorology on meteorology, in order to better understand chains and loops of interactions. For example, clouds modulate boundary layer outflow/inflow by changes in the radiative fluxes as well as alterations of vertical mixing and the water vapour modulates radiation. The temperature gradient influences cloud formation and controls turbulence intensity and the evolution of the atmospheric boundary layer (ABL). Similar feedback mechanisms exist for altered chemistry impacts on chemistry. For example, biogenic emissions affect the concentrations of ozone and secondary organic aerosols. The polymerisation of organic aerosols produces long chain secondary organic aerosol (SOA) with lower volatility. On a more general level, a number of chains and loops of interactions take place and should be properly simulated in an online coupled model. These may include: (a) A loop feedback starting with temperature that affects chemistry and thus chemical concentrations (Table 5.11); the changes in chemical concentrations will in turn affect radiative processes (Table 5.12), which will then affect temperature to close the loop (illustrated in Fig. 5.43). (b) A chain feedback starting with aerosol that affects radiation (Table 5.12) and thus photolysis and chemistry (Table 5.11). (c) A chain feedback starting with temperature gradients that affects turbulence mixing (Met-Met feedback); thus affecting surface-level pollutant concentrations (Table 5.11) and boundary layer outflow/inflow (Met-Met feedback). (d) A chain feedback starting with aerosols that affect cloud optical depth through influence of droplet number on mean droplet size (Table 5.12); the resulting changes in cloud formation will then affect the initiation of precipitation(Met-Met feedback). (e) A chain feedback starting with aerosol absorption of sunlight which results in changes in the temperature profile of the atmosphere and vertical mixing (Table 5.12) and thereby changes in the cloud droplet formation, which affects cloud liquid water and thus cloud optical depth(Met-Met feedback).

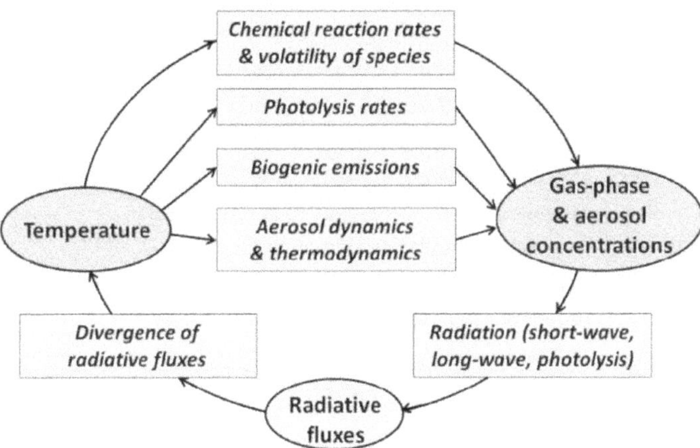

Fig. 5.43 *Conceptual model of impacts from temperature on concentrations and vice versa.*

Table 5.11 Meteorology's impacts on chemistry.

Temperature	Modulates chemical reaction and photolytic rates Modulates biogenic emissions (isoprene, terpenes, dimethyl sulfide, etc.) Influences biogenic emissions (isoprene, monoterpenes) Influences the volatility of chemical species Determines aerosol dynamics (coagulation, condensation, nucleation)
Temperature vertical gradients	Determines vertical diffusion intensity
Temperature & humidity	Affect aerosol thermodynamics (e.g. gas-particle partitioning, secondary aerosol formation)
Water vapour	Modulates OH radicals, size of hydrophilic aerosol
Liquid water	Determines wet scavenging and atmospheric composition
Cloud processes	Affects mixing, transformation and scavenging of chemical compounds
Precipitation	Determines the wet removal of trace gases and aerosol
Land surface parameterization (soil type and vegetation cover, soil moisture, leaf area)	Affects natural emissions (e.g. dust, BVOCs) and dry deposition
Lightning	Determines free troposphere NOx emissions
Radiation	Determines photolysis rates and influences many chemical reaction rates Determines isoprene emissions
Wind speed and direction	Determines horizontal transport and vertical mixing of chemical species Influences dust and sea-salt emissions
ABL height	Influences concentrations

Table 5.12 Chemical species' impacts on meteorology

Aerosols	Modulate radiation transfers (SW scattering/absorption, LW absorption, LW scattering by large particles like dust) Affect boundary layer meteorology (temperature, humidity, wind speed and direction, ABL height, stability) Extraordinary high concentrations can affect stability and wind speed Influence cloud formation, since they act as cloud condensation nuclei
Aerosolsphysical properties (size distribution, mass and number concentrations, hygroscopicity)	Influence cloud droplet and crystal number and hence cloud optical depth and hence radiation Modulate cloud morphology (e.g. reflectance) Influence precipitation (initiation, intensity) Affect haze formation and atmospheric humidity Influence scattering/absorption
Soot deposited on ice	Influences albedo
Radiatively active gases	Modulate radiation transfers

WRF-Chem package brief details:

The WRF-Chem is available from https://ruc.noaa.gov/wrf/wrf-chem. The WRF-Chem model package consists of the following components (inaddition to resolved and non-resolved transport) as well as some additional unlistedcapabilities:

Dry deposition, coupled with the soil/vegetation scheme

- Four choices for biogenic emissions:
  - No biogenic emissions included
  - Online calculation of biogenic emissions as in Simpson et al. (1995) and Guenther et al. (1994) includes emissions of isoprene, monoterpenes, and nitrogen emissions by soil
  - Online modification of user-specified biogenic emissions - such as the EPA
  - Biogenic Emissions Inventory System (BEIS) version 3.14. The user must provide the emissions data for their own domain in the proper WRF data file format
  - Online calculation of biogenic emissions from MEGAN
- Three choices for anthropogenic emissions:
  - No anthropogenic emissions
  - Global emissions data from the one-half degree RETRO and one-degree EDGAR data sets
  - User-specified anthropogenic emissions such as those available from the U.S. EPA NEI-05 and NEI-11 data inventories. The user must provide the emissions data for their own domain in the proper WRF data file format.
- Several choices for gas-phase chemical mechanisms including:
  - RADM2, RACM, CB-4 and CBM-Z chemical mechanisms
  - The use of the Kinetic Pre-Processor, (KPP) to generate the chemicalmechanisms. The equation files (using Rosenbrock type solvers) are currently available for RADM2, RACM, RACM-MIM, SAPRC-99, MOZART andNMHC9 chemical mechanisms
- Three choices for photolysis schemes:
  - Madronich scheme coupled with hydrometeors, aerosols, and convectiveparameterizations. This is a computationally intensive choice, tested with manysetups
  - Fast-J photolysis scheme coupled with hydrometeors, aerosols, and convectiveparameterizations
  - F-TUV photolysis scheme. This scheme, also from Sasha Madronich, is fasterthan the previous Madronich scheme option.
- Five choices for aerosol schemes:
  - The Modal Aerosol Dynamics Model for Europe - MADE/SORGAM
  - The Modal Aerosol Dynamics Model for Europe with the Volitity Basis Setaerosols – MADE/VBS

  - The Modal Aerosol Module (MAM) 3 or 7 bin schemes closely coupled to theCAM5 physics
  - The Model for Simulating Aerosol Interactions and Chemistry (MOSAIC - 4 or 8bins) sectional model aerosol parameterization
  - A bulk aerosol module from GOCART
- Aerosol direct effect through interaction with atmospheric radiation, photolysis, andmicrophysics routines. This is available for all aerosol options starting with version3.5
- Aerosol indirect effect through interaction with atmospheric radiation, photolysis, andmicrophysics routines. This feature is available for modal and sectional aerosoloptions starting with version 3.5.
- An option for the passive tracer transport of greenhouse gases
- Two options for a 10-bin volcanic ash aerosol scheme based upon emissions from asingle active volcano. One scheme includes SO2 degassing from the volcano while theother ignores SO2 degassing. Volcanic ash emissions can also be coupled to someaerosol modules (bulk and modal)
- A tracer transport option in which the chemical mechanism, deposition, etc. has beenturned off. The user must provide the emissions data for their own domain in theproper WRF data file format for this option. May be run parallel with chemistry
- A plume rise model to treat the emissions of wildfires

Emission Sources

Aanthropogenic emissions for use with WRF-Chem

Global-emissions data sets: Two data sources are used with WRF-Chem. The global emissions data comes from the RE analysis of the TROpospheric (RETRO) chemical composition over the past 40 years(http://retro.enes.org) and Emission Database for Global Atmospheric Research(EDGAR) (http://www.mnp.nl/edgar/introduction). Both RETRO and EDGAR provide global annual emissions for several greenhouse gases (e.g., CO2, CH4 and N2O) as well as some precursor gases on a 0.5x0.5 degree (RETRO) or a 1x1 degree (EDGAR) grid.A simple grid-mapping program called prep_chem_sources, developed at CPTEC, Brazil, will map the global anthropogenic emissions data to a WRF-forecast domain using a Mercator, polar stereographic or Lambert conformal projection.

The 4-km resolution NEI data set (2005 NEI emissions data for USA only)

Anthropogenic-emissions data is currently available for the contiguous 48 states of the United States, southern Canada and northern Mexico based upon the U.S. Environmental Protection Agency (EPA) National Emissions Inventory (NEI) 2005inventory. Area type emissions are available on a structured 4-km grid, while point type emissions are available by latitude and longitude locations and with stack parameters needed for plume-rise calculations (http://ruc.noaa.gov/wrf/WG11/anthropogenic.htm).

HYSPLIT: Hybrid Single-Particle Lagrangian Integrated Trajectory model (HYSPLIT) (88, 89) is a complete system for computing simple air parcel trajectories as well as complex transport, dispersion, chemical transformation, and deposition simulation studies. HYSPLIT has been developed by National Oceanic and Atmospheric Administration (NOAA) Air Resources Laboratory's(ARL). HYSPLIT model has many applications like back-trajectory analysis, which is useful to determine the origin of air masses and establish source–receptor relationships (92). HYSPLIT has applications like tracking and forecasting the release of radioactive material (87, 85, 95), wildfire smoke (97), wind-blown dust pollutants from various stationary and mobile emission sources, allergens (90), and volcanic ash (99, 98).

HYSPLIT model calculation method is a hybrid between the Lagrangian approach (using amoving frame of reference for the advection and diffusion calculations as the trajectories or air parcels move from their initial location), and the Eulerian methodology (using a fixed three-dimensional grid as a frame of reference to compute pollutant air concentrations). The HYSPLIT model has evolved throughout more than 30 years, from estimating simplified single trajectories based on radiosonde observations to a system accounting for multiple interacting pollutants transported, dispersed, and deposited over local to global scales.

The HYSPLIT model (version 4) is a complete system for computing rajectories complex dispersion and deposition simulations using either puff or particle approaches. It consists of a modular library structure with main programs for each primary application: trajectories and air concentrations. Gridded meteorological data, on a latitude-longitude grid or one of three conformal (Polar, Lambert, Mercator) map rojections, are required at regular time intervals. The input data are interpolated to an internal sub-grid centered to reduce memory requirements and increase computational speed. Calculations may be performed sequentially or concurrently on multiple meteorological grids, usually specified from fine to coarse resolution. Air concentration calculations require the definition of the pollutant's emissions and physical characteristics (if deposition is required). When multiple pollutant species are defined, an emission would consist of one particle or puff associated with each pollutant type. Alternately, the mass associated with a single puff may contain several species. The latter approach is used for calculation of chemical transformations when all the species follow the same transport pathway. Some simple chemical transformation routines are provided with the standard model distribution. The dispersion of a pollutant is calculated by assuming either a Gaussian or Top-Hat horizontal distribution within apuff or from the dispersal of a fixed number of particles. A single released puff will expand until its size exceeds the meteorological grid cell spacing and then it will split into several puffs. An alternate approach combines both puff and particle methods by assuming a puff distribution in the horizontal and particle dispersion in the vertical direction. The resulting calculation may be started with a single particle. As its horizontal distribution expands beyond the meteorological grid size, it will split into multiple particle-puffs, each with their respective fraction of the pollutant mass.

In this way, the greater accuracy of the vertical dispersion parameterization of the particle model is combined with the advantage of having an expanding number of particles represent the pollutant distribution as the spatial coverage of the pollutant increases and therefore a single particle can represent increasingly lower concentrations. Air concentrations are calculated at a specific grid point for puffs and as cell-average concentrations for particles. A concentration grid is defined by latitude-longitude intersections. Simultaneous multiple grids with different horizontal resolutions and temporal averaging periods can be defined for each simulation. Each pollutant species is summed independently on each grid. The routine meteorological data fields required for the calculations may be obtained from existing archives or from forecast model outputs already formatted for input to HYSPLIT. In addition, several different pre-processor programs are provided to convert NOAA, NCAR (National Center for Atmospheric Research) re-analysis, or ECMWF (European Centre for Medium-range Weather Forecasts) model output fields to a format compatible for direct input to the model. The model's meteorological data structure is compressed and in "direct-access" format. Each time period within the datafile contains an index record that includes grid definitions to locate the spatial domain, check-sums for each record to ensure data integrity, variable identification, and level information. These data files require no conversion between computing platforms. The modeling system includes a Graphical User Interface (GUI) to set up a trajectory, air concentration, or deposition simulation. The post-processing part of the model package incorporates graphical programs to generate multi-color or black and white publication quality Postscript printer graphics. The model can be run interactively on the Web through the ARL READY system (https://ready.arl.noaa.gov/HYSPLIT.php), or the code executable and meteorological data can be downloaded to a Windows or Mac PC. The web version has been configured with some limitations to avoid computational saturation of the ARL web server.

Methodology: HYSPLIT has the capability to generate the Lagrangian path of a particle as the trajectory and facilitates computation of the pollutant dispersion considering the pollutant particles either as a particle-ensemble or a puff. In the particle model, a consigned number of emitted particles will be advected within the model domain conforming to mean and turbulent wind components. In the puff model, the puff will expand and split as several puffs as and when the puff size exceeds the grid cube size with a share of puff mass. HYSPLIT configuration has several options for the choice of dispersion and the default is fixed as particle dispersion in the vertical and puff distribution in the horizontal directions to optimize and effectively reduce the computations of puff modeling. The vertical turbulent diffusion was treated using the Kantha and Clayson (2000) (96) approach where the boundary layer velocity variances were defined as a function of the surface layer parameters and the mixing height. The horizontal turbulence was assumed to be proportional to the vertical turbulence. The boundary layer stability functions were derived from heat and momentum fluxes using the meteorological model fields. In the HYSPLIT model (89), the sulphur module includes both the dry and wet removal of sulphur dioxide and sulphate particles considering the gas and aqueous phase oxidation of sulphur dioxide. The wet removable of gas is applied at all levels from the ground to the top of the cloud. The wet deposition can happen if the pollutant is within or below the cloud layer(as determined from humidity) with different washout coefficients which are dependent on its solubility and is governed by Henry's Law constant. Dry deposition is assumed to occur when the pollutant is close enough to the ground. In the 3D particle method, the particle within the surface layer are only deposited. The mass from dry deposition is computed by assuming a uniform vertical concentration distribution in the layer of deposition. If the pollutant is near the surface, there is dry deposition and the remaining part of the pollutant after deposition continue to be transported by the mean wind field at that height. In the 3D particle method, the dispersion carries the pollutant into the ground at given point and the pollutant with lesser mass after the deposition is reflected back and continues to transport and dispersion.

In the new HYSPLIT model, the advection algorithms have been updated to include temporal interpolation and although the dispersion of a pollutant can still be calculated by assuming the release of a single puff, the puff can now be defined with either a Gaussian or top-hat horizontal distribution. A three-dimensional particle dispersion routine has been added that computes air concentrations from the dispersal of an initial fixed number of particles. The particle approach includes a modification, following Hurley (1994), that combines both puff and particle methods by assuming a puff distribution in the horizontal and particle dispersion in the vertical direction. This hybrid calculation may be started with a single particle. In this way, the greater accuracy of the vertical dispersion parameterization of the particle model is combined with the advantage of having an expanding number of puffs represent the pollutant distribution as the spatial coverage of the pollutant increases.

Trajectories: The basis of any Lagrangian model is that the dispersion is computed following the particle or puff. That is, the advection of a particle is computed independently of the dispersion calculation. The time integrated advection of each particle can be viewed as a simple trajectory which only requires the three-dimensional velocity field.

Meteorological data

Meteorological data for trajectories and dispersion calculations are obtained as output fields from meteorological models. The minimum HYSPLIT required meteorological variables are U, V (the horizontal wind components), T(temperature), Z (height) or P (pressure), and the pressure at the surface, P_0 and rainfall (wet deposition processes). Most of the datasets have ground level($\leq$10m) winds, otherwise these winds are estimated using a logarithmic profile for neutral conditions. HYSPLIT has the ability to use different meteorological data sources (from different models) for input, the profiles of the meteorological data at each horizontal grid point are linearly interpolated to a terrain-following (σ) coordinate system. When the vertical velocity (W) field is missing,

HYSPLIT has an option to calculate W, based upon an assumption that the pollutant parcel is transported on some other surface.

Advection calculation

After acquiring the meteorological fields these are interpolated to the internal model grid, trajectories can be computed. In HYSPLIT, time interpolation of the meteorological fields is directly incorporated into the advection algorithm. The advection of a particle or puff is computed from the average of the three-dimensional velocity vectors for the initial-position P(t) and the first-guess position P'(t+Δt). The velocity vectors are linearly interpolated in both space and time.

The first guess position is $P'(t + \Delta t) = P(t) + V(P,t)\,\Delta t$

and the final position is $P(t + \Delta t) = P(t) + 0.5\,[\,V(P,t) + V(P',t + \Delta t)\,]\,\Delta t$

Trajectories are terminated if they exit the model top, but advection continues along the surface if trajectories intersect the ground. The integration time step (Δt) can vary during the simulation. The advection distance per time-step should be less than0.75 of the meteorological grid spacing.

Air Concentrations

Air concentrations is computed with two assumptions by Lagrangian model. In a puff model, the source is simulated by releasing pollutant puffs at regular intervals over the duration of the release. Each puff contains the appropriate fraction of the pollutant mass. The puff is advected according to the trajectory of its center position while the size of the puff (both horizontally and vertically) expands in time to account for the dispersive nature of a turbulent atmosphere. Air concentrations are then calculated at specific points (or nodes on a grid) by assuming that the concentrations within the puff have a defined spatial distribution.

In a Lagrangian particle model, the source can be simulated by releasing many particles over the duration of the release. In addition to the advective motion of each particle, a random component to the motion is added at each step according to the atmospheric turbulence at that time. In this way a cluster of particles released at the same point will expand in space and time simulating the dispersive nature of the atmosphere. Air concentrations are calculated by summing the mass of all the particles in a grid cell. In a homogeneous environment the size of the puff (in terms of its standard deviation) at any particular time should correspond to the second moment of the particle positions. Although a particle approach can more realistically represent dispersion in an in homogeneous atmosphere, the puff methods used in HYSPLIT permit the puffs to split into smaller components when they exceed certain dimensions, usually the dimensions of the meteorological grid. In this way, subgrid turbulence processes are modeled by the dispersion parameterizations while grid-scale processes are simulated by the puff-splitting process. However, one limitation of puff-splitting is that it can generate too many puffs, especially in the vertical direction if the mixing is strong. A novel approach, developed by Hurley (1994), uses particle dispersion in the vertical direction and puff dispersion in the horizontal, is incorporated in the HYSPLIT model. This permits the more accurate particle representation to be used in the vertical, where discontinuities can be large, and where the puff-splitting approach has its limitations, while using the puff representation in the horizontal to limit the number of particles that are required to adequately represent the horizontal distribution. It is necessary to compute the atmospheric stability, the corresponding atmospheric mixing, and the rate of pollutant dispersion. Boundary layer stability is estimated by two ways. The preferred method is touse the fluxes of heat and momentum provided by the meteorological model. Otherwise the vertical temperature and wind gradients at each grid point are used to estimate stability.

Particle and puff dispersion

Both the horizontal puff and vertical particle dispersion are formulated in terms of the turbulent velocity components which are a function of the diffusivities. In the particle model, the dispersion process is represented by adding a turbulent component to the mean velocity obtained from the meteorological data. Puff dispersion is treated in two domains, when the puff is smaller than the meteorological model grid size and when it is larger. In the latter case it is assumed that the meteorological model is capable of resolving turbulent motions on that scale. Gaussian and "top-hat" puffs are treated almost identically.

Deposition

There are three different removal mechanisms available which are dry deposition, wet depletion, and radioactive decay. Dry deposition can be either explicitly defined by a deposition velocity, a gravitational settling velocity for particles, or it may be computed using the resistance method. Wet removal for soluble gases is computed from its Henry's Law constant and particle wet removal is computed from a scavenging ratio within the cloud and by an explicit scavenging coefficient for pollutants below the cloud base.

Dry removal

Dry removal is computed only when the bottom of the puff or the particle center position is within the surface layer, defined as the dispersion model's second meteorological data level (usually around 75 m).

Wet removal

Wet deposition is divided into two processes, those in which the polluted air is continuously ingested into a cloud from a polluted boundary layer and those in which rainfalls through a polluted layer. The dominant process is difficult to determine during a model simulation using only the data available from most meteorological archives. Therefore, for particulate pollutants, the simplifying assumption of a scavenging ratio is assumed for pollutants located within a cloud layer and the scavenging coefficient is used for pollutant removal in rain below a cloud layer.

Volcanic eruptions

HYSPLIT has another application of emergency response situations, dealing with the forecasting of the movement of debris from volcanic eruptions, which may penetrate well into the stratosphere. HYSPLIT is also capable to handle the volcanic eruptions. In the volcanic ash simulation, an emission point would be represented by several particles or puffs, each corresponding to a different size range of ash. The puffs representing each range need to be treated independently because they will have different atmospheric transport pathways due to gravitational settling. The sizes range from diameters of 0.3 to 30 μm, but they all have the same density (2.5 g cm-3), typical of a mix of pumice, basalt, and similar materials. The release amount for each size range is adjusted according to what fraction of the total mass is represented by that size in the particle distribution. Most of the mass (99%) resides in the 3 to 30 μm range, but the larger particles fall quickly, and only the smallest particles are subject to long-range transport.

Source Code: The HYSPLIT model is available on both Linux and as well as windows platforms. The source code both HYSPLIT model and as well as the other supporting software (e.g., graphical tools) can be downloaded from https://ready.arl.noaa.gov /HYSPLIT.php. The input datasets are available for Global as well as USA (CONUS) regions separately. The datasets can be downloaded from https://ready.arl.noaa.gov/ archives.php. There is an option to run web based HYSPLIT, which is accessible from the HYSPLIT web page.

Further Reading

"AIR QUALITY MODELING": www.envirocomp.org/aqm (ref 29-84)

Problems

1. Estimate the concentration of carbon monoxide at the downwind edge of a city. The city may be considered to consist of three parallel strips, located perpendicular to the wind. For all of the strips the wind velocity u equals 3 m/s. The properties of each of the strips are described in the following table:

Name of strip	Length, km	Emission rate q, g/s km^2	Mixing height H, m
Upwind suburbs	5	100	400
Downtown	2	500	500
Downwind suburbs	5	100	400

Assume that the fixed box model applies to each of the strips. The background concentration b in the air entering the upwind suburbs is 1 mg/m^3.

2. Suppose the following atmospheric altitude versus temperature data have been collected

Altitude (m)	Temperature (°C)
0	20
100	18
200	16
300	15
400	16
500	17
600	18

   (a) What would be the mixing depth?
   (b) How high would you expect a plume to rise if it is emitted at 21°C from a100 m stack and it rises at the dry adiabatic lapse rate? Would you expect the plume to be looping, coning, fanning, or fumigating?
3. For the temperature profile given in Problem 2, if the maximum daytime surface temperature is 22 °C, and a weather station anemometer at 10 m height shows winds averaging 4 m/s, what would be the ventilation coefficient? Assume stability class C and use wind at the height halfway to the maximum mixing depth.
4. A large poorly controlled copper smelter has a stack 150 m high and a plume rise of 75 m. It is currently emitting 1000 g/s of SO_2. Estimate the ground level concentration of SO_2 from this source at a distance 5 km directly downwind when the wind speed is 3 m/s and the stability class is C.
5. The management of the smelter in Problem 4 has been informed that the concentration calculated at that location and for those conditions is twice the allowable. They propose to remedy this situation by installing a higher stack. How high must this stack be so that the estimated concentration will be exactly one-half that in Problem 4?

Questions

Short Questions

1. Write a short note on global air circulation and biomes.
2. Brief about Ground level wind patterns and velocities.
3. Summarize the phenomena of Temperature Inversion.
4. Debrief Wind Rose.
5. Explain the different types of plumes.
6. List out the different models employed in air pollution diffusion.
7. Deduce the Holland's equation for buoyant plumes.

Essay Questions

1. Explain about the horizontal movements of winds with suitable diagrams.
2. How do the Coriolis force and pressure gradient force combine to produce gradient winds?
3. Bring out the significance of Cold and Warm front in frontal weather conditions.
4. How does Air pollution influence the meteorological factors based on their movement?
5. Explain the meteorological aspects of air pollution dispersions in detail.
6. Describe the Gaussian plume model. Give its advantages and limitations

References

1. Frederick K. Lutgens and Edward J.Tarbuck, 1992. The Atmosphere.
2. G. Tyler, 1995. Miller Junior Environmental Science: Working with Earth. Wadsworth Publishing Company, Behnuti, California.
3. O. G. Sutton, 1953. Micrometeorology. McGraw Hill, New York.
4. H. Lamb, 1932. Hydrodynamics, Dover, New York.
5. C. Junge, 1983. Chemical and Engineering News. August 16, 1971. p. 40. Cited Manhan. Environmental Chemistry. 3rd Edition. Willard Grant Press, Boston. Chapter 11.
6. G. C. Holzworth, 1972. 'Mixing heights' wind speed and potential for urban air pollution. USEPA Report. AP-101. Washington DC.
7. P. V. Portels and P. J. Lewis, 1987. Meteorology in atmospheric pollution. E. E. Picket (Ed.) Hemisphere, Washington DC.
8. W. B. Peterson, 1978. User guide for PAL: A Gaussian's plume algorithm for point, area and line sources. USEPA Research Triangle Park NC.
9. G. A. Briggs, 1969. Plume rise U.S. Atomic Energy Commission critical review series. T 10-25075.
10. P. E. Church, 1949. Dilution of waste stack gases in the atmosphere. Industrial engineering chemistry.
11. S. R. Hanna, 1971. A simple model for calculating dispersion from urban area sources. J. Air Pollution Control Asso. 21:774-777.
12. F. Pasquill, 1961. The estimation of dispersion of wind borne materials. Meteorological magazine. 90:30.
13. F. A. Gifford, 1961. Use of routine meteorological observations for estimating atmospheric dispersion. Nuclear safety. 2:47-5 1.

14. D. B. Turner, 1970. Workbook of Atm. dispersion estimates. USEPA Report AP-26.US. Govt. Printing Office, Washington DC.

15. D. O. Martin, 1976. J. Air Pollution Control Association 26. No. 145.

16. H. C. Perkins, 1974. Air pollution. McGraw Hill, New York.

17. G. Briggs, 1972. Atmos. Emissions 6:507.

18. J. Z. Holland, 1963. Meteorological survey of the Oakridge area. ORA 99. US Atomic Energy Commision, Oakridge team.

19. H. Moses and J. E. Carson, 1968. Stack design parameters influencing plume rise. J. Air Pollution Control Asso. 18:454.

20. T. L. Montego many and M. Corn, 1967. J. Air Pollution Control Assoc. Vol. 17:5 12.

21. J. A. Tickvart, 1988. Memorandum on stock-structure relationships. Research Triangle Park. NC USEPA office of air quality planning and standards.

22. Rober B. Wilson, 1993. Review of development and application of CRSTER and MPTER models. Atm. Env. 27B: 41–57.

23. USEPA 1989. User's guide to the complex dispersion model plans algorithms for unstable situations (CTDMPAUS). Vol. 1. EPA 600/8-89-041, New York.

24. USEPA 1990(a). User's guide to urban air shed model. Vol. 1. 8EPA 450/04-90-007.Research Triangle Park NC. USA.

R. E. Morris and T. C. Myers, 1990(b). Users guide for urban air shed model. USEPA report. EPA 450/4-90-007A and B.

25. R. D. Scheffe, 1993. A review of the development and application of urban air shed model. Atmos. emission. 278:23-29.

26. USEPA 1993. Reactive plume model IV RPM-IV. User's guide. EPA-454393012. Research Triangle Park NC.

27. P. E. Benson, 1979. CALINE 3 – A versatile dispersion model for predicting air pollutant levels near highways and arterial-streets, Interim report FHWA/CA/TL-79/23, Washington DC.

28. L. L. Schulman and S. Scire, 1980. Buoyant line and point sources (BLP) dispersion model user's guide doc. p-7304B. Concord, Mass Env. Research Tech.

29. Beryl and, M.Y., 1975, Contemporary problems of atmospheric diffusion and pollution of the atmosphere. Gidrometezdat, Leningrad, translated into English by NERC, US EPA.

30. Boubel et al. (1994) Fundamentals of Air Pollution, 3rd edition. Academic Press.

31. Briggs, G.A., 1965, A plume rise model compared with observations J.Air Poll. Control Association 15:433.

32. Bosanquet, C.H. (1936) The Spread of Smoke and Gas from Chimmneys. Trans. Faraday Soc. 32:1249.

33. Carmichael, G.R., and Peters, L.K., 1979, Numerical simulation of the regional transport of SO_2 and sulfate in the eastern United States, Proc. 4 th Symp. on turbulence, diffusion and air pollution, AMS 337.

34. Chamberlain, A.C., 1953, Aspects of travel and deposition of aerosol and vapour clouds A.E.R.E., HP/R 1261, H.M.S.O.

35. Deardorff, J.W., and Willis, G.E., 1975, A parameterization of diffusion into the mixed layer J. Appl. Met 14:1451.

36. Deardorff, J.W., 1970, Convective velocity and temperature scales for the unstable planetary boundary layer and for Rayleigh convection. J. Atm. Sci. 27, 1211-1213.

37. Dyun, D.W. and J.K.S. Ching, 1999, Science algorithm of the EPA Models-3 Community Multiscale Air Quality (CMAQ) Modeling System., EPA-Dep./600/R-99/030.

38. Egan, B.A., Rao, K.S., and Bass, A., 1976, A three dimensional advective-diffusive model for long-range sulfate transport and transformation 7 th ITM, 697, Airlie House.

39. Eliassen, A., and Saltbones, J., 1975, Decay and transformation rates of SO_2 as estimated from emission data, trajectories and measured air concentrations Atm. Env. 9:425.

40. Eschenroeder, A.Q. and J.R. Martinez, 1970, "Mathematical Modeling of Photochemical Smog", American Institute Aeronautics and Astronautics (Proceedings), Eight Aerospace Sciences Meeting, New York, Jan 19-21.

41. Finzi, G. and G. Nunnari (2005) Air Quality Forecast and Alarm Systems. Chapter 16A of Zannetti, P., Ed. Air Quality Modeling, Vol. II. www.envirocomp.org/aqm

42. Fisher, B.E.A., 1975, The long-range transport of sulfur dioxide, Atm.Env.9,: 1063.

Friedlander, S.K. and J.H. Seinfeld, 1969, A Dynamic Model of Photochemical Smog, Environ., Science Technol., 3, 1175 (1969).

43. Gifford, F.A., 1957a, Relative atmospheric diffusion of smoke plumes J. Met. 14:410

44. Gifford, F.A., 1957b, Further data on relative atmospheric diffusion J. Met. 14:475.

45. Haugen, D.A., 1959, Project Prairie grass, a field programme in diffusion Geographical research paper 59, vol III, G.R.D.A.F.C., Bedford, Mass.

46. Hay, J.S., and Pasquill, F., 1957, Diffusion from a fixed source at a height of a few hundred feet in the atmosphere J. Fluid Mech. 2:299.

47. Högstrom, U., 1964, An experimental study on atmospheric diffusion Tellus, 16:205.

48. Holzworth, G.C., 1967, Mixing depth, wind speed and air pollution potential for selected locations in the U.S.A., J.Appl.Met. 6:1039.

49. Isaksen, I.S.A., and Rohde, H., 1978, A two-dimensional model for the global distribution of gases and aerosol particles in the troposphere Rep. AC-47, Dep. of Meteor. Univ Stockholm, Sweden.

50. Junge, C.E., 1963, Air chemistry and radioactivity, academic press, New York, London.

51. Johnson, W.B., 1980, Interregional exchange of air pollution: model types and applications 10th ITM, 3, Amsterdam.

52. Klug, W., 1968, Diffusion in the atmospheric surface layer: comparison of similarity theory with observations Quart. J.R. Met.Soc. 94:555.

53. Kolmogorov, A.N., 1941. The local structure of turbulence in incompressible viscous fluid for very large Reynolds numbers, C.R. Acad. Sci. U.R.S.S., 30, 301.

54. Lamb, R.G. and J.H. Seinfeld, 1973. Mathematical modeling of urban air pollution - General theory. Envir. Sci. Technol. 7, 253-261.

55. Levy, H., 1971, Normal atmosphere: large radical and formaldehyde concentrations predicted Science 173:141.

56. Liu, M.K. and J.H. Seinfeld, 1974, On the Validity of Grid and Trajectory Models of Urban Air Pollution, Atmos. Environ., Vol. 9, pp. 555-574.

57. Monin, A.S., 1959, Smoke propagation in the surface layer of the atmosphere Atmospheric diffusion and air pollution, ed. Frenkiel and Sheppard, advances in Geophysics, 6:331, Academic Press.

58. Monin, A.S., 1955, The equation of turbulent diffusion, Dokl. Akad. Naak., 105, 256.

59. Moore, D.J., 1967, Physical aspects of plume models Atm.Env. 1:411.

60. Nieuwstadt, F.T.M., and Dop, H. van, 1982, Atmospheric turbulence and air pollution modeling, D.Reidel Publish. Comp.

61. Ogura, Y., 1959, Diffusion from a continuous source in relation to a finite observation interval' Atmospheric diffusion and air pollution, ed. Frenkiel and Sheppard, advances in Geophysics 6,149, Academic Press.

62. Pasquill, F., 1962, Atmospheric diffusion Van Nostrand, New York.

63. Peters, L.K., and Jouvanis, A.A., 1979, Numerical simulation of the transport and chemistry of CH_4 and CO in the troposphere, Atm.Env. 13:1443.

64. Record, F.A., and Cramer, H.E., 1958, Preliminary analysis of Project Prairie grass diffusion measurements J.AirPoll.Cont.Ass 8:240.

65. Reynolds, S., Roth, P., and Seinfeld, J., 1973, Mathematical modeling of photochemical air pollution Atm.Env 7.

66. Reynolds, O., 1895, On the dynamical theory of incompressible viscous fluids and the determination of the criterion Phil. Transactions of the Royal Soc. of London. series A, 186:123.

67. Roberts, O.F.T., 1923, The theoretical scattering of smoke in a turbulent atmosphere Proc. Roy. Soc. A, 104, 640.

68. Rohde, H., 1972, A study of the sulfur budget for the atmosphere over northern Europe Tellus, 24:128.

69. Rohde, H., 1974, Some aspects of the use of air trajectories for the computation of large scale dispersion and fallout patterns Adv. in Geophysics 18B: 95, academic press.

70. Roth, P.M., P.J.W. Roberts, J.K. Liu, S.D. Reynolds and J.H. Seinfeld, 1974, Mathematical Modeling of Photochemical Air Pollution - Part II. A Model and Inventory of Pollutant Emissions, Atmos. Environ., Vol. 8, pp. 97-130.

71. Seinfeld, J.H., 1986, Atmospheric chemistry and physics of air pollution, John Wiley & Sons.

72. Shir, C.C. and L.J. Shieh, 1974, A generalized urban air pollution model and its application to the study of SO_2-distribution in the St. Louis Metropolitan area, J. Appl. Met. 19, 185-204.

73. Sklarew, R.C. et al., 1971, A particle-in-cell method for numerical solution of the atmospheric diffusion equation and application to air pollution problems; Systems, Science and Software, Ca-Reg 35R-844, Vol I.

74. Smith, F.B., 1957, The diffusion of smoke from a continuous elevated point source into a turbulent atmosphere J.Fluid Mech. 2:49.

75. Stewart, N.G. et al., 1958, The atmospheric diffusion of gases discharged from the chimney of the Harwell Pile IntJ. Air Poll. 1:87.

76. Sutton O.G. (1932) A theory of Eddy Diffusion in the Atmosphere. Proc. Roy. Soc. A, 135:143

77. Taylor, G.I., 1915, Eddy motion in the atmosphere Phil. Transactions of the Royal Soc. of London. Series A, 215:1.

78. Taylor, G.I., 1921, Diffusion by continuous movements proc. London Math. soc. 20:196.

79. Turner, D.B., 1964, A diffusion model for an urban area J.Appl.Met. 3:83.

80. Warneck, P., 1988, Chemistry of the natural atmosphere, Int.Geoph.Series 41, Academic Press.

81. Willis and Deardorff, 1978, A laboratory study of dispersion from elevated source within a modeled convection mixed layer, Atm. Env. 12, 1305-1311.

82. Yamartino, R. J. (1977) A new Method for computing pollutant concentrations in the presence of limited vertical mixing. APCA Note Book 27(5):467

83. Zannetti, P. and R. Sire (1999) MONTECARLO – A new, fully-integrated PC Software for the 3D simulation and Visualization and Air pollution dispersion using Monte Carlo Lagrangian Particle (MCLP) techniques. AIR POLLUTION 99, Stanford, CA, WIT Publications.

84. Zimmermann, P.H., 1988, Moguntia: a handy global tracer model 17th ITM. 593, Cambridge.

85. Bowyer, T. W., R. Kephart, P. W. Eslinger, J. I. Friese, H. S. Miley, and P. R. J. Saey, 2013: Maximum reasonable radioxenon releases from medical isotope production facilities and their effect on monitoring nuclear explosions. J. Environ. Radioact., 115, 192–200, doi:10.1016/j.jenvrad.2012.07.018.

86. Chen, B., A. F. Stein, P. Guerrero Maldonado, A. M. Sanchez de la Campa, Y. Gonzalez-Castanedo, N. Castell, and J. D. de la Rosa, 2013: Size distribution and concentrations of heavy metals in atmospheric aerosols originating from industrial emissions as predicted by the HYSPLIT model. Atmos. Environ., 71, 234–244, doi:10.1016/j.atmosenv.2013.02.013.

87. Connan, O., K. Smith, C. Organo, L. Solier, D. Maro, and D. Hébert, 2013: Comparison of RIMPUFF, HYSPLIT, ADMS atmospheric dispersion model outputs, using emergency response procedures, with 85Kr measurements made in the vicinity of nuclear reprocessing plant. J. Environ. Radioact., 124, 266 -277, doi:10.1016/j.jenvrad.2013.06.004.

88. Draxler, R. R., and G. D. Hess, 1998: An overview of the HYSPLIT_4 modeling system for trajectories, dispersion, and deposition. Aust. Meteor. Mag., 47, 295–308.

89. Draxler, R.R. and Hess, G.D. 2014: Description of the HYSPLIT_4 Modeling System. NOAA Technical Memorandum ERL ARL-224, Air Resources Laboratory, Silver Spring, MD. NOAA ARL READY Website, http://www.arl.noaa.gov/documents/reports/arl-224.pdf.

90. Efstathiou, C., S. Isukapalli, and P. Georgopoulos, 2011: A mechanistic modeling system for estimating large-scale emissions and transport of pollen and co-allergens. Atmos. Environ., 45, 2260–2276, doi:10.1016/j.atmosenv.2010.12.008

91. Escudero, M., A. Stein, R. R. Draxler, X. Querol, A. Alastuey, S. Castillo, and A. Avila, 2011: Source apportionment for African dust outbreaks over the Western Mediterranean using the HYSPLIT model. Atmos. Res., 99 (3–4), 518–527, doi:10.1016/j.atmosres.2010.12.002.

92. Fleming, Z. L., P. S. Monks, and A. J. Manning, 2012: Review: Untangling the influence of air-mass history in interpreting observed atmospheric composition. Atmos. Res., 104–105, 1–39, doi:10.1016/j.atmosres.2011.09.009.

93. Gaiero, D. M., L. Simonella, S. Gassó, S. Gili, A. F. Stein, P. Sosa, R. Becchio, J. Arce, and H. Marelli, 2013: Ground/satellite observations and atmospheric modeling of dust storms originated in the high Puna-Altiplano deserts (South America): Implications for the interpretation of paleo-climatic archives. J. Geophys. Res., 118, 3817–3831, doi:10.1002/jgrd.50036.

94. Hurley, P. 1994. PARTPUFF - A Lagrangian particle-puff approach for plume dispersion modeling applications, Jnl. appl. Met., 33, 285-294.
95. Jeong, H., M. Park, H. Jeong, W. Hwang, E. Kim, and M. Han, 2013: Radiological risk assessment caused by RDD terrorism in an urban area. Appl. Radiat. Isot., 79, 1–4, doi:10.1016/j.apradiso.2013.04.018.
96. Kantha, L. H., and C. A. Clayson, 2000: Small Scale Processes in Geophysical Fluid Flows. International Geophysics, Vol. 67, Academic Press, 750 pp.
97. Rolph, G. D., Draxler, R. R., Stein, A. F., Taylor, A., Ruminski,M. G., Kondragunta, S., Zeng, J., Huang, H.-C., Manikin, G., McQueen, J. T., and Davidson, P. M. 2009: Description and verification of the NOAA Smoke Forecasting System: The 2007 fire season. Wea. Forecasting, 24, 361–378, doi:10.1175/2008WAF2222165.1.
98. Stein, A.F., R.R. Draxler, G.D. Rolph, B.J.B. Stunder, M.D. Cohen, and F. Ngan, 2015. NOAA's HYSPLIT atmospheric transport and dispersion modeling system, Bulletin of the American Meteorological Society. 96, 2059–2077. doi: http://dx.doi.org/10.1175/BAMS-D-14-00110.
99. Stunder, B. J. B., J. L. Heffter, and R. R. Draxler, 2007: Airborne volcanic ash forecast area reliability. Wea. Forecasting, 22, 1132–1139, doi:10.1175/WAF1042.1.
100. Ackermann, I. J., H. Hass, M. Memmesheimer, A. Ebel, F. S. Binkowski, and U. Shankar, 1998: Modal aerosol dynamics model for Europe: Development and first applications, Atmos. Env., 32, No.17, 2981-2999.
101. Binkowski, F. S. and U. Shankar, 1995: The regional particulate matter model, 1. Mode desription and preliminary results, Journal of Geophysical Research, 100, 26191-26209
102. Chang, J. S., F. S. Binkowski, N. L. Seaman, J. N. McHenry, P. J. Samson, W. R. Stockwell, C. J. Walcek, S. Madronich, P. B. Middleton, J. E. Pleim and H. H. Lansford, 1989: The regional acid deposition model and engineering model. State-of-Science/Technology, Report 4, National Acid Precipitation Assessment Program, Washington D.C.
103. Dudhia, J, 1989: Numerical study of convection observed during winter monsoon experiment using a mesoscale two-dimensional model, J. Atmos. Sci. 46, 3077-3107.
104. Fast, J. D., Gustafson Jr.,W. I., Easter, R. C., Zaveri, R. A., Barnard, J. C., Chapman, E. G., and Grell, G. A, 2006: Evolution of ozone, particulates, and aerosol direct forcing in an urban area using a new fully coupled meteorology, chemistry, and aerosol model, J. Geophys. Res., 111, D21305, doi:10.1029/2005JD006721.
105. Grell, G. A., S. E. Peckham, R. Schmitz, S. A. McKeen, G. Frost, W. C. Skamarock, and B. Eder, 2005: Fully coupled "online" chemistry within the WRF model, Atmos. Environ., 39, 6957– 6975.
106. Guenther, A. B., P. R. Zimmerman, P. C. Harley, R. K. Monson, and R. Fall, 1993: Isoprene and monoterpene emission rate variability: Model evaluations and sensitivity analyses. J. Geophys. Res., 98D, 12609-12617.

107. Guenther, A., P. Zimmerman, and M. Wildermuth, 1994: Natural volatile organic compound emission rate estimates for U.S. woodland landscapes. Atmos. Env., 28, 1197-1210.
108. Joseph, J. H., W. J. Wiscombe, J. A. Weinmann, 1976: The delta-Eddington approximation for radiative flux transfer. Journal of Atmospheric Science, 33, 2452-2458.
109. Madronich, S., 1987: Photodissociation in the atmosphere, 1, actinic flux and the effects of ground reflections and clouds. Journal of Geophysical Research, 92, 9740-9752.
110. Middleton, P., W. R. Stockwell, and W. P. L. Carter, 1990: Aggregation and analysis of volatile organic compound emissions for regional modeling. Atmos. Env., 24A, 1107-1133.
111. Odum, J. R., T. Hoffmann, F. Bowman, D. Collins, R.C. Flagan and J.H. Seinfeld, 1996: Gas/particle partitioning and secondary organic aerosol yields, Environ. Sci. Technol., 30, 2580– 2585.
112. Schell B., I. J. Ackermann, H. Hass, F. S. Binkowski, and A. Ebel, 2001: Modeling the formation of secondary organic aerosol within a comprehensive air quality model system, Journal of Geophysical research, 106, 28275-28293.
113. Schoenemeyer, T., K. Richter, and G. Smiatek, 1997: Vorstudie uber ein raumlich und zeitlich aufgelostes Kataster anthropogener und biogener Emissionen fuer Bayern mit Entwicklung eines Prototyps und Anwendung fur Immissionsprognosen. Abschluss bericht an das Bayerische Landesamt fur Umweltschutz. Fraunhofer-Institut fuer Atmosphaerische Umweltforschung, Garmisch-Partenkirchen.
114. Simpson, D., A. Guenther, C. N. Hewitt, and R. Steinbrecher, 1995: Biogenic emissions in Europe. 1. estimates and uncertainties. J. Geophys. Res., 100D, 22875-22890.
115. Skamarock, W. C., Klemp, J. B., Dudhia, J., Gill, D. O., Barker,D. M., Duda, M. G., Huang, X.-Y., Wang, W., and Powers, J.G., 2008: A description of the Advanced Research WRF version 3, National Center for Atmospheric Research Tech. Note, NCAR/TN-475+STR, 113 pp.
116. Stockwell, W. R., P. Middleton, J. S. Chang, and X. Tang, 1990: The second generation regional acid deposition model chemical mechanism for regional air quality modeling. J. Geophys. Res., 95, 16343-16367.
117. Stockwell, W.R., F. Kirchner, M. Kuhn, and S. Seefeld, 1997: A new mechnism for regional atmospheric chemistry modeling, J. of Geophys. Res., 102, 15847-25879.
118. Weseley, M. L., 1989: Parameterization of surface resistance to gaseous dry deposition in regional numerical models, Atmos. Env., 16, 1293-1304.
119. Zaveri, R. and L. K. Peters, 1999: A new lumped structure photochemical mechanism for large-scale applications, J. Geophys. Res., 104, 30387–30415.

CHAPTER 6

Air Pollution Sampling and Analysis

6.1 Planning a Sampling Program for Emission Testing

The ambient air quality is not only influenced by the rate at which pollutants are emitted by various sources but also by certain physical parameters like meteorological conditions, wind speeds and directions, temperature, humidity, etc., which influence the concentration of pollutants by dispersion or diffusion. So for proper monitoring of ambient air quality, both physical and chemical parameters of the ambient air are to be measured [1]. The detailed procedures for sampling air pollutant emissions from various sources, ambient air and determining air pollutants for assessing ambient air quality are described in this chapter.

The primary purpose of industry performance monitoring is to provide information to the plant operator. This allows the operation of a plant to an adequate level of pollution control and assessment of compliance with the discharge licence. In addition, the data can be useful in the design and implementation of clean technology systems and to assist with identifying waste minimisation opportunities. Most often this will involve sampling from a stack or other designated discharge point and determining the characteristics of the emission gas stream.

Performance monitoring should provide timely, reliable and accurate information regarding the composition and rate of emission of waste to the environment. This will enable non-compliance with licence conditions to be promptly acted upon. These objectives can only be achieved if sampling and analysis methods are of a high standard.

Those engaged in sampling and analysis of air emissions require special expertise. It is most important that sampling personnel are trained and familiar with methods of obtaining representative samples, their handling and preservation, requirements of the analysis and any sampling and analysis limitations and importantly, safety issues.

A detailed knowledge of the operation of the source(s) and any associated air pollution control devices is also important in obtaining a representative sample of the gas stream. The choice of measurement methods needs to be compatible with the objectives of the monitoring program.

6.1.1 Emission sampling

The sample(s) analysed must be representative of the gas stream being tested. Therefore, the objective of sampling is to ensure that the sampled gas stream is representative of either the total or a known portion of the source emissions. Unfortunately there is no single sampling method to cope with the variable and complex nature of source emissions.

Health and Safety Precautions

The risks associated with stack sampling must be carefully assessed prior to commencement of sampling. Appropriate risk management steps must be put in place. The risks may arise from any number of hazards associated with the tasks to be done and the physical conditions prevailing.

Some of the hazards involved include:

- working at height or on temporary platforms,
- exposure to toxic, corrosive or hot gases,
- electrical hazards,
- trip hazards from cables,
- noise or heat from plant equipment,
- objects falling from work platform or into the duct, and
- flammability hazards.

6.1.2 Designing a sampling program

Planning a source emission test program requires that the objectives be clearly defined to ensure that the data will meet the stated objectives. Designing an adequate sampling program requires a good knowledge and understanding of the system to be sampled. The data user, sampling team and staff from the analysing laboratory should be involved in the planning stage if the program is to be successful. The following information should be acquired prior to sampling and testing:

- detailed information on the process conditions,
- the process conditions under which the test is required,
- the location of the sampling plane,
- provision of access holes and safe working platform,
- selection of the number of sampling points,
- safe access to the area,
- suitable sampling equipment,
- the availability of sufficiently sensitive and specific methods of analysis, and
- the distance from and capacity of the analytical laboratory.

There are wide variations in process conditions in most industrial plants, which may produce different characteristics and quantities of waste emissions. Strategies for sampling vary depending on whether the process is continuous, cyclic or batch and whether the results are to reflect peak, normal or any other designated plant operating conditions.

During the design stage of the sampling program personnel involved in emission sampling and analysis will need to consider the physical and chemical variability of the discharge. Sampling and analytical method validation will also be required to ensure the integrity of data where non-validated methods are used. Even in situations where validated methods exist, they may not be applicable to some emission sources.

In all test situations consideration must be given to:

- fluctuations in velocity, temperature or pollutant concentration due to uncontrollable variation in the process,
- moisture content (particularly wet stack gases),
- expected gas composition and likely interfering compounds,
- high vacuum, high pressure or high temperature gas streams, and
- corrosive or very reactive components.

6.1.3 Preliminary site inspection

Prior to testing, physical inspection of the source to be tested is used to establish the location of sampling access holes and determine accessibility and work platform requirements (including power and safety). Discussion with site personnel should include plant operating conditions and plant safety requirements for testing equipment and testing personnel.

The inspection should also determine if access holes are pre-existing, and that access hole covers can be removed for sampling (not corroded shut). Preliminary determinations of temperature, velocity, pressure and moisture content can then be made. If the sampling plane fails to meet the requirements necessary for obtaining a representative sample then alternative sampling locations should be sought.

Following the inspections, and after considering the test objectives and plant operating conditions, a decision can be made on the sampling equipment and test procedures to be employed.

6.1.4 Numbers of samples and sampling times

A manual sampling program shall consist of a minimum of two test runs per pollutant with minimum sampling times. The minimum sampling times do not apply to variable or batch processes. These cases may require sampling during an entire cycle or taking sufficient samples to characterize the gas stream to meet the objective of the sample program.

6.1.4.1 Manual sampling

Virtually all stack tests require, as a minimum, the measurement of the following parameters to enable the calculation of mass emission rates of the waste in the discharge:

- gas velocity,
- gas pressure and temperature,
- gas composition and density,
- moisture and,
- volumetric flow rate.

To minimize the number of variables, tests should be conducted when there is constant flow through the duct. This should remain constant over the period of the test.

Sample conditioning is generally required to successfully transfer the analyze to the collection medium. This often incorporates filtration, heating, cooling or condensation to maintain the integrity of the sample.

6.1.5 Sampling emissions for particles

To obtain a representative sample of the gas stream containing solid or liquid particles, the sample must be withdrawn from the emission source at a rate such that the velocity and the direction of the gas entering the sampling nozzle is the same as the gas in the duct at the sampling point (iskinetic sampling). Generally, multi-point sampling at selected points across the sampling plane is performed. The number of points is dependent upon the cross-sectional area of the stack and distance of the sampling plane from flow disturbance within the stack.

The degree to which a sample represents the particles in the total gas flow depends on:

- homogeneity of the gas velocity within the sampling plane,
- sufficient sampling points used across the sampling plane, and
- the maintenance of isokinetic sampling conditions.

Australian Standard 4323.2 describes an isokinetic gravimetric method for the determination of total solid particulate matter emitted from stationary sources. Solid particulate matter for this method is defined as particles remaining after oven conditioning at 105 ± 5°C for a minimum of one hour. For some industries, consideration must be given to the temperature of the discharge and the effect that conditioning at 105 ± 5°C may have on the collected particulate matter.

6.1.6 Sampling emissions for gases

Multi-point sampling is generally not required for sampling of gaseous emissions. However, in some situations, notably after the junction of several different streams, stratification of the gas stream will persist for some distance downstream.

A survey of a suitable constituent of the gas stream such as carbon dioxide or oxygen should be performed to determine the degree of stratification. In cases where stratification does not exist, single point sampling at one quarter the diameter across the stack, should be representative of the gaseous emission.

If stratification exists, the gaseous emission determination will require multi-point sampling techniques, unless an alternative sampling plane can be found.

Sampling emissions for organic compounds

The four environmental issues normally associated with stack emissions of organic compounds are:

- toxicity and potential for detrimental environmental or human health effects,
- odorous properties of the volatile species,
- potential to participate in photochemical reaction to produce oxidants,
- greenhouse gas or ozone-depleting potential.

Emissions of organic compounds to the environment may be in various phases or combinations of phases (solid, liquid or gas). Sampling techniques are therefore governed by the phase(s) of the compounds. To establish the most appropriate sampling method for determination of organic compounds from emission sources information on the composition and expected concentration is required. Details of the process operation are also required to establish whether the emission is intermittent, cyclic or continuous. Selection of an appropriate test method will be based on factors including:

- the chemical composition of the organic compounds emitted,
- the expected concentration range,
- the chemical and physical properties (boiling point, reactivity, solubility etc.), and
- the characteristics of the discharge (temperature, moisture etc.).

When the type and concentration of organic compounds in the discharge are unknown it is recommended that preliminary sampling and analysis be undertaken to identify and determine approximate concentrations of organic compounds beforehand to facilitate the selection of an appropriate test method. Some of the information can be obtained from literature surveys, plant personnel or previous experience with similar industrial processes.

6.1.7 Sampling organic emissions with solid sorbents

For multi-phase sampling of organics the common practice is to utilise a filter for solids and a solid sorbent cartridge for vapour-phase organics. This approach is used for sampling of semi-volatile organic compounds in USEPA Method 23 and USEPA SW-846 Method 0010.

EPA Method 4230 is generally applicable to hydrocarbons in the boiling point range 36 to 126°C and involves collection of a representative sample of stack gas on a solid sorbent tube. The method should not be used to sample very volatile organic compounds because low molecular weight hydrocarbons such as methane and propane breakthrough the absorption trap and are not measured.

6.1.8 Sampling organic emissions with bags/containers

If the discharge contains only very volatile gaseous organic compounds, gas samples may be collected in an inert sample container, then analysed by gas chromatographic techniques. With this sampling and analysis procedure very volatile organic compounds can be identified and quantified. Sampling using this technique with subsequent GCFID (gas chromatograph – flame ionisation detector) or GC-MS (gas chromatography – mass spectrometry) analysis is generally used for:

- screening of emissions to identify unknown species, and
- organic compounds that are non-reactive volatile species.

For container samples, analysis must be completed within 48-hours of sample collection. This procedure is generally not applicable to higher molecular weight hydrocarbons due to the potential for loss in the sample container. A recovery evaluation is a mandatory requirement of container sampling. The recovery evaluation will include a spike of similar concentration to the sample and must remain in the container for the same duration that the collected sample was in the container.

6.1.9 Detection limits

If it is known or likely that pollutant concentrations in the stack discharge will be close to the method detection limit (for example, Class 3 indicators) then attempts must be made to increase the pollutant concentration in the collected samples. There are a number of ways to increase the pollutant concentration above the detection limit including:

- increasing the sample volume
- concentrating the sample, and
- using more sensitive analytical techniques.

The equation used for determining in-stack detection limits (ISDL) is:

$$ISDL = A \times \frac{B}{C}$$

$$= \frac{0.04mg / Lx0.25L}{3.0E - 6mg / L} \qquad .....(6.1)$$

where A = analytical detection limit
B = amount of analyte
C = volume of stack gas sampled

E.g., the licence limit for chromium is 0.03 mg/m^3. The ISDL is predetermined to be 0.1 of the licence limit, 0.003 mg/m3 (3.0E-6 mg/L). The sample will be prepared in 250 ml volumetric flask and the analysis method has a detection limit of 0.04 mg/L. What sample volume is required to obtain an ISDL of 0.003 mg/L?

$$C = A \times \frac{B}{ISDL} \qquad .....(6.2)$$

$$= 3333 \text{ L} = 3.33 \text{ m}^3$$

6.1.9.1 Emission monitoring

Semi-continuous emission monitoring

Instrumental gas analysers are often used in a semi continuous (batch) monitoring mode to measure emissions for short periods of time (typically from 1hour to several days). These systems are not permanently attached to the stack and generally stack gases are withdrawn from the discharge through a gas conditioning system to the instrumental analyser.

Because of the variety of parameters measured and techniques available in gas analysers it is not possible to specify performance specifications for all types of instrumental analysers. However, users must operate equipment according to the manufacturer's instructions and be familiar with the characteristics of their analyser for their particular application. A requirement for instrumental analysers is that the following performance characteristics are assessed before use:

- response time,
- zero and span drift,
- detection limit,
- effect of interfering substances, and
- effect of temperature and pressure on instrument stability.

6.1.10 Continuous emission monitoring

A continuous emission monitoring system (CEMS) is the total equipment necessary for the determination of a gas or particulate matter concentration or emission rate using instrumental analyser

There are two types of systems for continuous emission monitoring:

- extractive, and,
- *in-situ*.

Extractive systems involve continuous withdrawal of a sample from the gas stream. For gaseous components the extracted sample gas must be conditioned to prevent condensation and filtered to remove particle contaminants. Some conditioning systems include dilution with dry air to reduce the dew point of the sample gas to prevent condensation. An extractive system typically includes the following components: in-stack sampling probe, coarse in-stack filter, sample transport tubing, sample pump, moisture removal (dilution, refrigerant or diffusion), fine filter, analyser calibration system and data recorder.

In-situ monitoring means monitoring pollutants inside the stack under actual stack conditions. The monitoring path may be across the stack or part thereof.

6.1.11 Sampling checklist

- Conduct a preliminary site visit.
- Select an approved method and sampling equipment.
- Observe all safety precautions.
- Ensure plant is operating according to test requirements.
- Adhere to requirements of sampling method.
- Transport samples to laboratory as soon as practicable ensuring complete sample identification and description.
- Emission monitoring systems must be evaluated for performance.

6.2 Planning a Ambient Air Quality Sampling Program

There are several possible reasons why ambient air quality are to be monitored. Frequently these are requirements of Govt approvals which are needed to meet the requirements of a Neighborhood Environment Improvement Plan.

Many times the following are reasons for monitoring:

- to determine the concentrations of pollutants in the environment or entering the environment from a specific source,
- to measure the efficiency of pollution control equipment,
- to allow process operators to control their processes within prescribed limits and to achieve optimum process efficiency,
- to provide data for emission inventories, and
- to assess the impacts of improvement strategies on the local environment.

The critical steps in any monitoring program (either source emission or ambient air) include, but are not restricted to:

- determining the objectives of the program,
- assessing a suitable sampling location or monitoring site,
- deciding on the measurements to be made (if not pre-stipulated),
- selecting an approved measurement method and appropriate equipment,
- obtaining representative samples,
- ensuring that the integrity of the sample is maintained prior to its presentation to the analytical device,
- documenting key steps in the sampling and analytical processes, including:
  - labelling of samples to be analysed,
  - recording process conditions related to emission source samples,
  - maintaining logs of instrument and equipment operating conditions,

- analysing the samples or sampled air accurately and precisely following approved analytical methods, and
- reporting results accurately and completely, and where appropriate providing informed interpretation of the results.

Most industrial operations have variable process conditions which change the characteristics and quantities of pollutants discharged to the environment. With care, these difficulties can be minimised and acceptable results produced. For large emission sources continuous emission monitoring may be required to provide a detailed record of emissions over time. In other circumstances, periodic testing may be sufficient, but this may also be needed to provide a check on the results produced from continuous monitoring.

For all monitoring or testing, a documented quality system is required. This would include:

- written procedures for the work undertaken,
- an outline of the training provided to staff who will operate the procedures,
- a description of the processes employed to check the quality of results, and
- procedures to be used in the event that the quality control checks identify a problem with the sampling and analysis.

As per US EPA the following guide lines provide general direction on appropriate selection of sampling locations, and sites, sampling and preservation of samples, storage and transport of samples, and analytical and quality assurance procedures for the testing and monitoring of source emissions and ambient air quality.

6.2.1 Determination of physical parameters of air

6.2.1.1 Wind velocity

The instrument which is used for measuring wind velocity is the anemometer. Rotational and pressure plate type anemometers are used for measuring wind velocity.

Rotational type anemometer: The most common type of anemometer is the cup anemometer (Fig. 6.1) which is fixed at an elevated point where the wind rotates its cups. The movement of wind over a selected period of time is recorded and the wind velocity is calculated in m/sec or km/h.

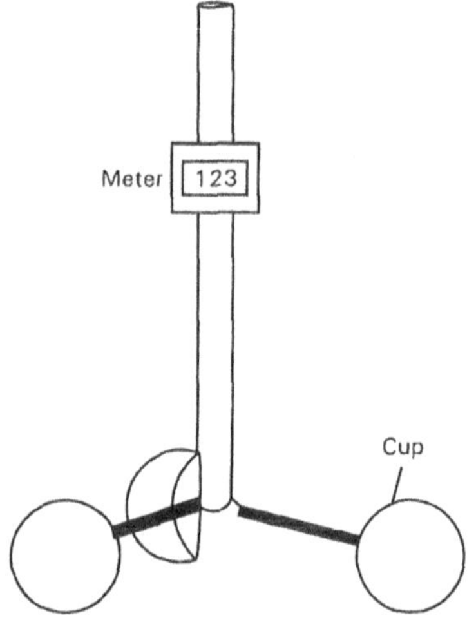

Fig. 6.1 *Cup type anemometer*

Pressure-plate anemometer: It has a round pressure-plate of about 30 cm diameter (Fig. 6.2).When wind pressure falls on the face of this plate, it gets transmitted through a shaft provided at its back which moves the indicator on a graduated scale, from which it is possible to know the velocity of wind at any specific time.

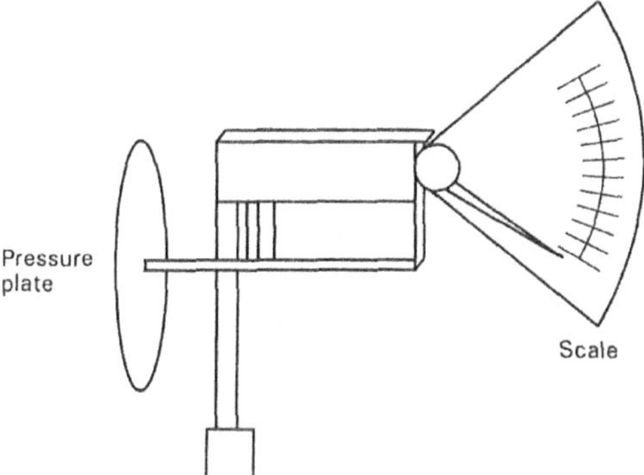

Fig. 6.2 *Pressure-plate anemometer*

6.2.1.2 Direction of wind

Wind direction is an important feature which can be measured from the wind direction recorder fixed on most of the anemometers. The recorder is set in an appropriate direction with the help of a magnetic compass and the direction is expressed on a 16-point scale of geographical directions (Fig. 6.3).

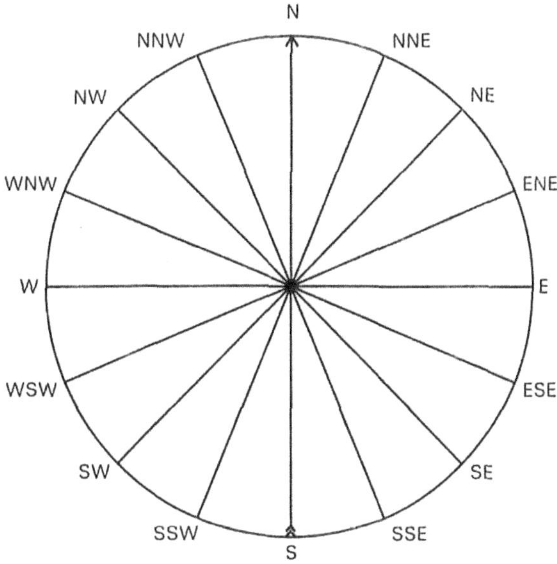

Fig. 6.3 *Sixteen point scale of geographical directions in a wind direction recorder*

6.2.1.3 Atmospheric pressure

Pressure falls with increasing altitude, being reduced by one-thirtieth for every 274 m rise in altitude.

Barometers are used to measure the atmospheric pressure. They work on the principle of rise or fall in the mercury column along a glass tube. Atmospheric pressure is conveniently measured by modern barographs and dial type barometers. Atmospheric pressure is expressed in millibars or mm Hg.

6.2.1.4 Temperature

Temperature fluctuations are more pronounced in air than in water and can be caused by other environmental factors such as moisture, wind, sunshine, etc., as discussed in Chapter 5. Monitoring of diurnal and annual ranges of atmospheric temperature is of high importance in many ecological investigations.

A minimum-maximum self registering type of thermometer with a U-shaped capillary tube is shown in Fig. 6.4. A pear-shaped reservoir is present at the left limb of the capillary tube and the lower half of the U-shaped tube is filled with mercury while the remaining part is filled with creosote. Creosote fills the left-elongated reservoir completely while it fills the right pear-shaped reservoir partly. Near the two limbs of the U-shaped tube, temperature scales will be fixed. The left one is able to read minimum temperature while the right is able to read maximum temperature. Indicators made up of iron will be present in the capillary tube above the mercury level, which move about with the rise or fall in mercury level. In the early morning hours, the indices in the two limbs will be set on the mercury using a horseshoe magnet and then the thermometer will be kept in a shady place. After a lapse of 24 h, the temperature will be recorded in the

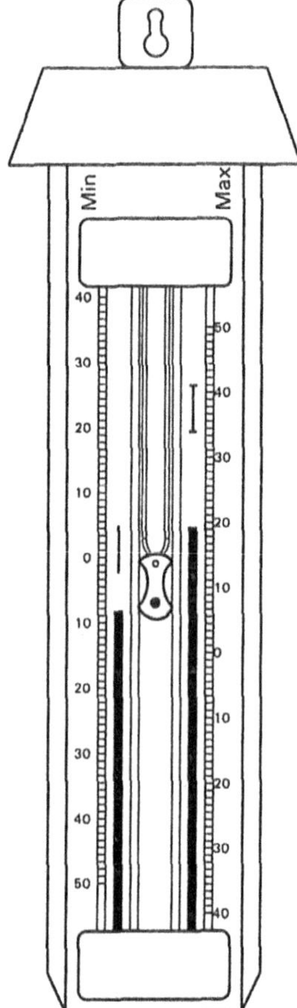

Fig. 6.4 *Minimum maximum thermometer*

two limbs of the instrument which is indicated by the lower end of each index, with minimum temperature on the left limb and maximum temperature on the right limb.

6.2.1.5 Relative humidity

Absolute humidity is defined as the amount of water vapour present in air, expressed as weight of water per unit weight of air. The amount of water vapour held by air varies with temperature and pressure. Relative humidity (RH) is defined as the percentage of water vapour which is actually present as compared with what would be expected at saturation under the said conditions of temperature and pressure. RH is an important environmental parameter as it modifies the effects of temperature and influences the rate of evapo-transpiration from the surfaces of organisms.

Dial-type hygrometers, hydrographs and electrical sensing devices are used to measure the RH. The conventional psychrometer (Fig.6.5) with wet and dry bulb thermometers was widely used for RH measurement. The bulb of the wet-bulb thermometer is maintained with saturated water by covering it with a cotton wick dipped in the water. By using a handle provided at the top of the instrument, the set of thermometers is swung in air until the readings in the two thermometers become constant. The temperatures in the two thermometers are then recorded. The difference in the two readings is referred to as depression in the wet-bulb thermometer which can be used to calculate RH (9%). If the two thermometers have the same reading, it indicates 100% RH.

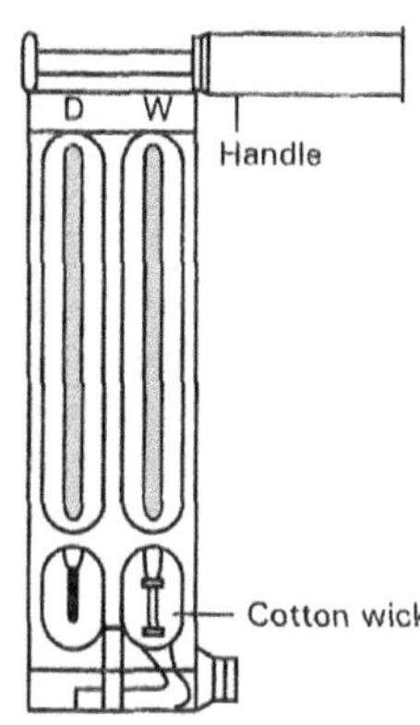

Fig. 6.5 *Psychrometer*

6.2.2 Experimental methods for sample collection and estimation of air pollutants from ambient air /emission sources

One of the most difficult tasks in the analysis of ambient air quality for various pollutants like SPM, SO_x, NO_x, etc., is to obtain a genuine and representative sample of the ambient air. This is because the concentrations of pollutants in the ambient air are likely to be extremely small and vary from place to place. In general ambient air sampling devices consist of a sample collector, a flow meter, and a pump to draw the air sample into the system. The type of pollutant and the method of analysis to be adopted for pollutant analysis dictate the type of sampling system to be used. Mainly there are two types of collectors–one to collect particulate pollutants and one for gaseous pollutants.

6.3 Particulate Sampling

The following methods are used for sampling and estimation of particulate matter.

6.3.1 Gravitational method

For collecting dust particles of 1μ (micron) or larger (40μ) in the atmosphere, clean glass jars are kept in the area where dust fall is to be determined and after a few hours or days, the dust is collected from each jar and then weighed. The average weight of dust in each jar is estimated and the dust fall is expressed as weight of dust per unit area per unit time.

Containers, generally conical plastic jars, 10-15 cm in diameter (Fig. 6.6), open at the top are used. The jars are kept in strategic locations throughout a community or in the vicinity of particulate sources under study. Grit and dust fall into the jars which sometimes have water to hold the dust. After a one month exposure, the jars are collected and brought into the laboratory where their contents are analysed. In most cases only the total particulate matter is determined, and the results are expressed in term of tons per square kilometre per month. Monthly isopleth maps can be constructed showing the variation of dust fall throughout the area.

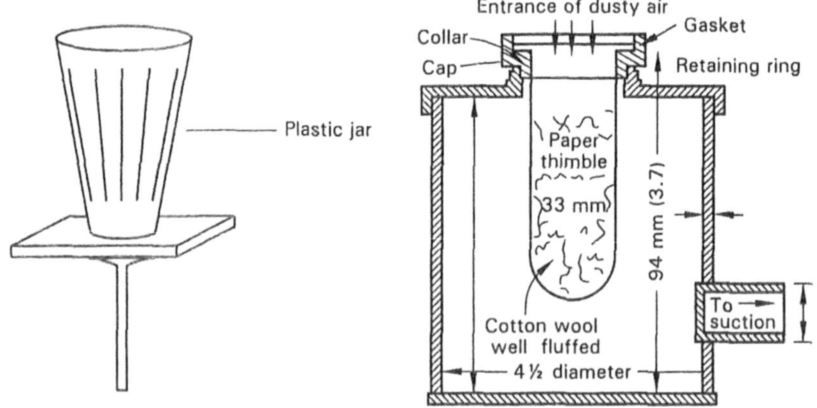

Fig. 6.6 *Dust fall collector*

Fig. 6.7 *Thimble filter*

For a rough comparison of the total fallout in one area with that of another, some fairly approximate methods are used. A funnel is placed in the neck of a large bottle mounted about 1.5 m above the ground in some open space; the dust falling into the funnel falls into the bottle or is washed by rain. At the end of a fixed period the collected material is weighed and subsequently analysed to estimate the relative proportions of soluble organic matter. These values determined as mg/cm^2/day are later scaled up to units of tons/sq km/month.

As scaling up of the small measurements may be inherently inaccurate due to large variations, month by month measurements and simple procedures which permit observation of long-term trends have to be taken up.

Other methods for evaluating the character and extent of dust fall involve the use of trays and glass slides coated with an adhesive, or sticky paper tape.

6.3.2 Filtration

A fibrous filter which is woven with randomly oriented fibres acts as a target for the collection of fine particles. When airborne particles enter a filter and flow around the fibres, they are subjected to aerodynamic forces which result in their collection on the fibres. The principal mechanisms operating are gravitation, inertia, interception and diffusion.

Many types of filters are available for removal of particulate matter from gas streams. The chief variations are in the filter media material and in the shape of the membrane. One common type is the thimble filter shown diagrammatically in Fig. 6.7. The filter paper thimble is filled with well-fluffed cotton which operates at a sampling rate of 2 cfm. By determining the change in weight of the dried filter after exposure, dust concentration in the gas stream is determined. The paper thimble cannot be used with high temperature gases; an alundum thimble is useful in this case.

6.3.2.1 High volume air sampler

The most commonly used particulate matter sampler for ambient air sampling is the high volume sampler (Fig. 6.8). The various designs available commercially work on the principle of collecting a large volume of air through suitable filter media. One specific unit has a specially fired glass-fibre filter, web mounted on a rectangular aluminium filter holder. The capacity of the sampler is 60-70 cu ft/min.; 84,000 cu ft of air may get filtered in 24 h. The collected particulate matter is dried and weighed, and then removed by washing for subsequent size separation and composition analysis.

High volume air samplers can be generally run as long as 24 h without an appreciable increase in airflow resistance. The multiple advantages are: relatively high air flow rates at low pressure drop, high particles storage capacity, and low moisture regain. High collection efficiency and low cost make the high volume sampler with glass fibre medium a very popular sampling device for ambient air quality monitoring.

The main difficulty in particulate analysis is the lack of a suitable universal filter material. Different filter materials serve very well for specific applications, but none can be used for all applications. In practice two types of fibrous filters are in use for collecting particles for weight or size determination, or chemical analysis. These are the usual what man filters composed of cellulose fibres and filters consisting of microfine filters made of glass or synthetic plastics. The glass fibre medium can filter at the rate of 70 cu ft/ min. The filter is over 99% efficient for the collection of particles as fine as 0.3 microns in diameter. Fibre filters composed of polystyrene were found to be very good for pollutant analysis due to their low background level of inorganic materials. However, they are not useful for organic analysis. Glass fibre filters have good weighing qualities and are therefore very useful for determining particulate concentration. However, metals silicates, sulphates, and other species are readily released from fine glass fibres, introducing error into analyses for inorganic particulate constituents.

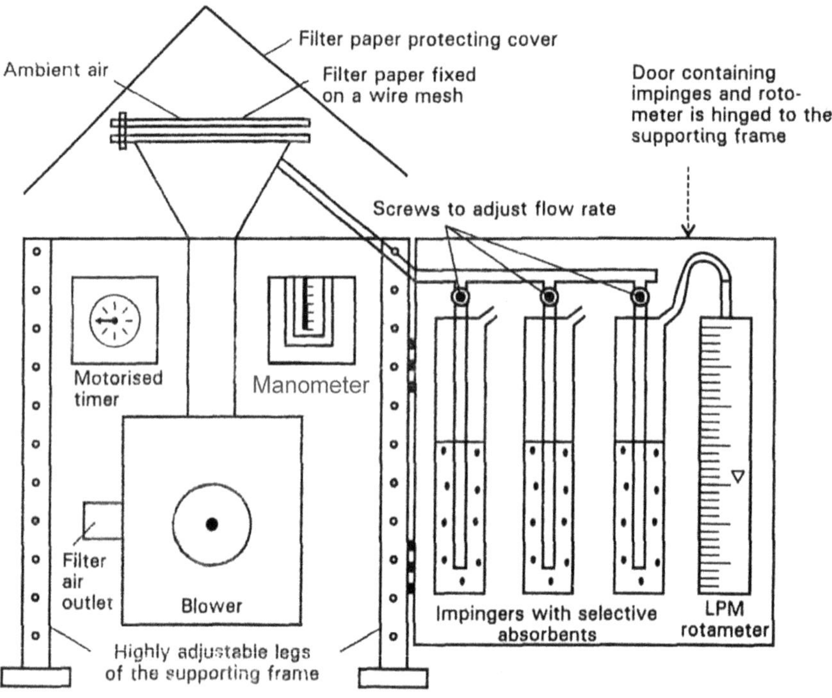

Fig. 6.8 *High volume sampler*

The weight of material collected is determined by weighing the filters before and after collection. Subsequent treatment of the filter depends upon the analyses to be carried out. Collected particulate matter could be extracted with a suitable solvent or volatilized by heating. The filter and the material collected can then be incinerated or "digested" by oxidizing acid mixtures before carrying out elemental analysis.

Data obtained for particulate matter should be carefully handled. Many reactions are likely to take place on the filter and during the process of removing the sample from the filter which can bring about serious misinterpretation of the data. For example, volatile particulate matter may be lost from the filter and due to chemical reactions on the filter, the material analysed may not be the material which was collected.

Electron microscopes can be used for evaluation of fine particles as magnifications of several thousands can be achieved. The enlarged image can either projected onto a fluorescent screen for viewing or on to photographic film for a permanent record.

6.3.3 Inertial collectors

6.3.3.1 Impactors

If a high velocity jet of dust-laden air gets directed on to a flat surface, the particles impact on the surface, lose their kinetic energy and thus get collected. The cascade impactor (Fig. 6.9) utilizes this principle for collecting dust in 3–4separate size fractions. As the air flows through a series of slits of progressively smaller area, the velocity of the gas stream passing through the slits correspondingly gets increased at each stage. The efficiency of impaction being directly proportiona1 to the particle velocity, the cascade impactor collects larger particles in the initial stages (where the gas velocity is lower) and relatively finer particles in the later stages. The collecting surfaces are collodin coated glass slides which can be removed for counting the sizing under a microscope.

A known volume of air is drawn through the slits using a suction pump in the impactor and after a considerable volume of air has passed through the cascade impactor, the suction is stopped and the particles collected in the cascade impactor are weighed. The dust particles can also be counted by observing them through a microscope; and we can then express the weight of a certain number of particles per unit volume of air.

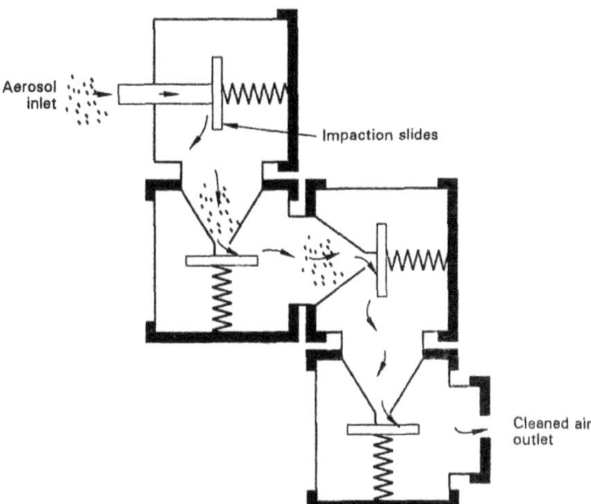

Fig. 6.9 *Cascade impactor for particulate sampling*

The wet impinger is essentially a single stage impactor. In this case the gas flows through a glass tube terminating in a fine nozzle submerged in water. The aerosol particles impinge on the bottom of the impinger flask and get trapped in the liquid, which is usually water. The Greenburg-Smith or standard impinger samples at the rate of 1 ft^3/min of air, while the midget or micro-impinger samples at the rate of one-tenth cubic foot per minute and may be operated with a hand pump. Both devices are highly efficient for the removal of particles around 10microns in diameter.

Greenburg-Smith impinger (Fig. 6.10) is one of the best devices for collecting particulates. It uses a high velocity air jet that impinges on a glass plate immersed in water or alcohol. The particles are collected in the surrounding fluid.

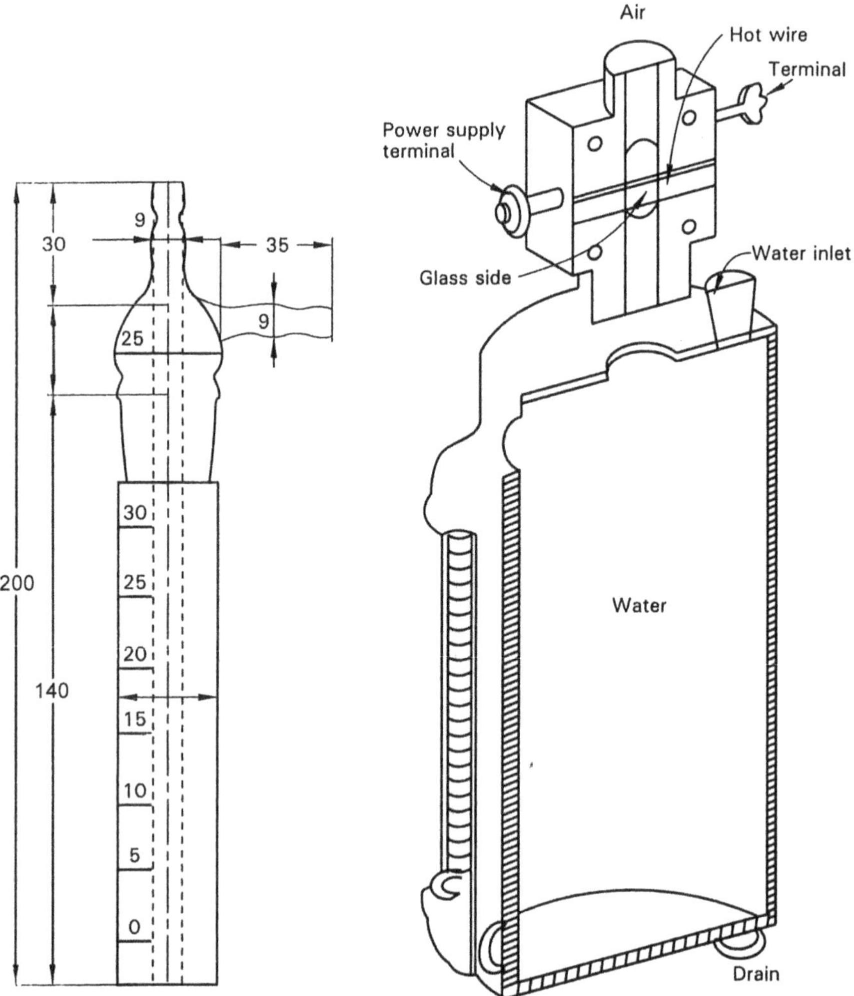

Fig. 6.10 *Greenburg - Smith impinge*

Fig. 6.11 *Thermal precipitator*

The Andersen air sampler consists of six sieves stacked one above the other. The top sieve has a number of relatively large holes, while the other plates have equal number of holes of progressively smaller diameter. Beneath each perforated plate is a petri dish containing agar. The air, which is sampled at the rate of 1 ft^3/min, passes sequentially through the six stages from top to bottom, impacting the relatively large particles on the dish beneath the first stage; the finer particles are collected in the subsequent stages. The Andersen air sampler can be used for analysing inert dust by simply substituting an adhesive material for the agar. A seventh stage can be added in the form of a 5 cm high efficiency filter connected to the exit of the sixth stage. The petri dishes can be weighed before and after sampling, and the resulting data can be used to make a size-frequency analysis of the aerosol.

6.3.3.2 Cyclones

High efficiency cyclones can be used to collect dust. As the efficiency of collection varies inversely with the cyclone diameter, cyclones as small as 2.5 cm in diameter are required for the efficient collection of sub-micron particles.

6.3.3.3 Electrostatic precipitators

An electrostatic precipitator consists of an ionizing electrode charged with a high negative potential and a collector to be maintained at positive potential produced by a special transformer and rectifier. A known volume of air is allowed to pass through the precipitator where the incoming particles become negatively charged by the electrode and adhere to the positively charged collecting tube. The collected particles are weighed and their number can also be counted accurately by using a microscope. This method of collection of particulate matter is very efficient as high flow rates with small pressure drops can be used. Sampling upto 8 ft^3/min can be carried out with about 99% efficiency for the removal of particles down to sub-micron sizes.

6.3.4 Thermal precipitators

In thermal precipitators the particles in the sampled stream move past a very hot wire and are repulsed to the nearby cold plate where they are collected. However, it can operate only at very low sampling rates–approximately 50 millilitres per minute–and is useful only for R&D work. Figure 6.11 shows a thermal precipitator of the type supplied by Casella Company Ltd., London, England.

6.3.4.1 Sampling in ducts and stacks

An accurate determination of the components of a flowing gas stream requires are representative and reproducible sample to be obtained for analysis. In the case of particulate matter, wide differences exist in the concentrations at different portions of the stream and so the sampling devices must meet the following requirements.

(a) The sampler must be made of non-corrosive material.
(b) Temperature in the sampling line must be held above the dew point of normally gaseous components.
(c) The sample must be obtained isokinetically (rate and direction of flow of gases into the sampler must be the same as in the gas stream being analysed).

(d) The long axis of the sampling head must be parallel to the direction of gas flow. The collector must be least 95% efficient. Occasionally, lower efficiencies can be accepted if the reproducibility of samples is high and the samples are representative of the total gas stream.

(e) The sample must be of sufficient size for required analysis.

(f) The total gas flow must be known. The volume of gas flow is usually determined by traversing the cross-section of the duct or stack with a Pitot tube to get gas velocity. Calculated velocities based on the Pitot readings can be averaged and multiplied by the cross-sectional area to give the gas volume. The temperature and static pressure of the gas stream could also be measured or determined so that the volume may be referred to a standard condition, e.g. 20 °C and 76 cm Hg.

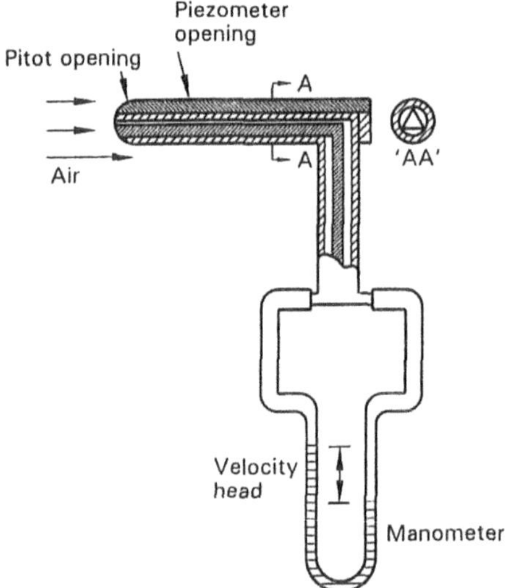

Fig. 6.12 *Pitot static tube*

The most difficult criterion for getting representative sampling is the maintenance of isokinetic sampling conditions. After the duct or stack has been traversed with a Pitot tube (Fig. 6.12), the average stack velocity could be calculated. The suction device that will draw the gas sample through the sampling apparatus should be then set so that velocity at the probe nozzle is the same as that of the main gas stream at the sampling point. A sampling nozzle can be then chosen for a suitable gas flow rate. A gas flow rate of 0.5 to 1 ft^3/min for combustion effluents at the selected velocity is recommended for good results.

6.3.5 Determination of particle size

The distribution of particles by size of any particulate sample either from the stack or ambient air is one of the important characteristics to be determined.

Using high-powered microscopes, the collected material can subsequently be identified by size and type and appropriate size distribution patterns evaluated.

For determining the number and size of particles in a sample of dust, particles suspended in a suitable liquid get pumped through a current carrying orifice and the particles will then be collected. In passing through the orifice, a particle changes the resistance of the circuit. The change in the resistance produces a voltage pulse having a magnitude proportional to the volume of the particle. By electronically counting the pulse above a certain threshold voltage, the instrument yields a count of all particles above a selected size. Comparatively small samples could be used, but the particles must be insoluble in the liquid carrier and have but few fibrous particles.

Bahco micro-particle classifier, which is a combination air centrifuge-classifier is a very widely used unit for the classification of particle size and was specified as the standard instrument for particle-size measurement by the ASME Power Test Code.10 g of the sample is needed for calibration of each specific unit.

Screen analysis by separating the particulate matter into size groups by a graduated series of metal-cloth screens is one of the simplest methods for classification of particles larger than 40μ.Separation of dust particles finer than 40 μ into various size ranges is usually done by air elutriation. The Roller particle six analyser is used in this method and allows the separation of a 25g dust sample into six size ranges in a few hours. Particle counts and size determination could also be made microscopically. A very powerful tool for the examination of aerosols is the electron microscope having a resolving power of the order of 0.02 μ.

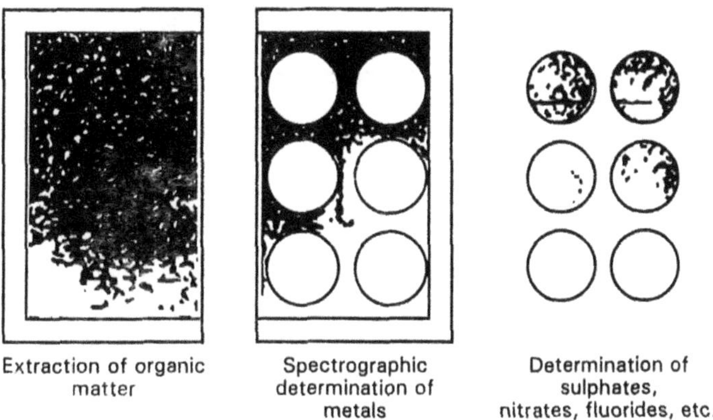

Fig. 6.13 *Division of high volume particulate sample analysis*

In another type of unit, a fixed volume of air is drawn through a white filter medium, with subsequent study of the "darkness" of the stain. This provides comparative information on the amount of coloured suspended matter in the atmosphere.

Particles larger than 5 to 7 μm rarely enter the lungs, and only particles 0.25 to 5μ in diameter are believed to be effectively retained in the lungs.

Haze density readings, as measured by a nephelometer unit, (Fig. 6.14) provide comparative information on the effect of suspensions on light transmission. Air is drawn at a fixed rate, e.g. $0.01 m^3/min$ through an optical haze to a chamber illuminated by a light beam. Some of the light is scattered by the fine particles present and the intensity of this scattered radiation can be measured by a photo-multiplier tube mounted at right angles to the primary light beam.

The instrument's response reflects changes in dust concentrations only if the suspensions examined are similar with respect to particle nature and size distributions. Calibration in absolute units is very difficult.

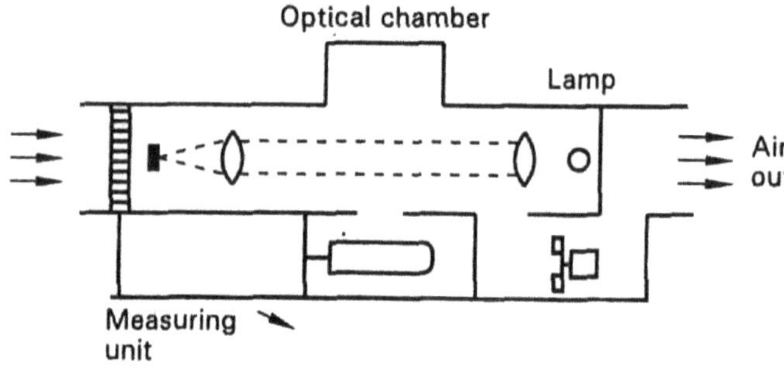

Fig. 6.14 *Greenburg Smith impinge*

6.3.6 Chemical analysis of particulate matter

A variety of analytical techniques are available for determining the concentrations of components of interest in collected particles.

Chemical methods are most satisfactory for the analysis of some elements present in particulates.

6.3.6.1 Sulphates and nitrates

These ions are often analysed in the context of acid rain related studies. Until recently, analysis was done by rather laborious procedures involving reduction of nitrate to nitrite which was determined spectrophotometica llyafter diazotization and azo coupling reactions. A number of colorimetric reagents are available for sulphate and a barium chloride turbidi metric procedure is also in use. In advanced laboratories these methods are supplemented by ion chromatography, in which the anions are separated by ion exchange and detected most commonly by conductivity after suppression of the conductivity of the eluent.

6.3.6.2 Arsenic

Arsenic can be analysed by the colorimetric silver diethyldithiocarbamate method. Zinc is used to reduce the arsenic in the acidified sample to gaseous arsine, AsH_3, which gets swept into an absorbing solution of silver diethyldithiocarbamate dissolved in pyridine. A soluble red complex is formed and its concentration can be monitored spectrophoto metrically by measuring absorbance at 535 nm.

6.3.6.3 Lead

The lead in particulate matter can be determined by dithizone method. The sample is collected on a 0.45 mm membrane filter. If it is desired to trap gaseous lead compounds, a tube filled with I_2 crystals is used after filtration. The filtered sample is digested with HNO_3, H_2SO_4, and $HClO_4$ to solubilize the lead. Iodine crystals are dissolved in acid KI and reduced with sulphite:

$$I_3^- + H_2SO_3 + H_2O \rightarrow 3I^- + SO_4^{2-} + 4H^+$$

Again the lead is solubilized. Acidic Pb^{2+} from either source is made basic with NH_3 and extracted with a chloroform solution of dithizone. After careful removal of excess dithizone from the chloroform phase, the lead dithizone complex is determined spectrophotometrically by measuring the absorbance at 510 nm.

6.3.6.4 Chemical microscopy

Chemical microscopy has been an extremely useful technique for the characterization of atmospheric particles. Either visible or electron microscopy can be used. Particle morphology and shape would tell the experienced microscopist a great deal about the material being examined; reflection, refraction, microchemical tests, and other techniques are used to further characterize the materials being examined. Microscopy is also used for determining levels of specific kinds of particles and for determining particle size.

Metal ions like lead, cadmium, mercury and zinc have been most commonly determined **by atomic absorption** after acid digestion to extract the metals of interest; any technique capable of measuring metal ions in the required concentration range could be used provided the possible interferences present have been accounted for.

Among metals analysed quantitatively by atomic absorption are barium, chromium, iron, lithium, manganese, beryllium, lead, cadmium, nickel, calcium, magnesium, copper and zinc. Mercury can be analysed by a technique called flameless atomic absorption. In this technique, mercury solution is reduced to the elemental state and is swept as a vapour through a gas cell by involving a carrier gas.

Non-destructive methods include X-ray fluorescence (both wavelength dispersive and energy dispersive) and neutron activation analysis. Both methods involve costly instrumentation but can give data on a range of elements from a single analysis.

Alternatively, the sample may be dissolved in oxidizing acids and the solution analysed by methods such as atomic absorption spectrometry, inductively coupled plasma (ICP) emission spectrometry, ICP-mass spectrometry, or anodic stripping voltametry.

X-ray fluorescence (Fig. 6.15) depends on fluorescent X-rays emitted from the inner shells of atoms bombarded by energetic X-radiation, gamma rays, or protons. Electrons get displaced from the inner shells of the atoms. As outer electrons fall into the vacancy created, fluorescent X-rays possessing a very characteristic energy for the particular atom are emitted. The wavelength of the emitted radiation can be used for qualitative analysis of the elements while the intensity of radiation from a particular element can be used for quantitative analysis.

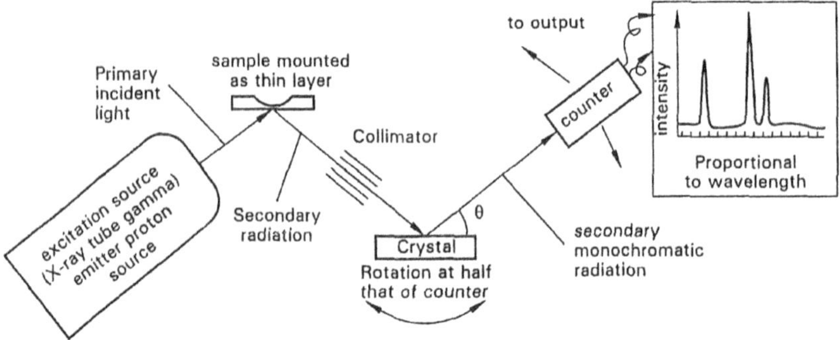

Fig. 6.15 *X-ray fluorescence spectrophotometer*

Energy-selective detectors of the semiconductor type have been developed for measurement of fluorescent X-rays of different energies without the need for wavelength dispersion by electronically distinguishing the energies of a number of lines incident on a detector simultaneously.

The wavelength-dispersive X-ray fluorescence is superior to the energy-dispersive technique except for speed and convenience. The wavelength-dispersive technique can be used for the analysis of over 20 pre-selected elements within approximately two minutes. One more advantage of this technique for multi-element analysis is that sensitivities and detection limits do not vary greatly across the periodic table as they do with methods such as neutron activation analysis or atomic absorption. Proton-excited X-ray emission has been especially found to be sensitive. X-ray fluorescence is commonly used for the multi-element analysis of atmospheric particulate matter.

Mass spectrometry is also important in lead analysis because it can be used to determine the source of the element. Lead has four stable isotopes; ^{204}Pb, ^{206}Pb, ^{207}Pb, ^{208}Pb and a given area contains unique proportions of each. By determining the isotope ratios ^{204}Pb: ^{206}Pb, ^{207}Pb: ^{206}Pb and ^{206}Pb: ^{208}Pb from the mass spectrum of a sample, its origin may be inferred. On the basis of such ratios, which are nearly identical for tetraethyl lead in gasoline and atmospheric aerosols, the overwhelming contribution of the former to urban particulates can be determined. Atomic absorption analysis is also an important tool in the study of lead concentrations.

Emission spectroscopy is one of the techniques by which metal can be analysed in particulate matter. This technique is able to measure the distinct light emission lines which are produced by metals highly energized in an electrical discharge. The electromagnetic radiation emitted in the ultraviolet and visible wavelength regions will have very narrow wavelength lines whose intensity and wavelength can be recorded on strips of photographic film after passage through a monochromator. It is possible to compare these lines with similar lines from a standard. The wavelength pattern of the lines observed provide a qualitative analysis, and the intensities of selected lines provides a quantitative analysis of metals in the sample. Quantitative analysis by emission spectroscopy is not very precise. Certain elements like beryllium, lead, iron, cadmium, nickel, vanadium, manganese, chromium, copper, zinc, barium, and tin are analysed by emission spectroscopy under controlled conditions.

6.3.6.5 Particulate organics

The important class of polycyclic aromatic hydrocarbons which have several known carcinogens can be extracted from particulate matter in the collecting filter with an organic solvent. Analysis generally is carried out by fluorescence or ultraviolet absorption spectroscopy (Fig. 6.16(a)). Analysis of these compounds, like analysis of many particulate species gets complicated by losses resulting from incomplete recovery, volatilization, and photo-oxidation. PAH compounds occur in urban atmosphere at levels of about 200 $\mu g/m^3$.These will adsorb on the particulate matter [1].

6.3.6.6 Polynuclear aromatic hydrocarbons

These compounds have long been of interest as some, most notably benzo (a) pyrene, are carcinogenic in experimental animals. The structures of some of the more commonly encountered compounds are given in Fig. 6.16(b). These compounds may be separated by gas-liquid chromatography, with detection by flame ionization detection (after cleanup) or electron capture. Nowadays, the most popular method is high performance liquid chromatography; if both UV absorption and fluorescence detection are used a considerable number of compounds can be determined with high specificity.

Organic components of aerosols can be determined by first obtaining the sample in solution using solvent extraction techniques which do not decompose the compounds of interest, followed by standard analytical techniques for trace concentrations, like thin layer chromatography, liquid chromatography, gas chromatography, etc. Detection of the separated compounds can be achieved by ultraviolet spectroscopy, which is more sensitive but only useful for compounds that fluoresce in ultraviolet light. Complete identification needs the use of mass spectroscopy but this is not always necessary when pure standards are available for comparison with unknowns in the chromatographic separation step. The appropriate sample preparation for polychlorinated aromatic compounds is also done by extracting the different fractions and by concentrating (vacuum expiration) the species of interest.

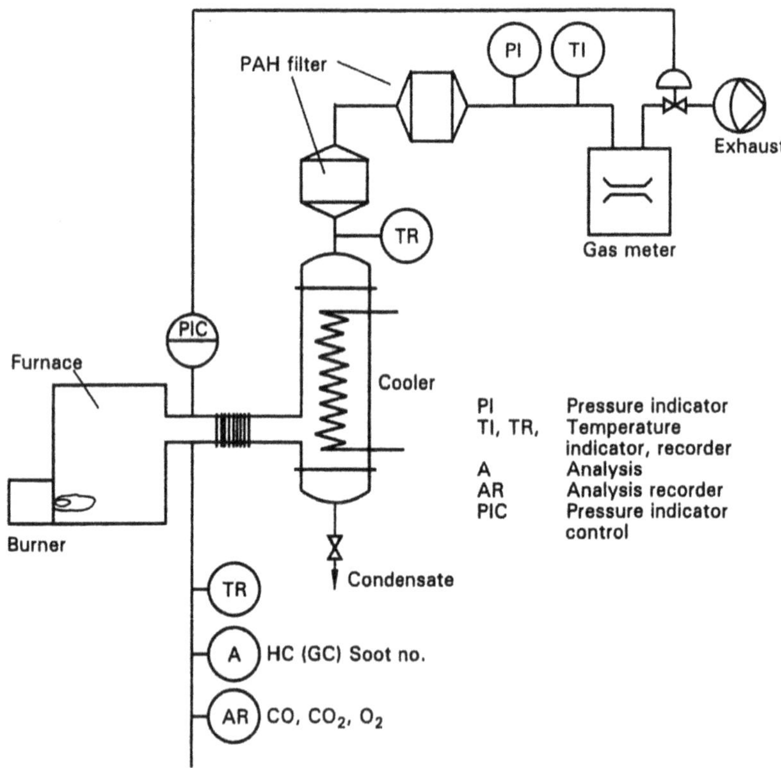

Fig. 6.16(a) *Diagram of a PAH sampling system with the total flow method*

When collecting the wanted components on filters and during preparation of samples chemical transformations can occur which lead to errors. E.g. Hartung found nitro-derivatives of the PAH as artefacts in samplings of diesel exhaust. PAH measurement depends on the reactivity and on the temperature of PAH, on the concentrations of the interfering components (e.g, oxides of nitrogen and on the sampling time.)

Analysis of the prepared samples is generally carried out gas-chromatographically with separating capillaries and mass-selective detectors or, in chlorine compounds, with electron-capture detectors.

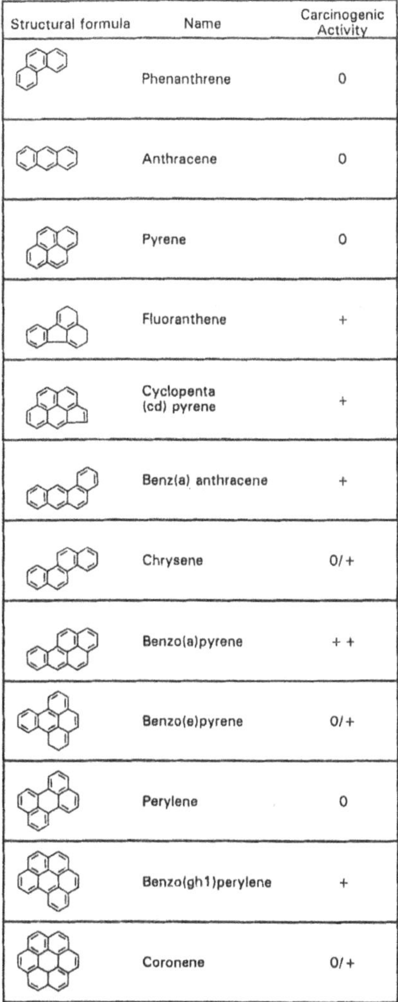

Structural formula	Name	Carcinogenic Activity
	Phenanthrene	0
	Anthracene	0
	Pyrene	0
	Fluoranthene	+
	Cyclopenta (cd) pyrene	+
	Benz(a) anthracene	+
	Chrysene	0/+
	Benzo(a)pyrene	+ +
	Benzo(e)pyrene	0/+
	Perylene	0
	Benzo(gh1)perylene	+
	Coronene	0/+

Fig. 6.16(b) *Structures of commonly encountered polynuclear aromatic hydrocarbons*

6.4 Sampling and Collection of Gaseous Air Pollutants

The following are the main criteria for adopting a sampling procedure for monitoring gaseous pollutants

(a) The sample collected should be representative with reference to time, location and physico-chemical conditions of the pollutant.

(b) The volume of the sample collected should be appropriate in accordance with the analytical method adopted and the concentration of the pollutant present.

(c) The frequency of sampling and the duration of sampling time (4h or 8h) should be fixed keeping in mind the expected flow or 24h average or variations of the pollutant levels.
(d) The rate of sampling should correspond to maximum collection efficiency.
(e) Sampling should not be cause physicochemical change in the pollutant.

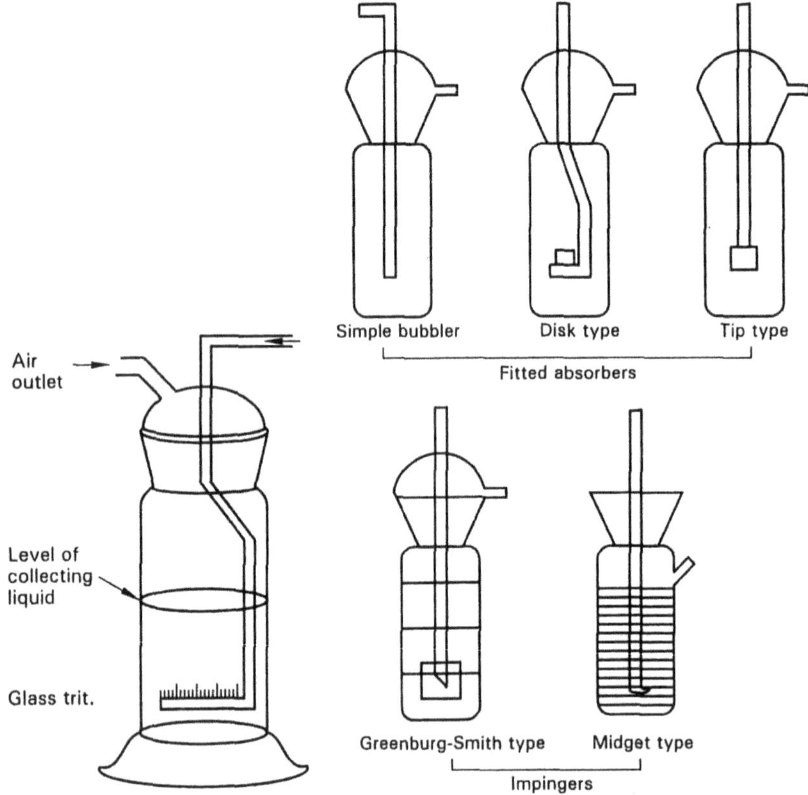

Fig. 6.17 *Scrubber*

Fig. 6.18 *Gas absorbing devices*

The common methods adopted for sampling of gaseous pollutants are

6.4.1 Grab sampling

In this method the sample is collected by filling an evacuated flask or an inflatable bag. Plastic bags can be used for sampling and storing for analysis. Moisture condensation or diffusion through the walls of the bag will create some losses in the sample in this method. Using containers made of glass or stainless steel these losses can be minimized. These sample collectors can be initially evacuated and opened for sucking the air sample or filled with water and then used as a collector by draining away the water and sucking the ambient air.

6.4.2 Absorption in liquids

Using a liquid absorption medium for collecting gaseous samples is a versatile technique under use. By properly selecting the specific absorption liquid, the desired pollutant from air can be separated by its solubility in the media or a chemical reaction. Different types of collectors are available for this purpose. The most widely used collector is the impinger. In the impinger the gas stream is impinged at high velocity and to a flat surface which may be the bottom of a collector or a specially designed plate. The midget impinger can handle flow rates of about 31/min while the Greenburg-Smith type impingers can be used with flow rates of 30 1/min. The pollutant is estimated according to the method of selection of the liquid. The various types of gas absorbing devices like impingers, and glass scrubbers are shown in Fig. 6.18.

6.4.3 Adsorption on solid materials

In this method a packed column containing a finely divided solid adsorbent is used as a medium for the gaseous pollutant which is sucked from the air passed through the thin column. The pollutants are concentrated and retained on the surface of the adsorbent. The common adsorbents used are activated charcoal or silica gel. The gases are desorbed by heating to the temperature of volatility of the trapped material or by sucking with inert gas or applying vacuum. The adsorbed gas can also be dissolved in a liquid solvent. Organic vapours trapped by this technique are analysed by gas chromatography.

Figure 6.19 shows a typical charcoal packed column through which air is pumped at the rate of 11 min^{-1}. After adsorption the charcoal is washed with a suitable solvent such as CS_2 which is then analysed by gas chromatography. Apart from charcoal, Al_2O_3, silica gel, and molecular sieves can also be used as adsorbents.

6.4.4 Freeze out sampling

In this method the air sample containing gaseous pollutants is passed through a series of cold traps, maintained at progressively lower temperatures. The pollutants are condensed in these traps which are brought to the laboratory, and after defreezing the gaseous pollutants are estimated by gas chromatography or IR or UV spectrophotometry, mass spectrophotometry or by wet chemical methods. The various coolants used for freeze out traps and the temperatures attained are given in Table 6.1.

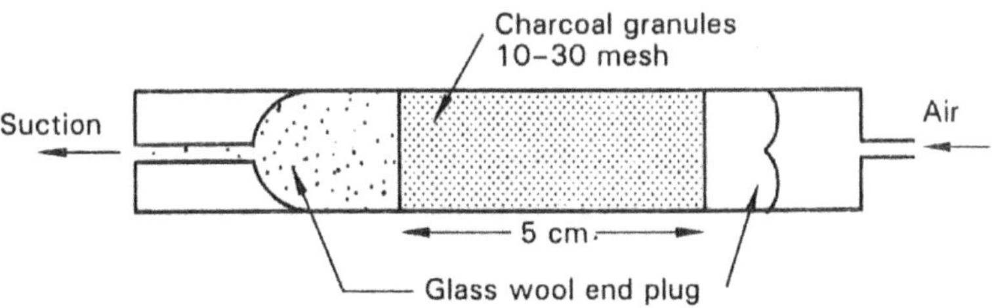

Fig. 6.19 *Charcoal adsorption tube*

Table 6.1 *Different types of coolants for freeze out*

Coolant	Cooling temp. °C
Ice-water	0
Ice-salt	–21
Dry ice and acetone	–79
Liquid air	–147
Liquid oxygen	–183
Liquid nitrogen	–196

This method can be used only if special facilities are available for obtaining the coolants, and cannot be a routine technique.

Undesirable aspects of this approach include solubility loss of some components in the residual films of liquid, introduction of liquid vapours which may have to be removed before testing, and the inconvenience of transporting and releasing large volumes of liquid. Dry sampling methods are thus often the preferable alternative, but it must be emphasized that many samples have a natural liquid vapour content and condensation can yield a liquid film capable of selectively dissolving some sample components.

In the dry sampling approach, gas gets drawn through a perfectly dry vessel by some form of pump or aspirator until the volume of the vessel gets displaced at least six times. If piping or tubing is needed to connect the sampling area to the sampling vessel, it too must be flushed thoroughly with the test gas. The flushing process can be avoided if the sample vessel is evacuated prior to exposure to the test atmosphere, but the slightest preliminary leak in the evacuated vessel leads to a major sampling error. In the laboratory, the gas samples could be transferred into the measuring apparatus by liquid displacement (Fig. 6.20).

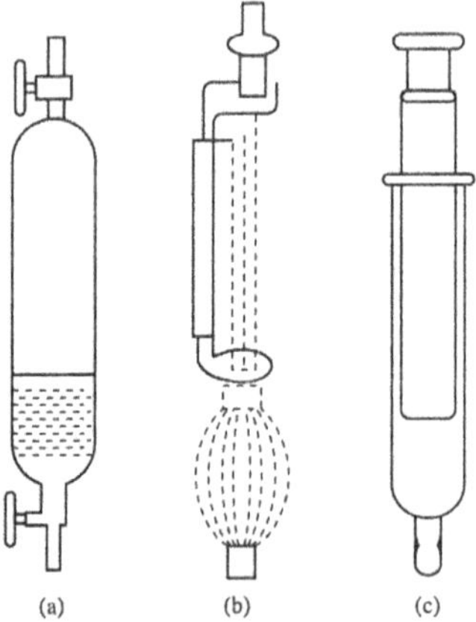

Fig. 6.20 *Methods of gas sampling* **(a)** *liquid displacement using gas sampling tubes* **(b)** *rubber aspirator bulb* **(c)** *syringe (or any other type of gas tight pump)*

6.4.5 Stack sampling of gaseous pollutants

In any stack sampling technique, obtaining a representative sample is the most important aspect for reliable results. The designer for air pollutant management needs data, neither too high nor too low compared to the true value. So the objectives of any stack sampling operation is to examine total quality and quantity, the physico-chemical characteristics of the pollutant emitted from any stack and evaluate how any control device is working. Often this information is useful for distinguishing the appropriate emission control device and also to arrive at the factors for emission inventory. The actual purpose for which the stack gas is sampled determines the type of equipment used for sampling. Further specific stack conditions like high temperature of effluent gases, moist, dusty or corrosive atmosphere require certain specific stack sampling equipment.

Basically it should be kept in mind that there should not be any change in the physical or chemical nature of the stack sample while sampling. The flow patterns and pollutant concentrations at the sampling point should truly represent the stack atmosphere which depend mainly on precise measurement of pressure, density, moisture, humidity and fixing suitable sampling points in the stack. In some stacks, it is not possible to ensure uniform flow, so concentration patterns and hence, multiple samples are required to obtain a representative sample That is, the actual sampling must be performed at a number of traverse points in the stack. Two sets of traverse readings, 90° apart, are taken in the same plane for circular stacks. The number of traverse points required on each of the two perpendiculars may be estimated from Table 6.2. According to the sampling point criterion, different numbers of traverse points are required depending upon the location of downstream and upstream disturbances, and the greater of the two values of the traverse points calculated should be chosen. The location of the traverse points in a circular stack is at the centre of the annular rings of equal area shown in Fig. 6.24. For fixing the traverse points, the distances are to be taken from the inside wall of the stack and the traverse point should be selected within 25 mm from the inside wall of the stack. Therefore attention must be paid to the selection of sampling point or points required. The usual practices in determining stack sampling points are: (1) The sample site should be located at *least* 8 *times* the stack or the duct diameters downstream. (2) The sample site should be located at least two diameters upstream from any source of flow disturbance such as bends, fittings, or constrictions since any flow disturbance would cause a non-uniform particle concentration pattern.

The sampling sites for a rectangular stack can be located by estimating equivalent diameter as per the following equation:

$$D_{eq} = 4 \times \frac{\text{Cross - sectional area of flow}}{\text{Wetted perimeter}} = \frac{2LW}{L+W} \qquad \text{.....(6.3)}$$

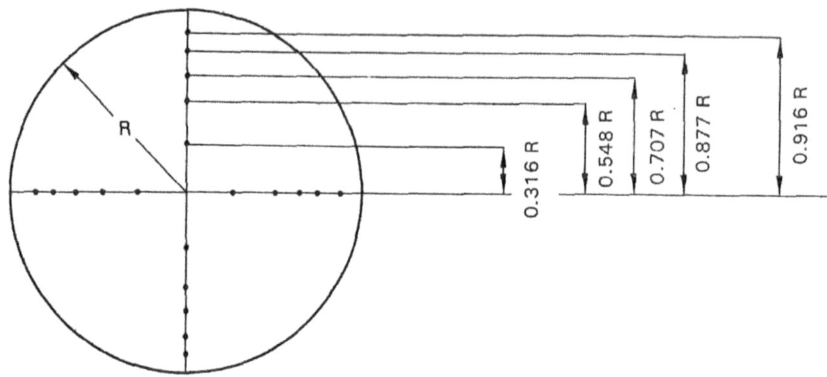

Circular stack (10-point traverse)

Rectangular stack (measure of centre of least 9 equal areas)

Fig. 6.21 *Fixation of traverse points and sampling points on the stack*

Table 6.2 *Minimum number of traverse points required [2]*

Inner diameter of stack, m	No. of points
D < 0.3	4
0.3 < D < 0.6	8
0.6 < D < 1.2	12
1.2 < D < 2.4	20
2.4 < D < 5	32

Table 6.3 *Location of traverse points in circular stacks*

Traverse point number for a diameter	Number of traverse points for a diameter										
	4	**6**	**8**	**10**	**12**	**14**	**16**	**18**	**20**	**22**	**24**
1	6.7	4.4	3.3	2.5	2.1	1.8	1.6	1.4	1.3	1.2	1.1
2	25.0	14.7	10.5	8.2	6.7	5.7	4.9	4.4	3.9	3.5	3.2
3	75.0	29.5	19.4	14.6	11.8	9.9	8.5	7.5	6.7	6.0	5.5
4	93.3	70.5	32.3	22.6	17.7	14.6	12.5	10.9	9.7	8.7	7.9
5		85.3	67.7	34.2	25.0	20.1	16.9	14.6	12.9	11.6	10.5

Table 6.3 ***Contd...***

Traverse point number for a diameter	Number of traverse points for a diameter										
	4	6	8	10	12	14	16	18	20	22	24
6		95.6	80.6	65.8	35.5	26.9	22.0	18.8	16.5	14.6	13.2
7			89.5	77.4	64.5	36.6	28.3	23.6	20.4	18.0	16.1
8			96.7	85.4	75.0	63.4	37.5	29.6	25.0	21.8	19.4
9				91.8	82.3	73.1	62.5	38.2	30.6	26.1	23.0
10				97.5	88.2	79.9	71.7	61.8	38.8	31.5	27.2
11					93.3	85.4	78.0	70.4	61.2	39.3	32.2
12					97.9	90.1	83.1	76.4	69.4	60.7	39.8
13						94.3	87.5	81.2	75.0	68.5	60.2
14						98.2	91.5	85.4	79.6	73.9	67.7
15							95.1	89.1	83.5	78.2	72.8
16							98.4	92.5	87.1	82.0	77.0
17								95.6	90.3	85.4	80.6
18								98.6	93.3	88.4	83.9
19									96.1	91.3	86.8
20									98.7	94.0	89.5
21										96.5	92.1
22										98.9	94.5
23											96.8
24											98.9

6.4.6 Isokinetic sampling for particulates

Generally the gases inside the stack are at high temperature and high velocity. So in stack sampling for particulates, for getting a representative sample, one must maintain isokinetic flow into the sampling probe. The problem is illustrated in Fig. 6.22. If the gas velocity inside the sampling probe is the same as the gas velocity from which the sample is taken then the sampling conditions are called isokinetic ($V_n = V_s$) and the measured concentration V is equal to true concentration (C_t). If the velocity in the nozzle exceeds that in the stack then the gas streamlines will bend into the nozzle and the inertia of the particles will carry some of them past the nozzle even though the gas through which they are carried will be collected and the measured concentration will be less than the true concentration. If the velocity in the nozzle is less than the stack then the gas stream will bend away from the nozzle and the inertia will carry some particles into the nozzle away from the gas that accompanied them and the measured concentration will be higher than the true one. There are a number of methods practised to adjust the flow in the nozzle to keep its velocity same as the flow in the stack and thus produce isokinetic conditions. Some of the problems associated with stack sampling are the high temperature and high velocities of gases causing difficulties in collecting the sample and requiring a number of additional measurements like moisture content, pressure, temperature, the flow rate of the gases and their composition [3].

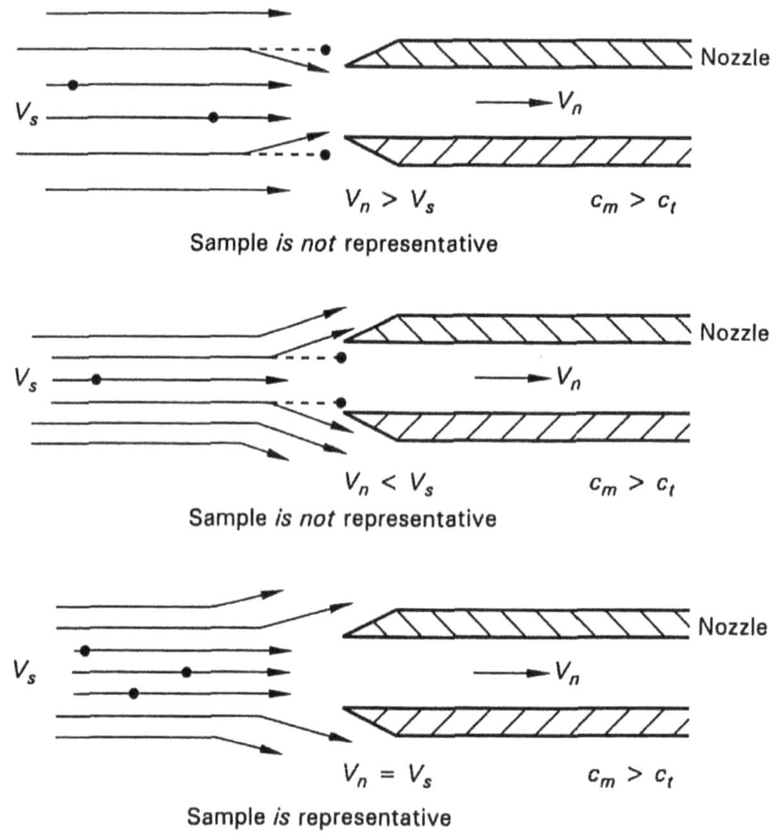

Fig. 6.22 *Isokinetic sampling problem for particulates*

In actual practice a part of the gas stream is diverted through a sampling *train* (Fig.6.23) for collecting a representative sample of the stack gases. The procedure for obtaining gas samples is similar to that of particulate emission sampling. The sample can be withdrawn from the stack at a constant rate independent of the flow rate in the stack. To prevent losses resulting from condensation, the sampling line may be heated. Tubing and the sampling probe are of inert material to avoid reaction with the gas to be measured. In a sampling train a nozzle attached to one end of a sampling probe for collecting gas samples at different transverse points is pushed into the gas stream. The sample is passed through a particulate or gas collector which has a flow measuring device, and a prime mover such as a vacuum pump or an ejector. The end of the probe is sharp-edged, pointing inward from the outside edge and the traversing probe is usually made of stainless steel with glass or teflon lining.

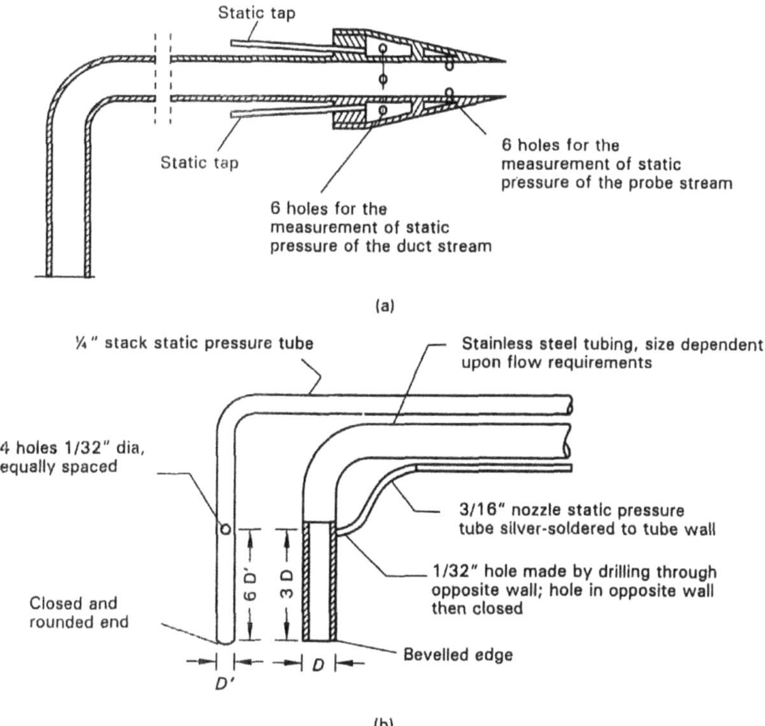

Fig. 6.23 *Stack gas sampling*

Hot gases can be sampled by probes that have a circulating coolant system around them to prevent combustion of particulate matter inside the probe. Besides selecting a proper sample nozzle, care has to be taken in choosing the other components of the sampling train, depending upon the nature of the gas to be sampled. Figure 6.24 shows a typical sampling train arrangement for SO_2.

Similar to devices for ambient air collection, a number of devices can be used for collection of particulate matter and gases in stack sampling. These include filtration, wet or dry impingement, impaction, electrostatic and thermal devices for collecting particulates, and absorption, adsorption and freeze-out devices for collecting gases. Flow measurements can be made with devices such as wet test meters, dry gas meters, capillary flowmeters, rotameters or orifice meters. The important factor is that they must be accurately calibrated. Commonly used suction devices include vacuum pumps and water or steam ejectors. Figures 6.25(i) and 6.25(ii) show some of the common sampling system components.

After drawings known volume of the sample pollutant gas from the stack into a sample collection device where gas of interest is removed and concentrated, the concentration of the remaining gas is monitored with a flowmeter located downstream of the collection device. The quantitative calculation of the pollutants is carried out at different points and the rate and the duration of the sampling time and the rate at which the sampling is collected is important for determining the amount of constituent gases.

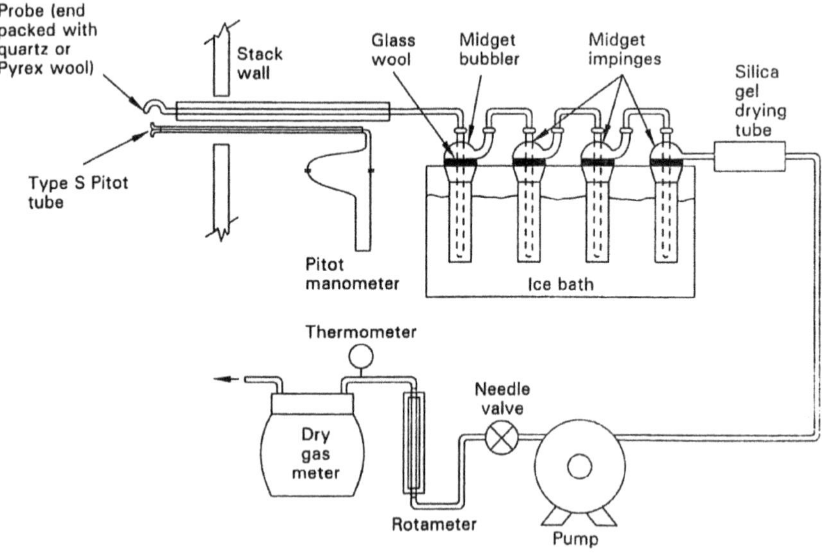

Fig. 6.24 *USEPA Method 6 sampling train for SO_2 [4]*

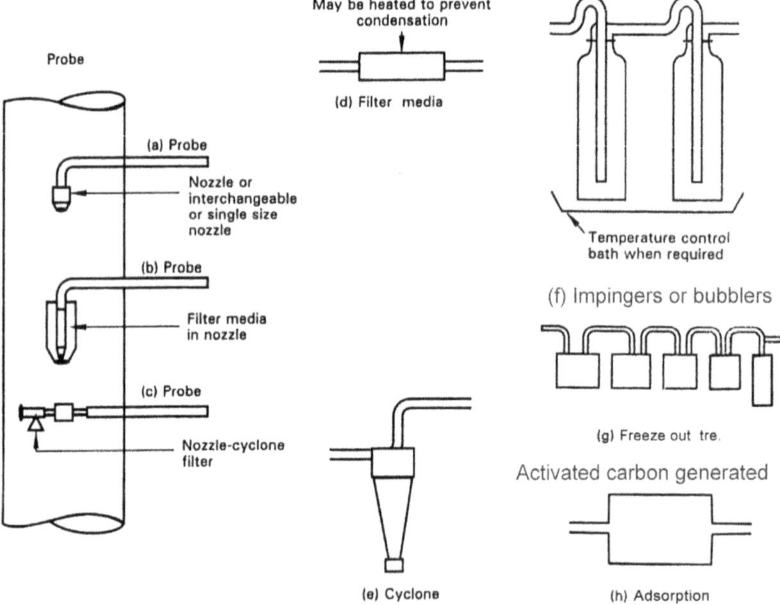

Fig. 6.25(a) *Common sampling system components*

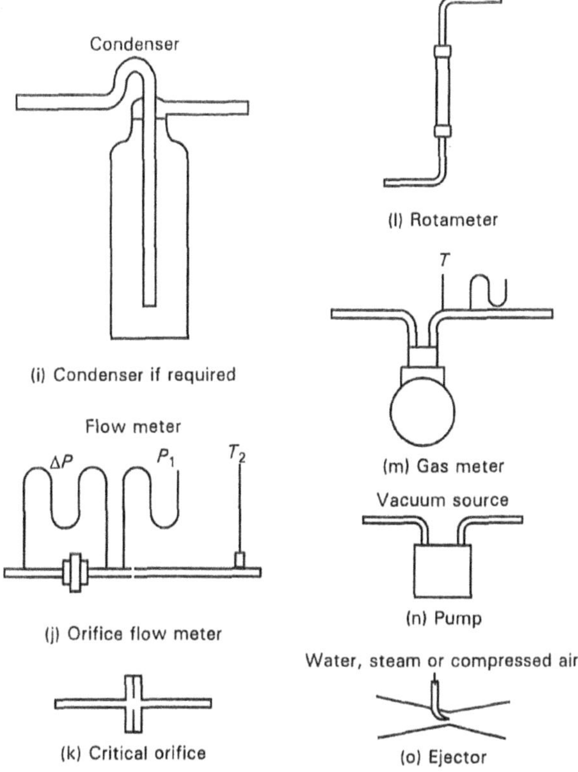

Fig. 6.25(b) *Common sampling system components*

6.4.7 Analysis of gaseous air pollutants

Continuous analysis of air quality, though desirable, is not always possible and conventional laboratory analysis of grab samples or discrete samples is important. For most of the gas pollutants, absorption of the gas in a specific chemical solution or solvent and analysing the absorbed gas by titrimetric or colorimetric procedures is the normal practice.

6.4.8 Sulphur dioxide

The methods for the estimation of sulphur dioxide are based on chemisorption, coulometric, conductometric, colorimetric, idometric or turbidmetric procedures.

6.4.8.1 Chemisorption method

A simple method for the estimation of SO_2 is the lead peroxide candle method involving the reaction of lead peroxide with sulphur dioxide through chemisorption:

$$SO_2 + PbO_2 \rightarrow PbSO_4$$

On a ceramic support of candle shape, the lead peroxide is coated as a paste and protected from the weather. The total amount of sulphate formed over the time of exposure gives a measure of the integrated sulphur dioxide concentration over that period of time. As the collection efficiency is a function of humidity, temperature and wind speed, this method is satisfactory for only rough estimates.

Another adaptation of the chemisorption technique is the use of chemically impregnated crystals, which undergo a colour change because of reaction with the compound being measured. Tubes having specific reactants for a wide range of pollutants are available commercially. The tubes are open-ended at the test site and a known volume of air is drawn with a hand pump through the absorbent. The length along the tube where the characteristic colour change takes place provides a measure of the concentration. Such methods are useful in industrial plants where concentrations are higher than in ambient air.

6.4.8.2 Electrometric/coulometric method

This method is suitable for continuous monitoring of air. In this method the air sample is bubbled through an electrolyte solution having 0.1M of potassium bromide in 2N sulphuric acid. The SO_2 in air brings about a decrease in bromine concentration in solution which is indicated by electrode current and could be measured electrometrically which corresponds to SO_2 content of air.

Figure 6.26 is a schematic representation of one type of electrometric unit for monitoring sulphur dioxide levels. Air is drawn into the apparatus through a filter to remove dust, and then over a heated silver wire to eliminate substances such as hydrogen sulphide, ozone and chlorine. The purified gas is then bubbled through an aqueous solution of bromine. Some bromine is reduced by the sulphur dioxide content and this is continually replaced by an electrolytic generation process. The current flow during the electro-generation step can be correlated with the SO_2 gas admitted to this purified stream. If care is taken in the calibration and control of flow rates, and if sorption of SO_2 in the sample line and dust filters can be kept to a minimum, an overall accuracy of better than ±15% may be achieved using this procedure.

In another type of bromocoulometric analyser, air containing SO_2 is drawn continuously into an electrolytic cell containing acidified bromine solution and two sets of electrodes. The indicator reference set of electrodes is used to detect the bromine concentration while the other set comprising the generating electrodes and the auxiliary electrode, is used to generate any bromine necessary to maintain the proper balance. Bromine will oxidize SO_2 in the air sample resulting in decrease in bromine concentration as shown:

$$2H_2O + SO_2 + Br_2 \rightarrow H_2SO_4 + 2HBr$$

The oxidation-reduction potential of the reagent changes due to decrease in the concentration in bromine resulting in development of a voltage between the indicator electrode and the reference electrode, which is sensed by the other electrode system causing an electric current to flow between them. Simultaneously sufficient bromine is generated to maintain the original concentration. The current flow necessary to maintain the proper balance is equal to the SO_2 concentration in the air stream. By using this analyser SO_2 in the range of 0.01–10 ppm can be measured.

$$2Br^- \rightarrow Br_2 + 2e^-$$

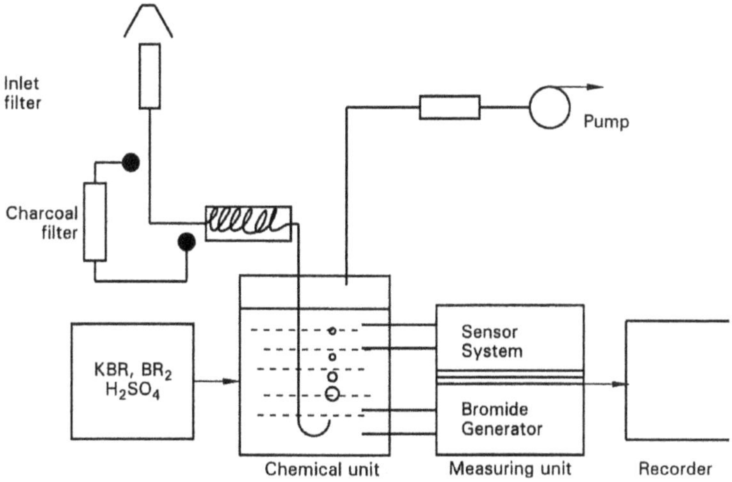

Fig. 6.26 *Electrometric SO_2 monitor*

6.4.8.3 Conductometric method

A known volume of air is allowed to bubble through a dilute solution of sulphuric acid and hydrogen peroxide. Sulphur dioxide present in the air gets oxidized to sulphuric acid, thus increasing the electrical conductivity of the solution which can be measured (Fig.6.29) and estimated as SO_2 content per unit volume of air.

$$SO_2 + H_2O_2 \rightarrow H_2SO_4$$

6.4.8.4 West-Gaeke method (Pararosaniline method) [5]

In this method sodium tetrachloromercurate is used as absorber solution. A measured volume of air is passed through 0.1 N sodium tetrachloromercurate which absorbs SO_2 present in air quantitatively (collection efficiency 95%) at concentrations as low as 0.002 ppm and disallows its oxidation to SO_3. The absorbed solution is then allowed to react with formaldehyde and bleached pararosaniline causing the formation of a reddish purple pararosaniline and sulphuric acid which can be measured photometrically and SO_2 estimated. SO_2 in air in a wide range of concentrations (from 0.002 to 5 ppm) can be estimated by this method.

$$SO_2 + H_2O_2 \longrightarrow H_2SO_4$$

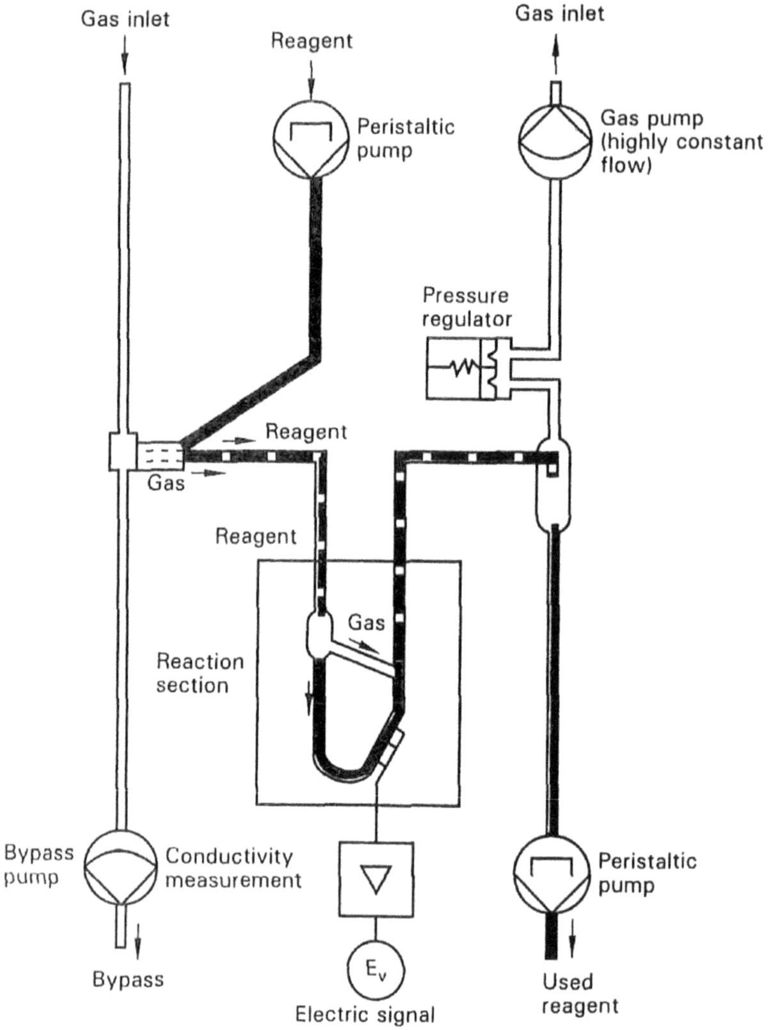

Fig. 6.27 *Flow diagram of conductometric gas analyser*

The sampling should be carried out at a flow rate of 200 ml–2.01/min for 30 min to 24 h. The duration and rate of sampling depends upon the concentration of SO_2 expected in the ambient air.

$$SO_2 + HgCl_4^{2-} + H_2O \rightarrow HgCl_2SO_3^{2-} + 2H^+ + 2Cl$$

$$SO_2 + H_2O + HCHO \rightarrow HOCH_2 - SO_3H + H^+$$

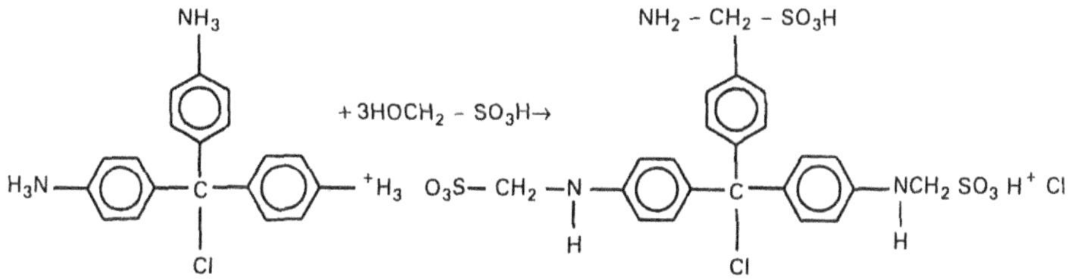

NH_3 $NH_2 - CH_2 - SO_3H$

$+ 3HOCH_2 - SO_3H \rightarrow$

H_3N — C — $^{+}H_3$ $O_3S - CH_2 - N$ — C — $NCH_2SO_3H^+Cl$

Cl   H   Cl   H

Fig. 6.28

Procedure

Pump 30-60 l of air through 10 ml of the scrubbing solution (13.6 g $HgCl_2$ and 7.5g KCl in 1 litre water) in a small impinger, at a rate of 1-2 l min^{-1}. Then add 1 ml of dilute pararosaniline reagent solution (4 ml of 1% aqueous solution with 6 ml of conc. HCl and diluted to 100ml with water) and 1 ml of 0.2% HCHO solution (5 ml of commercial 40% solution diluted to 1 litre with water). After 20–30 min., measure the absorbance at 580 nm. Use a dilute solution of Na_2SO_3, counter checked iodometrically as the standard.

After the collection of a 1-h or 24-h average sample in the field, the solution should be returned to the laboratory. Saiphonic acid is to be added to reduce nitrogen dioxide, which otherwise would destroy some of the color reagent. The standard curve, for obtaining the unknown concentration, can be prepared with sodium metabisulphite in the range of 2 to 25 μg of SO_2. The concentration of SO_2 in μg of SO_2/m^3 can be calculated as follows

$$SO_2 = g/m^3 = \frac{\mu g \text{ of } SO_2 \times 10^3}{\text{Total volume of air sampled } (L)}$$

When high concentrated $HgCl_2SO_3^{2-}$ is used more than one site on the pararosanilinedye may react, and Beer's law will no longer be obeyed. This limits the method to solution concentrations of SO_2 between 0.05 and 5 mg/dm^3. The range of atmospheric concentrations may be extended by appropriate dilutions of concentrated samples from 0.002 to 5ppm.

Nitrogen dioxide at levels exceeding 2 ppm constitutes a major interference, with the original West-Gaeke method. It is possible to get rid of the interference by the addition of sulphamic acid, (H_2NSO_3H) which acts as a reducing agent converting NO_2 to nitrogen gas.

Method of Dasgupta et al. (1980) [6]: The only drawback of the West-Gaeke method is the handling of relatively high concentrations of toxic and costly $HgCl_2$. Dasgupta et.al. developed a method (Fig. 6.29) based on absorption of SO_2 into HCHO buffered at pH 4. Sulphite liberated from the compound by base is added to acidic pararosaniline and the absorbance measured at 580 nm. This method is comparable to the West-Gaeke method in absorption and recovery efficiency, sensitivity and precision. No interferences are observed due to O_2, NO_2 and transition metal ions, except Mn (II).

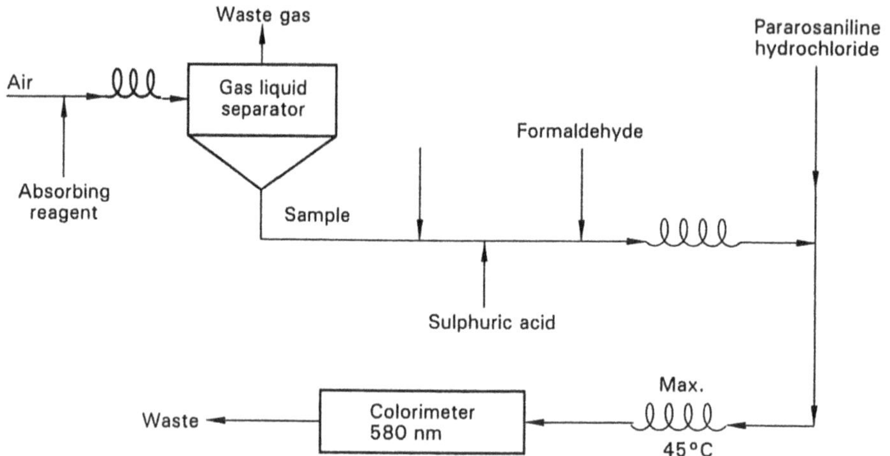

Fig. 6.29 *Sulphur dioxide monitor using a modified West-Gaeke method*

Buffered HCHO absorber is prepared by diluting HCHO solution (37%, 530/μL) and potassium hydrogen phthalate (204 mg) to 1 litre. This buffer shows a pH of 4.2 at 25°C.

Midget bubblers, filled with 15 ml of the buffered HCHO absorber, can be used for sampling. The usual sampling rate is 250 ml/min. The contents of the fritted tube and side arm are washed into the bubbler with 2 ml water. Then NaOH (1 ml, 4.5M) is added from a pipette and the contents of the bubbler are then transferred into pararosaniline reagent (5 ml of 0.30% in 1.2 M HCl) kept in a 25 ml culture tube. The contents of the tube then are made up to the mark (25 ml) with water and the tube capped and inverted thrice. The absorbance of the resulting colour is measured at 580nm after a period of 15 min.

The method is less susceptible to NO_2 interference than the West-Gaeke method. However, with fairly high levels of NO_x or O_3, sulphamic acid/sodium sulphamate addition is necessary as in the West Gaeke method. For elimination of Mn (II) interference addition of 0.1 millimole CDTA (trans 1, 2-cyclohexane dinitrilo tetraacetic acid) to the absorber solution is recommended.

6.4.8.5 Flame emission method

In the flame photometric sulphur analyser, gaseous sulphur compounds are burned in a reducing hydrogen air flame and the emission of the S_2 species at 394 nm is measured. The technique has a very fast response as polluted air may be passed directly to the flame. The sensitivity is similar for all volatile sulphur compounds and in urban air the level of total volatile sulphur, as given by a flame photometric analyser, approximates closely to the sulphur dioxide level.

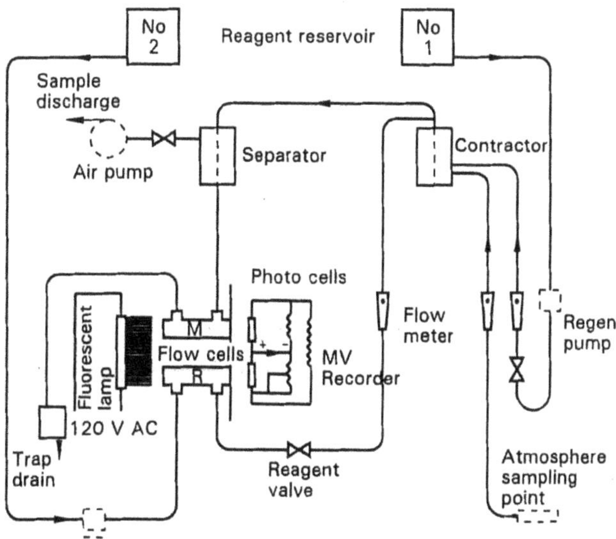

Fig. 6.30 *Schematic flow diagram for a continuous sulphur dioxide analyser using the West-Gaeke method*

6.4.8.6 Pulsed fluorescent analyser

Pulsed fluorescent analyser is specific for measurement of sulphur dioxide at levels down to 1 ppb. Fluorescence, excited by radiation in the region of 214 nm, is measured (Fig.6.31) and a wide linear range of response is found. Instruments covering the range of 1–5000 ppm and 0–0.5 ppm of SO_2 are available.

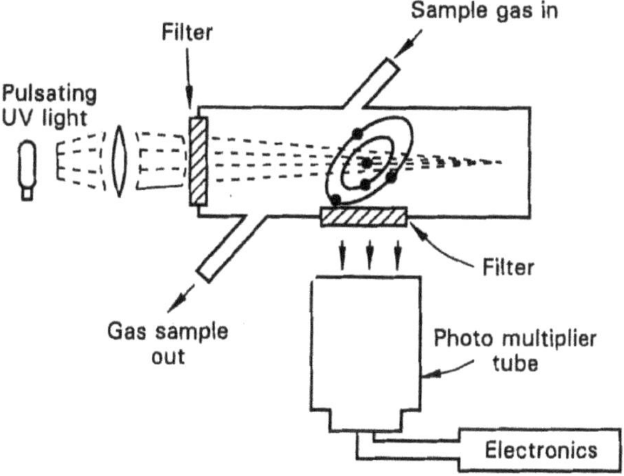

Fig. 6.31 *Pulsed fluorescent analyser for sulphur dioxide*

6.4.8.7 Permeation tubes as standards for gaseous air pollutants

Permeation tubes are made from inert plastic tubes having the liquefied gas to be analysed (Fig. 6.32). The gas permeates the walls of the tube at a constant rate. The permeation rate for a gas has been found to be reproducible and depends upon temperature. It is possible to calibrate permeation tubes gravimetrically to determine the rate of gas loss. Other calibration techniques such as volumetric, colorimetric, and coulo metric techniques can be used. It is possible to maintain a permeation tube at a constant temperature by suspending it in a condenser through which the thermostated water gets circulated, or by placing the tube inside a holder immersed in a constant temperature bath.

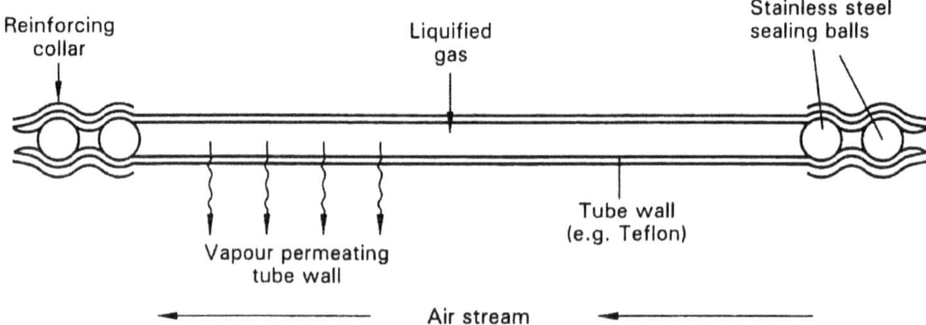

Fig. 6.32 *A permeation tube which is used for atmospheric pollutant standards*

6.4.8.7.1 *Oxides of nitrogen*

The most common method of measuring the oxides of nitrogen is the colorimetric method of Griess-Saltzman [7]. In this method NO_2 reacts with sulphanilic acid to form the diazonium salt which is coupled with N-(1 napthyl) ethylene diamine dihydrochloride to get a pink-coloured dye complex. Within 15 minutes at room temperature itself the colour development is complete. The developed colour is measured at 550 nm. This method is also applicable to NO analysis or a combined analysis, by oxidizing NO to NO_2 by passing it through potassium permanganate solution.

Analysis must be completed within an hour of sample collection. The method can be used to determine atmospheric concentrations of NO_2 from 40 to 1500 $\mu g/m^3$. High concentrations of SO_2 may bleach the reagent, giving low results, and the stoichiometry of the process is not completely clear, but it can be calibrated using permeation tubes and is easily modified for automated analysis.

Nitric oxide, NO, can be determined by passing a parallel stream of air through an oxidation tube containing supported chromic oxide to convert the NO to NO_2, prior to collection and conversion to the azo dye. The nitric oxide concentration can then be known by difference.

A more stable technique for analysing field samples of 24 h is the Jacobs-Hochheiser method. In this method NO_2 is passed through a dilute solution of 0.01 N sodium hydroxide to form the nitrite ion. This is then reacted with sulphanilamide and N-(1 napthyl) ethylene diamine in phosphoric acid to form the highly coloured azo dye which can be determined calorimetrically.

The Jacobs-Hochheiser [8] method was adopted by USEPA as a standard method as it permits upto 2 weeks between sample collection and analysis. Polluted air is passed through aqueous NaOH solution, in which NO_2, is converted to sodium nitrite and some sodium nitrate. The collected sodium nitrite after bringing it to laboratory, is treated with H_2O_2 to remove SO_2 interference, acidified, and used to diazotize sulphanilamide. The latter is then coupled to N-ethylenediamine dihydrochloride as in the Griess-Saltzmananalysis.

The phenol-disulphonic acid method is another simple spectrophotometeric method that can be used for NO_x concentrations between 15 ppm to 1500 ppm by volume and as such is very suitable for measuring stack gas emissions. The absorbing solution consists of 3% hydrogen peroxide and 0.1 N sulphuric acid. The peroxide oxidizes nitrogen compounds to nitric acid. Analysis can be carried out using phenol-disulphonic acid which is treated with ammonium hydroxide to obtain a yellow compound which can be measured by a spectrophotometer.

The NO_2 is collected in an impinger containing the absorbing reagent (sodium hydroxide-sodium arsenite). The sampling train is similar to that used for SO_2 measurement (Fig. 6.29). In the absorbed air sample, hydrogen peroxide (H_2O_2) is first added to eliminate the interference of SO_2 by converting it into sulphates. Colour is developed by adding sulphaniline solution with phosphoric acid and N (l-napthyl) ethylenediamine dihydrochloride. The intensity of the colour can be measured after 10minutes by obtaining optical density at 540 nm.

The standard curve can be made with sodium nitrite, ranging from 0.04 to 2.0μg NO_2/ml. The concentration of NO_2 in the air can be calculated by the following formula.

$$\mu g\ NO_2\ /\ m^3 of\ air = \frac{\mu g\ NO_2\ in\ absorbent \times 10^3}{volume\ of\ air\ sampled\ (L) \times 0.82} \qquad .....(6.4)$$

where the value 0.82 is a factor for collecting efficiency.

The sampling rate is to be usually kept at 200 ml/min, and the sample is to be collected for 24 h for air containing NO_2 in the range of 20 to 750 $\mu g/m^3$.

Thus manual wet chemical procedures for NO_x are based upon the conversion of NO_2 to nitrite, which is then determined typically by diazotization and azo dye formation. This, however, is found to be very problematic when reliable sources of NO_2 calibration gas are not available, since the collection efficiency of NO_2 is found to be variable, and the stoichiometry of the NO_2 to nitrite conversion is uncertain. This finding led to further development and introduction of improved procedures such as that involving the use of arsenite in the collection solution to enhance the formation of nitrite.

6.4.8.7.2 Chemiluminescence method (Fontjin et al., 1970) [9]

Instrumental analysers not subject to these problems have now been available for some years. The currently favoured technique for determination of oxides of nitrogen is based upon the chemiluminescent reaction of nitrogen oxide and ozone to give electronically excited nitrogen dioxide which emits light in the 600–3000 nm region with a maximum intensity near 1200 nm.

$$NO + O_3 \rightarrow NO_2^* + O_2$$

$$NO_2^* \rightarrow NO_2 + h\nu \text{ (light)}$$

In the presence of excess ozone generated within the instrument, the light emission varies linearly with the concentration of nitrogen oxide from 1 ppb to 10^4 ppm. The apparatus is shown in Fig. 6.33.

The method is free of interferences for measurement of NO and may be used to measure NO_x by prior conversion of NO_2 to NO in a tube heated to about 650°C. Dependent upon which converter is used, some interference in the NO_x mode is likely to occur from compounds such as peroxyacetyl nitrate and nitric acid. Some instruments incorporate two reaction chambers, one running permanently in the NO mode, the other analysing NO_x after NO_2 to NO conversion. Thus pseudo-real-time NO_2 concentration scan be measured.

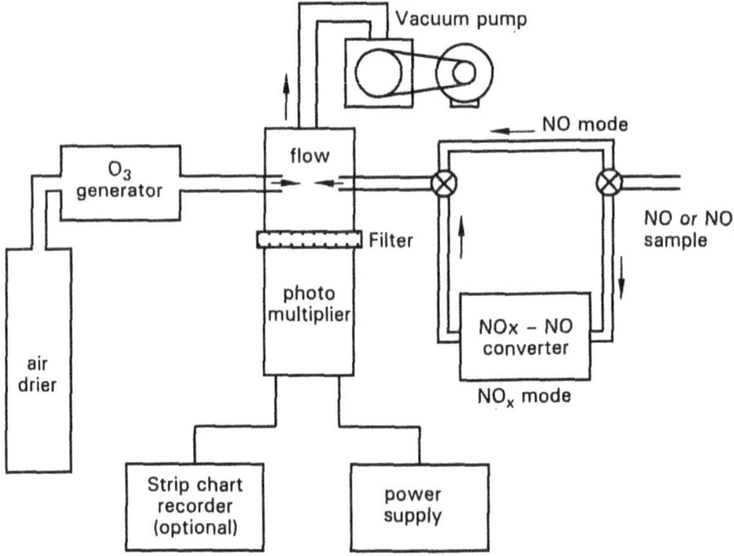

Fig. 6.33 *Chemiluminescent analyser for oxides of nitrogen*

6.4.8.7.3 Carbon monoxide

For estimating carbon monoxide levels in the range of 1–50 ppm in ambient air, non-dispersive IR spectrophotometry is widely used. However in this method carbon dioxide and water vapour which have absorption bands in the same region will interfere. The interference of water vapour can be eliminated by passing the sample through a drying agent. By interposing a carbon dioxide cell between the sample and reference cells and detecting, the interference of carbon dioxide can be nullified.

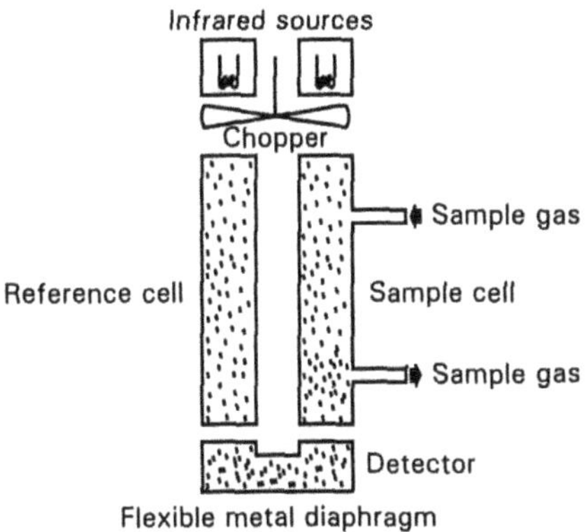

Fig. 6.34 *Non-dispersive infrared analyser*

In non-dispersive IR spectrophotometry the intensity of CO stretching frequencies in the air sample can be continuously monitored with reference air containing no CO. This is achieved without dispersion of the IR radiation by using cells containing CO at a reduced pressure as detectors for two beams which are chopped at a frequency of about 10 Hz and passed respectively through sample and reference cells. Absorption of radiation by the CO causes a differential pressure between the two detector cells which is sensed by a flexible diaphragm between them and used to generate an electrical signal (Fig. 6.34).

The non-dispersive infrared spectrophotometer is one of the versatile and standard methods for continuous monitoring (Fig. 6.34).These instruments have rapid response and high sensitivity. This instrument consists of a reference cell filled with a non-absorbing gas such as nitrogen and a sample cell which is continuously flushed with an air sample containing CO which absorbs at 4.6 μm. There are two IR sources to pass through the sample cell and reference cell and a detector. The detector has two compartments which are separated by a thin diaphragm and filled with CO. The infrared light source is allowed to pass through the sample and reference cells alternatively by a chopper. While the reference cell which has no absorbing gas allows the entire energy to pass to one compartment of the detection cell, the sample cell containing CO allows energy to pass through it varying proportionally to the concentration present in it. Since more IR energy passes through the reference cell side of the detector compartment, the CO present in it expands and pushes the diaphragm in. An electrical signal is generated based on the mechanical movement of the diaphragm and is recorded after amplification.

Non-dispersive IR instruments can be used to estimate CO even in the ranges of 1-25-100 ppm and the results are controlled by operator pressures and amplification of electrical signals.

The main interference comes from water vapour which also has got IR absorption. It can be eliminated by drying the air sample before it enters the instrument.

An analyser for continuous determination of carbon monoxide at levels down to 1ppm uses an electrochemical cell. Gas diffuses through a semipermeable membrane into the cell and at an electrode CO is oxidized to CO_2 at a rate proportional to the concentration of the CO in air. The response time is fairly short and interferences from other air pollutants at normally encountered levels are minimal. Analysers based upon electrochemical cells are also available for measurement of sulphur dioxide and oxides of nitrogen.

6.4.8.7.3.1 Gas chromatographic method for CO

A gas chromatographic technique for the measurement of CO levels in ambient air has very high sensitivity though it is not useful for continuous monitoring (Robbins et al., 1968) [10]. In this method CO is catalytically reduced to methane (Fig. 6.35) and then estimated it by GC using a Flame Ionization Detector (FID). For estimating methane, CO

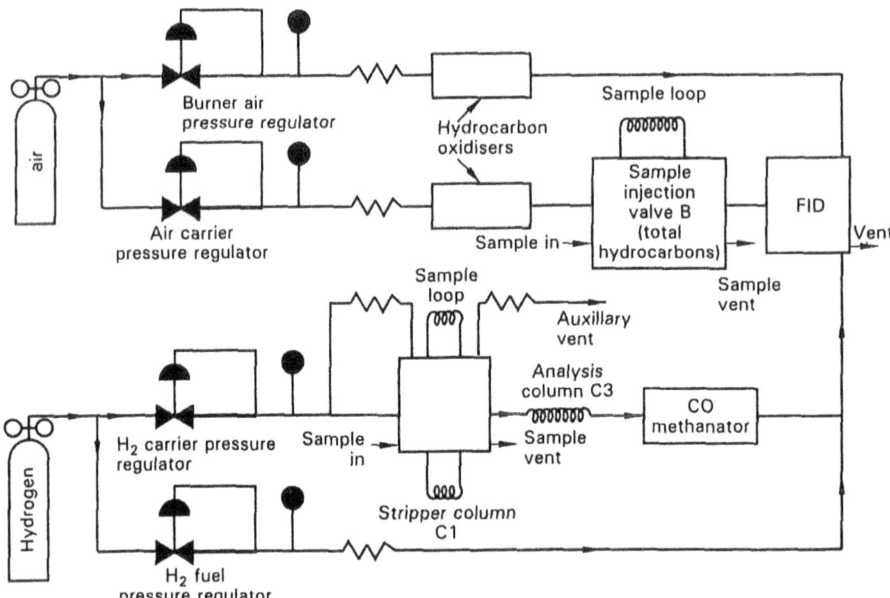

Fig. 6.35 *Analyser giving sequential measurements of methane, carbon monoxide, and total hydrocarbons*

and total hydrocarbons, this method is used in commercial instruments. A small air sample is flushed using hydrogen carrier gas from a sample loop through a stripper column (Cl) packed with an adsorbent porous polymer, which will allow movement of methane and carbon monoxide only while the heavier compounds are obstructed. Later these compounds are removed by back-flushing. When the methane and carbon monoxide are allowed to pass through a chromatographic column containing mole cularsieves (C_3), separation of these compounds takes place and finally they are passed to the FID via a catalytic methanetor. This method permits estimation of CO at ppb levels. By appropriate modification of the instrument (C_2) hydrocarbons can also be allowed to pass through the stripper column, separated on a second column of porous polymer and finally fed to the FID. By passing 10^3 of ambient air the total hydrocarbons can be determined using the FID.

6.4.8.7.4 Hydrocarbons

The commonly used methods for determining the hydrocarbon content of the atmosphere and of process streams are the spectrographic, chromatographic, and flame ionization methods. All three methods find use for the comparatively high concentrations in process streams. Both dispersive and non-dispersive infrared instruments are used. However, the non-dispersive instrument is normally used in process streams. It gets sensitized for the particular hydrocarbon of interest so that quantitative analyses can be made.

Hydrocarbon pollutants are normally determined as **Total hydrocarbons** but individual chemical species can also be determined. The most abundant hydrocarbon is methane. Methane is not a hazardous air pollutant and hence the total hydrocarbon content is reported on a non-methane basis.

Analysis of total hydrocarbons or non-methane hydrocarbons is complicated by the great complexity of the mixture present and as a rule no attempt is made to analyse every compound individually.

The unburnt hydrocarbons present in engine exhaust gases consist mainly of alkanes, alkenes and aromatic compounds together with their products of partial oxidation (formaldehyde, acetaldehyde, etc.). Since the concentration of these hydrocarbons reaches a peak with early morning traffic flows, the quality control limit specifies a mean value (0.24 ppm) for a 3-h period (6 to 9 a.m.). At later hours, sunlight promotes a series of photochemical reactions that lead to the formation of secondary pollutants such as nitrogen dioxide, ozone, aldehydes, ketones, peroxycyl nitrates (PAN, e.g. $CH_3COO_2NO_2$), and alkyl nitrates as discussed in chapter 1. These secondary products give rise to photochemical smog.

With such a complex mechanism, the concentrations of individual components vary widely during the day. Hence the values such as those recorded in Table 6.4 are mainly indicative of relative magnitudes.

Table 6.4 *Concentrations of specific C_2–C_6 hydrocarbons measured in high traffic urban air (ppb)*

Compound	Arithmetic mean	Standard deviation
Ethane	54.2	28.1
Ethane	42.3	9.3
Acetylene	10.3	6.0
Propane	12.2	5.8
Propane	9.9	7.1
Isobutane	40.7	10.2
N-butane	16.6	12.8
1-butene	11.0	8.0
2-methylpropane	1.2	1.6
Trans but-2-ene	18.9	17.0
Cis but-2-ene	8.7	7.2
N-pentane	11.2	7.1
2-methylbutane	81.2	65.7
2,2 dimethylpropane	58.8	16.6
1-pentane	10.3	4.9
Trans-pent-2-ene	19.7	12.4
Cis pent-2-ene	13.6	17.7
2-methylbut-2-1-ene	0.4	0.3
3-methylbutane	38.3	17.7
n-hexane	10.1	4.5
2,2 dimethylbutene	47.9	16.9
2-methylbutane	21.2	17.5
3-methylpentane	40.2	20.8
1-hexane	10.4	8.3
Trans-hex-3-ene	5.1	5.4
Cis hex-3-ene	6.3	7.1

The universal choice nowadays for total hydrocarbon analysis is the flame ionization analyser, based upon use of ambient air as an oxidant in an air/hydrogen flame ionization detector; the presence of hydrocarbons leads to enhanced conductivity of the flame.

In the flame ionization detector, the sample gas is injected into the flame created by burning hydrogen gas in air or oxygen. When hydrogen alone burns very few ions are generated. When hydrocarbons are present, the flame produces a complex ionization in which a large number of ions are present. An electric field is set up near the flame by making the burner jet positive with respect to a wire loop. The electric field induces ionmigration in such a manner that a small ionization current is established between the electrodes and this current is proportional to the concentrations of the ions in the flame. The current is amplified and displayed on an output meter. Since the response of the detector is approximately proportional to the number of the carbon atoms present in the hydrocarbons being consumed in the flame the data are usually expressed in the units of the gas used in calibration (ppm of C as methane). The FID probe does not respond to other gases like CO, CO_2, H_2O, SO_2, or NO_x.

The response of a flame ionization detector is approximately proportional to the number of carbon atoms in the compound. The flame ionization detector has a very wide range of linear response and is found to be sensitive to extremely low levels of hydrocarbons, ranging down to 1×10^{-4}μg per litre. Due to the extremely high sensitivity of the detector it sometime becomes possible to analyse hydrocarbons without a pre-concentration step. A schematic diagram of a flame ionization detector is depicted in Fig. 6.36.

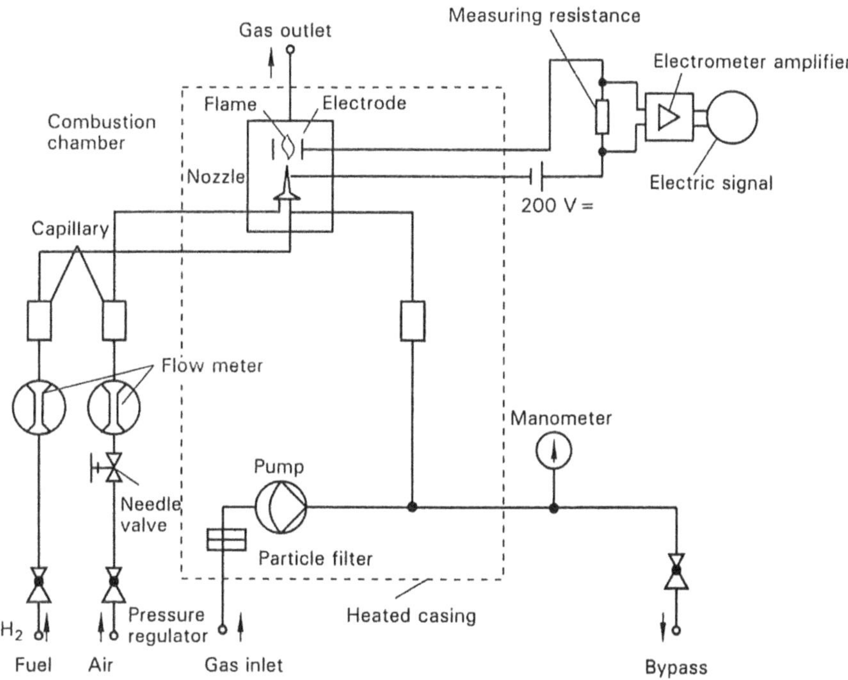

Fig. 6.36 *Flame ionization detector for the analysis of hydrocarbons*

Sensitivity per carbon atom of individual hydrocarbons varies little, whilst oxygenated and halogenated compounds cause a lower response. Water vapour, carbon dioxide, and carbon monoxide do not significantly affect the reading. The instrument is calibrated with methane and results are reported as ppb C, or parts per billion carbon since 25 ppb butane, 50 ppb ethane, and 100 ppb methane all give the same response. Methane is determined in such instruments by selectively scrubbing all other hydrocarbons from the air stream prior to analysis and the non-methane hydrocarbon concentration is derived by difference.

6.4.8.7.4.1 Gas chromatographic methods for hydrocarbons

The determination of specific hydrocarbons in ambient air normally requires a pre-concentration stage in which air is drawn through an adsorbent such as a porous polymer or activated carbon, or a tube where freeze out of the compounds by reduced temperature occurs, followed by injection into a gas liquid chromatography. Excellent separations of many compounds have been achieved, the best results coming from the use of capillary columns. Detection may be by flame ionization, or by passage into a mass spectrometer, the latter techniques allowing a more positive identification of individual compounds.

In gas chromatography (Fig. 6.37) each component can be identified on the basis of the time it takes to pass though the column. With careful calibration, the area of intensity of each recorded signal peak becomes a measure of the amount of the component present.

The diversion of the stream leaving the column into a mass spectrometer unit (Fig.6.38) allows positive identification of the separated components. The individual molecular species are first separated from the carrier gas and are then admitted via a "leak" into a high-vacuum system. In the ionization chamber, the organic materials are exposed to accelerated electrons which on impact convert molecules into charged ionsor fragments. All the charged particles are accelerated through electrostatic and magnetic fields and become separated on the basis of their mass/charge ratio. The separated components are sequentially focused on a detector to yield a characteristic mass spectrum. To identify the initial molecular species, the spectrum is compared with that of known pure compounds.

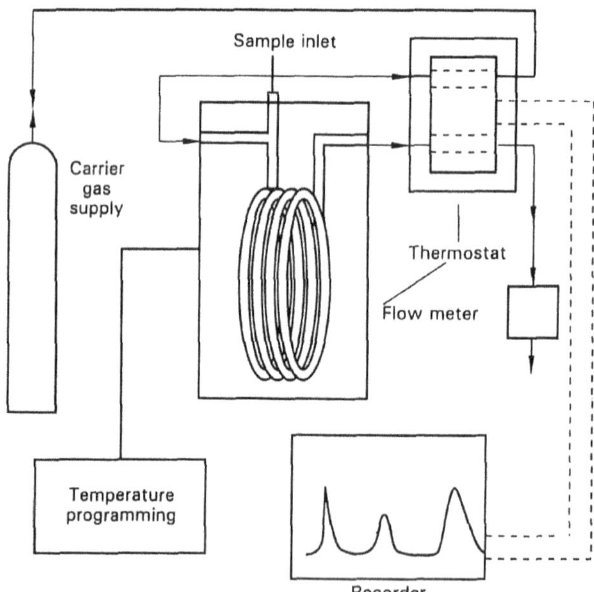

Fig. 6.37 *Gas chromatograph*

Both non-dispersive infrared analysis and catalytic oxidation which are used for the analysis of carbon monoxide can be used for hydrocarbon analysis. Dispersive infrared analysis in which selection of a measuring wavelength is achieved with a monochromator can be used for the quantitative analysis of hydrocarbons.

At one time, use of non-dispersive infrared spectrum overlapped with that of hexane, most notable in the regions of the C-H stretching vibration. Since not all hydrocarbons absorb at the same length, and absorption intensities differ, the sensitivity per carbon atom in the molecule is similar for alkanes other than methane but is substantially less for alkanes and aromatics and almost nil for ethylene.

Chemiluminescence is also used for the analysis of reactive hydrocarbons. Chemiluminescence from the reaction of atomic oxygen with hydrocarbons is used. Benzene at concentrations as low as 0.3 ppm can be detected.

6.4.8.7.5 Ozone and other oxidants

Oxidants are those substances that oxidize a reference material or solution under certain conditions. These include ozonides, peroxides of hydrocarbons and substances such as peroxyacyl nitrates. The standard method for the net determination of total oxidants is the neutral phosphate-buffered KI colorimetric technique of Saltzman [7]. In this method, a sample of air at a known flow rate is passed through a solution of 1 % neutral buffered potassium iodide and the liberated iodine is measured colorimetrically at 352 nm.

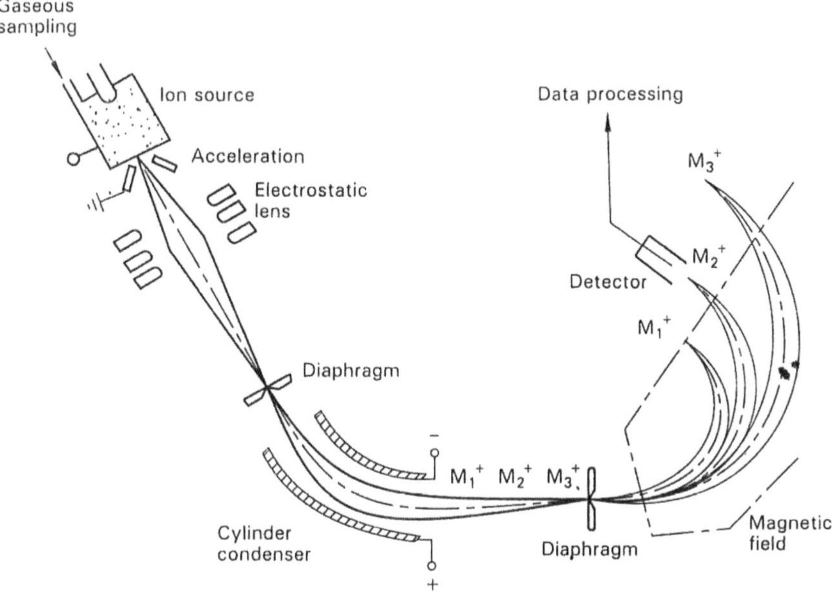

Fig. 6.38 *Schematic diagram of a mass spectrometer*

The overall reaction with ozone is assumed to be as low with a stoichiometry of I_2/O_3 equal to 1.0.

$$2KI + O_3 + H_2O \rightarrow O_2 + 2KOH + I_2$$

$$I_2 + I^- \rightarrow I_3^-$$

6.4.8.7.5.1 The tri-iodide can be measured colorimetrically

Since many reducing agents can react with I_3, the method gives a net oxidant level unless they are removed. One method is scrubbing with $CrCO_3$, but this also oxidizes NO to NO_2, necessitating a correction factor.

The method responds to all oxidants although not at the same rate or to the same extent. For example, the response to ozone and peroxides has been fast but the response to NO_2 is only 10% of the response to ozone. The value obtained has been therefore a complex composite of all the oxidants present.

Some researchers, however, have found a stoichiometry for I_2/O_3 of 1.5 and observed an enhanced formation of iodine and a reduced oxygen formation with increasing pH and postulated a more complex mechanism. The feasibility of iodide oxidation to iodate has been shown and the possibility of catalysis of this reaction by a glass frit. Hence, in addition to the lack of specificity of the reagent the stoichiometry of the reaction with ozone is in doubt. The use of a borate, rather than a phosphate buffer, has been shown to overcome many of these problems.

Approximate chemical filters improve the selectivity of methods based upon the oxidation of neutral buffered halide. Sulphur dioxide produces a negative interference equal to the equivalent molar concentration of oxidant and may be eliminated by incorporating a chromic acid paper absorber in the sampling line. The chromic acid does, however, oxidize nitrogen oxide to nitrogen dioxide causing further interference in the presence of NO.

Another colorimetric method widely used in continuous analysers for measuring total oxidants is that proposed by Mast and Saunders (1962) [11]. In this method, the air sample is passed through a sensing cell containing a thin film of the KI solution which acts as a source. The iodine liberated at the cathode of the cell produces an external current flow which is proportional to the iodine (or oxidant) concentration which can be measured with a microammeter. This method is sensitive enough to detect ozone even less than 0.01 ppm, and can be precisely used in the best working range of 0.01 to 0.20 ppm.

6.4.8.7.5.2 Chemiluminescence method

The EPA reference method used for ozone monitoring was based upon the chemilumin escent reaction between ozone and ethylene [11, 12] and later improved as an analytical technique. Chemiluminescence from this reaction takes place over a range of 300 to 600 nm having a maximum at 435 nm. The intensity of the emitted light is directly proportional to the level of ozone. Ozone concentrations ranging from 0.003 to 30 ppm can be measured. The corresponding chemiluminescence reaction is shown below.

$$C_2H_4 + O_3 \rightarrow O_2 + C_2H_4O^*$$

$$C_2H_4O^* \rightarrow C_2H_4O + h\nu \text{ (300–600nm)}$$

Emission is centred on 435 am and hence no interference from the reaction of ozone and nitrogen oxide occurs. The method is sensitive to as little as 1 ppb of ozone.

The UV absorption of ozone at 254 nm may be used for its determination at levels down to 1 ppb. Interference from other UV absorbing air pollutants such as mercury and hydrocarbons may be minimized by taking two readings. The first reading is of the absorbance of an air sample after catalytic conversion of ozone to oxygen and ozone content of the air. Available instruments perform this procedure automatically and give a read-out in digital form. Although truly continuous measurement of ozone levels is not possible, the response is fast and readings may be taken at intervals of less than one minute. The instrument is shown diagrammatically in Fig. 6.39.

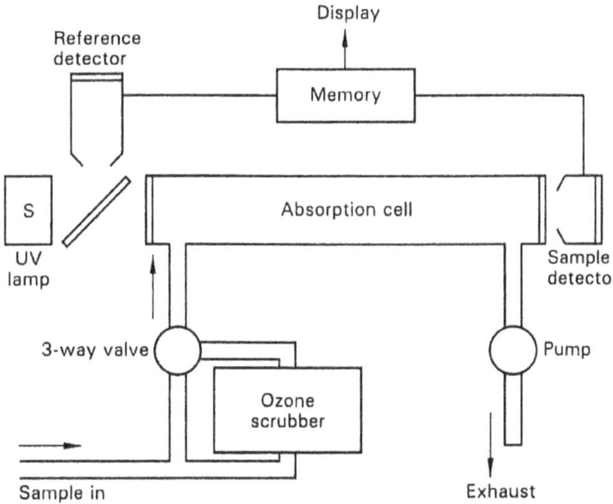

Fig. 6.39 *A UV photometric analyser for ozone*

The reaction of O_3 with substances such as Rhodamine B or ethane produces products in excited electronic states. Their decay to the ground state is accompanied by emission of characteristic visible radiation. The measurement of the level of luminescence by means of a photomultiplier determines the amount of reaction and hence, the amount of ozone present. The chemilum inescent method described earlier for NO can also be adapted to determine O_3.

The reaction of ozone and nitric oxide to produce excited nitrogen has been used as a method for the analysis of nitric oxide. It is also used for the analysis of ozone. Reducers are needed to prevent the quenching of the chemiluminescent reaction. Emission lies in the 600–3000 nm range, and a photomultiplier detector can only be used in the lower wavelength region of that range. When low levels of ozone down to 0.01 ppm have to be measured, the detector has to be cooled to –20°C.

6.4.8.7.5.3 Peroxyacyl nitrate (PAN)

Peroxyacyl nitrate is a product of atmospheric photochemical reactions and is characteristic of photochemical smog (Table 6.5).

$$CH_3 - \overset{C}{\underset{O}{\|}} - O - O - NO_2$$

Chemical analysis of PAN may be determined by long path IR measurements, the greatest sensitivity being achieved when Fourier transform methods are used. The most sensitive and specific routine technique involves gas chromatographic separation and detection of specific peroxyacyl nitrates by electron capture. The detection limit of below 1 ppb permits it to be used as a direct atmospheric monitor under circumstances of high pollution. Peroxyacyl nitrate, aldehydes and ketones and hydrocarbons may be analysed by using gas chromatography or electron capture of flame ionization detectors. The latter type of detector gives a response proportion to the number of carbon atoms in a compound and is used in the reference method for total hydrocarbons. Gas chromatographs using mass spectrometers as detectors make analysis of specific compounds in the complex smog mixture more feasible.

Table 6.5 *Typical concentrations of pollutants in photochemical smog*

Pollutant	Concentration range (ppm)
Carbon monoxide	200–2000
Total hydrocarbons	20–50
Aromatics	10–30
Aldehydes	5–25
Alkenes	2–6
Nitric oxide	1–15
Nitrogen dioxide	5–20
Peroxyacyl nitrates	1–4
Ozone	2–20

6.4.8.7.6 Hydrogen sulphide

Procedure: Into a set of five 25 ml standard flasks, pipette 1.5, 2.0, and 2.5ml stock solution Na_2S. To each flask add the following solutions: 2 ml $NaClO_4$, 1 ml N, N-diethyl-p-phenylenediamine and 1 ml ferric ammonium sulphate. Make up each to25 ml with chloroform. After 10 min but before 20 min, measure the absorbance of these solutions in a 1-cm or 4-cm cell at 670 nm. Draw the calibration curve of sulphidein ppm. Take in a standard wash bottle with a fritted glass disperser, 50 ml zinc acetate solution.

This is the absorbent. Pump 500 litres air at the rate of 12 litres per minute through the absorbent. After the pumping is stopped, add the following solutions in order: 2 ml $NaClO_4$, 1 ml N, N-diethyl-p-phenylenediamine and 1 ml ferric ammonium sulphate. Shake the solution well and after 10 minutes extract the solution 2 to 3 times each time with 5 ml chloroform. Transfer the extracts into a 25 ml standard flask. Make up the volume to the mark with chloroform. Measure the absorbance in the same cell used for the standards at 670 nm. Determine the concentrations of sulphide in the given sample of air from the calibration curve.

Note: In all absorbance measurements, the blank solution should contain all the reagents except the substance under investigation.

The various commonly adopted methods for the analysis of air pollutants in ambient air are summarized in Table 6.6.

Table 6.6 *Summary of commonly employed methods for the measurement of air pollutants*

Pollutant	Measurement technique	Sample collection period	Response time (continuous technique)[a]	Minimum detectable concentrations
Sulphur dioxide	Absorption in H_2O_2 titration Absorption in tetrachloromercurate/spectrophotometry Flame photometric Analyser Gas phase fluorescence	24h 15 min		2 ppb 10 ppb 0.5 ppb 0.5 ppb
Oxides of Nitrogen	Chemiluminescent reaction with ozone		1 s	0.5 ppb
Total hydrocarbons	Flame ionization analyser		0.5 s	10 ppb C
Specific hydrocarbons	Gas chromatography/flame ionization detector	b		<1 ppb
Carbon monoxide	Catalytic methanation FID Electrochemical cell Non-dispersive infrared	c	25 s 5 s	10 ppb 1 ppm 0.5 ppm
Ozone	Chemiluminescent reactions with ethane UV adsorption		3 s 30 s	1 ppb 1 ppb
Peroxyacetyl nitrate	Gas chromatography/electron capture detection	c		1 ppb
Particulates	High volume sampler β - guage	24 h min/h		5 μgm^{-3} 5 μgm^{-3}

(a) Time taken for a 90%response to an instantaneous concentration change; **(b)** Grab samples of air collected in on inert container and concentrated prior to analysis; **(c)** Instantaneous concentrations measured on a cyclic basis by flushing the contents of a sample loop into the instrument.

Questions

Short Answer Questions

1. Brief the importance of preliminary site inspection prior to testing.
2. Write a short note on continuous emission monitoring.
3. List out the key points in a sampling checklist.
4. How is a cascade imparter used for particulate sampling?
5. Outline the principle of an X-ray fluorescence spectrophotometer.
6. Write a note on Isokinetic sampling for particulates.
7. Explain the working of a Pulsed fluorescent analyzer.
8. Write a short note on Non-dispersive infrared analyzer.
9. List out the typical concentrations of pollutants in photochemical smog.
10. Summarize the commonly employed methods for measurement of air pollutants.

Essay Questions

1. Discuss how the physical parameters of air are determined.
2. Explain about Detection limits. Derive the equation for determining in-stack detection limits.
3. Describe the gravitational method employed for particulate sampling.
4. With a suitable sketch, bring out the working of a high volume air sampler used for fibre filtration.
5. Chemical methods have higher efficiency in analyzing certain elements. Justify
6. Explain about sampling and collection of gaseous air pollutants in detail.
7. Analyze the methods employed for estimating Sulphur dioxide.
8. Outline the working of a Chemiluminescent analyzer used for oxides of nitrogen.
9. Explain the Gas chromatographic methods used for hydrocarbons.

References

1. Manahan, S. E. 1979. Environmental Chemistry.3rd edition. Willard Grant Press, Boston, Massachusetts.
2. Marrow, N. L., R. S, Brief and R. R. Bertrand, 1972. Sampling and analysis of air pollution sources. Chem. Engg.79:84-98.
3. Maugh, J. 1972. Envi. Sci. Tech. 8(6):584-5.
4. Harris, J. C., D. J. Larsen, C. E. Rechsteiner and K. E. Thrun, 1984. Sampling and analysis method for hazardous waste combustion. EPA 600/8-84/002 PB84-155545USEPA.
5. West, P.W. and G.C. Gaeke, 1956. Fixation of sulphur dioxide as sulfitomercurate (III) and subsequent colorimetric determination. Analyt. Chem. 28: 1816.

6. Dasgupta, P. K. et al., 1980. Analyt. Chem.52:1912-22.
7. Saltzman, B. E. 1954. Analyt. Chem. 26, 1949.
8. Jacobs, M. B. and S. Hochheiser, 1958, Analyt. Chem.30, 426.
9. Fontjin, A. A. J. Sabadell and R. J. Ronco, 1970. Analyt. Chem. 42, 575.
10. Robbins, K. C. K. M. Borg and E. Robinson, 1968. J. Air Pollut. Conts. Ass. 18,106.
11. Mast, G. M. and H. E. Saunders, 1962. Research and development of instrumentation of ozone testing, Instr. Soc. Am. Trans. 1:320-318.
12. Air quality criteria for photochemical oxidants. U.S. National air pollution control administration. Dept. of Health, Education and Welfare. Publ, No, A.B. 63. 1970.

CHAPTER 7

Air Pollution Control: General Source Correction Methods and Particulate Control Technologies

7.1 Sources Correction Methods

Unlike water pollution control, air pollution control is more difficult since air is more diffusive and disperses more in space. Polluted water can be collected at a place for treatment unlike polluted air. Therefore any control technology for air pollutants will be more effective at the source or generating point rather than after dispersion into environment. The following source corrections [1] in various industrial operations can help control air pollution.

(a) Raw material changes

(b) Operational changes that involve modification of the process or modification or replacement of process equipment.

7.1.1 Raw material changes

Sulphur dioxide is one of the main air pollutants produced by burning of fuels. All fuels used by humans contain some sulphur. Fuels like wood have very little sulphur, whereas most coals have 0.5-3% sulphur. Oils generally have more sulphur than wood but less than coal. If we burn these fuels, the sulphur in them will mostly forms sulphur dioxide,

$$S + O_2 \rightarrow SO_2$$

In natural gas most of the sulphur is in the form of H_2S which is easily separated from the other constituents of the gas. In oil and also in oil shales and tar sands, the sulphur is chemically combined with the hydrocarbon compounds; normally it cannot be removed without breaking chemical bonds. In oils the sulphur is concentrated in the higher-boiling fraction of the oil, so the same crude oil can yield a low-sulphur petrol and a high-sulphur heavy fuel oil. In coal, much of the sulphur is also in the form of chemically bound sulphur, but some coals have a large fraction of their sulphur in the form of small crystals of iron pyrite. When the fuel is burned, almost all of the sulphur in the fuel, whether chemically bound or pyretic, will be converted to sulphur dioxide and carried along with stack gas. A small fraction will be captured in the ash, and some will be converted to SO_3.

Coal cleaning: Pyritic sulphur can be removed by grinding the coal to a small through size so that the pyrites will be mostly present as free pyrite particles. Gravity methods can then used to separate the low-density coal from the high-density pyrites. This widespread approach is particularly suited for coals in which a substantial fraction of the sulphur is present as pyrites. Unfortunately, pyrite particles are generally quite small, so that very fine grinding is needed to separate them from the rest of the coal.

Solvent-refined coal: It is also possible to dissolve coal in strong enough solvents and then to treat the solution by the same kind of catalytic hydrogenation processes that can be used to remove sulphur from petroleum products. The mineral materials do not dissolve, so it will be rejected by filtration or settling. When the solvent is then removed for reuse, the remaining product will be a very clean-burning combustible solid, free of ash and sulphur, called solvent-refined coal.

Hydrocarbon fuels: Sulphur in hydrocarbon fuels is also normally converted to SO_2 during combustion and then emitted to the atmosphere. Large oil-burning facilities can have equipment to capture that SO_2, but autos, trucks, and airplanes cannot have such equipment. The only way to limit the SO_2 emissions from these sources is to limit the amount of sulphur in the fuel. Crude oils vary in their sulphur content; low-sulphur crudes are called "sweet", high-sulphur crudes, "sour". If the fraction of the crude oil going to petrol or diesel fuel has too high a sulphur content, most of that sulphur is removed by catalytic hydrodesulphurization,

$$\text{Hydrocarbon containing S} + H_2 \xrightarrow[\text{Hydrocarbon } + H_2S]{\text{Ni or Co catalyst promoted with Mo or W}}$$

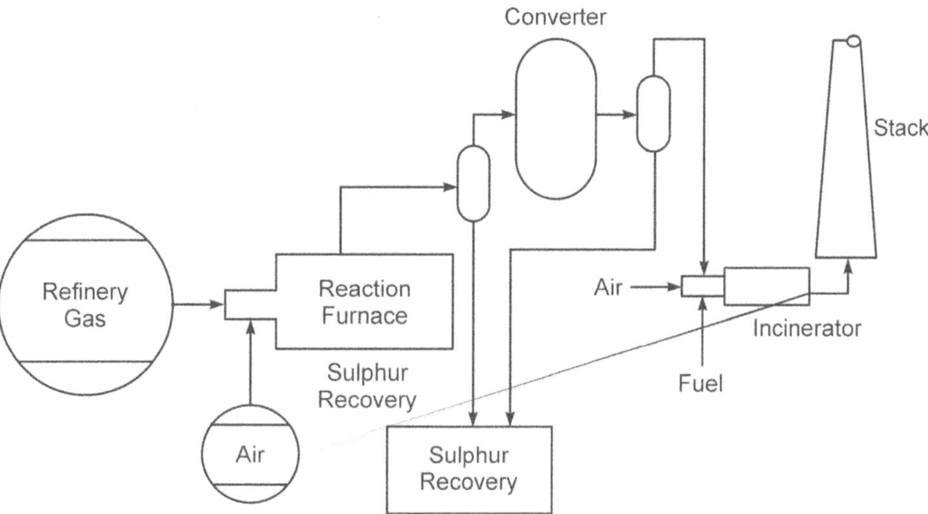

Fig. 7.1 *Claus process for removal of H_2S*

The resulting gas stream, a mixture of H_2 and H_2S, is one of the streams treated in a refinery for H_2S removal by the Claus process shown in Fig. 7.1. Some petroleum streams in refineries are treated over these catalysts to remove both sulphur and nitrogen because those elements interfere with the catalysts used for subsequent processing. The resulting gas streams contain both H_2S and NH_3.

Substitution: Replacing petrol as a motor fuel with compressed natural gas or propane can reduce the emissions of VOCs, because those fuels can be handled, metered, and burned with fewer VOC emissions than petrol. The petroleum industry is working hard to improve the burning properties, handling, and use of petrol, to make it as low-emission a fuel as compressed natural gas and propane, so that petrol can keep its dominant position in the auto fuel market.

7.1.2 Process changes or modification or replacement of process equipment

7.1.2.1 Process change

This technique involves a modification of an existing process (or). The introduction of a new process. Since 90's the process change has been known as "pollution prevention". Consider the example of painting operations in automobile industry. Large quantities of volatile organic compounds and hazardous air pollutants were released as a result of these operations. After the passage of 1990 CAAA many operations have substituted water based, low HAP paints for oil based paints for reducing emissions. Some operations have implemented processes such as electro-deposition, dip tank and powder coating.

7.1.2.2 Process modification (PM)

PM is a popular technique to control air pollution. Major efforts are underway in all industries to modify processes to reduce pollution. For example, new oxy furnaces are being developed by Libbey Owens Ford in Toledo in US to reduce ozone emissions during glass manufacturing. Changes industrial process to reduce raw materials and fuels also lead to reduction in air emissions. The promotions to use fluorescent lights and to conduct energy audits by electric utilities are examples of less electricity demand.

This results in less fuel use at the power plant and hence less air pollutants are emitted. The use of wind energy, geothermal energy, hydroelectric power and solar energy is increasing and is helping in reducing air pollution.

The USEPA has initiated a major effort in the area of pollution prevention. The purpose is to "prevent" Pollution rather than control it at the point of release. The goal is not to produce solid or hazardous waste.

Change in Fuel: This technique involves the use of less polluting fuel to reduce air pollution. Use of low sulfur fuel instead of high sulfur fuel by electric utilities is an example of this method.

Use of oil with low ash content or natural gas fir a dryer at an asphalt plant to reduce particulate matter is another example of this method. Introduction of compressed Natural gas, propane, ethanol and oxygenated fuels for automobiles have helped in the reduction of air pollutants in the U.S.

The use of natural gas in North America and Europe for winter heating brought significant improvement in air quality in most cities. Nuclear power plants are relatively pollution free when compared to the coal fired power plants. However, they have been subjects of controversy in their overall environmental impact.

Improve Dispersion: This approach is based on the concept that dilution of air contaminants before they reach ground will lower the concentrations to which the population is exposed. The use of this approach for industry is discouraged by the US EPA. However, local and state agencies use the concept to develop air pollution control strategies for their area. Before the 1970 CAAA the most widely approach to air problems was based on the motto-Dilution is the solution to pollution. This is evident from the from the tall stacks build by Tennessee Valley Authority and others in 60's and 70's.

Another form of this approach is practiced in the form of intermittent control or an air quality prediction system. This approach attempts to control source emission rate during the periods of high ground level concentrations. Production curtailments, a plant shutdown, fuel switching, or other strategies achieve the reduction in source emission rate. The critical periods are determined from weather related data in the area. The prediction systems are based on observations, predictive equations or a combination of observations and predictions.

Good Operating Practices: Release of unnecessary air pollutants could be avoided by maintaining good housekeeping in the plant and performing proper maintenance. For example, liquid chemicals spread over the floor evaporates rapidly and will cause an increase in emission rate for that chemical.

7.1.3 Operational changes

Radical changes in the chemical and petroleum refining industries have resulted in minimizing the release of materials to the atmosphere. The volatile substances can be recovered by condensation and the non-condensable gases can be recycled for additional reactions. Hydrogen sulphide, which was once flared in refineries, is now recycled and used in the Claus process to recover elemental sulphur. In the Claus reaction sulphur is recovered from H_2S as follows.

$$2H_2S + SO_2 \rightarrow 2H_2O + 3S\downarrow$$

The required SO_2 can be obtained by burning part of the H_2S in air.

$$2H_2S + 3O_2 \rightarrow 2H_2O + 2SO_2$$

The recovered sulphur is an added value for the industry.

7.1.3.1 Modification of the combustion process

One standard way of burning large amounts of coal is to grind the coal to about 10-50μ size and blow it with hot air into a large combustion chamber. There the small coal particles decompose and burn in the one to four seconds that they spend in the furnace, transferring most of the heat generated to the walls of the furnace as radiant heat. The furnace walls are made of steel tubes in which fluid is heated. The hot gases leaving the furnace then pass over banks of tubes and transfer much of their remaining sensible heat to the fluid being heated.

Fluidized bed combustion is an alternative way to burn coal. In it, coal is burned in gravel-sized pieces by injecting them into a hot fluidized bed of limestone particles instead of as a finely dispersed powder in air. A fluidized bed is a dense bed of solid particles suspended in air, such beds are widely used in chemical engineering, e.g. in fluidized bed catalytic cracking. The coal spends a much longer time in the bed than it would in a pulverized coal furnace, because more time is needed to get complete combustion of the much larger particles.

In such a fluidized bed combustor, SO_2 is formed in the presence of a large number of limestone particles and has a high probability of reacting with one of them in the combustion bed. Here the temperatures are much higher than in the dry processes and most of the limestone can be converted to CaO, as the reaction of SO_2 with CaO is rapid enough to provide adequate SO_2 control. The limestone in the bed is steadily replaced, and the material withdrawn has largely been converted to $CaSO_4$. Here again, a dry powder waste is produced instead of a wet scrubber sludge.

The fluidized bed has tubes full of water and steam projecting into the bed. The heat transfer between the hot bed in which the coal is burned and the tubes is much better than that between the flames and the walls of an ordinary coal-fired boiler. For this reason fluidized bed combustors are smaller and operate at lower temperatures than ordinary coal-fired boilers. This saves on some costs and greatly reduces the formation of nitrogen oxides. These boilers, however, have other problems, so that they are not yet a clear winner over conventional boilers.

A second combustion modification alternative, is to convert the coal to a synthetic fuel gas and then burn that in combination gas turbine, steam turbine power plants (Fig. 7.2).This seems complex and costly, but it has the advantage that in the synthetic fuel gas the sulphur is present as H_2S; since no other acid gas is present, the sulphur can be easily removed from the gas by absorption. The second and more important advantage is that modern gas turbine, steam turbine plants have a much higher thermal efficiency than typical coal-fired steam plants. If some problems with this technology can be solved, it may offer a more efficient and economical way of converting coal to electricity than the systems currently used even though it is much more complex.

In petroleum refiners, hydrocarbon vapours are released into the atmosphere from storage tanks due to temperature changes, direct evaporation, and displacement during filling. These losses can be minimized by designing the tanks with floating roof covers or by pressuring the tanks.

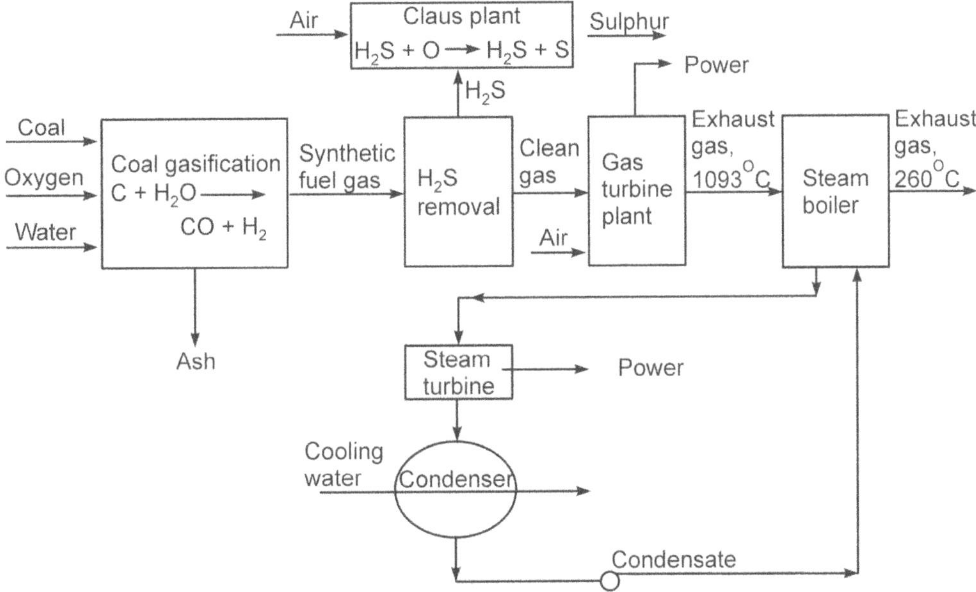

Fig. 7.2 *Equipment modification for combustion process in gas turbine or steam turbine plants*

Replacement of the open hearth furnace in steel industry and development of alternative power sources for automobiles in place of the internal combustion engine are additional examples of equipment alternatives.

The SO_2 concentrations in off-gases from the smelting of metal sulphide ores depend on which process is used and vary with time within the batch smelting cycle. However, they generally range from 2-12% SO_2. Such gases can be economically treated in plants that produce sulphuric acid by the following reactions:

$$SO_2 + 0.5O_2 \xrightarrow{\text{Vanadium catalyst}} SO_3 \quad \text{.....(7.1)}$$

$$SO_3 + H_2O \rightarrow H_2\ SO_4 \quad \text{.....(7.2)}$$

Reaction I is an equilibrium reaction which does not go to completion. Further more, this reaction is exothermic, so that the per cent conversion at equilibrium is higher at low temperatures than at high temperatures. For this reason, the reaction is customarily carried out in three or four separate catalyst beds with intercoolers between them (Fig. 7.3).

The preheated feed gas enters the fist catalyst bed at about 420 °C. The exothermic reaction heats it. All of the catalyst beds except the last are small enough such that the gases do not come close to their equilibrium conversion. The gases leave each of the first three beds and are cooled before entering the next bed. As shown in Fig. 7.3, the equilibrium conversion decreases rapidly with increasing temperature. The exit gas from the final bed is at about 425 °C and close to equilibrium at this temperature. In this way, about 98% of the incoming SO_2 can be converted to H_2SO_4. In a simple acid plant, the remaining 2% of the SO_2 is vented to the atmosphere.

In response to legal pressure to reduce this SO_2 emission, plants have been developed that are, in effect, two plants in series; i.e. equilibrium in Reaction I is obtained, the SO_3 is removed by Reaction II, and then the "tail gas" is again passed over a catalyst so that it will again come to equilibrium according to Reaction I. The resulting gas is again contacted with water so that the SO_3 is absorbed according to Reaction II. Such plants are called double contact or double absorption plants. They can typically convert over 99.7% of the incoming SO_2 to H_2SO_4. Because the feed material is quite cheap; the extra cost of the second reaction and absorption step is not repaid by increased acid production and is truly a pollution control expenditure. Fig. 7.3 shows simplified flow diagrams for single- and double-absorption sulphuric acid plants.

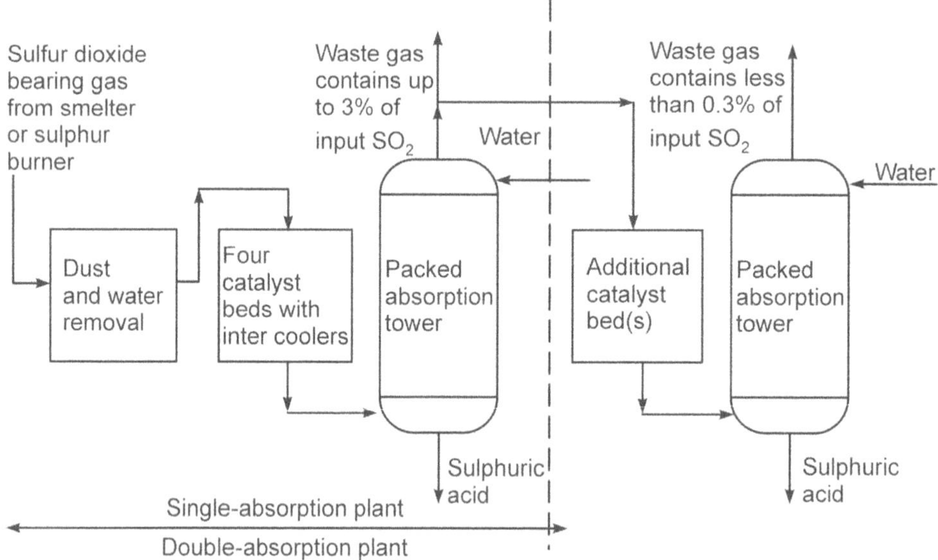

Fig. 7.3 *Single- and double-absorption plant for waste gas collection*

In addition to the above mentioned source correction methods, air pollutants emissions from industrial operations can also be reduced by proper equipment maintenance, good housekeeping practices and cleanliness in the facilities and premises. Often changes in the design of local exhaust hoods and proper installation can minimize the emission of pollutants to the atmosphere.

Chemical process plants often have excessive leakage around ducts, piping, valves and pumps. Many such leaks can be prevented by checking the seals and gaskets routinely. Floors, decks, storage bins and soils, loading areas, and material transfer conveyors must be kept clean to reduce dust pollution. Replacing open-hearth furnaces with controlled basic oxygen furnaces or electric furnaces can reduce smoke, carbon monoxide and metal fumes and, at the same time, conserve energy. Such changes, coupled with various combinations of gas cleaning devices, can be very effective in reducing air pollution. Smelting and paper industries are notorious for emissions of highly objection able sulphurous materials. These emissions are being curtailed by major process changes such as hydro-metallurgical separations of ores and avoiding the use of sulphides in papermaking.

Rotary kilns are a major source of dust generation in cement plants. Some degree of dust control may be achieved in the kiln operation by optimizing the operating conditions. In the steel industry; a radically different process has been proposed to lower sulphurous emissions during combustion. In this process the sulphur bearing fuel is mixed with limestone and air which are injected into a molten iron bath. The combustibles in the fuel are partially oxidized to carbon monoxide within the molten iron bath. Other examples involving process change include:

- Reduction of the formation of nitric oxides in combustion chambers by low excess air combustion in two stages, flue gas re-circulation and water injection.
- Washing the coal before pulverization to reduce the fly ash emissions and
- Substitution of bauxite flux for fluorine-containing fluorspar in the open hearth method.

7.2 Particulate Control Technologies

7.2.1 Principles of particulate control

In the most general terms, the control of particulate air pollutants involves passing the gas stream having the particles through a chamber and allowing forces to act on the particles to take them out of the gas stream. This method is dependent on several factors including the nature of the plant operation; whether it is cyclic or continuous, and whether the emission is likely to vary at different times of the day.

Particulate removal devices operate basically on the principle that a gas stream containing particles is passed through a region where the particles are acted on by external forces or caused to intercept obstacles, thereby separating them from the gas stream. When acted upon by external forces, the particles acquire a velocity component in a direction different from that of the gas stream. In- order to design a separation device based on particulate separation by external forces, one must be able to compute the motion of a particle under such circumstances.

A preliminary selection of suitable particulate emission control systems is generally based on knowledge of four items:

(i) particulate concentration in the stream to be cleaned,
(ii) the size distribution of the particles to be removed,
(iii) the gas flow rate, and
(iv) the final allowable particulate emission rate.

Once the systems that are capable of providing the required efficiencies at the given flow rates have been chosen, the ultimate selection is generally made on the basis of the total cost of construction and operation. The size of a collector, and therefore its cost, is directly proportional to the volumetric flow rate of gas that must be cleaned.

The operating factors that influence the cost of a device are:

(a) the pressure drop through the unit,
(b) the power required, and
(c) the quantity of liquid needed (if a wet scrubbing system).

7.2.2 Devices that remove particles

Devices that remove particles from gas streams rely on one or more of the following physical mechanisms:

1. Sedimentation. The particle-containing gas stream is introduced into a device or chamber where the particles settle under gravity to the floor of the chamber. Devices of this type are called settling chambers.
2. Migration of charged particle in an electric field. The particle-containing gas stream is introduced into a device in which the particles are charged and then subjected to an electric field. The resulting electrostatic force on the particles causes them to migrate to one of the surfaces of the device, where they are held and collected. Devices of this type are called electrostatic precipitators. Electrostatic precipitators can treat large volumetric flow rates of gas at relatively low pressure drops with very high removal efficiencies. However, electrostatic precipitators are expensive and are relatively inflexible to changes in process operating conditions
3. Inertial deposition. When a gas stream changes direction as it flows around an object in its path, suspended particles tend to keep moving in their original direction due to their inertia. Particulate collection devices based on this principle include cyclones, scrubbers, and filters.

   Fabric filters tend to have very high efficiencies but are expensive and are generally limited to dry, low-temperature conditions. Scrubbing can also achieve high efficiencies and offers the auxiliary advantage that gaseous pollutants can be removed simultaneously with particles. However, scrubbers can be expensive to operate, owing to their high pressure drop and to the fact that they produce a wet sludge that must be treated or disposed of.
4. Brownian diffusion. Particles suspended in a gas are always in Brownian motion. When the gas stream flows around obstacles, the natural random motion of the particles will bring them into contact with the obstacles, where they adhere and are collected. Because we know that Brownian motion is more pronounced the smaller the particle, we expect that devices based on diffusion as the separation mechanism will be most effective for small particles.

The key parameter that influences the choice of which device to employ in a particular case is the particle diameter Dp As we will see, the physical mechanisms above vary greatly in their effectiveness depending on the size of the particle.

Particles as large as sand grains are brought into direct physical contact; they generally will stick together by electrostatic and van der Waal's bonding forces.

Electrostatic and van der Waal's forces are, in general, proportional to the surface area of the particle. Most of the particles we are used to are large enough so that gravity or inertia will overcome electrostatic or van der Waal's forces; and unless they are wet, sand grains do not stick together. But gravity and inertia forces are proportional to the particle mass, which is proportional to $d^3{}_p$, whereas the surface area is proportional to $d^2{}_p$ (which is the diameter of the particle). Thus, as the particle size decreases, $d^3{}_p$ goes down much faster than $d^2{}_p$ so that the ratio of electrostatic and van der Waal's forces to inertia and gravity forces becomes larger. As a result, when particles of small size are brought into intimate contact, they would not fragment into individual particles but rather would aggregate into agglomerates that are the size of ordinary sand. With a microscope one can see that these agglomerates are really masses of much smaller particles held together by the forces described above.

For this reason, the basic strategy of control for particulate pollutants is to agglomerate them into larger particles that can be easily collected. This can be done by forcing the individual particles to contact each other or by contacting them with drops of water. The drag forces that the surrounding air or gas exerts on such particles can be understood from Stokes' law.

7.2.3 Stokes' law

Figure 7.4 shows the forces acting on a spherical particle settling through a fluid under the influence of gravity. Newton's law for the downward acceleration of the particle can be written as [2]

$$ma = \rho_{part}\left(\frac{\pi}{6}\right)d_p^3 g - \rho_{fluid}\left(\frac{\pi}{6}\right)d_p^3 g - F_d \quad \text{.....(7.3)}$$

where $\rho_{part,}$ and ρ_{fluid} are the density of the particle and air respectively. d_p = particle diameter in cm; g = acceleration due to gravity. *ma* represents the downward acceleration of the particle. The gravity, buoyant force, and drag force acting on the particle are given by the three terms on the right hand side of the equation. With increasing speed the drag force increases and becomes zero for zero speed. The particle accelerates rapidly; as it accelerates, the drag force increases as the velocity increases, until it equals the gravity force minus the buoyant force. At this terminal settling velocity, the sum of the forces acting is zero, so the particle continues to move at a constant velocity. To find this velocity, we set the acceleration to zero in Equation (7.3) which simplifies to

$$F_d = \left(\frac{\pi}{6}\right) d_p^3 g \left(\rho_{\text{part}} - \rho_{\text{fluid}}\right) \qquad .....(7.4)$$

which is the relation between the drag force F_d and velocity V.

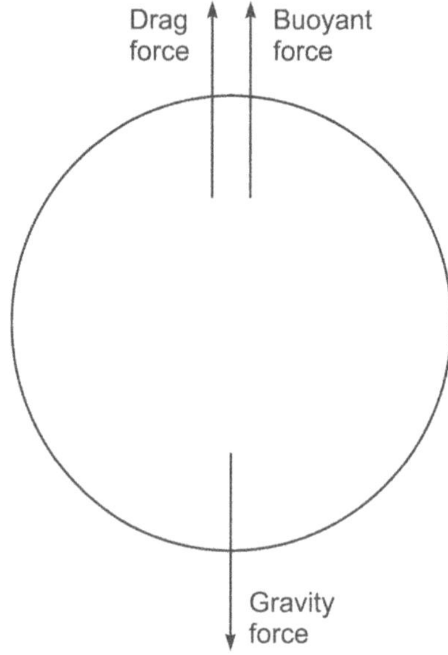

Fig. 7.4 *Forces acting on a spherical particle in a fluid*

The relation between drag force F_d and velocity V_t is also given by Stokes' Law as

$$F_d = 3\,\pi\mu\, d_p V_t \qquad .....(7.5)$$

where μ = the viscosity of the fluid. If we substitute Equation (7.4) into Equation (7.5) and solve for V_t, we find

$$V_t = \frac{g d_p^2 (\rho_{\text{part}} - \rho_{\text{fluid}})}{18\mu} \qquad .....(7.6)$$

which is commonly referred to as Stokes' Law which gives the terminal settling velocity of the particle in terms of the particle diameter, density of the fluid and viscosity of the medium.

When the application of Stokes' law was verified for a range of conditions, it was observed the Stokes' Law cannot be applied for very large and very small particles as then its assumptions break down. Fig. 7.5 which is a logarithmic plot of settling velocity in air as a function of particle diameter for spheres with a specific gravity of 2 depicts this behaviour of particles. For the range of values in which Stokes' law applies, the result is a straight line with slope 2 on log paper.

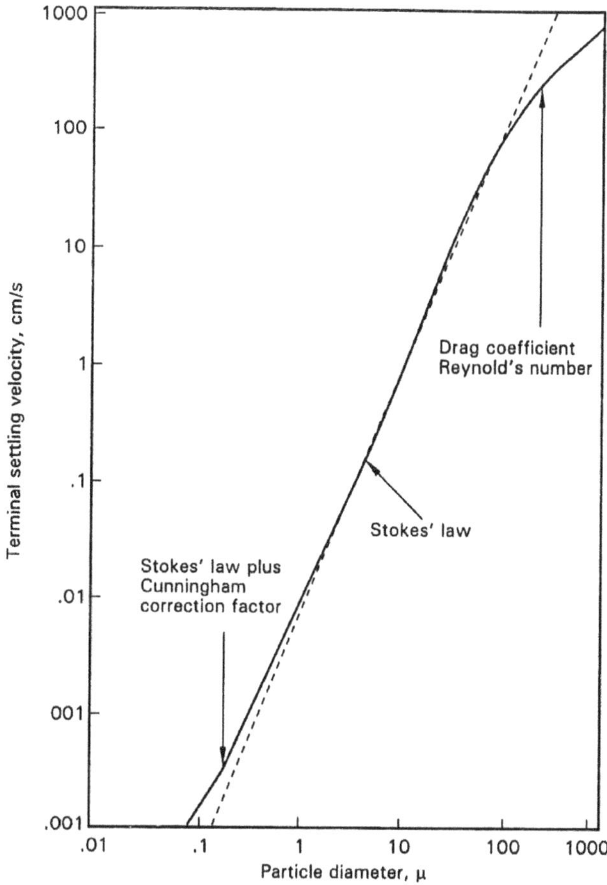

Fig. 7.5 *Relation between terminal settling velocity and particle diameter*

Problem 7.1

Estimate the terminal settling velocity in air of a particle with diameter 5μ. Given g = 9.81m/s^2, specific gravity of the particle = 2.0, density of air = 1.184 kg/m^3 at 25 °C, viscosity of air 1.85 × 10^{-5} kg/ms.

Solution: Substituting these values in Equation (7.4):

We have

$$V_t = \frac{(9.8 \text{ m/s}^2)(5\times10^{-6})\text{m}^2(200 \text{ kg/m}^3) - (1.184 \text{ kg/m}^3)}{(18)(1.85\times10^{-5}\text{kg / ms})}$$

$$= \frac{48.70992\times10^{-9}}{33.3\times10^{-5}} = 1462.76\times10^{-7}\text{ m / s} = 14.63\times10^{-5}\text{ m / s} = 0.01463 \text{ cm/s}$$

7.2.3.1 Particles too large for Stokes' law

For larger and larger particles, the flow of the fluid around the sphere does not obey the approximation that the terms involving velocities squared are negligible and the Stokes' drag equation, which is based on that assumption, no longer holds good. As the particle sizes further increase the flow of fluid around the particle becomes turbulent and the principal assumptions of Stokes' law will not be applicable.

For larger particles, the experimental data can be correlated by a non-dimensional relationship using a new parameter, called the drag coefficient C_d, as given in Equation (7.7)

$$C_d = \frac{F_d}{(\pi/4)d_p^2 \rho_{\text{fluid}}(v^2/2)} \quad \text{.....(7.7)}$$

Introducing Reynolds number (given by the Equation (7.8)) which is a dimensionless ratio of inertial forces to viscous forces acting on a mass of fluid in the same flow:

$$R_e = \frac{d_p V_t \rho_{\text{fluid}}}{\mu} \quad \text{.....(7.8)}$$

And substituting for F_d [Equation (7.5)] in Equation (7.7), we have

$$C_d = \frac{24}{R_p}$$

which is another form of Stokes' drag term.

It is observed experimentally that the Stokes' law is obeyed by particles whose Reynolds number is less than 0.3. For higher values of Reynolds number, for example 0.3 $< R_e >$ 1000, the following equation holds good.

$$C_d = \frac{24}{R_e}(1 + 0.14R_e^{0.7}) \quad \text{.....(7.9)}$$

7.2.3.2 Importance of drag coefficient for larger size particles

Problem 7.2

If a particle with diameter 200 μ is falling in air, using Stokes' law calculate its falling velocity and Reynolds number.

Solution: If the Stokes' law is applicable the velocity of the particle will be proportional to square of the diameter.

From Problem 7.1 we have the terminal settling velocity of 5 μ particle as 14.63 × 10^{-5}m/s.

So we have

$$V_t = 14.63 \times 10^{-5}\,\text{m/s} \times \left(\frac{200}{5}\right)^2 = 2.3408\ \text{m/s}$$

Substituting the value of V_t in the equation for Reynolds number Equation (7.8) we have

$$R_e = \frac{d_p V_t \rho_{fluid}}{\mu} = \frac{(200\times10^{-6}\text{m})(2.3408\text{ m/s})(1.20\text{ kg/m}^3)}{1.8\times10^{-5}\text{kg / ms}}$$

$$= \frac{56.16\times10^{-5}}{1.8\times10^{-5}} = 31.2$$

The value of R_e in the above problem is greater than 0.3 which indicates that Stokes' law cannot be applied reliably and we have to use the relation in which drag coefficient is used.

If we assume R has a value of 20, Equation (7.9) reduces to

$$C_d = 24/20(1+0.14\times20^{-0.7}) = 2.57 \qquad .....(7.10)$$

Solving Equation (7.6) for velocity by substituting Equation (7.8), we find

$$V_t = \left(\frac{F_d}{C_d \pi/8 d_p^2 \rho_{fluid}}\right)^{1/2} \qquad .....(7.11)$$

Here the terminal settling velocity is $F_d = \text{mg}$

$$= (\pi/6) d_p^3 \rho_{part} g \qquad .....(7.12)$$

So $$V_t = \left(\frac{4/3\, d_p \rho_{part} g}{C_d \rho_{fluid}}\right)^{1/2}$$

Substituting values of the particle's diameter, density and C_d, we will have

$$V_t = \left(\frac{4/3(200\times10^{-6}\text{m})(2000\text{ kg/m}^3)(9.81\text{ m/s}^2)}{(2.57)(1.20\text{ kg/m}^3)}\right)^{1/2} = 1.30\text{ m/s} \qquad .....(7.13)$$

Substituting the value of V_t in Equation (7.6) we get

$$R_e = \frac{d_p V_t \rho_{fluid}}{\mu} = \left(\frac{(200\times10^{-6}\text{m})(1.30\text{ m/s})(1.20\text{ kg/m}^3}{1.8\times10^{-5}\text{kg / ms}}\right) = 17.3 \qquad .....(7.14)$$

Substituting the values in Equation (7.7) we get

$C_d = 2.82$

Substituting the new value of C_d in Equation (7.11), we get

$$V_t = \left(\frac{4}{3}\frac{(200\times10^{-6}\text{m})(2000\text{ kg/m}^3)(9.81\text{ m/s}^2)}{(2.82)(1.20\text{ kg/m}^3)}\right) = 1.24\text{ m/s} \qquad .....(7.15)$$

Repeating these substitutions, we get R_e= 16.5 and C_d = 2.90 for which V_t = 1.22 m/s. As we get nearly the same value, we can stop the iteration at this stage taking V_t as1.22 m/s. This value calculated involving the drag coefficient for large particles agrees well with the experimentally observed value.

It can be seen that the velocity of the larger size particles (> 50 μ) is nearly twice the value obtained involving drag coefficient which agrees well with the experimental values.

Thus for particles larger than 50 μ, serious errors will result if V_t calculations are carried out using only Stokes' law and as such we have to use the relation based on experimental drag coefficient. It can been seen in Fig. 7.5, where the experimental settling velocities, labeled *Drag coefficient Reynolds number*, are significantly less than those shown by the extrapolated straight line, which represents Stokes' law.

Problem 7.3

Calculate the settling velocity of a fine silica particle (with a particle size of 2 μ) emitted from a talc factory at 25 °C, present in still air with specific gravity 2.0.

Solution: From Stokes' law we have

$$V_t = \frac{g(\rho_p - \rho_g)d_p^2}{18\mu}$$

where ρ_p = density of particle = 2000 kg/m^3

ρ_a = density of air = 1.184 kg/m^3 $\rho_p >> \rho_a$ (ρ_a can be neglected)

d_p = diameter of particle = 2×10^{-6}m

μ = viscosity of medium in which the particle is settling = 1.85 × 10^{-3}kg/ms

$$V_t = \frac{9.81 \text{ m/s}^2 \times 2000 \text{ kg/m}^3 \times (2\times10^{-6})^2 \text{m}^2}{18\times1.85\times10^{-5} \text{kg/ms}} = 2.378\times10^{-4} \text{m/s}$$

As Stokes' law is applicable the settling velocity = 2.378 ×10^{-4}m/sec.

Problem 7.4

A spherical particle with diameter 200 μ and specific gravity 4.0 is settling in still air.

(a) What is the terminal settling velocity of this particle according to Stokes' law?

(b) What is the terminal settling velocity of this particle, according to Stokes' law taking the Cunningham correction factor (2.57) into account?

Solution:

(a) $$V_t = \frac{9.81 \text{ m/s}^2 \times 4000 \text{ kg/m}^3 \times (200\times10^{-6})^2 \text{m}^2}{18\times1.85\times10^{-5} \text{ kg/ms}} = \frac{1.5696\times10^{-3}}{3.3310^{-4}} = 4.7 \text{ m/s}$$

(b)
$$V_t = \left(\frac{F_d}{C_d \pi / 8 d_p^2 \, \rho_{\text{fluid}}} \right)^{1/2}$$

$$= F_d = \pi / 6 \, d_p^3 \, \rho_{\text{fluid}}$$

So
$$V_t = \left(\frac{4/3 \, d_p \, \rho_{\text{part}} g}{C_d \, \rho_{\text{fluid}}} \right)^{1/2}$$

$$= \left(\frac{4/3 \times 200 \times 10^{-6} \text{m} \times 400 \text{ kg/m}^3 \times 9.81 \text{ m/s}^2}{2.57 \times 1.184 \text{ kg/m}^3} \right)^{1/2} = 0.583 \text{ m/s}$$

7.2.3.3 Particles too small for Stokes' law

Stokes' law assumes that the fluid in which the particle is moving is a continuous medium. However, real gases, liquids, and solids are not truly continuous but are made up of atoms and molecules. As long as the particle we are considering is much larger than the spaces between the individual gas molecules or atoms, the fluid interacts with the particle as if it were a continuous medium. When a particle becomes as small as possible or smaller than the average distance between molecules, then its interaction with molecules changes. When a particle has a large number of molecular collisions per unit time, most of the molecules bounce off the particle in a mirror-like way, with angle of reflection equaling the angle of incidence. If the number of collisions is small, then some significant fraction of the colliding gas molecules are adsorbed onto the surface of the particle and remain long enough to "forget" what direction they came from. In this case, their direction of leaving is diffuse, meaning random, subject to some statistical rules.

The effect of the change from specula to diffuse reflection is to lower the drag force, which causes the particle to move faster. The most widely used correction factor for this change is

$$F_d = \frac{F_{d-\text{stokes}}}{1 + A\lambda / d_p} \qquad \text{.....(7.16)}$$

where A = an experimentally determined constant

λ = mean free path (the average travel distance of a gas molecule between successive collisions)

$F_{d\text{-stokes}}$ = the drag force computed according to Stokes' law

The $[1 + A(\lambda/d_p)]$ term used here is commonly called the Cunningham correction factor. It is only applicable for values of λ/d_p of order of magnitude one. For much larger values of λ/d_p, more complex formulae are used [3]. Although formulae for the a precise calculation of λ are reasonably well-known, the A in Equation (7.16) is not a universal constant for all particles but varies from one kind of particle to another [4]. Most workers use the value found by Millikan for oil droplets settling in air, A = 1.728.

7.2.3.4 Importance of cunningham correction factor

Problem 7.5

Estimate the settling velocity of a small particle with diameter 0.1 μ and specific gravity 2.0 in still air.

Solution: By combining Equation (7.15) and (7.11) we have

$$V_t = V_{\text{Stokes}}(1 + A\lambda / d_p) \qquad .....(7.17)$$

where V_{Stokes} is the terminal settling velocity calculated from Equation (8.4)

$$V_{\text{Stokes}} = \frac{(9.81 \text{ m/s}^2)(10^{-7}\text{ m})^2(2000 \text{ kg/m}^3 - 1.2 \text{ kg/m}^3)}{(18)(1.85\times10^{-5}\text{kg/ms})} \qquad .....(7.18)$$

$$= \frac{19608.22}{33.3} = 588\times10^{-9}\text{ m/s} = 5.88\times10^{-7}\text{ m/s}$$

Here the mean free path depends on temperature *T*, pressure *P* and mass *M* of the medium. For air at one atmosphere pressure and room temperature, λ= 0.07 m and so the correction term is

$$1 + A\lambda/d_p = 1 + 1.728\ (0.07\ \mu/0.1\ \mu) = 2.21 \qquad .....(7.19)$$

So $V_t = 5.88 \times 10^{-7}$ m/s (2.21) = 1.29×10^{-6} m/s .....(7.20)

Table 7.1 *Extent of errors in the application of simple Stokes' law and Stokes' law with Cunningham correction factor [3]*

Formula	Permissible error	
	1%	10%
Stokes' law: Equation (8.4)	$16 < d_p < 30$ μ	$1.6 < d_p < 70$ μ
Stokes' law with simple form of Cunningham correction: Equation (8.9)	$0.36 < d_p < 30$ μ	0.1 m $< d_p < 70$ μ

Thus it is evident that for small particles, the estimated settling velocity will be 2.21times faster than the value obtained from Stokes' law. However, the estimated values are only rough estimates as it involves the value of A which is uncertain. However we can infer that such smaller particles will have a smaller drag force. So they should be settled by electrostatic force in the settling chamber. Table 7.1 gives the extent of errors in the application of Stokes' law and Stokes' law with Cunningham correction factor for various particle sizes as suggested by Fuchins.

7.2.4 Collection efficiency

The definition of overall efficiency above is based on particle number. We can also define overall efficiencies based on other particle properties, such as surface area and volume (or mass). For example, the collection efficiency based on particle mass η_{in} is defined as

We define the *collection efficiency* $\eta(D_p)$ of a device for particles of diameter D_p as

$$\eta(D_p) = 1 - \frac{\text{number of particles of diameter } D_p \text{ per m}^3 \text{ of gas out}}{\text{number of particles of diameter } D_p \text{ per m}^3 \text{ of gas in}} \qquad \text{.....(7.21)}$$

The *overall efficiency* of the device based on particle number is

$$\eta = 1 - \frac{\text{number of particles per m}^3 \text{ of gas out}}{\text{number of particles per m}^3 \text{ of gas in}} \qquad \text{.....(7.22)}$$

These efficiencies can be expressed in terms of the particle size distribution functions at the inlet and outlet sides of the device.

$$\eta(D_p) = \frac{n_{in}(D_p)dD_p - n_{out}(D_p)dD_p}{n_{in}(D_p)dD_p}$$

$$= 1 - \frac{n_{out}(D_p)}{n_{in}(D_p)} \qquad \text{.....(7.23)}$$

and

$$\text{h} = \frac{\int_0^{\infty}\left[n_{in}(D_p) - n_{out}(D_p)\right]dD_p}{\int_0^{\infty} n_{in}(D_p)dD_p}$$

$$= \frac{\int_0^{\infty}\eta(D_p)n_{in}(D_p)dD_p}{\int_0^{\infty} n_{in}(D_p)dD_p} \qquad \text{.....(7.24)}$$

$$\eta_{in}(D_p) - 1 - \frac{\text{number of particles of diameter } D_p \text{ per m}^3 \text{ of gas out}}{\text{number of particles of diameter } D_p \text{ per m}^3 \text{ of gas in}} \qquad \text{.....(7.25)}$$

and the overall efficiency is

$$\eta_{in} = \frac{\int_0^{\infty}\left[(\pi/6)\rho_p D_p^3 n_{in}(D_p) - (\pi/6)\rho_p D_p^3 n_{out}(D_p)\right]dD_p}{\int_0^{\infty}(\pi/6)\rho_p D_p^3 n_{in}(D_p)dD_p}$$

$$= \frac{\int_0^{\infty}\left[\eta(D_p)D_p^3 n_{in} - (D_p)\right]dD_p}{\int_0^{\infty} D_p^3 n_{in}(D_p)dD_p} \qquad \text{.....(7.26)}$$

The overall collection efficiency by mass is usually the easiest to measure experimentally. The inlet and outlet streams may be sampled by a collection device, such as a filter, that collects virtually all of the particles.

A term that is sometimes used to express collection efficiency is the *penetration*. The penetration is based on the amount emitted rather than captured; penetration based on particle mass is just $P_m = 1 - \eta_m$. Alternatively, the penetration can be defined on the basis of particle number, $P = 1 - \eta$.

We have called the relationship between collection efficiency and particle size simply the collection efficiency. Other terms that are used for this quantity are the *grade efficiency* or the *fractional efficiency*. An important point on the collection efficiency curves is the size for which h = 0.5. The particle size at this point is called the *cut size* or the *cut diameter*.

7.3 Particulate Control Equipment: Types of Equipment in Use

As stated in the beginning there are several different classes of particulate control equipment that operate based on different mechanisms for different types of particulates.

The selection of a specific particulate control system for cleaning industrial effluent gases depends on the nature of effluent gaseous discharge, the extent of purification required and gaseous discharge characteristics which affect the performance of the purification system. Particle sizes range generally from 100 µm down to 0.1 µm and even less. To remove particulate matter from gas streams, various types of control equipment are available which can be selected based on the following information:

- Nature of the gas phase and particulate matter
- The physical and chemical characteristics of particulates
- The particulate size and concentration in the gas
- Volume of particulates to be handled
- Temperature and humidity of gaseous medium
- Factors like toxicity and inflammability must be taken into consideration when evaluating column efficiency
- Quality of the treated effluent, i.e. efficiency of the removal of particulate matter

The simplest particulate control device is a **settling chamber**, a large chamber in which the gas velocity is slowed, allowing the particles to settle out by gravity.

A cyclone operates by causing the entire gas stream to flow in a spiral pattern inside a tapered tube. Because of the centrifugal force, particles migrate outward and collect on the wall of the tube. The particles slide down the wall and fall to the bottom, where they are removed. The clean gas generally reverses its flow and exits out of the top of the cyclone.

An electrostatic precipitator utilizes the electrostatic force on charged particles in an electric field to separate particles from the gas stream. A high voltage drop is established between two electrodes, and particles passing through the resulting electric field acquire charge. The charged particles migrate to and are collected on an oppositely charged plate while the clean gas flows on through the device. Periodically, the plates are cleaned by rapping to shake off the layer of dust that has accumulated.

A variety of **filters** operate on the principle that the particulate-laden gas is forced through an assemblage of collecting elements, such as a fiber or a filter mat. As the gas passes through the assemblage, particles accumulate on the collectors.

Wet collection devices called **scrubbers** operate on the basis of the collision of particles with droplets of water that can easily be separated from the gas because of their large size.

Some general statements can be made about the nature of the various types of particulate gas-cleaning equipment. Mechanical collectors such as settling chambers or cyclones are typically much less expensive than the others but are generally only moderately efficient in particle removal. Since they are much better for large particles than for small ones, they often are used as pre cleaners for the more efficient final control

The sizes of particles removed by various particulate control devices and their efficiency are listed in Table 7.2.

Table 7.2 *Sizes of particles removed by various particulate control devices and their efficiency are listed below*

Device	Minimum particle size (μ)	Efficiency (%)
Venturi	> 0.5	< 99
ESP	> 1	< 99
Bag filter	< 1	> 99
Spray chamber	> 10	< 80
Cyclonic spray chamber	> 3	< 80
Impingement scrubber	> 3	< 80
Centrifugal settler	> 5	< 80
Gravitational settler	> 50	< 50

7.3.1 Gravitational settling chambers

7.3.1.1 Gravitational settling

It is perhaps the most obvious means of separating particles from a flowing gas stream. A settling chamber is, in principle, simply a large box through which the effluent gas stream flows and in which particles in the stream settle to the floor by gravity. Gas velocities through a settling chamber must be kept low enough so that settling particles are not re-entrained. The gas velocity is usually reduced by expanding the ducting into a chamber large enough so that sufficiently low velocities result.

Although in principle settling chambers could be used to remove even the smallest particles, practical limitations in the length of such chambers restrict their applicability to the removal of particles larger than about 50μ m. Many times settling chambers are used as pre -cleaners to remove large and possibly abrasive particles, prior to passing the gas stream through other collection devices.

Settling chambers offer the advantages of

(i) simple construction and low cost,

(ii) small pressure drops, and

(iii) collection of particles without need for water.

The main disadvantage of settling chambers is the large space that they require.

A settling chamber is, as noted above, simply a horizontal chamber through which the particle-laden gas flows and to the floor of which the particles settle.

A gravity settler is simply a long chamber through which the contaminated gas passes slowly, allowing time for the particles to settle by gravity to the bottom. Solid or liquid particles suspended in the gas reach a terminal free falling velocity which is given by Stokes' law for small particles, and is proportional to the product of the square of the particle diameter and the density difference between the particle and the carrier gas, and inversely proportional to the viscosity of the carrier gas as discussed in detail earlier. In practice, this implies that the method can only by applied for removing particles greater than 100 μm in diameter. As most of the troublesome particles have sizes smaller than 50 μm, these devices are normally used as pre-cleaners prior to passing the gas stream through high efficiency collection devices.

The cross-sectional area of (WH) of the settling chamber is much larger than that of the duct through which the pollutant gas stream enters this settling chamber. The gas stream, containing the particulate matter is allowed to flow at a low velocity in the settling chamber allowing sufficient time for the particles to settle down (Fig. 7.6).

Simple and horizontal flow gravity settlers [Figs. 7.6(a) and 7.6(b)], which have a cross-sectional area much larger than that of the duct, bring the dust laden gas stream in to the settling chamber.

The gravitational settling chambers usually operate with velocity between 0.5 and 2.5 m/s, although for best operating results, the gas flow should be uniformly maintained at less than 0.3 m/s. Some settling chambers have simply enlarged conduits and some have horizontal shelves and baffles, about 2.5 cm apart. The collection efficiency of the chamber depends upon the type of flow also. For both laminar and turbulent flows, the design considerations are discussed in the following pages.

7.3.1.2 Collection efficiency

In analyzing the performance of a settling chamber, the key feature is the nature of the gas flow through the device. We can distinguish three basic idealized flow situations:

1. laminar flow,

2. plug flow (velocity uniform across the cross section) with no vertical mixing of particles,
3. plug flow with complete vertical mixing of particles

Laminar flow is characterized by a parabolic-type velocity profile; such a flow would only be realized for Reynolds numbers below that for transition to turbulent flow.

In a laminar flow, the time required for a particle at height y above the floor of the chamber to settle is y/vt, where is the particle's setting velocity, and vertical mixing of particles is absent in laminar flow. (The effect of Brownian motion is generally neglected relative to the steady downward movement due to settling.)

The second flow category above, plug flow with no vertical mixing of particles, is, in a sense, an approximation to laminar flow in that vertical mixing of particles is still ignored, but a flat velocity profile is assumed and the particles all settle at their settling velocities.

The third category, plug flow with thorough vertical mixing, is the model for turbulent flow. In a turbulent flow settling chamber the gas velocity is assumed to be uniform across the chamber due to the turbulent mixing. Moreover, the turbulent mixing in the core of the chamber overwhelms the tendency of the particles to settle and maintains a uniform particle concentration vertically across the chamber. Removal by settling can be assumed to occur in a thin layer at the bottom of the chamber.

The collection efficiency of any particulate collection system represents the actual performance, and it is expressed in terms of weight of the material caught or retained by the collector compared to the total amount entering the collector. This is also known as **gravimetric efficiency** and is usually expressed as a percentage.

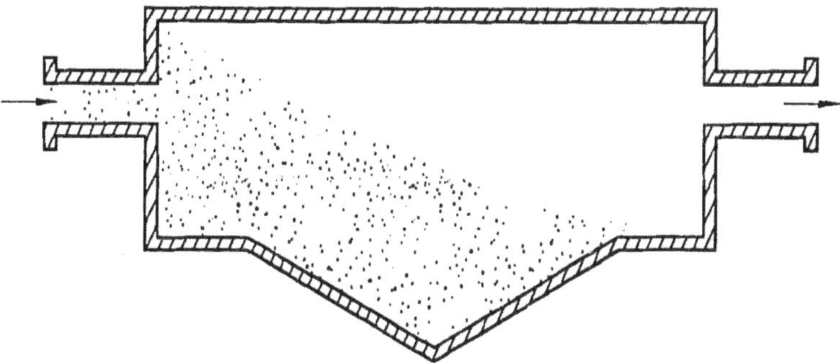

Fig. 7.6(a) *Simple settling chamber*

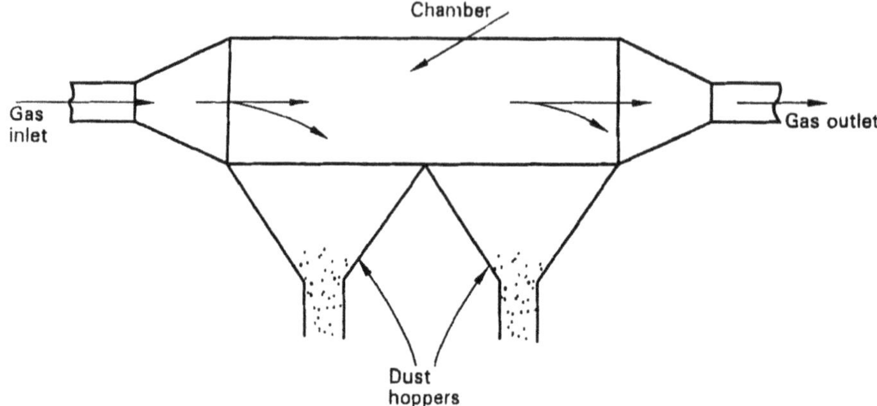

Fig. 7.6(b) *Horizontal flow settling chamber*

$$\eta_T = 100 \times \frac{\text{Weight of material collected}}{\text{Total amount entering collector}} \qquad \text{.....(7.27)}$$

For any particulate system the efficiency varies with particle size and is expressed as fractional efficiency. Fractional efficiency is the efficiency of the collector for a particular particle size for given conditions like flow rate, temperature and particle composition.

7.3.1.3 Fractional efficiency

The fractional efficiency is given by

$$\eta_T = 100 \times \frac{\text{Fraction in range } i \text{ collected}}{m_i} \times 100 \qquad \text{.....(7.28)}$$

where m_i = amount in range *i* entering the collector.

The overall efficiency η_T can then be calculated over *n* number of size fractions as

$$\eta_T = \frac{\sum_{i=1}^{n} m_i n_i}{M} \text{percent} \qquad \text{.....(7.29)}$$

where *M* = total amount entering the collector.

Problem 7.6

A gas cleaning equipment has the following fractional efficiencies for various sizes < 10 μ; weight fraction, *w* = 0.02, efficiency 16%; 10 – 30 μ; *w* = 0.3, efficiency 52%; 30 – 50 μ; *w* = 0.4, efficiency 92%; > 50 μ; *w* = 0.2, efficiency 98%; > 20 μ: *w* = 0.16, efficiency 100%. Calculate its overall collection efficiency.

Solution: $\eta_T = \sum_{i=1}^{n} m_i n_i = \sum_{i=1}^{n} m_i \mu_i$

$$= (0.02 \times 16) + (0.3 \times 52) + (0.4 \times 92) + (0.2 \times 98) + (0.16 \times 100)$$

$$= 0.32 + 15.6 + 36.8 + 19.6 + 16 = 88.32\%.$$

So the overall collection efficiency of the device is 88.32%.

7.3.1.4 Design considerations for gravitational settling chamber

In air pollution engineering the behaviour of the gravity settler can be modeled with two types of conditions.

(a) Block flow or plug flow model and

(b) Back mixed or mixed model

Block flow or plug flow model

1. In this model the gas fluid entering the settling chamber is assumed to be totally unmixed.
2. The horizontal velocity of the gas in the settling chamber is equal to V_{avg} everywhere in the chamber.
3. The horizontal component of the velocity of the particles in the gas is always equal to V_{avg}.
4. The vertical component of the velocity of the particles is equal to their terminal settling velocity due to gravity V_t.
5. The particle after settling down will not be entrained.

Back mixed or mixed model

In this model it is assumed that there is total mixing in the entire device or in its entire cross-section perpendicular to the flow. Each of the above models with assumptions leads to simple calculations explaining the behaviour of the gravity settler. In actual practice the observed behaviour of a variety of gravitational settling chambers falls within these two models and so they can be set as limits between which any gravitational settler works.

The average horizontal gas velocity in a settling chamber for either block flow or mixed flow can be written as

$$V_{avg} = \frac{Q}{WH} \qquad .....(7.30)$$

where Q is the volumetric flow rate, W is the width and H is the height of the chamber.

Block flow

Assuming block flow, consider a particle entering the settling chamber at height H from the bottom. The time required for the particle to travel the length of the chamber is

$$t = \frac{L}{V_{avg}} \quad .....(7.31)$$

During this travel time the particle will settle by gravity at a distance given by

$$\text{Vertical settling distance} = tV_t = V_t \frac{L}{V_{avg}} \quad .....(7.32)$$

If this vertical settling distance is greater or equal to H then the particle will settle at the bottom of the settling chamber. If all the particles are of the same size, then there will be some distance above the bottom of the settling chamber below which all the particles will settle down and above which more of them will be collected.

If we assume uniform distribution of all the particles at the inlet and they do not interact with each other then the fractional collection efficiency is given by

$$\text{Fractional efficiency } \eta = \frac{LV_t}{HV_{avg}} \quad .....(7.33)$$

By substituting Stokes' law (Equation (7.4)) for V_t and ignoring the air density, it being very small compared to particle density, we have

$$\eta = \frac{L_g d_p^2 \rho_{part}}{HV_{avg} 18\mu} \quad .....(7.34)$$

Mixed flow

In the mixed flow model, it is assumed that the gas flow is totally mixed in z direction also. Consider a section of a settling chamber with length dx. If V_t is the vertical distance which an average particle falls due to gravity in passing through the section and H is the height of the section, then [5]

$$\text{Fractional collection efficiency, } \eta = V_t dt/H \quad .....(7.35)$$

The change in concentration as the gas passes through this section is

$$dc = -c\,(\text{fractional collection}) = -cV_t\, dt/H \quad .....(7.36)$$

The time taken by particle to pass through this section is

$$dt = \frac{dx}{V_{avg}} \quad .....(7.37)$$

Combining Equations (7.36) and (7.37), we have

$$\frac{dc}{c} = \frac{V_t}{HV_{avg}} dx \quad .....(7.38)$$

which can be integrated within the limits, $x = 0$ at the inlet and $x = L$ at the outlet.

$$ln\left(\frac{c_{out}}{c_{in}}\right) = \frac{V_t L}{HV_{avg}} \quad \text{.....(7.39)}$$

$$\eta = 1 - \left(\frac{c_{out}}{c_{in}}\right) = 1 - \exp\left(\frac{V_t L}{HV_{avg}}\right) \quad \text{.....(7.40)}$$

Substituting Stokes' law, Equation (7.6), for V_t we have

$$\eta = 1 - \exp\frac{L_p d_p^2 \rho_{part} g}{HV_{avg} 18\mu} \quad \text{.....(7.41)}$$

Substituting Equation (7.34) in Equation (7.41), we have

$$\eta_{mixed} = 1 - \exp(-\eta_{block\ flow}) \quad \text{.....(7.42)}$$

Problem 7.7

A gravity settler has the following dimensions, H = 2 m, L = 10 m and V_{avg} = 1 m/s. Assuming Stokes' law is applicable for both block and mixed flows, estimate the relation between the particle diameter and efficiency of collection.

Solution: For a particle size 5 μ, from Equation (7.39) for block flow we have

$$\eta = \frac{Lgd_p^2 \rho_p}{HV_{avg} 18\mu}$$

$$= \frac{(10\ \text{m})(9.81\ \text{m/s}^2)(5\times10^{-6}\text{m})^2(2000\ \text{kg/m}^3)}{18(1.8\times10^{-6}\text{kg / ms})(2\ \text{m})1\ \text{m/s}} = 15.15\times10^{-4} \text{ or } 1.515\times10^{-3}$$

For mixed flow, from Equation (7.40), we have

$$\eta_{mixed} = 1 - \exp(-1.515\times10^{-3}) = 2\times10^{-3}$$

For particles of other diameters we can calculate using the relation that the block efficiency is proportional to the square of the particle diameter. These values for different particle sizes are presented in Table 7.3.

Table 7.3 *Collection efficiency for different particle sizes*

Particle diameter, μ	η(block flow)	η(mixed flow)
5	0.001515	0.000303
10	0.0303	0.0298
30	0.273	0.239
50	0.26	0.53
57.45	1.0	0.63
80	–	0.86
100	–	0.95
120	–	0.99

As can be seem from Table 7.3, the estimated collection efficiencies for smaller particles are equal for both block flow and mixed flow systems while they deviate for larger particles. Block flow always gives values higher than the mixed flow. The block flow gives 100% efficiency for particle sizes of 57 μ while for mixed flow the efficiency will increase nearly to 100% for particles of size 100 μ

Thus gravitational settling chambers will be useful for collecting particles greater than 100 μ and so can be used as pre-cleaners before using any high efficiency cleaning systems.

For increasing the efficiency of settling chambers for smaller particles, we have to increase L or make H smaller or lower V_{avg} which requires larger cross-sectional areas as per Equation (7.34). To achieve greater efficiency by decreasing H, Howard type of settling chambers are designed (Fig. 7.7). In Howard settling chambers, horizontal trays or shelves are fixed in the chambers at about 1 to 3 cm height intervals. The increase in the efficiency of the Howard type gravitational chamber is proportional to number of trays as can be understood by the following equations.

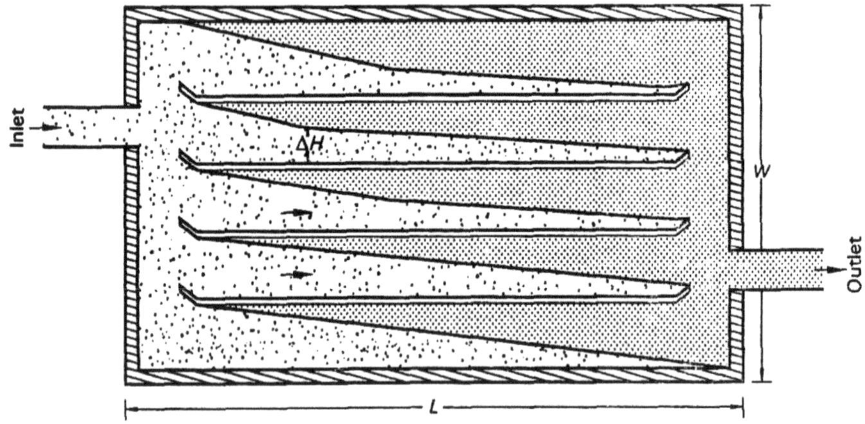

Fig. 7.7 *Howard settling chamber with five trays*

The hydraulic diameter for the air flowing between the trays of a settling chamber having n number of trays in a settling chamber of length L, width W and height H is:

$$D_h = \frac{2W\Delta H}{W + \Delta H} \qquad .....(7.43)$$

The relation between the Reynolds number and hydraulic diameter is given by

$$R_e = \frac{V_h D_h \rho}{\mu} \qquad .....(7.44)$$

where V_h, the velocity inside the chamber, is given by

$$V_h = \frac{Q}{nW\Delta H} \quad .....(7.45)$$

where Q = volumetric flow rate of the gas stream.

Substituting for V_h and D_h in the Equation (7.44), we get

$$R_e = \frac{2Q\rho}{n\mu(W + \Delta H)} \quad .....(7.46)$$

The spacing between the trays, ΔH, is given by

$$\Delta H = \frac{H}{n} \quad .....(7.47)$$

Provided there is no dust layer initially present on the tray surface, substituting for ΔH in Equation (7.47), we have

$$R_e = \frac{2Q\rho}{\mu(nW + H)} \quad .....(7.48)$$

If a layer of dust H_d is initially present, then

$$\Delta H = \frac{H}{n} - H_d \quad .....(7.49)$$

With this modification,

$$R_e = \frac{2Q\rho}{\mu(nW + H - nH_d)} \quad .....(7.50)$$

Assuming that V_h = V_t when the particle with diameter d_p attains 100% settling efficiency. We can then substitute

$$V_h = \frac{Q}{WL} \quad .....(7.51)$$

For V_t in Stokes' law (Equation (7.11)) we set

$$V_t = \frac{4}{3}\left(\frac{d_p\rho_{\text{part}}g}{C_d\rho_{\text{fluid}}}\right)^{1/2} \quad .....(7.52)$$

and solving for dp we have

$$d_p = \frac{0.75}{g}\frac{C_dQ^2\rho_a}{L^2W^2\rho_p} \quad .....(7.53)$$

d_p is now the diameter of the spherical particle that can be removed 100%. This diameter corresponds to v_0, the settling velocity for 100% removal. Particles of other diameters less than d will be removed in the ratio of their terminal settling velocities. Expressing Q in terms of the flow-through velocity v_h, depth H, and W yields

$$d_p = \frac{0.75}{g}\frac{C_d v_h^2 H^2 \rho_a}{L^2 \rho_p} \qquad \text{.....(7.54)}$$

From fluid dynamics, the fractional removal of solids R is given by

$$R = 1 - x_0 + \int_0^{x_0} \frac{v_p}{v_0} dx \qquad \text{.....(7.55)}$$

where x_0 is the fraction of the particulates of settling velocity equal to or less than v_0, and V_p is the settling velocity of the particles of diameter less than d_p. The formulas above apply to spherical particles only. For non spherical particles, the volume shape factor β should be used.

A procedure for determining β is as follows: Soak a sample of particulates to saturation with water and record the resulting volume and weight. Evaporate the water and dry to constant weight. The loss in weight represents the volume of water in the voids; hence the average volume of a particulate v_p is equal to the volume of the soaked sample minus the volume of water evaporated (represented by the loss in weight) divided by the number of particles. The number of particles must have to be counted under a microscope using a particulate counter. Since the volume of the particle is also equal to the volume of the equivalent spherical particle,

$$V_p = \frac{4\pi}{3}\left(\frac{d}{2}\right)^3 = \beta\, d_p^3 \qquad \text{.....(7.56)}$$

where d is the average diameter of the equivalent spherical particle and d_p is the average "diameter" of the particle. From Equation (7.56), β may be obtained.

For calculation of the collection efficiency of the settling chamber, the above design concept is useful. However, several factors hinder settling at high particle concentrations non-uniform gas velocity over the settling height and width, particle re-entrainment and turbulence cause deviations. As a general rule, chamber velocities below 3 m/s are satisfactory for avoiding re-entrainment of most of the materials.

Problem 7.8

A sample of particulates of average diameter 80 μm is soaked with water in a graduated cylinder to the 1 mL mark. The loss in weight after evaporating to dryness is 0.02 g. Determine β if the number of particles in the sample is 6.0 (10^6).

Solution:

$$0.02 \text{ g} = 0.02 \text{ cc}$$

$$V_p = \frac{1.0-0.01}{5.0(10^6)} = 2.0\ (10^{-7})\text{cm}^3$$

$$V_p = \frac{1.0-0.02}{6.0\times10^6} = \frac{0.98}{6.0\times10^6} = 1.6\times10^{-7}\,\text{cm}^3$$

$$\beta = \frac{1.6(10^{-7})}{\left[80(10^{-6})(10^2)\right]^3} = 0.3125$$

Problem 7.9

What is the terminal settling velocity of the particle that is removed 100% in a settling chamber from the dust released from an industry if the horizontal velocity is 0.3 m/s, the temperature is 77 °C, the specific gravity of the particle is 2.0 and the chamber length and depth equal 7.5 m and 1.5 m, respectively?

Solution:

$$d = 1.24\beta^{0.333} d_p = \frac{0.75}{g}\frac{C_d v_h^2 H^2 \rho_a}{L^2 \rho_p}$$

$$d_p = \frac{1}{1.24(0.90^{0.333})}\left(\frac{0.75}{9.81}\right)\left[\frac{C_d(0.3^2)(1.5^2)(1.2)}{7.5^2(2.0)(1000)}\right] = 1.37(10^{-7})C_d$$

ρ_a= 1.2 kg/m^3 and μ= 2.1(10^{-5}) kg/ms at 77 °C

$$V_t = \sqrt{\frac{4}{3}g\frac{\rho_p d}{C_d \rho_a}} = \sqrt{\frac{4}{3}(9.81)\frac{(2.0)(1000)(1.24)(0.9^{0.333})d_p}{C_d}} = 161.56\sqrt{\frac{d_p}{C_d}}$$

$$V_t = 161.56\sqrt{1.37(10^{-7})} = 0.060 \text{ m/s}$$

Problem 7.10

In a cement factory a settling chamber with the following dimensions was designed to remove particulate matter from an effluent gas for a flow rate of 8 m^{3}s at 75 °C (75 × 2.1 × 10^{-5}kg/min/sec).

Width W = 6 m, Length L = 10 m, Height H = 2 m, Specific gravity = 2. Estimate the cut size of the particle that will be removed with 100% efficiency for a block flow model.

Solution:

As per Stokes' law

$$\text{Terminal settling velocity } V_t = \frac{g(\rho_p-\rho_g)d_p^2}{18\mu} = d_p^2 = \frac{V_t\times18\mu}{g(\rho_p-\rho_g)}$$

Horizontal velocity $V_h = Q/W \times W$, where Q = volumetric flow, W = width, H = height of the gravitational settling chamber. So

$$V_h = \frac{8 \text{ m}^3/\text{s}}{6 \text{ m} \times 2 \text{ m}} = 0.66 \text{ m/s}$$

as $$\frac{L}{V_h} = \frac{H}{V_t}$$

$$V_t = \frac{HV_h}{L}$$

$$d_p^2 = \frac{HV_h \times 18\mu}{Lg\left(\rho_p\right)}$$

We can write Stokes' law, Equation (7.4) as

$$d_p^2 = \frac{18V_h H_\mu}{gL(\rho_p - \rho_g)}$$

Ignoring p_g, it being very very small when compared to ρ_p, we have

$$d_p^2 = \frac{18 \times 0.66 \text{ m/s} \times 6.0 \text{ m} \times 1.8 \times 10^{-5} \text{kg / ms}}{9.81 \text{ m/s}^2 \times 10 \text{ m} \times 2000 \text{ kg/m}^3} = \frac{24.9 \times 10^{-5}}{2.0 \times 10^3}$$

$$= \frac{129.0 \times 10^{-5} \text{kgm}^3 / \text{s}^2}{196200 \text{ kg/s}^2\text{m}}$$

$$d_p^2 = 6.57 \times 10^{-8} \text{m}^2$$

$$d_p = 8.06 \times 10^{-5} \text{m}$$

Problem 7.11

In a talcum factory particulate matter (specific gravity 2.0) is emitted along with flue gases (bulk gas velocity of 1 m/s). For achieving 100% efficiency for 80 μ size particulate matter with a settling chamber of height 3 m, estimate the length of the gravity settler. What will be the length if 4 trays are used for achieving the same efficiency?

Solution: By Equation (7.32), the efficiency of the gravity settler is given by

$$\eta = \frac{L \times V_t}{H \times V_{\text{avg}}}$$

From Stokes' law

$$\text{Terminal settling velocity } V_t = \frac{g(\rho_p - \rho_g)d_p^2}{18\mu}$$

$H = 3$ m

$V_{avg} = 1$ m/sec

$$V_t = \frac{9.81 \text{ m/s}^2 \times 2000 \text{ kg/m}^3 \times (80\times 10^{-6})^2 \text{m}^2}{18\times 1.85\times 10^{-5} \text{ kg/ms}} = 0.377 \text{ m/s}$$

$$L = \frac{\eta \times HV_{avg}}{V_t} = \frac{1.0\times 3\times 1}{0.377} = 7.90 \text{ m}$$

If four trays are installed, considering the bottom floor as one tray, $\Delta H = 4/3$ m =1.33m.

The length of the chamber with four trays will be $L/4 = 2.22$ m.

Problem 7.12

In a cement factory a settling chamber was installed as a pre-cleaner with 6 trays including the bottom surface with a spacing in between of 0.3 m. The dimensions of the settling chamber are Width $W = 2$ m, Length $L = 6$ m for an effluent gas velocity of 6m^3/s at 75°C ($\mu = 2.1 \times 10^{-5}$kg/ms). Estimate the collection efficiency for particles of 50μ and 125μ for laminar flow and mixed flow. The specific gravity of the particulate matter is 2. Estimate the residence time and the distance settled for each particle.

Solution:

$Q = 6$ m^3/sec, $n = 6$ trays $\Delta H = 0.3$m, $L = 6$m, $W = 2$m, $\rho_p = 2000$kg/m^3.

For a particle size of 125 μ, horizontal velocity inside the chamber

$$V_h = \frac{Q}{nW\Delta H} = \frac{6 \text{ m}^3/\text{s}}{6\times 2 \text{ m}\times 0.3 \text{ m}} = 1.666 \text{ m/s}$$

$$\text{Residence time } \frac{6 \text{ m}}{1.66 \text{ m/s}} = 3.615 \text{ s}$$

For a laminar flow condition the particle settles by a distance y which is related to settling velocity by the equation

$$\frac{y}{V_t} = \frac{L}{V_{avg}}$$

By Stokes' law

$$\text{Terminal settling velocity } V_t = \frac{g(\rho_p - \rho_g)d_p^2}{18\mu}$$

$$V_t = \frac{9.81\ \text{m/s}^2 \times 2000\ \text{kg/m}^3 \times (125\times10^{-6})^2\ \text{m}^2}{18\times1.85\times10^{-5}\ \text{kg/ms}} = \frac{3.06\times10^{-4}}{33.3\times10^{-5}} = 0.918\ \text{m/s}$$

$$y = \frac{L}{V_h}\times V_t = \frac{6.0\times0.918\ \text{m/s}}{1.666\ \text{m/s}} = 3.30\ \text{m}$$

Efficiency of collection is

$$\eta = \frac{L\times V_s}{V_h\times\Delta H} = \frac{6\times0.918}{1.66\times0.3} = \frac{5.508}{0.498} = 11.0 \text{ or } 89\%$$

For mixed flow, the collection efficiency for the particles can be calculated as

$$\eta = 1-\exp\frac{(V_t L)}{V_{\text{avg}}} = \frac{0.918\times6}{1.666} = \frac{5.508}{1.666}\left[1-\exp(3.31)\right] = 1-27.38 = 26.38$$

7.4 Cyclone Separators and Electrostatic Precipitators

7.4.1 Cyclone separators

Cyclone separators are gas cleaning devices that utilize the centrifugal force created by a spinning gas stream to separate particles from a gas. In a cyclone the particulate laden gas is accelerated through a spiral motion which imparts centrifugal force to the particles. The centrifugal force forces out the particles from the spinning gas and makes them fall on the wall of the cyclone. The particles will finally slide down.

The cyclone separator sketched in Fig. 7.8 is merely a gravity settler made in the form of two concentric helices and utilizes the centrifugal force generated by the spinning gas to separate the particulate matter from the polluted gas stream.

A simple cyclone collector consists of a cylinder with a tangential inlet and an inverted cone attached to the base. Gas enters the cyclone through the tangential in let which imparts a whirling motion to the gas. Suspended particles will be forced towards the wall on which they collect and slide down into the conical collector. Near the bottom of the cone, the gas turns abruptly upward and forms an inner spiral, which leaves through the pipe or duct extending into the center of the cyclone body.

There are three types of cyclones which are in industrial use. They are

1. high-throughput cyclone
2. conventional cyclone
3. high-efficiency cyclone.

High-through put cyclones process large volumes of waste air as input but operate at low efficiencies. The conventional cyclones are in between high-throughput and high-efficiency cyclones. Cyclones can also be arranged to operate in multiples to produce higher efficiencies. The efficiency of removal is, of course, not only a function of the size of the particle but of the other variables such as the airflow.

In gravitational settlers particulates of diameters up to 60 to 70 μm settle at Reynolds numbers equal to or less than 1, the Stokes law range. In centrifugal settling, particulates also settle at the Stokes law range, and $C_D = 24/R_e$ may be substituted in the equations derived for the terminal settling velocity [Equation (7.6)].

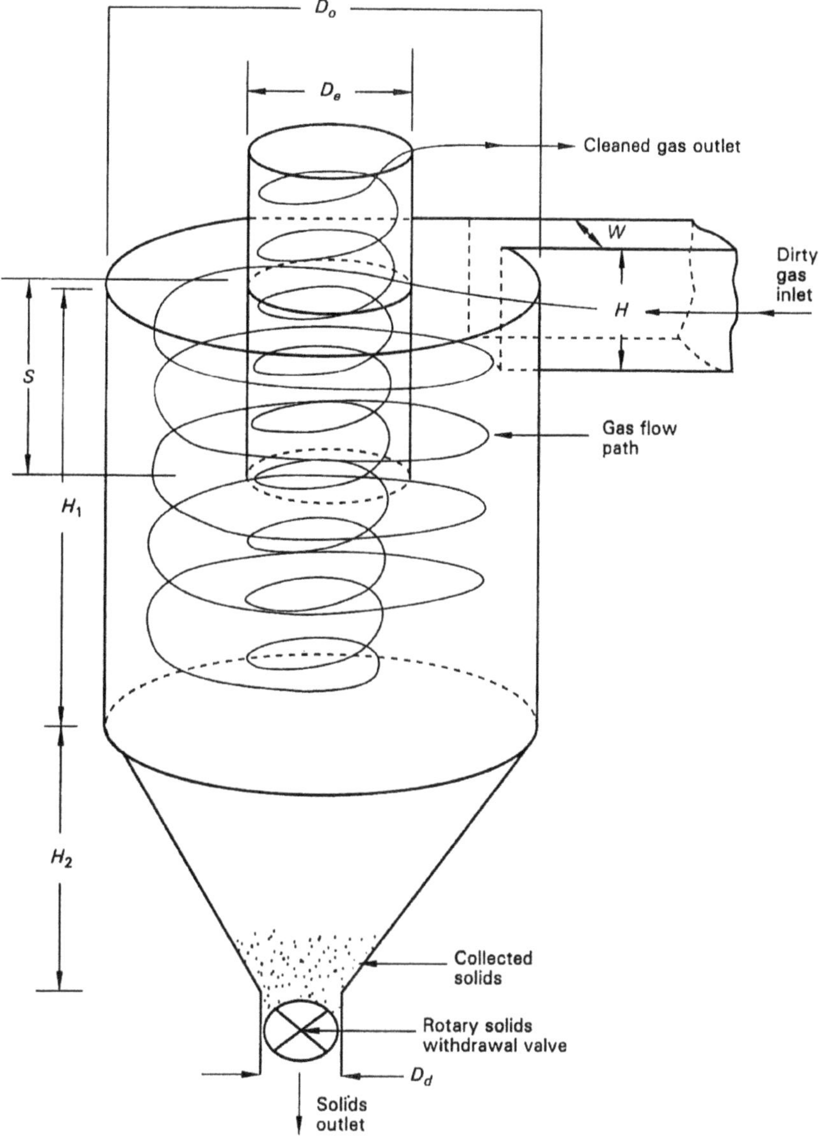

Fig. 7.8 *Cyclone separator*

The *g* in the equations above should be substituted by the g in a centrifugal field. In the centrifugal field, two forces are acting on a particle: the *inertial force*, which keeps the particle moving in a straight line as soon as it enters the body of the cyclone, and the *sweeping force* of the air, which tends to move the particle in the direction of flow. The vector sum of these forces produces a curved trajectory for the particle, which eventually bumps the side of the cyclone. The radial component of this "trajectory" force is the centrifugal force, and the body force acting on the particle in radial motion is the centrifugal force *mg*, where g is the centrifugal acceleration.

7.4.1.1 Cyclone dimensions

Cyclone collection efficiency increases with increasing

1. particle size,
2. particle density,
3. inlet gas velocity,
4. cyclone body length,
5. number of gas revolutions, and
6. smoothness of the cyclone wall.

On the other hand, cyclone efficiency decreases with increasing

1. cyclone diameter,
2. gas outlet duct diameter, and
3. gas inlet area.

For any specific cyclone whose ratio of dimensions is fixed. The collection efficiency increases as the cyclone diameter is decreased. The design of a cyclone separator represents a compromise among collection efficiency, pressure drop, and size. Higher efficiencies require higher pressure drops (i.e., inlet gas velocities) and larger sizes (i.e., body length

Table 7.4 shows the dimensions/characteristics of various types of standard cyclones.

Table 7.4 *Standard cyclone design dimensions*

Dimensions	Conventional	High-throughput	High-efficiency
Cylinder diameter, D	D	D	D
Cylinder length, L_1	2D	1.5D	1.5D
Cone length, L_2	2D	2.5D	2.5D
Total height, $L_1 + L_2$	4D	4	4D
Outlet length, $h + L_3$	0.675D	0.875D	0.5D
Inlet height, h	0.5D	0.75D	0.5D
Inlet width, b	0.25D	0.375D	0.2D
Outlet diameter, D_e	0.5D	0.75D	0.5D
Dust exit diameter, D_d	0.25D	0.375D	0.375D

The dimensions required to specify a tangential-entry, reverse-flow cyclone are shown in Figure 7.9. In classic work that still serves as the basis for cyclone design, Shepherd and Lapple determined "optimum" dimensions for cyclones. All dimensions were related to the body diameter Dc. A common set of specifications is given on the right-hand side of Figure 7.9. Other standard cyclone dimensions are given by Licht (1984) and Cooper and Alley (1986). The number of revolutions that the gas makes in the outer vortex can be approximated by

$$n_r = \frac{1}{H_c}\left(L_c + \frac{Z_c}{2}\right) \qquad .....(7.57)$$

where the dimensions are shown in Figure 7.9.

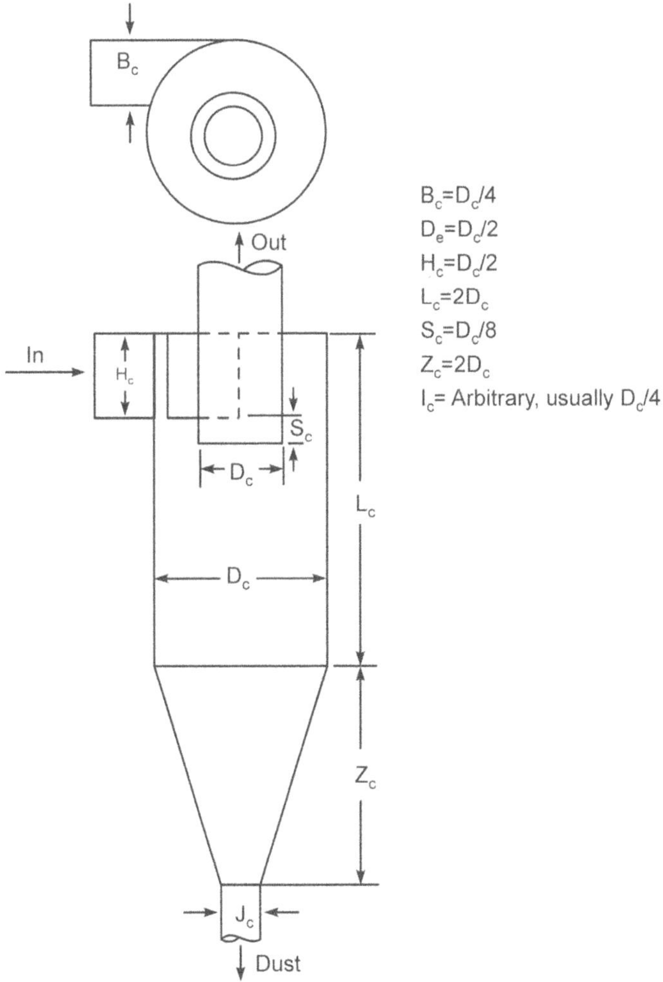

Fig. 7.9 *Geometric specifications for the design of a cyclone separator*

The dimensions given on the right-hand side of the figure are those of the classic design of Shepherd and Lapple

To find the value of the centrifugal acceleration, consider a particle and its component motion in the radial direction. Since the particle is forced to undergo a radial motion, by Newton's law there is an unbalanced force in this direction. This unbalanced force is the centripetal force that tends to force the particle toward the center of rotation and is simply V_t^2 / r multiplied by the mass *m* of the particle (mV_t^2/r), where V_t is the tangential velocity of the particle and r is the radius at the location of the particle. mV_t^2 / r is also equal to *mg*. The *g* in this expression is the centrifugal acceleration. Hence solving for g, we have

$$g = \frac{V_t^2}{r} \qquad .....(7.58)$$

Assuming no slippage between particle and air, v_t may be assumed equal to the velocity of the waste air revolving around the cyclone interior.

The gas entering the cyclone travels in a circular path and also radially between the inlet on the periphery and the exit on the axis. As the particle and gas have the same tangential velocity, their relative velocity is simply equal to the radial velocity of the gas which results in a drag force on the particle towards the centre. By equating the radial drag force with the centrifugal force, the equilibrium radius of the rotation of the particle can be then obtained from Equation (7.58).

$$3\pi\mu g d_p V_r = \pi / 6 d_p^3 (\rho_p - \rho_g) V_t^2 / r \qquad .....(7.59)$$

where d_p is the particle diameter, V_r = radial velocity of the gas at radius *r*, V_t = tangential velocity of gas or particle of radius *r*.

By rearranging the above equation for V_r and substituting V_t^2 / r for g

$$V_r = \frac{d_p^2 (\rho_p - \rho_g) V_t^2}{18 \mu r} \qquad .....(7.60)$$

The magnitude of the centrifugal force is normally expressed in terms of the number of times it exceeds the gravitational force.

Problem 7.13

A particle of 1 μ size is travelling in a stream of gas with a velocity of 5 m/s in a circular path of radius 0.2 m. Calculate the ratio of centrifugal force to the gravitational force acting on it. The value of g = 9.81 m/s^2.

Solution:

$$\frac{\text{Centrifugal force}}{\text{Gravity force}} = \frac{mV_t^2 / r}{mg} = \frac{(5 \times 5 \text{ m}^2/\text{s}^2 / 0.2 \text{ m})}{9.81 \text{ m/s}^2} = 12.74$$

Even for modest velocities and particles of normal size, the centrifugal forces are at least twice the gravitational forces. The ratio of centrifugal force to gravitational force varies from 5 for low velocities and small-sized cyclones to 2500 in high pressure cyclones.

As the airflow is introduced from the inlet into the cyclone chamber, it is subjected to a sudden increase in the cross-sectional area of flow. As the air negotiates the expanding cross-section and moves along with the curvature of the cyclone, an outer vortex flow is formed. It is in this outer vortex that removal of particulates occurs. This vortex acts as the settling chamber or the settling zone. The dimensions of this settling zone are depth = $(D_o/2 - D_c/2)$, length = $[\pi(D_o - D_c)/2]$, and width = $(L_1 + L_2/2)$. The average cross-sectional area flow A_h as the air spirals through the outer vortex is

$$A_h = \left(\frac{D_o}{2} - \frac{D_c}{2}\right)\left(L_1 + \frac{L_2}{2}\right) \qquad .....(7.61)$$

Calling the inlet as Q, v_t, can therefore be calculated. Substituting the result in Equation (7.60) and simplifying yields

$$g = \frac{\{4Q/[(D_o - D_c)(2L_1 + L_2)]\}^2}{r} \qquad .....(7.62)$$

Since the air has to go out, an inner vortex is also formed. This vortex is the outlet zone of this sedimentation basin.

The overflow area of the settling zone may be considered as the average of the outside and the inside lateral areas of the outer-vortex settling chamber. This area is

$$A_s = \frac{\pi}{2}\left(L_1 + \frac{L_2}{2}\right)(D_o + D_c) \qquad .....(7.63)$$

and hence the overflow velocity is

$$V_0 = \frac{Q}{(\pi/2)(L_1 + L_2/2)(D_o + D_c)} \qquad .....(7.64)$$

Equating the overflow velocity to the terminal settling velocity, substituting $C_D = 24/R_e$, and solving for d, we obtain

$$d = \sqrt{\frac{72\mu}{\pi g \rho_p (2L_1 + L_2)(D_o + D_c)}} \qquad .....(7.65)$$

By equating the overflow velocity to tangential velocity we get an equation relating the gas flow rate Q and cross-sectional area of the cyclone A_l by the equation

$$V_t = \frac{Q}{A_1}\left(\frac{D_c}{2r}\right)^n \qquad .....(7.66)$$

where D_c = diameter of the cyclone.

The exponent n is a dimensionless constant which has a value of 1 for ideal gas while for real gases the value varies from 0.5 to 1.0 based on the radius of the cyclone and temperature of the gas.

Substituting Equation (7.64) in Equation (7.6), we have

$$V_r = \frac{d_p^2(\rho_p - \rho_g)}{18\mu g} \frac{Q}{A_i} \left(\frac{D_c}{2r}\right)^n \qquad \text{.....(7.67)}$$

In general the collection efficiency of any cyclone increases with increase in radial velocity V_r though factors like particle-particle interaction modify it to a certain extent.

It can be seen from Equation (7.64) that V_r increases as the size of the particle increases and cross sectional area of the inlet decreases resulting in an increase in efficiency.

The fractional collection efficiency for a cyclone can be obtained by substituting the values of the inlet stream height W_i, in the radial direction (corresponding to H values in gravity settlers). The length of the flow path is equal to $N\pi D_o$ where N is the number of turns the gas makes traversing the outer helix of the cyclone before it enters the inner helix. D_o is the outer diameter of the cyclone. (This corresponds to L in the gravity settler in the Stokes' law Equations (7.33) and (7.34)).

We have

For block flow

$$\eta = \frac{N\pi D_o V_r}{W_i V_i} \qquad \text{.....(7.68)}$$

For mixed flow

$$\eta = 1 - \exp\left(\frac{N\pi D_o V_r}{W_i V_c}\right) \qquad \text{.....(7.69)}$$

Substituting for V_r from Equation (7.68) in Equation (7.67) we have

$$\text{Block flow } \eta = \frac{N\pi d_p^2(\rho_p - \rho_g)V_c}{W_i 18 W_i \mu} \times \frac{D_o}{r} \qquad \text{.....(7.70)}$$

Ignoring ρ_g which is very small compared to ρ_p, we have

$$\eta = \frac{N\pi d_p^2 \rho_p V_c}{9\mu W_i}$$

$$\text{For mixed flow} = 1 - \exp\left(\frac{N\pi\, d_p^2(\rho_p - \rho_g)V_c}{9\mu W_i}\right) \qquad \text{.....(7.71)}$$

Equations (7.68) and (7.69) contain the term N which represents the number of turns the gas makes around the cyclone before it leaves the collecting areas near the cyclone wall. A value of $N = 5$ is found to give good correlation with the experimental data.

The efficiency of collection of various particle sizes (η) can be determined from an empirical expression and graph developed by Lapple [6, 7].

$$d_{0.5} = \left(\frac{9\mu \, W_i H}{\rho_p - Q_g N} \right)^{1/2} \quad .....(7.72)$$

or

$$d_{0.5} = \left(\frac{9\mu \, W_i}{2 \, N V_i \rho_{\text{part}}} \right)^{1/2} \quad .....(7.73)$$

where $d_{0.5}$ = cut diameter, the particle size for which the collection efficiency is 50%, μ = dynamic viscosity of gas, P_a.s, w = width of entrance in m, Q_g = gas flow rate m^3/s which is equal to $2\pi V_C$, ρ_p = gas density, kg/m^{3}s, N = effective number of turns made in traversing the cyclone.

When the ratio of the actual particle size (d_p) to cut size d_{pc} or $d_{0.5}$ is plotted against collection efficiency a generalized curve (Fig 7.10) for collection efficiency can be obtained for various particles.

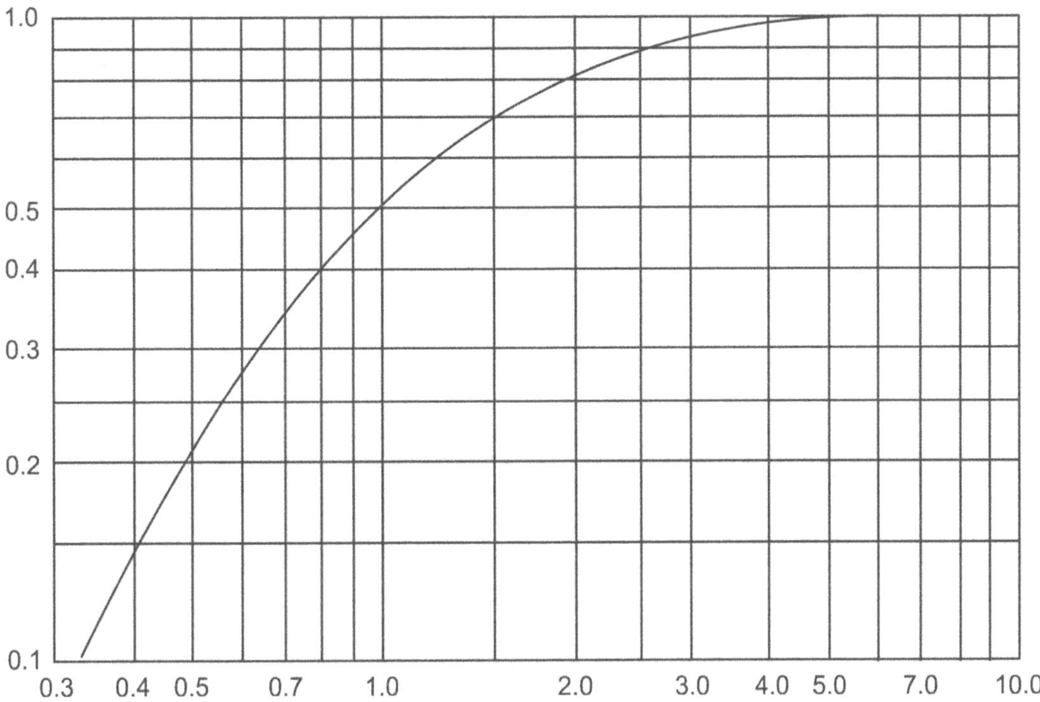

Fig. 7.10 *Empirical collection efficiency as a function of particle size for reverse flow collectors*

The value of N can be obtained by the following equation:

$$N = \frac{\pi}{H}(2H_1 + H_2) \qquad \text{.....(7.74)}$$

where H_1 and H_2 are the heights of the cylinder and the cone.

Pressure drop

In general as the diameter of the cyclone is reduced, the efficiency of collection is increased. However, the pressure drop also increases. This increases the power requirements for moving the gas through the collector. Since an efficiency increase will result, even if the tangential velocity remains constant, the efficiency is increased without increasing the power consumption by using multiple cyclones in parallel.

The pressure drop ΔP may be estimated according to the following equation, [8]:

$$-\Delta P = \frac{K\rho_g V_i^2 (HW)}{2D_e^2} \qquad \text{.....(7.75)}$$

where K = a constant, which averages 13 and ranges from 7.5 to 18.4, ΔP = pressure drop, H, W, and D_e= cyclone dimensions, V_i= inlet gas velocity, and ρ_g= gas density.

Since the pressure drop is proportional to the square of the inlet velocity, it is obvious that high velocities cause not only re-entrainment but also excessive pressure drop.

In the design of cyclones the pressure drop $-\Delta P$ across the unit is an important consideration. The pressure drop is used to size the blower used for forcing the waste air into the unit using the following empirical equation [9].

$$-\Delta P = \frac{0.0027QF}{CD_e^2 bh(L_1/D)^{1/3}(L_2/D)^{1/3}} \frac{1}{0.01C_i + 1} \qquad \text{.....(7.76)}$$

where $-\Delta P$ is the pressure drop of water ($-\Delta P = P_1 - P_2$,); Q the waste air flow; F is 0.8 for high-throughput cyclones, 1.0 for conventional cyclones, and 2.2 for high-efficiency cyclones; C = 0.5 with no guide vanes from the inlet to the cyclone chamber = 1.0 with guide vanes straight from the inlet to the inner edge of the outer vortex; and C_i is the inlet dust loading in grains/m^3.

High-efficiency cyclones

Small diameter cyclones are more efficient than conventional cyclones [10] (Table 7.5). Theoretically the collection efficiency for particulate matter of a given cyclone design increases as the ratio of centrifugal force to drag force increases.

$$\text{Ratio} = \frac{\text{centrifugal force}}{\text{drag force}}$$

$$= \frac{(\text{particle diameter})^2 (\text{particle density})(\text{particle tangential velocity})}{(\text{gas viscosity})\ (\text{cyclone diameter})}$$

$$= \frac{(d_p)^2(\rho_p)V_o}{(\mu)D_c} \quad .....(7.77)$$

The efficiency ranges of various types of cyclones are given in Table 7.75 Cyclones are quite efficient for particles larger than 10 μm and not very efficient for particles1μm or less in diameter. Thus, they are employed only for coarse dusts. Cyclones are used in the controlling of wood dust, paper fibers, and buffing fibers emission. Multi clones are frequently used as pre-cleaners for fly ash control devices in power plants.

Table 7.5 *Efficiency ranges of cyclones*

Particle size range (microns)	Efficiency range % weight collected	
	Conventional	High-efficiency
Less than 5	Less than 50	50–80
5-20	50–80	80–95
15-40	80–95	95–99
40 and above	95–99	95–99

Problem 7.14

Estimate the cut diameter for a cyclone with inlet width of 0.15 m, $Q_g = 18.3$ m/s, $N = 5$.

Solution:

$$d_{0.5} = \left(\frac{9(0.15\text{ m})(1.87\times10^{-5}\text{ kg/ms})}{2\pi(5)(18.3\text{ m/s})(2000\text{ kg/m}^3)}\right)^{1/2}$$

$$= 4.63 \times 10^{-6}\text{m} = 5\ \mu$$

Problem 7.15

Determine the efficiency of a standard cyclone having the following characteristics for particles 10 μm in diameter with a density of 800 kg/m^3.

Cyclone outer diameter = 0.50 m

Gas flow rate = 4.0 m^3/s

Gas temperature = 25 °C

Solution: From the standard cyclone dimensions we can estimate the following: (See Fig. 7.8)

$W_i = (0.25)\ (1.5) = 0.375$ m

$H = (0.50)\ (1.50) = 0.75$ m

$H_1 = H_2 = (2.00)\ (1.5) = 3.0$ m

The number of turns N is then

$$N = \frac{\pi}{H}(2H_1 + H_2)$$

$$N = \frac{\pi}{H}[2 \times 3 + 3.0]$$
$$= 37.7$$

The cut diameter is then

$$d_{0.5} = \left(\frac{9(1.85 \times 10^{-5}\ \text{kg/ms})\ (0.375\ \text{m})\ (0.75\ \text{m})}{2\pi(800\ \text{kg/m}^3)\ (37.7) \times 4.0\ \text{m}^3/\text{s}} \right)^{1/2} = 2.41 \times 10^{-6}\ \text{m}$$

The ratio of particle sizes is

$$\frac{d}{d_{0.5}} = \frac{10\ \mu\text{m}}{2.41\ \mu\text{m}} = 4.15$$

Problem 7.16

In a standard conventional cyclone, the airstream is 21.0m^3/s and the diameter of the cyclone is 2.0 m. The temperature is 650 °C and the specific gravity of the particle is 2.0. What are the diameter and the terminal settling velocity of the particle that is removed 100%? Assume that β equals 0.90.ρ_a = 0.59 kg/m^{3}and μ = 4.0 (10^{-5}) kg/ms at 650 °C.

Solution:

$$d = 1.24\beta^{0.333} d_p = \sqrt{\frac{72Q\ \mu}{\pi g\ \rho_p (2L_1 + L_2)(D + D_e)}}$$

$$d_p = \left[\frac{1}{1.24(0.90^{0.333})} \right] \sqrt{\frac{72Q\ \mu}{\pi g\ \rho_p (2L_1 + L_2)(D + D_e)}}$$

$$= 4.0 \sqrt{\frac{Q\mu}{g\rho_p (2L_1 + L_2)(D + D_e)}}$$

From Table 8.3,

$L_1 = 2D = 2(2) = 4$m; $L_2 = 2D = 4$m; $D_e = 0.5D = 1.0$ m;
$r = 0.5 + (1.0 - 0.5)/2 = 0.75$ m.

$$g = \frac{\left[\frac{4Q}{(D - D_e)(2L_1 + L_2)} \right]^2}{r} = \frac{\left[\frac{4(21)}{(2.0 - 1.0)(2(4) + 4)} \right]^2}{0.75} = 65.33\ \text{m/s}^2$$

$$d_p = 4.0 \sqrt{\frac{21(4.0)(10^{-5})}{65.33(2.0)(1000)(2(4) + 4)(2 + 1)}} = 53.5 \times 10^{-6}\text{m}$$

$$v_0 = \frac{Q}{(\pi/2)(L_1 + L_2/2)(D + D_e)} = \frac{21}{(\pi/2)(4 + 2)(2 + 1)} = 0.74\ \text{m/s}$$

Problems

1. We are passing a gas stream through a cyclone with D_{cut} = 10 μ. We now must treat twice as much gas as the same cyclone, so the average gas velocity will be increased by a factor of 2. We are told by the cyclone manufacturer that we should have no special trouble with re-entrainment or disturbance of the flow patterns if we double the flow rate. What will the cut diameter be at the new flow rate?
2. A particle stream has 33% particles of diameter 1 μ, 33% of 5 μ, and 34% with diameter 10 μ. We now pass this stream through a cyclone whose efficiency is d_{cut} = 5 μ. What fraction by weight of the particles will the cyclone collect?
3. In any cyclone separator is a slow moving boundary layer adjacent to the outer wall of the cyclone body. Using the nomenclature and ideas of boundary layer theory, estimate the thickness of this layer for a cyclone with V_i = 183 m/s, D_o= 0.3 m, N = 5.

A conventional cyclone with diameter 1.0 m handles 3.0 m^3/s of standard air. Using N_e = 6 determine the cut size and the efficiency as a function of particle diameter for particles of density (a) 1500 kg/m^3, and (b) 2500 kg/m^3

7.4.2 Electrostatic precipitators (ESPs)

7.4.2.1 Principle

Electrostatic precipitators are one of the most widely used particulate control devices, ranging in size from those installed to clean the flue gases from the largest power plants to those used as small household air cleaners.

The basic principle is to give particles an electrostatic charge and then put them in an electrostatic field that drives them to a collecting wall. A gas stream containing suspended particulates is allowed to pass between two electrodes electrically insulated from each other and between which there is a considerable difference in electric potential. The high-voltage electrode usually has a small cross-section and some curvature, e.g. a wire. The other electrode could be a plate or a surface of only slight curvature. The high voltage on the electrode of small cross-section ionizes the gas and aerosol particles. A corona is formed of ions which attach to the particulates making them ionic; they are then attracted to the large collecting electrode. There they are allowed to flow down or are drained by gravity and collected at the bottom. The electrostatic precipitator is like a gravity settler or centrifugal separator, but as the electrostatic force drives the particles more powerfully to the collecting plate, it is more effective on smaller particles than the gravity settlers and cyclone separators.

In general, in all these particulate collecting systems–gravitation settling chambers, cyclone separators or ESPs–a driving force acts on the particle to push it to the collecting wall against the resistance offered by the viscous medium which is proportional to the particle diameter. For gravity settlers and cyclone separators this force is proportional to the cube of the diameter of particles of constant density. Further the ratio of the driving force on the particle to push it to the wall to the resistance force offered by the viscous medium is equal to the square of the particle diameter and so this ratio falls rapidly as the diameter decreases. However, in electrostatic precipitators though the resistance force offered by the viscous medium is Stokes' drag, the force driving the particle to the wall is electrostatic which is proportional to the square of the particle diameter. So the ratio of the force driving the particle to the wall to the viscous drag force is proportional to the particle diameter. Thus though it will be more difficult in an ESP to collect smaller particles than larger particles, this hindrance is proportional to $1/d_p$ rather than $1/d^2_p$ as in the case of centrifugal or gravitational collectors.

7.4.2.2 Corona formation

A gas when subjected to a voltage gradient will conduct electricity as it is ionized due to electrons being knocked out by incident background radiation. Then these free electrons move towards the positive electrode while the positive gaseous ions attach to particulate matter and migrate to the negative collecting electrode. The initially formed free electrons in the presence of the electric field are accelerated in the direction of the field towards the positive wire electrode; this acceleration continues until these electrons with sufficient energy collide with the gaseous atoms and produce more free electrons which also get accelerated in the electric field leaving behind a large number of positive ions. This process continues until each free electron produces thousands of free electrons resulting in an abundance of electrons which is accompanied by emission of light and sometimes sound also. This is known as positive corona (Fig 7.11).

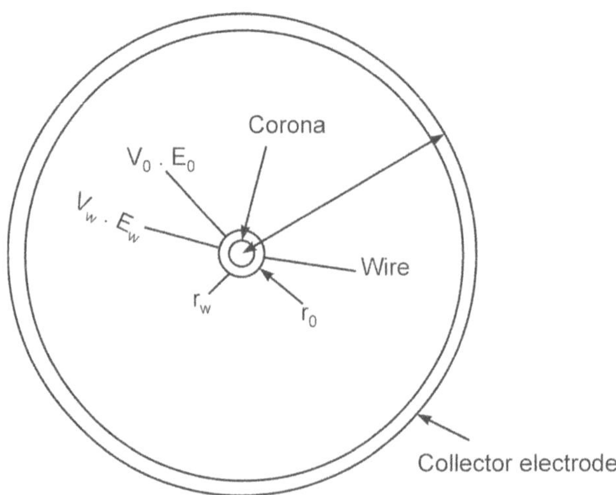

Fig. 7.11 *Formation of corona around wire in a cylindrical ESP*

7.4.2.3 Negative corona

However if the electrons migrate to a lower electric field region with gaseous molecules which will absorb them, it results in the formation of a large number of negative ions.

This is known as a negative corona. These negative ions attach to particulate matter making them negatively charged. They will migrate to the collecting electrode maintained at a positive potential relative to the other electrode. If the free electrons cannot be absorbed by gas molecules they will move to the positive electrode resulting in a spark which indicates the breakdown of the gases. Nitrogen does not absorb electrons while gases like SO_2, O_2, H_2O vapour, CO_2 absorb electrons.

An electrostatic precipitator has four principal parts: (1) a source of voltage, (2) high voltage ionizing electrodes and collecting electrodes, (3) a means for disposal of the collected material and (4) an outer housing to form an enclosure around the electrode. A simple precipitator is shown in Fig. 7.12. This is a single stage precipitator in which ionization and precipitation take place simultaneously.

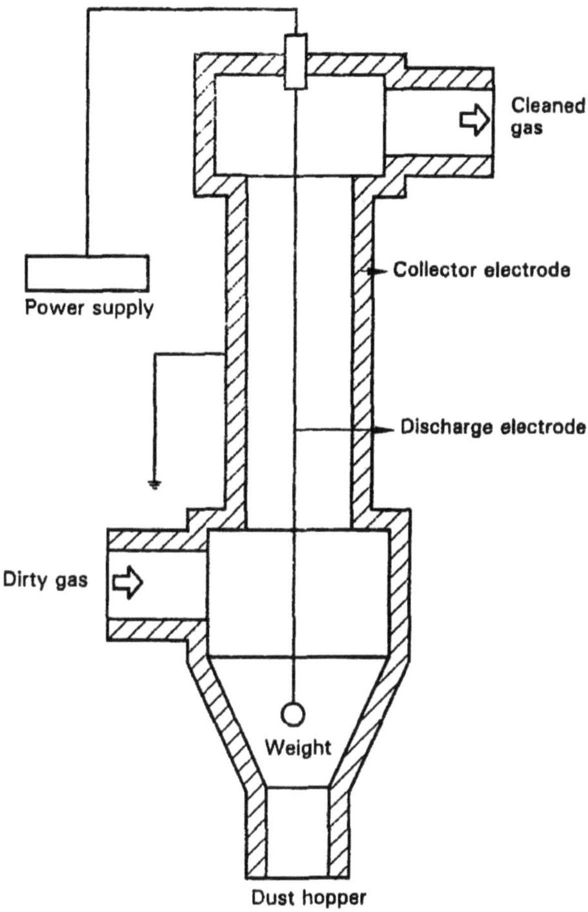

Fig. 7.12 *Electrostatic precipitator*

In this the particle-laden gas flows upward entering from the bottom of the cylindrical ESP. As the gas passes through the cylindrical portion of the ESP, the dust particles are ionized and under the influence of the applied field move towards the collecting electrode where they are collected by rapping while the cleaned gas escapes from the top. The discharge electrode (properly aligned with the cylindrical axis of the ESP) is a wire suspended from the top by an insulator and is held straight by a weight at the lower end. Either a steady or pulsed D.C. voltage of 40-50 KV is maintained on the discharge electrode using a large power supply.

A group of these collectors arranged in parallel in a single line is called the parallel plate single stage collector (Fig. 7.13).

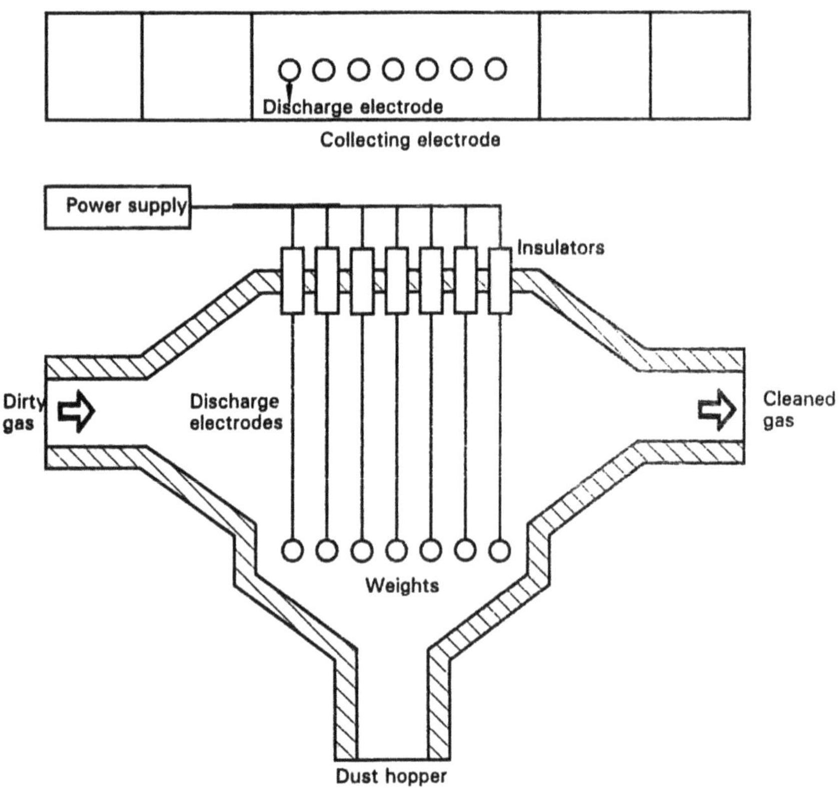

Fig. 7.13 *Parallel plate single stage electrostatic precipitator*

For most industrial applications the two separate steps of charging and collecting are carried out simultaneously in the same part of the ESP. The charging function is done much more quickly than the collecting function, and the size of the ESP is largely determined by the collecting function.

The parallel plate two stage electrostatic precipitator is shown in Fig. 7.13. The gas passes between the plates which are electrically grounded. Between the plates are rows of wires, held at a voltage of typically 40,000 volts. The power is obtained by transforming ordinary alternating current to a high voltage and then rectifying it through some kind of solid-state rectifier. This combination of charged wires and grounded plates produces both the free electrons to charge the particles and the field to drive them against the plates. On the plates the particles lose their charge and adhere to each other and the plate, forming a "cake". The cleaned gas then passes out of the far side of the precipitator.

Solid cakes are removed by rapping the plates at regular time intervals with a mechanical or electromagnetic rapper that strikes a vertical or horizontal blow on the edge of the plate. Some of the cake is always re-entrained, thereby lowering the efficiency of the system. If the collected particles are liquid, e.g. sulphuric acid mist, they run down the plate and drip off. For liquid droplets the plate is often replaced by a circular pipe with the wire down its center.

In a two stage cylindrical ESP (Fig. 7.14) the corona electrode in the form of a thin wire is fixed at the bottom connected to a collecting plate electrode of large diameter. Both are maintained at the same voltage by a simple power supply. Since the corona electrode is smaller in diameter, for the same voltage it produces a corona while the collecting plate electrode being larger in diameter cannot induce any corona. In some designs, the corona electrode and the collecting plate electrode (high voltage) will be maintained at different voltages by a separate power supply.

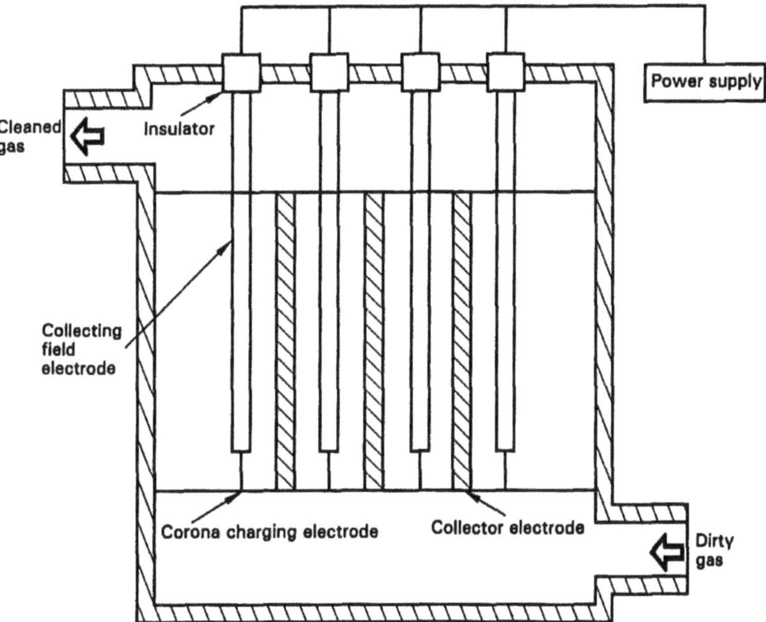

Fig. 7.14 *Cylindrical two stage electrostatic precipitator*

In a single stage ESP operated by negative corona significant levels of ozone are produced. As the ozone level produced along with the clean gas is objectionable when discharged into indoor air single stage ESPs are not used for air-conditioning applications but are widely used for industrial applications. In a double stage ESP, the positive corona produces less ozone than negative corona; so these are used in air-conditioning applications.

7.4.2.4 Theory for design considerations

In an ESP as the electrons flow toward the positive plate, they encounter particles and can be captured by them, thus charging the particles. Then the same electric field that created the electrons which is driving them toward the plate also drives the charged particles toward the plate. By capturing electrons in their path, particles larger than 0.15μ get charged by the electric field.

For particles larger than about 0.15 μ, the dominant charging mechanism is field charging. This is practically equivalent to the capture of any electron by any particle lying in its path. However, as the particles become more highly charged, they bend the paths of the electrons away from them. As time elapses the charge grows with time, and attains a steady state value of

$$q = 3\pi\left(\frac{\varepsilon}{\varepsilon+2}\right)\varepsilon_0 d_p^2 E_0 \quad \text{.....(7.78)}$$

where q is the charge on the particle, ε is the dielectric constant (which is a dimensionless number that is 1.0 for a vacuum, 1.0006 for air, and 4 to 8 for typical solid particles) of the particle; ε_0 is the permittivity of free space which has a value (SI units) of $8.85 \times 10^{-12} C/(V.m)$, d_p is the particle diameter, and E_0, is the local field strength.

The particle experiences an electrostatic field

$$F = qE_p \quad \text{.....(7.79)}$$

where E_p is the local electric field strength effecting the force.

If we substitute for q from Equation (7.78), we find

$$F = 3\pi\left(\frac{\varepsilon}{\varepsilon+2}\right)\varepsilon_0 d_p^2 E_0 E_p \quad \text{.....(7.80)}$$

where E_0 represents the field strength at the time of charging, and E_p represents the instantaneous field strength. For all practical purposes $E_0 = E_p = E$ and if the particles' resistance to being driven to the wall by electrostatic forces is given by Stokes' drag force, Equation (7.6) can be reset equating resistance to the electrostatic force and the resulting velocity of the particle can be written as

$$V_t = \frac{d_p \varepsilon_0 E^2\left(\frac{\varepsilon}{\varepsilon+2}\right)}{\mu} = w \quad \text{.....(7.81)}$$

This is called as the drift velocity which is designated as w.

Problem 7.17

Estimate the drift velocity of a particle of size 5 μ of a material with dielectric constant which reaches equilibrium charge in an ESP at a place where the field strength is 500 KV/m.

Solution:

$$V_t = \frac{d_p \varepsilon_0 E^2 \left(\dfrac{\varepsilon}{\varepsilon + \varepsilon_1} \right)}{\mu} = w$$

$$w = \frac{(5\times10^{-6}\,\text{m})(8.85\times10^{-12})(C/Vm)(5\times10^{5}\,Vm)^2\, 6/8 \times Nm \times CV}{(1.8\times10^{-5}\,\text{kg/ms})(Ns^2/\text{kg/m})}$$

$$= \frac{829.7\times10^{-8}}{1.8\times10^{-5}}$$

$$= 0.461 \text{ m/s}$$

Equation (7.81) indicates that the drift velocity is proportional to the square of E, which is approximately equal to the wire voltage divided by the wire-to-plate distance. So either by raising the voltage or lowering the wire-to-plate distance, an unlimited drift velocity can be achieved but this will be limited by sparking. Sometimes an ionized conduction path may be formed between the wire and the plate; which acts as a good conductor resulting in a continuous standing spark. By sensing the sudden increase in current the power supply stops the current flow into the wire to avoid the burnout of the transformer. The current is shut off for a fraction of a second which prevents the lightning stroke, and then the field is reestablished. As one raises the values of E, the frequency of sparks increases. These sparks are energetic events that disrupt the cake on the plate, thus reducing the collection efficiency; so a large number of sparks is bad. Experimentally it has also been found that setting the voltage low enough to have zero sparks results in too low an E for optimum efficiency. Most ESP control systems are set for about 50 to 100 sparks per minute, which seems to be the optimum balance between the need to increase E and the need not to have too many sparks. Also it is a common practice to subdivide the power supply of a large precipitator into many sub-supplies so that each part of the precipitator can operate at the voltage optimum for its local conditions, and during the fraction of a second in which the system is shut down to neutralize a spark, only a small part of the whole ESP is shut down.

If we compare the drift velocity (w) of an ESP with the terminal settling velocity V, in a cyclone for the same sized particle, it can be observed that though w is only about five times faster than V_1, the collection efficiency of an ESP is much higher than that of a cyclone. This can be understood from the fact that the drift velocity in an ESP is proportional to d_p (Equation (7.81)) while the settling velocity in a cyclone is proportional to d^2_p (Equation (7.80)). In a cyclone to achieve high drift velocities the gas velocity should also be high which results in less time for the particle to be exposed to the centrifugal force in a cyclone. However, in an ESP, it can be seen from Equation (7.81) that the gas velocity does not exist in the equation of the drift velocity and the particle approaches the collecting plate under the influence of electrostatic forces and is independent of gas velocity. So the ESP can be made large enough so that the particle spends enough time under the influence of electrostatic forces for a high probability for its collection. Normally in an ESP the gas velocity is in the range of 0.3–3 m/s and the particle spends nearly 1 to 10 seconds before it is captured.

For a tube type of ESP with diameter *D* and length *L* equating airflow velocity and drift velocity, substituting $C_D = 24/R_e$ and solving for d_p the diameter of the particle which can be removed with 100% efficiency, we have

$$d_p = \frac{6\mu}{\pi\varepsilon_0}\frac{Q}{E^2DL}$$

For a plate length of ESP with length *L* and height *h*, we have

$$d_p = \frac{3\mu}{\varepsilon_0}\frac{Q}{E^2Lh} \qquad .....(7.82)$$

The volumetric flow though the section is given by

$$Q = HV_{avg} \qquad .....(7.83)$$

E is force/charge. If the force is in new tons and the charge in coulombs and if both the numerator and denominator are multiplied by metres, the result is new ton-metre per coulomb-metre or volt per metre. Hence *E* has also the unit of volts per metre. So, if the voltage drop between electrodes is *V* volts and the distance between the discharge and the collector electrodes is Z_0 metres, the average *E* between electrodes is

$$E = \frac{V}{Z_0} \qquad .....(7.84)$$

The other consideration in the design of ESPs is the power requirement. Power is consumed whenever electric current is flowing through a wire: in the absence of current flow, no power is consumed. In ESPs, current flow is required to produce the corona: the greater the corona discharge, the greater is the current flow. In addition, the A C current from the power plant is converted to DC by a rectifier. Hence, from electricity, the power consumption P_c is

$$P_c = I_cV \qquad .....(7.85)$$

where I_c is the corona current.

Problem 7.18

To design a plate-type ESP the following values are given. The airstream is 330 m^3/s and the plates are spaced 30 cm apart. The voltage drop between discharge and collector plates is 7000 V. The aspect ratio, which is the ratio of the length of the plate to its width, is 1.2. The plate width is 10 m and there are a total of 80 channels. The temperature is 650 °C and the specific gravity of the particles is 2.0. (a) What are the diameter and terminal settling velocity of particles that are removed 100%? (b) What is the power requirement assuming that the corona current is 2.3A? Assume that β is 0.90.

Solution:

(a) $$d = 1.24\beta^{0.333} d_p = \frac{3\mu}{\varepsilon_0}\frac{Q}{E^2 LW}$$

$$d_p = \left(\frac{1}{1.24(0.90^{0.333})}\right)\left(\frac{3\mu}{\varepsilon_0}\frac{Q}{E^2 LW}\right) = 0.835\left(\frac{3\mu}{\varepsilon_0}\frac{Q}{E^2 LW}\right)$$

$\mu = 4.0(10^{-5})$ kg / ms and $\rho_a = 0.59$ kg/m^{3}at 650 °C

$Q = \frac{330}{80(2)} = 2.06\ \text{m}^3/\text{s}$ $\quad \varepsilon_0 = 8.85(10^{-12})\ \text{C}^2/\text{Nm}^2$ $\quad W = 10\ \text{m}$

$L = 1.2(10) = 12\ \text{m}$ $\quad E = \frac{70{,}000}{0.302} = 4.67(10^5)\ \text{V/m}$

$$d_p = 0.835\left(\frac{3\mu}{\varepsilon_0}\frac{Q}{E^2 LW}\right) = 0.835\left(\frac{3(4)(10^{-5})(2.06)}{8.85(10^{-12})[4.67(10^5)]^2(12)(10)}\right)$$

$$= 8.91(10^{-7})\ \text{m}$$

$$V_0 = \frac{Q}{LW} = \frac{2.06}{12(10)} = 0.017\ \text{m/s}$$

(b) $P_c = I_c V = 2.3(70{,}000) = 161{,}000$ W or 161 KW

Since an ESP is also a gravity settler, substituting the drift velocity for V_t and substituting from Equations (7.30) and (7.42) in Equations (7.28) and (7.34), we have

$$\eta = \frac{wA}{Q}\text{(block flow)} \quad .....(7.86)$$

$$\eta = 1 - \exp\frac{wA}{Q}\text{(mixed flow)} \quad .....(7.87)$$

Equation (7.86) is also called the "theoretical laminar flow equation", which would hold if we had block flow of gas with no mixing. Equation (7.87) is the Deutsche-Anderson [11] equation, the most widely used equation for design, analysis, and comparison of ESPs, where

η = fractional collection efficiency
A = collection area of plates, m^2
w = migration velocity or drift of particles, m/s
Q = gas flow rate, m^3/s

Problem 7.19

Calculate the relation between particle diameter and efficiency of collection for an ESP with *A/Q* value 0.4 min/0.30 m and drift velocity of 0.040 m/s.

Solution:

$$\eta = 1 - \exp\frac{wA}{Q}$$

$$= 1 - \exp\,(-(0.040 \text{ m/s})\,(0.4 \text{ min}/0.30 \text{ m})\; 60 \text{ s/min}$$

$$= 1.05$$

$$\eta = d_p^{\;1.05}$$

Problem 7.20

A coal fired power plant has an ESP working with 85% efficiency. To meet the statutory regulations, the company is proposing to upgrade the ESP with 99% efficiency. Estimate to what extent the collection area of the ESP has to be increased.

Solution: Percentage increase in efficiency $p = 1 - \eta_{\text{existing}} = 0.15$

$$0.15 = \exp\left(\frac{wA_{\text{existing}}}{Q_g}\right)$$

$$P_{\text{new}} = 1 - \eta_{\text{new}} = 0.01$$

$$= \left(\exp\frac{wA}{Q}\right)$$

$$\frac{\log 0.15}{\log 0.01} = 0.45 = \frac{-w\text{A}85/Q}{w\text{A}99/Q}$$

or $$\frac{A_{\text{new}}}{A_{\text{existing}}} = \frac{1}{0.45} = 2.5$$

So for achieving 99% efficiency, the collection area should be increased by 2.5 times. If the system obeys Deutsche-Anderson equation exactly, then for increasing the efficiency from 85% to 99%, the collecting area should be increased by 2.5 times, while for increasing the efficiency from 99% to 99.9%, the area should be increased by three times. As the drift velocity is proportional to the particle diameter (down to the very small particles, where diffusion charging becomes important), the big particles, which contain most of the mass, are removed first and as the percentage efficiency by weight increases, the remaining particles become smaller and smaller and harder and harder to collect. In view of this, the Deutsche-Anderson equation was further modified [12].

$$p-1-\eta=\exp-\left(\frac{wA}{Q}\right)^k \quad .....(7.88)$$

where k is an arbitrary exponent, which is typically about 0.5.

Problem 7.21

A electrostatic precipitator operating at 96% efficiency has two identical sections operating in parallel, each handling one-half of the gas. If the total gas flow is maintained constant, but unequally distributed so that ¾ of the gas goes through one of the sections, and ¼ through the other, estimate the overall collection efficiency.

Solution:

For the existing situation we calculate

$$p=0.05=\exp\frac{wA}{Q_g}$$

$$\frac{wA}{Q}-\ln 0.04=3.218$$

For the new situation we have

$$p_1=\exp(-3.218)\frac{1/2}{1/4}=0.08$$

$$p_2=\exp(-3.218)\frac{1/2}{3/4}=0.026$$

$$Q_1P_1={}^3\!/_4\,Q(0.08)=0.06Q$$

$$Q_2P_2={}^1\!/_4\,Q(0.026)=0.00665Q$$

$$(Q_1+Q_2)P=0.06+0.0066=0.0666$$

$$P=0.0666$$

$$\eta=1-0.0666=93.34\%$$

From the above problem it can be understood that an uneven distribution in an ESP can affect the efficiency as the particle escapes with high velocity and spends less time under the influence of electrostatic forces.

Problem 7.22

A thermal power plant installed an electrostatic precipitator with 5000 m^2 of collector plate area. The ESP is 95% efficient in treating 200 m^3/s of flue gas. Estimate how large the plate area should be to achieve 98-99% efficiencies for the electrostatic precipitator.

Solution: Flow rate of the flue gas = 200 m^3/s

Surface area of the collecting plate = 5000 m^2

Drift velocity $w = \frac{Q}{A} \ln(1 - \eta)$

$$w = \frac{200 \text{ m}^3 / \text{s}}{5000 \text{ m}^2} \ln(1 - 0.95)$$

$= 0.1198$ m/s

To achieve 98% efficiency, the area required would be

$$A_{98} = \frac{Q}{w} \ln(1 - \eta)$$

$$A_{98} = \frac{200 \text{ m}^3/\text{s}}{0.1198 \text{ m/s}} \ln(1 - 0.98)$$

$= 6530.69 \text{ m}^2$

To achieve 99% efficiency, the area required would be

$$A_{99} = \frac{200 \text{ m}^3/\text{s}}{0.1198 \text{ m/s}} \ln(1 - 0.99)$$

$= 7687.9 \text{ m}^2$

It can be seen from the above calculations that to increase the collection efficiency from 95 to 98%, the additional plate area required is 1530.7m^2 while for even a 1% increase, i.e. 98 to 99%, an increase of 1157.2 m^2 is required.

Problem 7.23

A cement plant installed a parallel type electrostatic precipitator with 7 channels to handle 25,000 m^3/h of flue gas. Determine the plate length of the ESP to achieve efficiency of (1) 90% (2) 98% and (3) 99.9%. v_p, velocity of particle towards plate is 0.1 m/sec, size of each plate is L m long × 2 m height, and spacing, s is 0.15 m.

Solution: Gas flow rate Q, 25,000 m^3/hr = 6.94 m^3/sec

No. of channels = n = 7

No. of plates = 7 + 1 = 8

Gas flow rate/channel = 0.99 m^3/sec

Area of collecting electrode/channel = 2 × L = 2 L m^2

$$V_g = \frac{Q}{H \times S} = \frac{0.99 \text{ m}^3/\text{s}}{2 \times 0.15} = 3.3 \text{ m/s}$$

Efficiency $\eta = 1 - \exp \frac{-wA}{Q_g}$

$$\eta = 1 - \exp \frac{-w2L}{Q_g}$$

$$\eta = 1 - \exp\frac{-w2L}{V_g \times HS}$$

$$\eta = 1 - \exp\frac{-0.1 \text{ m/s}}{3.3 \text{ m/s}} \times \frac{2L}{0.15}$$

$= 1 - \exp(-0.404\ L)$ for 90% efficiency

$0.90 = 1 - \exp(-0.404\ L) = 5.70$ m $\quad\therefore\quad L = 5.70$ m

Similarly $99\% = 1 - \exp(-0.404L) = 11.40$m $\quad\therefore\quad L = 11.40$ m.

and $99.9\% = 1 - \exp(-0.404L) = 17.40$ m $\quad\therefore\quad L = 17.40$ m.

Problem 7.24

A cement industry is emitting particulates of 0.6 μm with flue gas at the rate of 12 m^3/s. The drift velocity is 1.5 m/s. Design a parallel plate ESP with an efficiency of (i) 95% (ii) 99% (iii) 99.9%, if the dimensions of the plates and spacing between the plates are 5 m, 1 m and 0.15 m respectively. Estimate the number of channels for achieving 95%, 99% and 99.9% efficiency, if the ESP is cylindrical in shape, of diameter 0.65 m, and length 2 m. Estimate the number of plates required for 95%, 99% and 99.9% efficiency.

Solution:

(i) The efficiency of the ESP can be calculated by using the equation

$$\eta = 1 - \exp\frac{-wA}{Q_g}$$

$$\frac{-wA}{Q_g} = \ln(1-\eta)$$

$$A = \frac{-Q}{w}\ln(1-\eta)$$

For 95% efficiency $A = \dfrac{-12 \text{ m}^3/\text{sec}}{1.5 \text{ m/sec}} \ln(1-0.95) = 29.96 \text{ m}^2$

For 99% efficiency $A = \dfrac{-12 \text{ m}^3/\text{sec}}{1.5 \text{ m/sec}} \ln(1-0.99) = 36.84 \text{ m}^2$

For 99.9% efficiency $A = \dfrac{-12 \text{ m}^3/\text{sec}}{1.5 \text{ m/sec}} \ln(1-0.99.9) = 55.26 \text{ m}^2$

When length, height and spacing of plates are 5 m, 1 m, and 0.15 m.

Area of each channel = 2 × 5 m × 1 m = 10 m^2

$$\text{Number of channels} = \frac{\text{Total area of the electrode}}{\text{Area of channels}}$$

(i) For 95% efficiency $\dfrac{29.96\text{ m}^2}{10^2} = 2.99 \simeq 3$

(ii) For 99% efficiency $\dfrac{36.80\text{ m}^2}{10\text{ m}^2} = 3.68 \simeq 4$

(iii) For 99.9% efficiency $\dfrac{55.26\text{ m}^2}{10^2} = 5.53 \simeq 6$

So number of plates required for 95%, 99% and 99.9% efficiency are 3, 4, 6, respectively.

Volume of channel = area of each plate × spacing

$$V = \frac{A}{2} \times S = h \times L \times S$$

Volume of ESP for 95% efficiency = 3 × 1 m × 5 m × 0.15 m = 2.25m^3

For 99% efficiency = 4 m × 1 m × 5 m × 0.15 m = 3.0 m^3

For 99.9% efficiency = 6 m × 1 m × 5 m × 0.15 m = 4.5 m^3

(ii) If the ESP is cylindrical in shape

When the diameter and length of cylinder are 65 cm and 2 m, the area of each electrode = 3.14 × 0.65 × 2 = 4.08 m^2

So the number of plates required for the cylindrical ESP:

95% efficiency = 7.34 so seven plates are required

99% efficiency = 9.03 so nine plates are required

99.9% efficiency = 13.54 so fourteen plates are required

Problem 7.25

A cement plant was emitting flue gas at the rate of 20,000 m^3/h. Assuming inlet gas velocity of 2 m/s, design a tubular ESP with 0.20 m diameter with 7 cylinders to achieve an efficiency of a) 90% b) 95% c) 99.9%.

Solution:

$$\eta = 1 - \exp\left(\frac{-V_p A}{Q}\right)$$

$$\eta = 1 - \exp\left(\frac{-V_p (T_1 DL)}{Q}\right)$$

$$\text{A} = \frac{\text{Q}}{\text{V}_\text{p}} \mu_{\text{gc}} (1 - \eta)$$

$$\pi \text{DL} = \frac{\text{Q}}{\text{V}_\text{p}} \mu_{\text{gc}} (1 - \eta)$$

As μ_{gc} has a value of 1.0, we have

$$L = \frac{Q}{\pi D V_p}(1-\eta)$$

Assuming the diameter of each cylinder to be 20 cm, inlet gas velocity V_g to be 2 m/s, and length of the tubular electrode as L m.

Volume of flow = 20,000 m^3/hr 20000/(60 × 60) = 5.55 m^3/s.

Gas flow for each tube = Total flow/no of tubes = 5.55/7 = 0.79 m^3/s.

For 90% efficiency

$$L = \frac{0.79 \text{ m}^3/\text{s}}{3.14 \times 0.20 \text{ m} \times 2.0 \text{ m}} \ln(1-0.90)$$

$$= 1.45 \text{ m}$$

For 95% efficiency

$$L = \frac{0.79 \text{ m}^3/\text{s}}{3.14 \times 0.20 \text{ m} \times 2 \text{ m/s}} \ln(1-0.95) = 1.89 \text{ m}$$

For 99.9% efficiency

$$L = \frac{0.79 \text{ m}^3/\text{s}}{3.14 \times 0.20 \text{ m} \times 2.0 \text{ m/s}} \ln(1-0.99) = 2.90 \text{ m}$$

Problem 7.26

Determine the length of the ESP to achieve 99% efficiency with the following conditions. The electrostatic precipitator is a cylindrical type with 0.25 m diameter. The volumetric flow rate of flue gas is 0.04 m^3/s. While the drift velocity of the particle is 0.12 m/sec, calculate the electrode collection area for a 99% and 99.99% collection efficiency and give the percentage increase in the surface area.

Solution: Q = 0.04 m^3/sec, D = 0.25 m V_p = 0.12 m/sec

Area of collecting surface $A = \pi DL$

$$= \pi \times 0.25 \times L$$

For 99% efficiency

$$\eta = 1 - \exp\left(\frac{-WA}{Q}\right)$$

$$\eta = 1 - \exp\left(\frac{-W\pi DL}{Q}\right)$$

$$A = \pi DL = \frac{Q}{V_p}\ln(1-\eta)$$

$$A_c = \frac{-0.04}{0.12}\ln(1-0.99) = 1.52 \text{ m}^2$$

Area of the collecting electrode = 1.52 m^2 = 22/7 × 0.25

Since $A = \pi DL$

$$L = \frac{A}{\pi D}$$

$$L = \frac{1.52}{3.14 \times 0.25} = 1.94$$

For 99.99% efficiency

$$A_1 = \frac{-Q}{V}\ln(1-\eta)$$

$$= \frac{0.04}{0.12}\ln(1-0.999) = 2.28 \text{ m}^2$$

$$L = \frac{2.28}{3.14 \times 0.25} = 2.9 \text{ m}$$

Percentage increase in collection area

$$= \frac{2.28 - 1.94}{1.94} = 17\%$$

Problem 7.27

A paper mill flue gas control ESP (parallel plate, single channel) is 5.0 m high, 8 m long with a plate to plate spacing of 0.15 m. With a flue gas flaw rate of 80 m^3/min, 95% collection efficiency was obtained. Calculate the bulk velocity of the gas, drift, velocity, and outlet loading. Determine the collection efficiency if the flow rate is increased to 200 m^3/min and a revised collection efficiency, if the plate spacing is decreased to 50 cm.

Solution: The bulk velocity of the flue gas

$$v_g = \frac{\text{volumetric flow}}{\text{sec tion area of passage}}$$

$$= \frac{80/60 \text{ m}^3/\text{s}}{5.0 \times 0.15 \text{ m}} = 1.78 \text{ m/s}$$

Outlet loading = inlet loading (1 – η)

= 120 (1 – 0.95)

= 6.0/m^3

Collection efficiency, if the flow rate is increased to 200 m^3/min

$$\eta = 1 - \exp\frac{-WA}{Q} = 0.95$$

$$\frac{-WA_c}{Q} = \ln(1-\eta) = \ln(1-0.95) = -2.99$$

$$W = \frac{2.99 \times 80 \text{ m}^3/\text{min}}{2 \times 5 \times 8 \text{ m}} = 2.99 \text{ m/min} = 0.05 \text{ m/s}$$

Revised collection efficiency

$$\eta = 1 - \exp\left(\frac{-WA}{Q_g}\right)$$

$$\eta = 1 - \exp\left(\frac{-2.99 \times 2 \times 5 \times 8}{200}\right) = 0.698 = 69.8\%$$

Problem 7.28

A cement plant installed two ESPs, of collecting surfaces of 2500 m^2 each to handle 60 m^3/s flow rates. Calculate the total removal efficiency and the efficiency of total treatment system if one ESP is shut down. The drift velocity is 0.15 m/s.

Solution: Volumetric flow in each precipitator, $Q = 60$ m^3/s

Collecting surface area of the precipitator, $A = 2500$ m^2

Drift velocity $V_p = 0.15$ m/s

$$\eta = 1 - \exp\frac{-WA}{Q}$$

$$= 1 - \exp\left(\frac{0.15 \times 2500}{60}\right) = 0.998 \text{ i.e., } 99.8\%$$

If one precipitator is shut down and the total gas volume is treated by only one ESP, we have

$$\eta = 1 - \exp\left(\frac{-0.15 \times 2500}{2 \times 60}\right) = 0.9561 = 95.61\%$$

Problems

1. Before the installation of an electrostatic precipitator, the stack gas of a power plant contained 6.0 g particulates per m^3 of gas. The gas flow rate is 350 m^3/min and the new precipitator can remove 2500 kg particulates/day.

(a) What is the emission rate of particulates before and after pollution control in kg/day?

(b) What is the efficiency of the electrostatic precipitator?

(c) Will the new system meet emission standards?

2. A cylindrical electrostatic precipitator of diameter 0.3 m is used for separating pulverized coal fly ash particles from a furnace gas stream. If the volumetric flow rate of the gas is 0.05 m^3/s, what will be the length of the precipitator for obtaining a collection efficiency of 99.90%? What per cent change in electrode collection area is required to increase the collection efficiency from 99.9 to 99.95%?

7.4.2.5 Industrial ESPs

Electrostatic precipitators can easily remove more than 98% of the particles passing through them, including particles of sub-micrometer size. Some have efficiencies even greater than 99.9%. They can handle large flue gas flow rates with little pressure drop, and they have relatively low operation and maintenance costs. They are quite versatile, operating on both solid and liquid space. Area requirements increase non-linearly with collection efficiency [13] in accordance with the Deutsche-Anderson equation [Equation (7.77)].

There are also two general types of precipitators, depending on the physical arrangement of the electrodes. In pipe-type precipitators, the high-voltage ionizing electrodes hang axially in collecting electrode pipes. The collected material is removed from the inside of the pipes by a continuous flow of liquid down the pipes, or by periodic rapping of the pipes. The plate-type precipitator (Fig. 7.12) may have either a horizontal or vertical gas flow arrangement. The collecting electrodes are in the form of curtains, which act as ducts with high-voltage electrodes equidistantly located in the ducts between the collecting electrode curtains.

Because of the very large scale of these units, and the very high voltages used, these processes are all carried out automatically. In actual practice, the processes are not as simple as in the above description. For example, the discharge of newly arrived particles at the plate, where there is already a deposited cake acting as a partial insulator, might be very slow. Indeed a charge can build up on the particle layer which is greater than the charge on the central wire, and "back arcing" will occur, which strips some of the deposited particles and re-entrains them into the gas stream.

One operational problem of ESPs of particular note is the resistivity of the fly ash [14]. Fly ash is a generic term used to describe the particulate matter carried in the effluent gases from furnaces burning fossil fuels. ESPs often are used to collect fly ash. The strongest force holding fly ash to the collection plate is electrostatic and is caused by the flow of current through the fly ash. The fly ash acts like a resistor and, hence, resists the flow of current. This resistance to current flow is called the resistivity of the fly ash. It is measured in units of ohm.cm. If the resistivity is too low, not enough charge will be retained to produce a strong force and the particles will not "stick" to the plate. Conversely, and often more importantly, if the resistivity is too high, there is an insulating effect. The layer of fly ash breaks down locally and a local discharge of current from the normally passive collection electrode occurs. This discharge lowers the spark-over voltage and produces positive ions that decrease particle charging and, hence, collection efficiency.

Table 7.6 *Drift velocities of certain particulates in industrial practice [15]*

Particles	Drift velocities m/s
Pulverized coal (fly ash)	0.44
Paster mills	0.25
Open heart furnace	0.19
Second blast furnace	0.41
Gypsum	0.52–0.64
Hot phosphorus	0.09
Acid mist (TcO_2)	0.19–0.25
Flash roaster	0.25
Multiple hearth roaster	0.26
Portland cement (dry process)	0.19–0.23
Portland cement (wet process)	0.33–0.37
Catalyst particles	0.25

To calculate an appropriate value of the effective drift velocity in the Deutsche-Anderson equation (Equation 7.77) the knowledge of particle size distribution, the dielectric properties of the material, and the sensitivity coherency of the particle is necessary. Often this information is not available for all new materials, so trials have to be made. Some typical effective drift velocities (w) values of certain particles encountered in industrial practice are given in Table 7.6.

In spite of these problems, modern electrostatic precipitators operating under favorable conditions can achieve efficiencies of 99% or better in the collection of fine particles, down to and below 0.01 μm. They could also be built to the efficiency needed allowing for considerable cost saving.

Table 7.7 *Advantages and Disadvantages of ESPs*

	Advantages	Disadvantages
1.	Pressure drop and hence power requirement is small compared to the other devices; economical and simple to operate	Relatively high initial cost and large space requirement
2.	99%+ efficiency obtainable. Very small particles can be collected, wet or dry.	Sensitive to variable particulate loadings or flow rates
3.	Can handle both gases and mists for high voltages volume flow.	Safeguarding operating personnel from high is necessary.
4.	Few moving parts; can be operated at high temperatures and pressures.	Collection efficiency can deteriorate gradually.

7.5 Gas Filtration for Particulate Removal

7.5.1 Major types of gas filters

A major class of particulate air pollution control devices relies on the filtration of particles from gas streams. A variety of filter media is employed, including fibrous beds, packed beds, and fabrics. Fibrous beds used to collect airborne particles are typically quite sparsely packed, usually only about 10% of the bed volume being fibers. Packed bed filters consist of solid packing in, say, a tube and tend to have higher packing densities than do fibrous filters. Both fibrous and packed beds are widely used in ventilation systems. Fabric filters are frequently used to remove solid particles from industrial gases, whereby the dusty gas flows through fabric bags and the particles accumulate on the cloth.

Two main types of gas filters are in use for the removal of particulate matter from industrial gases: (a) Packed bed or depth filters (b) Fabric elastic filter.

Fabric filters are used for cleaning industrial gases with a particulate loading gravity ofl g/m^3 while depth filters are used only for gases containing particles of the order of 1 mg/m^3.

7.5.2 The performance efficiency

The performance efficiency of any filter is determined from the following criteria

(a) Pressure loss: It is expressed in mm of water and gives the required fan power. Air is a major factor that determines the operational cost.

(b) Collection efficiency: The collection efficiency (mass percentage of dust collected) determines the performance of a fibre and is measured instantaneously or cumulatively since the efficiency varies with dust adhered to the filter which varies with filtering time.

(c) The life time or dust holding capacity of any filter is determined by actual experimentation under operational conditions and this determines the major portion of the initial costs. It is the maximum capacity of the filter to hold the dust per unit area of the filter medium for a certain range of pressure drop. Normally filters will have 0.3–2 kg of dust per square meter of filter paper.

7.5.3 Depth filters

For low dust concentrations depth filters are very useful and their simple construction offers easy maintenance. Two types of depth filter are in use.

(i) Throw away type in which the filter medium is packed (like in a filter cigarette) and can be replaced after a certain period of time.

(ii) Renewable type in which the filter medium is replaced or renewed.

(i) ***Throw away panel filters:*** These are made in the form of mats packed closely and fixed in a metal or glass or plastic frame of 50 cm square and 2–5 cm thick which can be operated at gas velocities of 0.3 to 2 m/s and will have pressure losses of 5 to 50 mm H_2O. Filters with mineral, ceramic or metallized pebbles can be used for high temperature applications.

(ii) ***Renewable medium filters:*** In this a compressed medium will be used which can be periodically renewed with specific velocity. Some partially renewable medium filters are also in use where a portion of the medium in the filtering zone will be replaced periodically by a fresh one. Both of these have advantages over the throw way type from the economic point of view and reduced filtration pressure drop and flow rate.

7.5.4 Fabric filters

In fabric filtration, the dirty air is passed through a woven or felted fabric that filters out the particulate matter. The initial process of removal when the fabric is still clean is by direct interception, impaction, diffusion, and electrostatic attraction. After a dust mat or cake has been formed, submicron particles are removed by a sieving action. The principle of operation is similar to that of an ordinary vacuum cleaner used in homes.

Many industries utilize fabric filters for collection of dust particulates from the effluent gases. Fabric/cloth/surface filters particularly when high degree of cleaning of gases is required and the size of the particulate is very small the fabric collector are the preferred choice. Further using fabric filters one can recover large quantities of valuable process materials in a dry state. When the particle effluent gases from various industrial processes flow through a cloth tube or envelope, the particles larger in size than the fabric interstices will be deposited by simple sieving action (Fig. 7.15).

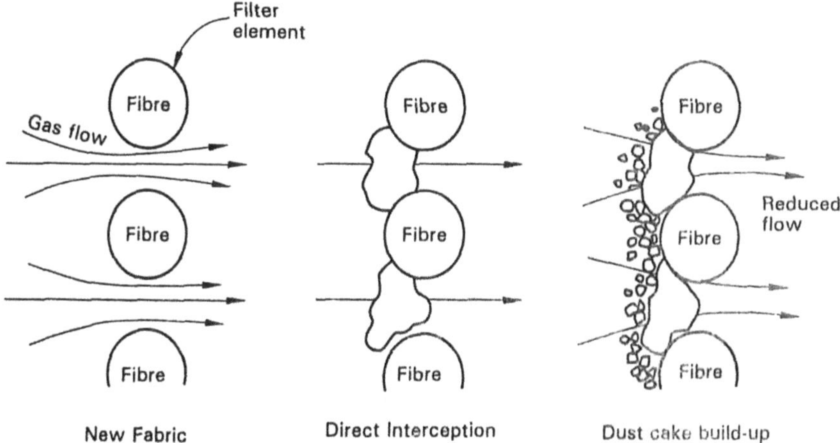

Fig. 7.15 *Sieving Mechanism of Filtration*

The porous structure of the granular or fibrous material of the filter will retain particulates of larger size than the fabric interstices and allows the gas to pass through the voids of the filter.

The gas entering through the inlet pipe strikes a baffle plate, which causes the larger particles to fall into a hopper due to gravity. The carrier gas then flows upward into the tubes and then outward through the fabric leaving the particulate matter as a "cake" on the inside of the bags. The filter efficiency during pre-coat formation is low, but increases as the pre-coat is formed, until a final efficiency of over 99% is achieved. This pre-coat acts as a part of the filter medium, which helps in further removal of the particulates. Thus, dust becomes the actual filtering medium. The bags, in effect, act primarily as a matrix to support the dust cake. The cake is usually formed in a matter of minutes or sometimes even seconds.

7.5.5 Theory of filtration

When a gas fluid containing suspended particulates flows through a filter medium, enters a cloth, but sometimes a paper or porous metal, or bed of sand the solid particles in the stream deposit on the face of the filter medium forming the filter cake (Fig. 7.15) while the cleaned gas, free from solids, flows through both the cake and filter medium. The flow of the gas is horizontal, it will undergo a small change in velocity due to pressure drop, which makes the gas expand, as it leaves behind its contained particles. For most filters of air pollution, the combined effect of these changes is negligible. In most industrial filters, both for gases and liquids, the flow velocity in the individual pores is so low that the flow is laminar and so the well-known relations for laminar flow of a fluid in a porous media, can be used.

In depth filters there are three possible mechanisms through which the particulates from gas streams are separated. They are (a) inertial impaction (b) direct interception and (c) diffusion.

7.5.6 Inertial impaction

When a particle-laden gas flows through cylindrical fibre in a filter the gas flow must bend to flow around the fibres, just as the wind bends to flow around a building or a river bends to flow around a rock in its middle. The contained particles, which are much denser than the gas, are carried by their inertia, which makes them travel straight. Thus some of them hit the target rather than following the gas around it.

Inertial impaction is the mechanism by which large particles, traveling on a collision course with a filtering element, impact with the element. Inertia keeps the larger particles on a collision course even though the smaller particles and gas stream diverge around the element.

Separation by impaction and interception is also referred to as inertial separation. Impaction is important when the particle has appreciable mass but negligible size. The impaction parameter, I, is the ratio of the diameter of the tube of gas streamlines, containing particles, and the diameter of the fibre or droplet on which the particles impact. This is shown in Fig. 7.16(a).

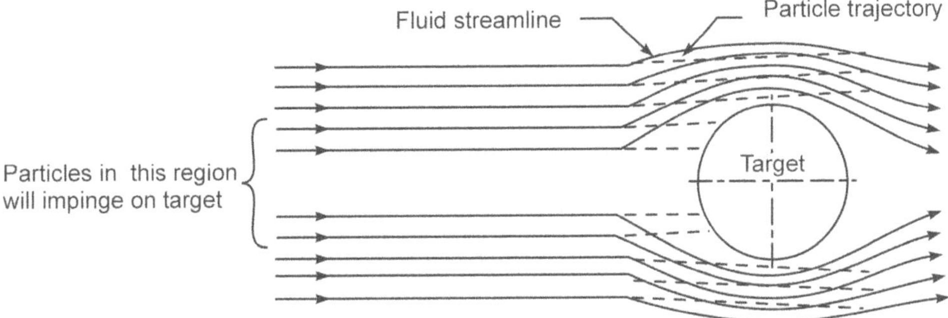

Fig. 7.16(a) *Separation by impaction*

7.5.7 Direct interception

Direct interception is the mechanism which brings the particles of appreciable size, which are able to follow the streamlines, into contact with the element since the air streams have a tendency to converge as they flow around the element [Fig. 7.16(b)].

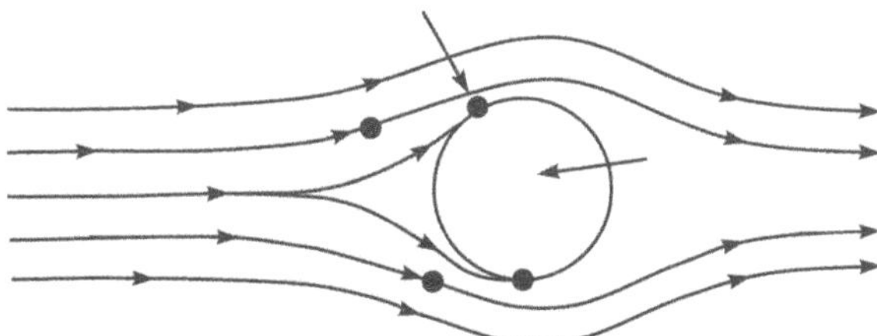

Fig, 7.16(b) *Separation by interception*

The interception parameter, R, is the ratio of the diameter of the particle to the diameter of the body. For both impaction and interception, a separation number is used to depict the performance of the separation equipment and is defined as the ratio of the stopping distance and the diameter of the fibre or droplet.

$$\text{Separation number, } S = \frac{u_t v_0}{g\,D} \qquad .....(7.89)$$

where u_t is the terminal velocity of the particle, v_0 the relative velocity between the gas and body and D the diameter of the body (fibre or droplet). Figure 7.17 gives a typical plot for the inertial collection efficiency and the separation number, S. It can be seen that the efficiency of collection increases with increase in relative velocity between the gas and droplet and with a decrease in droplet size.

The collection efficiency by interception of a cylindrical target may be calculated by the following expression if potential flow is assumed:

$$\eta_{\text{inter}} = \left(1+\frac{d_p}{d_f}\right) - \frac{1}{\left(1+\frac{d_p}{d_f}\right)} \qquad .....(7.90)$$

For a viscous flow situation, Strauss [16] recommends Equation (8.80)

$$\eta_{\text{inter}} = \frac{1}{2.002 - \ln\ R_{ef}}\left\{\left(1+\frac{d_p}{d_f}\right)\ln\left(1+\frac{d_p}{d_f}\right) - \frac{2(d_p/d_f)+(d_p/d_f)^2}{2+2(d_p/d_f)}\right\} \qquad .....(7.91)$$

The combined mechanisms of impaction and interception usually account for more than 99.9% of the collection of particles larger than 1 μm.

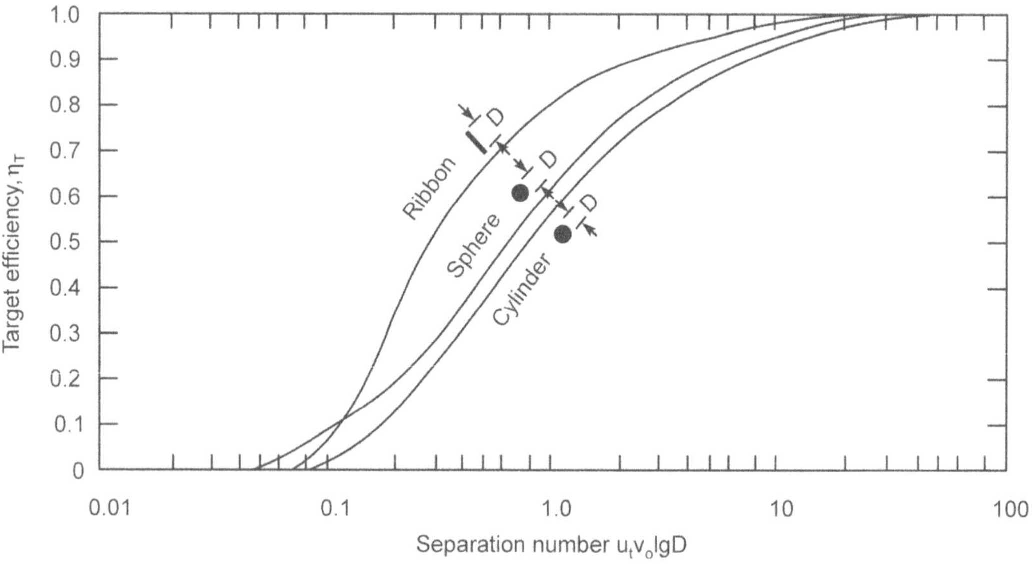

Fig. 7.17 *Inertial efficiency vs separation number plots for ribbon, spherical and cylindrical impacting bodies [4]*

7.5.8 Diffusion

The third mechanism for collecting particles in depth filters is diffusion or Brownian motion, the same mechanism that causes diffusion in gases and liquids. If a particle is large, then in any short period of time it will experience many collisions with the surroundings gas molecules that are hitting it from all sides, and the resulting net force will be quite small relative to the mass of the particle. If the particle is small enough that it can only expect a few collisions per second and its inertia is small because of its small size, then the force of an individual collision is enough to make it move. Subsequent collisions, whose directions are random, will move it in other directions, so that its time-series path will be a series of short jumps in one direction and then another. One must use a microscope to observe such behavior, because the particle size at which it becomes important is too small for our eyes to distinguish.

For very small particle sizes, separation due to diffusion of particles becomes significant. Particles possess diffusivity, which can be calculated by the following equations.

$$D_M = \frac{CKT}{3\pi\ \mu d_p} \qquad \text{for } d_p > 0.1 \text{ micron} \qquad .....(7.92)$$

$$D_M = \frac{4\ KT}{3\pi\ \mu d_p^2 P} - \left[\frac{8RT}{\pi M}\right]^{1/2} \qquad \text{for } d_p > 0.1 \text{ micron} \qquad .....(7.93)$$

where C is the Cunningham correlation factor, K the Boltzmann constant, d_p the particle diameter, μ the viscosity of gas at temperature T Kelvin and M the molecular weight of the gas. There are a number of reported equations for finding the contribution of diffusion to the efficiency.

One such equation is

$$\eta_D = 0.775\text{Pe}^{-0.6}\left[\frac{C_D R_e}{2}\right]^{0.4} \qquad .....(7.94)$$

where η_D is the diffusion efficiency, Pe is the Peclet number and represents a product of Reynolds number (R_e) based on fibre or droplet size and Schmidt number (Sc).

The combined efficiency due to impaction, interception and diffusion η_{ICD} can be evaluated if the individual efficiencies are known.

$$\eta_{ICD} = 1 - (1 - \eta_I)(1 - \eta_C)(1 - \eta_D) \qquad .....(7.95)$$

where η_{ICD} is the combined efficiency (also called target efficiency η_T) and $\eta_P \eta_C$, and η_D are the contributions due to impaction, interception and diffusion.

A widely-used expression for determining the combined efficiency or target efficiency is

$$\eta_{ICD} = 6\text{Sc}^{-2/3}\,R_e^{-1/2} + 3R^2\,R_e^{1/2} \qquad .....(7.96)$$

where R is the ratio of the particle diameter to the diameter of the fibre or droplet. Equation (8.85) gives the single-fibre target efficiency and should be corrected for the presence of other fibres in mats or fibre filters. The following empirical expression gives the target efficiency when the void fraction of the filter or spray is known.

$$\eta = \eta_0[1 + 4.5(1 - \varepsilon)] \qquad .....(7.97)$$

where η is the new target efficiency, η_0 the target efficiency for a single fibre or droplet and ε the void fraction in the filter or spray-tower. It may be noted that η is larger than η_0 due to the increased local-gas velocity as well as the change in flow pattern of the gas around the fibre or droplets, because of the presence of other fibres or droplets.

7.5.9 Classification of fabric filters

Classified according to the methods of cleaning, there are three types of baghouses: shaker, reverse-air and pulse-jet baghouses. The filtering element is usually in tubular form as shown in Fig. 7.18, called *filter bag*. Filter bags may range from 1.8 to 9 m in length and may have diameters of around 20 cm. Several of these bags are grouped together and put into a compartment, and several of these compartments are put together and assembled inside a structure called a *baghouse*. The capacity of a compartment is determined by the area of the fabric filter. Table 7.8 shows some design guidelines to size compartments.

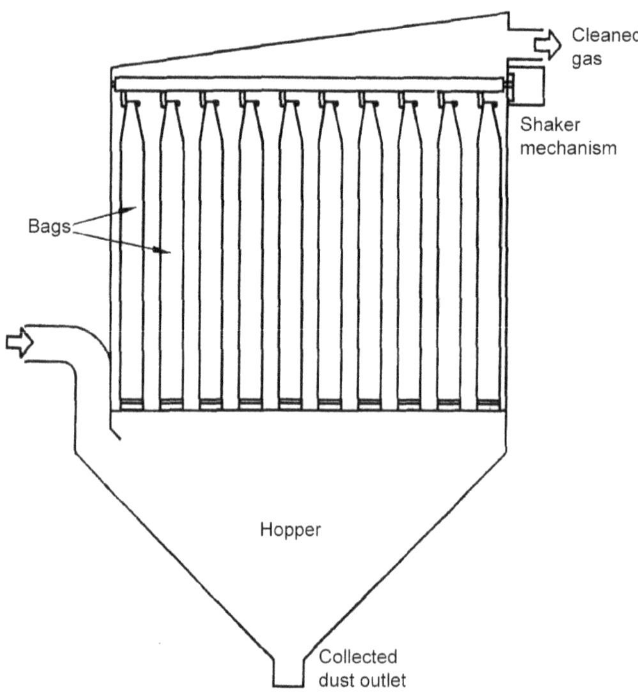

Fig. 7.18(a) *Vertical cross-section of a simple bag house*

In the reverse-air and shaker baghouse, the filters in each compartment are bundled in bag headers at the bottom and the individual filter bags are suspended from the end sat the top. Dirty air is introduced at the bottom and flows through the fabric from the inside of the tubes. The cakes are therefore collected at the inside of the bags. When it is time for the filters to be cleaned, one compartment is isolated from the rest. In the reverse-air system, air is blown through the bag in the reverse direction from the outside toward the inside. In the shaker system, the bags are shaken. In both systems, during cleaning, chunks of the previously collected dust fall down to the bottom of the hopper below. As soon as the compartment has been cleaned, it is put back to operation, and the next compartment in-line for cleaning is now, in turn, isolated, and another compartment cleaning begins. The cleaning proceeds around a cycle.

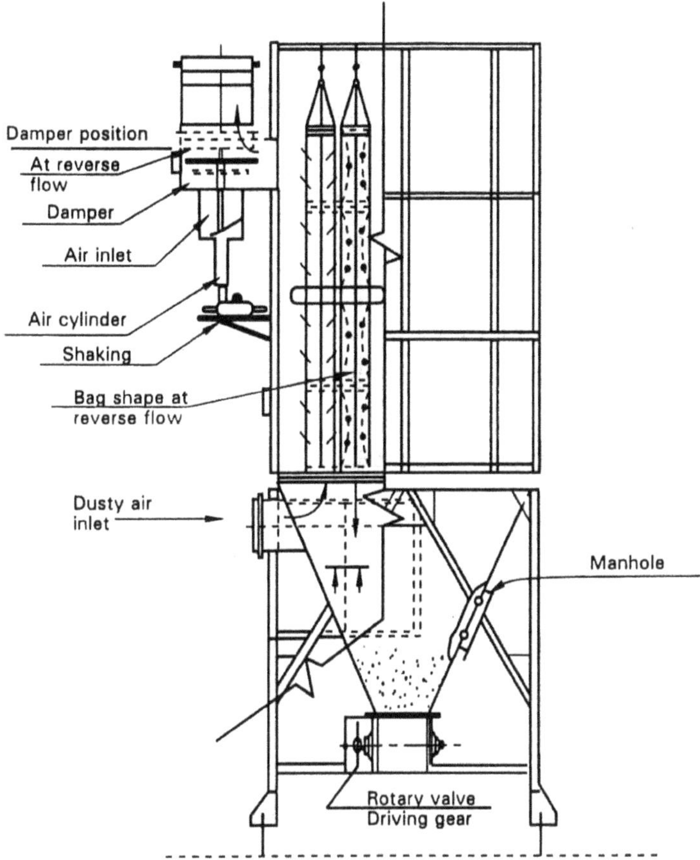

Fig. 7.18(b) *Bag filter with mechanical shaking of middle ring and reverse flow*

In the pulse-jet baghouse, bags are suspended on tube-bag headers at the top and mounted on cage supports. The cage is to prevent bag collapse. Dirty air flows from the outside toward the inside of the bag, thus the necessity of the cage support. Clean, filtered air then flows up through the tube header, thence into the outlet. The bags are cleaned by pulsing a blast of high-pressure air from a blow pipe at 4.17 to 5.16 $\times$ 10^3 t or of every few minutes. There is no need to isolate a compartment, as cleaning is done at the same time that filtration is going on. In fact, no compartment is needed.

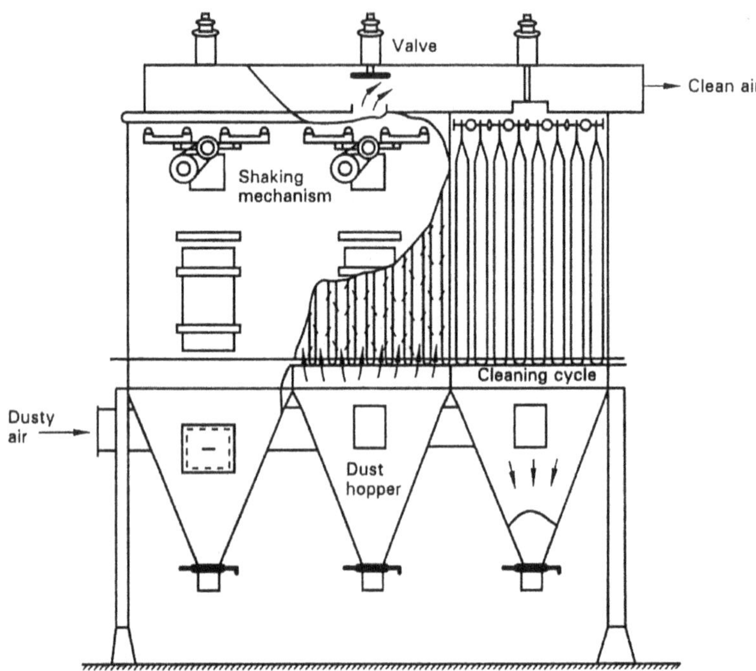

Fig. 7.18(c) *Typical bag filter with mechanical shaking of upper ends*

Table 7.8 *Fabric filter area vs number of compartments*

Filter area (m^2)	Number of compartments
1–200	1–2
201–600	3–4
601–1000	5–7
1001–1600	8–10
1601–2200	11–13
2201–3000	14–16
3001–4100	17–20
> 4100	> 20

Problem 7.29

A baghouse is to be constructed to control emissions from a grain elevator. The filter bags to be used measure 0.4 m in diameter and 6.0 m in length. The system is used to control 25 m^3/s of waste-air flow. Determine the number of bags required.

Solution: From the air/cloth ratio (Table 7.10), use an air/cloth ratio of 3.0 cfm/ft^3 = 0.0152 m/s.

Then

$$\text{No. of bags} = \frac{25}{0.0152(\pi)(0.4)(6)} = 216.6, \text{ say 216 bags.}$$

In fabric filters, the open spaces in woven fabric are usually many times the size of the individual particle. Because of this, the collection efficiency in the initial stages is rather poor. After a short period of operation, the captured particles form a bridge across the cloth openings and a particle layer forms which improves the collection efficiency greatly. Fabric filters usually provide collection efficiencies exceeding 99% and even efficiencies of 99.99% are not uncommon. Careful maintenance is necessary to prevent a deterioration of efficiency. Periodic mechanical cleaning is necessary and this usually does not remove all the interstitial layer so that collection efficiencies remain high throughout the useful life of the filter.

Bag houses have found a wide variety of applications. Examples include the carbon black industry, cement crushing, feed and grain handling, gypsum, limestone crushing, and sanding machines. Bag houses application to boiler flue gas is finding wide acceptance currently, in contrast to past practice. This is because of the better thermal properties of the bag material. Cotton and wool fibre bags, for example, cannot be used for sustained temperatures above 90–100 °C. Glass fibre bags, however, can be used for sustained temperatures up to 260 °C. Bag house sizing is based on the ratio of filter area to gas flow rate. Note that this air to cloth ratio is a velocity with an inward direction. An average value would be about 0.01 m^3/s. m^2 for a woven fabric in a conventional bag house.

The cleaned compartment is put back into service. Bag life varies between one and five years. Two years is considered normal.

Fabric filters may be classified according to the filtering media: woven fabric or felt cloth. Woven fabrics have a definite long range repeating pattern and have considerable porosity in the direction of gas flow. These open spaces must be bridged by impaction of interception to form a true filtering surface.

Felted cloth consists of randomly oriented fibres, compressed into a mat and needled to some loosely woven backing material to improve mechanical strength. Each one of the randomly oriented fibres acts as a target for impaction and interception. This type of filter usually requires substantially lesser cloth area for a given loading than woven type filters and they cannot be used in high humidity service, especially if the particles are hygroscopic because of the high tendency for clogging.

Operating temperature and the corrosiveness or abrasiveness of the particles are the main factors to be considered before selecting the type of fabric filter to be used. Various filter media are prepared from fibre aggregates, woven fibres, loosely packed beds and paper filters, using both natural and synthetic materials as filament fibres of 1 mm or less in diameter.

Table 7.9 *Main characteristics of some commonly used fibre filter materials*

Fibre	Relative strength	Specific gravity	Normal moisture content (%)	Max. usable temp °C	Relative resistance to attack by			Other properties
					Acid	Base	Organic solvent	
Cotton	Strong	1.6	7	180	Poor	Medium	Good	Low cost
Wool	Medium	1.3	15	210	Medium	Poor	Good	Fair resistance to abrasion
Paper	Weak	1.5	10	180	Poor	Medium	Good	–
Polyamide (nylon)	Strong	1.1	5	220	Medium	Good	Good	Easy to clean, excellent resistance to abrasion
Polyester (dacron)	Strong	1.4	0.4	280	Good	Medium	Good	Easy to clean
Acrylonitirile (Orlon)	Medium	1.2	1	250	Good	Medium	Good	–
Vivylidene chloride	Medium	1.7	10	210	Good	Medium	Good	–
Polyethylene	Strong	1.0	0	250	Medium	Medium	Medium	–
Tetrafluoroethylene (teflon)	Medium	2.3	0	500	Good	Good	Good	Expensive
Polyvinyl acetate	Strong	1.3	5	250	Medium	Good	Poor	–
Glass	Strong	2.5	0	550	Medium	Medium	Good	Poor resistance to abrasion
Graphitized fibre	Weak	2.0	10	500	Medium	Good	Good	Expensive
Asbestos	Weak	3.0	1	500	Medium	Medium	Good	–
"Nomex" nylon	Strong	1.4	5	450	Good	Medium	Good	Poor resistance to moisture

The general characteristics of some commonly used fibres are presented in Table 7.8 along with recommendations for maximum operating temperatures, as well as an indication of chemical resistances. Note that polypropylene and teflon have excellent chemical resistances to acids and alkalis. In addition, teflon has a high temperature of operation of 204.7 °C. Glass has the highest temperature of operation (287.8 °C) but has poor resistance toward alkali, although it has high resistance toward acids.

The **advantages** of fabric filters which make them the best choice are:

1. Very high efficiencies
2. Retention of finest particles
3. Collection of particulates in dry form
4. Relatively low pressure drop

The **disadvantages** are:

1. Their large size
2. High construction costs
3. Their application only to process temperatures generally below 285 °C

7.5.10 Design aspects of fabric filters

Variables to be established in the design of bag houses are filtering area, cleaning cycle, pressure drop, and compressor requirement for pulse-jet bag houses. The filtering area is calculated using *filtration velocities* or *air/cloth ratios* that must be determined experimentally for a particular type of dust. Table 7.9 shows some air-to-cloth ratios for some types of dust.

The cloth shaker-fabric collector is applied to a wide variety of dust producing operations in the steel making, rock products and metal mining, foundry, fertilizer, cement, metalworking, woodworking, chemical, and pharmaceutical industries. It is normally rated at air-to-cloth ratios of 1:5:1 to 4:5:1. The usual filter media is specially woven cotton sateen.

Table 7.10 *Air/cloth ratios for various types of dust*

Dust	Maximum air/cloth ratio (15.52×10^{-4} m^2/s)
Leather, paper, tobacco, wood	3
Alumina, clay, coke, charcoal, cocoa, lead oxide, mica, soap, sugar, talc	2
Aluminium oxide, carbon, fertilizer, graphite, iron ore, lime, paint pigments	2
Asbestos, limestone, quartz, silica	2
Cork, feeds and grain, marble, oyster shell	3
Bauxite, ceramics. chrome ore, feldspar, flour, flint, glass, gypsum, plastics	2

7.5.11 Design of cleaning cycle

For the shaker and reverse-air bag houses, the compartments are cleaned around a cycle. Imagine that compartment one has just been cleaned and immediately put back to service. Then a second compartment is now in line for the next isolation and cleaning. Let t_r, the *run time* be the time from when the first compartment is put into service and the time when the second compartment is isolated for cleaning. Also let t_c, be the time it takes to clean a compartment. Call $t_r + t_c$ the run clean time of the cycle. If there are N compartments in a cleaning cycle, there will be N of these times; and the total length of run-clean times is $N(t_r + t_c)$ [i.e., $N(t_r + t_c)$ is the cycle time]. Let t_f be the total filtration time of compartment one before it is cleaned again. During the $N(t_r + t_c)$ run-clean times around the cycle, compartment one is running, except for its cleaning time. Therefore, its filtration time t_f which is equal to the filtration time of any compartment is

$$t_f = N(t_r + t_c) - t_c \qquad \text{.....(7.98)}$$

The value of t_c can range from 1 to 5 min.

From the discussion above it can be deduced that the air/cloth ratio is highest during cleaning. Hence, if the design value is not to be exceeded, the air/cloth ratio should be based on $N - 1$ compartments running any time.

Problem 7.30

In Problem 7.29, determine the number of compartments and the number of bags in each compartment to be used. Design the cleaning cycle.

Solution: Total filter area = $216(\pi)(0.4)(6) = 1643\ m^2$; hence from Table 8.7, number of compartments = 12.

To maintain design conditions, when one compartment is isolated for cleaning, there should be 245 bags in use. Hence

$$\text{No. of bags / compartments} = \frac{216}{12} = \text{say 18 bags.}$$

Assume that a study on a similar baghouse treating a similar waste has established the following values: $t_r = 9$ min and $t_c = 2.5$ min. Then

$$t_f = N(t_r + t_c) - t_c = 12(9 + 2.5) - 2.5 = 135.5 \text{ min}$$

7.5.12 Determination of pressure drop

As filtration proceeds, dust continues to build up on the filter, forming a cake. This process forms two resistances in the path of airflow: the filter cloth and the cake. The equation for the pressure drop $-\Delta P$ is as follows: ($-\Delta P$ is $P_1 - P_2$ or ΔP is $P_2 - P_1$)

$$-\Delta P = \frac{\mu c \bar{\alpha}}{2A^2 t} V^2 + \frac{\mu R_m}{At} V \qquad \text{.....(7.99)}$$

where μ is the dynamic viscosity of air, c the mass of cake collected per unit volume of air filtered, $\bar{\alpha}$ the average specific cake resistance, $\mathcal{V}$ the volume of air filtered. A the filter cloth area, t the filtration time, and R_m the resistance of the filter cloth. But $\mathcal{V}/At = V$, the filtration velocity or air/cloth ratio. Substituting in Equation (7.73) and dividing all throughout by V,

$$\frac{-\Delta P}{V} = \frac{\mu c\ \alpha\ Vt}{2} + \mu R_m \qquad .....(7.100)$$

But $c = (L_iQ_0 - L_eQ_0)/Q_0 = L_i - L_e$, where L_i and L_e are the inlet and exit dust concentrations, respectively, and Q_0 is the waste airflow. Substituting in Equation (7.100) yields

$$\frac{-\Delta P}{V} = \frac{\mu\bar{\alpha}}{2}V(L_i - L_e)t + \mu\ R_m \qquad .....(7.101)$$

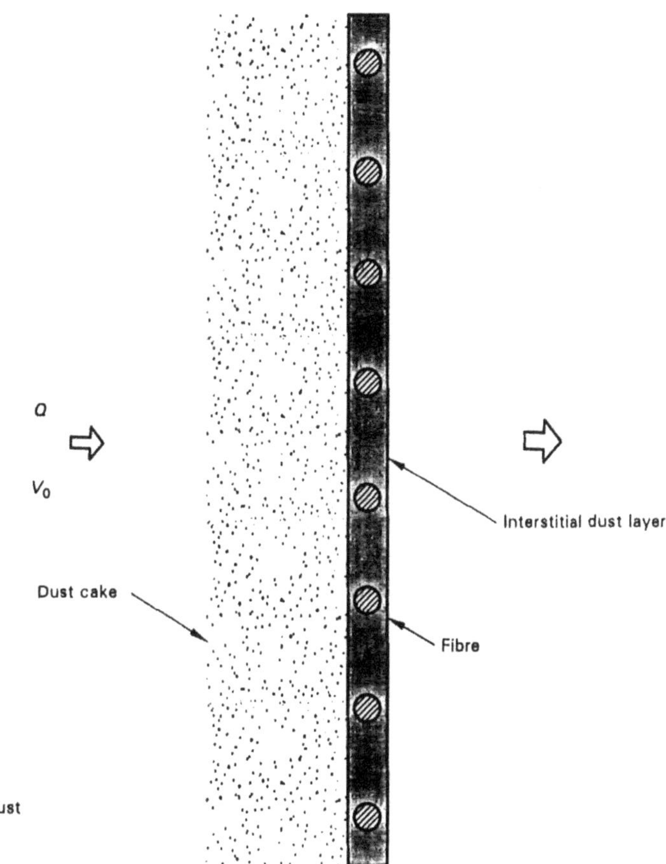

Fig. 7.19 *Single layer filter with interstitial dust layer and with dust cake attached*

In the above equation, the plot of $V(L_i - L_e)t$ and $-\Delta P/V$ is a straight line with a slope of $\mu\,\bar{\alpha}/2$ and intercept of μR_m; the equation requires only two experimental points to determine the parameters. $\bar{\alpha}$ is $\bar{\alpha}_0(\Delta P)^s$ where $\bar{\alpha}_0$ is a proportionality constant and s is a compressibility coefficient.

7.5.13 Compressor power

In addition to the power needed to force the waste air through the baghouse, compressed air is also needed for pulse-jet bag houses. In pulse-jet baghouses, the bags are cleaned by pulsing a blast of high-pressure air from a blow pipe at 4.17 to 5.16 × 10^4 t or at intervals of every few minutes. Typically, the flow rate of compressed air is about 0.5 to 1.0% that of the filtered air when both are corrected to the same temperature and pressure. This high-pressure air is provided by a compressor, is compressed inside the unit, and expelled at the discharge port at a much higher pressure in to the blow pipe (Figure 7.19).

Air compression is of three different types: isothermal, adiabatic, and polytropic. In the process of compression, the air is heated and the resulting temperature may become excessive. For this reason the compressed air is cooled by after coolers and intercoolers. In the design of the machine, channels are provided in the body to enable a coolant such as water to be circulated. Units that provide coolant between stages are called intercoolers and those that provide coolant at the final discharge port are called aftercoolers. The cooling process can become so effective that the air exit temperature may approach that of the inlet. The air is then said to be subjected to an *isothermal* process.

7.6 Particulate Liquid Scrubbers

7.6.1 Principle

Liquid scrubbers which are known as wet collectors usually utilize water, to assist in removing particulate matter or liquid or gaseous contaminants from a polluted gas stream. The primary collection mechanisms are impaction and centrifugal force.

The underlying principle of operation of the liquid scrubbers is that a differential velocity between the droplets of collecting liquid and the particulate pollutant allows the particle to impinge onto the droplet. Since the droplet-particle combination is still suspended in the gas stream, an inertial collection device is placed downstream to remove it. Because the droplet enhances the size of the particle, the collection efficiency of the inertial device will be higher than it would be for the original particle without the liquid drop.

As fine particles have a tendency to adhere to a liquid drop when in contact by creating an atmosphere for attaining intimate contact, the drop and the fine particulate can be efficiently collected between them. When one has to collect a set of 0.5 μ particles, cyclones will not work at all. However, by introducing a large number of 50μ diameter drops of liquid into the gas stream to collect the fine particles, one can pass the stream through a cheap, simple cyclone and collect the drops and the fine particles stuck on them. This idea is the basis of almost all scrubbers for particulate control.

When the particulate matter to be collected is wet, corrosive, or very hot, the fabric filter will not work while liquid scrubbing may be highly effective. Typical scrubbing applications include control of emission of talc dust, phosphoric acid mist, foundry dust, and open hearth steel furnace fumes.

As shown in Fig. 7.20 a scrubber will have three main components.

(a) A gas liquid contactor
(b) A gas liquid separator and
(c) A gas solid separator

In many scrubbers the gas liquid separator will be a simple cyclone and the liquid-solid separator will be a gravity settler.

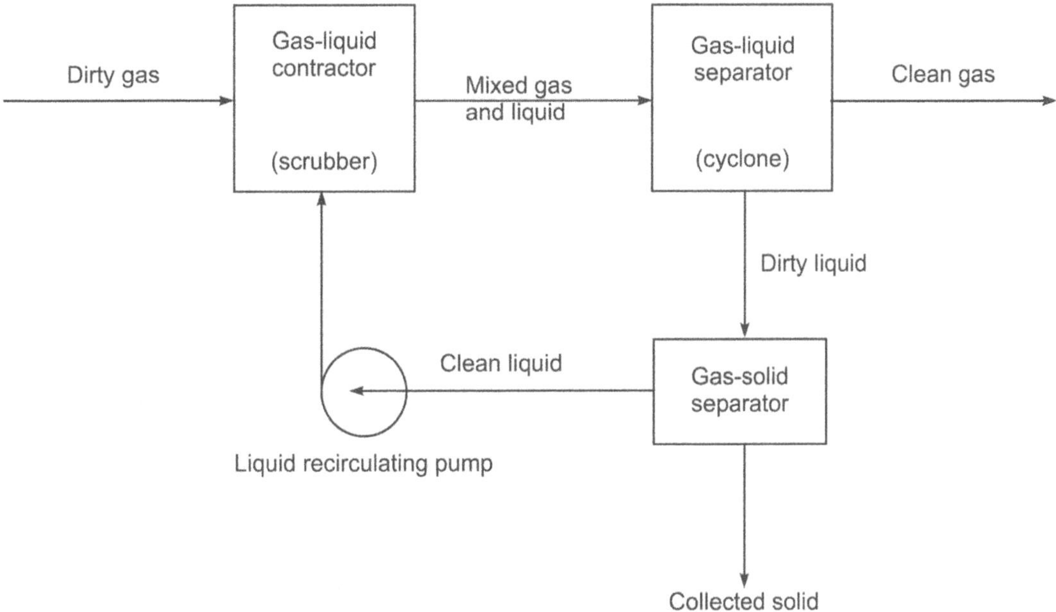

Fig. 7.20 *Components of a scrubber*

These scrubbers work on the same principle of the removal of particulate matter from the atmosphere by the rain drop in the process of rain.

7.6.2 Commercial wet scrubbers

Wet collectors are commercially available in a wide variety of designs, shapes, and sizes and the contact between the scrubbing liquid and the particles is achieved by a variety of ways. The following types are categorized according to their most dominant characteristics.

Liquid scrubbers vary in complexity. Simple spray chambers are used for relatively coarse particle sizes. For high efficiency removal of fine particles, the combination of a venturi scrubber followed by a cyclone can be selected.

The following are the major types of wet collector in use:

1. Spray towers
2. Centrifugal scrubbers
3. Packed beds and plate column scrubbers
4. Venturi scrubbers

The advantages and disadvantages of wet scrubbers are given in Table 7.11.

Table 7.11 *Advantages and disadvantages of wet scrubbers*

Advantages	Disadvantages
Simultaneous removal of gases and particulates can be effected	Relatively costly due to high energy usage
Effective performance over a wide loading range can be achieved	Disposal of wet sludge a big problem
Equipment occupies only a moderate amount of space compared to dry collectors such as baghouses	Corrosion problems due to wet sludge
Explosive dust-air mixtures are reduced	Visible wet plume, reduction in buoyancy and dispersion of pollutant
Temperature and moisture content of gas have no influence	Very small particles (sub-micron sizes) cannot be collected
Corrosive gases may be neutralised by proper choice of scrubbing liquid	High water usage

7.6.3 Water requirements

Collector water requirements represent a continuing operating cost which must be evaluated when selecting specific equipment. The water requirements for various types of scrubbers are presented in Table 7.12.

Table 7.12 *Water requirements for wet scrubbers*

Wet collector	Water rate (gal/1.000 ft^3)
Spray tower	10
Scrubber	2 – 5
Packed bed/plate column	0.5 – 1.5
Orifice type	10 – 40
Venturi scrubber	5 – 15

Problem 7.31

A fabric filter consists of 100 row of parallel fibers arranged in series which have individual collection efficiency of 9.5%. They are spaced far enough apart that the flow field becomes completely uniform between one row and the next (i.e., the rows do not interact). What is the collection efficiency of the entire filter?

Solution: Here we calculate

$$\eta_{\text{overall}} = 1 - p_{\text{overall}} = 1 - (p_{\text{individual}})^n = 1 - (1 - 0.095)^{100} = 0.999 = 99\%$$

7.6.4 Collection mechanisms of scrubbers

In actual application impaction, interception and diffusion mechanisms are not independent so all three mechanisms must be combined for calculating the fibre efficiency. A reasonable estimate for the combined efficiency of collection η_f is

$$\eta_t = 1 - (1 - \eta_{\text{impact}})(1 - \eta_{\text{inter}})(1 - \eta_D) \quad .....(7.102)$$

In addition to the above mentioned three collection mechanisms, electrostatic forces between particles and fibres increase the collection efficiency. The generation of electrostatic charges in filter fabrics may be due to friction between gas and fabric at high gas velocities (1.5 to 2.0 m/s). The increase in collection efficiency on account of electrostatic forces becomes greater with increase in the strength of the electric charge.

The resistance to airflow is directly proportional to the air velocity and collected dust load. The initial resistance is based upon a specified rated face velocity. Face velocity is determined by dividing the airflow rate by the face area of the filter.

$$\text{Face velocity} = \frac{\text{Flow rate}}{\text{Flow area}} \quad .....(7.103)$$

The initial resistance of most filters is based upon rated flow velocity of 225 m^3/min. Initial resistances at other face velocities can be approximated by the equation (7.102)

$$U_t = \frac{Q}{A} \quad .....(7.104)$$

The relative efficiencies of different mechanisms are given in Table 7.13.

Table 7.13 *Comparison of collection mechanisms*

	Impaction and interception	**Diffusion**
Increasing particle size causes efficiency to	Increase	Decrease
Increasing gas velocity causes efficiency to	Increase	Decrease
Increasing target diameter causes efficiency to	Decrease	Decrease

Problem 7.32

In a fabric filter a bag house is to be designed to handle effluent gas flowing with the velocity of 600 m^3/min. The filtering velocity is 4 m/min. Each bag is 20 cm in diameter and 40 m high. The baghouse is to be square in cross section, with 30 cm spacing between bags and 20 cm clearance from the walls. Calculate (a) the number of bags required, and (b) the width of the baghouse.

Solution:

Total gas flow, $Q = 600\ m^3/min$

Let the filtering velocity be 4 m/min (i.e., 2 $m^3/m^2/min$)

$$\text{Cloth area for filtering} = \frac{600\ m^3/min}{4\ m/min} = 150\ m^2$$

Area of each bag $A = \pi DH$

$$= \pi \times 0.2 \times 4$$

$$= 22/7 \times 0.2 \times 4 = 2.51\ m^2$$

$$\text{Total number of bags} = \frac{150\ m^2}{2.51} = 59.76 = 60$$

7.6.5 Spray towers

In spray towers the water is introduced through nozzles, which direct the spray to wards the bottom of a circular or rectangular chamber (Fig, 7.21). The polluted gas enters at the

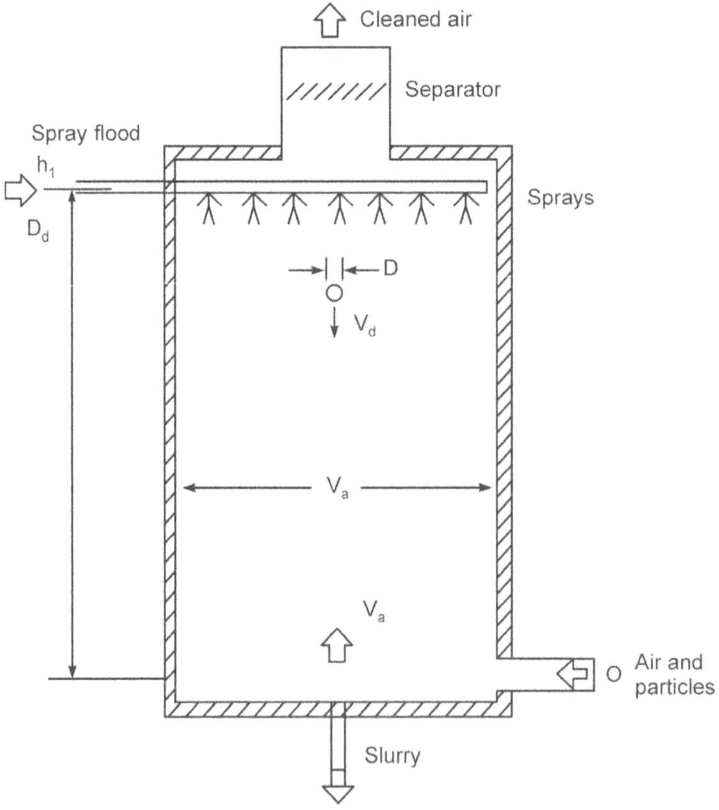

Fig. 7.21 *Spray tower*

bottom, and as it flows upward the dust particles collide with the liquid drop lets generated by the nozzle spray. The liquid droplets and their captured dust particles impinge on the water eliminator and are removed from the gas stream. This is the simplest type of counter flow gas scrubber with a moderate contact between phases and so is used only for removing coarse dust when high efficiency is not necessary.

The particle collection can be done by the mechanism of inertial impaction and interception on the droplets. Spray towers cause very little pressure loss and can handle large volumes of gases. The towers are very effective in removing particles larger than10μm. Stairmand [17] observed that collection efficiency by inertial impaction is independent of particle size, when droplets fall under gravity through air. The maximum efficiency occurs if droplets have a diameter of 800 μm. Figure 7.22 gives the target efficiencies for different droplet diameters. Thus the efficiency of a spray tower depends upon the droplet size, flow velocity of the gas, liquid gas ratio and droplet trajectories.

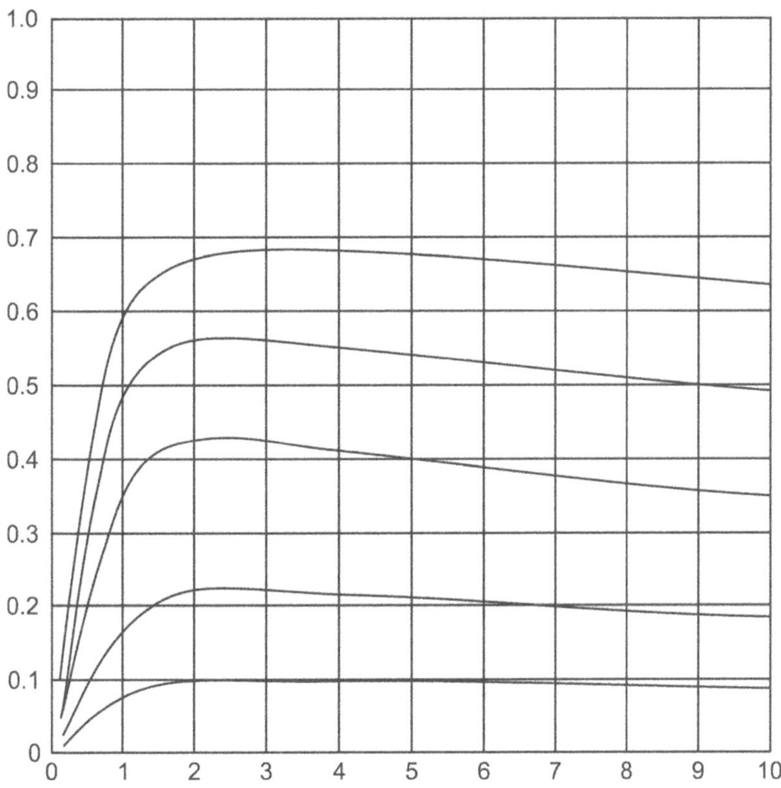

Fig. 7.22 *Target efficiency and droplet size for 2–10 μm diameter particles*

For identical spherical droplets falling through a slow upward moving gas stream, the overall collection efficiency is given by

$$\eta_{sc} = 1 - (1 - \eta_d)^n \qquad .....(7.105)$$

where n = number of collecting droplets encountered by a group of particles and η_d = individual drop collection efficiency.

For spray towers the value of h can be calculated from the expression

$$n = \frac{1.5 m_s L}{\rho_L d}\left(\frac{1}{Q} + \frac{1}{A_{sc}(V_{td} - V_p)}\right) \qquad .....(7.106)$$

where m_s = mass flow rate of spray liquid, L = height of spray tower, d = average droplet size, Q = volumetric flow rate of gas, ρ_L = spray liquid density, A_{sc} = tower cross sectional area, if it is circular $A_{sc} = D^2Sc/4$, V_{td} = terminal free falling velocity of the droplet, and V_p = particle velocity (taken to be the same as that of the gas velocity = Q/A_{sc}). The efficiency of a conventional spray tower varies from 94% for 5 μm particles and 99% for particle size of 25 μm. If water is the spray liquid this equation becomes

$$\eta_d = \frac{0.0015 m_s^2}{d} \times \frac{1}{Q} + \frac{1}{A_{sc}(V_{td} - V_p)} \qquad .....(7.107)$$

7.6.6 Centrifugal scrubbers

The wet centrifugal collector utilizes centrifugal force and the wet impingement principle in collecting light to heavy loading of all sizes of granular dust (Fig. 7.23). Dry dust particles

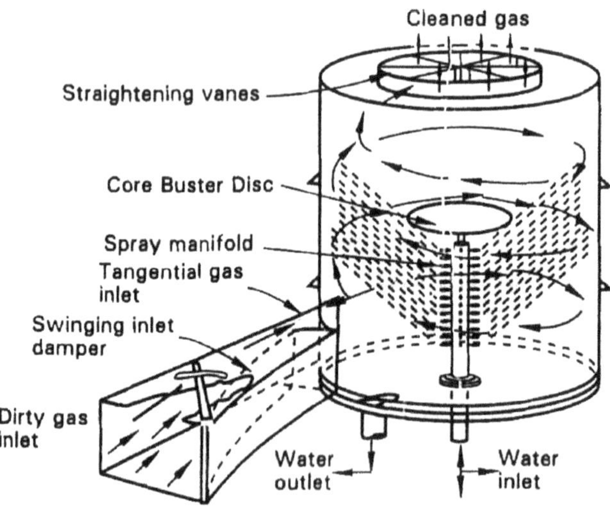

Fig. 7.23 *Centrifugal scrubbers*

impinge against the wetted peripheral surfaces, thus reducing re-entertainment of the particle into the gas stream. As the incoming gas is spun down the inlet cones, the water and collected solids are separated from the gas stream.

The collection efficiency of the centrifugal scrubber is given by

$$\eta = 1 - \frac{C_n V_L}{C_n V_0} \quad \text{.....(7.108)}$$

$$= 1 - \exp\left(-\frac{\Delta N_p}{C_{nv} - N_d P_d D^3} + \frac{6Q_s - \rho_d}{\pi Q_p} \right)$$

where C_nV_L and C_nV_0 are the number density of particulates per unit area in the bottom and top of the centrifugal tower, ΔN is the rate at which the particulates are collected by the drops. D is the diffusion of the particle cm^2/s, N_d is the number of drops collecting the particulates, Q_s is the gas volumetric flow rate.

The collection efficiency for particles smaller than those recovered in spray towers can be increased through the use of centrifugal scrubbers. The simplest type of centrifugal scrubber can be constructed by inserting banks of nozzles inside a conventional dry cyclone. The spray acts on the particles in the outer vortex, and the droplets loaded with particles are thrown outward against the wet inner wall of the cyclone. In the absence of the spray, the efficiency will be the same as that for a dry collector. The particle diameter (d_{pc}) for centrifugal spray scrubbers is between 2 and 3 μm. Commercial scrubbers have operating efficiencies of 97% or better for particles larger than 1 μm. The centrifugal acceleration involved may range from 25 to 300 g. In a centrifugal field of 100 g, for example, the most effective droplet size is 100 μm. Smaller droplets can be entrained in the gas stream whereas the bigger droplets are found to have smaller inertial impaction parameters.

7.6.7 Packed beds and plate column scrubbers

Packed towers are very efficient absorption devices involving a continuous contact of two phases. These use a variety of packing materials ranging from specially designed ceramic packing to crushed rock. The liquid is distributed over the packing, which provides high interfacial surface area, and flow down the packing surface in the form of thin films or sub divided streams.

7.6.7.1 Packed scrubber

In a packed scrubber, fibre glass, saddles, coke or broken stone, etc. are used as the collection material. (Fig. 7.24) The polluted gas stream moves upward in a counter current-flow packed scrubber and comes in contact with the scrubbing liquid stream which is moving downward over the packing in a film. The gas stream passes through the packing pore spaces and captures the particles by the inertial impaction. Because of the good mass transfer characteristics of the packing, efficient collection of fine particles by diffusion is also possible. Smaller packing increases the efficiency of collection but its shape does not appear to affect the collection efficiency. Sometimes packing towers encounter plugging problems, which can be reduced by employing sprays to wash the packing or by using low density spheres, etc. agitated by gas flow. Chemical foam packing is being employed now which drains slowly from the scrubber with captured particles and is replaced with fresh material. The Impingement baffle plate scrubber is shown in Fig. 7.25.

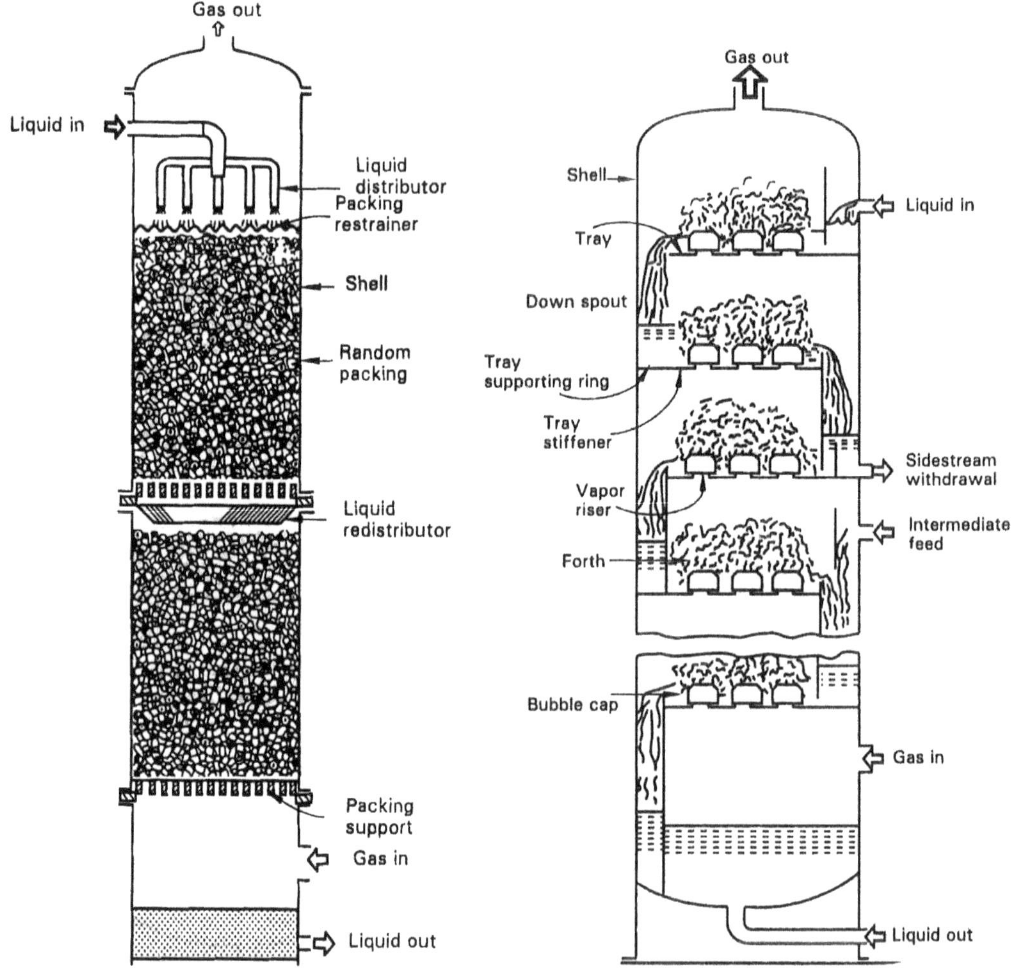

Fig. 7.24 (a) *Packed tower*

Fig. 7.24 (b) *Bubble cap plate tower*

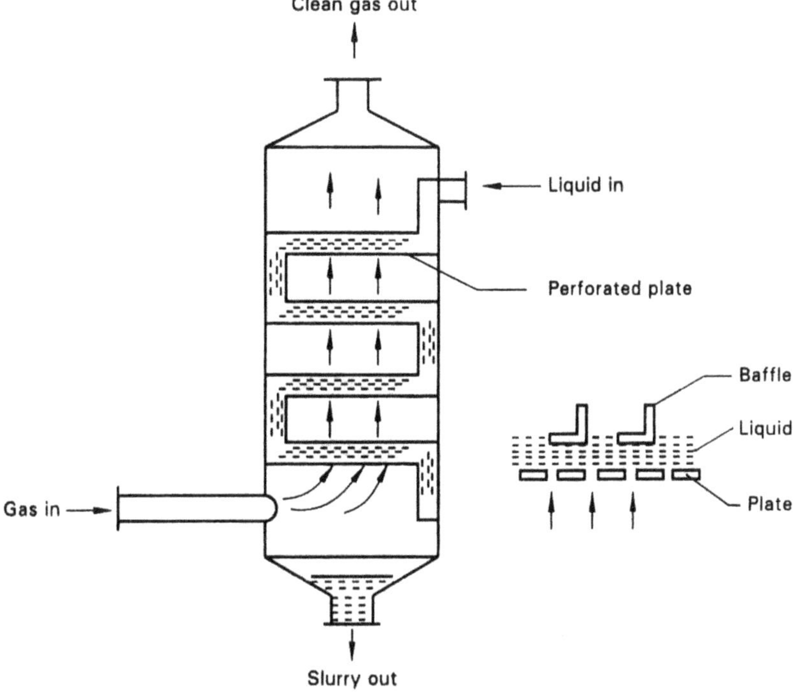

Fig. 7.25 *Impingement baffle plate scrubber*

7.6.8 Venturi scrubbers

Venturi scrubbers are employed when high collection efficiencies are required and when most of the particles are smaller than 2 pun in diameter. There are a number of instances, in fact, where a venturi scrubber is the only practical device for a gas-cleaning application. If the particles to be removed are sticky, flammable, or highly corrosive, for example, electrostatic precipitators and fabric filters cannot be used. Venturi scrubbers are also the only high-efficiency particulate collectors that can simultaneously remove gaseous species from the effluent stream.

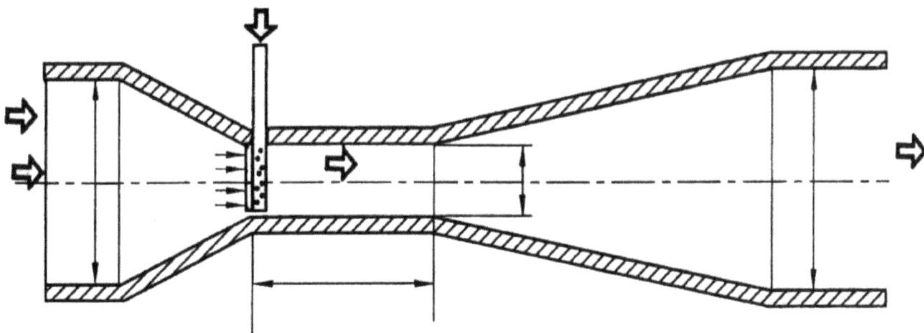

Fig. 7.26 *Cross-section of a venturi scrubber*

The distinguishing feature of a venturi scrubber is a constricted cross section or throat through which the gas is forced to flow at high velocity. A typical venturi configuration is shown in Figure 7.26.

The configuration includes a converging conical section where the gas is accelerated to throat velocity, a cylindrical throat, and a conical expander where the gas is slowed down. Liquid can be introduced either through tangential holes in the inlet cone or in the throat itself. In the former case, the liquid enters the venturi as a film on the wall and flows down the wall to the throat, where it is atomized by the high-velocity gas stream. In the latter, the liquid is injected perpendicular to the gas flow in the throat, atomized, and then accelerated. Gas velocities in the range 60 to 120 m s -l are achieved, and the high relative velocity between the panicle-laden gas

Venturi scrubbers offer high performance collection of fine particles, usually smaller than 2 to 3 μm in diameter. They are particularly suitable when the particulate matter is sticky, flammable or highly corrosive. The high performance of the venturi scrubbers is achieved by accelerating the gas stream velocities, of the order of 60–120 m/s.

Due to the high speed action, the feed liquid is atomized in a uniform fashion across the throat through several low pressure spray nozzles directed radially inward as shown in Figs. 7.27.

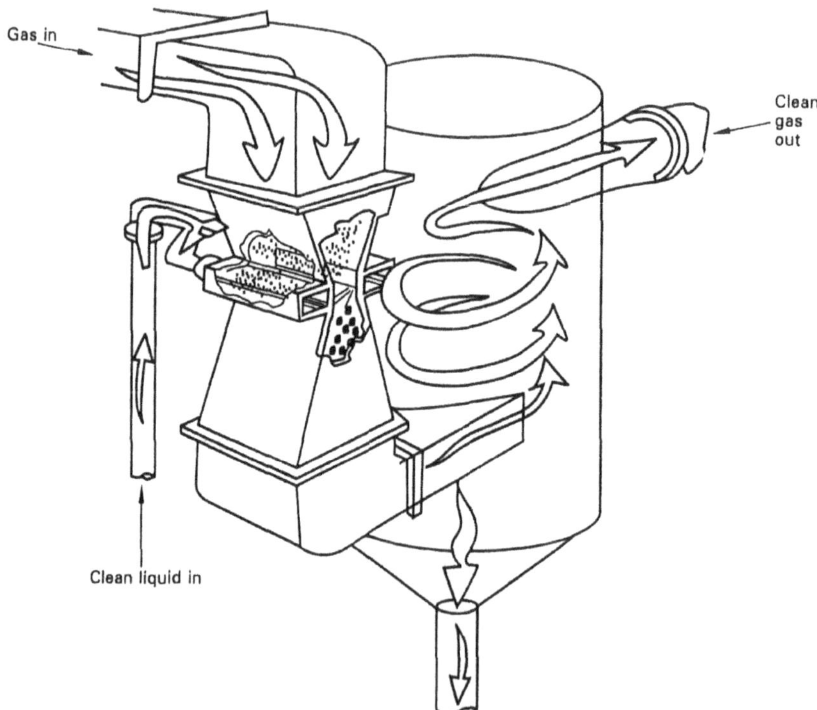

Fig. 7.27 *Schematic of a typical down flow venturi scrubber, with liquid injection at the throat and the discharge passing to a large cyclone separator*

The droplets accelerate in the throat section (7.25) and due to the velocity difference between the particles and the droplets the particles are impacted against the slow moving droplets. The acceleration continues to some extent into the diverging section of the venturi. The gas liquid mixture is then directed to a separation device such as a cyclone separator where particulate matter is separated from the gas stream. The affecting mechanisms for collection of particulates in the scrubber are inertial impaction, diffusion, electrostatic phenomenon and condensation and agglomeration, the principal mechanism is inertial impaction. The application of venturi scrubber is more often in Kraft mill furnaces, metallurgical furnaces, sulphuric acid concentrators, etc. for removing mists and dusts from gases.

The collection efficiency of the venturi scrubber with the impaction parameter (ψ) is given by John stone, Field, and Tassler [18]:

$$\eta_{sc} = 1 - \exp(-k_L \sqrt{\psi}) \quad .....(7.109)$$

where n = scrubber efficiency

K_{sc} is empirical parameter whose magnitude will be based on the throat geometry which may be of the order 0.1 to 0.2.

L = liquid flow rate (in $m^3/1000\ m^3$ of gas)

The inertial impaction parameter (ψ) relates the particle and droplet sizes and relative velocities which is given by

$$\psi = \frac{C\rho_p V_r (d_p)^2}{18 D_d \mu} \quad .....(7.110)$$

where V_r = is the relative velocity of the gas to liquid at venturi throat in m/s

ρ_p = particle density, kg/m^3

D_p = diameter of particle, m

D_d= diameter of droplet, m

μ= dynamic viscosity of gas, Pa. s

The Cunningham correction factor accounts for the fact that very small particles do not obey Stokes' settling equation. They tend to "slip" between the gas molecules. Thus, the drag coefficient (C_D) is reduced and the particles fall faster than otherwise would be expected. This is particularly true for particles less than 1 mm in diameter. The Cunningham factor may be approximated with the following equation: [19]

$$C = 1 + \frac{6.21 \times 10^{-4} (T)}{d_p} \quad .....(7.111)$$

where T = absolute temperature, K

The diameter of the droplet obtained from gas atomization in the venturi can be evaluated from the equation

$$Dd = 16400/V_r + 1.45\ L^{1.5} \quad .....(7.112)$$

The co-flow venturi scrubber shown in Fig. 8.25 has a rectangular cross-section, chosen for ease of fabrication. Liquid is injected into the throat of the venturi; the discharge passes immediately to a large cyclone separator. The venturi design is widely used because it saves fan power.

Pressure drop in venturi scrubbers: The cross flow and counter flow scrubbers have very small pressure drops (and very poor efficiencies). Venturi scrubbers have higher pressure drops and higher efficiencies. The power cost of the fan that drives the contaminated gas through a venturi is much more important than the cost of the scrubber.

Application of different types of venturi scrubbers

Based on the particle size, its concentration, desired efficiency of collection, costs, space available, and maintenance factors, different types of venturi scrubbers are used and efficiency parameters for different types of scrubbers are given in Table 7.14.

Table 7.14 *Scrubber efficiency parameters [20-23]*

Scrubber	Aerosol	Correlation parameter	
		α	β
Venturi scrubber	Talc dust particulates	2.97	0.362
	Phosphoric acid mist	1.33	0.647
	Foundry cupola dust	1.35	0.621
	Open hearth steel furnace fume	1.26	0.569
Venturi and cyclonic spray scrubber	Raw gas (lime dust and soda fume)	1.47	1.05
	Black liquor recovery furnace fume	1.75	0.620
	Ferrosilicon furnace fume	0.870	0.459
Venturi pipeline and cyclonic spray scrubber	Pre-washed gas (soda fume)	0.915	1.05
	Hot fume solution for scrubbing	0.740	0.861
Venturi condensation scrubber with			
(i) mechanical spray generator	Copper sulphate	0.390	1.14
(ii) hydraulic nozzles	Copper sulphate	0.562	1.06
Cyclone	Talc dust	1.16	0.655

7.7 Summary of Particulate Emission Control Techniques

The summary of particulate emission control techniques, including minimum particle sizes, ranges of efficiency, and advantages and disadvantages of each type of unit are presented in Table 7.15 which help in understanding which method to be adopted to meet a particular gas cleaning need.

Table 7.15 *summary of particulate emission control techniques*

Device	Minimum particle size (μm)	Efficiency (%) (mass basis)	Advantages	Disadvantages
Gravitational settler	> 50	< 50	Low-pressure loss Simplicity of design and maintenance	Much space required Low collection efficiency
Cyclone	5-25	50-90	Simplicity of design and maintenance Little floor space required Dry continuous disposal of collected dusts Low-to-moderate pressure loss Handles large particles Handles high dust loadings Temperature independent	Much head room required Low collection efficiency of small particles Sensitive to variable dust loadings and flow rates
Wet collectors Spray towers Cyclonic Impingement Venturi	> 10 > 2.5 > 2.5 > 0.5	< 80 < 80 < 80 < 99	Simultaneous gas absorption and particle removal Ability to cool and clean high-temperature, moisture-laden gases Corrosive gases and musts can be recovered and neutralized Reduced dust explosion risk Efficiency can be varied	Corrosion, erosion problems Added cost of wastewater treatment and reclamation Low efficiency on Submicron particles Contamination of effluent stream by liquid entrainment Freezing problems in cold weather Reduction in buoyancy and plume rise

Problem 7.33

For removing the fly ash particles (d_p = 10 m, r_p = 700 kg/m^3) of effluent gas stream in a thermal power plant, a venturi scrubber with the following dimensions is used.

Throat area =1.00m^2; Gas flow rate = 82.4 m^3/s

Gas temperature = 100 °C while $P_g = 25.2 \times 10^{-6}$ Pas

Liquid flow rate =0.13 m^3/s; Coefficient $k = 200$

Droplet diameter = 100 mm.

Derive an expression for the collection efficiency of the scrubber as a function of particle size.

Solution: We know

$$\eta_{sc} = 1 - \exp(-k_L \sqrt{\psi})$$

where ψ is

$$\psi = \frac{C \rho_p V_r (d_p)^2}{18 D_d \mu}$$

The correction factor from Equation (8.99)

$$C = 1 + \frac{6.21 \times 10^{-4} (T)}{d_p}$$

For estimation of ψ the gas velocity at the throat should be determined

$$v_g = \frac{Q_g}{A_t}$$

$$v_g = \frac{6.21 \times 10^4 \times 473 \text{ K}}{1.0 \text{ m}^2}$$

$$= 82.4 \text{ m}^3/\text{s}$$

Substituting these values in the equation for ψ can be calculated in terms of d_p in mm.

$$\psi = \frac{(1)(700 \text{ kg/m}^3)(82.40 \text{ m}^3/\text{s}) 1 \times 10^{-12} \mu\text{m}^2/\text{m}^2 (d_p)^2}{(18)(100 \times 10^{-6} \text{ m})(25.2 \times 10^{-6} \text{ Pa.s})}$$

$$= 1.27 (d_p)^2$$

$$\eta = 1 - \exp[(1 - 200)(1.38 \times 10^{-3})(1.27) d_p]$$

$$= 1 - \exp[-0.35(d_p)]$$

Problems

1. A cross flow scrubber is collecting 90% of the particles passing into it. But due to experimental limitations it is necessary to double the gas flow rate. If we hold everything else constant, what is the anticipated collection efficiency at the new flow rate?
2. A cross flow scrubber is collecting 90% of the 3 μ particles entering into it. The water drops are all the same diameter, 400 μ. If new spray nozzles make all the drops 200 μ in diameter are installed and Q_L is not changed, what is the new collecting efficiency?
3. If in a bulk chemical plant 90% of the particles from a gas stream are collected in a cross flow scrubber. It is proposed to increase the collection efficiency to 95% by

increasing the water flow to the scrubber, holding everything else constant. By what percentage must we increase the water flow to the scrubber?

4. A crossflow wet scrubber has $Q_G = 28.31$ m^3/min, $Q_L = 0.2831$ m^3/min, $\underline{A} = 0.12$ m^2, and an inlet liquid velocity of zero. The gas has the same properties as air, and the liquid the same properties as water. (a) Estimate the pressure drop of the scrubber (b) Estimate the cut diameter.

Questions

Short Answer Questions

1. Outline the clause process employed for removing H_2S.
2. Brief the principles employed for particulate control.
3. Write a note on Stoke's law and give its significance.
4. How is Stoke's law applied for particles which are too large?
5. How is Stoke's law applied for particles which are too small?
6. Determine the settling velocity of a small particle with diameter 0.25 μ and specific gravity 2.0 in still air.
7. Write a brief note on Gravitational settling chambers.
8. Mention the advantages and disadvantages of settling chambers.
9. Brief the significance of pressure drop in a cyclone separator.
10. List out the advantages and disadvantages of Electrostatic precipitators.
11. Write a short note on Direct interception.
12. Outline the components of a scrubber.
13. List out the advantages and disadvantages of wet scrubbers.
14. Write a short note on Collection Mechanisms of Scrubbers.
15. Summarize the particulate emission control techniques.

Essay Questions

1. Explain the physical mechanisms employed to remove particles from gas streams.
2. Using Stokes' law, calculate the particles falling velocity and Reynolds number of the particle falling in air which has a diameter of 225 μ.
3. Describe the collector efficiencies of a settling chamber.
4. Explain the two types of conditions in which the gravity settler can be modelled.
5. Outline the working principle of a Cyclone separator. Any one.
6. Discuss the working principle of an Electrostatic precipitator. Explain about Corona formation.

7. Bring out the Electrostatic precipitators employed in Industrial applications.
8. Explain the different types gas filters used in industrial scale for particulate removal.
9. Describe how fabric filters are classified with suitable diagrams.
10. Explain the working of Spray towers used for particle collection.
11. Explicate the working principle of Packed beds and plate column scrubbers.
12. Illustrate the working of Centrifugal scrubbers.

References

1. Air pollution engineering manual. USEPA report. PB 228 – 132 10. US Govt. Printing Office, Washington DC. 1967.
2. H. Lamb, 1962. Hydrodynamics. Dover, New York.
3. N. DeNivers. Fluid mechanics for chemical engineering. 2nd edition. McGraw Hill, New York.
4. R. H. Perry and H. Chilton, (Eds). Chemical engineers workbook, 5th edition. McGraw Hill, New York.
5. L. Schiller and A. Naumann, 1933. Z. Ver Dent Ing. 77:318.
6. C. E. Lapple, 1951. Chem. Engg. 58:144–151.
7. C. E. Lapple, Air pollution manual. (Ed.) J. A. Danielson. US Dept. Health Education &Welfare Pub. No. 999-AP 90.
8. C. B. Shepard and C. E. Lapple, 1939. Ind & Eng. Chem. 31:72.
9. Cooper, C.D., and Alley, F.C., Air Pollution Control: A Design Approach, Waveland Press, Inc., Long Grove, Illinois, USA, 2011
10. Hesketh, 1974. Understanding and controlling air pollution. Ann Arbor Science. Ann Arbor.
11. L. J. Stairmanch, 1991. The design and performance of cyclone separators. Trans. Inst. Chem. Eng. 25:356.
12. W. Dentson, 1922. Ann. Phys(Lec_zig). 68:335.
13. D. A. Lloyd. Electrostatic precipitator handbook. Adam Hilger, Bristol, England.
14. White, H.J., Industrial Electrostatic Precipitation, Addison-Wesley Publishing Company, Massachusetts 1963
15. T. G. Ramsdell, 1968. Design criteria for precipitation for modern control. Proc. amm. power center. 30:444–449.
16. A. Renord. Proof of concept testing of ESP retrofit technology for low and high resistivity fly ash. Proceedings of the 7th particulate control symposium. R. F. Altman (Ed.). 1:17–41.

17. S. J. Olesby and G. B. Nichols. 1978. Electrostatic precipitation. Marcel Dekker, New York.
18. W. Strauss, 1966. Industrial gas cleaning. 1st edition. Pergammon Press, London.
19. C. J. Stairmand, 1950. Trans. Inst. Eng. London. 28:130.
20. H. F. Johnstone, R. S. Field and M. C. Tassler, 1954. Gas absorption and aerosol collection in venturi. Ind. Engg. Chem. 46:1601–1608.
21. H. Hesketh, 1977. Fine particles in gaseous media. Ann Arbor Science. Ann Arbor.
22. K. T. Semasn, 1960. Correlation of dust scrubber efficiency. J. of air pollution control assoc. 10:200

Suggested References

1. Adachi, M., Kousaka, Y., and Oku Y Ama, K. "Unipolar and Bipolar Diffusion Charging of Ultrafine Aerosol Particles," J. Aerosol Sci., 16, 109—123 (1985).
2. Adamc'zyk, Z., and Van De Ven, T. G. M. "Deposition of Brownian Particles onto Cylindrical Collectors," J. Colloid Interface Sci., 84, 497-518 (1981).
3. Balasubramanian, M., and Meisen, A. "A Note on the Diffusional Deposition of Aerosol Particles in Packed Beds," J. Aerosol Sci., 6, 461-463 (1975).
4. Bhatia, M. V., and Cheremisinoff, P. N. "Cyclones, in Air Pollution Control and Design Handbook: Part I, P. N. Cheremisin off and R. A. Young, Eds., Marcel Dekker, New York, 281-316 (1977).
5. Bickelhaupt, R. E. "A Technique for Predicting Fly Ash Resistivity," U.S. Environmental Protection Agency Report No. EPA-600/7-79-204, Research Triangle Park, NC (1979).
6. Brock, J. R., and Wu, M. "Field Charging of Aerosol Panicles," J. Colloid Interface Sci., 45, 106-114 (1973).
7. Calvert, S. "Particle Control by Scrubbing" In Handbook of Air Pollution Technology, S. Calvert and H. M. Englund, Eds., Wiley, New York, 215-248 (1984).
8. Cooper, C. D., and Alley, F. C. Air Pollution Control: A Design Approach, PW S Engineering, Boston (1986).
9. Crawford, M. Air Pollution Control Theory, McGraw-Hill, New York (1976).
10. Davison, S. W., and Gentry, J. W. "Differences in Diffusion Charging of Dielectric and Conducting Ultrafine Aerosols," Aerosol Sci. Technol., 4, 157-163 (1985).
11. Dennis, R., and Klemm, H. A. "Fabric Filter Model Format Change," Vol. 1, "Detailed Technical Report," U.S. Environmental Protection Agency Report No. EPA-600/7-79-043a, Research Triangle Park, NC (1979).
12. Friedlander, S. K. Smoke, Dust and Haze: Fundamentals of Aerosol Behavior, Wiley, New York (1977).

13. Gebhart, J., ROTH, C., and Stahlhofen, W. "Filtration Properties of Glass Bead Media for Aerosol Particles in the 0.1-2 Size Range," J. Aerosol Sci., 4, 355-371 (1973).
14. Han, H., and Ziegler, E. N. "The Performance of High Efficiency Electrostatic Precipitators, ' Environ. Prog. , 3, 201-206 (1984).
15. Happel, J. 'Viscous Flow Relative to Arrays of Cylinders," Aiche J., 5, 174-177 (1959).
16. Happel, J., and Brenner, H. Low Reynolds Number Hydrodynamics, Prentice-Hall, Englewood Cliffs, NJ (1965).
17. Kuwabara, S. "The Forces Experienced By Randomly Distributed Parallel Circular Cylinders or Spheres in A Viscous Flow At mall Reynolds Numbers," J. Phys. Soc. Jpn., 14, 527-532 (1959).
18. Lassen, L. "Die Anlagerung Von Zerfallsprodukten Der Naturlichen Emanationen An Elektrisch Geladene Aerosole (Schwebstoffe), Z. Phys. , 163, 363-376 (1961).
19. Leith, D., and Licht, W. The Collection Efficiency of Cyclone Type Particle Collectors—A New Theoretical Approach, "In Aiche Symposium Series, No. 126, Vol. 68, 196-206 (1972). LICHT, W. Air Pollution Control Engineering, Marcel Dekker, New York (1980).
20. Licht, W. "Control of Particles by Mechanical Collectors," In Handbook of Air Pollution Technology, S. Calvert and H. M. Englund, Eds., Wiley, New York, 319-329 (1984).
21. Liu, B. Y. H., and Kapadia, A. "Combined Field and Diffusion Charging of Aerosol Particles in the Continuum Regime," J. Aerosol Sci., 9, 227-242 (1978).
22. Liu, B. Y. H., and Pui, D. Y. H. "On Unipolar Diffusion Charging of Aerosols in the Continuum Regime," J. Colloid Interface Sci., 58, 142-149 (1977).
23. Liu, B. Y. H., and Y EH, H. C. "On the Theory of Charging of Aerosol Particles in an Electric Field," J. Appl. Phys., 39, 1396-1402 (1968).
24. Lio, B. Y. H., Whitby, K. T., and Yo, H. H. S. "Diffusion Charging of Aerosol Panicles at Low Pressures," J. Appl. Phys., 38, 1592-1597 (1967).
25. Marlow, W. H., and Brock, J. R. "Unipolar Charging of Small Aerosol Particles," J. Colloid Interface Sci., 50, 32-38 (1975).
26. Mccabe, W. L., and Smith, J. C. Unit Operations of Chemical Engineering, 3rd Ed., Mcgrawhill, New York (1976).
27. Michael, D. H., and Norey, P. W. "Particle Collision Efficiencies for a Sphere," J. Fluid Mech., 37, 565-575 (1969).
28. Nielsen, K. A., and Hill, J. C. "Collection of Inertialess Particles on Spheres with Electrical Forces, Ind. Eng. Chem. Fund., 15, 149-156 (1976a).
29. Nielsen, K. A., and Hill, J. C. "Capture of Particles on Spheres by Inertial and Electrical Forces, Ind. Eng. Chem. Fundam., 15, 157-163 (1976b).

30. Oh, S. H. , Macdonald, J. S., Vaneman, G. L., and Hegedus, L. L. "Mathematical Modeling Of Fibrous Filters for Diesel Particulates–Theory and Experiment," SAE Paper No. 8101 13, Society of Automotive Engineers, Warren dale, PA (1981).
31. Paretsky, L., Theodore, L., Pfeffer, R., and Squires, A, M. "Panel Bed Filters for Simultaneous Removal of Fly Ash and Sulfur Dioxide: II. Filtration of Dilute Aerosols By Sand Beds," J. Air Pollut. Control Assoc., 21, 204-209 (1971).
32. Perry, R. H., and Chilton, C. H. , Eds., Chemical Engineers' Handbook, 5th Ed., McGraw hill, New York (1973).
33. Rajagopalan, R., and Tien, C. "Trajectory Analysis of Deep Bed Filtration with the Sphere in Cell Porous Media Model," Aiche J., 22, 523-533 (1976).
34. Rosenhead, L. Laminar Boundary Layers, Oxford University Press, Oxford (1963).
35. Slinn, W. G. N. "Precipitation Scavenging," In Atmospheric Sciences And Power Production
36. 1979, Div. of Biomedical and Environmental Research, U.S. Dept. of Energy, Chap. I L (1983).
37. Smith, W. B., and Mcdonald, J. R. "Development of a Theory for the Charging of Particles by Unipolar Ions," J. Aerosol Sci., 7, 151-166 (1976).

CHAPTER 8

Air Pollution Control Technologies for Gaseous Pollutants

8.1 Control of Gaseous Pollutants

The principal gaseous pollutants of concern in air pollution are SO_x, NO_x, CO, the hydrocarbons, and other organic and inorganic gases

There are two general methods by which gaseous pollutants can be removed from the industrial gaseous effluent stream.

(a) Sorption of gaseous pollutants by absorption in a liquid

(b) Adsorption on a solid

(c) Decomposition or conversion of the pollutant, from toxic form to non-toxic form stable form, chemically either by combustion or by catalytic incineration

8.2 Removal of Pollutants by Absorption

8.2.1 Principle

Absorption is an operation involving selective mass transfer of a soluble vapor component to a solvent liquid in a device that promotes intimate contact between the gas and the liquid. The separation principle involved is the preferential solubility of a gaseous component in the liquid. So-called physical absorption takes use of the remarkable solubility of a gaseous pollutant in the solvent, and involves only the dissolution process. For example, water is, by itself, quite efficient for removing soluble acidic gaseous such as HCl, HF and the soluble basic gas NH_3.

Gases of very little solubility such as SO_x, H_2S and NO_x, can be absorbed readily in an alkaline solution such as NaOH. Thus, when water is used as the solvent, it may contain added species, such as acids, alkaline, oxidants, or reducing agents to react with the gas being absorbed and enhance its solubility

The equilibrium vapor pressure can be decreased almost zero by adding a chemical reactant to the absorbing liquid. This in effect ties up or chemically changes the solute gas, since the driving force for absorption is the difference between the partial pressure of the soluble gas in the gas mixture and its equilibrium vapor pressure just above the surface of the liquid. Due to chemical reactions involved, such an operation is called chemical absorption.

Both in the physical and chemical absorption, the mechanism by which the species is removed from the gas consists of three steps that occur in series: (1) diffusion of the pollutant molecules through the gas to the surface of absorbing liquid, (2) dissolution into the liquid at the interface, and (3) diffusion of the dissolved species from the interface into the bulk of the liquid.

In the chemical absorption, the dissolved species reacts with the liquid species during its diffusion and/or residence in the bulk. Figure 8.1 gives the scheme of the absorption process.

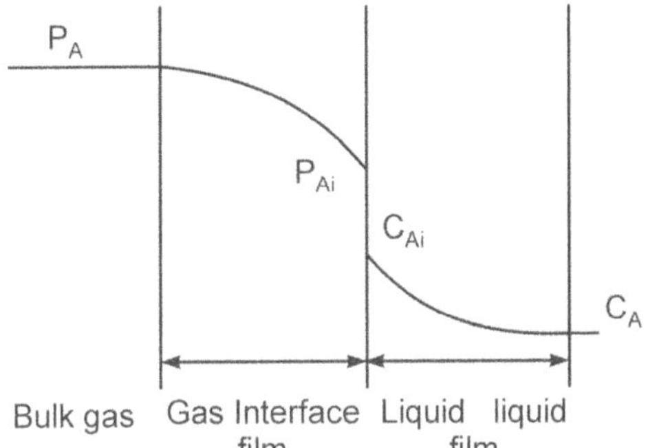

P_A : Partial pressure of pollutant A in the gas

P_{Ai} : Partial pressure of A at the interface

C_{Ai} : Concentration of dissolved A at the interface

C_A : concentration of A in the liquid bulk

Fig. 8.1 *The scheme of absorption process: two film model*

The frequently encountered examples of the application of gas absorption in air pollution control include:

- Control of sulfur dioxide and nitrogen oxides from combustion sources
- Removal and recovery of ammonia in fertilizer manufacture
- Removal of hydrogen fluoride from glass furnace exhaust
- Control of odorous gases from rendering plants
- Recovery of water-soluble solvents such as acetone and methyl alcohol.

8.2.2 Removal of pollutants by absorption of gases by moving liquid drops

Moving Liquid drop absorption is one of the most versatile techniques used for reducing the pollutant levels from various industrial gaseous effluents. In this process due to a concentration gradient the pollutant from the gaseous phase is transferred across the gas-liquid interface and the concentration decreases in the direction of mass transfer. Thus the basic collection mechanism is diffusion.

8.2.3 Theory of absorption of gases by moving liquid drops

In this process, one constituent of a mixture in gaseous phase diffuses into a liquid due to a concentration gradient. This is a process in which mass transfer of the pollutant gas takes place from points of higher concentration to points of lower concentration by diffusion. The removal of the pollutant gas takes place in three steps:

1. Diffusion of the pollutant gas to the surface of the liquid.
2. Transfer across the gas/liquid interface (dissolution).
3. Diffusion of the dissolved gas away from the interface into the liquid.

When a special drop of absorbing liquid passes through a mixture of gaseous pollutants, a boundary layer is formed (Fig. 8.2) next to the surface of the drop across which the concentration of the pollutant gas varies from $C_m v$ ($C_m v$ is the concentration of the pollutant gas at a distance from the drop) to $C_m v_0$ at surface of the drop.

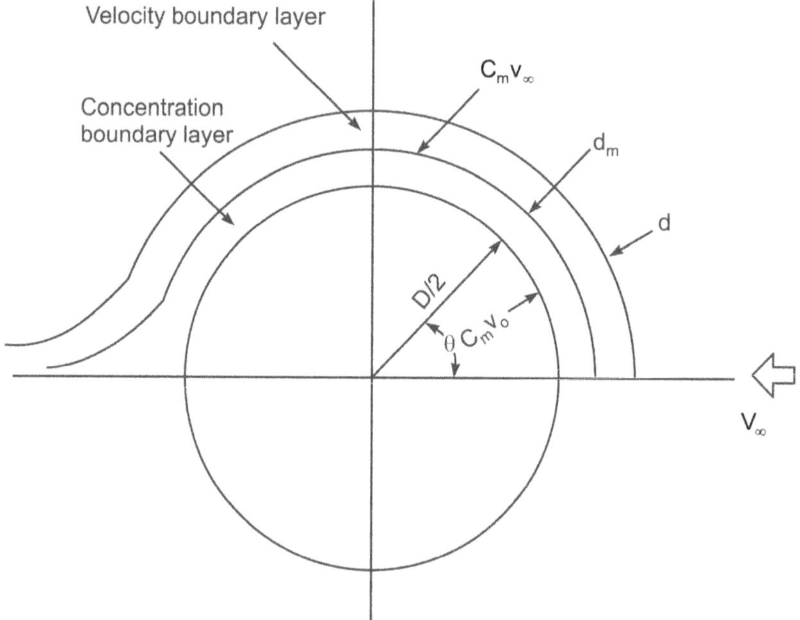

Fig. 8.2 *Concentration and velocity boundary layers on a spherical drop*

The variation of concentration of the pollutant gas across the boundary layer is given by

$$C_m v - C_m v_0 = (C_m v_\infty - C_m v_0)\left[\frac{3}{2}\frac{y}{\delta_m} - \frac{1}{2}\left(\frac{y}{\delta_m}\right)^3\right] \quad \text{.....(8.1)}$$

where δ_m is the concentration boundary layer

The collection efficiency of the absorption system is given by

$$\eta_d = 3D\frac{C_m v_\infty - C_m v_0}{C_m v_\infty}\frac{1}{\sqrt{\nu D v_\infty}} Sc^{1/3} \qquad .....(8.2)$$

where D = diffusivity coefficient

$$D = \frac{\nu}{Sc} \qquad .....(8.3)$$

where ν is velocity of the boundary layer, D is the diameter of the drop, v_∞ is the velocity of air relative to the velocity of drop, Sc is the number given $Sc = \nu/D$.

The equation can be written as

$$\eta_d = 3\times\frac{C_m v_\infty - C_m v_0}{C_m v_\infty}\times\frac{1}{Sc^{2/3} R_e^{1/2}} \qquad .....(8.4)$$

Or in terms of attachment number σ, it is given by

$$\eta_d = \frac{3\sigma}{Sc^{2/3} R_e^{1/2}} \qquad .....(8.5)$$

where

$$\sigma = \frac{C_m v_\infty - C_m v_0}{C_m v_\infty} \qquad .....(8.6)$$

The value of the concentration of the pollutants at the surface of the drop $C_m v_0$ is related to the value of the pollutant absorbed in the drop by Henry's law which states that for dilute solutions a linear relationship exists between the concentration of the gaseous pollutant in solution and partial pressure given by

$$P'_A = H'C'_A \qquad .....(8.7)$$

where

P'_A = equilibrium partial pressure of pollutant gas, kPa

H' = Henry's law constant, kPa m^3/g

C'_A = concentration of pollutant gas in the liquid drop, g/m^3

In terms of mole fraction unit Henry's law can be expressed as

$$P_A = H' x_A \qquad .....(8.8)$$

where x_A is the mole fraction of the pollutant gas A in the liquid. The values of Henry's constant of various gases in water are presented in Table 8.1 where smaller values of H′ indicate higher solubilities of the gas for a given partial pressure at equilibrium.

The validity of Henry's law extends up to partial pressures of 1 at m or more which are used in air pollution control systems using absorption by liquids.

The mass volume concentration of the pollutant vapour in the air ($C_m v$) in terms of Henry's constant is given by

$$C_m v = \frac{M_p H x_A}{RT} = \frac{HM_w}{RT} \times \frac{C'_m}{1+(M_w / M_p - 1)C'_m} \quad \text{.....(8.9)}$$

where M_p= molecular weight of pollutant gas and M_w= molecular weight of solvent, R = gas constant, C'_m is the mass fraction of the pollutant in the drop.

The mass volume concentration of the pollutant gas in the solvent phase $C'_m v = \rho_d C'_m$, where ρ_d is the density of the solution in the drop which is equated to solvent density. By substituting we have

$$C_m v = \frac{HM_w}{RT} \times \frac{C'_m v}{\rho_d + (M_w / M_p - 1)C'_m v} \quad \text{.....(8.10)}$$

If Henry's constant is known for a given concentration of the pollutant in gaseous phase one can calculate the concentration of the pollutant to be absorbed using the above equation.

Equation (8.7) indicates that the partial pressure of the gas increases as the liquid accumulates more pollutant or else it will come out of solution. However, when the liquid is removing pollutant from the gas phase, the partial pressure is decreasing as the gas is cleaned which is the reverse of what is to happen as per Equation (8.7). Bypassing the gas and liquid in opposite directions this problem can be solved; such flow is called *countercurrent flow*. In this manner, the high concentration gas is absorbed into a liquid with a high pollutant concentration. The lower concentration gas is absorbed by the liquid with no pollutants in it.

8.2.4 Scrubbing medium

The required characteristics for achieving high scrubbing efficiencies of any scrubbing process are: (a) High gas solubility and high selectivity between solute species; (b) Low vapour pressure; (c) Non-corrosiveness; (d) Easy regeneration and recovery; (e) Low cost; (f) Low viscosity at adsorption temperature; (g) Non-flammability; (h) Chemical stability.

Table 8.1 *Henry's constant (H') for various gases in water [1, 2, 3, 4]*

Substance	Formula	Molecular weight	Henry's law constant N/m^2	Diffusion coefficient	
				Air 10^{-5} m^2/s	Water 10^{-9} m^2/s
Acetylene	C_2H_2	26.0	1.35×10^9	1.7	2.0
Air		29.0	7.29×10^9		2.0
Ammonia	NH_3	17.0	303,000	2.2	2.0
Argon	Ar	79.8	3.61×10^9	1.7	1.4
Bromine	Br_2	159.8	7.47×10^6	1.0	1.3
Carbon dioxide	CO_2	44.0	1.65×10^8	1.5	2.0

Table 8.1 ***Contd...***

Substance	Formula	Molecular weight	Henry's law constant N/m²	Diffusion coefficient	
				Air 10^{-5} m²/s	Water 10^{-9} m²/s
Carbon monoxide	CO	28	5.87×10^9	2.0	2.0
Carbonyl sulphide	COS	60.1	2.63×10^9	1.3	1.5
Chlorine	Cl_2	70.9	6.82×10^7	1.2	1.5
Ethane	C_2H_6	30.0	3.07×10^9	1.5	1.4
Ethylene	C_2H_4	28.0	1.16×10^9	1.6	1.5
Gasoline			7.33×10^9		0.7
Helium	He	4.0	1.27×10^{10}	7.0	6.0
Hydrogen	H_2	2.0	7.07×10^9	7.5	4.8
Hydrogen cyanide	HCN	27.1	6.4×10^5	1.5	1.8
Hydrogen sulphide	H_2S	34.1	5.52×10^7	1.7	1.6
Krypton	Kr	83.8	2.20×10^9	1.5	2.0
Methane	CH_4	16.0	4.19×10^9	2.2	1.8
Methyl chloride	CH_3Cl	50.5	1.49×10^5	1.3	1.5
Neon	Ne	20.2	8.4×10^9	3.2	3.0
Nitric oxide	NO	30.0	2.91×10^9	2.0	2.4
Nitrogen	N_2	28.0	8.40×10^9	2.0	1.9
Nitrous oxide	N_2O	44.0	2.27×10^8	1.5	1.8
Oxygen	O_2	32.0	4.44×10^9	2.0	2.5
Ozone	O_3	48.0	4.64×10^8		2.0
Phosphine	PH_3	34.0	5.71×10^9	1.6	
Propylene	C_3H_6	42.0	5.73×10^8		1.1
Sulphur dioxide	SO_2	64.1	4.85×10^6	1.3	1.7
Water	H_2O	18.0		2.6	
Xenon	Xe	131.3	1.28×10^9	1.3	1.7

Various solvents used as a scrubbing medium for the removal of various industrial gaseous effluents are given in Table 8.2. One important precaution in working with the scrubbing solvent is that it should not again become a source of water pollution.

Table 8.2 *Scrubbing solvents for various pollutant gases*

Pollutant	Scrubbing solvent
Sulphur dioxide	Sodium hydroxide, sodium sulphite, magnesium oxide, calcium carbonate, calcium oxide and calcium hydroxide solutions
Oxides of nitrogen	Ammonium bicarbonate, ammonium bisulphate, calcium hydroxide, magnesium hydroxide, and sodium hydroxide
Hydrogen sulphide	Sodium hydroxide, potassium hydroxide solutions
Hydrogen chloride	Water, ammonia, calcium and magnesium hydroxide solutions
Chlorine	Solutions of sodium hydroxide, sodium sulphite, sodium thiosulphite, and water

Table 8.2 ***Contd...***

Pollutant	Scrubbing solvent
Phosgene	Sodium hydroxide and water
Ammonia	Sulphuric acid, nitric acid
Mercaptans	Sodium hypochlorite solution

8.3 Gaseous Pollutant Control Systems based on Absorption in a Liquid

8.3.1 Criteria for low cost and effective absorption equipment

To achieve the high performance of absorption and low cost of both capital and operation, the effective absorption equipment tends to meet the following criteria.

- large contact surface between gas and liquid;
- enough contact time between two phases;
- highly turbulent mixing of gas and liquid;
- steady operation and appropriate flexibility;
- low pressure drop;
- simple device easy for maintenance;
- Anti-corrosion and anti-congestion while necessary.

There are two common types of towers, plate and packed, make effective pollution control devices. They utilize simple mechanical methods of achieving good contacting between the gas and liquid phases to provide favorable overall mass transfer. The three general types of towers are packed, bubble cap and perforated-plate towers.

In these systems intimate contact between the gas phase and liquid phase is created by dispersion of the liquid in the gas. The pollutant gas enters the absorber in an upward direction and the liquid solvent passes downward (Fig. 8.3(a)). For achieving intimate contact between the pollutant gas and solvent drops, bubble caps or sieve trays or packing is used in the interior of the column. By the time the pollutant gas reaches the top of the column the entire pollutant is removed and the cleaned gas will pass through the outlet. The pollutant liquid solvent is then allowed to pass through a stripper whereby keeping temperatures higher or pressure lower than in the absorber, the dissolved pollutant gas will come out due to the decrease in solubility. The solvent after removing the pollutant gas is sent back to the absorber column.

Structures such as spray chambers and towers or columns which were discussed earlier for particulate matter removal can be employed to absorb pollutant gases also in a scrubber, which is a type of spray chamber in which liquid droplets are used to absorb the gas. In towers, a thin film of liquid is used as the absorption medium. Regardless of the type of device, the solubility of the pollutant in the liquid must be relatively high. If water is the solvent, this generally limits the application to a few inorganic gases such as NH_3,

Cl_2, and SO_2. Scrubbers are relatively inefficient absorbers but have the advantage of being able to simultaneously remove particulate matter also. Towers are much more efficient absorbers but they become plugged by particulate matter.

Plate towers: In plate towers the scrubbing liquid is introduced from the top of the cascades from one stage to another which present a stepwise contact between the two phases. (Fig. 8.3(a)) The scrubbing liquid after absorption is finally collected from the bottom of the tower. At each tray the polluted gas flows in the opposite direction of the falling scrubbing liquid through small holes of the tray where it forms bubbles during which absorption takes place efficiently.

8.3.2 Packed towers

The typical schematics of packed towers are displayed in Figure 8.3b. within the tower, a pack material is filled, supported on the plate. The liquid enters at the top of the tower and distributed by the distributor such as a spray system, overflow weirs, or a perforated distribution plate. The liquid is scattered to form the film on the surface of packing and flow downward through the voids between the packings.

The gas flow upward through the packing. The wetting surfaces of packing thus compose of the contact sites of the gas and liquid. There are many types of packings, some of which the commercially used are Raschig rings, Berl saddles, Pall rings, Intalox saddles, and cannon packings. In general, it's required for the packings to have large surface per unit volume, and low pressure drop of gas through it.

The packing in the tower can be installed as an entire column. However, when the liquid flows through the packing, it tends to flux onto the tower wall, and makes the packing in the center not fully wetted. To overcome such a wall effect, sometimes it's necessary to add intermediate packing support plates to divide the packing into several layers, as displayed in Figure 8.3(b).

Usually, the ratio of tower diameter to the packing size is required to be larger than 8 to avoid the wall effect. When the multi layers of packing are installed, each layer should keep below 3-5 meters high. Or based on the recommended factors for specific packing types, for example, the packing height is suggested to be 3 times as much as the tower diameter for Rasching ring, and 5-10 times of the tower diameter for Pall rings and saddles.

In packed towers a wide variety of packing materials are used ranging from specially designed ceramic packing to crushed rock and liquid s distributed over the packing. This provides high interfacial surface are, and liquid flows down the packing surface in the form of thin films or sub-divided streams [Fig. 8.3(b)]. These systems by providing continuous contact between the gaseous phase and the scrubbing liquid phase act as a very efficient absorption device.

In venturi scrubbers the absorbing liquid is introduced at the throat of the venturi where the gas with high velocity disappears into the liquid and stream. The resulting large interfacial area makes the venturi scrubber an extremely efficient absorber. However, the energy required for achieving high efficiency is considerable.

8.4 Design consideration for a Countercurrent Absorption Column

For a counter current flow absorption column (Fig. 8.3) the mass balance equation is:

$$(G_{m1}y_1)+(L_{m2}x_2)=(G_{m2}y_2)+(L_{m1}x_1) \quad .....(8.11)$$

$$(G_{m1})(y_1)-(G_{m2})(y_2)=(L_{m1})(x_1)-(L_{m2})(x_2) \quad .....(8.12)$$

where G_{m1}, G_{m2} = total gas flow (air plus pollutant) into and out of the column respectively, kg mole/h; y_1, y_2 = mole fraction of pollutant in the gas phase at inlet and outlet of column, respectively; L_{m1}, L_{m2} = total liquid flow (solvent plus absorbed pollutant) out of and into the column respectively, kg mole/h; x_1, x_2 = mole fraction of the pollutant in the liquid phase out of and into the column, respectively

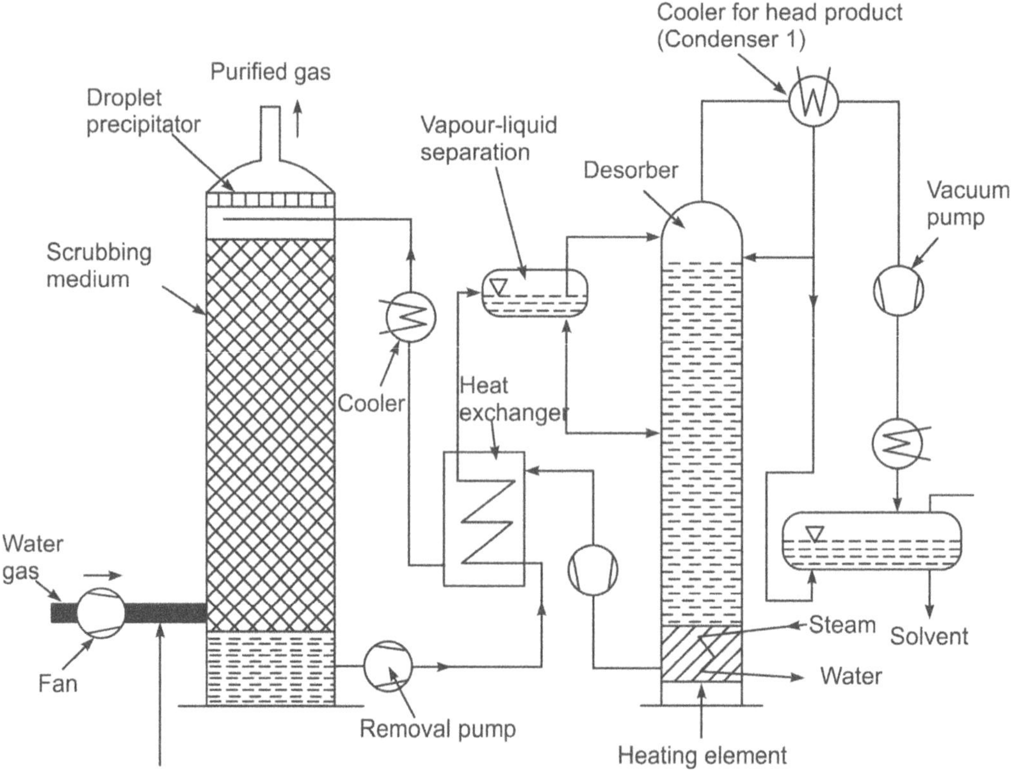

Fig. 8.3(a) *Physical absorption system with solvent recovery*

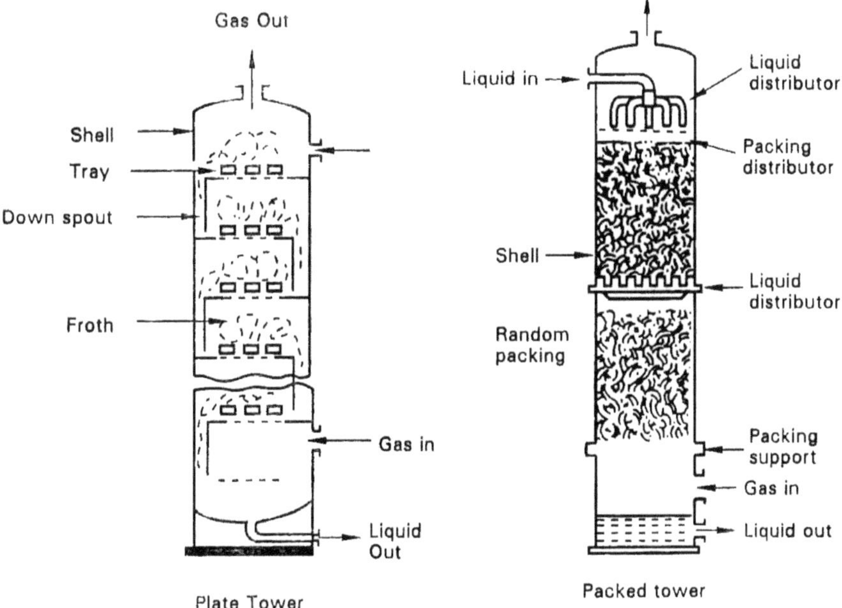

Fig. 8.3(b) *Gaseous pollutant absorption systems: plate towers and packed towers*

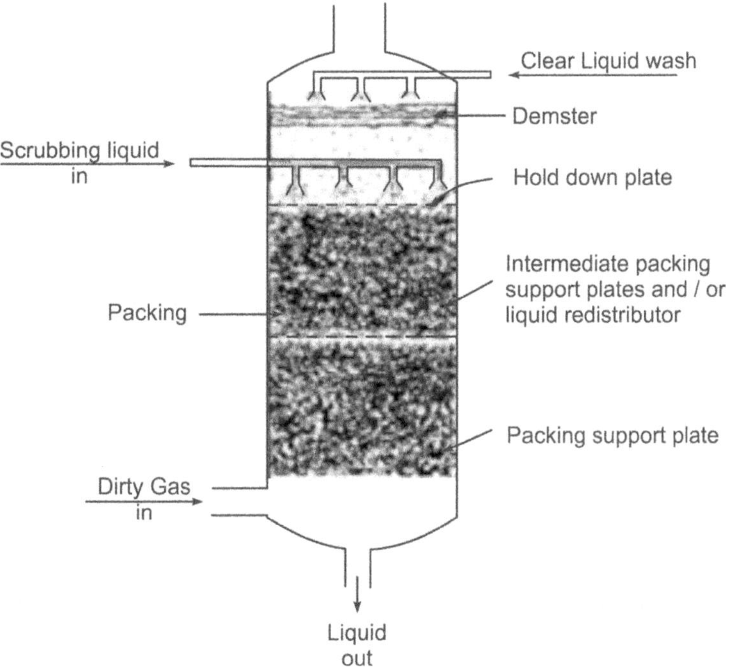

Fig. 8.3(c) *Counter current packed tower absorber*

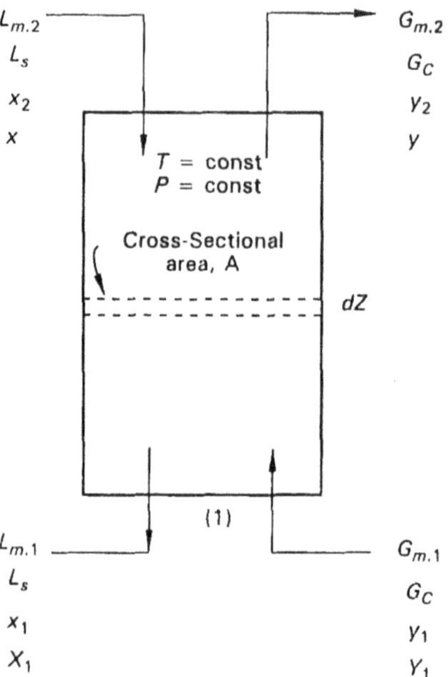

Fig. 8.4(a) *Design of gas absorption equipment*

Three variables of interest in the design of a packed tower are the gas flow rate, the liquid flow rate, and the height of the tower which are also interrelated.

8.4.1 Height of the column

If d_z is differential height of the absorber, [Fig. 8.4(a)], the total interfacial area open to mass transfer is given by

Area for mass transfer = (a) ($A\ dZ$) .....(8.13)

where a = area per unit volume of packing; A = cross-sectional area of column.

The rate of mass transfer of a gas, i, into solution (N_i) can be described by the following differential equation:

$$N_i = \frac{dC}{dt} = K_y(y - y') \quad .....(8.14)$$

where K_y = overall mass transfer coefficient for gas; y, y'= mole fraction of gaseous pollutant and equilibrium mole fraction, respectively, then

Rate of mass transfer = (N_i) (A) ($a\ dZ$) .....(8.15)

which is equal to the concentration of mass from the gas phase when it passes through the differential height

$dZ = d(G_m y)$ .....(8.16)

which can be expanded by using the terms: the mass flow rate per unit area G'_m and the mole ratio:

$$Y = \frac{y}{1-y} \qquad .....(8.17)$$

$$G_c Y = G_m y \qquad .....(8.18)$$

where G_c is the mass flow of the carrier gas without the pollutant. Equating the mass transfer (Equation (8.14)) with the mass loss (Equation (8.16)) and making substitutions from Equations (8.15, 8.17, and 8.18) yields

$$K_y a(y - y')dZ = G'_m \frac{dy}{1-y} \qquad .....(8.19)$$

or $dZ = G'_m \dfrac{dy}{(1-y)K_y a(y-y')}$

The overall driving force $(y - y')$ at any location in the tower may be written in the form

$$Y - y' = (1 - y') - (1 - y) \qquad .....(8.20)$$

We can define the log-mean value of $(1 - y')$ and $(1 - y)$ as

$$(1-y)_{LM} = \frac{(1-y') - (1-y)}{\ln \dfrac{(1-y')}{(1-y)}} \qquad .....(8.21)$$

Multiplying the numerator and denominator in Equation 8.11 by $(1 - y)_{LM}$, we obtain

$$dZ = \frac{G'_m}{K_y a(1-y)_{LM}} \times \frac{(1-y)_{LM} dy}{(y-y')(1-y)} \qquad .....(8.22)$$

Although G'_m, $K_y a$, and $(1 - y)_{LM}$, vary along the absorption column, the first term of this equation is reasonably constant which is termed as the overall height of a transfer unit (H_{og}) which is related to the height of the tower by Equation 8.17.

$$Z = H_{og} \int_{y_2}^{y_1} \frac{(1-y)_{LM}}{(y-y')(1-y)} dy \qquad .....(8.23)$$

The integral is called the number of transfer units (N_{og}). The height of the tower is computed from the following equation:

$$Z_t = (H_{og})(N_{og}) \qquad .....(8.24)$$

For dilute solutions which obey Henry's law, the number of overall gas transfer units may be calculated as follows:

$$N_{og} = \frac{\ln\left[\frac{(y_1 - mx_2)}{(y_2 - mx_2)}(1-A)+A\right]}{1-A} \qquad .....(8.25)$$

where

y_1, y_2= mole fraction of pollutant in the gas phase at inlet and outlet of tower, respectively

m = slope of equilibrium curve defined by Henry's law = y^*/x^* in mole fraction units (m has no units)

x_2= mole fraction of pollutant in the liquid phase entering the tower

$A = mQ_g/Q_l$

Q_l = liquid flow rate, kg mole/hm^2

Q_g = gas flow rate, kg mole/hm^2

The height of a single overall mass transfer unit (H'I'U) may also be expressed as the sum of the gas and liquid HTUs.

$$H_{og}= H_g+ AH_l \qquad .....(8.26)$$

in which H_g and H_l are complex functions of the flow rate, surface area of the packing and viscosity of the liquid and air diffusion of the pollutant gases.

In actual practice H_{og} of an overall transfer unit is determined using data obtained from small pilot-plant scale absorbers filled with the packing to be used in the large-scale equipment. The experimental conditions like flow rates, concentrations, and temperatures which are expected to prevail in the large plant is simulated in the pilot plant also. If the pilot plant data is not available H_{og} can be estimated from individual phase transfer unit heights according to the relation,

$$H_{og} = H_g + \frac{mG}{L} H_l \qquad .....(8.27)$$

The heights of gas and liquid transfer units, in Equation 9.21, are correlated by

$$H_g = \frac{\alpha (G')^{\beta}}{(L')^{\gamma}} \sqrt{Sc_G} \qquad .....(8.28)$$

$$H_l = \phi \left(\frac{L'^{\eta}}{\mu_L} \right) \sqrt{Sc_L} \qquad .....(8.29)$$

where H_g and H_l are in meters, G' and L' are in kg/m^{2}sec. μ_L is in kg/m sec and the gas and liquid Schmidt numbers are $Sc_G = (\mu_G/\rho_G D_G)$ and $Sc_L = (\mu_L/\rho_L D_L) = D_G$ and D_L are the gas and liquid diffusivities respectively of the solute absorbed. Values of the constants α, β, γ for several packing materials are given in Table 8.3 while those of ϕ and η are given in Table 8.4.

Table 8.3 *Values of α, β, γ for several packing materials [4]*

Packing (cm)	α	β	γ	Range of (kg/m^{2}s)	
				G'	L'
Raschig rings					
1.5	0.620	0.45	0.47	0.27–0.68	0.68–2.0
2.5	0.608	0.39	0.58	0.27–1.1	0.54–0.68
2.5	0.557	0.32	0.51	0.26–0.81	0.68–6.1
3.75	0.830	0.38	0.66	0.27–0.95	0.68–2.0
3.75	0.689	0.38	0.40	0.27–0.95	2.0–6.1
5	0.894	0.41	0.45	0.27–1.1	0.68–6.1
Berl saddles					
1.25	0.540	0.30	0.74	0.27–0.95	0.68–2.0
1.25	0.367	0.30	0.24	0.27–0.95	2.0–6.1
2.5	0.461	0.36	0.40	0.27–1.1	0.54–6.1
3.75	0.652	0.32	0.45	0.27–1.36	0.54–6.1

Table 8.4 *Values of φ and η for several packing materials [4]*

Packing (cm)	ϕ	η	Range of L''(kg/m^{2}s)
Raschig rings			
1	0.000321	0.46	0.54–20
1.25	0.000718	0.35	0.54–20
2.5	0.000235	0.22	0.54–20
3.75	0.00261	0.22	0.54–20
5.0	0.00293	0.22	0.54–20
Berl saddles			
1.25	0.00146	0.28	0.54–20
2.50	0.00129	0.28	0.54–20
3.75	0.00137	0.28	0.54–20

8.4.2 Solvent rate and column diameter

After evaluating the feed and effluent specifications the minimum solvent rate which is the rate below which separation is impossible even with a column of infinite height can be calculated [Fig.8.4 (b)]. From the data on the mole fraction of the pollutant in the solvent (x_2) and effluent gas (y_2) at the top of the column as well as that in the feed gas (y_1) one can draw a straight line from (x_2, y_2) to intercept the equilibrium line at y_B the slope of which represents the ratio of the minimum solvent rate to the feed gas rate.

The normal operating solvent rate should be 30 to 70% above the minimum rate.

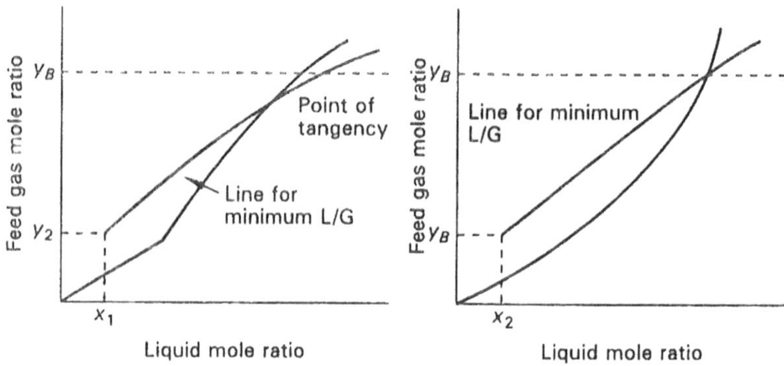

Fig. 8.4 (b) *Calculation of minimum solvent rates for two different shapes of equilibrium line*

$$\Delta P_{\text{flood}} = 0.307 F_P^{0.7} \quad \text{.....(8.30)}$$

From Figure 8.4(b) one can calculate Q_G value from which required column diameter of the column can be deducted.

8.5 Removal of Gaseous Pollutants by Adsorption on Solids

8.5.1 Introduction

This is a mass transfer process in which the gas is bonded to a solid and is a surface phenomenon. The gas (adsorbate) penetrates into the pores of the solid (adsorbent) but not into the lattice itself. The bond may be physical binding where electrostatic forces hold the pollutant gas or chemical bonding by reaction with the surface.

The adsorption process is endothermic while the desorption process is exothermic and both are reversible. Gases with higher molecular weights and lower boiling points are preferentially adsorbed and these displace gases which are less preferentially adsorbed effecting separation. For example organic vapour in air can be preferentially adsorbed by activated carbon than air.

Some of the important adsorbents and their respective properties are listed in Table 8.5.

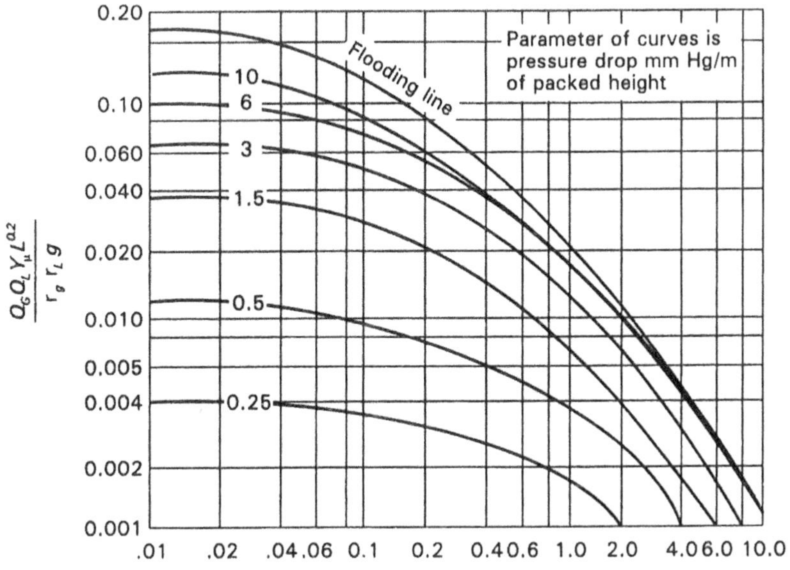

Key:

Q_L = Liquid rate, kg/s m$^2 \dfrac{L}{G}\left(\dfrac{P_g}{P_L}\right)^{\frac{1}{2}}$

Q_G = Gas rate, kg/s m^2

r_L = Liquid density, kg/m^3

r_g = Gas density, kg/m

F_p = Packing factor, m^{-1}

m_L = Viscosity of liquid, mPas

Y = Ratio, (density of water)/(density of liquid)

g = Gravitational constant, 9.81 m/s^2

Fig. 8.5 *Generalized correlation for the calculation of pressure drops in packed columns*

Table 8.5 *Typical adsorbents and their properties [5]*

Absorbent	Internal porosity	External Porosity, η(%)	Bulk density (kg/m^3)	Surface area (m/kg × 10^{-3})
Active carbons	55–75	35–40	160–480	600–1400
Fuller's earth	50	40	470–630	120–240
Iron oxide	22	37	1400	20
Magnesia	75	45	400	200
Silica gel	70	40	400	200
Acid-treated clay	30	40	400	320
Activated alumina and bauxite	35	45	700–870	200–250
Molecular sieves	50	35	655–695	595–695
Bone char	50	18	640	100

Active carbon, molecular sieves, silica gel, and activated alumina are the most common adsorbents. Activated carbon has a large surface area per unit weight or volume of solid. For use in a fixed bed or air or gas purification, the particles must be so sized that they impose little difference to flow for a given sorption efficiency; the range of 4–20 mesh encompasses the predominant portion of carbon for such use.

Activated carbon is effective in adsorbing molecules of organic substances with less selectivity than is exhibited by other, more polar sorbents. Activated carbon is effective in adsorbing organic molecules even from a humid gas stream. The water molecules being highly polar exhibit strong attractions for each other, which compete with their attractions for the non polar carbon surface; consequently the larger, less polar organic molecules are selectively adsorbed. Molecular sieves are dehydrated zeolites. Sodium silicate is reacted with sulphuric acid to make silica gel. Activated alumina is a porous hydrated aluminium oxide. The common property of these adsorbents is a large "active" surface area per unit volume after treatment. They are very effective for hydro carbon pollutants. In addition, they can capture H_2S and SO_2. One special form of molecular sieve can also capture NO_2. With the exception of the active carbons, adsorbents have the drawback that they preferentially select water before any of the pollutants. Thus, water must be removed from the gas before it is treated. All of the adsorbents are subject to destruction at moderately high temperatures (150 °C for active carbon, 600°C for molecular sieves, 400 °C for silica gel, and 500 °C for activated alumina). They are very inefficient at these high temperatures. In fact, their activity is regenerated at these temperatures!

As soon as the adsorption column attains saturation, when air is passed through it will come out unchanged. The plot of adsorption capacity versus pressure of the adsorbate at a given temperature is called the adsorption isotherm. Fig. 8.6 shows how adsorption capacity increases with increasing partial pressure.

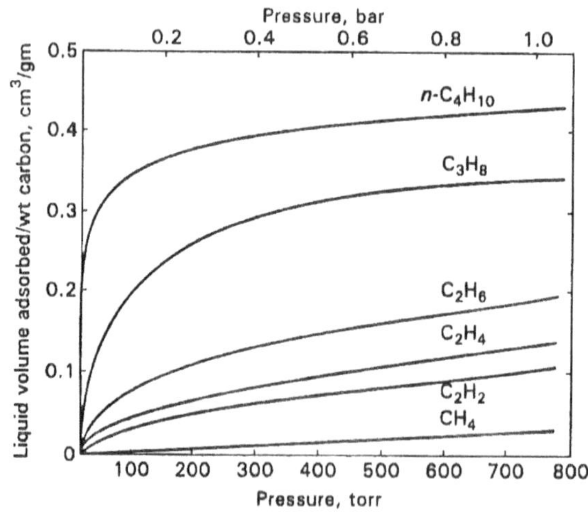

Fig. 8.6 *Adsorption isotherms of hydrocarbon vapours at 37.8 °C on air purification by activated carbon. Liquid volumes measured at boiling points of the hydrocarbons*

8.5.2 Gaseous pollutant control systems based on adsorption on solids

Certain solids have different affinities for different gases. If a gaseous mixture comes in contact with an adsorbent solid, some of the components of the gas stream will concentrate on the surface of the solid and separate from the main stream. There are basically three types of adsorption processes. The first type is physical adsorption or condensation of toxic or obnoxious gases and vapors on solids at temperatures above the dew point. The second type is chemical adsorption or chemisorption, where a chemical bonding occurs between the adsorbed pollutants and the solid adsorbent. The last type is ion exchange solids, where an exchange of ions takes place. The solids give up ions for the ions they adsorb from the fluid stream. This is mostly applied in water softening where the solids give up sodium ions to the water in exchange for magnesium and calcium ions, and this exchange decreases or removes water hardness. Of the three types of adsorption, physical adsorption is the most important in air pollution control.

An adsorption processing unit consists of a number of vessels containing the solid adsorbent. The polluted gas passes through one vessel until the adsorbent is saturated with pollutants, and then the gas is switched to another vessel. Regeneration of the adsorbent is accomplished by steaming the first vessel to drive out the adsorbed pollutants (Fig.8.7); steam and pollutants are collected, condensed and separated. The pollutants may be treated further to give some useful by-products, and the condensed steam maybe disposed of safely as waste water.

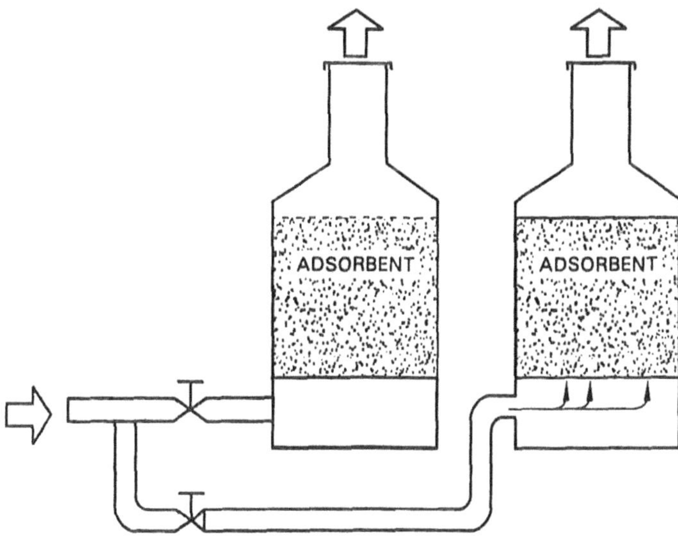

Fig. 8.7 *Adsorption system*

The amount of pollutants which may be picked up could reach 25% of the weight of the adsorbent. The effectiveness of the adsorbent is judged by the concentration of the pollutants in the exit air stream. When the values of inlet and outlet concentrations approach one another, it is high time to start the regeneration process. (Fig.8.8).The regeneration process is similar to adsorption process except that the driving force is acting in opposite direction. Various types of adsorption systems practised for industrial operations are shown in Fig. 8.9 (a) to (d).

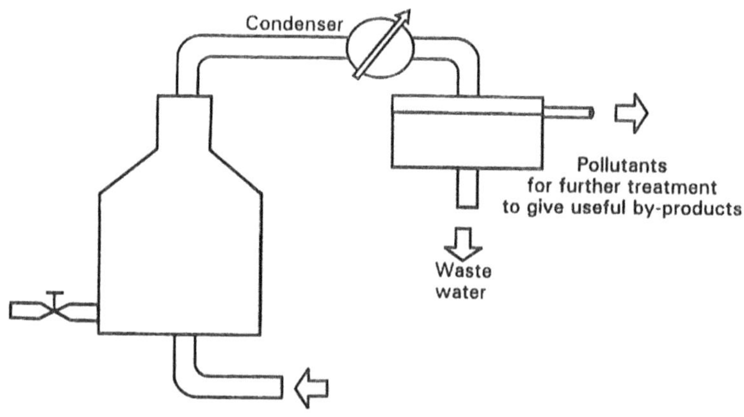

Fig. 8.8 *Regeneration of an adsorption unit*

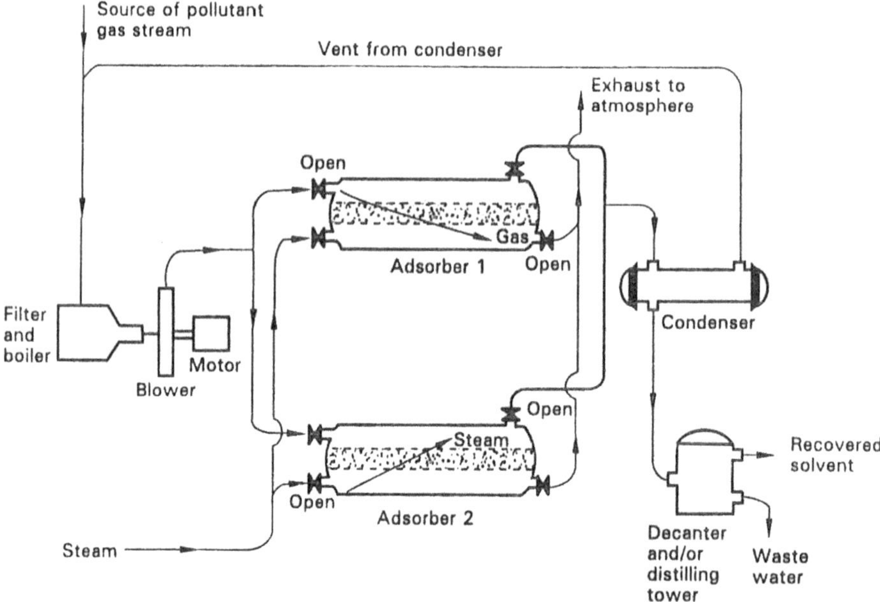

Fig. 8.9(a) *Two stage regenerative system. Adsorber 1 is adsorbing; adsorber 2 is steaming; after number 1 is saturated and number 2 is clean, their functions are reversed*

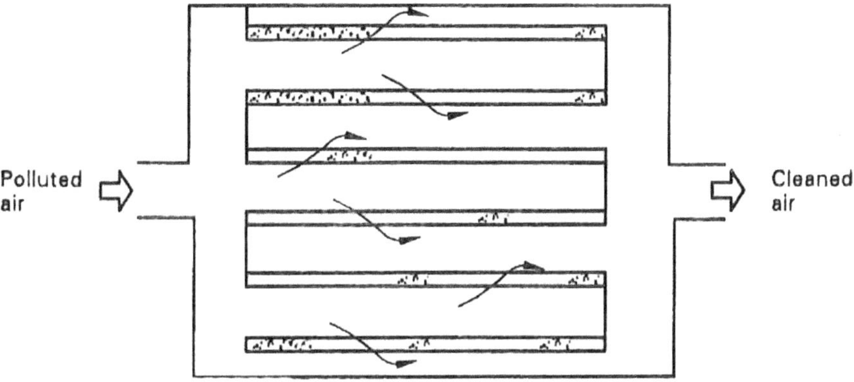

Fig. 8.9(b) *Adsorber with a number of adsorbent trays*

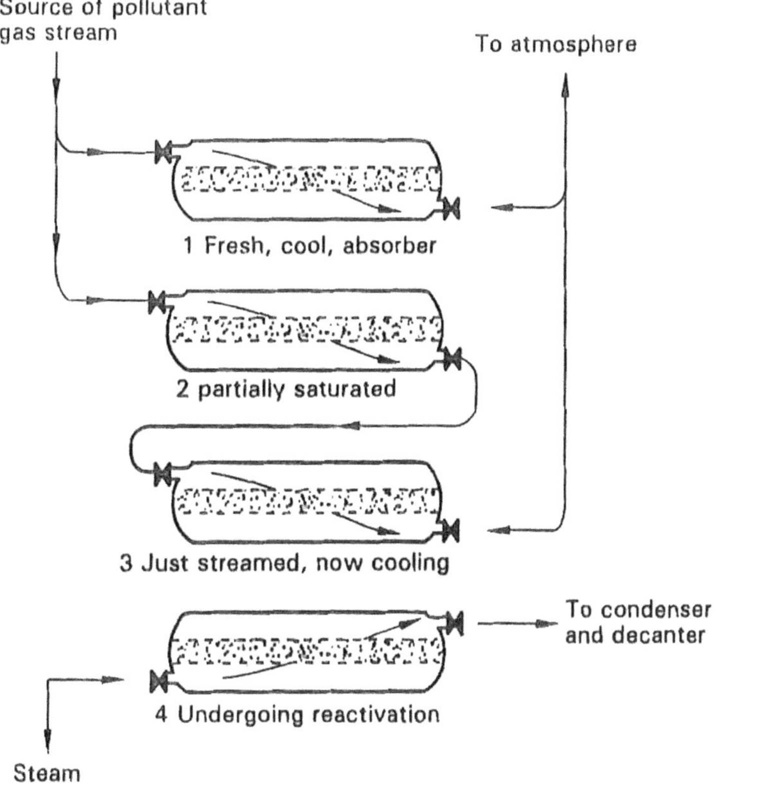

Fig. 8.9(c) *Four stage regenerative system*

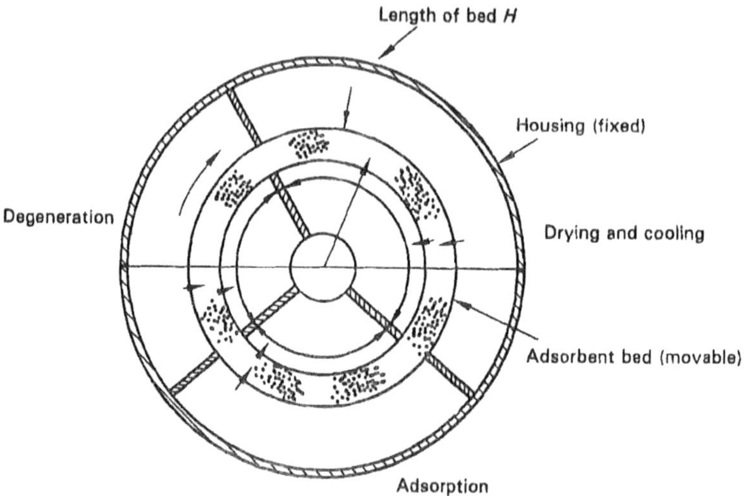

Fig. 8.9(d) *Cross-section of a rotating bed adsorber*

8.5.3 Fluidized adsorbers

When a gas is passed upward through a bed of granular adsorbent, the pressure drop imposed by the bed (Fig. 8.10) is in opposition to its own weight. If the gas velocity is

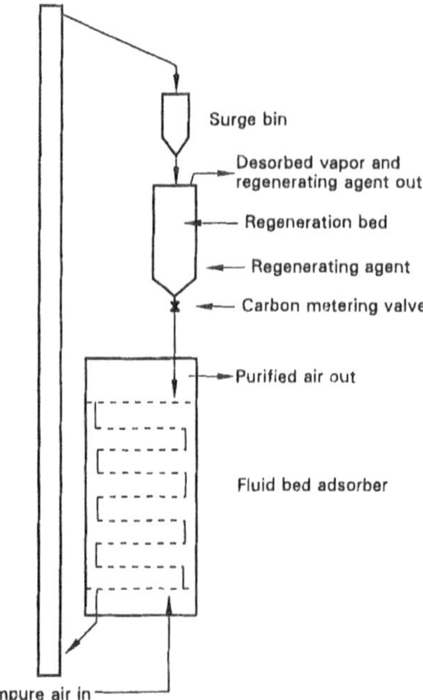

Fig. 8.10 *Fluidized bed solvent recovery system*

increased to a sufficient value, the pressure drop equals the weight of the bed and the solids begin to move. This motion is the beginning of fluidization. At higher gas velocities (about 76.2 m/min) the granular adsorbent particles may be maintained at constant motion. The pressure drop required for fluidization depends on the bed depth and the densities of the gas stream and of the solid particles.

8.5.4 Design of adsorption process

The design of the adsorption process depends mainly on the breakthrough curves which can be obtained by pilot-plant studies. The amount of pollutant adsorbed and the equilibrium pressure at constant temperature can be related to Langmuir adsorption isotherm [Fig. 8.11(a)]:

$$W = \frac{aC_g^{\cdot}}{1 + bC_g^{\cdot}} \qquad \text{.....(8.31)}$$

where

W = amount of gas per unit mass of adsorbent, kg/kg

a, b = constants determined by experiment

C_g = equilibrium concentration of gaseous pollutant, g/m3

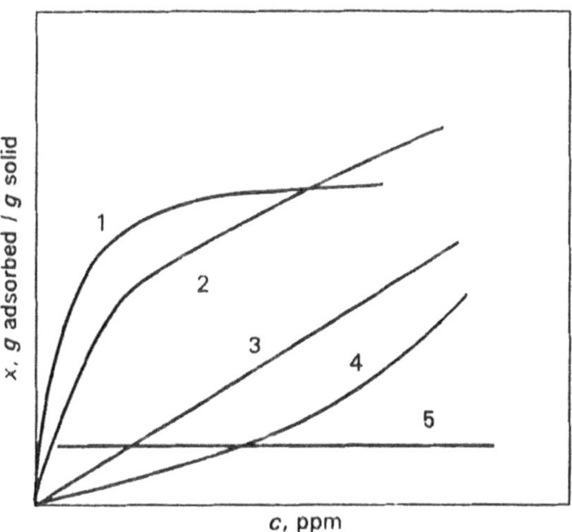

Fig. 8.11(a) *Adsorption isotherm*

Equation (8.22) can be rewritten for the analysis of experimental data as follows:

$$\frac{C_g^{\cdot}}{W} = \frac{1}{a} + \frac{b}{a}C_g^{\cdot} \qquad \text{.....(8.32)}$$

As per the Equation (8.30), a plot of (C'_g/W) versus C'_g should yield a straight line with a slope of (b/a) and an intercept equal to (1/a).

Key 1 = High loading (highly favourable)
2 = Favourable
3 = Linear
4 = LOW loading (unfavourable)
5 = Irreversible

When expressed as the volume of adsorbate adsorbed per unit mass of adsorbent V as follows

$$V = K_f G^{1/n} \qquad .....(8.33)$$

or expressed in terms of concentration as

$$\rho_e = K_F C\ 1/n\ (9.32) \qquad .....(8.34)$$

where K_F and n are constants and

$$n > 1\ (K_F \propto RT\ n\ be^{\Delta H/RT})$$

The expression in Equation 8.34 is known as the Freundich isotherm. For a binary or higher mixture in which the components compete for surface sites, the expression is more complex. The K_F and n values in the Freundich isotherm for selected adsorbates are presented in Table 8.6.

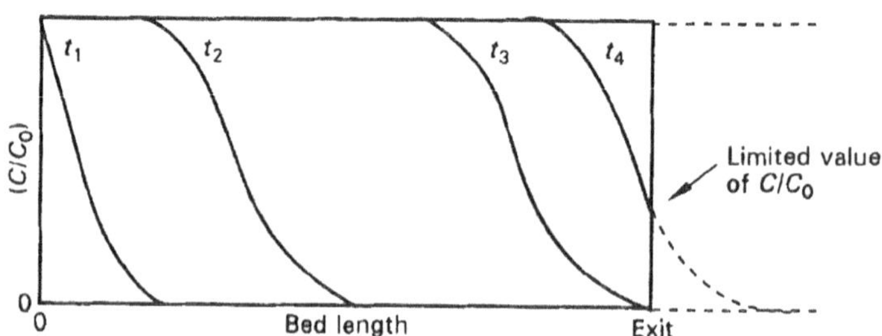

Fig. 8.11(b) *Concentration of pollutant in air stream as a function of bed length*

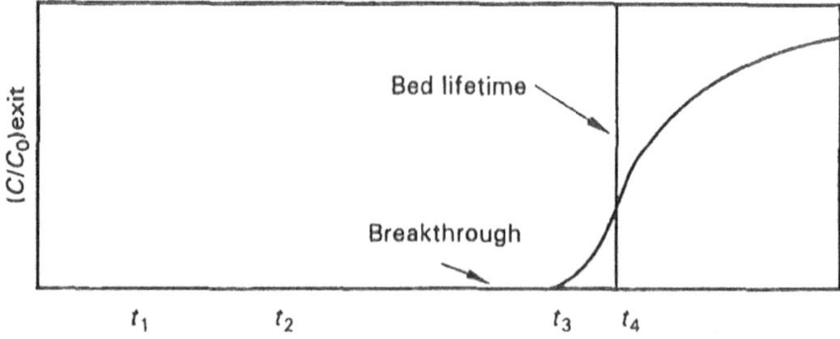

Fig. 8.11(c) *Adsorption profiles for a single contaminant in an air stream*

Table 8.6 *Freundich isotherm parameters for some adsorbates*

Absorbate	Temperature (K)	$K_F \times 100$	n	Partial pressure (Pa)
Acetone	311	1.324	0.389	0.69 – 345
Acrylonitrile	311	2.205	0.424	0.69 – 103
Benzene	298	12.602	0.176	0.69 – 345
Chlorobenzene	298	19.934	0.188	0.69 – 69
Cyclohexane	31	7.940	0.210	0.69 – 345
Dichloroethane	298	8.145	0.281	0.69 – 276
Phenol	313	22.116	0.153	0.69 – 207
Toluene	298	20.842	0.110	0.69 – 345
TCA	298	25.547	0.161	0.69 – 276
m-Xylene	298	26.080	0.113	0.69 – 6.9
p-Xylene	298	28.313	0.0703	69 – 345

Source: *U.S. Environmental Protection Agency (EPA), 1987, EAB control cost manual, 3d ed. (Research Triangle Park. N.C. USEPA)*

Notes: *The amount adsorbed is expressed in kg adsorbate/kg adsorbent. The equilibrium partial pressure is expressed in Pa. Data are for the adsorption on Calgon-type BPL activated carbon (4 × 10 mesh). Data should not be extrapolated outside of the partial pressure ranges shown.*

If an adsorbent adsorbs the bulk of a pollutant from air even when that material is in low concentration the adsorption process is called favourable curve (1 and 2) in Fig. 8.11(a). If high concentrations are to be present before significant quantities are adsorbed (curve 4), the profile is called unfavourable adsorption.

In absorption towers the collected pollutant is continuously removed by flowing liquid, whereas in adsorption systems the collected pollutant remains in the adsorption bed. As long as the bed has sufficient capacity, no pollutant is emitted [Fig. 8.11(b)]. At some point in time, the bed will become saturated with pollutant. Due to saturation the pollutant will begin to leak out of the bed; this process is called *breakthrough* [Fig. 8.11(c)]. The influent and effluent are equal when the bed capacity is exhausted and for *allowing continuous operation*, two beds can be provided so that when one is collecting pollutant, the other is regenerated. The concentrated gas released during regeneration is usually returned to the process as a recovered product. For the effective operation of the bed the important parameter is the length of time it can operate before break through occurs which can be calculated from the following equation [5].

$$T_B = \frac{Z_t - \delta}{v_f} \quad .....(8.35)$$

where

Z_t = height of bed, m

δ = width of adsorption zone, m

v_f = velocity of adsorption isotherm zone as defined by Equation (8.19), m/s

The height of the adsorption bed (Z_t) can be determined in the same manner as it was for absorption towers [Equation (8.16)], with a few exceptions. The value of N_{og} can be determined by integration of the following expression [5].

$$N_{og} = \int_{C_2}^{C} \frac{dC}{C - C_g^{\cdot}} \qquad \text{.....(8.36)}$$

where C_g= the equilibrium partial pressure given by Equation (8.16) and C is given by the operating line. The H_{og} in the equation can be modified by replacing H_l with H_g. The value of slope m in Equation (8.10) can be determined from Equation (8.17).

The width of the adsorption zone (Fig. 8.9) is a function of the shape of the adsorption isotherm.

The velocity of the adsorption isotherm zone can be calculated from the properties of the adsorption system by the following equation

$$v_f = \frac{(Q_g)(1 + bC_g^{\cdot})}{a\rho_s\rho_g A_c} \qquad \text{.....(8.37)}$$

where

ρ_s, ρ_g= density of the solid adsorbent packed and gas, kg/m^3

A_c= cross-sectional area of bed, m^2

Problem 8.1

An adsorption bed of thickness0.80 m and cross-section of 5 m^2 is operating with the following experimental conditions. The operating parameters for the bed are as follows: gas flow rate = 1.6 kg/s air, gas temperature = 30 °C, gas pressure = 101.325 kPa, ρ_g = 1.184 kg/m^3, bed density as packed = 140 kg/m^3, inlet pollutant concentration = 0.0040kg/m^3. Langmuir parameters = a = 18, b = 124, width of adsorption zone = 0.03 m. Calculate the breakthrough time for the adsorption bed.

Solution: The face velocity of the adsorption wave is

$$v_f = \frac{(1.6\text{ kg/s})[1 + 124(0.0040\text{ kg/m}^3)]}{18 \times 140\text{ kg/m}^3 \times 1.184\text{ kg/m}^3 \times 5\text{ m}^2}$$

$$= 1.60 \times 10^{-4}\text{ m/s}$$

The breakthrough time can be calculated from the equation

$$T_B = \frac{Z_t - \delta}{v_f} = \frac{0.8\text{ m} - 0.03\text{ m}}{1.6 \times 10^{-4}\text{ m/s}} = 13.33\text{ h}$$

8.5.5 Pressure drop across fixed adsorption beds and expansion of fluidized beds

As dirty air has to be sent by force through the bed, pressure drop determines the size and power requirement of the blower. To provide intimate contact between solute and adsorbent, the bed may also be fluidized or expanded. The formulas for head losses and the formula for bed expansion are given below. [6]

$$h = \frac{l(1-\eta)V_s^2}{\eta^3 g}\sum\frac{f_i' x_i}{d_{pi}},$$ head loss on stratified bed

$$h_{fb} = \frac{l_e(\gamma_p - \gamma_a)}{\gamma_a}\sum x_i(1-\eta_e),$$ head loss on expanded bed

$$\frac{l_e}{l} = (1-\eta)\sum\frac{x_i}{1-\eta_{e,i}},$$ bed expansion

$$f_i' = 150\times\frac{1-\eta}{\text{Re}} + 1.75$$

$$\eta_e = \left(\frac{V_B}{v}\right)^{0.22}$$

where h is the head loss in an unexpanded bed, h_{fb} the head loss in bed expansion, l the length of unexpanded bed, η the porosity of unexpanded bed, V_s the superficial velocity of air, g the gravity constant, x_i the mass fraction of particles of size d_{pi} or of particles in layer i, l_e the length of expanded bed, γ_p the specific weight of particle, γ_a the specific weight of air, η_e the porosity of expanded bed, V_B the upflow superficial velocity of air, and v the terminal settling velocity of particles.

Problem 8.2

An activated carbon adsorber with dimensions 4 m × 2 m in cross-section and 0.7 m deep is used to treat polluted air containing 5000 ppm of toluene at 1 atm and 37 °C. The *X/M* values of the bed is 5.0 kg of toluene per 50 kg of C. The bed has a bulk or packed density of 480 kg/m^3, η of 0.4 and a particle size of 4 by 10 mesh. Assuming that the unit absorbs 4.80 kg mol of toluene per hour, determine the pressure drop. Assume that $\beta = 0.52$.

Solution:

$$h = \frac{l(1-\eta)V_s^2}{\eta^3 g}\sum\frac{f_i' x_i}{d_{pi}}$$

$$f_t' = 150 \times \frac{1-\eta}{\text{Re}} + 1.75$$

Total moles of polluted air $= \dfrac{10^6}{5000} \times 4.8 = 960$ kg mol/h

$$= 960 \times 10^3 \times 22.4 \times \frac{1}{10^3} \times \frac{37+273}{273} = 24300 \text{ m}^3/\text{h}$$

$$V_s = \frac{24300}{2 \times 4 \times 60 \times 60} = 0.84 \text{ m/s}$$

No. 4 mesh = 4.699 mm opening; No. 10 mesh = 1.651 mm opening.

$$4 \text{ by } 10 \text{ mesh} = \frac{4.699 + 1.651}{2} = 3.175 \text{ mm} = d_p$$

$$\text{Re} = \frac{dV_s\rho_a}{\mu} \qquad \rho = 1.27 \text{ kg/m}^3 \qquad \mu = 1.8 \times 10^{-5} \text{ kg/ms}$$

$$d = 1.24\beta^{0.333} d_p = 1.24 \times 0.52^{0.333} \times 0.003175 = 0.0032 \text{ m}$$

$$\text{Re} = \frac{0.0032 \times 0.84 \times 1.27}{1.8 \times 10^{-5}} = 188.9$$

$$f_i' = 150 \times \frac{1-0.4}{188.9} + 1.75 = 2.226$$

$$h = \frac{0.7(1-0.4)0.84^2}{0.4^3 \times 9.81} \times \frac{2.226 \times 1}{0.003175} = 331.8 \text{ m of air}$$

$$331.8 \times 1.27 \times 9.81 = h_{\text{water}} \times \rho_{\text{water at } 37\,^{\circ}\text{C}} \times 9.81 = h_{\text{water}} \times 994 \times 9.81$$

$$h_{\text{water}} = 0.42 \text{ m} = 42 \text{ cm}$$

8.6 Gaseous Pollutant Control by Combustion

8.6.1 Methods of destruction of organic vapors by combustion

Destruction of organic vapour by combustion, called incineration, can be brought about by one of the following three methods depending upon the nature of the components in the vapour, their composition and concentration.

1. Direct flame combustion
2. Thermal incineration
3. Catalytic incineration

To incinerate is to destroy by combustion process, with or without the aid of a catalyst. Incineration is widely used in industry when the fumes from a stack contain toxic pollutants, especially those with offensive odors.

Combustion is an exothermic chemical reaction which usually takes place in air composed of mainly oxygen and nitrogen. As air contains 21% oxygen and 79% of nitrogen by volume so for every mole of oxygen supplied through air there are 0.79/0.21 = 3.76 moles of nitrogen and the common combustion reactions involving carbon, hydrogen and sulphur can be written as

$C + O_2 + 3.76\ N_2 \rightarrow CO_2 + 3.76N_2$

$H_2 + ½\ O_2 + 1.88\ N_2 \rightarrow H_2O + 1.88\ N_2$

$S + O_2 + 3.76\ N_2 \rightarrow SO_2 + 3.76\ N_2$

The above combustion reactions are the stoichiometric equations for combustion processes assuming complete combustion of fuel which occurs only when each fuel molecule reacts with oxygen. In view of this for many combustion processes higher volumes of air than the stoichiometric requirement should be supplied for achieving complete combustion. In the presence of 50% excess air the actual combustion reactions taking place can be written as

$C + 1.5\ O_2 + 5.64\ N_2 \rightarrow CO_2 + 0.5O_2 + 5.64N_2$

$H_2 + 1.5 \times ½\ O_2 + 2.82\ N_2 \rightarrow H_2O + 0.50\ O_2 + 2.82\ N_2$

$S + 1.5\ O_2 + 5.64\ N_2 \rightarrow SO_2 + 0.5\ O_2 + 5.64\ N_2$

Incomplete combustion may result in the absence of excess air due to less statistical probability for O_2 molecules to react with the carbon resulting in undesirable gaseous pollutant emissions.

For example in the absence of sufficient excess air the combustion reaction of carbon can be represented as

$C + O_2 + 3.76\ N_2 \rightarrow 0.8\ CO_2 + 0.2CO + 0.1\ O_2 + 3.76\ N_2$

The exact amount of air to be supplied to any fuel depends upon the characteristics of the fuel, temperature of the combustion process and the extent of turbulence and is given by the stoichiometric air-fuel ratio AFR.

The stoichiometrically required air-fuel ratio for any given combustion reaction can be given by

$$AFR = \frac{M_{air} \times n_{air}}{M_{fuel} \times n_{fuel}}$$

where M_{air}= molecular weight of air.

Since normal air has a composition of

21% O_2 and 79% N_2 $M_{air} = 0.21 \times 32 + 0.79 \times 28 = 28.97$

n_a = number of moles of air, i.e., the number of moles of O_2 and number of moles of N_2

M_{fuel} = molecular weight of fuel

n_{fuel} = number of moles of fuel

$$\text{AFR} = \frac{28.97 \times n_a}{M_f \times n_f}$$

8.6.2 The volume and composition of combustion products

For hydrocarbons fuels like CH_4, C_6H_6 or C_6H_{14} which can be written as C_xH_y where for methane, $x = 1$ and $y = 4$, etc., if one mol is in injected into the combustor then assuming complete combustion, the outlet gas will contain x mol of CO_2 and $y/2$ mol of H_2O. In addition, it will contain nitrogen and moisture that come in with the combustion air. The stoichiometric oxygen requirement is

$$n_{\text{stoichiometric oxygen}} = x + \frac{y}{4} \qquad \text{.....(8.38)}$$

which we abbreviate as n_{stoich} (Here we have $y/4$, because each H atom requires ¼ mol of O_2 but produces ½ mol of H_2O). The dry air flow to the burner is:

$$n_{\text{dry}} = n_{\text{stoich}} \left(\frac{1+E}{0.21} \right) \qquad \text{.....(8.39)}$$

where the 0.21 accounts for the fact that air is only 21% oxygen, and E is the fraction of excess air introduced to the burner. The total amount of air introduced is

$$n_{\text{total}} = n_{\text{dry}} (1 + X) = n_{\text{stoich}} \left(\frac{1+E}{0.21} \right) (1 + X) \qquad \text{.....(8.40)}$$

where X is the humidity, expressed as (mol H_2O/mol dry air).

$$n_{\text{total out}} = x + \frac{Y}{2} + n_{\text{stoich}} \left[\left(\frac{1+E}{0.21} \right) (1+X) - 1 \right] \qquad \text{.....(8.41)}$$

Problem 8.3

Methane is burned in air with 20 percent excess air and air moisture of 0.0116 mol/mol dry air. Calculate the stoichiometric AFR and actual AFR.

Solution: Here we choose as our basis 1 mol of methane so that all calculated flows are per mol of methane.

The combustion of methane as per stoichiometry can be given by the equation

$CH_4 + zO_2 + 3.76\ N_2 \rightarrow xCO_2 + yH_2O + 3.76\ N_2$

By balancing the above equation we have

$x = 1, y = 2$, and $z = 2$

So by stoichiometry

$CH_4 + 2O_2 + 7.52N_2 = CO_2 + 2H_2O + 7.52N_2$

Air-fuel ratio as per the stoichiometric equation

$$= \frac{28.97(2+7.56)}{16(1)} = 17.25$$

If 20% excess air is used the actual combustion equation is 100% + 20% air

$$CH_4 + 1.2 \times 2\ O_2 + 1.2 \times 7.52\ N_2$$

So the actual stoichio metric equation is

$$CH_4 + 2.4\ O_2 + 0.2\ N = CO_2 + 2H_2O + 0.02\ N_2$$

Actual AFR at 20% excess air

$$= \frac{28.97(2.4O_2 + 9.04)}{16} = 20.71355$$

8.6.3 Heat of combustion

In most industrial and commercial applications of combustion, the products of combustion leave the flame as hot gases and their heat can be transferred to heat something.

Fundamental aspects of the combustion process

In all combustion reactions the main components are fuel, oxidizer and ignition which can be represented as Fuel + Oxidizer + ignition = products of combustion. Excess air is an important factor for the low pollution content of combustion processes. Lack of air causes incomplete combustion. Apart from this factor, the following causes can impair the completeness of combustion and release various types of pollutants [Fig.8.12 (a)]: (a) inadequate mixing of fuel and combustion air (local lack of air) (b) insufficient pulverizing of solid fuels or atomizing of liquid fuels (c) sudden cooling of the flame gases through combustion chamber walls (d) too short residence time at high temperatures (e) flames burning in lifted condition, escaping of intermediate products from under the flame base.

During incomplete combustion the following pollutants can be formed and reach the environment in the form of emissions: carbon monoxide (CO), hydrocarbons (C_nH_m), oxidized hydrocarbons, and others such as odorous substances and soot (products of incomplete combustion: PIC).

Carbon monoxide (CO) is the intermediate product in the combustion process of carbon to CO_2, CO oxidation requires a so-called ignition temperature of at least 990 K and for a complete burn-up a certain residence time at this temperature is required. If combustion temperature and residence time in the flame are insufficient, or if there is a lack of air part of the CO can escape into the exhaust gas. Using oil furnaces as an example, the dependency of CO emission on air or oxygen excess during combustion is shown in Fig, 9.10(b). As can be seen, there is a dramatic increase of the exhaust gases' CO content when there is not enough air.

How high the minimum air excess has to be so that combustion is as complete as possible depends on the combustion temperature, on the residence time and on how well mixed the fuel-air mixture is. Thus combustion being an exothermic reaction the energy of the products of combustion is less than the combined energy of the reactants, i.e. fuel and oxidizer and the difference of energy is known as heat of combustion.

Table 8.7 lists heat of combustion values for many common fuels, as well as many other useful data about these fuels. It can be seen that the heat of combustion of methane is 13.9 kW/kg, (Table 8.7). When the water produced by combustion is condensed, its latent heat of condensation will add to the heat of combustion resulting in higher heating values. However, when water leaves the combustor as a gas, the heat of combustion will then have lower heating values which is less than the higher heating value by the amount of that latent heat of condensation. Table 8.7 shows lower heating values.

Table 8.7 *Combustion data for hydrocarbon fuels [8]*

Fuel	1	2 kcal/kg	3 kcal/kg	4	5		6 °C	7 °C
					Lean	Rich		
Acetone	58.1	124		4.97	59	233	561	2110
Acetylene	26.0		11519	7.72	31		305	
Benzene	78.1	93.89	9692	2.71	43	336	591	2288
n-butane	58.1	92.22	10919	3.12	54	330	430	2237
l-butane	56.1	93.33	10819	3.37	53	353	443	2302
Carbon monoxide	28.0	50.55		29.5	45	251	608	
Cyclohexane	84.2	85.55	10470	2.27	48	401	270	2232
n-decane	142.3	66.11	10633	1.33	45	356	231	2268
Ethane	30.1	116.66	11342	5.64	50	272	472	2227
Ethene	28.1	115.5	11264	6.52	41	610	490	2357
Ethyl alcohol	46.1	204.44		6.52			392	
n-heptane	100.2	75.55	10730	1.87	53	450	247	2196
n-hexane	86.2	80.00	10773	2.16	51	400	261	2221
Hydrogen	2.0	107.78	286511	19.5			571	
Hydrogen sulphide	34.1	131.66		12.24			290	
Isopropyl alcohol	60.1	158.89		4.44			455	
Methane	16.0	121.66	11945	9.48	46	164	632	
Methyl alcohol	32.0	262.78		12.24	48	408	470	
n-nonane	128.3	68.99	10673	1.47	47	434	233	
n-octane	114.2	71.66	10698	1.65	51	425	240	
n-pentane	72.1	85.55	10833	2.55	54	359	284	2232
Propane	44.4	101.66	11071	4.02	5	283	304	2232
Toluene	92.1	86.66	9778	2.27	43	322	567	2326

1–Molecular weight; 2–Heat of vaporization kcal/kg; 3–Heat of combustion gas-gas kcal/kg; 4– Stoichiometric mixture Vol%; 5–Flammability limit; % stoichiometric; 6– Spontaneous ignition temperature °C; 7–Adiabatic flame temperature °C

In the combustion of methane if 99% methane and 1% air are mixed, even when a spark is supplied the mixture will not burn as there is not enough air present, and the mixture is termed as being "too rich". Similarly, a mixture of 1% methane and 99% air also will not bum as there is not enough methane present, and the mixture is termed as being "too lean". Mixtures which will burn and which will not are called the lean limit or lower explosive limit (LEL) and the rich limit or the upper explosive limit (UEL).

Problem 8.4

What is the stoichiometric mixture, lean limit LEL and rich limit UEL for the combustion of methane in air?

Solution: For stoichiometric combustion of methane it is (2/0.21) = 9.52 mols of air per mol of methane. The resulting mixture will have 1 mol of methane in 10.52 total mols so the mol percent of methane is (1/10.52) × 100 = 9.5%.

Using the values in Table 8.7, we can estimate that the range of combustible mixtures for methane and air as

$$9.48\% \times 0.46 = 4.36 \text{ volume } \% \text{ to } 9.48\% \times 1.64 = 15.55 \text{ volume } \%.$$

Thus the ratio of stoichiometric air-fuel ratio to actual air-fuel ratio in the mixture is one of the important parameters which determine the combustion products. Higher ratio of stoichiometric AFR to the actual AFR > 100% makes the combustion process in complete producing intermediate hydrocarbons and carbon monoxide. In the presence of excess of air nitrogen reacts with O_2 leading to NO_x liberation. In view of this, optimum ratios of stoichiometric AFR to actual AFR have to be maintained to reduce emission of pollutants like CO and un burnt hydrocarbons.

Problem 8.5

In a polymer manufacturing company acetaldehyde fumes are discharged. Estimate the stoichiometric air-fuel ratio (AFR) in 20% excess air if these fumes are to discharged by combustion.

Solution: The stoichiometric combustion equation for acetaldehyde can be written as:

$$C_4H_4O + 3O_2 + 3 \times 3.76N_2 = 4CO_2 + 2\ H_2O + 3 \times 3.76\ N_2$$

The stoichiometric air fuel ratio (AFR)

$$= \frac{M_a(\text{mol. wt. of air}) \times n_a(\text{no. of moles of air})}{M_f(\text{mol. wt. of fuel}) \times n_f(\text{no. of moles of fuel})} = \frac{28.97 \times 14.28}{44 \times 1} = 9.40$$

In the presence of 20% excess of air, the AFR is 1.2 (9.40) = 11.28

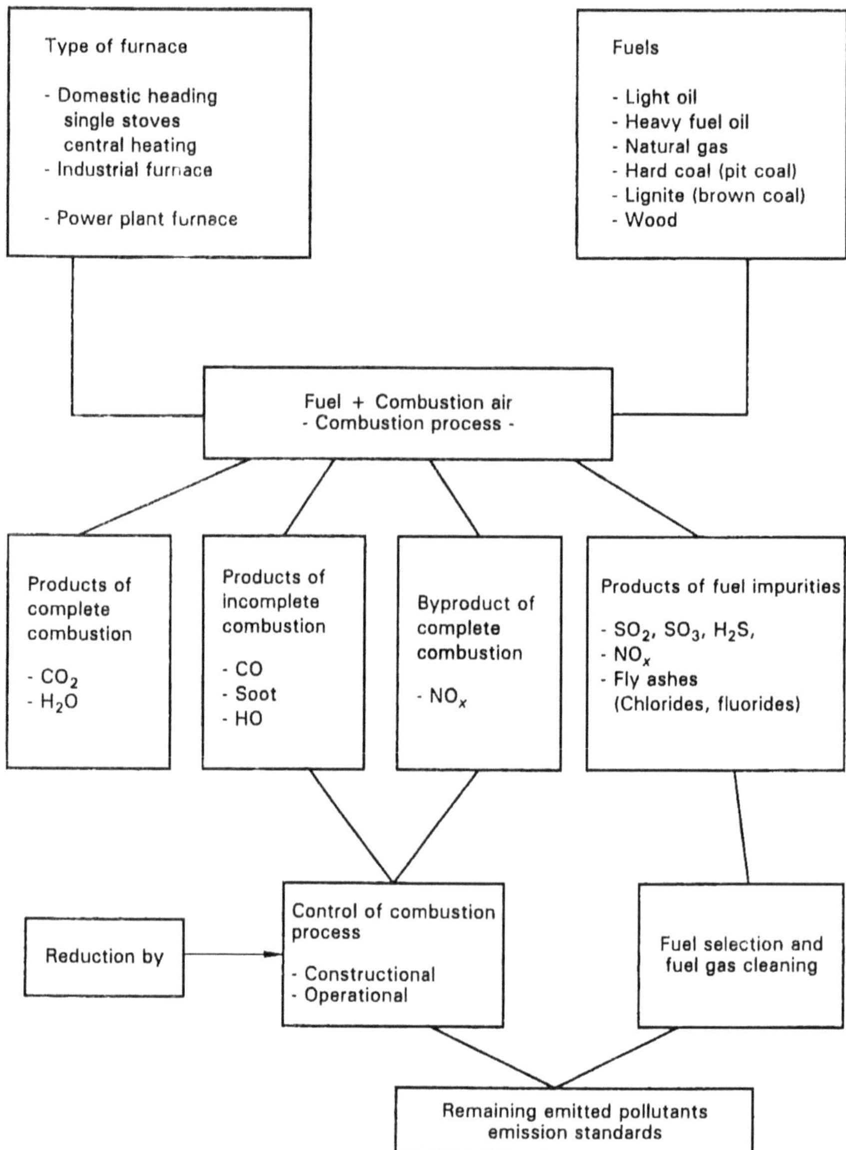

Fig. 8.12(a) *Parameters influencing pollutant emission spectrum of different types of furnaces*

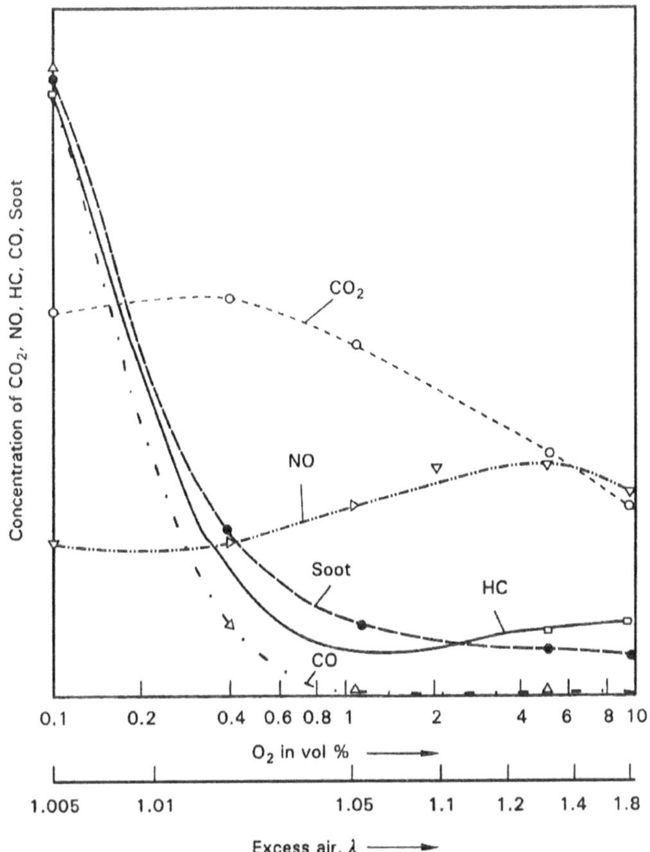

Fig. 8.12(b) *Typical course of pollutant emissions in the exhaust fuel gas of oil furnaces vs excess air*

Problem 8.6

In a pesticide factory which is emitting methyl mercaptan in the effluent gas, it is proposed to destroy it by burning in the presence of 200% excess air. Calculate the stoichiometric AFR and actual AFR.

Solution: The stoichiometric combustion reaction for burning of methyl mercaptan can be written as

$$CH_4S + 3O_2 + 3 \times 3.76N_2 = CO_2 + 2H_2O + SO_2 + 3 \times 3.76\ N_2$$

The stoichiometric AFR

$$= \frac{28.97 \times 14.28}{48 \times 1} = 8.62$$

In the presence of 200% excess of air the AFR = 3(8.62) = 25.85.

Problem 8.7

In an oil refinery an effluent gas with the following composition is discharged.

CH_4 = 8% C_2H_4O = 6% C_4H_8 = 4.8% H_2 = 43.5% CO = 25.2% CO_2 = 5% N_2 =7.5%

Calculate the stoichiometric air-fuel ratio and actual AFR in presence of 60% excess air and give wet and dry combustion products.

Solution: The stoichiometric combustion reactions of the various fuel constituents are given in the table.

The total number of moles of O_2 required for combustion of fuel = 0.97 moles

No. of moles of air required for complete combustion=0.97/0.21

Fuel	Moles /mole of gas	Stoichiometric combustion reaction	No. of moles of O_2 required for mole of fuel	Combustion products
CH_4	0.08	$CH_4 + 2O_2 = 2CO_2 + 2H_2O$	0.16	$CO_2 = 0.2$ $H_2O = 0.4$
C_2H_4O	0.06	$C_2H_4O + 3O_2 + 3 \times 3.76\ N_2 = 2CO_2 + 3H_2O + 3 \times 3.76\ N_2$	0.18	$CO_2 = 0.12$ $H_2O = 0.18$
C_2H_4	0.048	$C_2H_4 + 6O_2 = 4CO_2 + 4H_2O$	0.29	$CO_2 = 0.19$ $H_2O = 0.19$
H_2	0.435	$H_2 + ½\ O_2 = H_2O\ 2CO_2 + 3H_2O + 3 \times 3.76\ N_2$	0.217	$CO_2 = 0.0$ $H_2O = 0.435$
CO	0.252	$CO + ½\ O_2 = CO_2$	0.126	$CO_2 = 0.252$ $H_2O = 0.0$
CO_2	0.05	–	–	$CO_2 = 0.5$ $H_2O = 0.0$
N_2	0.075	$-2CO_2 + 3H_2O + 3 \times 3.76\ N_2$	–	$CO_2 = 0.0$ $H_2O = 0.0$
	1.00		0.9715	$CO_2 = 0.812$ $H_2O = 0.813$

The stoichiometric AFR volumes are = 4.63 moles/mole of fuel

The air: fuel ratio on the weight basis is 4.63 × 28.97/0.08 × 2 + 0.06 × 32 + 0.048 × 28 + 0.435 × 1 + 0.25 × 28 + 0.05 × 44 + 0.062 × 4 = 17.05

60% excess of air actual air AFR on volume basis = 4.63 × 1.6 = 7.866

Excess of oxygen is (equal to 0.9715 × 0.6) = 0.585 moles/mole

Oxygen present in the products = 0.585 moles/mole

Nitrogen present in the products=0.075 + 7.4 moles/mol × 0.79 = 5.921

So the combustion products in dry combustion are 0.812 CO_2 + 5.912 N_2

In wet combustion the combustion products are 8.13 H_2O + 0.812 CO_2 + 0.585 O_2 + 5.921 N_2 + 0.585 O_2

Problem 8.8

An oil refinery is emitting combustible gases which give products after combustion with the following composition

$N_2 = 85.7\%$ $CO_2 = 12.3\%$ $CO = 2\%$.

Estimate the stoichiometric air: fuel ratio and arrive at the fuel composition on volume basis in the presence of 30% excess air.

Solution:

As the combustion products contain only CO_2, CO, and N_2 the following combustion reaction can be written as:

$$C_xH_y + aO_2 + bN_2 = cCO_2 + dCO + eH_2O$$

Balancing the nitrogen, oxygen, carbon and hydrogen atoms,

$(x = 14, y = 17.28, a = 8, b = 22.79, c = 12.3, d = 2, e = 8.79)$

Thus the composition of the fuel can be written as $C_{14}H_{17.58}$. The theoretical or stoichiometric combustion equation is

$$C_{14}H_{17.58} + 8O_2 + 22.79\ N_2 = 12.3\ CO_2 + 2\ CO + 8.79\ H_2O + 22.79\ N_2$$

The stoichiometric AFR

$$\frac{m_a m_w}{m+m+w} = \frac{28.9\times(8+22.79)}{14\times12+17.58} = \frac{889.831}{185.58} = 4.79\ \text{kg/kg of fuel}$$

with 30% excess the AFR = 1.3 × 4.79 = 6.227 kg/kg of fuel.

8.6.4 Direct flame incineration

For destruction of gaseous pollutants by combustion one has to consider the safety aspects of combustion and improper combustion products. Many times the industrial gaseous effluents are not directly combustible and require addition of a fuel. However certain gaseous effluents from petroleum refinery and blast furnaces are flammable directly without the addition of any fuel.

Direct-flame incineration of vapour-air mixture can be carried out when the combined mixture has a heating value of 900 kcal/m^3 or more. The gases are reheated if the heating value is low but higher than 450 kcal/m^3. The equipment for flame combustion includes fume and vapour incinerators, after burners and flares. The factors affecting the efficiency and economy of flame combustion are the temperature of combustion, the time of contact of the gases, the concentration of oxygen in the gases and turbulence of the gases. The device for direct-flame incineration is often a conventional burner firing into an enclosure or even into the open, if practicable. Petrochemical plants and petroleum refineries make use of vertically oriented open-fired combustors. These are referred to as flares. Care should be taken that flashback of the flame up to the upstream channel does not occur. This can be achieved by setting the solvent-vapour-gas-mixture velocity on the upstream side well above the velocity of flame propagation. Figure8.13 shows a duct-type gas burner. The disposal of hydrocarbon gases released intermittently should preferably be by the use of flares.

Complete combustion normally takes place in the afterburner provided a temperature well above 500 °C is maintained there. The afterburner maintains continuous contact of the vapour-laden gas with the flame from an auxiliary burner. An adequate retention time of 0.2 s or more in the combustion zone is provided by proper design of the afterburner chamber.

Hydrocarbons can be burnt to CO_2 and H_2O and the exit gas stream is odorless. Direct flame incineration is the method of choice if two criteria are satisfied. First, the gas stream must have an energy concentration greater than 3.7 MJ/m^3. At this energy concentration, the gas flame is self-supporting after ignition. Below this point, supplementary fuel is required. The second requirement is that none of the byproducts of combustion be toxic. In some cases the combustion byproduct may be more toxic than the original pollutant gas. For example, the combustion of trichloroethylene will produce phosgene. Phosgene was used as a war gas in odd War I. Direct flame incineration has been successfully applied to varnish cooking, meat smokehouse, and paint bake oven emissions.

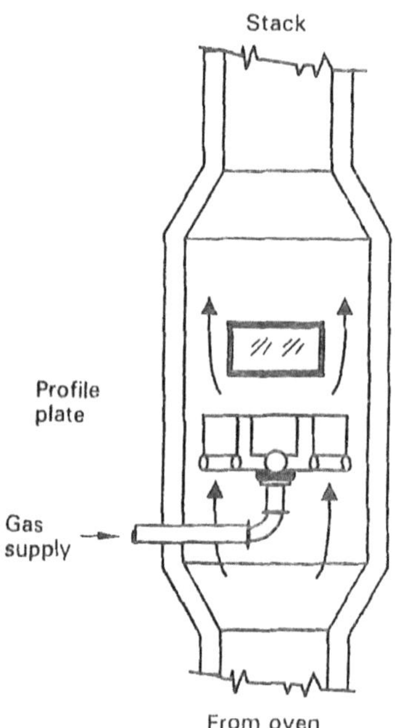

Fig. 8.13 *Duct-type burner*

8.6.5 Flares

For destroying fuel gases which cannot be tackled by any other method, flares are used. A flare is a combustion process in which a burner open to the atmosphere is kept at a height (5–100 m from gas) away from people and other material structures so that the heat liberated cannot do any damage. Figure 8.14 shows the flare burner which can be erected at any elevated place. The height required for an elevation is given by $H = \sqrt{Q/k} - 20D$ where H is the height of the flare burner which is kept above the structures, Q is the energy release rate of the flare in watts, D is the diameter of the flare and k is a constant. The contaminated gas stream containing pollutants which approach the lower limit of flammability, can be passed through a combustion chamber; a certain quantity of natural gas and fresh air to support combustion is fed to the chamber; the operating temperature is of the order of 1370 °C or higher.

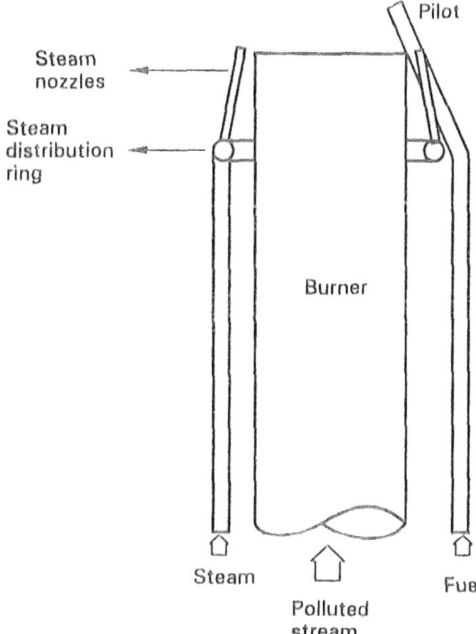

Fig. 8.14 *Sketch of a flare burner: this type is normally mounted several metres above ground*

In case of gases containing unsaturated hydrocarbons such as olefins and aromatics with low hydrogen to carbon ratio, and fuel containing in organics like Cl, F S, CN, a stream of steam through jets surrounding the tip of the flame (Fig. 8.15) is recommended for reducing smoke. The steam requirement is 0.05 to 0.3 kg steam per kg of gases depending on the types of compound present. The steam jet induces turbulence around the tip of the flame and delivers heat to the combustion process. Further the reaction between water and gases of the combustion gases results in smokeless flame.

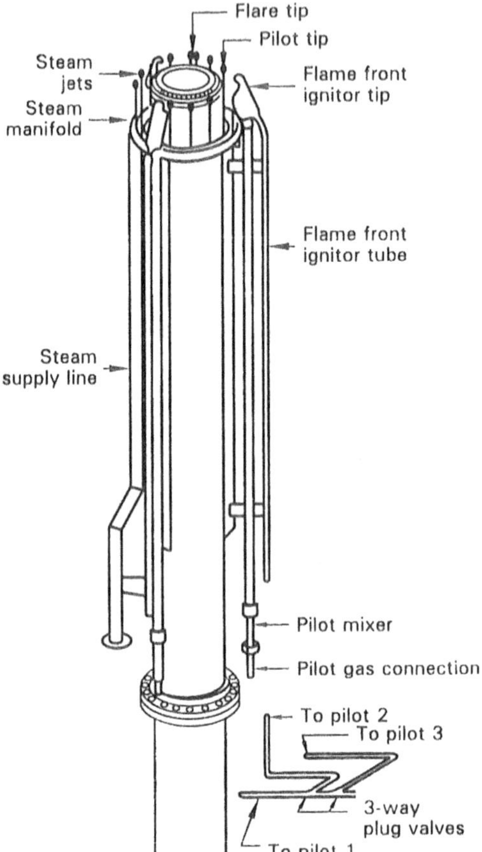

Fig. 8.15 *Steam injection type flare [7]*

8.6.6 Thermal incineration

When the combustible contaminants exist in very small concentrations, and thus cannot support combustion, thermal incineration becomes more economical to use than flame incineration, since the temperature needed here is of the order of 676 °C. A gas burner raises the temperature of the polluted stream to 540–820°C, causing thermal degradation of the fumes [Fig. 8.16(a)].

Volatile organic compounds (VOCs) are very serious air pollutants. Thermal incineration is one of the best ways to destroy them. This incineration process burns organic compounds and renders them harmless by giving off carbon dioxide and water. The proper length of residence time must be given so that complete combustion takes place. The temperature should be high enough so that total destruction of the toxic wastes occurs. Incinerators require residence times ranging from 0.5 to 1.0 second, with temperatures ranging from 650–871 °C. In general, destruction efficiencies are above

95%. One of the major advantages of thermal incinerators is the near total destruction of the hazardous waste without any residual waste being generated. Maintenance of the equipment and the initial cost are relatively reasonable. However, the main disadvantage is the need for fuels to support the high temperatures of combustion.

If the toxic stream contains halogenated compounds such as chlorine and fluorine, materials of construction must be carefully chosen because of the prevailing high corrosivity feed; materials such as stainless steel or fiber reinforced plastics are used. Many times, a scrubber is used prior to the release of the effluent gas, especially since any halogenated acid gases must not be released into the atmosphere.

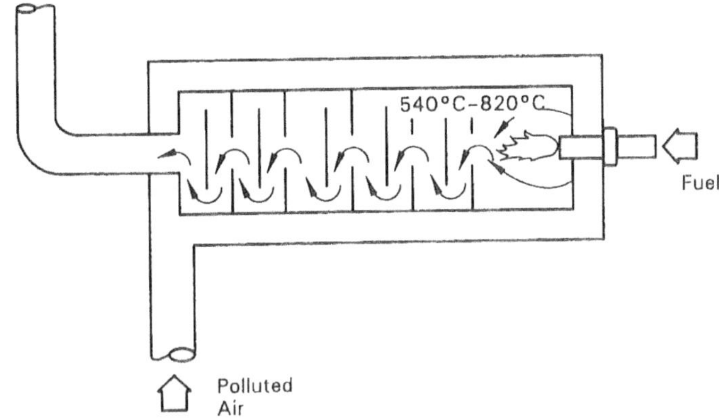

Fig. 8.16(a) *Thermal incinerator [7]*

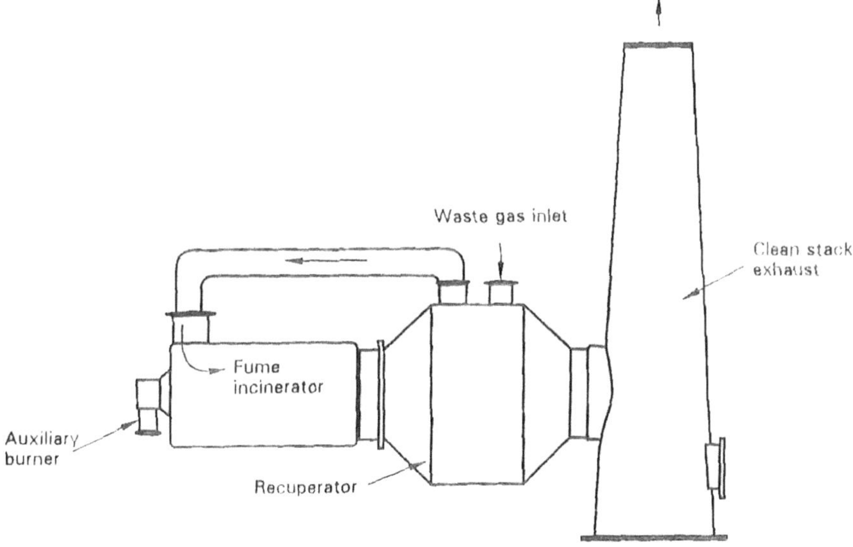

Fig. 8.16(b) *Thermal incineration with recuperation of heat*

8.6.7 Thermal incineration with recuperation

When the concentration of vapour in the gases is below the lower explosive limit, recuperation to increase the temperature of the vapour-gas mixture is practised. No additional fuel is necessary to support combustion with the use of recuperators. This scheme can be applied to solvent-gas mixtures from coating applications where the concentration of solvent is as low as 10–20 percent of the lower explosive limit. A properly designed and operated incinerator can completely destroy the organic vapour from the exit gases. Figure 8.16(b) shows a typical thermal-incinerator assembly incorporating a recuperator. It consists of an auxiliary burner, incinerator and recuperator.

8.6.8 Catalytic incineration

Some catalytic materials enable oxidation to be carried out in gases [9] that have an energy content less than 3.7 MJ/m^3. Conventionally, the catalyst is placed in beds similar to adsorption beds. Frequently, the active catalyst is a platinum or a palladium compound. The supporting lattice is usually a ceramic. Aside from expense, a major drawback of the catalysts is their susceptibility to poisoning by sulphur and lead compounds in trace amounts. Catalytic combustion has been successfully applied to printing press, varnish cooking, and asphalt oxidation emissions. A large number of hydrocarbons are catalytically decomposed to give innocuous compounds. Sulphur and metal bearing compounds poison the catalyst rapidly and render its use less attractive economically. Again, noxious odors are practically eliminated with this process. Typical catalytic incinerators are shown in Fig. 8.17(a) and (b).

Catalytic incineration is the least expensive and flame incineration is the most expensive. Catalytic incineration destroys organic fumes and readers them harm less by giving carbon dioxide and water as products of combustion. Here, the catalyst plays its role in speeding up the reaction by allowing the combustion to take place at a lower temperature, around 315 °C. The inlet stream of toxic wastes is heated in the recovery unit; it then moves on to a catalyst where it is converted into harmless substances. A

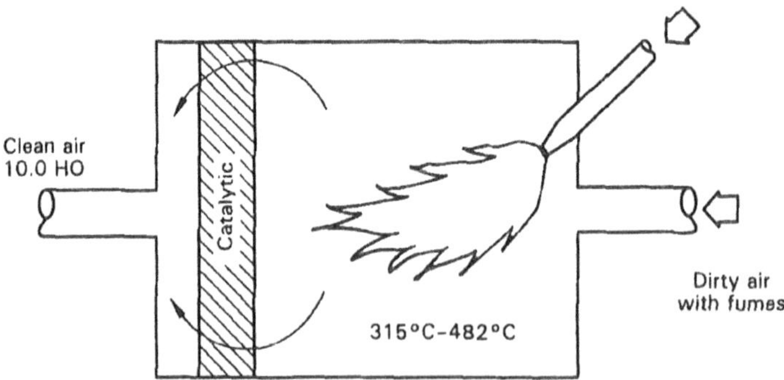

Fig. 8.17(a) *Catalytic incinerator*

plate-and-frame arrangement of the catalyst is the normal way of using it. This way the toxic waste will come into good contact with the catalytic substance, where it is burned and rendered harmless at a relatively low temperature. Catalysts used are normally platinum alloys, or copper chromite, or platinum itself along with some others like copper oxide, chromium, manganese, or nickel. They are usually placed on chemically inert materials like ceramics. The inlet temperatures of various catalytic processes are given in Table 8.8 [8]. Though the economy of the catalytic after burning process is low due to low operating temperatures the actual process and the type of heat values determine the operating costs.

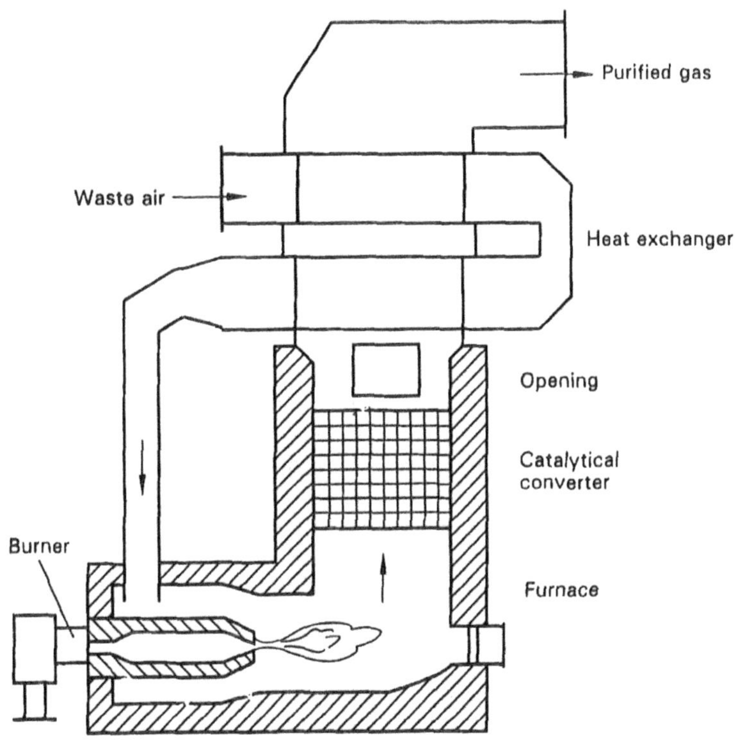

Fig. 8.17(b) *Catalytic after burning plant*

Table 8.8 *Inlet temperatures for catalytic combustion [8]*

Process	Contaminants in waste gas	Catalytic oxidation temp. °C
Asphalt oxidizing	Aldehydes, anthracenes, oil vapours hydrocarbons	315–371
Carbon black mfg.	Hydrogen, carbon monoxide, methane, carbon	649–982
Catalytic cracking units	Carbon monoxide hydrocarbons	343–427
Coke ovens	Wax, oil vapours	315–371

Table 8.8 ***Contd...***

Process	Contaminants in waste gas	Catalytic oxidation temp. °C
Formaldehyde mfg.	Hydrogen, methane, carbon monoxide, formaldehyde	343
Nitric acid mfg.	Nitric oxide, nitrogen dioxide	260–649
Lithography ovens	Solvents, resins	260–399
Octyl-phenol mfg.	Phenol	315–427
Phthalic anhydride mfg.	Maleic acid phthalic acid, napthaquinones, carbon monoxide formaldehyde	315–343
Polyethylene mfg.	Hydrocarbons	260–649
Printing presses	Solvents	315
Varnish cooking	Hydrocarbons	315–371
Wire coating and enameling ovens	Solvents, varnish	315–371

The key to the success of catalytic incineration is the fact that active sites on the catalyst must be made abundantly available to the oncoming toxic gas. Otherwise, the catalyst will become ineffective in doing the job it is meant to do. To maintain the catalyst in an active state and to achieve complete combustion, oxygen has to be maintained at 1% excess than required. Certain metal particulates may deposit on the catalyst and with time decrease its effectiveness. In this case, the catalyst must be reactivated by removing any impurities that may have deposited on it. This process of cleaning can be done by using steam on the catalyst or air at higher temperatures. A major advantage of the catalytic incineration process is the complete (96%) destruction of the toxic wastes.

These units are also very heavy when compared to thermal incinerators. Thus, for all practical purposes, they should be installed on ground levels. Fuel consumption is much lower than that needed for thermal incinerators, simply because operating temperatures are around 320 °C, instead of 870 °C.

8.6.9 Adsorption and incineration

A combination of adsorption by using activated carbon and catalytical incineration is the latest technology for destruction of gaseous pollutants. Initially the carbon adsorption unit will concentrate the organic components of the polluted air prior to the incineration process which can be used for toxic streams with low concentrations of organic matter. In this technique high destruction efficiency of the toxic components in the air stream can be achieved, without the generation of either solid or liquid waste. However the by product waste will have to be further treated and will add to the cost of the operation. Due to the catalysts activation of combustion the requirement for supplemental fuel is low when compared to the amount of fuel needed in a thermal incinerator and efficiency makes this process very attractive.

Problem 8.9

A chemical factory produces 20 m^3/sec of effluent flue gas at 200 °C. Estimate how much quantity of water can be heated for boiling at 30 °C.

Solution: The molecular weight of air = 29 which can be taken as the molecular weight of flue gas. From thermodynamics we know

$$C_p - C_v = R$$

where C_p is the specific heat at constant pressure and C_v is the specific heat at constant volume

Taking specific weight of air as 1.2 kg/m^3 the total weight of heated air flowing through is given as

Flow rate × density = 20 m^3/sec × 1.2 kg/m^3 = 24 kg/sec

$R = 8.314$ joules/mol. k

$C_p = 3.5 \times R = 3.5 \times 8.314 = 29.05$ joules

$d_r = 473.15 - 303.15 = 76.6$°C.

Heat lost = $m \times C_p \times d_r$

$$\frac{24 \times 10^4 \times 6.931 \text{ cal}}{29} \times 170 = 9.75 \times 10^5 \text{ cal/sec or } 975 \text{ kcal/sec}$$

Heat gained by water at 30 °C

$m \times s \times d_T$ where m = mass, s = density = 1, d_T= 70 K

$m = 9.73 \times 10^5/70 = 13928$ gm or 13.9 litres.

Problem 8.10

Estimate the air-fuel ratio for formaldehyde if it is burned in 100% excess air.

Solution: Air is supplied to the fuel to supplement oxygen requirement as it contains 21% O_2 and 79% nitrogen by volume. So for every mole of O_2 present in air supplied for combustion there are (0.79/0/21) 3.76 moles of nitrogen.

So the equation for the combustion of formaldehyde can be written as

$$CH_2O + xO_2 \rightarrow yCO_2 + zH_2O + 3.76\ N_2$$

So the stoichiometric equation for the burning of formaldehyde is

$$CH_2O + O_2 + 3.76\ N_2 \rightarrow CO_2 + H_2O + 3.76N_2$$

$$\text{AFR} = \frac{M_{\text{air}} \times n_{\text{air}}}{M_{\text{fuel}} \times n_{\text{fuel}}}$$

$$\text{AFR} = \frac{28.97 \times 4.76}{30} = 4.95$$

Problem 8.11

Find the stoichiometric AFR for methyl mercaptan.

Solution:

$CH_4S + xO_2 + 3.76 \times N_2 \rightarrow yCO_2 + zH_2O + kSO_2 + 3.76 \times N_2$

$1 = y;\ 4 = 2z \Rightarrow z = 2;\ 1 = k;\ 2x = 2y + z + 2k = 2 + 2 + 2 = 6 \ \therefore x = 3$

$\therefore$ Stoichio metric eqn. is: $CH_4S + 3O_2 + 11.3N_2 \rightarrow CO_2 + 2H_2O + SO_2 + 11.3\ N_2$

$AFR = 28.97\ (3 + 11.3)/48 \times 1 = 8.631$

Problem 8.12

Find AFR if 120% excess air is used for combustion of diethyl ketone, $C_5H_{10}O$.

Solution:

$C_5H_{10}O + xO_2 + 3.76 \times N_2 \rightarrow yCO_2 + zH_2O + 3.76 \times N_2$

$\therefore\ 5 = y;\ 10 = 2z;\ 2x + 1 = 2y + z \quad \therefore\ 2x = 10 + 5 - 1 = 14 \quad \therefore\ x = 7$

$C_5H_{10}O + 7O_2 + 26.3\ N_2 \rightarrow 5CO_2, + 5H_2 + 26.3\ N_2$

Stoichio metric $AFR = 28.97\ (7 + 26.3)/86\ (1) = 11.22$

$\therefore$ If 120% excess air is used. $AFR = 2.2 \times 11.22 = 24.68$

Problem 8.13

Find the air-fuel ratio if (a) 80% and (b) 150% mixture strengths are used for the combustion of C_2H_6O in a motor vehicle.

Solution: The stoichiometric equation of combustion is:

$C_2H_6O + 3O_2 + 11.28\ N_2 \rightarrow 2CO_2 + 3H_2O + 11.28\ N_2$

AFR (stoichiometric) = 28.9 (3 + 11.28)/46 ×1 = 8.96 kg/kg fuel.

(a) Mixture strength = stoichiometric AFR/actual AFR $\Rightarrow$ 0.80 = 8.96/AFR

$\therefore$ Actual AFR = 8.96/0.80 = 11.2 kg/kg fuel.

(b) Mixture AFR = 8.96/1.50 = 0.834 × 8.96 = 5.97 kg/kg fuel

Problems

1. Several industrial processes produce waste gases containing carbon monoxide. Normally these waste gases contain no oxygen and are mixtures of CO and N_2. If they contain enough CO, it is common practice to mix them with air and burn them, using the heat released to generate steam or for some other practical purpose.
   (a) Using the values in Table 9.7 estimate the lowest concentration of CO in the waste gas for which burning is possible. Here the condition we are seeking is

one where the amount of oxygen supplied is just the amount needed to burn up all the CO, with surplus oxygen, and the concentration of CO in the final mix of waste gas and air is at lower explosive limit. Using the LEL value in Table 9.7 is not exactly correct, because the lower explosive limit data in Table 9.6 is for CO in the presence of a mixture of N_2 and O_2 which has the ratio of 79/21 while in the case the ratio is much higher. But experimental data show that the result calculated this way is close to the observed value.

(b) List industrial processes that might produce such waste gases. These can be inferred by considering what kind of chemistry would produce such gases.

2. The explosive limit data in Table 9.7 (and all similar tables) are based on mixtures of the combustible material with air, which is a mixture of approximately 21% oxygen, 79% nitrogen. If we substitute pure oxygen for air, we normally find that the lean limit is unchanged, but the rich limit is greatly increased. Why? What are the safety consequences of this observation?

3. It is a rule of the thumb that for organic reactions near room temperature, raising the temperature by 10°C will double the reaction rate. If we take room temperature as 20 °C and assume that the rate of some reaction doubles when we raise the temperature to 30 °C and that $K = A_{exp}(-E/Z_t)$ is applied.

   (a) What is the value of E, the activation energy, for this reaction?

   (b) How much would the reaction double when we raise the temperature from1000to 1010 °C?

4. In tests of incineration of herbicides, the incinerator temperature averaged 1500°C, and the time spent by the material being burned at this temperature was 1.0 s. The destruction of dioxin (2,3,7,8 tetrachlorodibenzo-p-dioxin); averaged over three testing periods, was 99.93%. Assuming that this material burns according to the first order equation, estimate how long the material would need to be held at1500 °C to get 99.999% destruction.

5. We wish to treat an air stream from a paint dryer in a combustor to destroy 90%by weight of the combined hydrocarbons. The hydrocarbon vapor consists of components A, B, and C. Their weight percentages in the feed are 25, 25, 50. Each is believed to combust to carbon dioxide and water according to first order kinetics. The combustor will operate at 260 °C. At this temperature the first order rate constants ($1/s$) are 1.0, 2.0 and 3.5. How long must this gas be held at 815 °C to get 90% by weight conversion to carbon dioxide and water? Assume that the three components do not interact on burning (not a good assumption in the real world, but a satisfactory one for this problem) and that the heating up to 815 °C and cooling down from it are instantaneous.

Questions

Short Answer Questions

1. List out the scrubbing solvents for various pollutant gases.
2. Differentiate between plate and packed towers.
3. Outline the Two stage regenerative system.
4. Write a short note on Four stage regenerative system.
5. Brief about the Fluidized bed solvent recovery system.
6. Write a short note on how Flares are employed.
7. What is meant by Catalytic incineration?

Essay Questions

1. Explain the principle behind removal of pollutants by absorption.
2. Describe the theory of absorption of gases by moving liquid drops.
3. Outline the working principle of Physical absorption system with solvent recovery.
4. Explain how the pollutants are removed by adsorption on solids.
5. Discuss the 3 techniques employed for destruction of organic vapors by combustion.
6. Schematize the Parameters influencing pollutant emission spectrum of different types of furnaces.
7. Explicate the working of a Direct flame incineration system.
8. Bring out the working of a Thermal incineration system with recuperation process.

References

1. National Research Council "International Critical Tables" McGraw Hill, 1928, NewYork.
2. R.C. Reid and T. K. Sherwood, 1956. "The properties of gases and liquids and their estimation and correlation. McGraw Hill, New York.
3. R. H. Perry and C. H. Chilton (Eds), 1973. Chemical engineers handbook. 5th Edition. McGraw Hill, New York.
4. M. Crawford, 1976. Air pollution control theory. McGraw Hill, New York.
5. A. J. Buonicore and L. Theodore, 1975. Industrial control equipment for gaseous pollutants. Cleveland CRC Press.
6. E. Rosert Treybal, 1988. Mass transfer operations. McGraw Hill, New York.
7. W. Strauss, 1966. Industrial gas cleaning. Pergammon Press, London.

8. Lewis B. Anal and G. Von Elbe. Combustion flames and explosive gases. 2nd Edition.Academic Press, New York.
9. A. J. Buonicore and W. T. Davis, 1992. Air Pollution Engineers Manual. Van Nostrand Reinhold. New York.

Suggested References

1. Anna-Karin Hjaimarsson (1990), NOx control technologies for coal combustion, IEACR/24, June 1990.
2. Anthony J. Buonicore, Wayne T. Davis (1992), Air Pollution Engineering Manual, Van Nostr and Reinhold.
3. Faisal I. Khan, Aloke Kr. Ghoshal (2000), Removal of Volatile Organic Compounds from Polluted Air, Journal of Loss Prevention in the Process Industries, 13: 527-545.
4. Howard E. Hesketh (1996), Air Pollution Control: Traditional and hazardous pollutants, Technomic Publishing Company, Inc.
5. Levenspiel O. (1970), Chemical Reaction Engineering. New York: John Wiley.
6. Mitsuru T and Hermine S. (1993) FGD Performance and Experience on Coal-fired Plants. IEA CR/58. IEA Coal Research, London.
7. Rhoads T.W., Marks J.R., and Siebert P.C. (1990), Overview of industrial sources control for nitrogen oxides, Environmental Progress, 9(2): 126-130.
8. Richard C. Flatgan, John H. Seinfeld (1989), Fundamentals of Air Pollution Engineering, Prentice-Hall Englewood Cliffs.
9. Robert Jennings Heinsohn, Robert Lynn Kabel (1999), Sources and Control of Air Pollution, Prentice-Hall Inc., Simon & Schuster / A Viacom Company.
10. Rosana M. Alberici, Wilson F. Jardim (1997), Photo catalytic destruction of VOCs in the gas-phase using titanium dioxide, Apllied Catalysis B: Environmental, 14: 55-68.

CHAPTER 9

Air Pollution Control Technologies for Oxides of Sulphur

9.1 Introduction Control/Removal Technologies for SO_x

Sulphur dioxide is found in many industrial gases emanating from plants involved in roasting, smelting and sintering sulphide ores, or gases from power plants burning high sulphur coal or fuel oils or other sulphurous ores or other industrial operations involved in the combustion of sulphur-bearing fuels, such as fuel oil and auto mobiles. One of the more difficult environmental problems facing industry is how to economically control SO_2 emissions from these sources.

The following are the control technologies adopted for reducing SO_x emissions.

For controlling the emissions of SO, the following in-plant control measures or gaseous effluent treatment methods can be adopted: (i) Dilution of SO_x emissions by increased stack height (ii) Use of alternative fuels (iii) Fuel desulphurization (iv) Reduction of sulphur in the combustion process and (v) Flue gas desulphurization.

9.1.1 Reducing SO_x levels through dilution by increasing stack height

Natural dispersion at high elevations using stacks is one of the preliminary methods for controlling ground level concentrations of the pollutants. The minimum stack height prescribed for Indian conditions for any stack is 30 m. Minimization of gaseous discharges into atmosphere under stable atmospheric (inversion) conditions will greatly help in reduction of ground level concentrations.

9.1.2 Use of alternative fuels

This includes either switching to a fuel with reduced amount of sulphur or else a reduction of fuel-sulphur before its firing. Since SO_2 emissions are directly proportional to the sulphur content of fuel, and also to the amount of fired, a reduction in emissions can be achieved by switching to low sulphur fuels and to higher quality ones

Using liquefied natural gas instead of conventional high-sulphur fuels like coal and petroleum decreases SO_x gas emissions. However, its high cost limits its use in various industrial applications when compared to other alternative fuels. While using a low sulphur coal is a good option, acquiring low sulphur coal is a costly and time consuming option. Table 9.1 presents the sulphur content of various types of widely available Indian fuels which indicates that the burning of fuel is one of the main contributors to SO_2 emission and substituting of sulphur-containing fuel with alternative fuels can help in reducing SO_2 emissions.

Table 9.1 *Alternative fuels with varying sulphur content*

Solid fuels (coals), as received	**Fixed carbon**	**Volatile matter**	**Ash**	**Moisture**	**C**	**H**	**O**	**N**	**S**	**Higher heating value kJ/kg**
Anthracite	88	1	8	3	85	2	4	1	1	30,000
Semianthracite	75	10	12	3	78	4	4	1	1	30,000
Bituminous	60	25	10	5	78	5	7	1	3	28,000
Sub-bituminous	45	25	5	25	55	6	34	1	1	21,000
Lignite	22	35	8	35	42	7	42	1	1	16,000
Liquid fuels										**Higher heating value MJ/m^3**
No. 2 fuel oil	Specification for various grades of fuel oils are based on required									39,000
No. 4 fuel oil	Characteristics for firing in different types of burners, Fuel									40,400
No. 5 fuel oil	Composition varies with crude sources and refining									41,800
No. 6 fuel oil										42, 100
Gaseous fuels	**Composition**									**Higher heating value kJ/m^3**
Natural gas	86-95 % methane, remainder ethane, propane, CO_2, and N_2									37,200
Commercial propane	90-95 % propane, remainder propylene									93,100
Commercial butane	95%butane, remainder butylenes									119,200
Fuel gas from coal gasification	95 % butane, remainder butylene									3,700 to 13,000

9.1.3 Extraction of sulphur from fuels (fuel desulphurization)

Sulphur in coal is present in inorganic form as iron disulphide (FeS2) mostly in the form of pyrites; the organic form of sulphur is present as cystene, thiols, sulphides and some other cyclic compounds. Inorganic sulphur can be reduced by 30%, by washing. However, this results in loss of combustible material also. As organic sulphur is bound with carbonaceous material it can be removed only by chemical decomposition. Various cleaning methods have been developed for desulphurizing sulphur-bearing fuels prior to their firing. Sometimes this is associated with changes in the physical characteristic of the fuel, resulting in operational problems Sulphur removal process.

Both organic and inorganic sulphur can be removed by a solvent extraction [1] process (Fig. 9.1) which is also known as hydrodesulphurization process. The finely ground coal is slurried with anthracene oil in the presence of a small amount of hydrogen (about ½ %) to avoid repolymerization. To dissolve the coal the slurry is heated at 450 °C and the ash residue is removed by pressure filtration. The light fraction of coal filtrate is separated by flash evaporation while in subsequent operations the solvent is recovered by distillation. The liquid evaporator residue after cooling is pulverized its use as solid fuel, which can also be liquified at about 450 °C. This will have a higher heating value than raw coal, and contains less than 1% sulphur even when the starting material is coal of high sulphur content

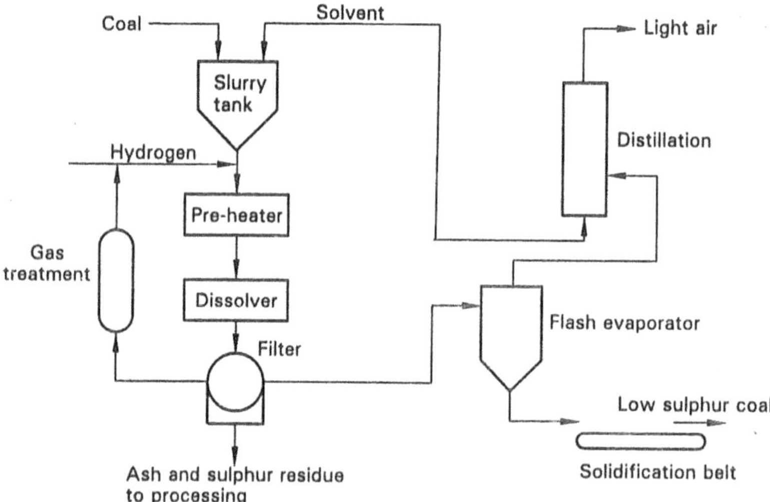

Fig. 9.1 *Solvent extraction process of coal*

9.1.4 Coal gasification

For India with abundant coal reserves with high sulphur content coal gasification (Fig. 9.4) is a promising method for which a number of technologically feasible processes are available for removing inorganic and organic coal.

In this process, coal is combined with oxygen and steam to produce a combustible gas, waste gases, char, and ash. The more than 70 coal gasification systems available or being developed can be classified by the heating value of the gas produced and by the type of gasification reactor used. High-Btu gasification systems produce a gas with a heating value greater than 900 Btu/scf (33,000 J/m^3). Medium-Btu gasifiers produce a gas having a heating value between 250 - 500 Btu/scf (9,000 - 19,000 J/m^3). Low-Btu gasifiers produce a gas having a heating value of less than 250 Btu/scf (9,000 J/m^3).

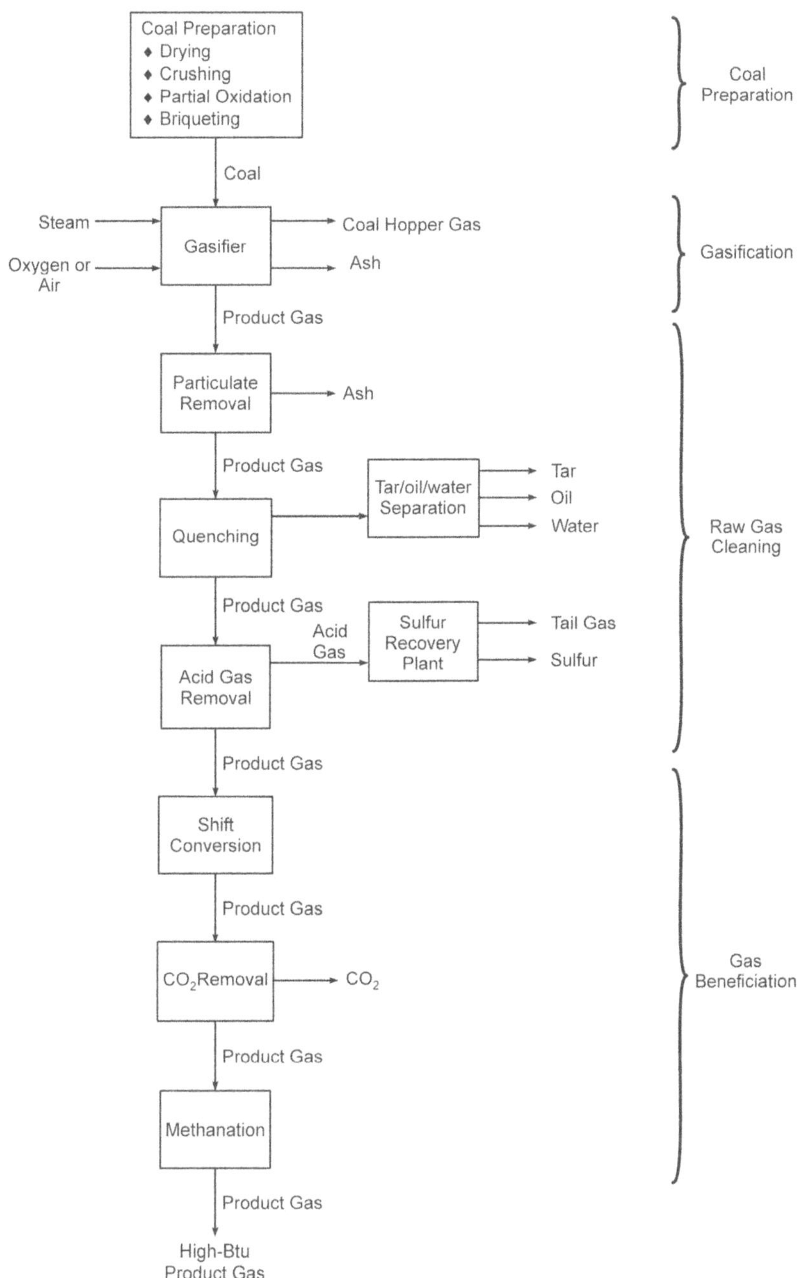

Fig. 9.2 *Schematic of various unit operations in Coal Gasification*

The majority of the gasification systems consist of 4 operations: coal pretreatment, coal gasification, raw gas cleaning, and gas beneficiation. Each of these operations consists of several steps. Figure 9.2 is a flow diagram for coal gasification.

Generally, any coal can be gasified if properly pretreated. High-moisture coals may require drying. Some caking coals may require partial oxidation to simplify gasifier operation. Other pretreatment operations include crushing, sizing, and briqueting of fines for feed to fixed bed gasifiers. The coal feed is pulverized for fluid or entrained bed gasifiers.

After pretreatment, the coal enters the gasification reactor where it reacts with oxygen and steam to produce a combustible gas. Air is used as the oxygen source for making low-Btu gas, and pure oxygen is used for making medium- and high-Btu gas (inert nitrogen in the air dilutes the heating value of the product). Gasification reactors are classified by type of reaction bed (fixed, entrained, or fluidized), the operating pressure (pressurized or atmospheric), the method of ash removal (as molten slag or dry ash), and the number of stages in the gasifier (1 or 2). Within each class, gasifiers have similar emissions.

The raw gas from the gasifier contains varying concentrations of carbon monoxide (CO), carbon dioxide (CO_2), hydrogen, methane, other organics, hydrogen sulfide (H_2S), miscellaneous acid gases, nitrogen (if air was used as the oxygen source), Sulphur, particulates, and water.

The following four gas purification processes may be required to prepare the gas for combustion or further beneficiation:

(i) particulate removal,
(ii) tar and oil removal,
(iii) gas quenching and cooling, and
(iv) acid gas removal.

The primary function of the particulate removal process is the removal of coal dust, ash, and tar aerosols in the raw product gas.

During tar and oil removal and gas quenching and cooling, tars and oils are condensed, and other impurities such as ammonia are scrubbed from raw product gas using either aqueous or organic scrubbing liquors.

Acid gases such as H_2S, SO_2, COS, CS_2, mercaptans, and CO_2 can be removed from gas by an acid gas removal process. Acid gas removal processes generally absorb the acid gases in a solvent, from which they are subsequently stripped, forming a nearly pure acid gas waste stream with some hydrocarbon carryover. At this point, the raw gas is classified as either a low-Btu or medium-Btu gas.

To produce high-Btu gas, the heating value of the medium-Btu gas is raised by shift conversion and methanation. In the shift conversion process, H_2O and a portion of the CO are catalytically reacted to form CO_2 and H_2. After passing through an absorber for CO_2

removal, the remaining CO and H_2 in the product gas are reacted in a methanation reactor to yield CH_4 and H_2O.

9.1.5 Liquefaction

Liquefaction is a conversion process designed to produce synthetic organic liquids from coal. This conversion is achieved by reducing the level of impurities and increasing the hydrogen-to-carbon ratio of coal to the point that it becomes fluid. There were over 20 coal liquefaction processes in various stages of development by various institutes.

These processes can be grouped into 4 basic liquefaction techniques:

(a) Indirect liquefaction, (b) Pyrolysis, (c) Solvent extraction, (d) Catalytic liquefaction

Indirect liquefaction involves the gasification of coal followed by the catalytic conversion of the product gas to a liquid.

Pyrolys is liquefaction involves heating coal to very high temperatures, thereby cracking the coal into liquid and gaseous products.

Solvent extraction uses a solvent generated within the process to dissolve the coal and to transfer externally produced hydrogen to the coal molecules.

Catalytic liquefaction resembles solvent extraction, except that hydrogen is added to the coal with the aid of a catalyst.

Figure 9.3 presents the flow diagram of a typical solvent extraction or catalytic liquefaction plant.

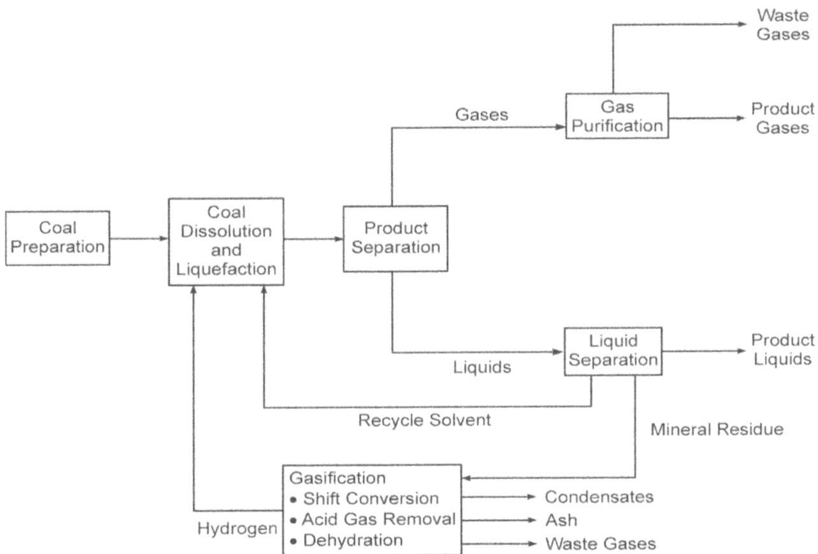

Fig. 9.3 *Schematic of Coal Liquefication*

These coal liquefaction processes consist of four basic operations:

(i) coal pretreatment,
(ii) dissolution and liquefaction,
(iii) product separation and purification, and
(iv) residue gasification.

Coal pretreatment generally consists of coal pulverizing and drying. The dissolution of coal is best effected if the coal is dry and finely ground. The heater used to dry coal is typically coal fired, but it may also combust low-BTU-value product streams or may use waste heat from other sources.

The dissolution and liquefaction operations are conducted in a series of pressure vessels. In these processes, the coal is mixed with hydrogen and recycled solvent, heated to high temperatures, dissolved, and hydrogenated. The order in which these operations occur varies among the liquefaction processes and, in the case of catalytic liquefaction, involves contact with a catalyst. Pressures in these processes range up to 2000 psig (14,000 Pa), and temperatures range up to 900°F (480°C). During the dissolution and liquefaction process, the coal is hydrogenated to liquids and some gases, and the oxygen and sulfur in the coal are hydrogenated to H_2O and H_2S

After hydrogenation, the liquefaction products are separated through a series of flash separators, condensers, and distillation units into a gaseous stream, various product liquids, recycle solvent, and mineral residue. The gases from the separation process are separated further by absorption into a product gas stream and a waste acid gas stream. The recycle solvent is returned to the dissolution/liquefaction process, and the mineral residue of char, un dissolved coal, and ash is used in a conventional gasification plant to produce hydrogen.

Coal suspended in tar at 100-250 bars at 450°C after catalytic hydrogenation can produce 75 percent desulphurization [2] by using nearly 20 kg of hydrogen per ton of coal to be purified of which 5 percent is transformed to hydrogen sulphide, an unwanted by-product (Figure 9.4). By scrubbing with mono ethanolamine or other amines impure natural gas can be freed of hydrogen sulphide.

A regenerative absorption process, most commonly used for the purification of refinery gases is the Girbotol process which uses monoethanolamine. The absorption of H_2S takes place in a 15-20% aqueous solution of the amine at a temperature of 30-40 °C according to the reaction:

$$RNH_3 + H_2S \rightarrow RNH_3HS$$

The solution is then sent to a regenerator where monoethanolamine is separated by heating with steam to obtain sulphur by the Claus process.

In the modified Claus process [3], one-third of the H_2S feed is oxidized to SO_2 which is then reacted over a catalyst with the remaining H_2S to form sulphur vapour.

$$1/3H_2S + \tfrac{1}{2}\, O_2 \rightarrow 1/3\ SO_2 + 1/3\ H_2O$$

$$1/3\ H_2S + 2/3\ H_2S \rightarrow S + 2/3\ H_2O$$

The sulphur is condensed and sent to a sulphur storage tank. Figure 9.5 shows schematically the process diagram of Claus process.

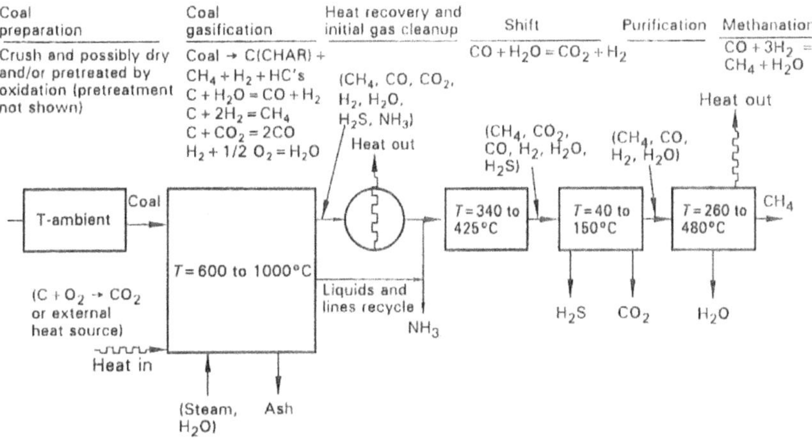

Fig. 9.4 *Coal gasification processes at various Temperatures (2)*

As the gases are required to be cooled in the absorption process there are significant heat losses. However the absorption processes are simple as they can be carried out at high temperatures; the dry methods of purification are more promising. Adsorption of H_2S on ferric oxide in a fluidized bed at 400 °C is one of the practical methods used on a commercial scale. After use the ferric oxide is regenerated by roasting it in air at 800°C producing SO_2 for sulphuric acid manufacture.

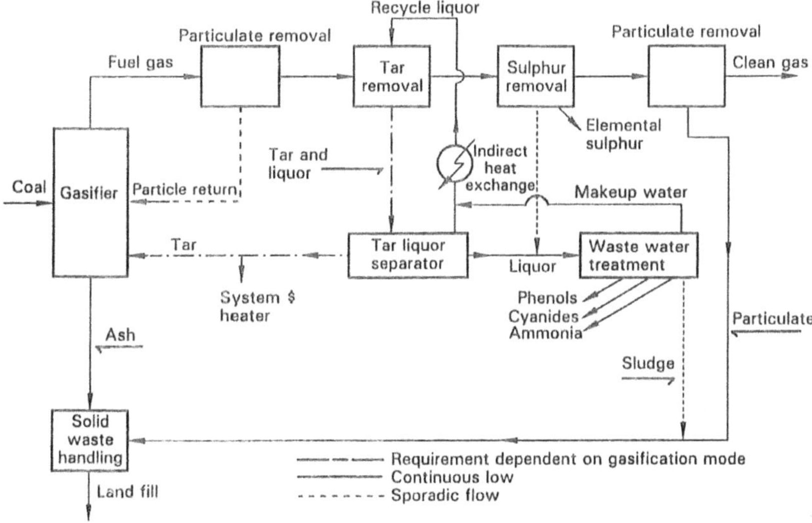

Fig. 9.5 *Fuel gas cleanup system*

9.1.6 Reduction of sulphur in combustion process

A number of technologies to prevent the production and release of SO_2 during combustion have been developed over the past decade, but very few have achieved wide commercial application to date.

Attempt to increase the desulphurization efficiencies have been fairly successful and one such improved process is the fluidized bed combustion process in which limestone and crushed coal together form the fluidized bed and oil is used as the fluidizing medium

The most developed are the Fluidized Bed Combustion (FBC) process and the integrated Gasification Combined Cycle (IGCC) system.

9.1.6.1 Fluidized bed combustion

This process involves the combustion of coal in a bed of inert material such as sand, with air being blown up from beneath the bed at high velocities. As velocity increases individual particles begin to be forced upwards until they reach a point at which they remain suspended in the air stream. The bed in this state behaves like a liquid and can be described as fluidized. Tubes containing water are immersed in the bed to absorb the generated heat (this water is converted to steam which is used to drive the steam turbine and thus produces electricity). The fluidized movement within the combustion chamber results in a greater heat transfer efficiency to the water filled tubes and therefore operating temperatures are lower than in a conventional system. SO_2 emissions can be controlled in this system by adding a sorbent (a substance used to absorb any SO_2 present, for example lime or limestone) to the bed of inert material. The limestone effectively absorbs the SO_2 as it is released from the coal and retains it within the ash, which is removed regularly. The low combustion temperatures allow efficient combustion to take place without causing the ash to soften, thereby allowing easy removal of the ash containing the absorbed SO_2.

Figure 9.6 is a schematic diagram of a simple coal-burning fluidized bed combustion boiler. It is formed by an enclosure consisting of water walls abutting boiler tubes in which water and steam flow. The pressure in this enclosure may be in the 1.25 atm range. A distributor plate with holes or bubble cap devices distributes the air flow uniformly over the base on the enclosure. The air then passes at superficial velocities in the range of 0.64.6 m/s through a bed of particles at temperatures of 760° -1040 °C. These particles, whose top size is 1.6-6.4 mm (1/6–1/4 in), consist of ash or other inert material, lime sorbent, and small quantities (<1%) of unburned coal or carbon. Coal and sorbent are fed to the bed by feeders extending through the air distributor or the waterwalls. From 60-70% of the heat released in burning the fuel with air is transferred to the waterstream in the tubes surrounding and submerged in the bed.

The air and combustion gases passing through the bed cause considerable agitation: particles are thrown from the bed into the empty volume above the freeboard. Larger, heavier particles fall back into the bed; smaller, lighter particles will come out of the enclosure by the gases. Much of the ash in the coal is eventually carried from the bed in the combustion gases. Larger ash particles may accumulate in the bed and require periodic removal.

Particles carried from the fluidized bed enclosure by the combustion gases can be captured by inertial separation apparatus such as a cyclone. These particles comprise ash, fragments of the limestone/dolomite sorbent, and typically 20–40%by weight unburned coal char or carbon. To obtain high combustion efficiencies, the particles captured in this primary cyclone can be recycled to the boiler bed or to a separate fluidized bed combustor where the burning of the char is. completed. Additional heat can be extracted from the combustion gases in the heat recovery section, which may be either an economizer, a boiler feed water heater, or an air preheater.

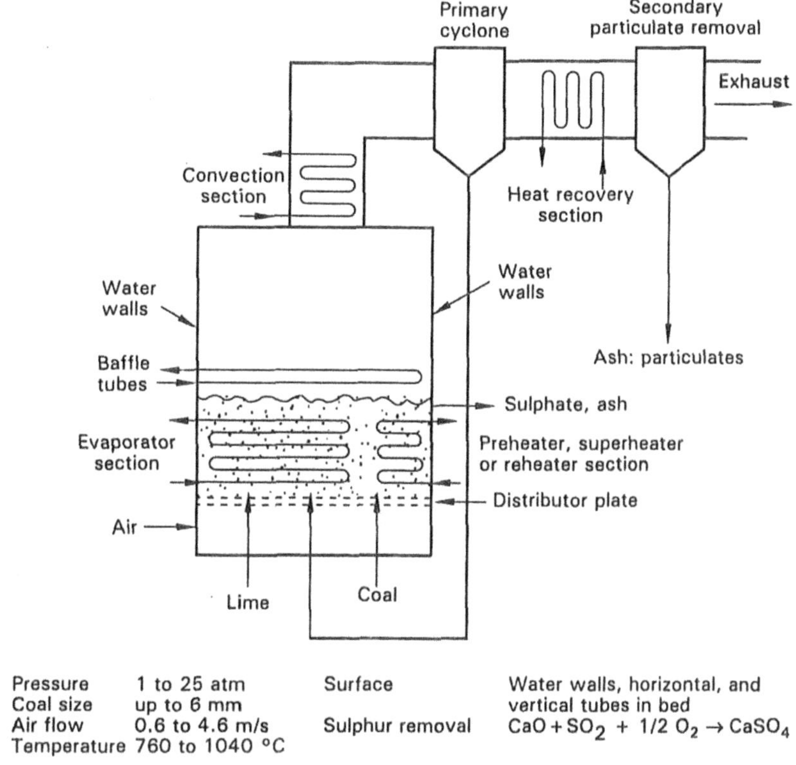

Pressure	1 to 25 atm	Surface	Water walls, horizontal, and vertical tubes in bed
Coal size	up to 6 mm		
Air flow	0.6 to 4.6 m/s	Sulphur removal	$CaO + SO_2 + 1/2\ O_2 \rightarrow CaSO_4$
Temperature	760 to 1040 °C		

Fig. 9.6 *Fluidized bed combustion boiler*

Final cleanup of the combustion gases is carried out in a secondary particulate removal system, using cyclones or an electrostatic precipitator.

Fluidized bed combustion has a high volumetric heat release rate of 5 MJ/s m^3, as compared to 0.2 MJ/s m^3 in a pulverized coal fired boiler. As the operating temperatures are lower in fluidized bed combustion process, the ash fusion is prevented which protects the corrosion of boiler tubes associated with the molten slag. The formation of nitrogen oxides by the nitrogen fixation reaction areas is reduced.

These advantages in addition to higher SO_2 removal efficiency make the fluidized bed combustion process attractive. However, the fluidized bed combustion process requires novel design for boilers and some special arrangement for the collection of solid particles and preparation of limestone, and regeneration of $CaSO_4$ which involve additional capital.

9.1.6.2 Integrated gasification combined cycle system

This process does not require the addition of a sorbent. Instead the coal is gasified under pressure with a mixture of air and steam. The resulting gas is expanded through a gas turbine to produce electricity. The waste heat from the gas turbine is then passed through a second steam turbine, the second stage of the combined cycle process, again producing electricity. As the coal is converted to gas the sulphur present is converted into hydrogen sulphide which can be easily removed and sold for use within the chemical industry. Gas cleaning can be integrated into the gas production process, and emissions can be reduced by more than 99%. Further very little waste is produced.

9.2 Flue Gas Desulphurization (FGD)

9.2.1 Classification of flue gas desulphurization systems

They can be broadly categorized as

1. Throw away or once through and
2. Recovery systems or regenerative system.

In the throwaway systems, the sulphur removed from the flue gas is rejected from the process in a waste sludge [3, 7].

In recovery systems, the sulphur-absorbing reagent is regenerated for recirculation to the flue gas contacting device and the sulphur removed from the flue gas is converted into its elemental form or into sulphuric acid.

This method is perhaps the most popular and practiced technology. The processes can be classified as dry methods, wet methods and wet and dry methods.

In dry methods SO_2 and SO_3 sorption take place purely physically (adsorption) on fixed sorption agents or through absorption and reaction processes (chemisorption).

Dry adsorption processes normally involve continuous moving bed operation, where by the adsorption agent is pulverized to meet the gas to be desulphurized. Later the agent is reactivated in the regenerator and is fed back into the adsorption process. These reactive additives can be added at different points in the furnace for different flue gas pattern (Fig. 9.7).

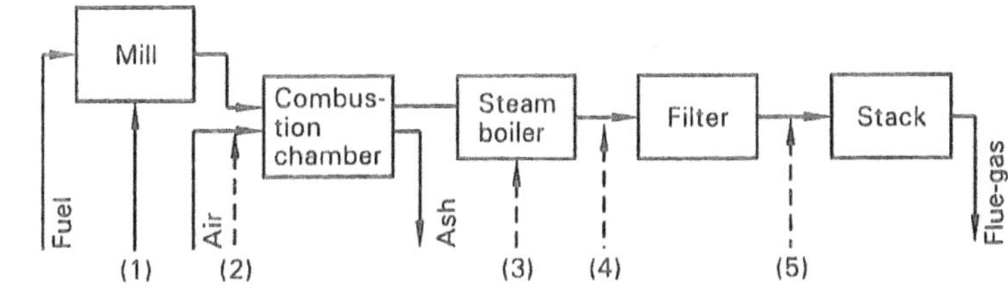

Sorbent addition	Process	Sorbents
(1)	Direct desulphurisation by addition to fuel	$Ca(OH)_2 \cdot (CaCO_3)$
(2)	Direct desulphurisation by addition to combustion air	$Ca(OH)_2 \cdot (CaCO_3)$
(3)	High temperature, non-regenerative	$Ca(OH)_2 \cdot (CaCO_3)$
(4)	Low temperature, non-regenerative	$Ca(OH)_2$
(5)	Low temperature, regenerative	Activated charcoal

Fig. 9.7 *Overview of sorbent dosing possibilities in furnaces [4]*

A number of parameters influence the binding of SO_2 and other acidic gases such as HCl and HF of which temperature is one important factor. The extent of SO_2 absorption for various additives and its dependence on temperature is shown in Fig.9.8 which indicates the effectiveness of the additives decreases significantly both at low and at high temperatures. Magnesium compounds exhibit low absorption efficiency and smaller temperature ranges while calcium based additives show higher absorption efficiency and wide temperature ranges.

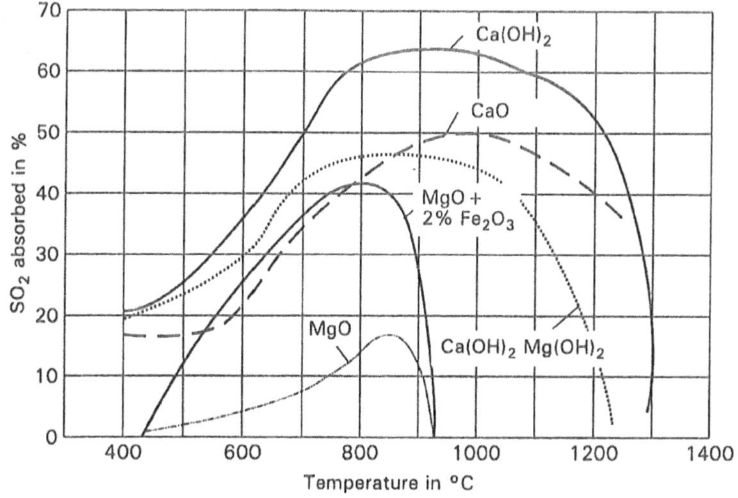

Fig. 9.8 *Temperature dependence of SO_2, absorption in different dry additives*

At low temperatures there will not be sufficient activation energy to initiate the absorption reactions, while at high temperatures the reaction equilibrium will be on the gas side, and surfaces of the additives particles sinter reducing the effective surface thereby causing reduction SO_2 absorption. It can be seen from Fig. 9.8 that MgO is less effective by itself as the compound $MgSO_3$ is not formed in the absence of a catalyst. Adding 2% of Fe_2O_3 as catalyst improves the adsorption of MgO significantly.

With raw limestone the reaction with SO_2 starts only at temperatures in excess of approx. 600 °C; with raw dolomite ($CaCO_3$) in excess of approx. 500 °C. In addition to temperature, other factors like residence time, mixing etc. also play an important role in the combustion process of the limestone. By impregnation with reagents, calcination, hydration as well as by pelletizing the sorbents also can be modified while SO_2 sorptionability will depend on the reaction temperature and reaction time.

Figure 9.9 presents the sequence of possible reactions likely to occur during calcination, hydration and sulphating (SO_2 absorption) of dolomite ($CaCO_3$, $MgCO_3$).

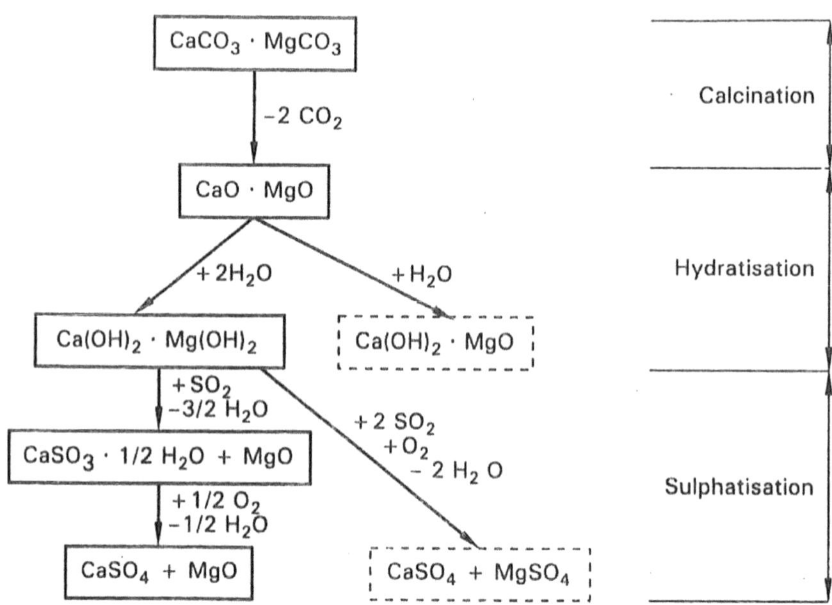

Fig. 9.9 *Reaction sequences of calcination, hydratisation and sulphatisation of dolomite*

As the specific surface increases with hydration the SO_2 absorption pertaining to burnt limestone or dolomite will also increase. As the CaO part is hydratized more than the MgO, calcination based compounds will have higher SO_2 sorption efficiency. The type of hydration strongly influences the sorption ability,

Based on the required application or dose inlet location the reaction additive has to be added at an appropriate point in the furnace. A wet particle surface favorable to SO_2 absorption can be further increased by pumping moisture and mixing with hydrated lime particles. As the SO_2 solubility increases exponentially with sinking temperatures it will be helpful if the flue gases are cooled as much as possible which can also be achieved by adding water. For additive preparation, conditioning the flue gas, re action and disposal of the used additive a separate equipment is required. Additives must beadded in hyper stoichio metric quantities in all dry additive processes for achieving high efficiency in desulphurization.

Though in dry additive processes no wastewater is produced, a number of solid residues will be generated which have some utilization potential (Fig. 9.10).

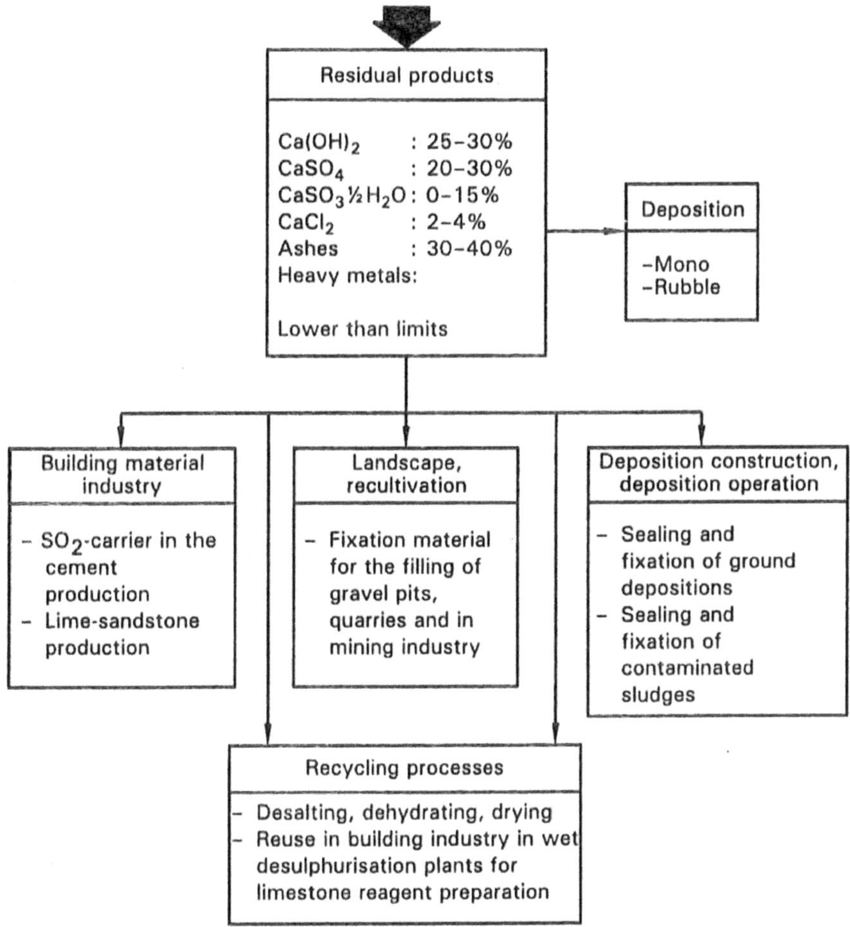

Fig. 9.10 *Utilization of disposal of residual products of dry flue gas desulphurization [9]*

The various FGD processes are summarized in Table 9.2

Table 9.2 *Summary of FGD Process*

Process Type	Process operations	Active materials	Key sulphur product
Non-generative process			
1. Lime or limestone			
2. Sodium	Slurry scrubbing	Cao, $CaCO_3$	$CaSO_3/CaSO_4$
3. Dual alkali	Na_2SO_3 solution	Na_2CO_3	Na_2SO_4
	Na_2SO_3 solution,	$CaCO_3/Na_2SO_3$	$CaSO_3/CaSO_4$
	Regenerated by CaO or	Or CaO/NaOH	
4. Magnesium promoted lime or limestone	$CaCO_3$	$MgO/MgSO_4$	$CaSO_3/CaSO_4$
Regenerative processes			
1. Magensium oxides	$Mg(OH)_2$ slurry	MgO	15% SO_2
	Na_2SO_3 solution	Na_2SO_3	91% SO_2
2. Sodium (Wellman-Lord)	Sodium citrate solution	H_2S	Sulphur
3. Citrate	Ammonia solution	NH_4OH	Sulphur (99.9%)
4. Ammonia	Conversion to SO_2		
Dry processes			
1. Carbon adsorption	Adsorption at 400° K, reaction with H_2S to S, reaction with H_2 to H_2S	Activated carbon carbon/H_2	Sulphur
2. Spray dryer	Absorption by sodium carbonate or slaked lime solutions	$Na_2CO_3/Ca(OH)_2$	Na_2SO_3/Na_2SO_4 or $CaSO_3/CaSO_4$

Presently practiced various commercial FGD process are described below

9.2.2 Catox process

Developed by Monsanto company in this process the flue gas is passed through a fixed catalyst bed where SO_2 in the presence of O_2, is converted to SO_3 (Fig.9.11) [5]. The SO_3 is then absorbed in recirculated sulphuric acid in an absorption tower. Since sulphuric acid is a salable product there is no solid waste product. Fly ash is first removed from the flue gas by a high temperature electrostatic precipitator and SO_2 is then catalytically oxidized to SO_3 which is then recovered as sulphuric acid. For achieving good combustion efficiency V_2O_5, [4] at 400–500 °C is used as a catalyst. In another improved efficiency process SO_2 and oxygen present in the stack gas are passed through the surface of an active carbon catalyst where they are absorbed and the catalyst catalyzes the oxidation of SO_2 to SO_3. The moisture present in the pores of active carbon reacts with SO_3 to form H_2SO_4. Complete combustion of SO_2 to SO_3 can be achieved by the combined effect of adsorption and catalysis of active carbon.

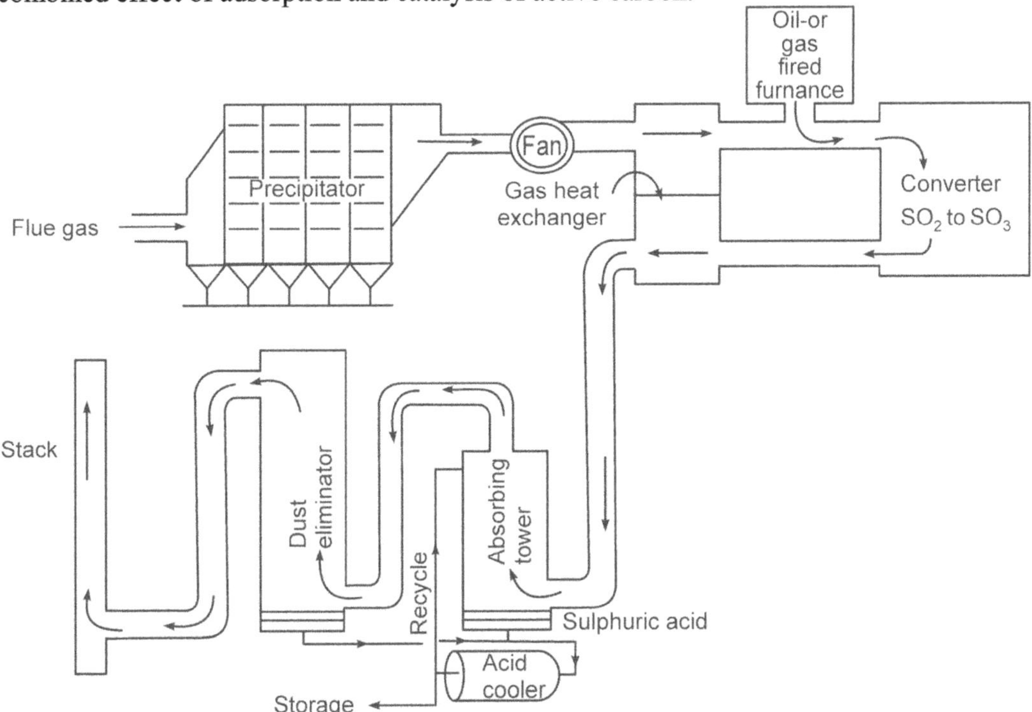

Fig. 9.11 *Catox process*

9.2.3 Adsorption of metal oxides

Many metal oxides and oxide mixtures (aluminium sodium oxide (alkalized alumina) and manganese oxide, bulk oxides of cobalt and copper) can be used as adsorbents or catalysts for SO_2 in flue gases. Compared to bulk metal oxides like sodium (or K), Sr, Cu supported on alumina are rated as superior catalysts.

9.2.4 Westvaco oxidation reduction process

In this process H_2S is produced in the regeneration train (Fig.9. 12) in the presence of a fluidized bed of high efficiency activated carbon, which acts as a catalyst and H_2S is reduced by H_2SO_4 to sulphur. The used carbon is then sent into sulphur regeneration where it will reduce H_2S and produce sulphur.

$$SO \xrightarrow{O_2/H_2O} H_2SO_4;\ H_2SO_4 + 3H_2S \rightarrow 4S + 4H_2$$

Part of the sulphur will then be recovered in a sulphur stripper by volatilization which is recovered by condensation of the molten product while the remaining sulphur reacts with hydrogen to form H_2S.

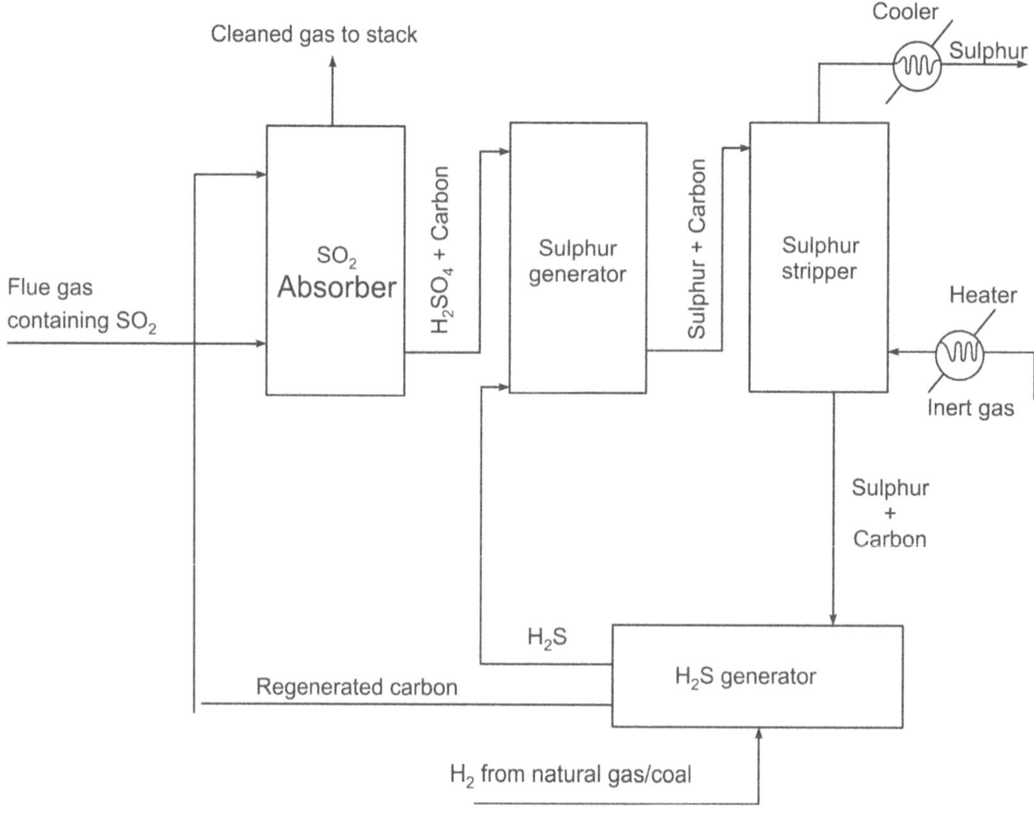

Fig. 9.12 *Westvaco process*

9.2.5 Lurgi's adsorption process

In this method stack gas passes to a dust collection unit for particulate removal and then proceeds to a cooler and then to the adsorption tower which is packed with carbon [6]. (Fig. 9.13). The SO_2 is adsorbed by carbon and is removed as a weak acid by spraying with water. The weak acid thus removed is contacted with the incoming polluted stream and the clean gas leaves the adsorption tower through the stacks.

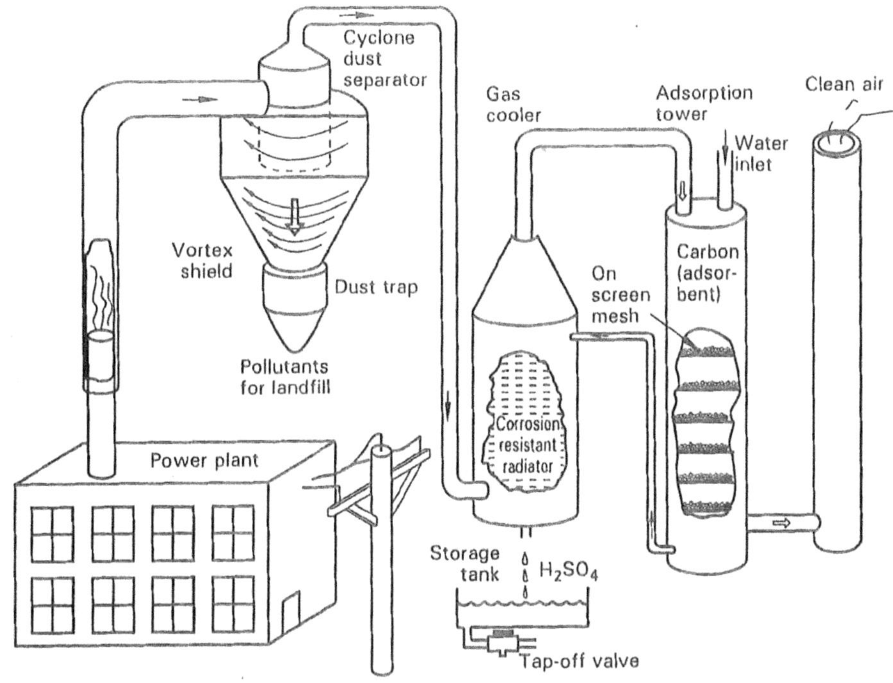

Fig. 9.13 *Lurgi's adsorption process*

9.2.6 Alkalized alumina process

In the alkalized alumina process, developed by U.S. Bureau of Mines, dust-free flue gas is fed to a reactor in which the porous form of sodium aluminate ($Na_2O.\ Al_2O_3$), adsorbent sorbs SO_2 at a temperature of ~ 315 °C. (Fig. 9.14). In this process, the SO_2 and O_2 in the flue gas react with the adsorbent [4].

$$Na_2O.\ Al_2O_3 + SO_2 + \frac{1}{2}\ O_2 \rightarrow Na_2SO_4 + Al_2O_3$$

The spent material is then contacted with a reducing gas such as hydrogen in a regenerator at about 680 °C to produce hydrogen sulphide.

$$Na_2SO_4 + Al_2O_3 + 2H_2 \rightarrow Na_2O.\ Al_2O_3 + H_2S + 3H_2O$$

By sending %S into a Claus unit sulphur can be produced as a byproduct:

$$2H_2S + SO_2 \rightarrow 3S + 2H_2O$$

The H_2S gas is sent to a Claus unit for the production of sulphur and the pellets of sodium aluminate are recycled. The maintenance of the granular strength of the sorbent is the major problem with the alkalized alumina process. Continued sorption-regenerated cycle of the sorbent results in uneconomically high attrition rates.

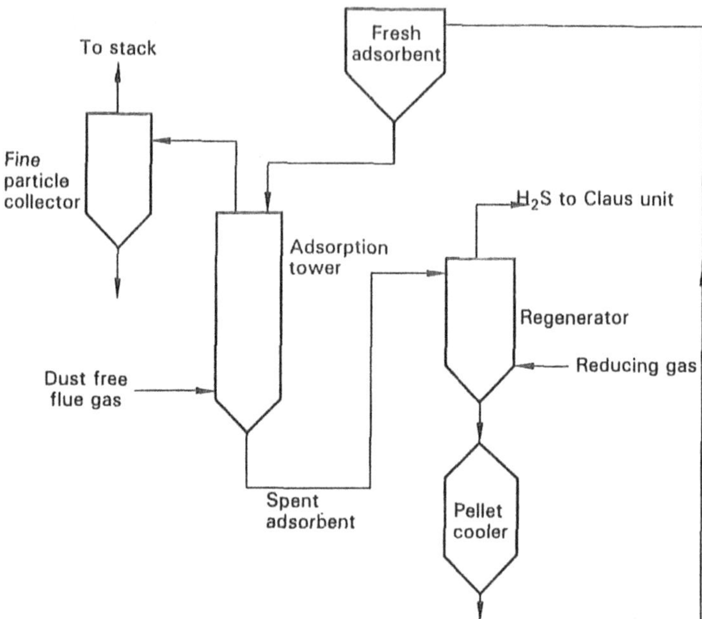

Fig. 9.14 *Alkalized alumina process*

9.2.7 Shell process with $CuO/CuSO_4$

In the Shell Claus off-gas treatment process **[7]**, an alumina bar kept between two electrodes parallel to the gas flow direction and copper oxide is impregnated in the alumina bar. In this process SO_2, reacts with copper oxide at a temperature of about 400°C to form copper sulphate.

$$SO_2 + \frac{1}{2} O_2 + CuO \rightarrow CuSO_4$$

NO_x present can also be removed simultaneously by the following reaction

$$2NO + 2NH_3 + \frac{1}{2} O2\ CuSO_4 \rightarrow 2N_2 + 3H_2O$$

Copper sulphate is then reduced at the same temperature in a hydrogen rich gas.

$$Cu\ SO_4 + 2H_2 \rightarrow Cu + SO_2 + 2\ H_2O$$

At about 400 °C flue gas is sent into one of the reactors where SO_2 reacts with CuO to form copper sulphate ($CuSO_4$). Both $CuSO_4$ and CuO act as catalysts in the reduction of NO_x with NH_3.When the $CuSO_4$ level in the reaction reaches a saturation level the flue gas is switched to a fresh reactor and the spent reactor is regenerated. In the regeneration cycle hydrogen (H_2) is used to reduce $CuSO_4$ to copper (Cu), yielding an SO_2 stream of sufficient concentration for conversion to sulphur or sulphuric acid.

The concentrated SO_2 is then sent to a Claus sulphur recovery plant where the effluent gas is heated with the addition of a reducing gas (H_2, CO) and passed through a reactor containing a cobalt-molybdenum catalyst. Di-isopropanolamine (DIPA) is used as an absorber solution. Through this process SO_2 levels can be reduced to less than 250 ppm.

9.2.8 Wet scrubbing methods

Wet flue gas desulphurization is mainly carried out in scrubbers. The principle is that alkaline or alkaline-earth absorption agents (solutions or suspensions) are dispersed in the gas to be desulphurized or are brought into homogeneous contact with it, so that the oxides of sulphur are washed out by absorption. Gas scrubbing with simple water would basically also be possible, but the tendency of the gases to be removed to solve in water is poor.

Solutions and slurries of compounds of calcium, magnesium and sodium are used as scrubbing media in the wet flue gas desulphurization method. Removal of fly ash prior to scrubbing in a pre-scrubber is a prerequisite in some of the processes while others, are capable of removing SO_2 and the ash simultaneously in the scrubber. In all the wet scrubbing methods the temperature of the gases is reduced to 30–50°C, resulting in a decrease of the gaseous dispersion in the atmosphere which necessitates the reheating of the effluent gases for effective dispersion.

The wet scrubbing methods for flue gas desulphurization fall into two broad categories: non-regenerative and regenerative. Non-regenerative means that the reagent used to remove the sulphur oxides from the gas stream is used and discarded. Regenerative means that the reagent is recovered and reused. In practical terms of the number and size of systems installed, the non-regenerative systems are more in number.

9.2.8.1 Non-regenerative systems

There are nine commercial non-regenerative systems. All have reaction chemistries based on lime (CaO), caustic soda (NaOH), soda ash (Na_2CO_3), or ammonia.

The SO_2 removed in a lime/limestone-based flue gas desulphurization system is converted to sulphite. The overall reactions are generally represented by:

$$SO_2 + CaCO_3 \rightarrow CaSO_3 + CO_2$$

$$SO_2 + Ca(OH)_2 \rightarrow CaSO_3 + H_2O$$

When using limestone and lime, respectively, part of the sulphite is oxidized with the oxygen content in the flue gas to form sulphate:

$$CaSO_3 + \tfrac{1}{2} O_2 + CaSO_4$$

Although the overall reactions are simple, the chemistry is quite complex and not well defined. The choice between lime and limestone, the type of limestone, and method of calcining and slaking can influence the gas-liquid-solid reactions taking place in the absorber.

Figure 9.15 shows the scrubber vessel as a spray tower with multiple sprays and a mist eliminator, some other designs use a packing with a very high open area in the tower or specialized bubbler designs. In the tower the SO_2 dissolves in the slurry and reacts with limestone producing CO_2 which enters the gas stream, and solid $CaSO_3$.Thelatter is almost entirely oxidized to $CaSO_4$, partly by the excess oxygen in the flue gases in the tower, partly in the effluent hold tank, or the thickener, or in an additional oxidizing vessel, sparged with air.

The slurry is recirculated from the hold tank. A side stream is taken to a thickener (also called a settler) and filtered to remove solids; ultimately the captured SO_2 must leave the system as $CaSO_3$ or $CaSO_4$, a solid waste. Finely ground fresh limestone is added to the effluent hold tank. The scrubber will operate at or near the a diabaticsaturation temperature of entering flue gas, which is about 52 °C. The cleaned flue gas is normally reheated to about 80 °C to restore plume buoyancy and to prevent acid corrosion of the ducts and stack downstream of the re heater, and then released to the stack. The water content of the waste slurry is reduced by thickening and then vacuum filtration. The filter cake is often mixed with dry fly ash from the same plant to make a solid waste stream that has a still lower water content. The resultant mixture is easier to handle and store as it goes to its ultimate destination, a landfill.

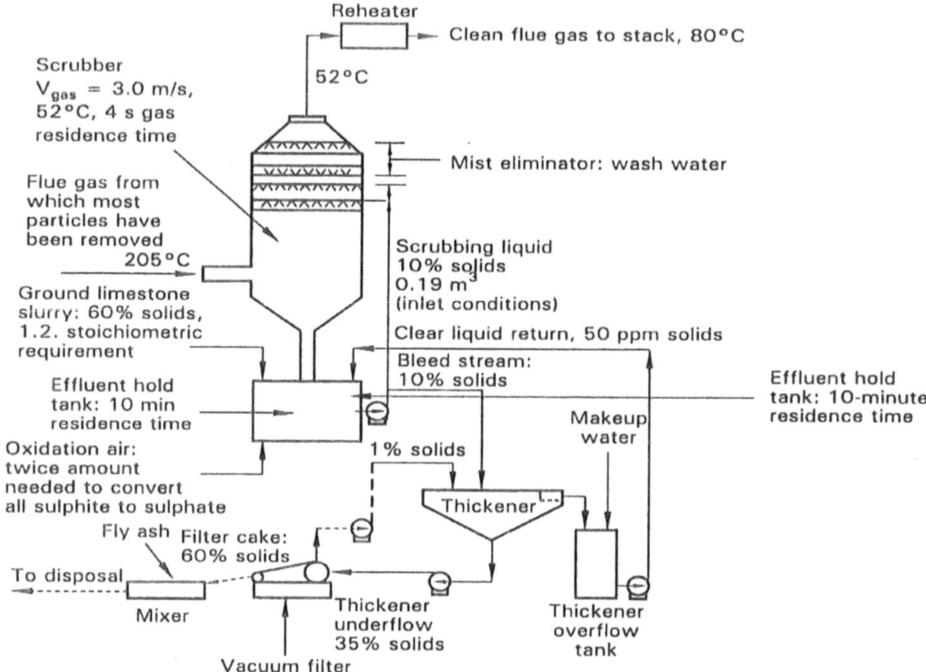

Fig. 9.15 *Flow diagram of a typical limestone wet scrubber [8]*

When the liquid is taken out of the scrubber, the alkali is regenerated with lime or limestone in the hold tank. The main reaction in the scrubber is

$$Na_2CO_3 + SO_2 \rightarrow Na_2SO_3 + CO_2$$

The overall reaction in the hold tank is

$$Na_2SO_3 + CaCO_3 + 0.5\ O_2 \rightarrow CaSO_4 + Na_2CO_3$$

which regenerates the sodium carbonate (or bicarbonate) in solution and precipitates the calcium as $CaSO_4$. All the liquid from the hold tank is sent to the thickener, where the calcium is removed, either as gypsum or as un reacted calcium carbonate (ox as calcium hydroxide). Some dissolved sodium carbonate or bicarbonate is added in the thickener overflow tank, and the clear liquid from it is used as the scrubbing liquid.

Because of the low solubility of $CaSO_4$, about 1800 ppm at 52 °C, most of the calcium is removed in the thickener, and the solution returning to the scrubber should be low enough in calcium so that no solids will form in the scrubber. The absence of solids makes the scrubber simple and trouble-free but the double alkali process adds the complexity of an entirely different set of chemicals and their control and requires much larger thickener.

9.2.8.2 Double alkali scrubber process

The many double alkali process variations involve the scrubbing of flue gases with a clear liquor containing dissolved sodium or ammonium salts, followed by treatment of the depleted liquor with lime or limestone in a reaction producing throwawaysliud for disposal and alkali liquor for scrubbing. (Fig. 9.16) In the United Stateg, primary development attention has been placed on sodium-based double alkali systems using lime regeneration. In Japan, emphasis has been on systems with limestone used as the input alkali and gypsum produced as a salable product.

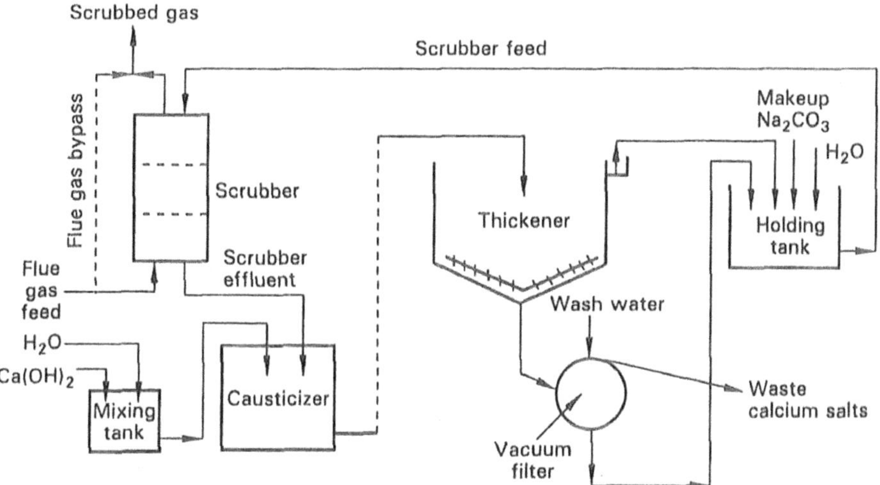

Fig. 9.16 *Double alkali process (one-variation) sodium scrubbing with lime regeneration-for desulphurization of stack gases*

9.2.9 Regenerative processes

In general regenerative processes are cost intensive and are Chosen only where waste disposal options are limited. Some of important regenerative processes are discussed.

9.2.9.1 Magnesium oxide scrubbing

In many respects, magnesium oxide scrubbing is similar to limestone scrubbing. The principal difference is that the spent magnesium sulphite and sulphate salts are regenerated, producing a concentrated stream of 10-15% SO, and regenerated magnesium oxide for reuse in the scrubber. Therefore, this system does not produce a solid waste problem, provided there is a firm market for these products.

Since the reactant recycled, it must be protected from contamination by particulates. It is therefore necessary that the process be applied in an oil-fired boiler or that particulate matter be sufficiently removed from the flue gas prior to passing it into this system.

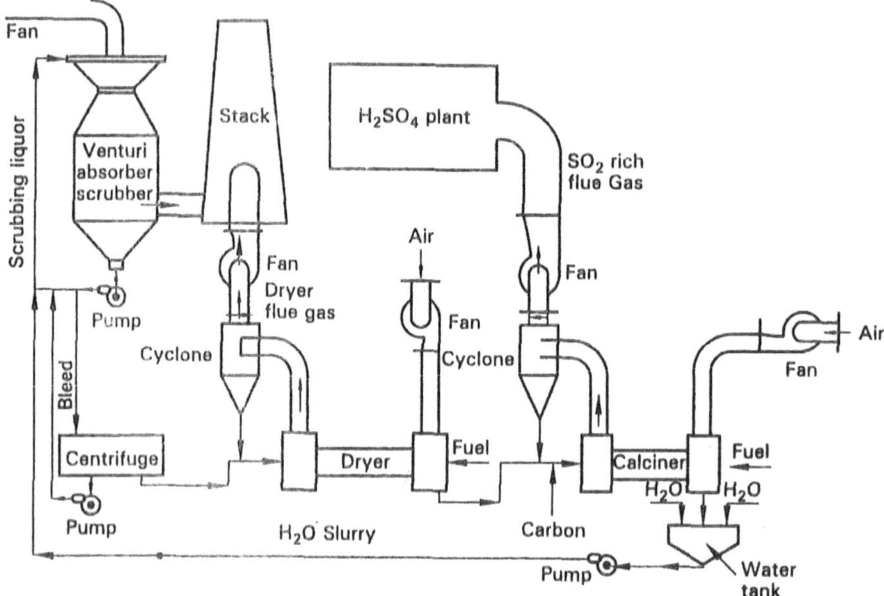

Fig. 9.17 *Magnesium oxide (MgO) slurry process for desulphurization of flue gas free of particulate matter.*

Magnesium oxide process is a regeneration /segregation process in which the flue gas is scrubbed with a slurry of magnesium oxide in a venturi scrubber which absorbs SO_2, and generates magnesium sulphite and sulphate as per the following reactions

$$MgO + SO_2 \rightarrow MgSO_3;\ MgSO_3 + SO_2 + H_2O \rightarrow Mg\ (HSO_3)_2$$

$$Mg\ (HSO_3)_2 + MgO \rightarrow 2\ Mg\ SO_3 + H_2O;\ Mg\ SO_3 + \tfrac{1}{2}\ O_2 \rightarrow Mg\ SO_4$$

$$MgSO_3 \rightarrow 800\ ^\circ C \rightarrow MgO + SO_2$$

The mixed sulphite/sulphate slurry along with unreacted MgO is separated from liquid phase in a centrifuge and the mother liquor is recycled into the absorber. The centrifuge wet solid cake is dried and contained for regeneration by adding dried carbon where $MgSO_4$ is reduced to MgO and SO_2.

$$MgSO_4 + \frac{1}{2}C \rightarrow MgO + SO_2 + \frac{1}{2}CO_2$$

This process was developed by the Chemical Construction company, CHRMICO U.S.A (Fig. 9.17)

9.2.9.2 Wellman-Lord (single alkali) process

In this process sulphur dioxide is removed by washing flue gas with an aqueous solution of sodium sulphite which is converted to bisulphite as the SO_2 from flue gases is absorbed by a saturated solution, of sodium sulphite. (Fig. 9.18)

$$Na_2SO_3 + SO_2 + H_2O \rightarrow 2NaHSO_3$$

Sodium bisulphite thus formed is sent to a double effect evaporator cum crystallizer where it decomposes into sodium sulphite at a temperature of around 100°C resulting in formation of SO_2 and steam as per the Equation

$$2\ NaHSO_3 - 100\ ^\circ C \rightarrow Na_2SO_3 + SO_2 + H_2O$$

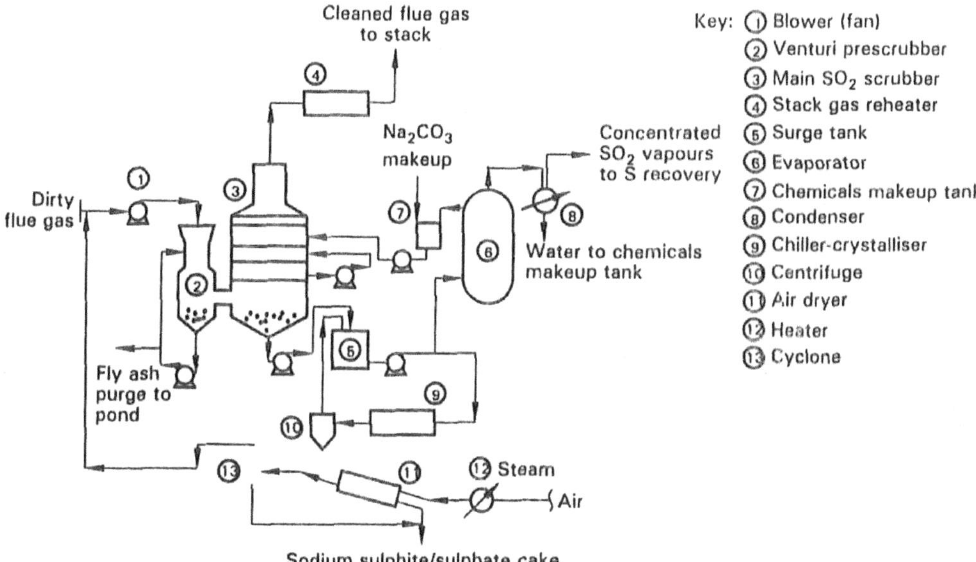

Fig. 9.18 *Schematic process flow diagram of the Wellman-Lord SO_2 scrubbing and recovery system. (Reprinted from U.S. Environmental Protection Agency (EPA), 1979)*

To keep the rate of accumulation in the scrubbing liquid low it is removed before scrubbing of SO_2, SO_2 and water vapour released from the evaporators are passed in to a condenser and the product goes to the dissolving tank for dissolution of Na_2SO_3, crystals and the rich SO_2 gas will thus be processed. Sodium sulphate is produced in this reaction which is removed and substituted by an equivalent amount of NaOH.

9.2.9.3 ASARCO process

Using demethylaniline as absorber the American Smelting and Refining Co (ASARCO) developed a technology for the control of SO_2 in flue gases (Fig. 9.19). The desulphurized gas is sent through sodium carbonate solution, and finally through dilute H_2SO_4 where the gas is scrubbed for traces of sulphur and dimethylaniline if any. The absorbent dimethylaniline with SO_2 is passed through a steam distillation column for treatment and removal. The ASARCO process is used for treating of gases with concentrations more than 3% of SO_2, The tail gas left into the stack after this scrubbing process consists of SO_2 of about 500 ppm.

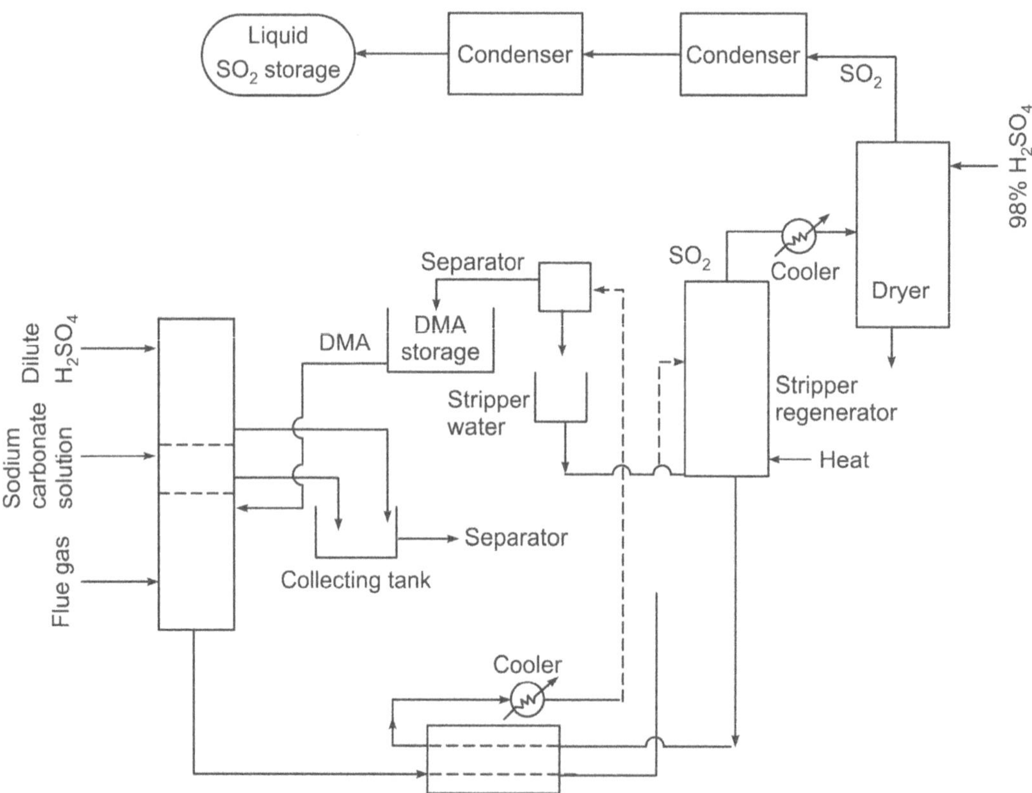

Fig. 9.19 *ASARCO process*

9.2.9.4 *COMINCO process*

The schematic diagram of the process is shown in Fig. 9.20 in which 30% aqueous ammonia is used as asorber for SO_2 and works with a removal efficiency of 95%. In this process a minimum of sulphur is produced by the absorption of SO_2 by 30% aqueous ammonia. The ammonium bisulphite produced is then added to 95% H_2SO_4 in a stripping column when ammonium sulphite is produced after stripping.

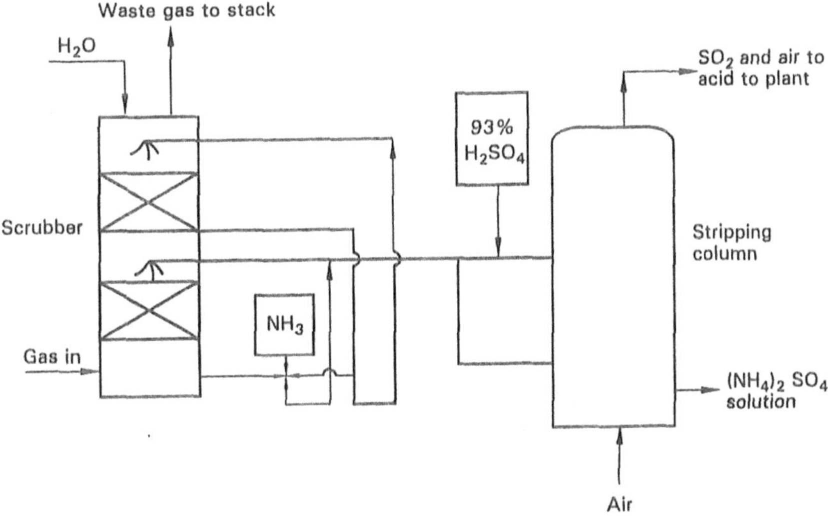

Fig. 9.20 *COMINCO process*

9 2.10 Wet-dry systems

In semi-dry processes an alkaline or alkaline-earth suspension is brought into contact with the flue gas for the removal of SO_2. A dry reaction product is formed in the hot flue gases, which is subsequently collected in particle precipitators.

The semi-dry processes were developed based on the knowledge that SO_2 absorption on limestone particles can be improved by making them moist. In these processes the absorption agent-generally a lime or sodium carbonate suspension, or caustic lye of soda–is sprayed into the hot flue gas in a superfine dispersion. Depending on the manufacturer of the process, the spray mist is either produced by spraying nozzles or with the help of a spinning-disk atomizer [4].When the spray mist contacts the flue gas the water of the absorbent vaporizes, and the SO_2 reacts with the absorption agent. The vaporization process is sustained until the ensuing reaction products take on the form of dry powder. This can then be removed from the flue gas with a standard dust collector (electrostatic precipitator or, when requirements are exacting, a fabric filter.)

The main advantage of the spray absorption process is that the water required vaporizes and there is no waste water. Generally, the flue gases also do not need to be heated again. Compared to dry process greater SO_2 absorption can be achieved with lesser sorbent.

As it is known that SO_2 absorption on limestone particles can be improved by making them moist the wet dry systems were developed for SO_2 removal. In wet dry system, for the SO_2 removal the flue gas will be intimately contacted by an alkaline or alkaline-earth suspension and the dry reaction product formed in the hot flue gases, will be subsequently collected in particle precipitators.

Lime or sodium carbonate suspension, or caustic lye of soda-is sprayed into the hot flue gas in a superfine dispersion spraying nozzles or with the help of a spinning disk atomizer. When the spray mist contacts the flue gas the water of the absorbent vaporizes, and the SO_2 reacts with the absorption agent and the possible reaction mechanism are summarized in Fig. 9.21. The vaporization process is sustained until the ensuing reaction products take on the form of dry powder which will be removed from the flue gas with a standard dust collector (electrostatic precipitator or a fabric filter). As the water required vaporizes no wastewater will be generated in this process. Further it is not necessary to heat the flue gases compared to dry process and even with smaller sorbent higher efficiency in SO_2 removal can be achieved.

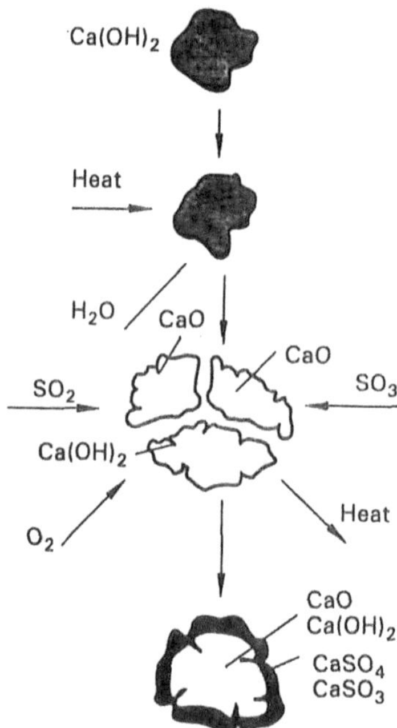

Fig. 9.21 *Possible mechanism of SO_2 absorption in the spray absorption process*

Spray dryers (Fig, 9.22) are the wet-dry systems in which the features of both wet and dry systems discussed earlier can be combined [9].

Spray dryers are widely used in the process industries. In treating SO_2 containing flue gas eaters the spray dryer chamber, usually from the side and/or top and flows out most often at the bottom or side or through an outlet tube that dips down in to the dryer vessel. The reagent slurry is dispersed as 10-50-μdrops, containing about 30percent by weight of solids. The resulting dry particles are small enough that most are carried along with the gas stream.

This is called a wet-dry system because part of it is like a wet lime scrubber and part is like dry sorbent injection. The freshly formed drops behave very much like the drops in a wet lime scrubber. The SO_2 dissolves in the water and reacts there with the dissolved $Ca(OH)_2$. As the water evaporates from the drops, the individual fine particles in them coalesce to form a single porous particle from each drop. This particle then behaves like the dry sorbent particles injected in the tests.

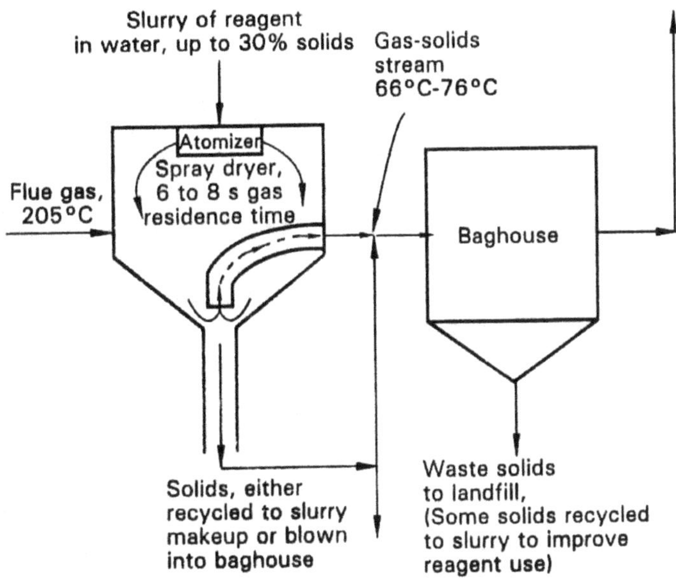

Fig. 9.22 *Spray dryers [10]*

Questions

1. Explain in detail with diagram the process of Desulfurization of coal.
2. For sulfur removal coal gasification is suitable? Justify.
3. Discuss the process of coal liquification for sulphur removal.
4. Describe the flue gas clean-ups system
5. Illustrate the reduction of sulfur using processes (i) Fluidised bed combustion (ii) Integrated gasification combined cycle system

6. Discuss the classification of Gas desulfurization
7. Discuss (i) Caltox process (ii) Westvaco process (iii) Lurg's adsorption process.
8. List various various wet scrubbing methods for sulphur removal.
9. Explain in detail the process of Magnesium oxide scrubbing (Regenerate process).
10. Discuss the mechanism of Wellman – lord (single alkali) process.
11. Discuss the mechanism of Spray and adsorption processes.
12. Write a short note on (i) Spray drying (ii) Scrubbing process

References

1. R. M. Timesom, 1966. Utilization of solvent refined coal in power plants. Chem. Engg. Press. 62(10):53.
2. Coal water gasification program. Final report EPRI GS-6808-. Electric Power Research Institute. Palo Alto California. 1990.
3. A.L. Kolix and F. C. Rienbenfelf, 1960. *Gas purification.* McGraw Hill, New York.
4. H. Michele and R. Wildner, 1984. *Chem. Engg. Tech. 56(115).*
5. H. Michele, 1986. *Chem. Engg. Tech.*
6. W. E. Miller, 1974. The catalyst process at Illinois power. *Chem. Engg. Pro.* 70(6): 49.
7. S. Katel, 1966. Removing sulphur dioxide from flue gases. *Chem. Engg. Pro.* 62(10):67.
8. F.M. Dantzenbug, J. E.Nadar and A. J. J.Vanginneken, 1971. *Chem, Engg. Pro* 67: 122.
9. D. S. Henzel, B. A. Laseki, E. O. Smith and D. O. Suenson, 1981. Limestone FGC scrubbers users handbook. U.S. EPA Report No. EPA-600/8/81-017. US Govt. Printing Office, Washington DC.
10. K. Maskes, 1985: *Spray drying: handbook.* John Wiley and Sons, New York.
11. S. M. Dalton, B. Toole-O. Neill, B.K. Gullet and G. J. Drurmm and, 1992. Summary of the SO_2 control systems. 1991 EPA / EPA/ DOE. *J. of Air waste ManualAssociation.*42: 1110–1117.

Suggested Reading

1. Taieb D, Brahim AB, Chimie CR (2013) Electro chemical method for sulphur dioxide removal from flue gases: Application on sulphuric acid plant in Tunisia. C R Chim 16: 39-50.
2. Ravi K Srivastava (2000) Controlling SO_2 emissions: A review of technologies. US Environmental protection agency.
3. Wang G, Bing L, Yang Z, Zhang J (2014) Selective catalytic reduction of Sulphur dioxide by carbon monoxide over iron oxide supported on activated carbon, School of Chemical Engineering, Qingdao University of Science and Technology Qingdao China.
4. Taylor MR, Rubin ES, Houn shell DA (2005) Control of SO_2 emissions from power plants: A case of induced technological innovation in the U.S. Technol Forecast Soc Change 72: 697-718.

5. Bhanarkar AD, Rao PS, Gajghate DG, Nema P (2005) Inventory of SO_2, PM and toxic metals emissions from industrial sources in Greater Mumbai, India. Atmos Environ 39: 3851-3864.
6. Luvsan ME, Shie RH, Purevdorj T, Lkhagvasuren B, Barkhasragchaa B, et al. (2012) The influence of emission sources and meteorological conditions on SO_2 pollution in Mongolia. Atmos Environ 61: 542-549.
7. Darmastuti Z, Bur C, Möller P, Rahlin R, Lindqvist N, et al. (2014) SiC–FET based SO_2 sensor for power plant emission applications. Sens Actuators B Chem 194: 511-520.
8. Cofala J, Syri S (1998) Sulfur emissions, abatement technologies and related costs for Europe in the RAINS model database. International Institute for Applied Systems Analysis.
9. Dehghani A, Bridganian H (2010) Flue gas desulphurization methods to conserve the environment. Petroleum & Coal 52: 220-226.
10. Rameshni M, Santo S (2005) Production of elemental sulphur from SO_2 RSR. Resources and energy 1-19.
11. Wendt A (2002) Optimization of WSA technology, Masters Thesis, Lund University.
12. ZhongY, Gao X, Huo W, Luo Z, Ni M, et al. (2008) A model for performance optimization of wet flue gas desulphurization systems of power plants. Fuel Processing Technology 89: 1025-1032.
13. Evans AP (2009) Operational Experience of Commercial, Full Scale, Ammonia Based Wet FGD for Over a Decade. Environmental Technologies 1-19.
14. Ma JX, Fang M, Lau NT (1999) Simultaneous catalytic reduction of Sulphur dioxide and nitric oxide. Catal Today 62: 127-130.
15. Nakazato T,Yunyi L, Kunio K (2004) Removal of SO_2 in Semi-Dry Flue Gas Desulphurization Process with a Powder-Particle Spouted Bed. Can J Chem Eng 82: 110-115.
16. Weiss (1999) LH Evaluating sulphur producing FGD processes. Chem systems Inc.
17. Zhu T, Kundakovic L, Dreher A, Stephanopoulos M (1999) Redox chemistry over CeO_2-based catalysts: SO_2 reduction by CO or CH_4. Catal Today 50: 381-397.
18. Kataoka S (1992) Coal burning plant and emission control technologies, Technical Note World Bank China Country Department Washington DC.
19. Happel J, Leon AL, Hnatow MA, Bajars L (1977) Catalyst composition optimization for the reduction of Sulphur dioxide by carbon monoxide. In d Eng Chem Prod Res Dev16: 150-154.
20. Paik SC, Kim H, Chung JS (1997) The catalytic reduction of SO_2 to elemental sulphur with H_2 or CO; Catalysis Today 38: 193-198.
21. Stephanopoulos MF, Zhu T, Li Y (2000) Ceria based catalysts for the recovery of elemental sulphur from SO_2 laden gas streams. Catal Today 62:145-158.
22. Puricelli S (2013) Regenerative SO_2 capture: recovering the future. Paper presented at Sulphur 2013, 29th international conference and exhibition, Florida, USA.

23. Processes for Stack SO_2 Recovery as Sulfur. J Air Pollut Control Assoc 22: 19: 687-692
24. Asphahani AI, Nicholas AF, Silence WL, Meyer TH (1989) High performance Catalysts for Catalytic Reduction of SO_2 with CO as a Reducing Agent. Ind Eng SO_2, HCl and Mercury Mitigation. Proceedings of the 18th American Waste to Energy.
25. Buecker B, Hovey L (2012) Circulating Dry Scrubbers: A New Wave in FGD? Mid Atlantic Regional air management SO_2/HCl. control technology 115: 1-176. systems.
26. Wang C, Lin S, Hwang W, Weng H (2002) Supported Transition-Metal OxideChem Res 41: 666-671.

CHAPTER 10

Air Pollution Control Technologies for Oxides of Nitrogen

10.1 Emission of Nitrogen Oxides

NO_x represent a family of seven compounds (Table 10.1.) NO_2 is not only an important primary air pollutant, but since it reacts in the atmosphere, it is also a secondary pollutant. NO_2 react with water in the atmosphere to form acid rain. Furthermore, NO_2 can react with air in the presence of ultraviolet light (UV) to form ozone and nitric oxide (NO). Then, NO can react with free radicals, created by UV irradiation of VOC, forming PAN (1).

Table 10.1 *Nitrogen Oxides (NO)*

Formula	Name	Properties
N_2O	Nitrous Oxide	Colorless, water soluble
NO N_2O_2	Nitro Oxide Dinitrogen dioxide	Colorless gas, slightly water soluble
N_2O_3	Dinitrogen trioxide	Black Solid, Water Soluble in water
NO_2 N_2O_4	Nitrogen dioxide Dinitrogen tetroxide	Red-brown gas, very water soluble, decomposes in water
N_2O_5	Dintrogen pentoxide	White solid, very water soluble, decomposes in water

10.1.1 NO_x sources

Mobile sources contribute for about 50% of the total NO emissions. Electric power plant boilers produce about 40% of the NO_x emissions from stationary sources (2). Additionally, substantial emissions arise from industrial boilers, incinerators, gas turbines, reciprocating spark ignition and Diesel engines in stationary sources, iron and steel mills, cement manufacture, glass manufacture, petroleum refineries, and nitric acid manufacture. Biogenic or natural sources of nitrogen oxides include lightning, forest fires, grass fires, trees, bushes, grasses, and yeasts. Nitrogen oxides can be grouped as follows accordingly with their formation processes during the combustion.

10.1.2 Thermal NO

Thermal nitrogen oxides, which are generally the most significant, are formed by the simple heating of oxygen and nitrogen, either in a flame or by some other external heating, e.g. a lightning bolt. The concentration of "thermal NO_x" is controlled by the nitrogen and oxygen molar concentrations and the temperature of combustion. Combustion at temperatures well below 1300 ^{0}C forms relatively low concentrations of thermal NO. During the combustion of fuels which may not contain any nitrogen, the oxygen reacts with the carbonaceous matter as well as the atmospheric nitrogen forming NO_x Between 20-90%.

10.1.3 Fuel NO_x

Fuels that contain nitrogen (e.g., coal) create "fuel NO" that results from oxidation of the already-ionized nitrogen contained in the fuel.

10.1.4 Conversion of organic nitrogen compounds in fuels to NO_x.

Some of the gaseous and liquid fuels contain little nitrogen, and the contribution of that nitrogen to the total NO in the combustion products is minimal. When the nitrogen content is less than 0.1%, in liquid fuels the percent conversion can approach 90%, but this percentage declines as fuel nitrogen content increases, reaching about 40-50% with fuel nitrogen content of about 0.5%. As the fuel nitrogen content increases above 1%, the efficiency of conversion of nitrogen to NO_x decreases further. About 20-25% of the organic nitrogen in coal is converted to NO_x under normal combustion conditions. With coal nitrogen content ranging from 0.7-2.8% it appears necessary to develop coal combustion techniques specifically oriented to minimizing conversion of fuel nitrogen to NO_x in order to minimize total NO_x emissions. Depending on the temperature and metal oxides various species of nitrogen are formed. Even though the primary oxide is NO it reacts quickly with oxygen or ozone to form NO_2. Water vapour also acts as a catalyst. In a combustion process since the time is very short at high temperatures formation of NO_x will not be complete to reach equilibrium.

However, some coals as well as the hydrocarbons derived from oil shale can contain up to 2% N. Most of the fuel nitrogen is converted in the flame to HCN, which then converts to NH or NH_2. The NH and NH_2 can react with oxygen to produce NO + H_2O, or they can react with NO to produce N_2 and H_2O. Thus the fraction of the fuel nitrogen that leaves the flame as NO is dependent on the NO/O_2 ratio in the flame zone. Low oxygen content of the gases in the high-temperature part of the flame significantly lowers the fraction of the fuel nitrogen converted to NO.

10.1.5 Prompt NO

Prompt refers to the nitrogen oxides that form very quickly as a result of the interaction of nitrogen and oxygen with some of the active carbon species derived from the fuel in flames. Prompt NO is formed from molecular nitrogen in the air combining with fuel in fuel-rich conditions, which exist, to some extent, in all combustion process. This nitrogen then oxidizes along with the fuel and becomes NO during combustion, just like fuel NO. Prompt NO species are not observed in flames of fuels with no carbon, e.g. H_2. They cannot be formed by simply heating oxygen and nitrogen; the participation of some active carbon species from the fuel is required. Fuel nitrogen oxide is formed by conversion of some of the nitrogen originally present in the fuel to NO_x.

10.1.6 Temperature wise formation of different kinds of NO_x

All three kinds of NO_x are present in many combustion gases.

Figure 10.1 shows estimates from coal combustion of the extent of formation of the thermal, fuel and prompt oxides of nitrogen and the formation mechanisms of NO_x emissions from coal combustion [3]. Below about 1300°C (2372 °F), the thermal NO mechanism is, negligible compared to the other two, while at the higher temperatures it is most important.

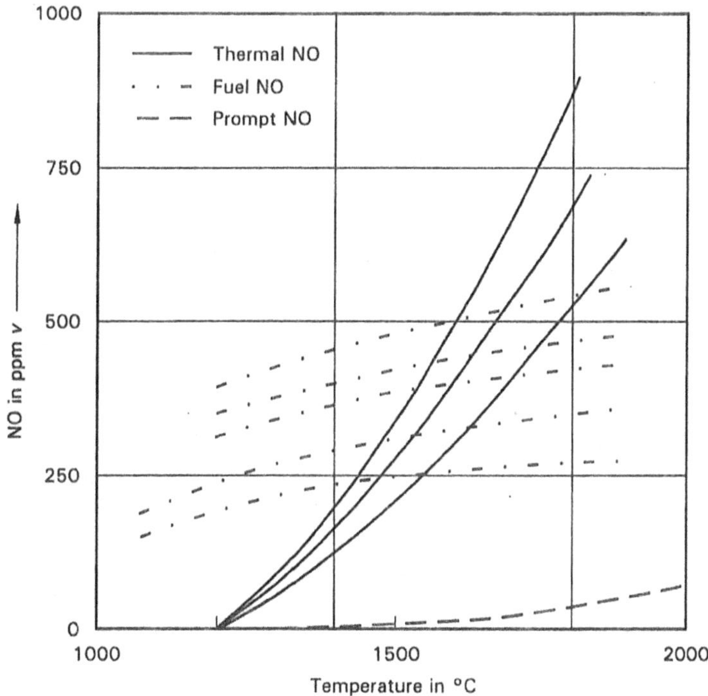

Fig. 10.1 *NO formation at different combustion temperatures*

The fraction of the fuel N that appears as NO_x in the exhaust gas is estimated to be 20-50%, depending on furnace conditions and, to some extent, the chemical nature of the N in the fuel. The thermal, prompt, and fuel NO products interact, as do the radicals involved. The NO_x emission factors for different fuels are presented in Table 10.2.

Table 10.2 *NO_x emission factors*

Source	Average emission factor
Coal	
Domestic and commercial	3.2 kg/ton of coal burned
Industries	
Fuel oil	5-7 kg/m^3 of oil burned
Domestic and commercial	7.0 kg/m^3 of oil burned
Industrial	10.5 kg/m^3 of oil burned
Utilities	
Natural gas	1.50 mg/m^3 of gas burned
Domestic and commercial	2.75 mg/m^3 of gas burned
Industrial	5.02 mg/m^3 of gas burned
Utilities	2.60 mg/m^3 of gas burned
Gas turbines	

10.1.6.1 Equilibrium concentrations of NO and NO_2 at various temperatures

The equilibrium reactions for the formation of NO and NO_2 at constant pressure are

$$\tfrac{1}{2} N_2 + \tfrac{1}{2} O_2 \leftrightarrow NO \quad .....(10.1)$$

$$NO + \tfrac{1}{2} O_2 \leftrightarrow NO_2 \quad .....(10.2)$$

The kinetics of the reaction and their thermodynamic parameters have to be evaluated to understand the conditions responsible for the formation of NO and NO_2.

Table 10.3 presents the equilibrium constants for the formation of NO and NO_2 at different temperatures.

Table 10.3 *Equilibrium constants for the formation of NO and NO_2 by Reaction (I) and Reaction (II) [4]*

Temp K	*Kp* of formation	
	NO	NO_2
300	7×10^{-31}	1.4×10^6
500	2.7×10^{-18}	4.9
1000	7.5×10^{-9}	0.11
1500	1.07×10^{-5}	0.011
2000	0.00040	0.0035
2500	0.0035	0.0018

Based on the relation between mole fraction and stoichio metric coefficient in the formation of NO the equilibrium constant for the Equation (10.1) can be written as

$$K(T)\frac{Y_{NO}^{vNO}}{Y_{O_2}^{vO_2}Y_{N_2}^{vN_2}}\left(\frac{P}{P_0}\right)^{vNO-NO_2-NO_2} \quad .....(10.3)$$

where Y= mole fraction and v = stoichiometric coefficient, P_0 is the atmospheric pressure and P is the actual pressure.

From the ratio of nitrogen to oxygen in the actual reaction in a given situation, the equilibrium concentration can be calculated from Equation (10.1) and using formation constants in Table 10.2. These typical data are given in Table 10.4 and graphically presented in Figs. 10.2 (a) and (b). NO levels at different temperatures and in the presence of varying amounts in air are shown in Fig. 10.3.

Table 10.4 *Calculated equilibrium concentrations of NO and NO_2*

Temperature	ppm NO	Starting with 78%	N_2 21% O_2	Starting with 78% N_2, 4% O_2
K		ppm NO_2	ppm NO	ppm NO_2
300	3.4×10^{-10}	2×10^{-4}	1.4×10^{-6}	4×10^{-5}
500	6×10^{-4}	0.04	3×10^{-4}	7.6×10^{-3}
1000	35	1.9	4.6	0.35
1500	785	6.8	580	1.3
2000	8100	13.2	3615	2.5
2500	20.660	20	11200	4.0

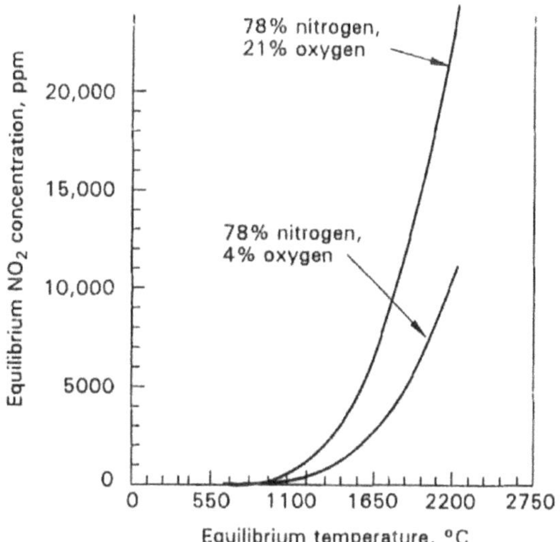

Fig. 10.2(a) *Calculated concentrations of NO for two different oxygen contents of the reacting gas*

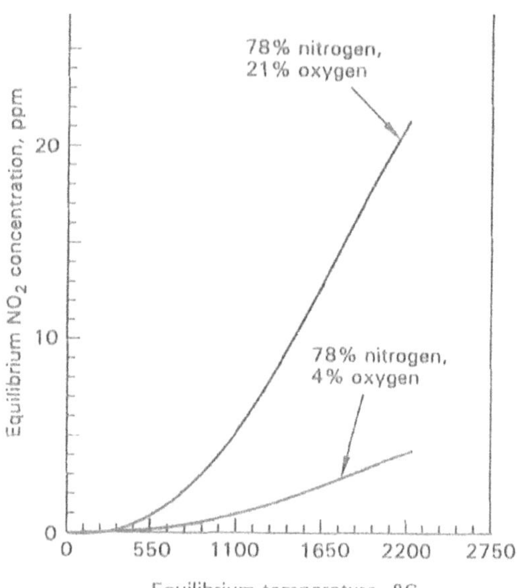

Fig. 10.2(b) *Calculated equilibrium concentrations of NO_2 for two different oxygen contents of the reacting gas*

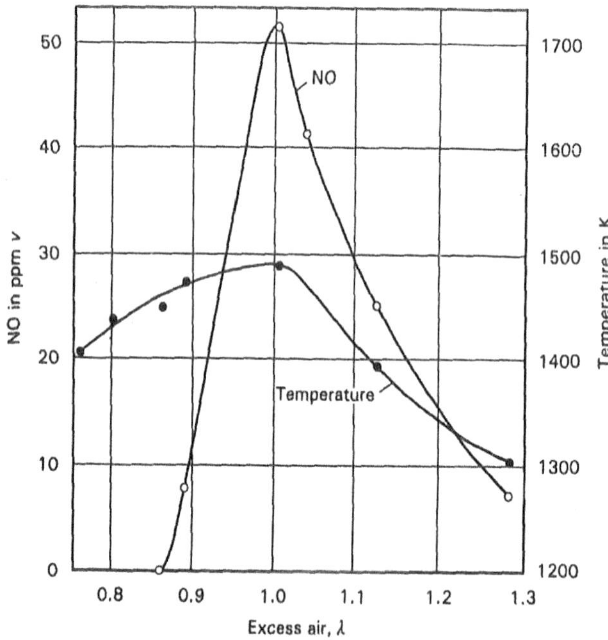

Fig. 10.3 *Influence of excess air on NO formation in an enclosed natural gas premix flume.*

From the equilibrium concentration of NO at different temperatures and different N:O ratios it can be inferred that NO formation increases rapidly within the vicinity of 2000 K which is the adiabatic flame temperature for almost all combustion reactions. Further the amount of NO formed also depends on available oxygen. Thus at low temperatures the equilibrium concentrations of NO_2 are much higher than NO whereas at high temperatures equilibrium NO is higher.

Based on the kinetics of the reaction for the formation of NO, a certain time is required for the NO concentration to reach equilibrium concentration and in many combustion processes the time the gas remains in the combustion process at high temperatures is far less the time than the time required for reaching equilibrium. As such the NO concentrations formed in combustion processes are less than the equilibrium concentration. In Table 10.5the times required to reach 500 ppm equilibrium concentration at different temperatures are presented.

Table 10.5 *Time in seconds required to achieve 500 ppm NO combustion concentration at different temperatures (Air 75% N_2, 3% O_2)*

Temperature K	Time (s)	Equilibrium (ppm) concentration	Equilibrium %
1588	1320	550	91
1801	16.2	1380	36
2033	1.1	2600	19
2255	0.11	4150	12

From the above data it can be inferred that NO_x release from any combustion process can be minimized by modifying the combustion process in the following ways

(a) decreasing the peak combustion temperature
(b) carrying out the combustion process with as less excess of air or oxygen as possible and
(c) minimizing the resident time at peak temperature.

10.2 Control Technologies for Nitrogen Oxide Emissions

It is relatively easy to remove SO_2 from combustion gases by dissolving SO_2 in water and reacting it with alkali. Aqueous SO_2 quickly forms sulphurous acid which reacts with alkali and then is oxidized to sulphates.

Collecting nitrogen oxides is not as easy this way because NO, the principal nitrogen oxide present in combustion gas streams, has a very low solubility in water. Unlike sulphur oxides which quickly react with water to form acids, NO must undergo a two-step process to form an acid:

$$NO + 0.5\ O_2 \rightarrow NO_2 \quad .....(10.4)$$

$$3NO_2 + H_2O \rightarrow 2HNO_3 + NO \quad .....(10.5)$$

The first reaction is relatively slow. It is fast enough in the atmosphere to lead to the formation of acid precipitation in the several hours or days that the polluted air travels before encountering precipitation, but slow enough that it does not remove significant quantities of NO in the few seconds that a contaminated gas spends in a wet limestone scrubber used for SO_2 control.

There are two possible approaches for controlling NO_x in combustion gases:

1. Modification of the combustion processes to prevent the formation of NO_x.
2. Treatment of the effluent gases to convert the NO_x to N_2.

Both approaches are used, separately or together.

10.2.1 Nitrogen oxide control by modification of combustion processes

As discussed earlier, the formation of thermal NO is increased by increasing any of the following three variables: peak temperature, time at high temperature, and oxygen content at the high temperature.

There are three ways to reduce the NO_x emissions from stationary sources:

1. By minimizing the resident time at peak temperature.
2. By reducing the peak temperature.
3. By minimizing the availability of O_2 for reaction with N_2.

The above conditions can be achieved by the following modified combustion process.

10.2.2 Reducing process temperature

This technique is based on the possibility to carry out the combustion at reduced / lower temperatures. This is achieved by avoiding the stoichio metric air to fuel ratio (the exact ratio of reactants that are needed for full combustion). This technique avoids the ideal stoichio metric ratio because this is the ratio that produces higher temperatures and therefore generates higher concentrations of thermal NO.

Figure 10.4 shows the variation of NO_x emission in a boiler when combustion is carried out using pulverized coal with varying percentages of excess air. It can be seen that NO_x decreased from 600 ppm at 25% excess air to as low as 175 ppm at 1.4% excess air. However, this is accompanied by lowering of carbon combustion efficiency which decreased from 99.5% at 25% excess air to around 92% at 2% excess air.

Thus the extent of excess air influences both the temperature and oxygen concentration of gases in the post combustion zone. In general boiler units operate in the presence of 10-20% excess air for achieving complete combustion of the fuel. However, excess air provides enough oxygen for reacting with nitrogen resulting in NO_x emission [5].

Combustion temperature may be reduced by:

(a) using fuel rich mixtures to limit the amount of oxygen available;

(b) using fuel lean mixtures to limit temperature by diluting energy input;

(c) injecting cooled oxygen-depleted flue gas into the combustion air to dilute energy;

(d) injecting cooled flue gas with added fuel; or

(e) injecting water or steam.

Burners are redesigned to exploit the above principles to reduce NO_x emissions

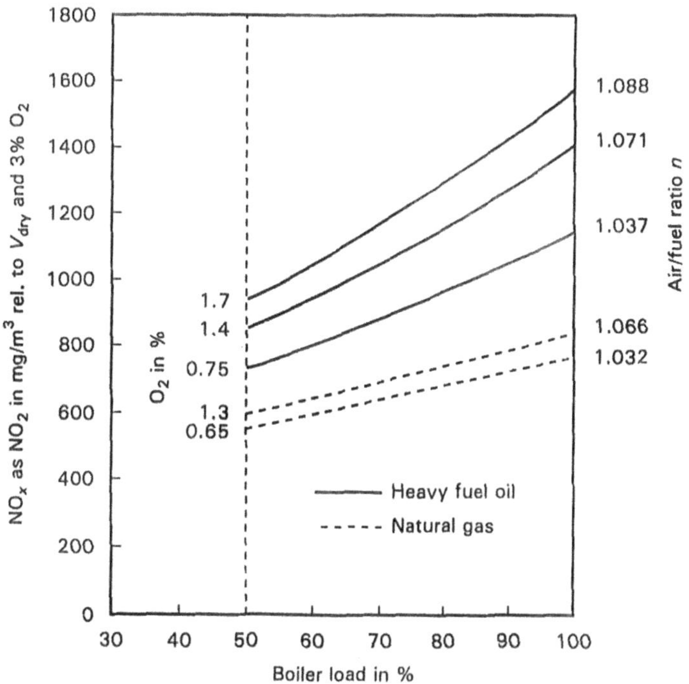

Fig. 10.4 *Dependency of NO_x emissions on boiler load and excess air*

10.2.3 Burner design modifications for NO_x control

The design of burner in the combustion chambers will have great influence on the NO_x formation. In general two types of furnaces are designed: (a) Tangential firing (b) Horizontal firing.

In the first type the furnace itself is used as a burner so that the flame and combustion products spiral upwards around the walls of the furnace resulting in low peak flame temperature and consequent reduction in NO_x emission.

In the horizontal type the flame is at right angles to the furnace walls due to which heat is concentrated resulting in higher peak flame temperature and increase in NO_x emissions. Figure 10.5 depicts the NO_x emissions for the two types of firing with variation of excess air. Larger discharges of NO_x emissions with the burner configurations produce more intense combustion and higher temperatures when compared to burners which have normal designs with larger combustion area.

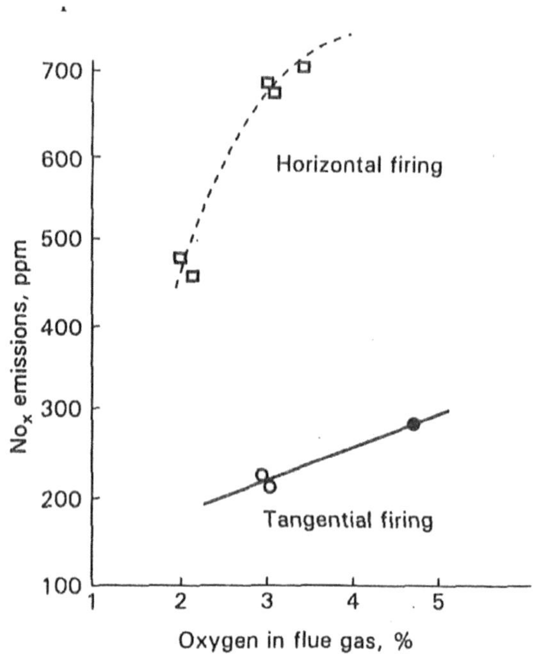

Fig. 10.5 *Effect of excess air on NO_x emission; for two types of boiler firing*

Fluidizer bed combustion operates with very high heat transfer rates with resultant low combustion temperatures less than 1050 ^{0}C. It is often used as a pollutant control technologies for control of NO_x emission. As it functions without good gas-solid contact within the bed, the excess air rate for combustion can be lowered resulting in low NO_x emissions.

10.2.4 Recirculation of the flue gases

Recirculating flue gases after cooling into the combustion chamber which can act as a thermal sink to reduce the overall combustion temperature was found to be effective in reducing NO_x emission. However, this method is effective only when the gas is mixed with primary air prior to combustion. Due to deficiency of oxygen, the N/O ratio and peak flame temperature are reduced resulting in low NO_x emission. Figure 10.6presents the effect of flue gas recirculation in furnaces. A 10%recirculation at 15% excess air results in high (nearly 30%) reduction in NO_x.

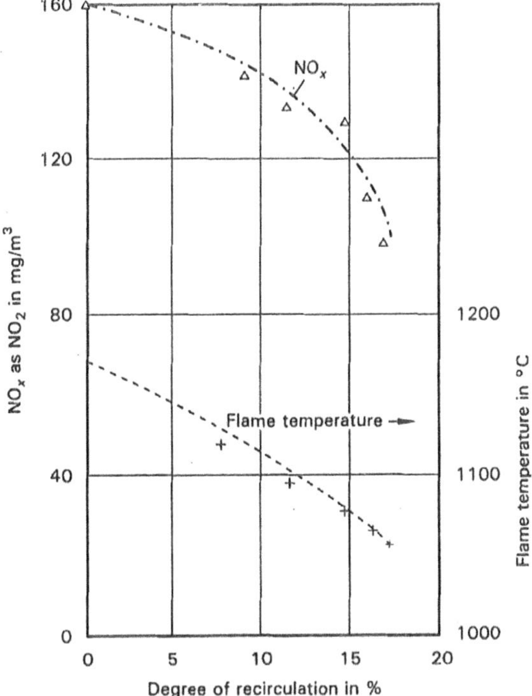

Fig. 10.6 *Reduction of NO_x emission by the gas recirculation in a light fuel oil furnace, 500 kW thermal power plant; dependency of flame temperature on degree of recirculation is also shown*

10.3 Combustion Process Modification

10.3.1 Two-stage combustion process

Most combustion modification schemes involve mixing part of the combustion air (90-95%) of the total air requirement with the fuel, burning as much of the fuel as that amount of air will burn, transferring some of the heat from the flames to whatever is being heated, then injecting the remaining air and finishing the combustion. This scheme is often called two-stage combustion or re-burning. In the first stage the maximum temperature is lowered because not all the fuel is burned, and the maximum temperature is reached when all the oxygen has been used up, so that there is not enough oxygen to form NO. In the second stage enough of the heat released in the first stage has been removed so that the maximum temperature reached-in the presence of excess oxygen is low enough that the formation of NO is less. In some re-burning schemes a small amount of additional fuel will also be added for the second stage, often a low-nitrogen content fuel like methane if the primary fuel has high nitrogen [6]. Under these conditions it was reported that reduction of NO_x by 35–40% for coal and oil firing and by 50% for natural gas can be achieved.

The advantage of this approach is that it is cheap. The disadvantages **are** that it requires a larger firebox for the same combustion rate and that it is difficult to get complete burning of the fuel in the second stage, so that the amount of unburned fuel and/or carbon monoxide in the exhaust gas is increased. A great deal of effort has gone into designing low-NO_x burners or burner systems that produce two-stage combustion and into installing them in large furnaces. These burners have reduced the NO emissions from furnaces, but are not enough to keep up with the more stringent new emission requirements.

10.3.2 Reducing residence time

Reducing residence time at high combustion temperatures can be done by ignition or injection timing with internal combustion engines. It can also be done in boilers by restricting the flame to a short region in which the combustion air becomes flue gas. This is immediately followed by injection of fuel, steam, more combustion air, or re circulating flue gas. This short residence time at peak temperature keeps the vast majority of nitrogen from becoming ionized. This bears no relationship to total residence time of a flue gas in a boiler.

10.3.3 Lowering the combustion air temperature by injection of water and stream

Generally in industrial operations, waste heat is used to preheat the air entering the combustion process which helps in appreciable energy savings. The added energy increases the flame temperature resulting in higher NO_x emissions. It was observed that a three-fold increase in NO_x emissions occurs when combustion air preheated from 25 °C to 300 °C. For controlling NO_x, of the primary flame zone by heat transfer to surrounding surfaces by injecting water or steam into the combustion zone is one of the methods practiced. Injecting water to the extent of 10–15% of the total combustion air can reduce the temperature level in the furnace to some extent.

10.3.4 Internal combustion reciprocating engines

Internal combustion engines use air-to-fuel ratio and ignition/injection timing to control maximum temperature and residence time. This can reduce the concentration of NO that is generated by reducing peak temperature. Valve timing adjustments can **reduce residence time at peak temperature** to control NO formation. Chemical reduction of NO is used in catalytic converters to reduce NO to N_2. Some stationary engines use both catalytic reduction and NO oxidation. A non-thermal plasma reactor was developed for treatment of diesel exhaust. A plasma ignition system allows greater freedom in the air-fuel ratio and the ignition timing of spark ignition engines.

10.4 Efficiency of NO_x Control Technologies used for Stationary Internal Engines

Table 10.6 *Presents efficiency of NO_x control technologies used for various stationary Internal Engines*

NO_x Abatement Method	Technique	Efficiency
1. Reduction of peak temperature	Air/fuel Ratio Timing of ignition/Type of ignition Pre-Stratified Combustion	20-97%
2. Reduction residence time at peak temperature	Valve Timing	NO-data
3. Chemical reduction of NOx	Selective Catalytic Reduction (SCR) Non-Selective Catalytic Reduction (NCSR) 80-90%	80-90%
4. Oxidation of NO_x with Subsequent absorption of NO_x	Non-Thermal Plasma Reactor (NTPR)	80-95%
5. Removal of nitrogen	Ultra-Low Nitrogen Fuel	NO-data
6. Use of sorbent	Sorbent in Exhaust Ducts	60-90%

10.5 Chemical Reduction of NO_X

This technique is based on the use of a reducing agent, which react with NO leading to N_2 formation. Examples include Selective Non-Catalytic Reduction (SNCR) that use ammonia or urea, Selective Catalytic Reduction (SCR), which uses ammonia, and Fuel Re burning (FR). Non-thermal plasma, in the presence of a reducing agent, is an emerging and innovative technology.

10.5.1 Selective non-catalytic reduction

(SNCR) technologies inject the reagent into a specific temperature zone. To initiate the reaction, it is important to properly mix the reagent with the flue gas, and the mixture must have adequate residence time at the required temperature. The acceptable temperature range for the reaction is 760 to 1100C (1400 to 2012F), although temperatures above 925C (1697F) are preferred. Below 870C (1598F), chemical enhancers, such as hydrogen, are needed to assist the reactions. As the temperature increases within this range, the ammonia or urea may react with available oxygen to form NO_x. This reaction becomes significant at temperatures above 1100C (2012F) and may become dominant as temperatures approach 1200C (2192F). Typical NO_x removal with SNCR ranges from 15 to 35%. NO_x removal can be pushed up to approximately 50% but with high ammonia slip, which can have undesirable consequences.

10.5.2 Selective catalytic reduction (SCR) of NO

Selective Catalytic Reduction of NO represent a key external combustion application which deserve great attention. The active catalytic component and temperature ranges may be classified as indicated:

Low temperature (175-250 °C): Platinum - based catalysts Medium temperature (300-450 °C): Vanadium - based catalysts

High temperature (350-600 °C): Zeolite - based catalysts

A more complete list of catalyst manufactures, general catalyst composition, catalyst support structures, and temperatures of operation are shown in Table 10.7.

Table10.8 present the solution for some problem for SCR.

Table 10.7 *SCR NO_x Catalyst Technologies*

Company	Catalyst Description	Operating Temperature (°C)
Babcock Hitachi	V/Ti/Metal plate	240-415
Carnet	Precious metal/metal monolith	225-275
Cormetech	V/T/Extruded monolin	200-450
Engelhard	Precious metal/Ceramic monolith	175-320
Engelhard	V/T/Ceramic monolith	300-440
Engelhard	Zeolite/Ceramic monolith	440-590
Hitachi Zosen	V/Ti/Extruded monolith	330-420
Hitachi Zosen	V/Ti/Wire mesh	330-420
JHI	V/Ti/Extruded monolith	200-400
JMI	V/Ti/Metal monolith	340-425
HKI	V/Ti/Extruded monolith	300-400
MIHI	V/Ti/Extruded monolith	200-400
Norton	Zeolite/Extruded monolith	220-520
Steuler	Zeolite/Extruded monolith	300-520
UBE	V/Ti/Extruded monolith	250-400

Table 10.8 *Problems of SCR and workable solutions*

Problems	Solution
Poisoning by SO_3	Use of TiO_2-V_2O_5 catalyst
Clogging by dust	Use of monolithic honeycomb type
Abrasion by coal dust	Hardening the part of gas entrance in catalyst
Catalytic Oxidation SO_2 to SO_3 by catalyst	Decrease of V_2O_5 loading and addition WO_3
Leak of ammonia	Decrease of NH_3/No_4 molar ratio
Pollution by aged catalyst	Recovery of heavy metal and TiO_2 from spent catalyst

10.5.3 Oxidation of NO

This technique is based on the possibility to fully oxidize NO to favor water to absorb them (i.e., it is based on the greater solubility of NO at higher N valence state). This is accomplished either by using a catalyst, injecting hydrogen peroxide, creating ozone within the air flow, or injecting ozone into the air flow. Notably, non thermal plasma, in the absence of a reducing agent, can effectively oxidize NO. A scrubber must be added to the process to absorb N_2 O_5, avoiding emissions to the atmosphere. Emissions of nitric acid can be controlled as well. In this case, nitric acid can be neutralized by basic liquids in the scrubber. Notably, nitric acid can be collected and then sold (also as a calcium or ammonia salt).

10.5.4 Removal of nitrogen from combustion

Reduction of NO_x formation can be achieved by removing nitrogen from the combustion either by: (a) using oxygen instead of air in the combustion process; or (b) using ultra-low nitrogen content fuel. Notably, the use of pure oxygen produces a rather intense flame that must be subsequently and suitably diluted. Using ultra-low-nitrogen content fuels with oxygen can nearly eliminate fuel and prompt NO_x

10.5.4.1 Combinations of these methods:

Many of these methods can be combined to achieve a higher NO_x reduction efficiency. For example, a fuel-rich cyclone burner can be followed by fuel re burn and over-fire air. Other control technologies that are intended to primarily reduce concentrations of sulfur also strongly affect the nitrogen oxide concentration.

10.5.5 Selection of proper method for industrial operations

The Destruction or Removal Efficiency (DRE) that each successful technology is capable of achieving is one of the key parameters for selecting the best available technology. Therefore effectiveness of reduction of NO and NO_2 is also expressed in terms of relative DRE; i.e., comparison of the NO generated in the presence of a NO reduction technology and in its absence.

In the case of combustion process, specific boiler types and combustion systems will be considered along with the specific NO_x technologies that can be applied. Notably, many new combustion systems incorporate NO_x prevention methods into their design and significantly reduce NO_x emissions compared to similar but older systems. As a result, the use of DRE (even a relative DRE) for NO_x may be inappropriate. In fact, the comparison between actual NO emissions from a new, well-designed system with those emitted by a similar older system, equipped with the same NO_x controlled technology, may be the best way of evaluating how effectively a new combustion system minimizes NO_x emissions.

10.6 NO_x Abatement System in Gas Turbines

Gas turbines use the Brayton Cycle with a burner to raise temperature of gas after compression and before expansion through the turbine. Turbines mainly use reducing peak temperature and reducing residence time approaches to limit NO emissions. Because addition of particles to air flow entering the turbine would accelerate erosion of turbine blades, sorbents could only be applied after the expansion in the turbine. NO_x reduction has been used to treat exhaust gases.

10.6.1 NO_x control by effluent gas treatment

In the modification of combustion processes the main objective is to reduce or suppress or prevent the emission of NO_x in the combustion chamber. However, sometimes these methods may not be practical or cost-effective to reduce emission levels, and in such cases effluent gas treatment methods like

(a) scrubbing,

(b) decomposition or reduction technologies

are effective alternatives for NO_x emission control

10.6.2 Scrubbing methods for effluent gas treatment for reduction of NO_x

The scrubbing techniques for NO_x control can be further subdivided into (i) absorption by liquids (ii) adsorption by solids.

10.6.3 Absorption by liquids

From power plants with 2000–15000 ppm levels of NO_x in the effluent gas, the following process can be adopted for NO_x control. Further these methods can also remove SO_2.

(i) Absorption by lime slurry (Figure 10.7) when gypsum and nitric acid can be recovered as byproducts [7].

(ii) Magnesium hydroxide scrubbing recovering concentrated NO as by-product [8].

(iii) Absorption by sulphuric acid which produces nitric acid and sulphuric acid [9].

(iv) Absorption by lime sodium hydroxide.

From nitric acid plants tail gas emissions vary from 0.1 to 0.69 volume percent with an average of 0.37%. Using sodium and calcium hydroxide solutions, a number of scrubbing technologies [10, 11] have been developed.

1. Nitrate and nitrite salts production by scrubbing first in water and then in sodium hydroxide, a two stage absorption process.
2. Ammoniacal solutions such as waste caustic ammonia liquor, ammonium bicarbonate and ammonium bisulphite as absorbing liquids.
3. Aqueous suspension of lime, as an absorbent where calcium and nitrate can be recovered for use as fertilizers.

With waste caustic ammonia liquor as the absorbing medium it was reported that the NO_x concentrations can be reduced to 100 ppm. NO_x can be used with ammonium bisulphate quickly at normal pressure and temperature to form ammonium sulphate which is useful as a fertilizer. A small residual amount of NO, some SO_2 and traces of NH_4 is present in the tail gas.

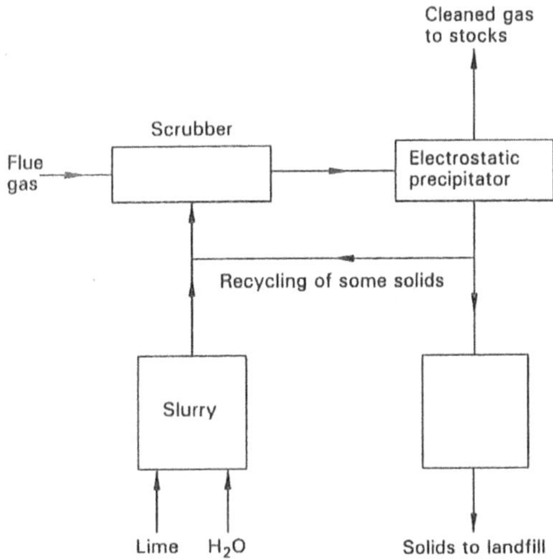

Fig. 10.7 *Lime slurry scrubbing sulphur dioxide and* NO_x

Aqueous suspension of lime can be used as a medium to reduce NO_x levels to 200ppm.whenuse lime, or limestone a slurry is prepared calcium hydroxide Ca $(OH)_2$ or calcium carbonate. It is used to remove the sulphur oxide from the flue gas stream as calcium sulphite and calcium sulphate and NO_x as calcium nitrite which is further converted to valuable calcium nitrate by treating with sulphuric acid as follows.

$$2H_2SO_4 + 3\,Ca\,(NO_2)_2 \rightarrow 2CaSO_4 + 4\,NO + Ca\,(NO_3)_2 + 2\,H_2O \qquad .....(10.6)$$

The two products precipitate from the solution. The lime slurry moves on to the scrubber, where the polluted gas makes intimate contact with it and reacts with the calcium. Then the slurry and products of the reaction, along with the gas prove on to a settling tank or a particulate removal unit, where residence time is essential for the particular matter to settle down. The solids are removed into a container, while some of the slurry is recycled to go through the process again. The clean gag moves up the stack. Fresh quantities of lime are added as needed [11]. The lime scrubbing process removes about 20% of the nitrogen oxides.

10.6.4 Magnesium hydroxide scrubbing

In power plants, large amounts of nitrogen oxides are released into the atmosphere. Tremendous volumes of flue gas must be dealt with in the cleaning process. For example, a 750MW gas fired plant releases almost 5tons of nitrogen oxides per hour, and flue gas volume is about 2million m^3/h.

In this process the effluent from a power plant containing nitrogen oxides and, of course, some sulphur dioxide, along with other pollutants and particulate matter, enters in to the spray tower where the particulate matter is removed from the bottom and taken to a settling pond. The water is recycled to the spray tower. (Figure 10.8)

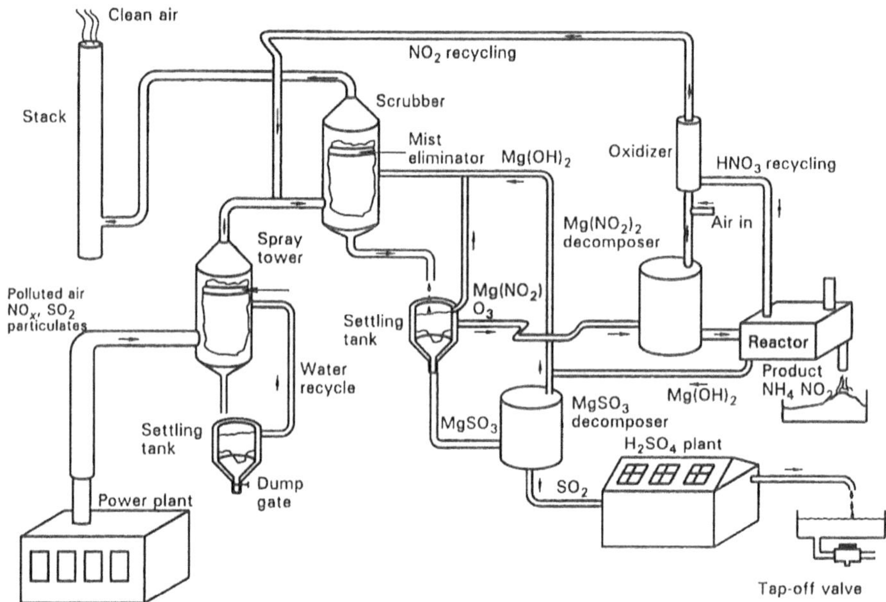

Fig. 10.8 *Magnesium hydroxide scrubbing for NO_x removal*

The exit gas from this unit enters a scrubber, where the magnesium hydroxide solution enters at the tog and the clean air leaves the scrubber at the top. The bottom stream is fed to a settler where the magnesium sulphite is taken to a unit for decomposing $MgSO_3$. Sulphur dioxide is given off and, taken to a sulphuric acid plant to give sulphuric acid as a by-product for sale. Magnesium nitrite leaves the settler into a nitrite decomposer where decomposition takes place, and the magnesium hydroxide enters into a react or which receives a stream of ammonia and some recycled nitric acid. The product is ammonium nitrate. It is stored for later sale.

A gas stream containing nitrogen oxides exits from the top of the magnesium nitrated is composer. This stream will combine with an air stream so that both are fed into the oxidizer, from which some recycled nitric acid is taken to the ammonia reactor. From the top, recycled nitrogen oxide is taken to be fed into the original scrubber. Limewater may be used in place of magnesium hydroxide. However, magnesium hydroxide scrubbing provides ease of regenerating oxides from the sulphites and nitrites of magnesium.

10.6.5 Tyco process

This process has been developed by Tyco Laboratories Inc. The basic chemistry of the chamber sulphuric acid process is applied for the purpose of removing both sulphuroxides and nitrogen oxides from a polluted air stream.

The flow chart for the process is given in Fig. 10.9. Concentrated sulphuric acid is used to scrub both the sulphuric acid mist and the nitrogen oxides.

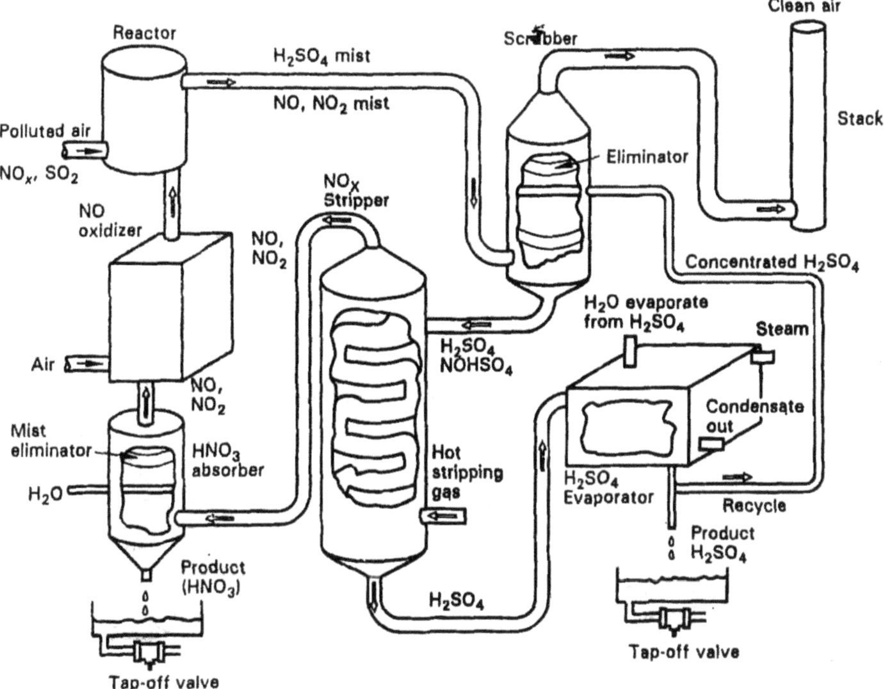

Fig. 10.9 *Tyco process for removing NO_x and SO_x*

This process gives a clean air stream free from nitrogen oxides and sulphur dioxide and it gives two by- ducts, namely, nitric acid and sulphuric acid. Both of these products help in improving the economic potential of such a process.

Polluted air containing sulphur dioxide and nitrogen oxides enters a reactor. The stream leaving this reactor consists essentially of sulphuric acid mist, NO, and NO_2 and it enters a scrubber. The scrubbing agent is concentrated sulphuric acid which scrubs the entering stream the acid mist and the nitrogen oxides; the exit gas is clean and harmless. The stream leaving the bottom of the scrubber consists of sulphuric acid and nitrogen thiosulphate. It enters a nitrogen oxide stripper where a hot stripping gas enters the unit, and NO and NO_2 leave the top of the stripper to enter a nitric acid absorber. The water entering the absorber will react with some of the NO and NO_2 to give nitric acid which leaves the bottom of the absorber. At the top of the absorber, the stream consists of nitrogen oxides, again NO and NO_2, which enter an oxidizer. The oxidizer also receives a stream of air or oxygen and transforms NO to NO_2, which enters the original first reactor, where it reacts again with the entering polluted stream.

Sulphuric acid leaving the stripper is concentrated in an evaporator. A certain portion of the acid is taken from the bottom of the evaporator as a marketable product (sulphuric acid) and another portion is recycled for use as a scrubbing agent in the scrubber, where the nitrogen oxides and acid mist are removed. Some thought has been given to combining the NO oxidizer and the NO_x stripper. Such a combination should help in reducing the equipment cost. The scrubber must have a very high efficiency (of the order of 99%) so that the nitric oxide balance can be kept in check since a large amount of NO_x is recycled (10 times of NO_x in original fuel).

10.6.6 Adsorption on solids

The solid adsorbents which show some capacity for oxidizing NO to NO_2 and for absorbing nitrogen dioxide [5]are activated carbon, silica gel, molecular sieves, ionex change resins, and certain metal oxides, particularly manganese and alkalized ferric oxides. However, when small quantities of oxygen are present in the flue gases, the sorption efficiency of activated carbon decreases drastically limiting its applicability as adsorbent. At the low concentrations of NO generally present in flue gas, other adsorbents also exhibit low capabilities.

The use of molecular sieves as possible adsorbents holds some promise, particularly for the treatment of nitric acid tail gas emissions. The process developed by Union Carbide [12] generates dry tail gas containing less than 50 ppm NO_x. In this process two beds operate batchwise: one adsorbs NO_x from the tail gas whilst the other is regenerated. The main disadvantage of using molecular sieves is the simultaneous adsorption of water vapour in the tail gas which decreases the efficiency of the bed.

The most suitable adsorbent for NO_x is the one which can be regenerated and at the same time does not preferentially react with water vapour or with CO_2 in the flue gas. The most promising adsorbent is a ferric salt. Researchers have studied the adsorption of nitric oxide on silica gel containing ferrous sulphate in the concentration range 50-100 ppm. Some conversion (2–4%) of ferrous sulphate was observed, but it was felt that this low conversion would not be a practical limitation.

10.6.7 Reduction/decomposition process for NO_x in flue gases (13)

There are two types of reduction processes, which are termed selective and non-selective. In selective reduction, the added reactant preferentially reduces NO_x and in non-selective reduction, the excess oxygen must be consumed first. Selective reduction is to be referred since it minimizes the amount of reactant required.

10.6.8 Reduction with natural gas

This is one of the simplest processes for removal of nitrogen oxides, where natural gas combines with nitrogen oxides and reduces them to harmless nitrogen and water. It is actually a combustion process and variations of burners been designed (Fig. 10 .10). The polluted air containing nitrogen oxides in noxious concentrations is fed into one conduit while a reducing gas such as natural gas is fed into another conduit. The two streams enter a cylindrical portion, and the resulting mixture passes onto an inverted cone section. This mixture is ignited, the ignition process converts the nitrogen oxides to nitrogen and water which may be discharged safely into the atmosphere.

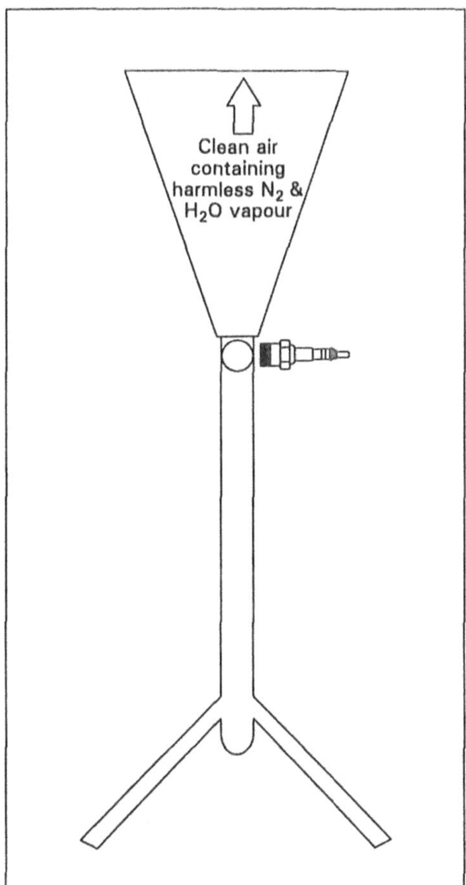

Fig. 10.10 *Reduction of NO with natural gas*

Thus in the non-selective reduction, two types of reactions take place. The first type involves the reaction of the fuel with oxygen and nitrogen dioxide, the latter being reduced to nitric oxide.

$$CH_4 + 2O_2 \rightarrow CO_2 + 2H_2O \quad .....(10.7)$$

$$CH_4 + 4\,NO_2 \rightarrow CO_2 + 2H_2O + 4\,NO \quad .....(10.8)$$

The second reaction reduces the NO but the reduction does not take place until all oxygen in the flue gas has been reacted with the fuel and reducing conditions havebeen achieved.

$$CH_4 + 4NO \rightarrow CO_2 + 2H_2O + 2N_2 \quad .....(10.9)$$

The polluted air must contain sufficient oxygen or another stream of oxygen must be fed in to support the combustion proems. The amount of fuel that is necessary for supporting the Combustion is very strongly influenced by the velocity at which the gaseous mixture enters. At 36.57 m/s the fuel ratio is high, but on the other hand, if the velocity is too low, the quantity of nitrogen oxides which escape unreduced also increases. Velocities above 6 m/s are usually used to help create some turbulent mixing and thus lessen the possibility of nitrogen oxides escaping the process without combustion.

While this process is simple, the economics must be proven, simply because of the cost of energy supporting the combustion. Of course, a catalyst may be added to the process, and this may reduce the cost to some degree. However, since there is no by product in such a process, it may be useful only for small operations. Whenever dealing with power plants, where volumes of stack gas are essentially huge, other processes must be studied for suitability, keeping in mind a saleable by-product recovery which will help offset the expense of pollution abatement.

10.6.9 Selective catalytic reduction for removal of NO_x from flue gases (14, 15)

The flue gas usually contains 2–6% oxygen and a few hundred ppm of NO_x, which consists of 90–95% NO and 5–10% NO_2. The only viable post combustion process for NO_x emission control is chemical conversion to N_2 by a reducing agent. Several gases can be used for this purpose including methane, hydrogen, carbon monoxide and ammonia. However, because the Rue gas contains much more than 0, than NO_x, the reducing agent must selectively react with NO_x rather than O_2 to minimize its consumption. To date, many commercial installations based on the selective reduction of NO_x by NH_3 can remove about 80% of NO_x in the flue gas. The governing equations for the ammonia based technology are as follows:

$$4\,NO + 4NH_3 + O_2 \rightarrow 4N_2 + 6H_2O \quad .....(10.10)$$

$$2\,NO_2 + 4NH_3 + O_2 \rightarrow 3N_{-2} + 6H_2O \quad .....(10.11)$$

$$4NH_3 + 5O_2 \rightarrow 4NO + 6H_2O \quad .....(10.12)$$

The first two reactions dominate when the flue gas is heated to about 1000 °C. Above 1000 °C the third reaction becomes significant leading to the unwanted formation of NO.

If the temperature is below 800 °C, the reaction rate is too low for practical use. Thus the process is temperature sensitive.

With the use of a suitable catalyst, the NO_x reduction can be carried out at 30–400 °C, a temperature normally available in a flue gas system. This process is 'called Selective Catalytic Reduction (SCR).

Selective catalytic reduction (SCR) is one of the most efficient methods for removal of NO_x pollutants. Here, anhydrous aqueous ammonia which is a selective reluctant is injected into the polluted flue gas through a bed containing the catalyst (Fig. 10.11). Both ammonia and nitrogen oxides combine to form an ammonium salt intermediate by the action of the catalyst. The intermediate compound then decomposes to given elemental nitrogen and H_2O. The nitrogen oxide reactions that take place at a small range of temperatures between 260-510 ^{o}C are shown in Fig. 10.11. type of catalyst determines this range. Operating below the optimum range will not give the energy necessary to initiate the desired reaction; while operating at a temperature above the maximum, oxidation of ammonia to either nitrogen oxides or ammonium nitrate and nitrite would take place. It is very important to operate in the proper temperature range. This is usually known from pilot plant tests and experience. The flue gas from industrial boilers is usually the input feed in this type of catalytic reduction process.

Thus SCR is potentially applicable to flue gas from fuel-lean firing combustion systems, that is, the flue gas is under oxidizing conditions. Non-selective catalytic reduction processes are applicable to fuel-rich firing combustion conditions.

The following main reactions take place at the catalyst during the reduction of, nitrogen oxides with ammonia:

$4\ NO + 2\ NH_3 + O_2 \rightarrow 4N_2 + 6\ H_2O$

NH3/NO molar ratio = 1:1 .....(10.13)

$6NO + 4NH_3 \rightarrow 5\ N_2 + 6\ H_2O$

NH_3/NO molar ratio = 2 : 3 .....(10.14)

NO_2 possibly present (generally no more than 5%) will also be reduced

$6\ NO_2 + 8\ NH_3 \rightarrow 7\ N_2 + 12\ H_2\ O$

NH_3/NO molar ratio = 3 : 4

$2\ NO_2 + 4\ NH_3 + O_2 \rightarrow 3\ N_2 + 6\ H_2O$

NH_3/NO molar ratio = 2 : 1

Undesirable side-reactions at the catalyst are e.g.

$4\ NH_3 + 3\ O_2 \rightarrow 2N_2 + 6\ H_2O$ additional NH_3 consumption

$2\ SO_2 + O_2 \rightarrow 2\ SO_3$

Excess ammonia reacts with sulphur trioxide (SO_3) to become ammonium hydrogen sulphate and ammonium sulphate:

$NH_3 + SO_3 + H_2O \rightarrow (NH_4)\ HSO_4$

$2\ NH_3 + SO_3 + H_2O \rightarrow (NH_4)_2\ HSO_4$

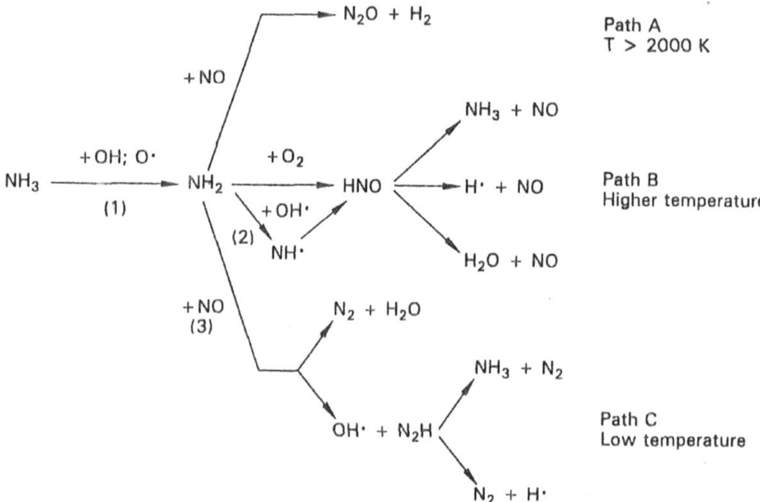

Fig. 10.11(a) *Simplified reaction mechanism of NO reduction or NO formation with NH_3*

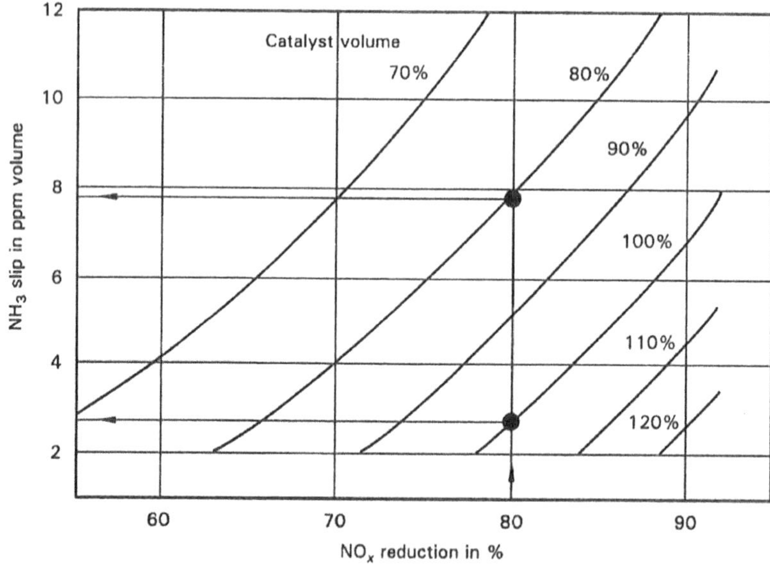

Fig. 10.11(b) *Correlation between NH_3 slippage, catalyst volume and NO_x reduction degree*

When temperatures fall below a certain range these ammonium salts can lead to considerable deposits on the components after the catalytic converter, e.g. the air pre heater. Reaction (XIII) represents the predominant reaction since approximately 95% of the NO_x in combustion flue gas is in the form of nitric oxide (NO). An NH_3: NO molar ratio of about 1:1 reduces NO_x emissions by 80–90% with a residual NH_3 concentration of less than 20 ppm.

The typical schematic of a SCR system in a power plant is depicted in Figure 10.12 in which the main component is a reactor, a catalyst, and a ammonia storage & injection system. For most of the reactions 520–720 °K is the optimum range for the catalytic reaction. Sometime due to increased pressure drop across the SCR reactor, some increase in boiler fan capacity may be necessary.

The NH_3: NO_x molar ratio and the flue gas space velocity are the major operating variables affecting the level of NO_x control achieved for a given boiler condition and SCR system. An NH_3 : NO molar ratio of 1: 1 can achieve about 90% NO_x reduction, with higher NH_3 rates resulting in higher undesirable NH_3 emissions.

Various shapes of catalysts exist including parallel-ridged plates, rings, pallets, and honeycomb plates. The catalyst compounds are either titanium, platinum, zeolites, some base metals or ceramics. Through research and experience, one determine the optimum combination to be used. Some problems in the early operation of this method included the poisoning of the catalysts by sulphur dioxide and other metallic components (such as arsenic) that may be in the flue gas. The conversion of compounds on the catalysts and downstream equipment also take place. Many improvements have been made in this process. Some catalysts such as zeolite and titania have been researched and developed to resist any poisoning by sulphur oxides. Selective reduction can also be carried out with H_2, CO, or NH_3 as the reactant gas. Table 10.9 gives the list of catalysts used in catalytic reduction.

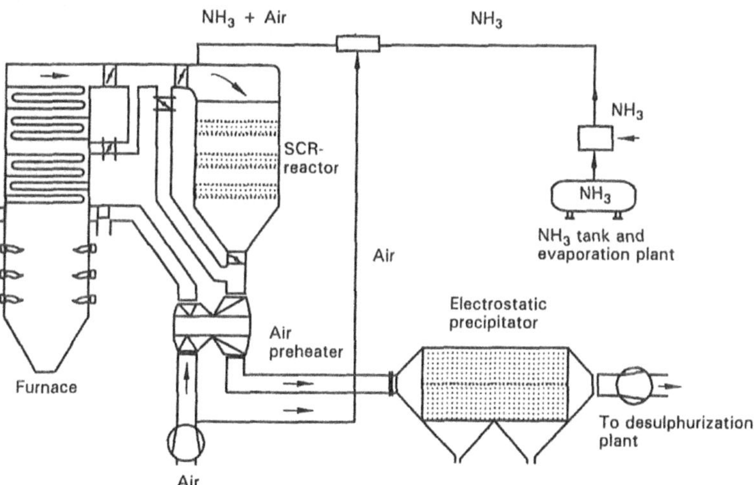

Fig. 10.12 *Construction of an SCR plant in the exhaust gas system of a power plant furnace*

Table 10.9 *Catalytic catalysts and reducing agents used in catalytic reduction of NO [13]*

Catalysts	Temp. range °C	Reducing agent
Cr-promoted Fe-oxide	250-240	Co or H_2
Fe-chromite	300-450	CO or H_2
Supported platinum	350	CO
	300	C_2H_5
Supported Cu-chromity	129-300	CO
Zn promored Co-chrumite	375-425	H_2
	370425	CH_4
Ba, pmmptad,Cu- hromite	225-525	C_1 to C_8 hydrocarbons

At a given temperature the following three factors are important for the design of catalyst converter: (i) the required degree of NO_x reduction (ii) the admissible or justifiable NH_3 slip (iii) the catalytic converter volume.

The basic assumption appears valid that under otherwise equal condition certain degree of reduction can either be attained by a-large catalytic converter volume and a small ammonia slip or by a small catalytic converter volume and a higher ammonia slip. The almost linear dependencies between reduction degree, NH_3 slip and catalytic converter volume shown in Fig. 11.12 are only valid up to a reduction degree of 80-85%. In excess of that, the required catalytic converter volume and/or slip rise disproportionately due to various factors such as irregularities in the flue gas stream (velocity, temperature, local NO_x, and NH_3 concentration).

Apart from the temperature, the molar ratio of NH_3/NO and the so-called space velocity are influential factors, pertaining to NO_x reduction. Depending on whether reaction (XII) or reaction (XIII) prevails at the catalytic converter, maximum reduction ensues at a NH_3/NO molar ratio of 1 or 0.66. Which reaction preferentially takes place depends, among other factor, on the material of the catalytic converter and on the oxygen content of the flue gases.

Questions

1. Discuss the various sources of NOx emissions in Air Pollution.
2. Outline the graphical representation of NOx formation at different combustion temperatures.
3. Discuss various factors responsible for NOx emissions.
4. Describe various control technologies used for NOx emissions.
5. Is it possible to reduce the dependency of NOx emissions on boiler load? If so justify.
6. Discuss and Outline the burner design for controlling the NOx emissions.
7. Describe the terms (i) Recirculate of the flue gases (ii) Two stage combustion process

8. Write a short note on (i) Selective non–catalytic reduction (ii) Selective catalytic reduction
9. Discuss the challenges in process of selective catalytic reduction
10. Describe effluent gas method to reduce NOx emissions.
11. Describe the process of $Mg(OH)_2$ scrubbing for NOx removal
12. Explain Tyco process
13. Discuss the methods adopted for reduction of NO in natural gas.

References

1. M. F, Hupa, F. Back man and S. Bastroem. 1989. Nitrogen oxide emissions in boilers. *J. Air pollution control assoc.* 39: 1496–1501.
2. Selective Catalytic Reduction Control of NO_x Emissions, SCR Committee of Institute of Clean Air Companies, November 1997
3. Nitrogen Oxides: Impacts on Public Health and the Environment, EPA 452/R-97-002
4. JANAF 1964. Thermo chemical tables. US Govt. Clearing House Number PB 168370.
5. D. Bienisstock, 1972. *Proc. of* 1972 *industrial conference.* October. Purdue University.
6. H. Brauer and Y. B. G. Varma, 1981. Air pollution control equipment. Springer Verlag, Berlin. 248.
7. J. Bagg, 1971. The formation and control of nitrogen oxides in air pollution. Air pollution control. Part I. (Ed.) W. Strauss. Wiley Inter Science, New York. 35–94.
8. S. M. Dalton, B. Toole-O. Neil, B. K. Gullert and C. J. Drumm and, 1991. Summary of 1991 EPRI / EPA / DOE IEACR/ 24 1990.
9. C. E. Blaskeslee and H. E. Burback, 1973. *J. of Air Pollution Control Assoc.* 23:37–42.
10. W. Bartork, 1971. Stationary sources and control of nitrogen oxide emissions. *Proc, of second international clean air conference*, (Eds) H. M. England and W. T. Beery. Academic Press, New York.
11. Atmospheric emissions from nitric acid manufacture processes. 1966. U.S. P.H.S. Publication No. 99–AP–27.
12. L. L. Form off, 1972. A new molecular sieve process for NO removal and recovery of nitric acid plant tail gas. *AlCME symposium series.* 68(116): 111.
13. C. S. Rao, 1978. Control of oxides of nitrogen from effluent gases. *Chem. Engg.* 13(4):102– 108.
14. H. R. L. Streight, 1958. Can. J. of Chemi. Engg. 36:3.
15. B. J. Mayland and R. C. Heinze, 1974. Che. Symposium series. 70(137):83.

Suggested References

1. Nitrogen Oxides: Impacts on Public Health and the Environment, EPA 452/R-97-002.
2. Selective Catalytic Reduction Control of NO_x Emissions, SCR Committee of Institute of Clean Air Companies, November 1997.
3. Nitrogen Oxides Control Technology Fact Book, Leslie L. S loss, Noyes Data Corp., 1992.

4. Nox Control Technology Data, EPA 600/2-91M9.
5. Bauer F-, Energy Engineering, 91 (1994) 17.
6. Alternate Control Techniques Document - NO_x Emissions from Cement, Manufacturing EPA-453/R-94-004.
7. The Economic Feasibility of using Hydrogen Peroxide for the Ehanced Oxidation and Removal of Nitrogen Oxides from Coal-Fired Power Plant Flue Gases, Hat wood J., Cooper C., AWMA Journal, March 1998.
8. Simultaneous SO, SO, and NO_x Removal by Commercial Application of the EBA Process, Hirano S., Aoki S., Izutsu M, and Yuki Y., EPRI-DOE-EPA Combined utility Air Pollution Control Symposium: The MEGA Symposium, Atlanta, August 1999.
9. Update on SNAP Technology for Simultaneous SOX and NOX Removal, Felsvang K., Boscak V., Iversen S., Anderson P., EPRI-DOE-EPA combined Utility Air Pollution control symposium: The MEGA symposium, Atlanta, August 1999.
10. Iwamoto, M. and Hamada, H. , Catal. Today, 10 (1991) 57.
11. Heck R.M. and Farrauto R.J., Catalytic air pollution control: commercial technologies, van Nostr and Reinhold, New York, 1995.
12. Zwinkels, M., Jaras, S., Menon, P., and Griffin, T., Catal. Rev-Sci. Eng. 35(1993) 319.
13. Li, Y. and Armor, J.N., J.Catal., 150 (1994) 376.
14. Li, Y., Battavio, P.J. and Armor, J.N. (1993) J. Catal., 1993, 142, 561.
15. Alternative Control Techniques Document - NO_x Emissions from Stationary Reciprocating Internal Combustion Engines, EPA 453/R-93-032.
16. Alternative Control Techniques Document - NO_x Emissions from Stationary Reciprocating Internal Combustion Engines. EPA-453/R-93032, July 1993.
17. Chess K., Yao S., Russell A. and Hsu H., Journal of the Air & Waste Management Association, 45 (1995) 627.
18. Anonimous, Oil & Gas Journal, 92 (1994) 134.
19. Manning M., Pipeline & Gas Journal, 222 (1995) 26.
20. Alternative Control Techniques Document - NO_x Emissions from Stationary Reciprocating Internal Combustion Engines. EPA-453/R-93032, July 1993.
21. Bartok, W., and Sarofim, A.F., Fossil Fuel Combustion, John Wiley & Sons, New York, 1991
22. Kuehn S.E., Power Engineering, 98 (1994) 23-27.
23. Advanced Reburning for SIP Call NO_x Control, EPRI-DOE-EPA Combined Utility Air Pollution Control Symposium: The MEGA Symposium, Atlanta, August 1999.
24. States' Report on Nitrogen Oxides Reduction Technology Options for Application by the Ozone Transport Assessment Group, OTAG, 1996.
25. Alternative Control Techniques - Nitric Acid and Adipic Acid Manufacturing Plants EPA 450/3-91-026.

CHAPTER 11

Control Technologies for Volatile Organics, Hydrocarbons and Hydrogen Sulphide

11.1 VOC Treatment Technologies

Volatile organic compounds (VOCs) are a large group of organic chemicals that include any compound of carbon (excluding carbon monoxide, carbon dioxide, carbonic acid, metallic carbides or carbonates, and ammonium carbonate).

VOCs are of great interest because they participate in atmospheric photochemical reactions that contribute to ozone formation. VOCs are emitted from a variety of sources, including motor vehicles, chemical manufacturing facilities, refineries, factories, consumer and commercial products, and natural (biogenic) sources (mainly trees).

VOCs which are organic carbon compounds bonded to H, S vapourize at significant rates even at room temperatures and participate in photochemical reactions when the gas stream is released to the ambient air. Almost all of the organic compounds used as solvents and as chemical feedstock are VOCs. However, a few organic compounds, such as methane, are not considered to be VOCs. Notably VOC compounds are not formed in industrial processes, they are lost.

Hydrogen sulfide is a very volatile gas and readily escapes into the air causing unpleasant odours that can be detected by the human sense in very low concentrations. In high concentrations, it is also very hazardous to humans. In concentrations as low as 10 ppm it can cause nausea, headache and conjunctivitis of the eyes.

Cost effective Control Technologies for the above volatile gaseous pollutants are very important in air quality management

11.1.1 Sources of Volatile Organic Compounds (VOC)

Generally in VOCs the number of organic carbon atoms is less than 12 particularly in photochemical oxidants like benzene, (which was discussed in detail in Chapter 1) and are typically organic solvents having boiling points less than 260 °C and vapour pressure greater than 0.007 atm (0.01psi).

Emissions of volatile organic compounds occur in various forms and from a variety of sources. In automobile transportation, one of the main VOC sources, mainly aliphatic and aromatic hydrocarbons are released from the exhaust pipes of vehicles. A large number of aromatic compounds are considered carcinogenic, e.g. benzene and different PAH (polycyclic aromatic hydrocarbons); others contribute to ozone formation in the troposphere, particularly alkenes and aromatic compounds. Diesel vehicles, but also vehicles with catalytic converters running on a too-rich mixture and engines burning motor oil substances produce an intense odour and are an emission problem. Moreover, many industrial processes also release strong odours. Odorous emissions are released during the production of food and semiluxuries. Odour-intensive substances are formed by protein degradation during the production of meat flavouring agents in spice factories. Emissions containing fat and water vapour develop in canned meat factories, e.g. when immersing the meat in hot fat or broth.

In starch factories strong-smelling vapours are produced and coffee and cocoa roasters give off the well-known odorants consisting of aldehydes, organic acids and other hydrocarbons. Further producers of odours are malt factories, beer breweries and smoking plants. Odour emitters processing high-protein material of animal origin are fish meal factories, bone processing plants and suet melting plants.

The following substances can be considered as odorants: ammonia and amines as alkaline nitrogenous substances, hydrogen sulphide and mercaptans as sulphurous compounds, carbonyl compounds, fatty and amino acids.

Odour emissions also emanate from intensive livestock breeding, from different waste disposal plants for sludge conditioning and composting. In animal carcass processing plants where carcasses of sick and condemned animals, of those killed in accidents and also slaughtering waste are processed small fragments having an intensive, nauseating odour are formed from protein particles as a result of rotting and decomposition processes.

In the industry adding organic solvents is a preliminary step to processing and treating many materials, e.g. in the surface treatment of paint shops, in the printing industry and in the chemical and pharmaceutical industry. Dry cleaners and metal degreasing plants, work with halogenated hydrocarbons. Volatile organic substances partially vapourize during the process and are frequently released, by vapourizing oil in hardening and electrolytic refining plants, in metals melting and in foundries during the production of cores, etc.

In plastic and textile processing, monomers (formaldehyde, styrol), solvents, decomposition products of thermal molding and propellants from foaming processes are released, depending on the task. For instance, during the spinning process in cellulose (viscose) fibre production the solvent carbon disulphide (CS_2) evaporates from the thread taking shape. The air required for this leaves the vent laden with carbon disulphide.

VOCs as fuels (propane, gasoline, jet fuel, diesel fuel) are used very widely because liquid hydrocarbon fuels are easy to produce, transport and store. Further they evaporate into the air, leaving either no residue (dry cleaning solvent, nail polish remover) or a thin layer of previously dissolved solid (paint, fingernail polish, floorwax, inks). The rate of evaporation is roughly proportional to the vapour pressure.

11.2 Control Measures for VOCs by Reduction of Leakages or Working Losses

11.2.1 Filling, breathing and emptying losses

From large storage tanks containing liquid VOCs, leakage of these vapours occurs when filling and emptying or due to changes in temperature and atmospheric pressure. Such losses are termed filling or displacement losses, emptying losses, breathing losses, or, collectively, called working losses and can be reduced by using special types of bends and seals.

11.2.2 Special seals for leakage control

Emissions of VOCs may occur as leaks at seals which become significant if not properly controlled. Figure 11.1 shows three kinds of seals. Figure 11.1 (a) shows a static seal which exists between the bottle cap and the top of a bottle of carbonated beverage. Figure 11.1 (b) shows a simple compression seal between a housing and a shaft in which a nut screws down over the body of the faucet to compress an elastomeric seal that is trapped between the body of the faucet and the stem of the valve. The compressed seal must be tight enough to prevent leakage of the high-pressure liquid inside the valve out along the edge of the stem, but not so tight that the valve cannot be easily rotated by hand. However, this type of seal often leaks and tightening the nut normally reduces the leak to a rate low enough that it becomes invisible (but does not become zero).

Figure 11.1(c) shows the seal that surrounds the drive shaft of an automobile at the point where the shaft exits from the transmission. The inside of the transmission is filled with oil. The elastomeric seal is like a shirt cuff turned back on itself with the outside held solidly to the wall of the transmission and the inside held loosely against the rotating shaft by a garter spring. If we set that spring loosely, then there is a great deal of leakage. If we set it tightly, then the friction and wear between the cuff and the shaft that rotates inside it are excessive. Setting the tension on that spring requires a compromise between the desire for low leakage and the desire for low friction and wear which leads to a low, but not a zero, leakage rate; a small amount of oil is always dripping out. Valves and pumps also have shafts that must rotate, and hence they also have the same kind of leakage problem. All of the pumps and valves in facilities that process VOCs will use seals. Some are more complex versions of the above types. For the seals to be made more and more leak-tight, the low-quality seals on pumps and valves should be replaced with more complex and expensive, higher-quality seals.

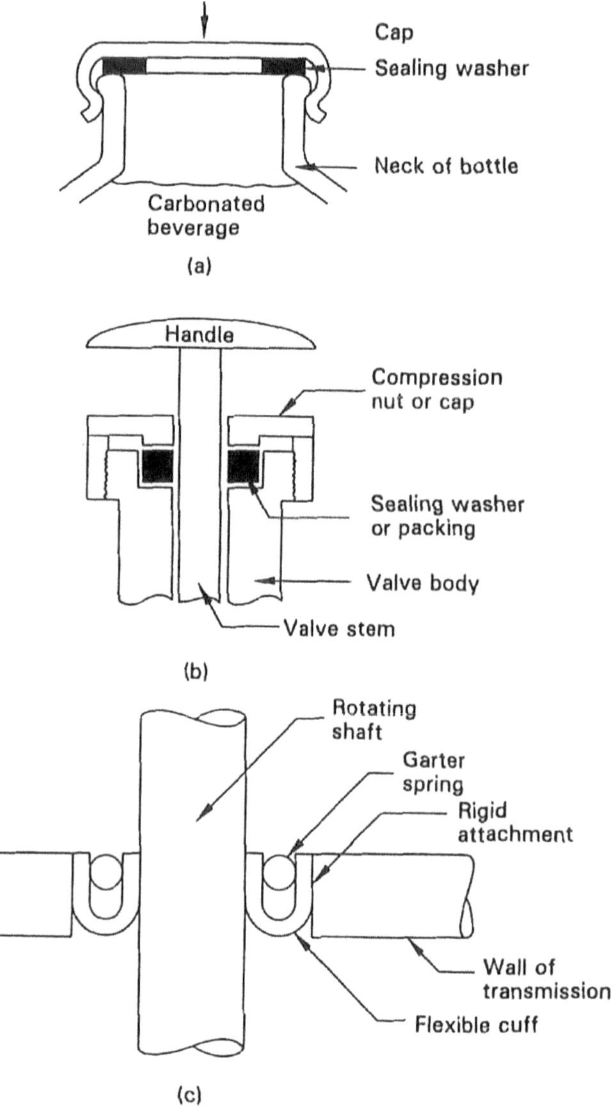

Fig. 11.1 *Three kinds of seals (a) a static seal, as exists between a carbonated beverage bottle and its bottle cap; (b) a packed seal, as exists between the valve stem and valve body of simple faucets, and as also exists on many simple pumps; (c) a rotary seal of the type common on the drive shafts of automobiles and some pumps*

11.3 Adsorption, Condensation and Combustion Techniques for Reducing/Controlling VOC Emissions

11.3.1 Adsorption techniques

Adsorption systems beds are generally used when the gas stream contains one to three volatile organic compounds, and it is economical to recover and reuse these compounds, or when a large number of organic compounds at low concentration, and it is necessary to pre-concentrate these organics prior to thermal or catalytic oxidation

Adsorbers remove VOCs from air or other gas streams in which the VOCs are present in low concentrations. VOCs can be recovered mainly by regenerating the adsorbents with steam and condensing the VOCs as liquids. The same principles apply to the throwaway adsorbers used in industrial face masks and to the adsorbers used to capture some of the fuel emissions in most modern autos.

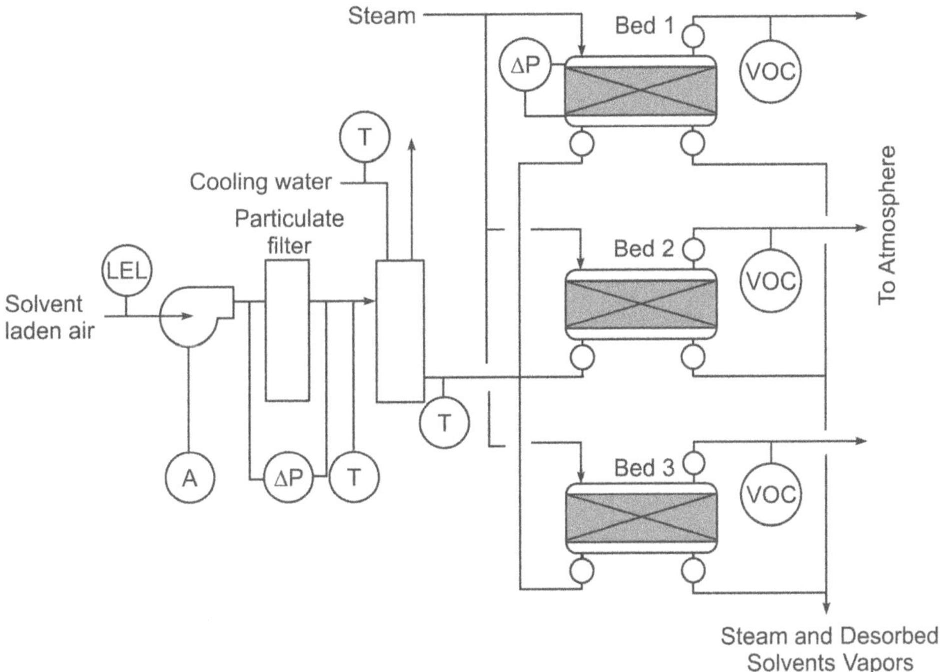

Fig. 11.2 *Multi-bed adsorber system for solvent recovery*

Figure 11.2 shows the diagram of a multi-bed adsorber system used for collection and recovery of organic solvent compounds. The VOC-Iaden gas is often cooled prior to entry into the adsorption system to improve the effectiveness of adsorption. As the gas stream passes through the bed, the organic compounds adsorb weakly onto the surfaces of the adsorbent (high surface area activated carbon, zeolite, or organic polymer). When the adsorbent is approaching saturation with organic vapor, a bed is isolated from the gas stream and desorbed. Low-pressure steam or hot N_2 is often used to remove the weakly adsorbed organics. The concentrated stream from the desorption cycle is treated to recover the organic compounds. After desorption, the adsorption bed is returned to service, and another bed in the system is isolated and desorbed.

In reality some of the adsorbable material appears at the outlet before the bed is filled. At time t_b (Fig. 11.3) the concentration in the outlet stream reaches the *breakthrough value* (typically 1% of the inlet value). The adsorbent bed is then switched to regeneration or, if it is a throwaway adsorber like those in workers' face masks, the adsorber cartridge is discarded and a new one is installed.

11.3.2 Breakthrough performance

Though it is assumed that an adsorbent fills up with adsorbed material uniformly in ideal conditions, in reality it will not happen. In many cases the pollutant adsorbed "breaks through" before the bed has reached its maximum capacity. The reason for this early breakthrough is that there is a finite resistance to mass transfer between the gas and solid, so some finite amount of time is needed for each particle to be loaded with adsorbate [1, 2, 3].

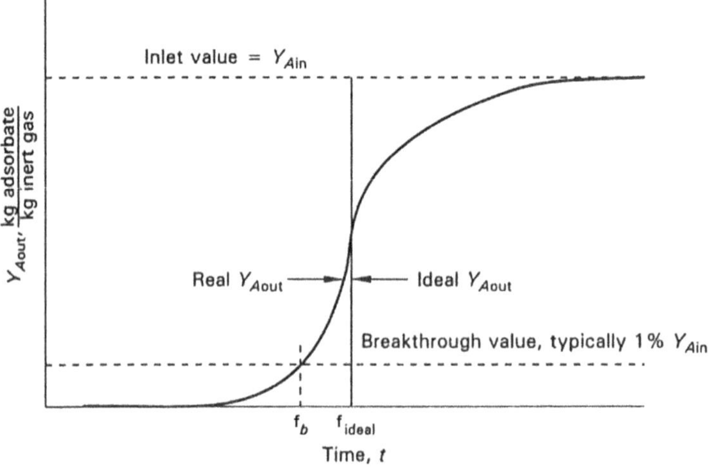

Fig. 11.3 *Ideal and real breakthrough curves for adsorption in a fixed bed*

11.3.3 Adsorption System for pre-concentration

An adsorption system used for pre-concentration prior to solvent recovery is shown in Figure 11.4. The gas stream containing the VOC passes through a rotary wheel containing zeolite or carbon-based adsorbents. Approximately 75-90% of the wheel is in adsorption service while the remaining portion of the adsorbent passes through an area where the organics are desorbed into a very small, moderately hot gas stream. The concentrated organic vapors are then transported to a thermal or catalytic oxidizer for destruction. The preconcentration step substantially reduces the fuel requirements for the thermal or catalytic oxidizer.

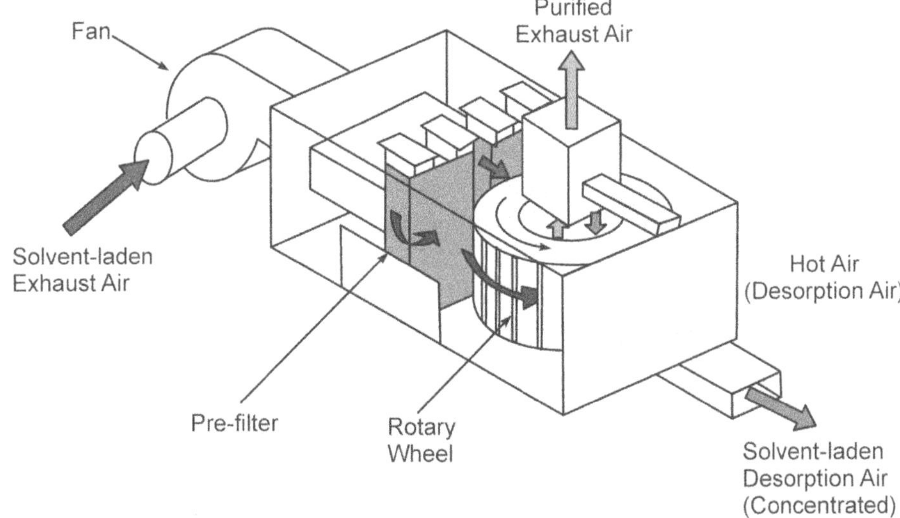

Fig. 11.4 *Pre-concentrator - type adsorption system (2)*

Adsorption systems are not recommended for gas streams that contain particulate matter and/or high moisture concentrations because the particulate matter and moisture compete with the gaseous pollutants for pore space on the adsorbent material.

The adsorption removal efficiency usually exceeds 95% and is often in the 98% to 99% range for both solvent recovery and preconcentrator type systems. In both types of units, the removal efficiency increases with reduced gas temperatures.

The suitability of an adsorption system for a particular situation should be considered on a case-by-case basis. However, as a general guideline, adsorption systems can be used for organic compounds having a molecular weight of more than 50 and less than approximately 200. In fact, the low molecular weight organics usually do not adsorb sufficiently. The high molecular weight compounds adsorb so strongly that is it is difficult to remove these materials from the adsorbent during the desorption cycle. Adsorption systems can be used for a wide range of VOC concentrations from less than 10 ppm to approximately 10,000 ppm. The upper concentration limit is due to the potential explosion hazards when the total VOC concentration exceeds 25% of the LEV.

11.3.4 Control by condensation

Condensation, refrigeration, and cryogenic systems remove organic vapor by making them condense on cold surfaces. In cases where pollutants have low vapour pressures, condensation is effective for removing a significant part of the vapor. The condenser works by cooling the feed gas to a temperature below the dew point of the feed gas. These cold conditions can be created by passing cold water through an indirect heat exchanger, by spraying cold liquid into an open chamber with the gas stream, by using a freon-based refrigerant to create very cold coils, or by injecting cryogenic gases such as liquid nitrogen into the gas stream. Cooling water is the most commonly used coolant although brine solutions are also used when low temperatures are required.

The concentration of VOCs is reduced to the level equivalent to the vapor pressures of the compounds at the operating temperature. Condensation and refrigeration systems are usually used on high concentration, low gas flow rate sources. Typical applications include gasoline loading terminals and chemical reaction vessels.

Although condensation can also occur by increasing the pressure without changing the temperature, this method is seldom used for the control of air pollution. Because condensation is a simple process and the design techniques for condensers are well—established. Condensation is also used as a pretreatment process for reducing the load and operating problems of other pollution control devices

The removal efficiencies attainable with condensation depend strongly on the outlet gas temperature. For cold-water-based condensation systems, the outlet gas temperature is usually in the 4 to 10 °C range, and the VOC removal efficiencies are in the 90 to 99% range depending on the vapor pressures of the specific compounds. For refrigerant and cryogenic systems, the removal efficiencies can be considerably above 99% due to the extremely low vapor pressures of essentially all VOC compounds at the very low operating temperatures of –56 °C to less than –130 °C. There are two widespread applications for condensation for VOC control. The first is at gasoline loading terminals while the second application is the condensation of VOC at the outlet of an adsorber

Condensation, refrigeration, and cryogenic systems are usually used on gas streams that contain only VOC compounds. High particulate concentrations are rare in the types of applications that can usually apply this type of VOC control system. However, if particulate matter is present, it could accumulate on heat exchange surfaces and reduce heat transfer efficiency.

11.3.4.1 Types of condensers

Two types of condensers are in use which are surface and contact condensers. Surface condensers are used when the coolant and pollutant form a **miscible mixture**. This mixture must be further separated (e.g. by distillation) for recovery of the pollutants.

If the coolant is **immiscible** with the pollutants, contact condensers are used because phase separation is a relatively easy operation. However, the coolant is not contaminated to the extent of causing water pollution problems.

Through condensation alone cannot remove all the volatile organic pollutant gases; a direct contact condenser using room temperature water as the cooling fluid is a valuable pre-cleaner before using the condensation process or adsorption unit.

11.3.4.2 Direct contact condensers

In a direct contact condenser a stream of water or other cooling liquid is brought into direct contact with the vapour to be condensed. This process is conducted in a chamber especially designed for the purpose; one such chamber is shown in Fig. 11.5. The liquid stream m_m, leaving the chamber contains the original cooling liquid plus the condensed substances. The gaseous stream m_2 leaving the chamber contains the noncondensable gases and such condensable vapours as did not condense; it is reasonable to assume that the vapours in the leaving gas stream are saturated. It is then the temperature of the leaving gas stream which determines the collection efficiency of the condenser.

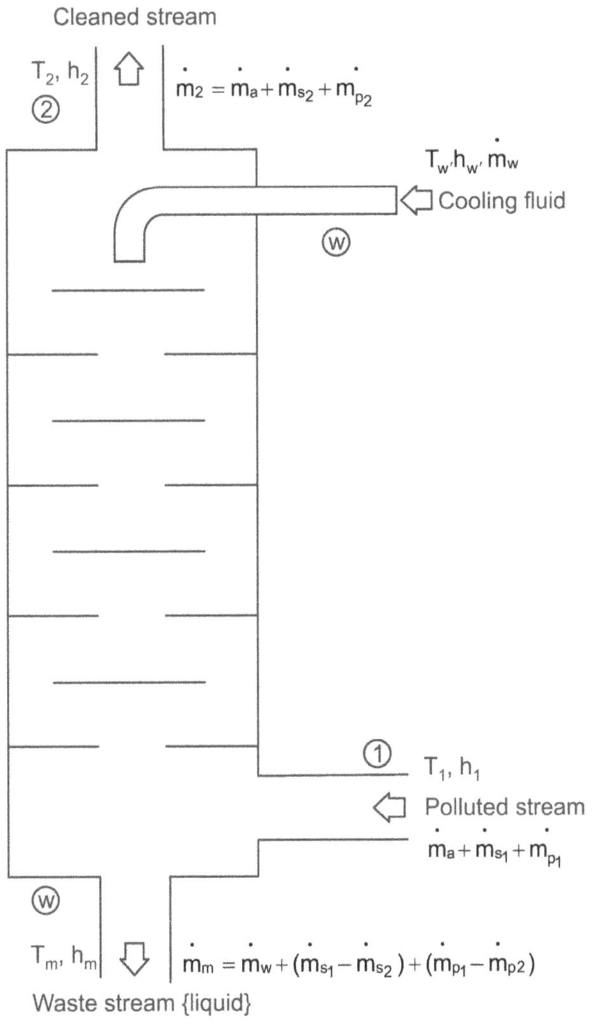

Fig. 11.5 *Direct contact condenser used as a pollution control device*

A variety of arrangements of direct contact condenser is possible; the one shown in Fig. 11.5 is a counterflow arrangement. Another common arrangement is the simple spray chamber. The cyclone scrubber and the venturi scrubber also function well as direct contact condensers.

11.4 Control by Oxidation/combustion

11.4.1 Biological oxidation

VOCs can be removed by forcing them to absorb into an aqueous liquid or moist media inoculated with microorganisms that consume the dissolved and/or adsorbed organic compounds. The control systems usually consist of an irrigated packed bed that hosts the microorganisms (bio filters). A presaturator is often placed ahead of the biological system to increase the gas stream relative humidity to more than 95%. The gas stream temperatures are maintained at less than approximately 40 °C to avoid harming the organisms and to prevent excessive moisture loss from the media.

Biological oxidation systems are used primarily for very low concentration VOC laden streams. The VOC inlet concentrations are often less than 500 ppm and sometimes less than 100 ppm. The overall VOC destruction efficiencies are often above 95%.

Biological oxidation systems are used for a wide variety of organic compounds; however, there are certain materials that are toxic to the organisms. In these cases, an alternative type of VOC control system is needed

11.4.2 Thermal and catalytic oxidation

Most VOCs when burnt produce substances which are harmless or less harmful. Thermal and catalytic oxidation are the most economical and practical ways of solving a VOC emission problem. For VOCs which include hydrocarbons (gasoline vapors, solvents, and aromatics), halogenated organics (solvents and vinyl chloride monomer), oxygenates (ketones, esters and aldehydes), and odorous compounds (amines, mercaptans, and others from the effluent treatment or food processing) the common oxidation products are carbon dioxide, water and an acidic component (HX [X–CI], Br, I) from halocarbons. SO_x from sulphur compounds; NO_x from amines, nitriles and nitrogen heterocycles; P_2O_5 from phosphorus compounds will also form.

Excess oxygen ensures that oxidation is complete and that partial oxidation products (which can be pollutants) are not formed. Scrubbing with a mild basic aqueous solution removes the acidic component in the tail streams.

11.4.2.1 Thermal oxidation

Thermal oxidation occurs by heating the polluted air to elevated temperatures (700 — 1000 °C). Thermal oxidation is a process whereby most of the VOCs are broken down and recombined with oxygen to produce water vapor and carbon dioxide. In a thermal oxidizer, the polluted air stream is heated to gas temperatures several hundred degrees

Celsius above the auto ignition temperature of the organic compounds that need to be oxidized. Due to these very high temperatures, thermal oxidizers have refractory-lined combustion chambers (also called fume incinerators) (Figure 11.6), which increase their weight and size considerably. The efficiency of oxidation and the design of most oxidizers is governed by the residence time (from a fraction of a second to more than two seconds), the combustion chamber temperature and the amount of turbulence the air stream sees. Thermal oxidizers usually provide VOC destruction efficiencies that exceed 95% and often exceed 99%. One of the main limitations of thermal oxidizers is the large amount of fuel required to heat the gas stream to the temperature necessary for high-efficiency VOC destruction. Termed regenerative thermal oxidizers (RTOs) use heat exchanger to recover some of the heat of the waste gas and to return it to the inlet gas stream. Therefore, these units, require less fuel to maintain the combustion chamber at the necessary temperature.

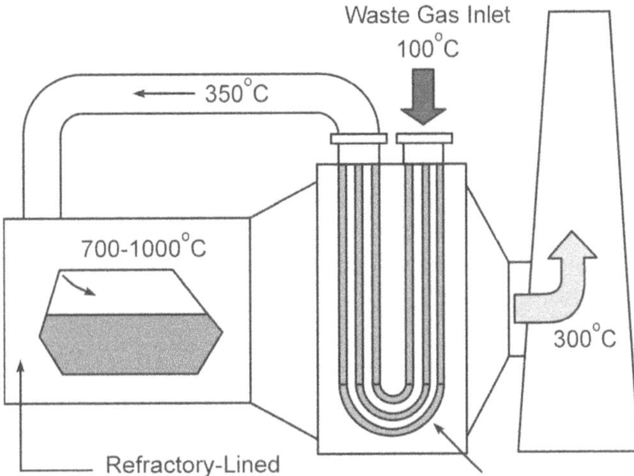

Fig. 11.6 *Thermal oxidizer with recuperative heat exchanger (I)*

Thermal oxidizers have the broadest applicability of all the VOC control devices. They can be used for almost any VOC compound. Thermal oxidizers can also be used for gas streams having VOC concentrations at the very low concentration range of less than 10 ppm up to the very high concentrations approaching 10,000 ppm.

Safety constraints impose to use thermal oxidizers for gas streams having VOC concentrations not exceeding approximately 25% of the lower explosive limit (LEL). This constrain has been introduced to be able to cope with possible short-term concentration spike that would exceed the LEL. The 25% LEL limit depends on the actual gas constituents and usually is in the 10,000 to 20,000 ppm range.

Thermal oxidizers handling VOC materials that contain chlorine, fluorine, or bromine atoms generate HCI, CIT HF, and HBr as additional reaction products during oxidation.

A gaseous absorber is used as part of the air pollution control system to collect these contaminants prior to gas stream release to the atmosphere.

11.4.2.2 Catalytic oxidation

Due to the presence of a catalyst, oxidation reactions can be performed at substantially lower temperatures (250-550 °C) than thermal oxidizers (700-1000 °C). Common types of catalysts include noble metals (i.e. platinum and palladium) and ceramic materials. VOC destruction by catalytic oxidizers usually exceeds 95% and often exceeds 99%. Due to the relatively low gas temperatures in the combustion chamber, there is no need for a refractory lining to protect the oxidizer shell. This minimizes the overall weight of catalytic oxidizers and provides an option for mounting the units on roofs close to the point of VOC generation. This placement can reduce the overall cost of the system by limiting the distance the VOC-Iaden stream must be transported in ductwork. (Figure 11.7)

Catalytic oxidizers are also applicable to a wide range of VOC-Iaden streams; however, they cannot be used on sources that also generate small quantities of catalyst poisons. Catalyst poisons are compounds that react chemically in an irreversible manner with the catalyst. Common catalyst poisons include phosphorus, tin, and zinc. Another potential operating problem associated with catalytic oxidizers is their vulnerability to chemicals and/or particulate matter that masks or fouls the surface of the catalyst. (Masking is the reversible reaction of a chemical with the catalyst and fouling is the coating of the catalyst with a deposited material). If the conditions are potentially severe, catalytic units are not installed.

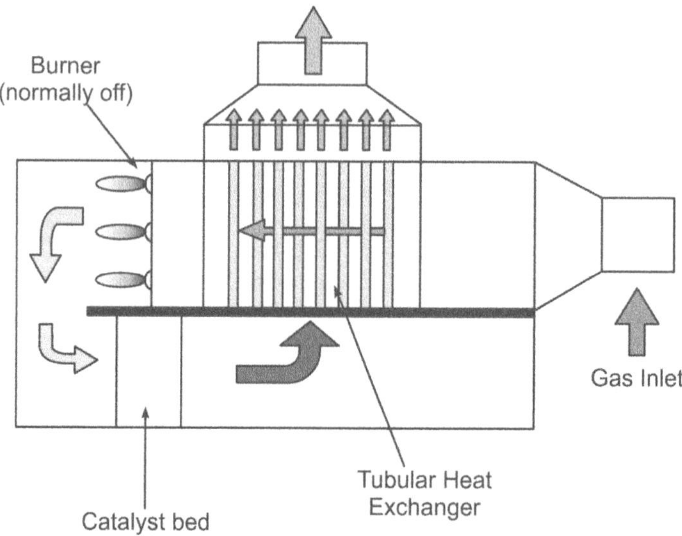

Fig. 11.7 *Scheme of a catalytic oxidation system (2)*

As with thermal oxidizers, catalytic oxidizers should not exceed 25% of the LEL, a value that is often equivalent to a VOC concentration of 10,000 to 20,000 ppm. The catalytic incineration method has become most popular because, in many case, it is more versatile and economic for the low concentrations of organic emission (i.e. < 5,000 vppm). The actual operating temperature and amount of preheat varies, depending on the organic molecule, space velocity, composition of feed (i.e., contaminants water vapor, and so forth), and organic concentration. Typical examples of operating temperatures are given in Table 11.1. One-way of comparing thermal versus catalytic abatement is to look at the energy required (air preheat temperature) to obtain quantitative removal of a given hydrocarbon. The operating temperatures shown in Table 11.1 are well below the corresponding temperatures necessary to initiate thermal (no catalytic) oxidation. The catalyst initiates reaction at lower temperatures. This demonstrates the major advantage of catalyzed processed, which is that they proceed faster than no catalytic reactions, allowing lower temperatures for the same amount of conversion. This translates directly into improved economic for fuel use and less expensive reactor construction materials, since corrosion is greatly reduced. Selection of the catalytic material for various organic pollutants has been the subject of many studies. Because metal oxides, precious metals, and combinations are used both for hydrocarbons and chlorinated hydrocarbons. As a rule, precious metals (especially platinum and/or palladium dispersed on carriers) are preferred because of their, resistance to deactivation, and ability to be regenerated.

Table 11.1 *Operating temperatures for catalytic destruction of organic compounds (2)*

Name of constituent	Chemical formula	Temperature rise 1,000 ppm (°c)	Operating temperature	Concentration before Treatment (ppm)
Styrene	COH@CHCH	138	250	310
Acetaldehyde	CH CHO	35	350	240
Benzene		103	210	380
Toluene	C6H5Cha	123	210	320
M-xylene	C6H5(CH3)	143	210	270
Phenol	C6H50H	101	300	380
Formaldehyde	HCHO	17	150	410
Acrolein	CH2CHCHO	51	180	500
Acetic acid	CH2COOH	26	350	590
Butyric acid	C3H COOH	66	20	370
Acetone	CH COCH 3	57	350	410
Methyl ethyl keton	CH COC H	74	220	380
Methyl isobutyl keton	CH COC H	116	250	270
Ethyl acetate	CH COOC H	68	350	350
Butyl acetate	CH COOC4Hg	108	350	480

Table 11.1 *Contd...*

Name of constituent	Chemical formula	Temperature rise 1,000 ppm (°c)	Operating temperature	Concentration before Treatment (ppm)
Methyl alcohol	CH OH	21	150	830
Ethyl alcohol	C2H50H		350	550
Isopropyl alchol	caH,0H	64	280	230
Butyl alcohol	C4H90H		260	330
Carbon monoxide	Co	9	150	4,000
Methyl cellosolve	HOCH CH OCH	55	300	110
Ethyl cellosolve	HOCH2CH20C2H	76	300	
Buty cellosolve	HOCH2CH20C4H9	118	300	50

11.4.2.3 Direct Fired Thermal Oxidizers (DFTOS)

The simplest Thermal Oxidizer is a Direct Fired unit (sometimes referred to as an After-Burner) that employs no heat recovery. In this system a fuel burner (mostly natural gas fired) raises the temperature of the pollutant-laden air to a predetermined combustion temperature. In order to achieve a high level of hydrocarbon destruction, the heated air is kept at the combustion chamber set point for a certain minimum time, called the residence or dwell time. In addition to temperature and dwell time, turbulence also plays an important role in making oxygen and hydrocarbon molecules to interact more vigorously. Since DFTOs employ no heat recovery, they are most often applied to very low volume air streams, usually with very high concentrations of VOCs, HAPs, and other pollutants. These oxidizers are characterized by specialized burners to insure mixing of combustion air and low volume, high concentration inert airstreams, which are often injected directly into the burner ports or directly in the flame cone via lances. This minimizes the risk of explosion and takes maximum advantage of the enriched fuel value of high VOC content exhaust streams.

11.4.2.4 Recuperative Oxidizers

A Recuperative Oxidizer is a Direct Fired unit that employs integral primary heat recovery. To minimize the energy consumption of the oxidizer, the hot air exiting the combustion chamber is passed over an air-to-air heat exchanger. The heat recovered is used to preheat the incoming pollutant laden air. The primary heat exchangers are usually supplied as either a plate-type or a shell and tube type heat exchanger. These heat exchangers can be designed for various heat transfer efficiencies, but the nominal maximum is 70%. Thus by the addition of a heat exchanger, the net heat load on the burner can be reduced by up to 70% of that required in a DFTO. The addition of the heat exchanger, because it is made of heat corrosion resistant alloy, substantially increases the cost of the oxidizer system. Also, the fan for moving the polluted gas through the oxidizer must be more powerful to overcome the additional pressure drop of the heat exchanger. In most cases, the savings in fuel will more than offset the additional up-front cost within the first two years of operation, however, even with 70% heat recovery, recuperative oxidizers can be expensive to operate, especially if the airflow is large and has dilute concentration levels, unless additional secondary heat recovery can be applied to the customer's process.

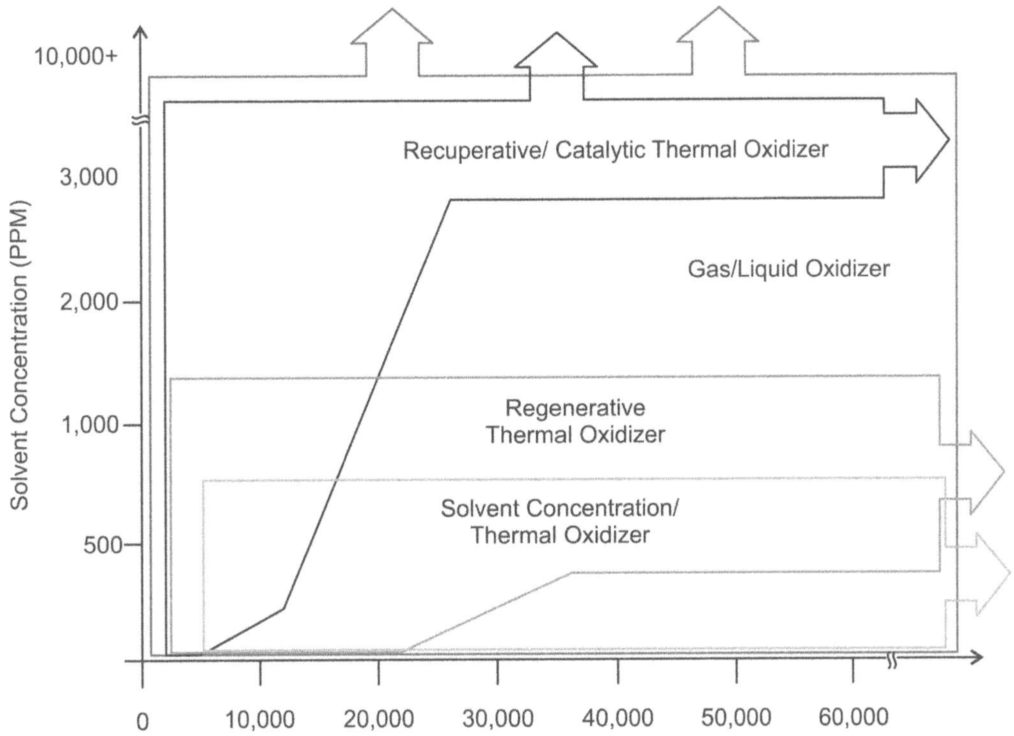

Fig. 11.8 *Chart for choosing equipment for specific applications*

Sources: *http://www.durrenvironmental*

11.4.2.5 Regenerative Thermal Oxidizers (RTOs)

A Regenerative Oxidizer is also a Direct Fired oxidizer that employs integral primary heat recovery. However, the RTO operates is periodic, repetitive cycle rather than a steady state mode. Instead of conventional heat exchangers, which indirectly transfer heat from hot side to cold side across the exchanger walls, RTOs use a store and release mechanism. The nature of a RTOs heat recovery process requires it to have at least two beds of appropriate heat recovery media. In many applications, the additional step of purging a bed before reversing the flow through it from inlet to exhaust is necessary to maintain very high destruction efficiencies. This purge step creates the requirement for an additional (or odd number) chamber making the RTO more complicated and more expensive than a recuperative oxidizer. RTO systems can utilize more than two beds (operating in parallel) in order to be capable of handling larger air volumes. The primary advantage of an RTO is lower operating costs due to high heat recovery and low fuel consumption. Depending on the mass of media included in an RTO, heat recoveries of up to 95% are common. Because of their capability for high heat recovery, RTOs are often operated in an "auto-thermal" or self sustaining mode, where the heat content of the VOCs being oxidized is enough to sustain the combustion chamber temperature at set point, requiring no external fuel input.

11.5 Selection of Suitable VOC control Technology-Chart

Choosing the right equipment for VOC control applications depends primarily on the exhaust air volume and the average concentration of VOCs. The chart displayed (Figure 11.8) provides general guidelines for choosing equipment to fit particular applications.

11.5.1 VOC Control Technologies–Specific applications to boilers, furnaces, flares

Control of volatile organics from coal-burning boilers, furnaces, etc. from unburned coal or products of incomplete combustion is one of the main air pollution problems. By proper control of sufficient excess air (Fig. 11.9) and adequate mixing between the burning coal, its decomposition products, and the air, this problem can be solved.

As can be seen in Fig. 11.9 for complete combustion with imperfect mixing, one must supply excess *air* in addition to that needed for stoichiometric combustion [4].

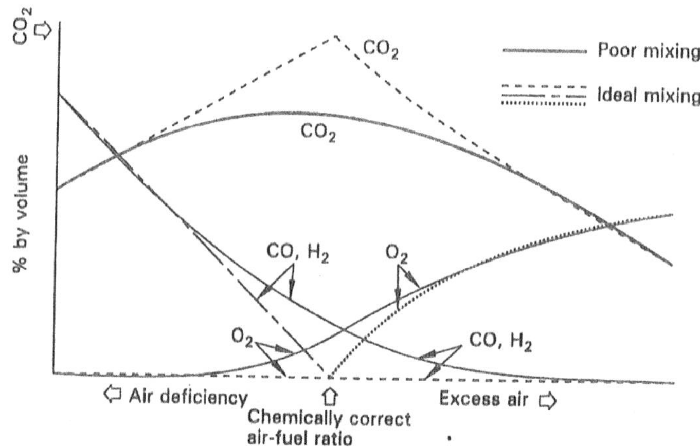

Fig. 11.9 *Effect of air-fuel ratio and quality of mixing on the composition of combustion gases*

The typical way of carrying out the combustion of VOCs by incinerator is shown in Fig. 11.10(a). The contaminated gas stream is mixed with fuel. If the contaminated gas stream does not contain enough oxygen to burn the fuel, additional air will also be supplied. The burning of the fuel takes place in a combustion chamber, where the high velocities of the inlet streams provide good turbulent mixing. From the combustion chamber the gases pass to an insulated retention chamber, where they remain long enough at high temperatures for the reactions to complete the destruction of the VOCs in the combustion chamber after which the hot gases are removed.

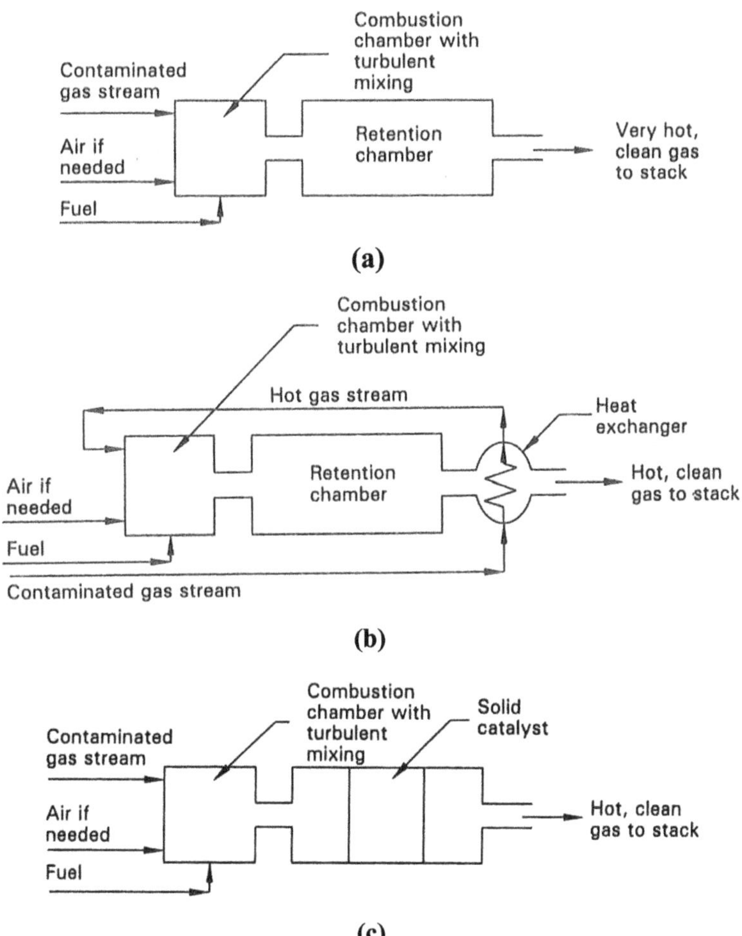

Fig. 11.10 *Arrangements for destroying VOC in a gas stream by incineration: (a) simple thermal incinerator, (b) thermal incinerator with regenerative heat recovery, and (c) catalytic incinerator*

The biggest drawback in this method is the high cost of the fuel. If the contaminated gas stream contained enough VOC to burn (i.e. the VOC concentration is above its Lower Explosive Limit) then no additional fuel would be needed, and the waste stream would be a valuable fuel source instead of an air pollution control problem. For air pollution control we almost always have to add fuel. One way to lower the fuel cost is to put a heat exchanger into the system, as shown in Fig. 11.10(b). There the hot, clean gas leaving the retention chamber transfers heat to the incoming contaminated gas stream, thus reducing the outlet temperature and reducing the amount of fuel needed to bring the mixture of fuel, air and the contaminated gas stream to the temperature at which the

VOC-destruction reactions proceed. Unfortunately, the hot-gas-to-cold-gas heat exchangers shown in Fig. 11.10(b) are expensive and often have severe corrosion problems. They are used in some situations, but not all. The second modification of the basic idea is to put an oxidation catalyst in the retention chamber.

The oxidation of a VOC is exothermic. However, because a VOC is present only as a dilute component, the reaction heat is insufficient to maintain me temperature required for oxidation. Therefore, additional heating is required, or a catalyst must be used that enables total oxidation at lower temperatures (Fig.11.10(c)).

The most commonly used catalyst is platinum metal as the active ingredient. The substrate for the catalyst is prepared by making porous ceramic pellets or a porous ceramic honeycomb with a high surface area. That ceramic is then immersed in a dilute solution of a platinum salt and subsequently dried, leaving the platinum salt behind on the ceramic support.

Most catalytic operations also require elevated temperatures to ensure a complete reaction at a suitably fast rate. Therefore, either the catalyst bed is heated, or the feed stream is preheated. Inlet gas streams are kept 50–150°C higher than the ignition temperatures. When the preheating uses an internal flame, the process is complex, involving both the products of combustion in the flame and processes occurring at the catalyst. The operating parameters (temperature and space velocity) for an application depend on the destructibility of the VOC.

In some cases, destruction of all VOCs and intermediates requires a higher operating temperature than that required to combust a single VOC (Fig. 11.11).

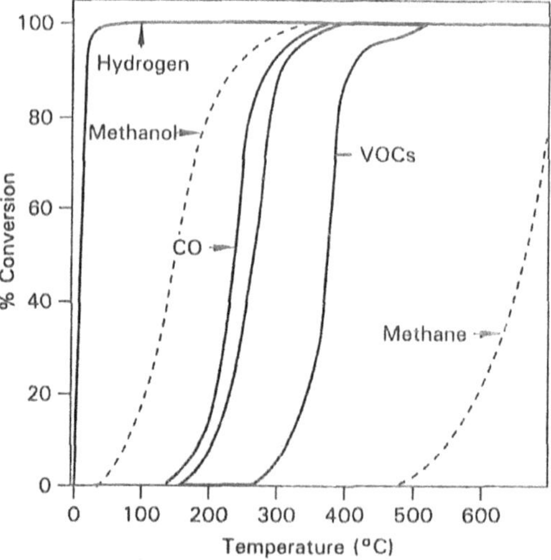

Fig. 11.11 *Conversion curves for VOC destruction using a catalytic incinerator. (Data from Brown Engineering, Seattle, Wash; Johnson Matthey, Catalytic System Division, Wayne, Pa.)*

The waste gases from several chemical, printing or related industries contain mixtures of halogenated and nonhalogenated VOCs. Converting each VOC requires a combination of catalysts. Further, scrubbing the effluent from the reactor is necessary to remove the acidic components. For specific applications, one or more catalyst types can be effective. The heat of combustion of values of the various VOCs, in gaseous effluents, explosive limits of combustion of VOCs, the thermodynamic equilibrium and the kinetics of the combustion of VOCs [4] reactions have to be taken into account in the application of the combustion process for the destruction of volatile organics. Care has to be taken for maintaining optimum parameter temperature and air-fuel ratio without which more harmful products may be discharged into the atmosphere.

Table 11.2 summarizes the various applications of thermal or catalytic oxidation for destroying or reducing VOCs.

Table 11.2 *Various applications of thermal /catalytic oxidation for VOCs*

Sources of pollution	Types of pollution	Oxidation catalysts
Chemical manufacture -plant operations -petrochemicals -storage	**Solvents** -hydrocarbons -aromatics ketones -esters -alcohols	**Metals/supported metals** Metal = Ag Pt Pd Ru Rh Ni
Chemicals use -coating processes -electronics industry -furniture manufacture -painting -dry cleaning -paper industry -printing	**Odor control** -amines -thiols/ thioethers -heterocyclics -aldehydes -acids -food by-products -smelter off gases	Support = SiO_2 Al_2O_3 Zeolite ThO_2 ZrO_2 Polymers
Environmental management -landfill -hazardous waste management	**Other effluents** -monomers -plasticizers **-CO** -Formaldehyde	**Metal oxides** MgO V_2O_5 Cr_2O_3 MoO_3 Pr_2O_3 Fe_2O_3 MnO_x NiO TiO_2 Co_3O_4 La_2MO_4 (M = Cu, Ni) $CoMoO_4$ Mixed oxides

Table 11.2 ***Contd...***

Sources of pollution	Types of pollution	Oxidation catalysts
Food/Beverage -breweries -bakeries -food processing -ovens **Energy** -power generation -stationary engines -diesel trucks		

11.6 Catalytic Converters for Automobiles for Reduction of VOCs from Transportation Sector

11.6.1 VOCs from automobiles

Air pollution generated from mobile sources is a problem of great interest. In the last sixty-years the world vehicle fleet has increased tremendously reaching nearly 1000 million (2). The environmental concern originated by mobile sources is due to the fact that the majority of engines employ combustion of fuels derived from crude oil as a source of energy. Burning of hydrocarbon ideally leads to formation of water and carbon dioxide, however, due to no perfect combustion control and high temperatures reached in the combustion chamber, the exhaust contains significant amounts of pollutants which need to be transformed into harmless compounds.

11.6.2 Automotive emission characteristics and regulations

Automobile emissions have a huge impact on human health and the well-being of our planet. Not only do they lead to elevated ozone levels, but the CO emitted by engines is toxic. Also, both CO_2 and ozone are greenhouse gases. After the link was made between automobiles and air pollution, and after overcoming strong opposition from the automobile industry, regulations to control automobile emissions were finally enacted in the U.S. when Congress passed the Clean Air Act in 1970. Since then, the Environmental Protection Agency (EPA) (US) has established emission standards that automobile manufacturers must abide by when producing cars, and these regulations have become more stringent over time as pollution abatement technology has improved.

Engine manufactures have explored a wide variety of technologies to meet the requirements of the Clean Air Act. Catalysts has proven to be the most effective passive system. The major worldwide suppliers of automotive catalysts are Engelhard, Johnson Matthey, Degussa, and Allied Signal. As the automobile engine has become more sophisticated, the control devices and combustion modifications have proven to be very compatible with catalyst technology, to the point where today engineering design incorporates the emission control unit and strategy for each vehicle.

The need to control engine emissions according to Clean Act Air, recent amendments have set up more stringent requirements for automotive emissions (2). New catalysts are required to last 100,000 miles for new automobiles after 1996. Furthermore, the amendments (contingent on tier 2 standards to be set by environmental protection agency) reduce no methane hydrocarbon (NMHC) emissions to a maximum of 0.125 g/mile by 2004 (down from 0.41 g/mile in 1991), carbon monoxide to 1.7 g/mile (down from 3.4 g in 1991), and nitrogen oxides to 0.2 g/mile (down from 1.0 g).

To meet these regulations new catalytic converters are developed by automobile industries

11.7 Control of Hydrogen Sulphide

11.7.1 Introduction

Sulfur occurs in several oxidative states, these include Hydrogen Sulfide, elemental sulfur and sulfate. Sulfate is typically the form of sulfur found in the aquifer. Sulfate has no odor and is highly soluble in water in various sulfate compounds. Sulfur reducing bacteria can reduce sulfate converting it into hydrogen sulfide in an environment devoid of oxygen. Since drinking water aquifers contains no oxygen, when a carbon source is introduced to a ground water containing sulfate, sulfide conversion can be produced by biological activity. The higher in the aquifer from which the water is withdrawn, the more prevalent will be the organic and biological sources necessary for this conversion. The chemical equation for sulfide conversion from sulfate is shown below:

$$SO_4^{-2} + 2C_{\text{organic}} + 2H_20 + \text{microbial activity} \rightarrow H_2S + 2HCO^{-3}$$

Hydrogen Sulfide is a very volatile dissolved gas and readily escapes as a gas into the air causing unpleasant odors. At pH of 7 both forms of Sulfide are present in equal concentrations. Downwardly adjusting pH to around 6 will result in converting all of the bisulfide ion, to the volatile form. Conversely, raising the pH to above 9, will convert the bisulfide ion to the sulfide ion. These relationships are shown in the figure 11.12.

11.7.2 Sources of H_2S

Hydrogen sulfide (H_2S) is a flammable, colorless gas with a characteristic odor of rotten eggs. Hydrogen sulfide occurs both naturally and from human processes. It is released from volcanoes, sulfur springs, undersea vents, swamps, salt marshes, and stagnant bodies of water, and is found in association with crude petroleum and natural gas. Hydrogen sulfide is also associated with municipal sewers and sewage treatment plants, landfill gases, manure handling operations, and pulp and paper operations. The levels of hydrogen sulfide in air are typically low. The amount of hydrogen sulfide naturally found in the air has been estimated as 0.11-0.33 ppb (0.15-0.46 ug/m^3). Lower levels (0.02-0.07 ppb; 0.03-0.1 ug/m^3) have been observed in some remote area. Concentrations closer to sources of hydrogen sulfide are higher. For example, annual H_2S concentrations of ~6 ppb and 24-hour concentrations as high as 74 ppb have been reported near pulp mills in Imatra, Finland (15). Hydrogen sulfide from manure at dairies has been measured at 20-40 ppb (27.8-55.6 ug/m^3; (16). Hydrogen sulfide is also a constituent of landfill gas emissions, accounting for up to 1% by volume (17, 18).

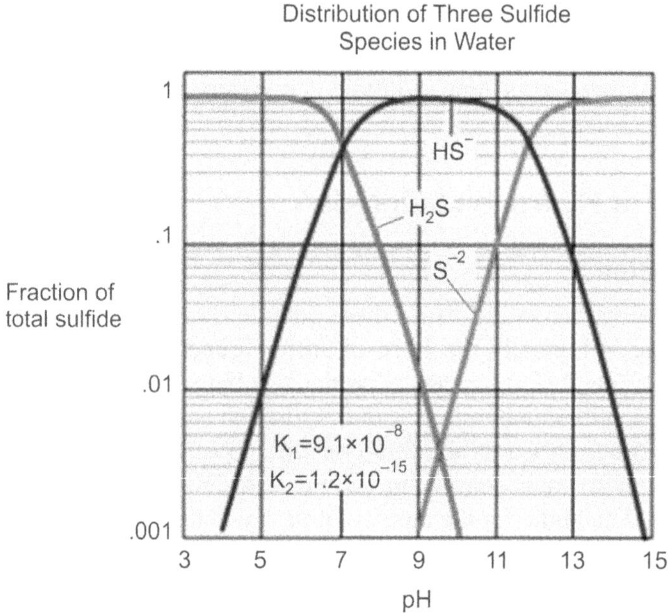

Fig. 11.12 *Distribution of Sulphide species in water at different pHs*

11.7.3 Threshold levels of hydrogen sulfide in ambient air

Concentrations that substantially exceed the odor detection threshold can result in annoying and discomforting physiological symptoms of headache and nausea (19). Several studies have been conducted to establish the ratio of discomforting annoyance threshold to detection threshold for unpleasant odors (20). The geometric mean ratio from these studies is 5; based on an odor detection threshold of 8 ppb, the odor annoyance threshold would be 40 ppb. CalEPA estimated that ~83% of the general population should be able to detect the odor of hydrogen sulfide at 30 ppb (41.7ug/m^3). The World Health Organization (WHO) recommends that in order to avoid substantial complaints about odor annoyance, 30-minute average hydrogen sulfide concentrations should not exceed 5 ppb (7 ug/m^3;(21).

11.7.4 H_2S Treatment technologies

H_2S rich gases are normally treated to recover sulphur by oxidizing H_2S to SO_2 and then reacting H_2S with SO_2 over an alumina based catalyst.

$$2H_2S + SO_2 = 3S + 2H_2O$$

The typical flow sheet for Claus process is shown.

Figure 11.13 Claus Process (1, 2)

H_2S is initially removed from a source, such as sour natural gas by dissolution in an alkanolamine solvent. However, for low concentrations of H_2S in the air, this method is not sufficiently effective for total removal of H_2S. Instead, the H_2S is either oxidized completely to SO_2, and then the system removes the SO_2 by scrubbing or capture, or it is captured by reaction with solid or liquid phase adsorbents. For stationary units, any liquid or solid base reacts with H_2S. Magnetite (Fe_3O_4), limestone ($CaCO_3$), lime (CaO), zinc carbonate ($ZnCO_3$), and zinc oxide (ZnO) are each effective, and Fe_3O_4, CaO and ZnO are used commercially. The recovery of H_2S from hydrocarbon processing using an amine as the solvent is a major application of regenerative absorption. Figure 11.13 shows a typical flow sheet.

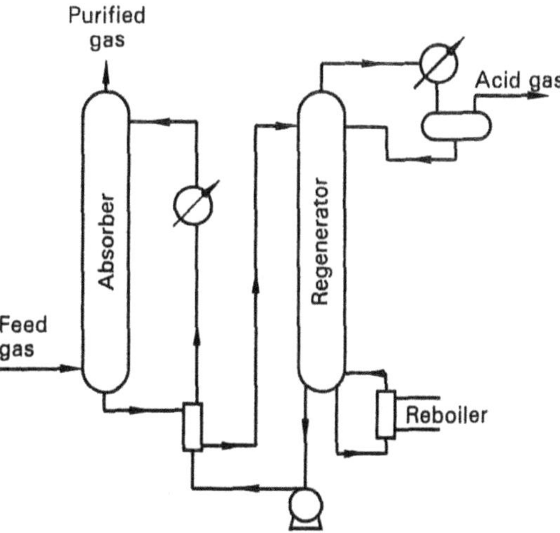

Fig. 11.13 *Flowsheet for recovery of H_2S using amine as a solvent (1, 2)*

For the capture of H_2S and volatile sulphur compounds (e.g. thiols, COS, and CS_2) released during the transportation of sour liquids, a portable system is used. The spent solution oxidizes to form ammonium sulphate, which is recovered from the spend absorbent and can be marketed as fertilizer. Alternate absorbents for portable units include basic hypochlorite (bleach) solutions, but the stability of the solution decreases with consumption of the base as the H_2S is adsorbed. A complicating factor is the use of basic fluid absorbent system is that CO_2 is also a weak acid and is absorbed to sequentially form bicarbonate (HCO_3^-) and carbonate (CO_3^{2-}) ions, thereby consuming the absorbent.

The absorption of SO_2 by dimethylaniline from off gases at copper, zinc, lead, nickel smelters is another application. The desorbed SO_2 produces H_2SO_4. The SO_2 concentration in the off-gases is approximately 10%.

For non regenerative or throwaway systems, irreversible reactions occur in the liquid phase. The reagents in the solvent are consumed and must be replenished for absorption efficiency. The resulting products are discarded. Typical examples are the absorption of acid gases by caustic solutions. Non regenerative systems are economical only when the reagents are inexpensive or the volume involved is small. Table 11.3 lists the major absorbents for H_2S the main advantage of using these solutions is to reduce high H_2S concentrations in gas streams.

Table 11.3 *The various absorbents for H_2S*

Liquid absorbents	Trade or Generic names	Source	Comments
Alkanolamines			
Monoethanolamine	MEA	Dow, USA	Moderately selective
Diethanolamine	DEA	Dow, USA	Moderately selective
Triethanolamine	TEA	Dow, USA	Selective
N-methyldiethanolamine	MDEA	Dow, USA	Selective
Diisopropanolamine	DIPA	Dow, USA	Selective
Aq. Potassium carbonate	Hot carbonate		Moderately selective
Aq. Tripotassium phosphate			Selective
Alkazid process			
Dialkylglycine sodium salt	DIK Solution		Selective
Sodium alanine solution	M solution		Moderately selective
Sodium phenolate solution	S solution		Not selective
Aq. Ammonia			
+ air/hydroquinone	Perox		Recovers as Sulphur
+ air	Am – gas	Am- gas Canada	Recovers as $(NH_4)_2$ SO_4
Aq. Sodium carbonate			
+ quinoline sulphonate salt and sodium metavanadate	Stretford	W.C.Holmes and Company UK	Recovers as Sulphur
+ naphthaquinone sulphonate salt	Takahax	Tokyo Gas Co. &Nittetsu Chemical Engineers Co.	Recovers as Sulphur
Aq. Zinc acetate			Precipitates as ZnS
Aq. Calcium acetate + heat to decompose product			$Ca(SH_2)$ soluble; CaS precipitates
Iron oxide + controlled Reoxidation	Dry box		Sulphur recovery, for ppm H_2S streams
Zinc oxide, Zno			Useful for ppm H_2S; only mod. Selective
Calcium oxide, CaO	Lime		Product releases H_2S on decomposition Oxidizes to sulphates
Copper Cu (or other metals)			Reagent not adsorbent; forms CuS and H_2 but requires expensive regenerations

Questions

1. Write a short note on VOC treatment technologies
2. Explain in detail the control measures for VOCs by reduction of leakages.
3. Discuss technologies used for reduction of VOC emissions.
4. Write a short note on methods (i) Adsorption (ii) Condensation and (iii) Combustion
5. Discuss types of condensers used in VOCs treatment
6. Write a short note on (i) Thermal oxidation (ii) catalytic oxidation (iii) Biological oxidation
7. Write a short note on VOCs treatment technologies and explain (i) Direct fire Thermal oxidisers (ii) Recuperative oxidisers (iii) regenerative thermal oxidisers
8. Write a short note on oxidation of VOCs and explain (i) thermal oxidation (ii) Catalytic oxidation
9. Short note on Catalytic converters used in automobiles.
10. Describe Claus process for treatment of H_2S removal.
11. Discuss various Adsorbents used for H_2S removal.

References

1. J. R. Fair, 1969. Sorption process for gas separations. Chem. Engg. July 14. 76(15): 90-110.
2. Air Pollution Control Engineering, de Nevers N., second Edition, 2000, McGraw Hill, New York.
3. K. E. Noli, V. Gouaris and W.S. Lauri, 1992. Adsorption technology for air water pollution. Lewis Chelsea, MI.
4. W. S. Smith and C. W. Grubus, 1960. Atmosphere emissions from coal combustion, an inventory guide. U.S. Department Health Education and Welfare. Public Health and Series Publication. NOAP-24:24.
5. R.H. Barnes, M.I. Saxtar, R. E. Barrett and A. Levay, 1979. Chemical aspects of afterburner systems. U.S. EPA report EPA– 600/7–79–096, US Govt. Printing Office, Washington DC.
6. Dilbeck, C. W.; Ganske, J. A. Chem. Educator 2008, 13, 1-5.
7. Medhurst, L. J. J. Chem. Educ.2005, 82, 278-281.
8. R. M. Heck and R. J. Farrauto, Catalytic Air Pollution Control: Commercial Technology, Van Nostr and Reinhold, New York, 1995.
9. D. A. Skoog, F. J. Holler and T. A. Nieman, Principles of Instrumental Analysis, 5th Edition, Harcourt Brace, Philadelphia, 1998, pp. 355-371.
10. H. H. Willard, L. L. Merritt, Jr., J. A. Dean and F. A. Settle, Jr., Instrumental Methods of Analysis, 6th ed., Van Nostr and New York, 1989, Ch. 4.

11. U.S. EPA website on air emissions: http://www.epa.gov/air/emissions/index.htm
12. California Air Resources Board website: http://www.arb.ca.gov/homepage.htm
13. www.epin.ncsu.edu
14. Buscom.co
15. www.engelhard.com
16. Parti-Pellinen K, Marttila O, Vilkka V, Jaakkola JJK, Jappinen P., Haahtela T 1996. South Karelia air pollution study: Effects of low-level exposure to malodorous sulfur compounds on symptoms. Arch. Environ Health 51(4):315-320.
17. Donham KJ, Knapp LW, Monson R and Gustafson K 1982. Acute toxic exposure to gases from liquid manure. J Occup Med 24:142-5.
18. ATSDR 1999. Toxicological profile for hydrogen sulfide. US Department of Health & Human Services. Public Health Service, Agency for Toxic Substances and Disease Registry.
19. ATSDR 2004. Draft toxicological profile for hydrogen sulfide update. US Department of Health & Human Services. Public Health Service, Agency for Toxic Substances and Disease Registry.
20. A moore JE 1985. The perception of hydrogen sulfide odor in relation to setting an ambient standard. Olfacto-Labs, Berkeley, CA: prepared for the CA Air Resources Board
21. CalEPA 1999. Determination of acute reference exposure levels for air borne toxicants: hydrogen sulfide acute toxicity summary. http://www.oehha.ca.gov/air/acute rels/pdf/7783064A. pdf
22. WHO, 1981. Hydrogen Sulfide. World Health Organization, Environmental Health Criteria No. 19. Geneva.

CHAPTER 12

Air Pollution by Automobiles and its Control Technologies

12.1 VOCs from Automobiles

Transport is the primary source of mobility in urban society. The problems of air pollution by vehicles is mostly confined to urban areas. Two-stroke petrol motor vehicles, scooters and such other two-and three-wheelers are the major contributors to the vehicular carbon monoxide and hydrocarbon pollution in India. The transport sector accounts for a large proportion (half or more) of the petroleum used in many countries. Air pollution generated from mobile sources is a problem of great interest. In the last sixty-years the world vehicle fleet has increased tremendously reaching nearly 1000 million. The environmental concern originated by mobile sources is due to the fact that the majority of engines employ combustion of fuels derived from crude oil as a source of energy. Burning of hydrocarbon ideally leads to formation of water and carbon dioxide, however, due to non perfect combustion control and high temperatures reached in the combustion chamber, the exhaust contains significant amounts of pollutants which need to be transformed into harmless compounds.

Air pollution caused by emissions from motorized vehicles is a serious environmental issue. The principal emissions from motorized road vehicles equipped with spark-ignition petrol engines are given in Table 12.1. For proper understanding of the air pollution control methods for automobiles it is necessary to examine the fuel parts of the following widely used engines: (a) Petrol or gasoline engines b) Diesel engines

12.1.1 Petrol engines

An internal combustion engine operates in four stages. In an air-tight cylinder in a block of engine 'head' a piston operates with oil lubrication. There is an inlet valve to let in the fuel and air mixture and an outlet valve that allows gases to escape and an electrically operated spark plug. The working is depicted in Fig. 12.1.

On the intake stroke (1), the piston moves down and a mixture of fuel and air is drawn into the cylinder past the open intake valve. With the compression stroke (2), the intake valve closes and the piston moves and compresses the, air-fuel mixture. On the power stroke (3), a spark from the spark plug ignites the heated, compressed mixture, which begins to burn, expands, and pushes the piston down. For the exhaust stroke (4), the exhaust valve opens, the spent, burned mixture exits with its pollutants, and the piston returns to the top of the cylinder. [2]

Table 12.1 *Air pollution from passenger vehicles [1]*

Transport	CO_2 Compounds (g/pass km)	Organic Compounds (g/pass km)	CO (g/pass k m)	NO_x (g/pass km)	SO_2 (g/pass km)
Truck, petrol					
Single occupancy	1.55	3.20	27.46	2.05	0.23
Average occupancy car	0.81	1.68	14.45	1.08	0.12
Car					
Single occupancy	1.12	2.57	20.36	1.61	0.14
Average occupancy	0.68	1.51	11.98	0.85	0.08
Vehicle ride share					
3-person car pool	0.37	0.86	6.79	0.54	0.05
4-person car pool	0.28	0.64	5.09	0.40	0.03
Bus (diesel)					
Transit	0.39	0.25	1.21	1.82	N/A
Rail intercity					
Diesel	0.43	1.12	0.6	0.9	0.51
Electric	0.26	-	0.05	1.1	2.07
Commuter (diesel)	0.53	1.04	1.44	4.10	0.63
Transit (electric)	0.37	-	0.06	1.48	2.89
Aircraft	0.57	0.5	0.52	1.08	0.08
Bicycle	0	0	0	0	0

The piston will be connected to a crankshaft by a connecting rod and the up and down movement of the piston will be converted to circular motion by the crankshaft which is also connected to a flywheel. With the piston at the top of the cylinder, the momentum of the fly wheel pushes the piston down, creating a vacuum. The inlet valve opens allowing the fuel-air mixture into the cylinder. The outlet valve will be closed at this time. As the piston reaches the bottom the inlet valve closes. In the upward movement of the piston, again deriving the movement from the flywheel, only the fuel-air mixture gets compressed since both the valves are closed. As the piston reaches the maximum compression position, the spark plug ignites. The fuel-air mixture suddenly explodes, pushing the piston down. This is the power stroke. The amount of energy released by this pushing is absorbed by the flywheel. As the piston goes up again deriving the energy from the flywheel, the outlet valve opens allowing the exhaust gases to go out of the cylinder. These four operations constitute the four aspects of the operation of a single cylinder. It should be observed that energy is generated and transmitted only in the ignition stroke. The other three movements of the cylinder are from the stored energy from the flywheel. If a single cylinder operation is selected, the movement of the crankshaft is jerky. If two cylinders are there, they can operate in a sequential way i.e. when one cylinder is having ignition stroke another may be having inlet suction stroke.

Similarly when the pistons are moving up one can be having a compression stroke while the other is having an exhaust stroke. In this process the crankshaft will receive two ignition strokes making the movement slightly smoother.

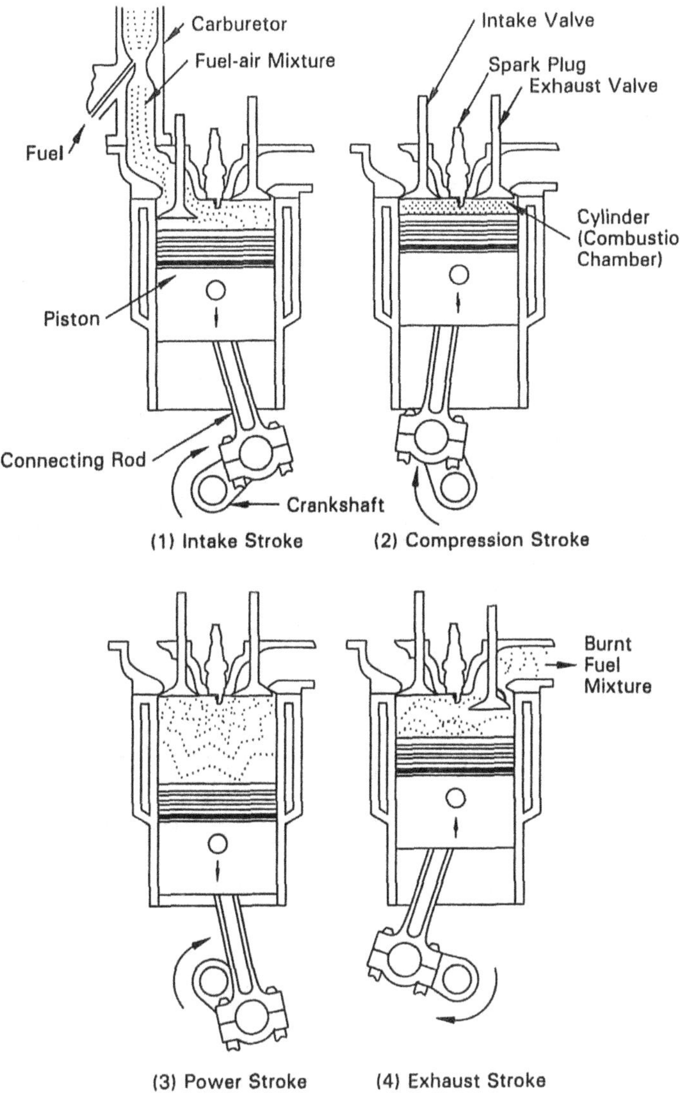

Fig. 12.1 *Combustion in an automobile petrol engine [2]*

The spark plugs on each of the cylinders operate with the help of a distributing mechanism in such a way that only at the end of the compression stroke of the compression cylinder the spark ignites. The spark plug operates at a very high voltage so that the electrical impulse can jump giving a spark which ignites the fuel mixture. The extent of combustion depends on the strength of the spark and the proportion of the fuel-air mixture. The fuel-air mixture will be prepared in the carburettor so that the mixture is obtained in a particular proportion. This will be part of the tuning of an engine. It is also essential that the spark plug discharges the spark exactly at the right time. Petrol-driven vehicles emit maximum amount of hazardous gaseous pollutants (CO, HC, etc.). On the other hand diesel-powered vehicles emit large amount of SPM and also aldehydes.

12.1.2 Diesel engines

As shown in Fig. 12.2 the diesel engine differs from the four-stroke petrol engine in three respects.

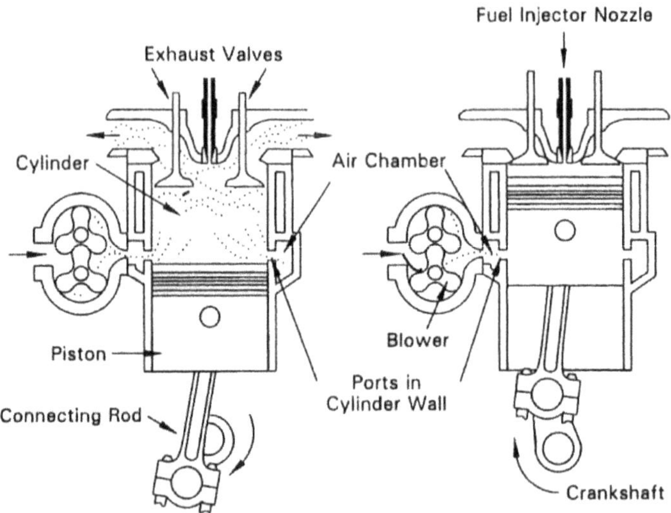

Fig. 12.2 *Combustion in a diesel engine*

First, the air supply is unthrottled; that is, its flow into the engine is unrestricted. Thus, a diesel normally operates at a higher air-fuel ratio than does a petrol engine.

Second, the fuel is injected directly into the combustion chamber, and a carburettor is not required. The power output is changed by the rate of fuel injection.

Third, there is no spark ignited system. The air is heated by compression. That is, the air in the engine cylinder is squeezed until it exerts a pressure high enough to raise the air temperature to about 540°C, which is enough to ignite the fuel oil as it is injected into the cylinder.

A well-designed, well-maintained, and properly adjusted diesel engine will emit less CO and hydrocarbons than the four-stroke engine because of the diesel's high air-fuel ratio. However, the higher operating temperatures lead to substantially higher NO_x

emissions. The exhaust emissions from the petrol engine consist of mainly CO, unburnt hydrocarbons, oxides of nitrogen and partial oxidation products of the aldehyde family. In addition, particulate matter in the form of lead compounds and carbonaceous matter are also emitted. On the other hand, exhaust emissions from diesel powered vehicles have low carbon monoxide and unburnt hydrocarbons, while nitrogen oxide is present in high concentrations. Besides these emissions certain smoke particles, oxygenated hydrocarbons including aldehydes and odour producing compounds are also formed.

12.2 Origin of Air Pollutants in Automobiles

12.2.1 Carbon monoxide

It is an intermediate product of the automobile petrol combustion. Its concentration depends mainly on the air-fuel ratio. The correct chemical ratio between mass of air to mass of petrol is 14.5: 1. If there is not enough air and insufficient time all the carbon in fuel cannot be burnt to carbon dioxide and some of it forms carbon monoxide. At low air-fuel ratios, very little oxidation takes place resulting in higher carbon monoxide emissions. The effect of the air-fuel ratio on CO formation is quite low in compression ignition engines, though it does influence CO formation in these engines also.

12.2.2 Nitrogen oxides (NO_x)

Chemistry of Combustion of hydrocarbons in automobiles and production of NO_x

Automobile engines convert the energy stored in chemical bonds into mechanical energy through the controlled combustion of gasoline (hydrocarbons, C_xH_y) in air:

$$C_xH_y + O_2 \rightarrow CO_2 + H_2O + \text{heat} \ldots\ldots\ldots \qquad \ldots..(12.1)$$

When there is the correct balance of O_2 and hydrocarbons in the combustion chamber (i.e., under "lean" conditions), complete combustion occurs and CO_2 and H_2O are emitted in the exhaust. When the mixture is not in balance, such as the case when there is too much fuel and not enough oxygen (i.e., under "rich" conditions), combustion is incomplete and the exhaust can contain CO and excess hydrocarbons. Since both nitrogen and oxygen are present in air used for combustion, the following reactions may also take place within the engine, especially when the combustion temperature is high:

$$O_2 + \text{heat} \rightarrow 2\,O\cdot \qquad \ldots..(12.2)$$

$$O\cdot + N_2 \rightarrow NO\cdot + N\cdot \qquad \ldots..(12.3)$$

$$N\cdot + O_2 \rightarrow NO\cdot + O\cdot \qquad \ldots..(12.4)$$

$$2\,NO\cdot + O_2 \rightarrow 2\,NO_2 \qquad \ldots..(12.5)$$

Nitrogen oxides are formed in the combustion chamber where the temperature is as high as 2500 °C. This high temperature makes O_2 and nitrogen combine resulting in the formation of nitric oxide (NO) which when exhausted into the atmosphere gets oxidized partially to NO_2. The formation of NO_x in a spark ignition engine is governed by two factors: [3-5]

(a) The availability of oxygen for the reaction

(b) High combustion temperature

NO formation in a compression ignition engine is also influenced by combustion pressure, temperature and time available for combustion and the availability of oxygen in the air-fuel ratio. On the other hand the maximum combustion temperatures are obtained at air-fuel ratios slightly richer than normal. Both nitric oxide and nitrogen dioxide are significant air pollutants. A high octane number fuel has been observed to result in lower NO_x emissions.

Thus nitrogen oxides ($NO_x = NO + NO_2$) are also present at high concentrations in car exhaust. Typical untreated car exhaust may contain CO concentrations of 1–2% by volume, unburned hydrocarbons levels between 500-1000 ppm and NO_x levels between 100-3000 ppm. The unit of ppm, or parts-per-million, are defined as follows:

1 ppm = 1 molecule reported per 1,000,000 molecules of other gases in air

1 ppb = 1 molecule reported per 1,000,000,000 molecules of other gases in air

12.2.3 Sulphur dioxide (SO_2)

It is formed during the combustion process. It depends on the quantity of sulphur compounds present in petrol and diesel, which further depends on the source of crude oil. The oxidation of sulphur produces SO_2.

12.2.4 Hydrocarbons (HC)

In spark ignition engines, the principal mechanism of HC formation has been attributed to the destruction of flame propagation radicals due to quenching in the engine combustion chamber. The thickness of the quenching layer depends on the air-fuel ratio and pressure. High hydrocarbon emissions can also result from either too lean or too rich local mixtures in the engine.

In compression ignition engines, there are two major sources which contribute to unburnt hydrocarbons:

(a) Fuel premixed to lower than the lean limit condition remaining unburnt.

(b) Emptying of fuel from the nozzle sac and hole volume, resulting in local fuel-rich conditions, the air-fuel ratio, injection timings, turbocharging, injection system design, ratio of injection and air swirl in the combustion chamber are some of the variables that influence the emissions of unburnt hydrocarbons.

12.2.5 Lead

It arises from the combustion of petrol. Alkyl lead compounds like tetraethyl lead (TEL) and tetramethyl lead (TML) are added to petrol during refining as an antiknock agent, to prevent the fuel from exploding spontaneously before its ignition by the spark plug.

The Bureau of Indian Standards (BIS) has specified a maximum amount of 0.56 and 0.8 g/l of TEL to be added to get octane number of petrol as 83 and 93. After combustion 70% of the lead is transmitted through exhaust as particulate lead halide. The size of lead particles varies from 0.001 to 10 microns. The fraction of lead escaping from petrol-driven vehicles increases as the vehicle mileage increases.

12.2.6 Soot

Soot emission in exhaust automobiles results from decomposition of hydrocarbon fuel, nucleation of carbon particles in the flame, growth of soot nuclei, agglomeration of particles and finally soot oxidation.

12.2.7 Composition of Auto Exhaust Pollutants and Corresponding Engine Conditions

Engine exhausts consist of a complex mixture, the composition depends on a variety of factors of engine such as: type of engine, driving conditions, vehicle speed, etc. The exhaust contains principally three types of pollutants, unburned or partially burned hydrocarbons (HC), carbon monoxide (CO) and nitrogen oxides (NO_x), mostly NO (Table 12.2) (5-6). In general, the emissions strongly depend on air-to-fuel (A/F) ratio, as exemplified in Figure 12.3.

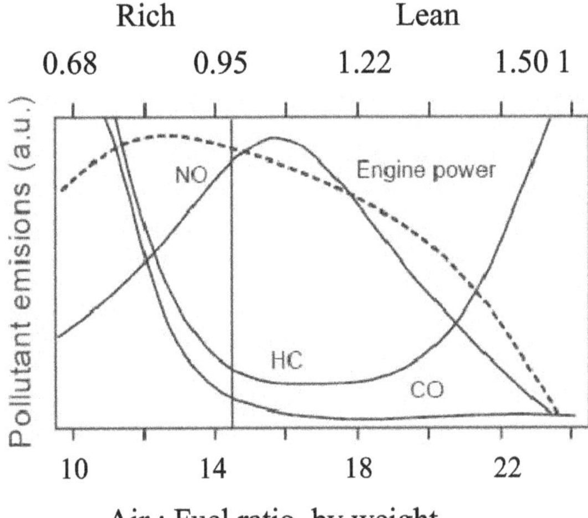

Fig. 12.3 *The effect of air-to-fuel ratio (A/F) on engine emissions (6)*

Tuning of the engine to rich feed gives the highest power output, which, however, occurs at expenses of high fuel consumption. Under lean-conditions lower combustion temperatures lead to lower NO emissions, however, at very high A/F engine misfire occurs, leading again to high HC emissions. Under any A/F conditions catalytic abatement of pollutants is needed to comply with the legislation limits. Only at stoichiometric conditions appropriate amounts of reducing and oxidising agents are present in the exhaust to carry out simultaneously the catalytic reactions of NO reduction and CO and hydrocarbon oxidation.

Table 12.2 *Exhaust pollutants and Engine conditions for their release (5)*

Exhaust components and conditions [a]	Diesel engine	Four stroke spark ignited-engine	Four-stroke lean-burn spark-ignited engine	Two-stroke spark-ignited engine
NO_X	350 - 10000 ppm	100 - 4000 ppm	∞ 1200 ppm	100-200 ppm
HC	50 – 330 ppmC	500 - 50000 ppmC	∞ 1300 ppmC	20000 - 30000 ppmC
CO	300 – 1200 ppm	0.1-6%	∞ 1300 ppm	1 - 3%
O_2	10 -15%	0.2-2%	4 - 12%	0.2 - 2%
H_2O	1.47 %	10-12%	12%	10 - 12%
CO_2	7 %	10-13.5%	11%	10 - 13%
SO_X	10 – 100 ppm [b]	15 - 60 ppm	20 ppm	∞ 20 ppm
PM	65 mg/m^3			
temperatures	rt – 650 °C	rt – 1100 °C	rt - 850 °C	Rt - 1000 °C
λ (A/F) [c]	~ 1.8 (26)	∞ 1 (14.7)	∞ 1.16 (17)	∞ 1 (14.7) [σ]

The mechanisms of formation of various automobile pollutants are summarized in Table 12.3

Table 12.3 *The mechanisms of formation of automobile pollutants*

Pollutant	Mechanism
Carbon monoxide	Incomplete combustion
Nitrogen oxides	Interaction of oxygen and nitrogen at high temperatures
Sulphur oxides	Interaction of sulphur and oxygen
Unburnt hydrocarbons	Wall quenching, incomplete combustion, scavenging
Particulate	Unburnt carbon particles

The types of pollutants emitted by petrol and diesel engines are similar but may differ in proportion due to differences in operating modes (Tables 12.4 and 12.5). In the automobile the exhaust pipe releases about 70% of the pollution; crankcase emission may account for the rest. While the petrol engine exhaust may be almost colourless or slightly grey, the diesel engine exhaust is dirty and may be bluish to black.

Table 12.4 *Emission factors for vehicles*

Types of vehicles	Avg. no. of km covered	Particulates tons	Sulphur dioxide tone	Nitrogen oxide (NO_x) tons	Hydrocarbon (HC) tons	Carbon monoxide tons
Petrol						
Two-wheelers	1000	0.20	0.02	0.07	10.0	17.0
Autos	1000	0.20	0.02	0.07	10.0	25.5
Cars/jeeps	1000	0.33	0.68	3.2	6.0	40.0
Diesel						
Cars/jeeps	1000	0.43	0.39	0.99	0.28	1.1
Trucks	1000	0.75	1.50	21	2.10	12.1
Buses	1000	0.75	1.50	21	2.10	12.7

Source: WHO official publication (40.61)

Table 12.5 *Average chemical composition of exhaust gases in ppm for diesel and petrol engines*

Pollutant	Idling		Cruising		Acceleration		Deceleration	
	Diesel	Petrol	Diesel	Petrol	Diesel	Petrol	Diesel	Petrol
CO	Traces	52000	20	8000	500	42000	Traces	52000
HC	250	750	65	3000	115	400	250	4000
NO_x	60	30	1500	850	850	300	10–50	12–30

When there is a lack of air, diesel engines are primarily prone to emissions of soot and odorous substances, but also to CO and hydrocarbon emissions. Figure 12.4 illustrates which components are to be expected in the exhaust gas of a diesel engine.

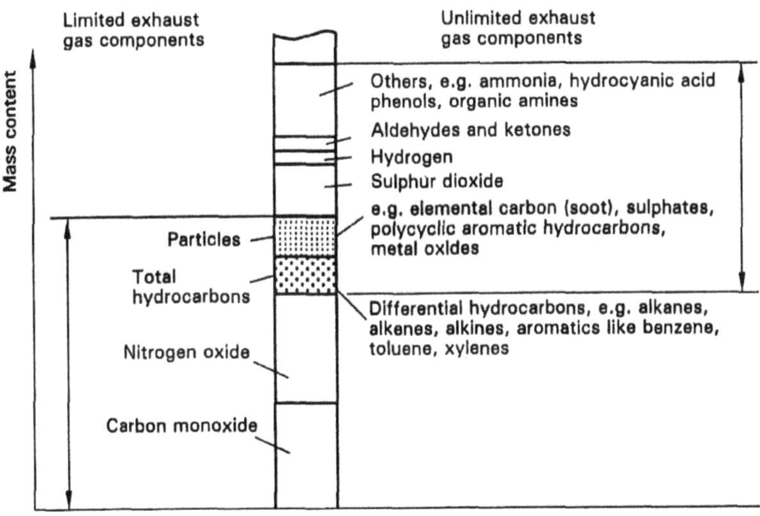

Fig. 12.4 *Content of limited and non-limited compounds in the exhaust gas of a diesel engine*

One distinguishes between the limited (CO, NO_x HC and particles) and the manifold non-limited components which can constitute up to a third of exhaust gas emissions (without CO_2 and H_2O). Apart from the air-fuel ratio, soot formation in diesel engines is dependent on pressure and temperature, on on the combustion process, on start of injection, injection ending and ignition delay. Particularly an inadequate mixing influenced by the factors listed above leads – at high temperatures and via cracking processes–to the formation of soot.

In diesel engines performance is not controlled via a throttle valve, as is the case in internal combustion engines, but via the amount of fuel injected. The amount of air sucked in is determined by the engine revolutions and the supply. Accordingly, when idling and when there is a partial load, vehicles generally operate with a great excess of air. If the injected amount of fuel is too high as against the amount of air sucked in (λ<1.2–1.4 depending on the factors listed above) at full load and revolutions in the low to medium range, then soot formation increases dramatically, (shown in Figs. 12.5 and 12.6) performance characteristics of a diesel engine. This is why diesel engines are prone to increased soot emissions at full loads, e.g. when driving up hill, with inadequate mixing and when large amounts of fuel are injected.

Motor vehicle emissions are highly variable. Apart from differences among vehicles, differences in operating conditions can cause emission from a vehicle to change by more than 100%. A consistent and replicable test procedure has to be prescribed for emission regulations or incentive systems to be enforceable.

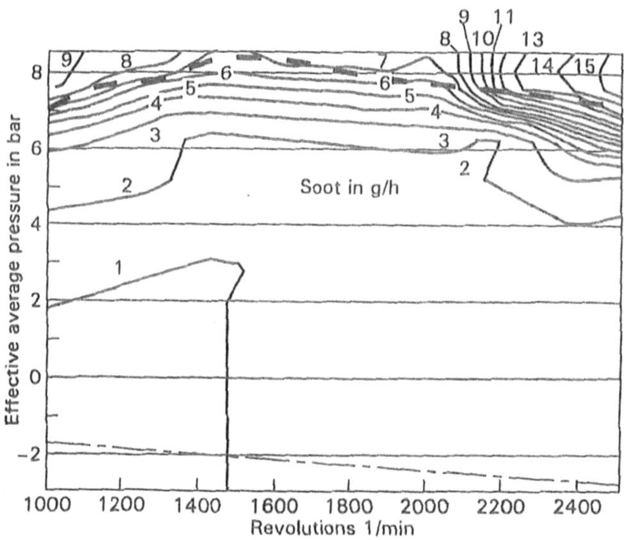

Fig. 12.5 *Engine performance curves of a diesel engine during stationary operating conditions sowing lines of constant soot emission*

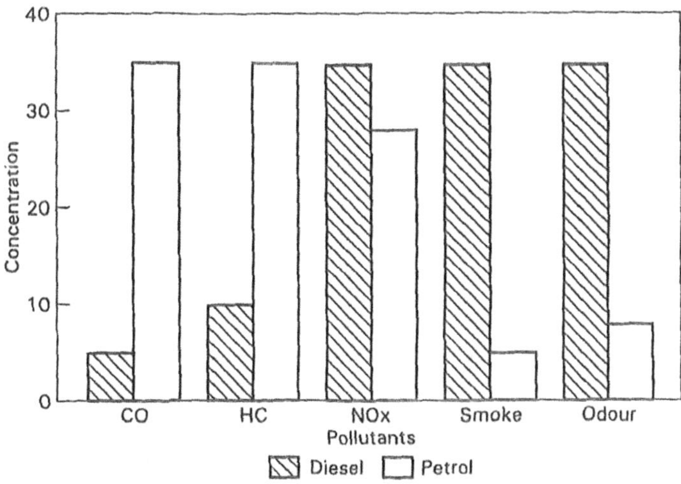

Fig. 12.6 *Petrol-diesel emission comparison*

Based on the driving speed, it was observed that NO_x emissions rise with increasing speeds owing to the increasing thermal load of the engines at higher speeds when thermal NO formation increases dramatically. The speed dependency of NO_x emissions illustrated in Fig. 12.7 does not apply at full loads. NO_x emissions would decrease again due to the enrichment of mixture. The inherent tendency to wait till the breakdown of the vehicle for maintenance instead of *preventive maintenance* has led to poorly maintained vehicles which emit excessive amounts of pollutants. Further, the boom of two-and three-wheelers has resulted in the emissions of enormous quantities of unburnt hydrocarbons into the environment. The driving pattern also plays a significant role in the release of air pollutants (Fig. 12.7).

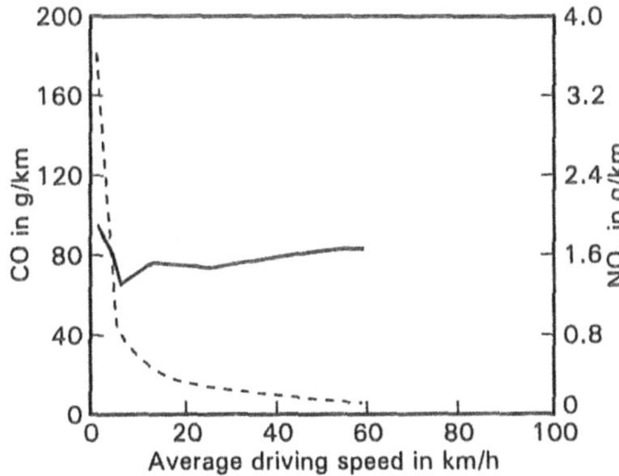

Fig. 12.7 *Typical nitrogen oxide emission vs driving speed on a plain roadway*

12.3 Control of Automobile Pollution at Source

In petrol vehicles the petrol vapour can escape from the fuel tanks and carburettor. This involves control of evaporative, crankcase and exhaust emissions by the following approaches

12.3.1 Evaporation emission control

Generally two basic techniques are used—vapour recovery system and the adsorption regeneration system.

Evaporation rates are very dependent on temperature; evaporation in the vicinity of the engine is especially high just after the engine is turned off, when the fans of cooling systems, and airflow due to vehicle movement are no longer operative. Carburettor losses are especially high during the hot soak period.

In vapour recovery system the crankcase is used as a storage tank for vapours from the fuel tank and carburettor. During hot soak period, the declining temperature in the crankcases cause a reduction in the crankcase pressure sufficient to draw in vapours from the carburettor and the fuel tank (via a vapour liquid separator). During engine operation, these stored vapours are sent to the air intake system via the positive crankcase ventilation.

A vapour recovery system that reduces this evaporative emission is shown in Fig. 12.8. In this approach, the crankcase space is used to temporarily store fuel vapours before they are recycled back into the combustion air intake by the PCV system. Another way to control evaporative emissions is with a canister of activated carbon, which collects vapours from the fuel tank and carburettor. When the engine is running, air is drawn through the canister to collect accumulated petrol vapours and return them to the intake manifold.

Evaporative emissions can also be reduced by using fuels with less tendency to vaporize. Petrol itself can be modified to be less volatile, but to achieve substantial improvements in evaporative emissions and still have the vehicle start easily and run smoothly, extensive modifications in the fuel intake and carburetion system are required. Instead, the mechanical systems just mentioned have been the preferred approach. It is appropriate to note that one fuel modification that has become fairly readily available is gasohol, which is a mixture of 90% unleaded petrol and 10% ethyl alcohol (ethanol).

The interaction between alcohol and petrol makes gasohol more volatile than either one alone, and evaporative emissions are correspondingly increased. While evaporative emissions, of hydrocarbons increase using gasohol, it appears that CO emissions are reduced while NO_x and exhaust emissions of HC may increase or decrease depending on the emission control systems used [5].

In the absorption regeneration system a canister of activated charcoal collects the vapours and retains them until they can be fed back into the intake manifold, to be burnt. The operating principle, however, remains the same as that of the evaporative emission control system.

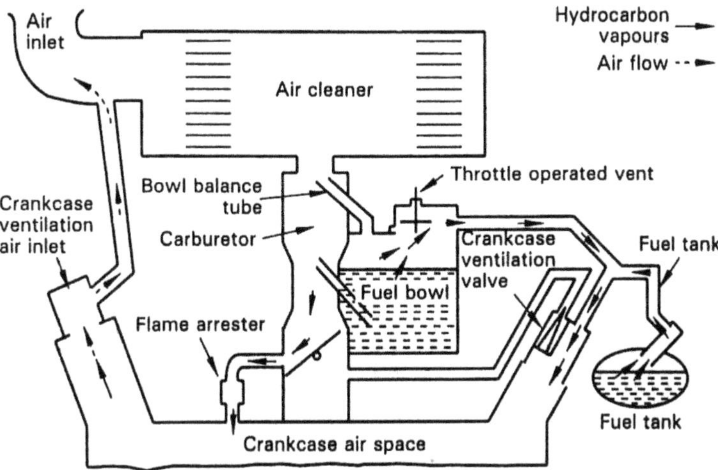

Fig. 12.8 *Use of crankcase air space and PCV system to reduce evaporative emissions [2]*

12.3.1.1 Crankcase emission control

Crankcase emissions occur when some of the air-fuel mixture in the cylinder is forced past the piston rings as blow-by. The primary objective of crankcase emission is to

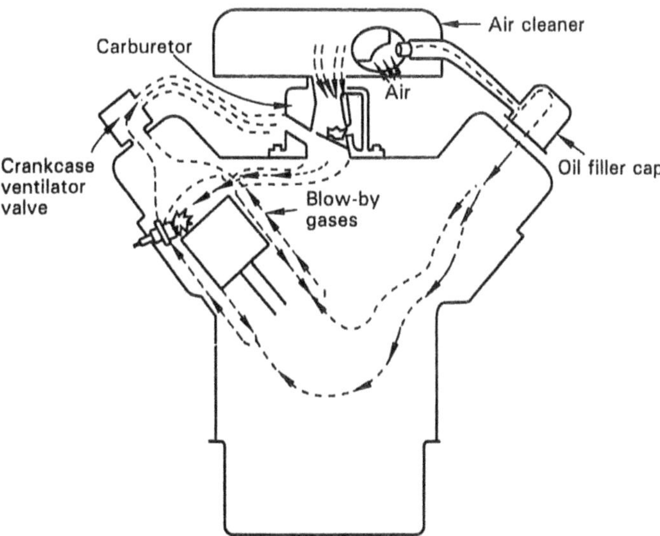

Fig. 12.9 *Positive crankcase ventilation (PCV) emission control system [2, 4]*

prevent blow-by gases from escaping into the atmosphere. All such control systems apply the approach, which involves recycling blow-by gases from the engine oil sump to the air intake system. Ventilation air is drawn down into the crankcase and then up a ventilator valve and hose and intake manifold. When the airflow through the carburettor is high, additional air from the crankcase ventilation system has little effect on engine operation, but during idling, the blow-by gases affect the air-fuel ratio causing rough idling. To check this, a valve called "Positive Crankcase Ventilation" (PCV) valve is used (Fig. 12.9). In PCV, atmosphere blow-by is not allowed and the gases are sent back to the engine and subjected once again to combustion. [4, 5]

12.3.1.2 Exhaust emissions control

12.3.2.1 Carburettor adjustment

Simple carburettor adjustment which regulates the air-fuel mixture results in reducing petrol consumption along with reduction in carbon monoxide emissions from a well-tuned healthy engine. Carburettor screw should be sealed after the pollution check and whenever the seal is broken for repairs and tune up, the vehicle should be tested again for emission level and sealed.

12.3.2.2 Effect of design and operating variables on emissions

The list of variables that effect internal combustion (automobile) emissions includes the following:

1. Air-fuel ratio
2. Load or power level
3. Speed
4. Spark timing
5. Exhaust back pressure
6. Valve overlap
7. Intake manifold pressure
8. Combustion chamber deposit build-up
9. Surface temperature
10. Surface-volume ratio
11. Combustion chamber design
12. Stroke-bore raio
13. Displacement per cylinder
14. Compression ratio.

The single most important factor in determining emissions from an internal combustion engine is the air-fuel ratio in the mixture as it enters the cylinder during the intake stroke. To analyze that ratio and its impact on emissions, let us begin with the stoichiometry of petrol combustion. While modern petrols are blends of various hydrocarbons, an average formulation can be represented as C_7H_{13}. We can write the following to represent its complete combustion in oxygen:

$$C_7H_{13} + 10.25O_2 \rightarrow 7CO_2 + 6.5\ H_2O \qquad .....(12.6)$$

If we want to show complete combustion in air, we can modify this reaction to account for the fact that about 3.76 moles of N_2 accompany every mole of O_2 in air. Thus, $10.25 \times 3.76 = 38.54$ mol of N_2 can be placed on each side of the reaction, yielding

$$C_7H_{13} + 10.25O_2 + 38.54N_2 \rightarrow 7CO_2 + 6.5\ H_2O + 38.54\ N_2 \qquad .....(12.7)$$

where any oxidation of nitrogen to nitrogen oxides has been neglected.

12.3.2.3 Stoichiometric ratio

For each mole of petrol, 10.25 mol of O_2 and 38.54 mol of N_2 are required. Using Reaction (6 and 7) we can determine the masses of each constituent as

$$1\text{ mol } C_7H_{13} = 7 \times 12 + 13 \times 1 = 97\text{ g}$$
$$10.25\text{ mol } O_2 = 10.25 \times 2 \times 16 = 328\text{ g}$$
$$38.54\text{ mol } N_2 = 38.54 \times 2 \times 14 = 1079\text{ g}$$

Considering air to be made up of only O_2 and N_2, the air-fuel ratio needed for complete oxidation of petrol is

$$\frac{\text{Air}}{\text{Fuel}} = \frac{328 + 1079}{97} = 14.5 \qquad .....(12.8)$$

This is known as the stoichiometric ratio for petrol.

If the actual air-fuel mixture has less air than what the stoichiometric ratio indicates is necessary for complete combustion, the mixture is said to be rich. If more air is provided than is necessary, the mixture is lean. A rich mixture encourages production of CO and unburned hydrocarbons since there is not enough oxygen for complete combustion. On the other hand, a lean mixture helps reduce CO and HC emissions unless the mixture becomes so lean that misfiring occurs. Production of NO_x is also affected by the air-fuel ratio. For rich mixtures, the lack of oxygen lowers the combustion temperature, reducing NO_x emissions. In the other direction, beyond a certain point lean mixtures may have so much excess air that the dilution lowers flame temperatures and reduces NO_x production.

Figure 12.10 shows the relationship between CO, HC and NO_x emissions, and the air-fuel ratio. Also shown is an indication of how the air-fuel ratios affect both power delivered and fuel economy. As can be seen, maximum power is obtained for a slightly rich mixture, while maximum fuel economy occurs with slightly lean mixtures. Before catalytic converters became the conventional method of emission control, most cars were designed to run slightly rich for better power and smoothness. When the first emission limitations were written into law, less rich mixtures were used, which had the desired effect of reducing CO and HC; at the same time, however NO_x emissions were increased. The difficulty automotive engineers have had in achieving control of all three pollutants, CO, HC and NO_x, all at the same time is apparent from the figure.

The air-fuel ratio is, of course, not the only factor that influences the quantity of pollutants created during combustion. Other influencing factors include ignition timing, compression ratio, combustion chamber geometry, and very importantly, whether the vehicle is idling, accelerating, cruising, or decelerating. Not all of the pollutants created in the combustion chamber pass directly into the exhaust system. Some find their way around the piston during the power and compression strokes, into the crankcase. This blow-by, as it is called, used to be vented from the crankcase to the atmosphere, but now it is recycled back into the engine air intake system to give it a second chance at being burned and released in the exhaust stream. The control method is called Positive Crankcase Ventilation (PCV), and the main component is a crankcase ventilator valve, as shown in Fig. 12.9. The PCV valve adjusts the rate of removal of blow-by gases to match the changing air intake requirements of the engine, making sure that the desired air-fuel ratio is not upset by these added gases.

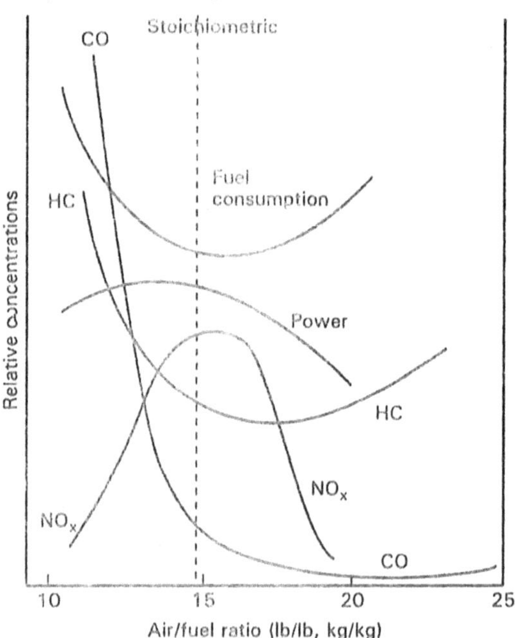

Fig. 12.10 *Effect of air-fuel ratio on emissions, power and fuel economy (5-7)*

12.3.2.4 Engine reconditioning

In improperly tuned vehicles, i,e, those requiring engine reconditioning, hydrocarbon emissions along with oxides of nitrogen from diesel exhaust result in more frequent occurrence of photochemical smog. So engines should always be kept properly tuned

12.3.2.5 Reduction in compression ratio

The use of low compression ratio results in high exhaust gas temperatures, and provides a better environment for oxidation of unburnt hydrocarbons in the exhaust system.

Modifications in engine design include in addition

(a) Use of leaner air-fuel ratio
(b) Retarding ignition timing
(c) Modification of combustion chamber configuration to reduce quench areas
(d) Pre-treatment of the mixture to improve vapourization and mixing of the fuel with air [5]

12.3.2.6 Exhaust gas recirculation (EGR)

In EGR method, a portion of the exhaust gas, which is inert is continuously recirculated through the engine so that the burning air-fuel mixture is diluted. The dilution decreases the oxygen concentration in the burning mixture thereby lowering the peak combustion temperature. The EG recirculation is regulated by an EG valve, and is generally 15–22% by the volume of the air-fuel mixture.

12.3.2.7 Fuel options for controlling emissions (8, 9)

Pollutant emissions can be reduced by 10–30% through cost-effective fuel modifications. This has major advantages–they can take effect quickly, can be targeted geographically or seasonally, and are easier to enforce since fuel is refined.

Major efforts have been made worldwide to remove lead from petrol, both to reduce lead emissions and to facilitate the use of vehicle emissions control technologies such as catalytic converters. Possible further changes to reduce emissions include reduced volatility, increased oxygen content, reduced aromatics and more widespread use of detergent additives. Conventional diesel fuel can be improved by reducing the sulphur and aromatic content and by using detergent additives.

The substitution of petrol and diesel by cleaner burning fuel has received attention in the last decade. The more important alternatives used are natural gas (generally compressed natural gas-CNG), liquefied petroleum gas (LPG), methanol, ethanol, and gasohol, biodiesel, hydrogen, etc.

Methanol has long been the fuel of choice in high performance racing cars. It has a much higher octane rating than petrol, which enables engines to be designed with higher compression ratios for increased power. It burns with a lower flame temperature than does either petrol or diesel fuel, so NO_x emissions are reduced. It burns more completely, as well, so that hydrocarbon and CO emissions are also reduced. Moreover, the hydrocarbons that are released are less photochemically reactive. As a result of these improved emission characteristics, the ozone-producing potential of methanol is perhaps half that of petrol [8].

Methanol is not without its problems. Its emissions are much higher in formaldehyde (HCHO), an eye irritant and a suspected carcinogen. The low volatility of pure methanol makes it difficult to start engines in cold conditions, and it is highly toxic. It burns without a visible flame, which can lead to especially dangerous conditions in an accident involving fire. Moreover, its energy content is only about half that of petrol, so either much larger fuel tanks are required or the driving range is considerably reduced.

Some of the problems of pure methanol can be overcome by using a blend of methanol and petrol. A mixture of 85% methanol and 15% petrol, called M85, eliminates the cold start problem and yields a visible flame as well. It also tastes much worse than methanol, which should discourage ingestion. The blend has an octane rating of 102 versus 87–92 for petrol, and an energy content of 68250 kWS per gallon compared with 1218000 kWS for petrol. While the increase in combustion efficiency does slightly offset the low fuel energy content, the driving range for M85 is still only about 60% of that for petrol. To help ease the transition to methanol, automobile manufacturers are introducing Flexible Fuel Vehicles (FFVs) designed to run on M85, petrol, or any combination of both. Eventually, if methanol becomes widely available, it would be possible to take advantage of the increased octane level with more efficient engines dedicated to M85 use.

Methanol derived from natural gas produces slightly less of the greenhouse gas CO_2 than petrol. On the other hand, if methanol is produced from coal, total CO_2 emissions released during coal processing and methanol combustion are double that of petrol. [8]

Hydrogen has the potential to be the cleanest burning motor fuel. Research and development work on hydrogen powered vehicles is progressing in several countries–-Canada, Germany, Japan, and USA. A prototype 60-passenger bus has been put into service in Vancouver, Canada.

Electric automobile: It is provided with lead-acid batteries and uses D-C current series not on with restive speed controls to give good start and maximum speed.

Advantages: a) Readily available refuelling b) Zero tail pipe emissions c) Quiet operation d) Reduced maintenance e) Reduced fuel cost f) Opportunity to fuel at home g) Readily available electricity distribution system

Disadvantages: Increased vehicle weight and initial vehicle cost, reduced driving range.

12.4 Exhaust Gas Treatment Devices-Catalytic Oxidation Converters

12.4.1 US clean air act and catalytic converters development

Automobile emissions have a huge impact on human health and the well-being of our planet. Not only do they lead to elevated ozone levels, but the CO emitted by engines is toxic. Also, both CO_2 and ozone are greenhouse gases. After the link was made between automobiles and air pollution, and after overcoming strong opposition from the automobile industry, regulations to control automobile emissions were finally enacted in the U.S. when Congress passed the Clean Air Act in 1970. Since then, the Environmental Protection Agency (EPA) has established emission standards that automobile manufacturers must abide by when producing cars, and these regulations have become more stringent over time as pollution abatement technology has improved.

Engine manufactures have explored a wide variety of technologies to meet the requirements of the Clean Air Act. Catalysis has proven to be the most effective passive system. The major worldwide suppliers of automotive catalysts are Engelhard, Johnson

Matthey, Degussa, and Allied Signal. As the automobile engine has become more sophisticated, the control devices and combustion modifications have proven to be very compatible with catalyst technology, to the point where today engineering design incorporates the emission control unit and strategy for each vehicle.

The need to control engine emissions according to Clean Act Air, recent amendments have set up more stringent requirements for automotive emissions (2). New catalysts are required to last 100,000 miles for new automobiles after 1996. Furthermore, the amendments (contingent on tier 2 standards to be set by environmental protection agency) reduce non methane hydrocarbon (NMHC) emissions to a maximum of 0.125 g/mile by 2004 (down from 0.41 g/mile in 1991), carbon monoxide to 1.7 g/mile (down from 3.4 g in 1991), and nitrogen oxides to 0.2 g/mile (down from 1.0 g).

To meet these regulations new catalytic converters are developed by automobile industries

12.4.2 Catalytic converters for destruction of auto exhaust pollutants

12.4.2.1 Catalytic converters with exhaust gas recirculation system

Emission control catalysts eliminate dangerous engine pollutants from a wide range of fuels, including gasoline, diesel, natural gas, and alternate fuels. Catalysts transform pollutants into harmless gases within a catalytic converter before releasing them into the environment. Historically, four generations of de-pollution catalysts/converters can be distinguished.

The first generation converter typically contained a simple oxidation catalyst based on Pd/Pt supported on Al_2O_3 an additional air supply system was introduced to promote the oxidation. This arrangement was sufficient to meet the initial standards. NO control was generally achieved by using an Exhaust Gas Recirculation system. By recirculating part of the exhaust gases back into the combustion chamber, the combustion temperature was lowered, resulting in reduced NO emissions.

12.4.2.2 Functioning of catalytic oxidation reactors

By using Pt, Pt-Pd catalyst loaded with CeO_2, Pt, Pd, Rd, MnO_2 or Fe_2O_3 80–90% of CO and hydrocarbons can be reduced (Figure 12.11). More than 600 different catalysts have been tried and the following were found better with certain advantages as well as disadvantages. Some of the catalyst systems are

(a) Alumina pellets in exhaust mufflers

(b) Copper oxide mufflers

(c) Mn catalyst

(d) Mufflers with stainless steel wool coated with alumina

(e) A catalyst after burner to reduce oxides of nitrogen

(f) Three-way catalyst

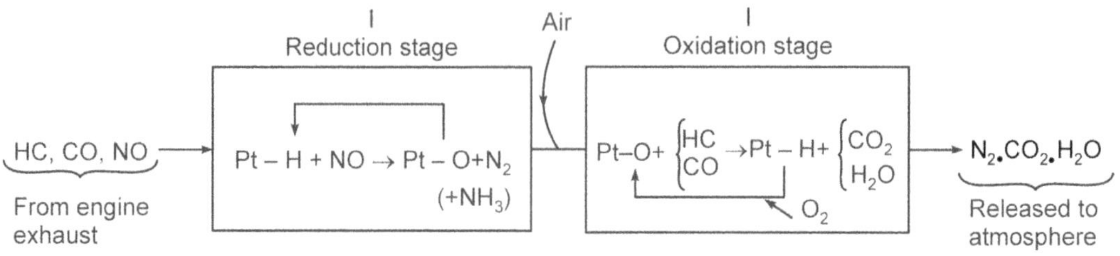

Fig. 12.11 *Catalytic Oxidation Reactor*

12.4.2.3 What is actually happening in a catalytic converter?

In a catalytic converter, hydrocarbons and CO are *oxidized* to CO_2 and H_2O:

$$C_xH_y + (1 + y/4)\, O_2 \rightarrow x\, CO_2 + y/2\, H_2O \quad .....(12.9)$$

$$CO + (1/2)\, O_2 \rightarrow CO_2 + y/2\, H_2O \quad .....(12.10)$$

and NO_x is *reduced* to harmless N_2 (note these reactions are intentionally unbalanced because the mechanism of catalysis is very complex)

$$NO\ (\text{or } NO_2) + CO \rightarrow N_2 + CO_2 \quad .....(12.11)$$

$$NO\ (\text{or } NO_2) + C_xH_y \rightarrow N_2 + CO_2 + H_2O \quad .....(12.12)$$

The metals and metal oxides present in the ceramic monolith act to adsorb both NO_x and O_2, and then break the normally very strong N–O or O–O bonds. On the surface, the resulting oxygen atoms react with CO or hydrocarbons, while N atoms recombine to form N_2. The whole system is catalytic, i.e., during a single catalytic cycle the metal species do not themselves permanently change their identity, but through their interaction with the reactants, serve to reduce the energy barrier (Activation Energy, E_a) to the reaction and speed the reaction up to form more benign gases.

Two conditions must be satisfied for the converter to work properly. First, the temperature of the catalyst must be high enough. This stems from the relation between the temperature and the rate of a chemical reaction, as described by the Arrhenius equation:

$$k = A \exp(-E_a/RT) \quad .....(12.13)$$

where k, is the rate constant of the reaction, A is the pre-exponential factor, and E_a is the activation energy for the reaction. The second requirement for the catalytic converter to work properly is that the engine combustion chamber must be operated close to the stoichiometric air to-fuel ratio. This is because there is a narrow range of air-to-fuel ratios in which both NO_x reduction and CO/hydrocarbon oxidation readily occur on the surface of the catalytic converter simultaneously. If there is too much O_2 present, the NO_x reduction does not occur. Conversely, if there is not enough O_2 present, the CO/hydrocarbon oxidation doesn't happen. This fine balance to keep the gas mixtures in the optimum range is continuously adjusted by an oxygen sensor and engine feedback system placed before the catalytic converter in the exhaust manifold.

The catalytic converter is normally located under the automobile between the engine and the muffler. When the car is first started (i.e. "cold start"), the catalytic converter is cold and the catalyst does not efficiently remove NO_x and CO from the exhaust. As the engine warms up, the exhaust warms the catalytic converter to a high enough temperature to initiate the catalytic reactions that remove pollutants. The catalytic reactions occur heterogeneously (i.e. when gaseous combustion products adsorb onto the surface of a ceramic monolith and react with the various metal atoms present there) and convert toxic gases into less harmful chemicals.

12.4.2.4 Study of efficiency of catalytic converter- UV/visible spectroscopic method for measurement of NO_x in the exhaust

The goals of this experiment are to measure quantitatively the concentration of NO_x in the exhaust of your vehicle using UV/VIS Spectroscopy, (10, 11) and to study the efficiency of your automobile's catalytic converter to decrease NO_x emissions to the atmosphere.

For this we need to sample exhaust from the tail pipe of the vehicle immediately at startup and then again after the car has had a chance to warm up for 10 minutes.

One or more exhaust samples can be collected to determine the amount of nitrogen oxides present in the gas stream using a chemical indicator.

12.4.2.5 The chemical indicator for NO_x in exhaust

NO_x at ppm levels can be determined using Ultraviolet-Visible (UV/VIS) Spectroscopy after it is converted into a brightly colored indicator dye in aqueous solution. In solution, NO and NO_2 undergo the following reactions to form nitrous acid (HONO):

$$4\ NO + O_2 + 2\ H_2O \rightarrow 4\ HONO \qquad .....(12.14)$$

$$2\ NO_2 + H_2O \rightarrow HONO + HNO_3 \qquad .....(12.15)$$

Nitrous acid reacts with sulfanilic acid and N-(1-naphthyl)-ethylenediamine to form an azodye:

The concentration of this purple-colored molecule can be easily determined from its absorption band at 550 nm using a UV/VIS spectrophotometer.

12.4.2.6 Ultraviolet-visible (UV/VIS) absorption spectroscopy

The absorption spectrum of a molecule is obtained by measuring the decrease in light intensity passing through the sample (compared to a "blank") as a function of wavelength: (Figure 12.12)

According to the Beer-Lambert law, the intensity of the transmitted light, I, is related to the initial intensity, I_o, as $I = I_0 \times 10^{-A}$, where A is called absorbance. For sufficiently diluted samples, absorbance is directly proportional to concentration (C) of the analyte and path length (l, in cm) of the sample holder, as shown below:

$$A = \varepsilon l C \quad .....(12.16)$$

The proportionality constant, ε is called molar absorptivity or molar extinction coefficient, and it is usually determined from the slope of a plot of absorbance against concentration at a known path length. Each molecule is characterized by a unique dependence of ε on the absorption wavelength. This plot can also be used to determine the concentration of an unknown solution.

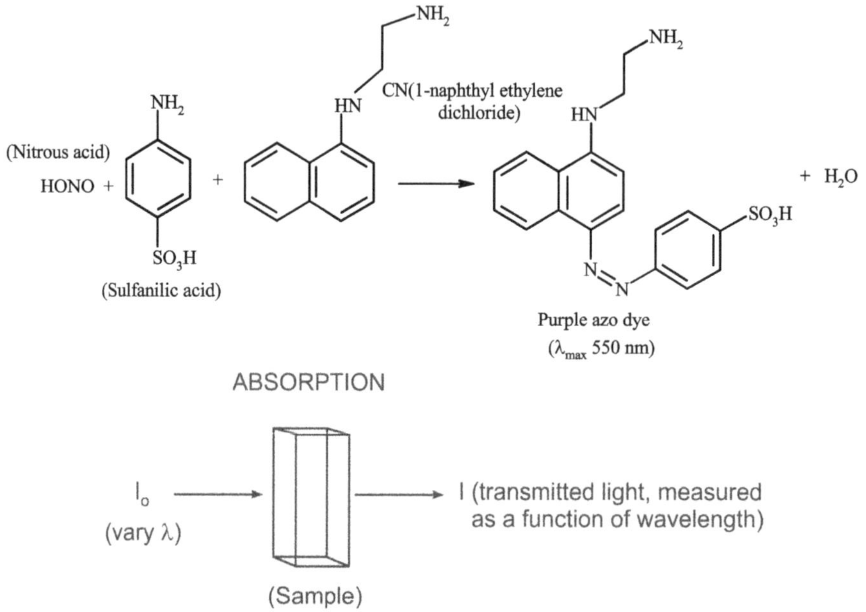

Fig. 12.12 *A chemical indicator (λ_{max} 550 nm) reaction for NO_x*

12.4.3 Dual way and three way catalytic converters

More stringent exhaust emissions standards, including tighter NO control, had been set in the USA by 1977-1981, which led to an implementation of the control of the exhaust emissions by developing the so-called dual-way catalyst (second generation) and, subsequently, three way catalyst (TWC, third and fourth generation). (6, 7)

The dual-way converter usually consisted of a primary bed containing a Rh/Pt-based catalyst for NO_x removal, followed by a secondary bed containing an oxidation catalyst equipped with an additional air supply to promote the oxidation reactions. By tuning the engine to run close to stoichiometric conditions, reduction of NO to N_2 could be achieved in the first bed and CO/HC oxidation to CO_2 and H_2O was carried out in the second bed. However, such devices could not withstand the progressively tightening demand for exhaust control in the USA in the early 80's, mainly due to the imperfect A/F control resulting from the use of carburettors. Further, NO can be easily reduced to NH_3 under rich conditions, which can then be back converted over the oxidation catalyst, making NO_x removal inefficient.

12.4.4 Three way catalytic converter (TWC)

Out of all the catalysts, the three-way catalyst is most recent and having more advantages, as there is no need for exhaust gas recirculation, after burner or other pollution control measures. The approach most favoured by American automobile manufacturers to achieve the emission standards dictated by the Clean Air Act has been the three-way catalytic converter. A three-way catalyst is able to oxidize hydrocarbons and carbon monoxide to carbon dioxide, while reducing NO_x to N_2. These catalysts are very effective in controlling emissions, and they have the advantage of allowing the engine to operate at near stoichiometric conditions where engine performance and efficiency are greatest. In fact, they must operate within a very narrow band of air-fuel ratios near stoichiometric point or else their ability to reduce all three pollutants at once is severely compromised, as shown in Fig. 12.13 (a,b).

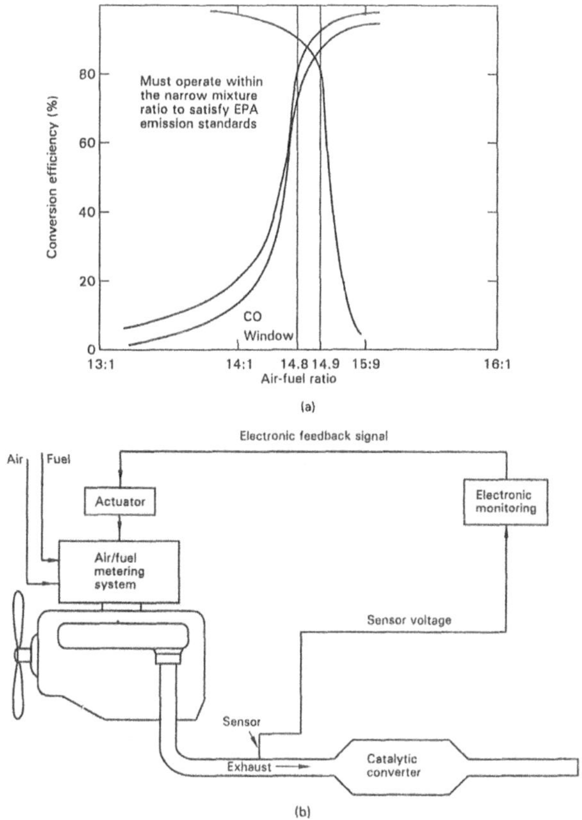

Fig. 12.13(a) *For a three-way catalyst to function correctly, the air-fuel ratio must be maintained within a very narrow band, as shown;* **(b)** *To maintain that ratio, a closed loop control system monitors the composition of exhaust gases and sends corrective signals to the air-fuel metering system*

These catalysts simultaneously oxidize hydrocarbons and carbon monoxide reducing the NO_x in the exhaust.

12.4.5 Reactions in 3-way catalysts reactors

The reactions in 3-way catalysts can be represented as:

$$CO + \text{oxidation catalysts} \rightarrow CO_2 + \text{reduced catalysts} \quad .....(12.17)$$

$$HC + \text{oxidation catalysts} \rightarrow CO_2 + H_2O + \text{reduced catalysts} \quad .....(12.18)$$

$$NO_x + \text{reduced catalysts} \rightarrow N_2 + \text{oxidation catalysts} \quad .....(12.19)$$

NO is removed by reduction using the CO, HC, and H_2 in the exhaust.

$$NO + CO \rightarrow ½\, N_2 + CO_2 \quad .....(12.20)$$

$$2NO + 5CO + 3H_2O \rightarrow 2NH_3 + 5CO_2 \quad .....(12.21)$$

$$2NO + CO \rightarrow N_2O + CO_2 \quad .....(12.22)$$

$$NO + H_2 \rightarrow ½\, N_2 + H_2O \quad .....(12.23)$$

$$2NO + 5H_2 \rightarrow 2NH_3 + 2H_2O \quad .....(12.24)$$

$$2NO + H_2 \rightarrow N_2O + H_2O \quad .....(12.25)$$

12.4.6 Latest design of a three-way catalytic converter

A typical latest design of a three-way catalytic converter is shown in Figure 12.13. Basically it is a stainless steel container which incorporates a honeycomb monolith made of cordierite or metal (12-14).

The active components of catalytic converters are precious metals such as Pt, Pd, and Rh dispersed on the high surface area ceramic monolith that is shaped like a honeycomb (Figure 12.13). The honeycomb support can be 10 inches in diameter, 7 inches long, have between 10-500 cells per square inch, and is held inside a metal housing (you will have an opportunity in class to take a catalytic converter apart). The honeycomb structure means that exhaust gases travel through the converter at the fastest possible rate while at the same time hitting a lot of catalyst covered surface on their way out. This design aims at maintaining fuel efficiency and preventing a loss of power that would happen if the exhaust gases build up pressure going through the converter

The choice and geometrical characteristics of the honeycomb monolith play a key role in determining the efficiency of the converter. In fact, high conversion must be achieved in the converter and therefore the catalyst works under conditions where severe mass and heat transfer limitations apply. Typically, both metal and ceramic monoliths are employed nowadays. The major advantage of the metallic substrate is that the wall thickness is limited by the steel rolling mill's capabilities, not strength. In a typical automotive 400 cell/inch^2 application, the frontal flow area in a ceramic monolith is 69% open (31% closed), while the metallic version has 91% open area. This is due to the higher wall thickness of ceramic monoliths (0.007 inch (0.178mm)) compared to metallic ones (0.002 inch (0.050mm)). However, even in this field there has been a strong improvement of the technology, cell densities as high as 900 cell/inch^2 or even higher are now commonly available on the market for both types of monoliths. Traditionally, cordierite monoliths have been employed quite extensively, primarily due to their lower production cost. However, a major advantage of the metal monoliths resides in their high thermal conductivity and low heat capacity which allow very fast heating of the CCCs during the phase-in of the engine, minimizing the light-off time.

12.4.7 Modern TWC with oxygen sensors

The advent of oxygen sensors (the so-called X sensor) based on zirconia type ceramics, led to the development of the modern TWC arrangement (Figure 12.14) The name is derived from the ability to simultaneously remove all of the three categories of pollutants, i.e. NO_x, CO and HC, which are present in the exhausts. Of these, NO_x removal/reduction

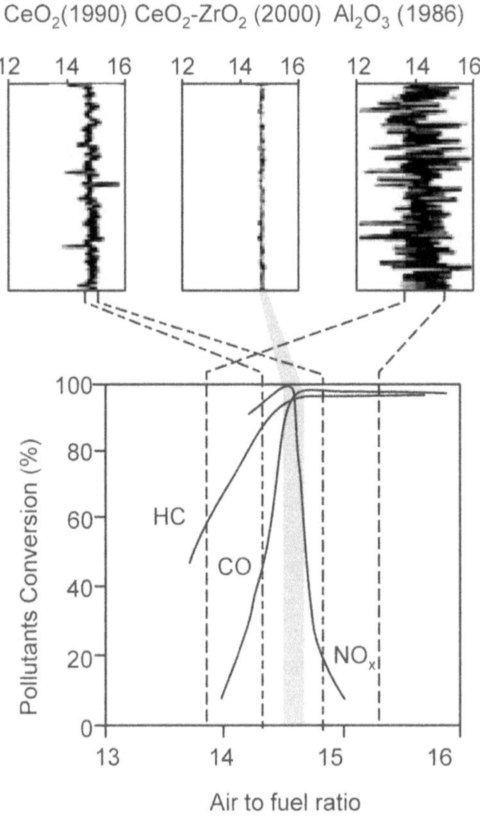

Fig. 12.14 *Effect of A/F ratio on the conversion efficiency of three-way catalysts. (7)*

is obviously most efficient under net reducing conditions, in a deficiency of O. In contrast, both CO and HC are best converted under net oxidising conditions, in excess oxygen. However, if one could maintain the A/F ratio close the stoichiometry, then in principle all the three pollutants could be converted simultaneously.

This is illustrated in Figure 12.14. In driving conditions, A/F is kept close to the stoichiometric point by a fuel injection device regulated by an electronic feedback system, which uses a signal from the sensor.

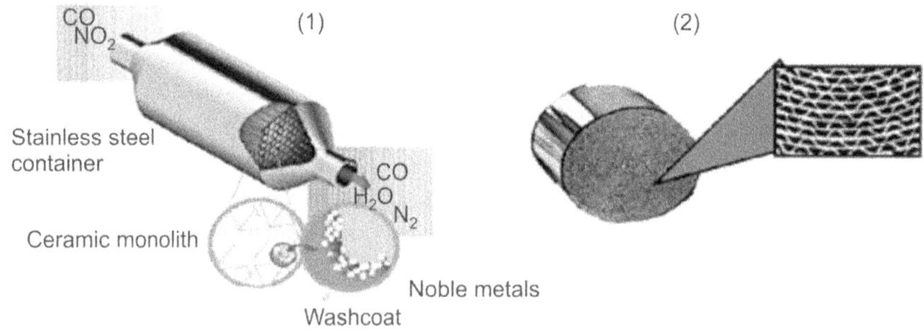

Fig. 12.15(a) *Typical car converter (I) monolith and (2) metallic honeycomb (6, 7)*

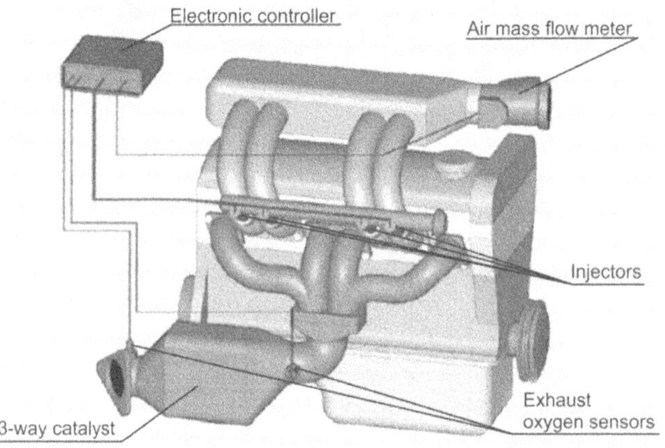

Fig. 12.15 (b) *Modern TWC Engine/oxygen sensor–control loop for the engine exhaust control (Courtesy of Mr. R. Crevatin, University of Trieste)*

To obtain an efficient control of the A/F ratio the amount of air is measured and the fuel injection is controlled by a computerised system which uses oxygen sensors (X) located at the inlet of the catalytic converter. The signal from the sensor is used as a feed-back for the control loop. A second additional sensor is mounted at the outlet of the catalytic converter (Fig. 12.15).

This configuration constitutes the basis of the so called engine on-board diagnostics (EOBD). By comparing the oxygen concentration before and after the catalyst, A/F fluctuations are detected. Extensive fluctuations of A/F at the outlet signal system failure. This EOBD arrangement implicitly assumes that only a narrow A/F window at stoichiometric point is the fingerprint of an effective TWC type of system. The location of the catalytic converter is another critical point which determines the conversion efficiency. TWCs typically feature a so-called light-off type conversion vs. temperature behaviour. This curves is characterised by conversion which steadily increases from O to 100 % conversion, the temperature of 50 % of conversion being indicated as the light-off temperatures.

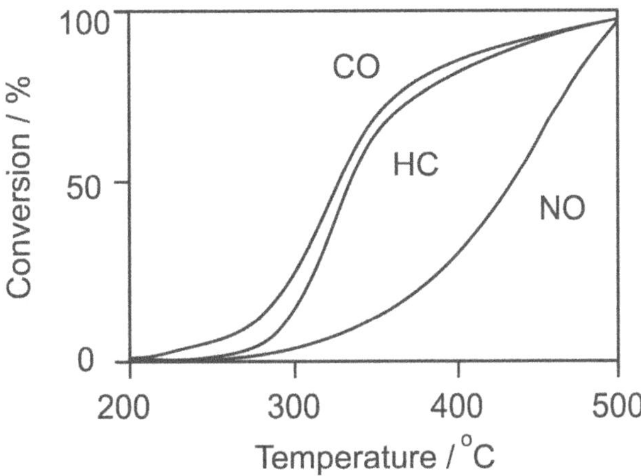

Fig. 12.16 *Light off type catalyst (6, 7)*

12.4.8 Light-off temperatures of three-way catalytic converters

TWCs are characterised by a light-off temperature around 250-350 °C (Fig. 12.16) (5). This means that an under-floor catalyst is heated above the light-off temperature within 90-120 s. Differently, when the catalyst is closely coupled to the engine (CCC close-coupled catalyst) the heating time typically drops down to 10-20 s. This dramatically affects vehicle emissions immediately after the start up of the engine.

The ULEV (Californian ultra low emissions vehicle) limit (0.064 g of HC emitted/km) is typically surpassed within 40 s after the engine start-up (9). This requires for an almost instantaneous heating up of the converter to achieve the required >95-98% conversion. As a consequence of this location of the catalyst, temperatures up to 1100 ^{0}C are easily met by a CCC

12.5 Automobile Air Pollution–Indian Scenario (15-18)

12.5.1 Types of automobiles on Indian roads

The three main types of automotive vehicles being used in India are: a) Passenger cars powered by four-stroke petrol engines; b) Motorcycles, scooters and autorickshaws powered mostly by small two-stroke petrol engines; c) Large buses and trucks powered mostly by four-stroke diesel engines

12.5.2 Growth rate of automobiles

The average annual growth rate of vehicle population from 1985—1988—89 has been of the order of 10.9% and 12.5% for public and private frieght transport respectively. In the case of intermediate public transport (IPT), three-wheelers registered the highest growth rate of 13.3% whereas buses and taxis registered growth rates of 9.8% and 7.2% respectively. (Figure 12.17)

The massive growth of 2-wheelers has put entire road systems on the brink. The auto rickshaw is the main mode of intermediate public transport (IPT) in urban areas, with a growth rate of 13.3%. Vehicles registered in five metropolitan cities—Mumbai, Delhi, Kolkata, Chennai and Bangalore account for 50% of cars, 16% of two-wheelers and 15% of diesel-driven vehicles in India. The number of vehicles per lakh of population in these cities are likely to go up to 33, 000 by the year 2020 from the present figure of 18,000 vehicles. Consequently the growth of vehicle population is likely to exert tremendous pressure on the meagre road space available in the urban areas.

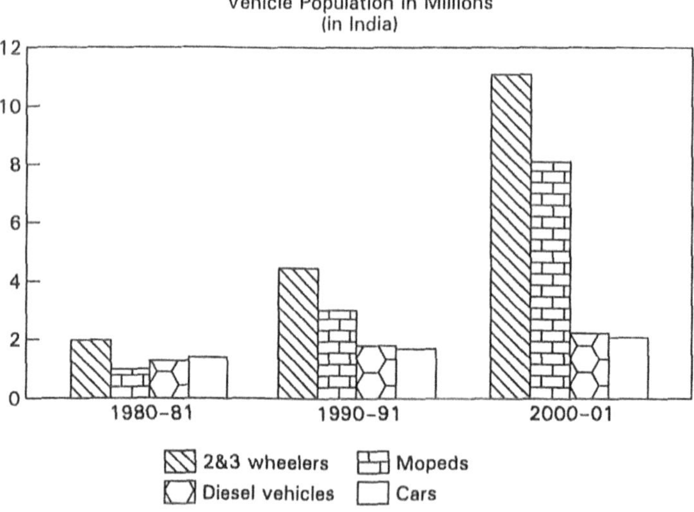

Fig. 12.17 *Vehicle population in millions [15]*

Since the liberalization of the sector in 1991 and allowing of 100 percent FDI through automatic route, Indian automobile sector has come a long way. Today, almost every global auto major has set up facilities in the country. Now, India is the one of the largest producers of automobiles in the world. The total production of Passenger Vehicles (PVs) was 669719 in the year 2001-02. It increased to 3233561 in the year 2012-13, recording an average annual growth of 34.8% during 2001-02 to 2012-13. The total production of Commercial Vehicles (CVs) was 162508 in the year 2001-02. It increased to 831744 in the year 2012-13, recording an average annual growth of 37.44% during 2001-02 to 2012-13. The production of Two Wheeler in India was 4271327 in 2001-02. It increased to 15721180 in 2012-13, recording an average annual growth rate of 24% during 2001-02 to 2012-13. The production of Three Wheeler in India was 212748 in 2001-02. It increased to 839742 in 2012-13, recording an average annual growth rate of 27% during 2001-02 to 2012-13 (Table 12.6).

Table 12.6 *Different types of Automobiles in India 2013-14*

Category	**2013-14**
Passenger Vehicles	30,87,973
Commercial Vehicles	6,99,035
Three Wheelers	8,30,108
Two Wheelers	1,68,83,049
Grand Total	**2,15,00,165**

12.5.3 Increased air pollution levels in Indian urban areas due to automobiles

The tremendous growth in these passenger vehicles is also greatly responsible for increased air pollution levels in various urban centers in India. In recent years, air pollution has acquired critical dimensions and the air quality in most Indian cities that monitor outdoor air pollution fail to meet WHO guidelines for safe levels (see Figures 3.17 to –3.21 in Chapter 3).

The levels of PM 2.5 and PM10 (Air-borne particles smaller than 2.5 micrometers in diameter and 10 micrometers in diameter) as well as concentration of dangerous carcinogenic substances such as Sulphur Dioxide (SO_2) and Nitrogen Dioxide (NO_2) have reached alarming proportions in most Indian cities, putting people at additional risk of respiratory diseases and other health problems.

According to World Health Organization (WHO), Delhi tops the list of most polluted cities. Among the world's 20 most polluted cities in the world, 13 are in India. India is in the group of countries that has the highest particulate matter (PM) levels. Its cities have the highest levels of PM10 and PM 2.5 (particles with diameter of 10 microns and 2.5 microns). At the level of more than 150 micrograms, Delhi has the highest level of airborne particulate matter PM2.5, considered most harmful. These figures are six times more than the WHO "safe" limit of 25 micrograms. Uncontrolled vehicular traffic seems to be the primary reason.

12.5.4 Reasons for air pollution from automobiles in India (as per Automobile and pollution in India Govt of India Report 2016-Jan)

Some of the reasons for pollution from Automobiles in India are as under:

(i) Absence of effective, efficient and well networked transport system in India. (ii) Lack of fast railway network across India which can run on cleaner technologies. (iii) Lack of fast and cost effective intra-city railway network particularly in major cities. (iv) Immense population in India with resultant huge number of vehicles in India. (v) Faulty traffic management systems and frequent jams. (vi) No stringent enforcement of various standards/norms for checking/curbing pollution. (vii) Bad road conditions. (viii) Adulterated fuels. (ix) Older vehicles. (x) Inferior technology used in the automobiles and (xi) Automobiles with faulty engines/equipment's or having bad maintenance.

12.5.5 Emission norms/standards in India

The first emission norms were introduced in India in 1991 for petrol and 1992 for diesel vehicles.

In 2002, the Indian government accepted the report submitted by the Mashelkar committee. The committee proposed a road map for the roll out of Euro based emission norms for India. It also recommended a phased implementation of future norms with the regulations being implemented in major cities first and extended to the rest of the country after a few years.

Based on the recommendations of the committee, the National Auto Fuel policy was announced officially in 2003. The roadmap for implementation of the Bharat Stage norms was laid out till 2010. The policy also created guidelines for auto fuels, reduction of pollution from older vehicles and R&D for air quality data creation and health administration.

Bharat stage emission standards are emission standards instituted by the Government of India to regulate the output of air pollutants from motor vehicles. The standards, based on European regulations were first introduced in 2000. Progressively stringent norms have been rolled out since then. All new vehicles manufactured after the implementation of the norms have to be compliant with the regulations. Since October 2010, Bharat stage III norms have been enforced across the country. In 13 major cities including Delhi (NCR), Mumbai, Kolkata, Chennai etc., Bharat stage IV emission norms have been in place since April 2010.

The Government plans to promote eco-friendly cars in the country i.e. CNG based vehicle, hybrid vehicle, and electric vehicles. The government has formulated a Scheme for Faster Adoption and Manufacturing of Electric and Hybrid Vehicles in India, under the National Electric Mobility Mission 2020.

12.5.6 Measures to tackle air pollution from automobiles in India

The emission norms have helped in bringing down the pollution levels in the India. But, it is still too less an impact, as the pollution levels are rising across India in unprecedented manner, particularly in the metro cities like Delhi.

Therefore, stringent and drastic measures are required immediately like

- Phasing out of old vehicles.
- Ban on inferior technologies used in automobiles/engines like diesel etc.
- Well maintained automobiles with heavy penalty on the ones in bad shape.
- Creation of effective and efficient transport system in India.
- Fast and cost effective intra-city railway network e.g. Metro (MRTS).
- Effective, fast and responsive traffic management systems and avoiding jams.
- Good road conditions with more flyovers and underpasses etc.
- Clean and good quality fuels like CNG.
- Superior and environment friendly technology to be used in automobile production.
- Stringent enforcement of various standards/norms for checking/curbing pollution.
- Faster implementation of the Bharat stage V and VI emission norms
- Curb on the huge number of vehicles in India by imposing heavy duties/taxes.
- Other measures like congestion tax, restricting entry/using of vehicles for example odd-even formula of Delhi etc.

Questions

Short Questions

1. Give a brief account of air pollution due to engines.
2. Brief about the effect of air-fuel ratio on engine emissions with a suitable graph.
3. List out the pollution formation mechanisms in an automobile.
4. Compare the pollutant concentration for a petrol and diesel engine.
5. List out the effect the internal combustion emissions from an automobile.
6. Bring out the significance of stoichiometric ratio in emissions.
7. Write a shot note on the catalyst systems used in a catalytic convertor.

8. Give the chemical reactions in a three-way catalytic convertor for emission control.
9. Brief about the chemical indicator for oxides of nitrogen in the exhaust.
10. How does the air-fuel ratio effect the conversion efficiency of a TWC.

Essay Questions

1. Explain the different emissions from an automobile in detail. Give the chemical reactions also.
2. Schematize and explain the crankcase emission control system.
3. Describe the basic techniques employed in evaporation emission control system.
4. Explain in detail about the fuel options for controlling emissions.
5. What are catalytic converters? How are they helpful in reducing HC, CO and NO_x emissions?
6. With suitable diagrams, explain the working of a Three Way Catalytic Convertor.
7. Discuss the modern TWC employed with oxygen sensors.
8. Air pollution levels in Urban India have been increased. Justify your answer.

References

1. Environmental Almanac. 1992. World Resources Institute.
2. R. C. Flagar and J. H. Seinfeld, 1998. Fundamentals of air pollution engineering. Prentice-Hall, Englewood Cliffs, NJ.
3. A. Gunther, 1980. Gasohol.
4. J. B. Heywood and R. Tabaicyzniski, 1976. Current development in spark ingnition engines in air history of the automobile internal combustion engine. SAE publications. SP-409.
5. J. D. Powell and R. P. Brennan, 1988. The automobile technology and society. Prentice-Hall, Englewood Cliffs. NJ.
6. Medhurst, L. J. J. Chem. Educ. 2005, 82, 278-281.
7. R. M. Heck and R. J. Farrauto, Catalytic Air Pollution Control: Commercial Technology, Van Nostrand Reinhold, New York, 1995.
8. California energy commission, 1989. News and comment. CEC Quarterly News Letter no. 23.
9. C. B. Moyer, 1989. Global warming benefits of alternative fuels. A Cinox Corporation, Mountain View, CA.Dilbeck, C. W.; Ganske, J. A. Chem. Educator 2008, 13, 1-5.
10. D. A. Skoog, F. J. Holler and T. A. Nieman, Principles of Instrumental Analysis, 5th Edition, Harcourt Brace, Philadelphia, 1998, pp. 355-371.

11. H. H. Willard, L. L. Merritt, Jr., J. A. Dean and F. A. Settle, Jr., Instrumental Methods of Analysis, 6th ed., Van Nostrand, New York, 1989, Ch. 4.
12. U.S. EPA website on air emissions: http://www.epa.gov/air/emissions/index.htm
13. California Air Resources Board website: http://www.arb.ca.gov/homepage.htm
14. www.epin.ncsu.edu
15. Vehicle population in millions in India. Survey of environment, The Hindu. 2001.
16. Estimated passenger travel demand. Tata Energy Research Institute report. 1993.
17. http://www.siamindia.com/technical-regulation.aspx?mpgid=31&pgidtrail=3
18. https://en.wikipedia.org/wiki/Bharat_Stage_emission_standards

Suggested References

1. Air Pollution Control Engineering, de Nevers N., second Edition, 2000, McGraw Hill, New York.
2. Sharma, Anuja and Roychowdhury, Anuja. (1996), "Slow Murder: The deadly story of vehicular pollution in India". Centre for Science and Environment, New Delhi.
3. Hadden, Susan G. (1987), "Statutes and Standards for Pollution Control in India". Economic and Political Weekly, 22(16).
4. Sharma H.C. (1994), "Environmental Pollution Compliance." CBS, Publisher, New Delhi, India.
5. Sankar, U. (2004). "Law and Institutions Relating to Environmental Protection In India." The India: Environmental Management Capacity Building Technical Assistance Project, the World Bank.
6. CPCB, (1995). "Air Pollution and Its Control." Parivesh Newsletter, Central Pollution Control Board." pp 20.
7. CPCB, (2000). "Air Quality Status and trends in India, National Ambient Air Monitoring Series: NAAQMS". Central Pollution Control Board, New Delhi.
8. World Bank, (1996). "India's Environment: taking stock of plans, programs and priorities. South Asia Regional Office, The World Bank, pp 319.
9. CPCB, (2010). "National Ambient Air Quality Status and Trends in India". Central Pollution Control Board, Ministry of Environment and Forest, New Delhi.
10. Hosamane, Sateesh N. and Dr. Desai. (2013). "Urban Air Pollution Trend in India- Present Scenario". International Journal of Innovative Research in Science, Engineering and Technology. Vol.2 Issue 8. pp 3738 -3747.

11. CPCB, 2007, Air quality trends, http://www.cpcb.nic.in/Trends.php.
12. The World Bank report, (2013) "India: green growth- overcoming India's environment challenges to promote development". http://www. world bank org/en/news/press-release/2013/07/17/india-green-growth-neccessary-and affordable-for-india-says-new-world

CHAPTER 13

Noise Pollution

13.1 What is Noise Pollution?

Noise is "unwanted or disturbing sound". Sound becomes unwanted when it either interferes with normal activities such as sleeping, conversation, or disrupts or diminishes one's quality of life. The fact that you can't see, taste or smell it may help explain why it has not received as much attention as other types of pollution, such as air pollution, or water pollution. The air around us is constantly filled with sounds, yet most of us would probably not say we are surrounded by noise. Though for some, the persistent and escalating sources of sound can often be considered an annoyance. This "annoyance" can have major consequences, primarily to one's overall health.

As noise travels through air and undergoes changes it is also considered as one form of air pollution and noise regulation laws were enacted as a part of US Clean Air Act. The 1990 Clean Air Act Amendments added a new title IV, relating to noise pollution. The U.S. Code designates the original title IV (noise pollution) as subchapter IV a code 7641.The Noise Pollution Clearinghouse in US is a national non-profit organization with extensive online noise related resources. The Noise Pollution Clearinghouse seeks to: Raise awareness about noise pollution Create, collect, and distribute information and resources regarding noise pollution Strengthen laws and governmental efforts to control noise pollution Establish networks among environmental, professional, medical, governmental, and activist groups working on noise pollution issues Assist activists working against noise pollution.

13.1.1 Definition of noise

Noise is defined as any sound, independent of loudness, that can produce an undesirable psychological effect in an individual or group. Thus, noise, is unwanted sound energy and is also considered as a pollutant when it exceeds certain limits. Noise has a short residence and decay time and hence does not remain in the environment for long periods, like air or water pollutants. By the time the average individual is spurred to action to reduce, or control, or at least, complain about sporadic environmental noise, the noise may no longer exist, either to be noticed or measured. In scientific terms, noise (or sound) is a pressure oscillation in the air or water or any medium which conducts and travels (radiates) away from the source. If noise can be controlled at the source, there is a saving in energy and energy costs which will have an impact on production and production costs.

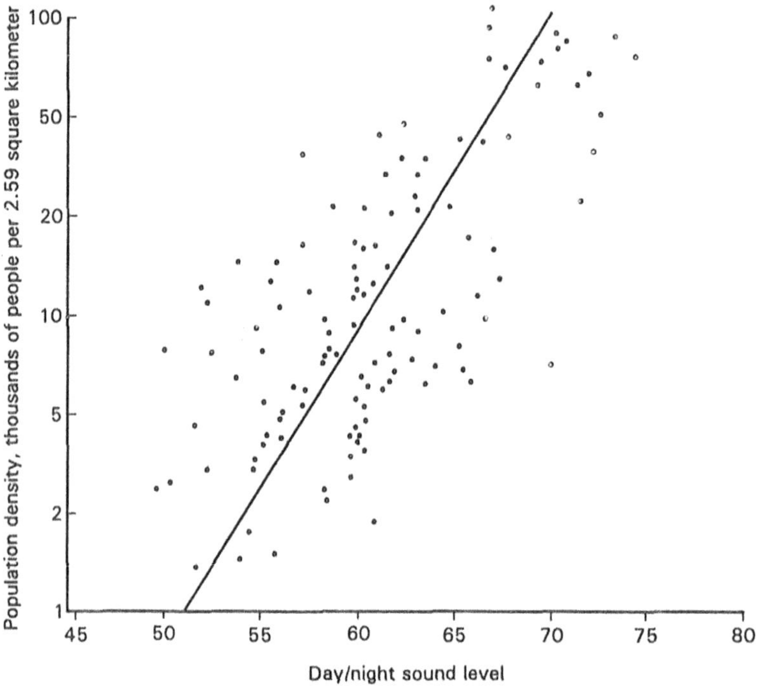

Fig. 13.1 *Day/night sound level as a function of population density*

Noise as a pollutant is a nuisance and cause for annoyance and affects the health of a person, his activities and mental abilities. Environmental noise pollution is not an entirely new phenomenon, but has grown steadily worse with time due to similar factors, responsible for air and water pollution, namely increasing population, urbanization, industrialization, technological change and usual relegation of environmental considerations to a position of secondary importance relative to economic ones.

Figures 13.1 and 13.2 show typical ambient day and night noise levels as a function of population density in a industrially developed society [1, 2] and in different environments respectively.

Noise annoys, distracts, disturbs and with sufficient exposure causes physiological effects leading to deafness. Annoyance results from interference with concentrated work, rest or sleep or with individual communication or speech. Noise in the work place reduces productivity, efficiency, accuracy and safety. The noise from industries is called Industrial Noise; from transport, Transport Noise; or present in the neighborhood due to natural and other causes, Ambient Noise.

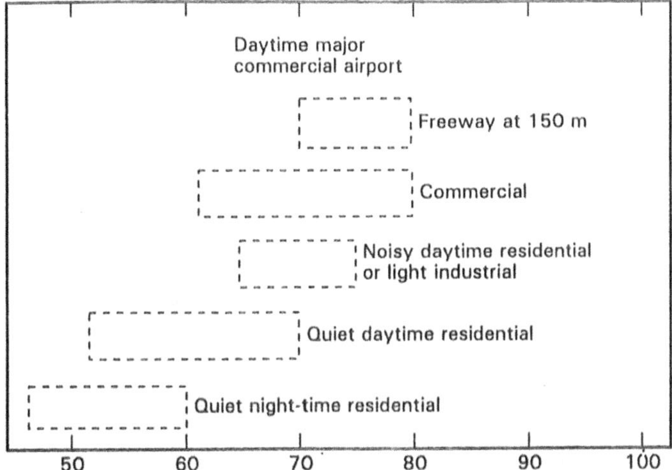

Fig. 13.2 *Representative ambient noise levels*

The extent to which noise contributes to the deterioration of our environment cannot be easily determined like that of pollution from other sources. It has been possible to assess impurities in the air or in water quantitatively but all persons are not affected to the same degree by the same noise. There is a vast variation in individual sensitivity to noise and people are affected differently when they are at home or outside or at work.

13.1.2 Comparison of air and noise pollution

There are a number of differences between air pollution and noise pollution.

1. The source of air pollution persists for some time, while the source of noise pollution can be immediately controlled or removed.
2. Air pollution is more serious than noise pollution, because air pollution adversely affects human health and can even be very fatal. The effects caused by noise pollution are not so fatal.
3. Noise pollution is local while air pollution is international.
4. Noise pollution does not contain harmful substances, while air pollution is due to various toxic substances such as CO_2, CO, SO_2, NO_x, particulate matter, organic matter, fly ash, etc.
5. Noise pollution cannot be fatal, even if not treated properly, but air pollution may cause havoc and so control measures are extremely necessary.

13.2 Characteristics of Sound

Sound is a form of energy giving the sensation of hearing and is produced by vibrations as longitudinal mechanical waves in materials including solids, liquid and gas and is transmitted by oscillation of atoms and molecules of matter. If we treat sound as an auditory sensation produced by these vibrating bodies, it gets characterized by the extent of vibrations like pitch, loudness or tonal quality. However, these characteristics cannot be objectively measured because they are psychological sensations and depend on the ear and judgment of the individual observer. In physical terms, sound may be defined as fluctuations in pressure of waves in a fluid medium. This definition gives rise to the objective characteristics of sound: frequency, intensity and waveforms. The pitch of a sound wave increases as its frequency increases but, there is no one-to-one ratio. Similarly, loudness varies with intensity. Loudness is defined as the observer's auditory impression of the strength of a sound. The loudness of a sound does not depend on the frequency or pitch. The human ear has a low sensitivity for low pitch and for extremely high pitch.

1. **The frequency (cycles/second)** of the waves determines the pitch of the sound.
2. **The amplitude or magnitude** (measured in decibel (dB) units) is a measure of loudness.

Sometimes sound intensity which is the amount of sound energy that flows through one unit area of medium in a unit time interval, is also measured in **watts per square metre (W/m^2).**

13.2.1 Sound pressure, power and intensity

There are three important parameters of noise, i.e., sound pressure, power and intensity.

13.2.1.1 Sound pressure

When sound is propagated through air, compression and rarefaction bands are formed, which are disturbances over existing conditions. While pressures are high in the compression band, they are lower in the rarefaction band. The differential of these pressures above and below existing atmospheric pressure will be responsible for the propagation of sound. No sound can be transmitted without these differences as the air will perceive no change.

Sinusoidal air waves propagated through the tube, depict the compression and rarefaction bands shown in Fig. 13.3 (a). The pressure differential Δp in a travelling sound wave is

$$\Delta p = \Delta p_m \sin (kx - \omega t) \qquad .....(13.1)$$

where Δp_m is the maximum pressure differential, k the angular wave number (the number of radians per unit of displacement x), and ω the angular frequency (the number of radians per unit time t) which when plotted appears as shown in Fig. 13.3(b).

In the figure, a differential strip of traveling air wave of thickness dx is isolated and depicted to be at a displacement s from the tip of x. This differential element oscillates to an amplitude of s_m that coincides with the middle of the rarefaction band (at minimum Δp). This means that, when s is at its maximum (s_m), the pressure differential is a minimum, indicating that if Δp uses the sine function, the displacement wave must use the cosine function and the displacement wave can be presented as

$$s = s_m \cos (kx - \omega t) \qquad \text{......(13.2)}$$

which when plotted appears as in Fig. 13.3(b).

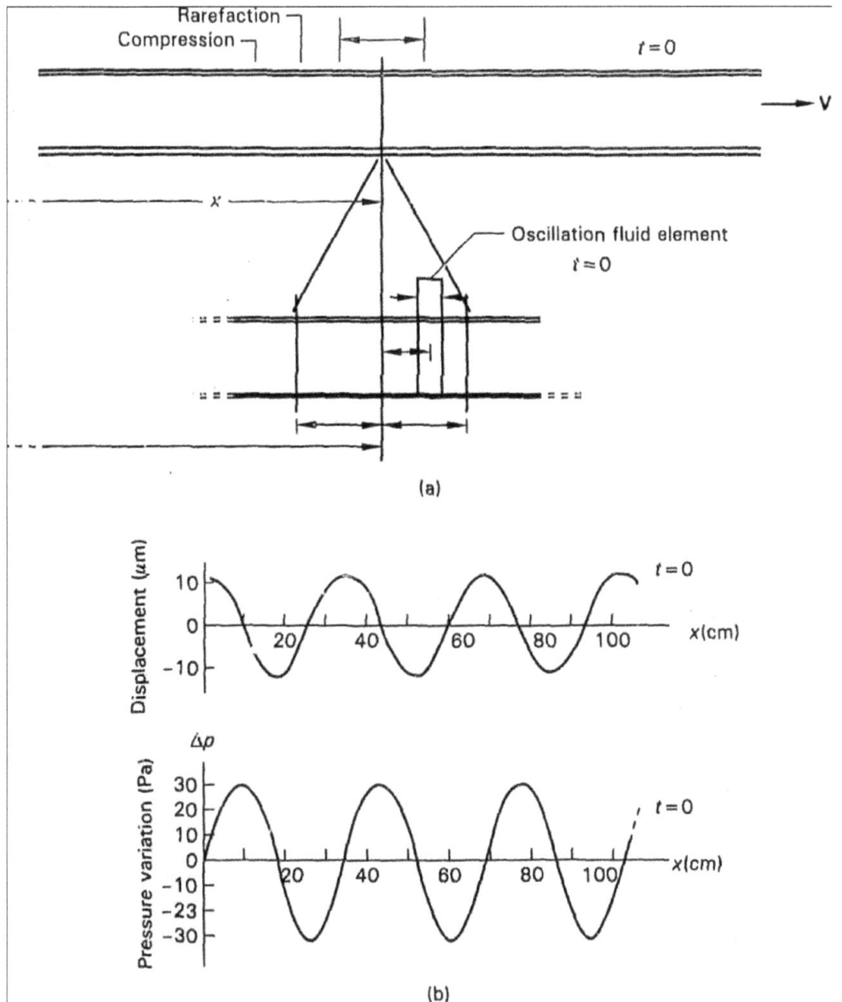

Fig. 13.3(a) *Sinusoidal wave sent through a tube;* **(b)** *Displacement and pressure waves*

From fluid mechanics the bulk modulus of elasticity of a gas can be written as

$$\Delta p = -\beta \frac{\Delta V}{V} \qquad \text{......(13.3)}$$

where β is the bulk modulus of elasticity, V the volume of a parcel of air, and ΔV the change in volume due to compression. V is equal to $A\Delta x$ and ΔV is equal to Δ $(A\Delta x)$ equals $A\Delta(\Delta x)$ = $A\Delta s$, where A is the cross-sectional area of the parcel of air through which sound, propagates. Substituting in Equation (13.3) and taking limits yields

$$\Delta p \lim_{\Delta x \to 0} -\beta\frac{\Delta s}{\Delta x} = -\beta\frac{\partial s}{\partial x} \qquad \text{......(13.4)}$$

Partially differentiating Equation (14.2) with respect to x and substituting the result in Equation (13.4), we have

$$\Delta p = \beta k s_m \sin(kx - \omega t) = \Delta p_m \sin(ks - \omega t) \qquad \text{......(13.5)}$$

From Equation (13.5),

$$\Delta p_m = (\beta k)s_m \qquad \text{......(13.6)}$$

The velocity of propagation of sound v is

$$v = \sqrt{\frac{\beta}{\rho}} \qquad \text{......(13.7)}$$

where ρ is the mass velocity of air. Therefore the differential pressure is

$$\Delta p_m = (v^2\rho k)s_m = (v\rho\omega)s_m \qquad \text{......(13.8)}$$

13.2.1.2 Sound power

Kinetic and potential energies constitute the energy of an oscillating sound wave. When a spring is compressed, it possesses potential energy, and when it is released against an object placed on it, the energy will be transferred to the object as kinetic energy. Neglecting the effect of friction, the resulting kinetic energy will be equal to the original potential energy. The air propagating the sound wave or noise is like a spring. In its compressed state, it also possesses potential energy. When it expands toward the rarefaction state, the compressed state potential energy will be released not to another object, however, but to air itself, as kinetic energy. The frictional energy of expansion may be neglected, and the kinetic energy is also therefore equal to the original potential energy. Hence, for any given parcel of air, the energy content of the travelling sound wave is equal to its potential energy plus kinetic energy equals twice its kinetic energy.

The kinetic energy dK of a differential mass of air dm propagating the sound wave is

$$dK = \frac{1}{2}dmv_s^2 \qquad \text{......(13.9)}$$

where v_s is the velocity of oscillation of the air mass but not the velocity of sound, which is v, v_s may be obtained by partially differentiating Equation (13.2) with respect to t and substituting the result for v_s in Equation (13.9) we have

$$dK = \frac{1}{2}(\rho A dx)(-\omega s_m)^2 \sin^2(kx - \omega t) \qquad \text{......(13.10)}$$

where $\rho A dx$ is dm.

The total energy dE is twice dK. Also, power is the rate of change of E with respect to time, dE/dt. Hence, from Equation (13.10) and the fact that dx/dt is v,

$$\frac{dE}{dt} = \frac{dK}{dt} = \rho A v \omega^2 s_m^2 \sin^2(kx - \omega t) \qquad \text{......(13.11)}$$

Taking the average over a whole number of wavelengths, the average power, $\frac{dE}{dt}$ is

$$\overline{\frac{dE}{dt}} = P_w = \rho A v \omega^2 s_m^2 \overline{\sin^2(kx - \omega t)} = \frac{1}{2} \rho A v \omega^2 s_m^2 \qquad \text{......(13.12)}$$

where $\sin^2(kx - \omega t) = \frac{1}{2}$ for a whole number of wavelengths.

13.2.1.3 Sound intensity

The average rate at which power is transmitted per unit cross-sectional area in the direction of travel is termed as the intensity I of a sound wave. So from Equation (13.12), we have

$$I = \frac{\frac{1}{2}\rho A v \omega^2 s_m^2}{A} = \frac{1}{2} \rho v \omega^2 s_m^2 \qquad \text{......(13.13)}$$

From Equation (13.8), Equation (13.13) may also be written as

$$I = \frac{\Delta p_m^2}{2\rho v} \qquad \text{......(13.14)}$$

$\Delta p_m / \sqrt{2}$ is termed as the *root mean square* (rms) pressure of the sound wave, Δp_{rms}. So, in terms of rms pressure, Equation (13.14) can be written as

$$I = \frac{\Delta p_{rms}^2}{\rho v} \qquad \text{......(13.15)}$$

13.2.1.4 Loudness

Loudness is expressed in a relative unit, called **sone.** One sone equals the loudness of 40 dB sound pressure at 1000 cycles per second (cps), while 40 dB sound at 5000 cps seems twice as loud and equals the value of 2 sones. Generally 85 dB, with loudness 10–50 sones, depending on the frequency may be regarded as the critical level for ear damage. Normally hearing is most acute in the frequency range 2000-5500 Hz. Perfect hearing lies in the range of about 16 (infra-audible) to 20,000 (Ultrasonic) Hz.

The noise value of sound waves depends upon the following important factors.

1. The frequency of sound waves.
2. The intensity of sound waves.
3. The time of exposure of sound waves.
4. Intermittence of sound waves.

The unit of loudness expressed by the mathematical exponent of number 10 is called a **bel**. One tenth of a **bel** is taken for quantitative measurements. Thus a **decibel** is considered to be the threshold of hearing. A whisper is about 20–30 decibels (db) and an average speaking voice is about 60 decibels. The loudness of a sound that a person can withstand without discomfort is about 80 decibels. Automobile horns may reach more than 90 decibels and a jet airplane at a distance of 30 m may have an intensity of about 140 decibels.

Loudness may be described as a listener's auditory impression which is not synonymous with sound pressure. For example, a sound of 20 Hz (Hertz) must have a sound pressure of 80 dB to be audible, but a sound of 2000 Hz is audible at a pressure of 20 dB.

In defining decibel, arbitrary reference values are chosen, depending on the type of decibel to be defined. For example, for power, the reference value is 10^{-12} W. For pressure and intensity, the references are, respectively, $2(10^{-5})$ Pa and 10^{-12} W/m^2. In terms of general definition, the decibel is

$$\text{decibel} \quad (dB) = 10 \log \frac{x}{y} \qquad \text{......(13.16)}$$

where x is any one of pressure, power, and intensity, and y relates to the respective reference values.

Using the respective decibels, the word *level* is now used. For example, for power, the decibel is the sound power *level* L_w, and for pressure and intensity, the decibels are sound pressure *level* L_p, and sound intensity *level* L_l, respectively. Using Equation (13.16) the respective definitions of the appropriate decibels follows.

$$L_w\,(dB) = 10\log \frac{p_w}{10^{-12}} \text{ sound power level} \qquad \text{......(13.17)}$$

$$L_l\,(dB) = 10\log \frac{l}{10^{-12}} \text{ sound intensity level} \qquad \text{......(13.18)}$$

L_w and L_l are the sound power and sound intensity levels, respectively.

In the definition of the sound pressure level, the rms value, Δp_{rms}, is used. Since the intensity I is proportional to the square of pressure, to be consistent with Equation (13.18), the definition for sound pressure level, L_p, should have the rms pressure and the reference pressure being squared, respectively.

Thus

$$L_p\,(dB) = 10\log\frac{\Delta p_{rms}^2}{(2\times10^{-5})^2} = 10\log\frac{\Delta p_{rms}^2}{(4.0\times10^{-10})}\ \text{sound pressure level} \qquad .....(13.19)$$

13.2.1.5 Relationships among intensity, pressure and power levels

Substituting Equation (13.15) in Equation (13.18) gives

$$L_I\,(dB) = 10\log\frac{\Delta p_{rms}^2}{\rho v\times10^{-12}} \qquad ......(13.20)$$

At 25 °C, ρ_{air} = 1.184 kg/m^3, β_{air} = 20.6 psi = 142038.51 N/m^2.

So $f = \sqrt{\dfrac{\beta}{\rho}} = \sqrt{\dfrac{142038.71}{1.184}} = 346.4$ m/s

Substituting the values of ρ and v in Equation (13.20) gives

$$L_I = 10\log\frac{\Delta p_{rms}^2}{4.1\times10^{-10}} \qquad ......(13.21)$$

Although Equation (13.21) holds at 25 °C, from Equation (13.19), $L_I = L_p$, approximately.

By definition, $P_w = IA$, where A is the area of air that is receiving noise normal to the direction of propagation. So

$$L_w = 10\log\frac{P_w}{10^{-12}} = 10\log\frac{IA}{10^{-12}} = 10\log\frac{I}{10^{-12}} + 10\log A = L_I + 10\log A \qquad ......(13.23)$$

Problem 13.1

The sound power from a loud speaker is 0.001 W. What is the sound power level? What are the sound intensity, the sound intensity level, the sound pressure, and the sound pressure level at a distance 10 m from the source?

Solution:

$$D_w = 10\log\frac{p_w}{10^{-12}} = 10\log\frac{0.001}{10^{-12}} = 90\,\text{dB}$$

Assume that the sound radiates from the source in all directions. Hence, the area of propagation at 10 m distance is the surface area of a sphere = $4\pi r^2 = 4\pi(10^2) = 1268$ m^2. Then

$$I = \frac{0.001\ \text{W}}{1268\ \text{m}^2} = 0.778\times10^{-6}\ \text{W/m}^2$$

$$L_I = 10\log\frac{0.78\times10^{-6}}{10^{-12}} = 58.9\,\text{dB}$$

$$L_I \approx L_p \qquad L_p = 58.9\text{ dB}$$

$$L_p = 10\log\frac{\Delta p^2_{rms}}{4.0\times10^{-10}} = 58.9$$

$$\Delta p_{rms} = \frac{58.9}{113.0} = 0.52\text{ N/m}^2$$

A scale showing some common sound pressure levels is given in Fig. 13.4.

13.2.1.6 Addition of sound pressure levels

Due to the logarithmic nature of sound pressure expressions when two types of sound pressures expressed in decibels are combined they do not add up or subtract in normal arithmetic fashion. For example if we add up 60 decibel noise (with reference 20 μPa) and another noise of 60 decibels we get a resultant noise of 63 decibels. If we have several levels of sound to be combined they should be combined two at a time with lower levels adding first and continuing two at a time with each successive pair until one number remains. Figure 13.5 gives the solution for the conversion of dB to the relative powers and adding and converting them to dB again.

Problem 13.2

Estimate the sound pressure level that results from combining the following three levels (all re: 20μPa): 65 dB, 81 dB, 71 dB.

Solution: Selecting the two lowest levels 65 dB and 71 dB, the difference between the values is 71–65 = 6.0. Using Fig. 13.5 draw a vertical line from 6.0 on the abscissa to intersect the curve. A horizontal line from the intersection to the ordinate yields about 1.0 dB. Thus, the combination of 65 dB and 71 dB results in a level of 72.0 dB. This, and the remainder of the computation is shown diagrammatically below.

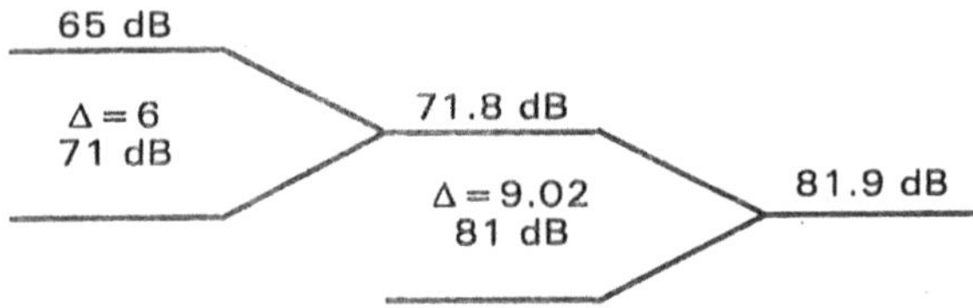

Rounding off to the nearest whole number yields an answer of 90 dB (re: 20 μPa)

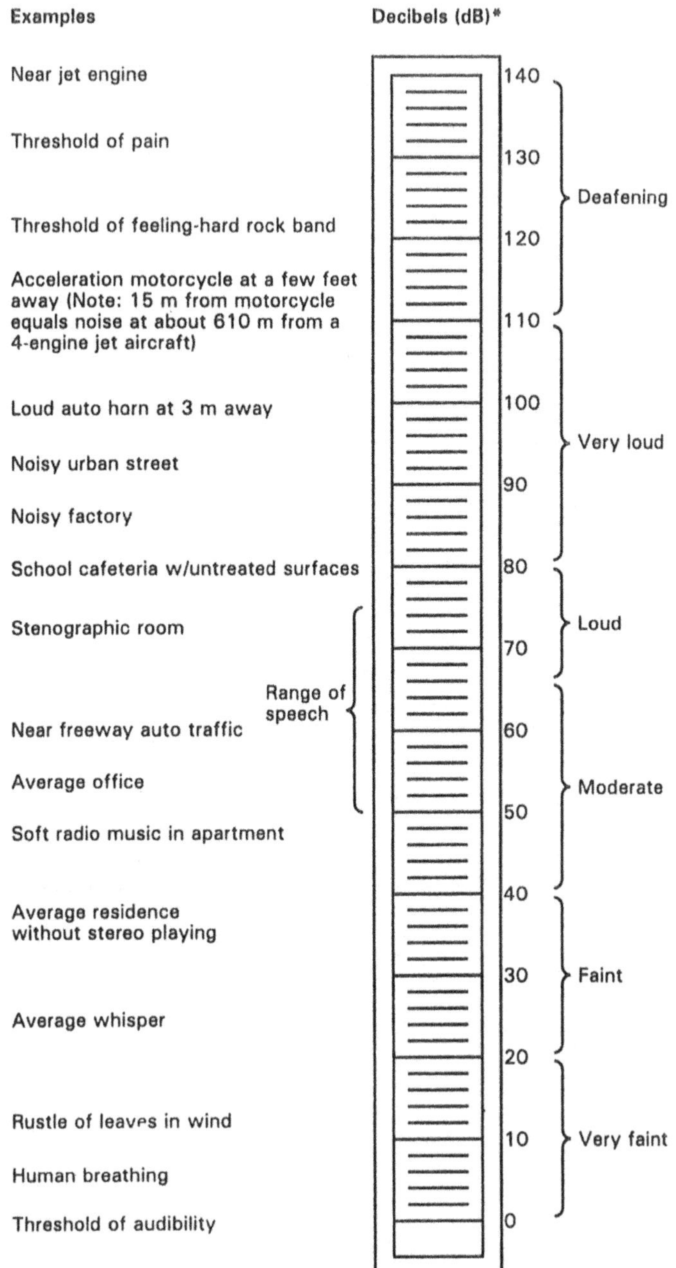

*dB are "average" values as measured on the A-scale of a sound-level meter

Fig. 13.4 *Scale showing common sound pressure levels*

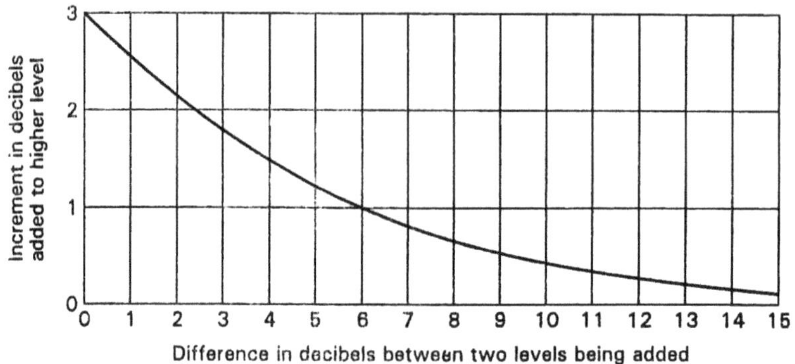

Fig. 13.5 *Noise levels log, addition*

13.2 1.7 Averaging of sound pressure levels

Due to the logarithmic nature of sound pressure levels expressed in dB, for measuring value of the sound pressure levels the following expression should be used.

$$L_p = 20\log\frac{1}{N}\sum_{j=1}^{N}10^{L_j/20} \qquad \text{......(13.24)}$$

where L_p = average sound pressure level dB x = 20μPa

N = number of measurements

L_j = the Jth sound pressure level dB x mPa J = 1, 2, 3–N.

The same equation holds good if the sound levels are expressed in dB(A) or if set 20 μPa is replaced by 10μPa.

Problem 13.3

Estimate the mean sound level from the following four readings: 25, 40, 72, and 85 dB(A).

Solution:

First estimate

$$\sum_{j=I}^{\Delta} = 10^{25/20} + 10^{40/20} + 10^{72/20} + 10^{85/20}$$

$$= 2.18 \times 10^4$$

Then compute the calculation

$$L_p = 20\log\frac{2.18\times10^4}{4} = 74.72\,\text{dBA} = 75\,\text{dBA}$$

However if arithmetic average is done directly, we get 55.5 dB(A).

13.2.1.8 Measurement of sound

The quantification of sound levels for the purpose of determining whether they are hazardous to health and welfare, or exceed local legally adopted limits, can be a complex task. Depending upon both the purpose of the measurement and the desired accuracy of the result, a wide range of measurement techniques and sound level descriptors are used. The instrumentation available to perform the assessment should be known to potential users so that their selection is compatible with the required task.

13.2.1.9 Sound level meter

A sound level meter can measure sound pressure levels in the range 20–40 dB(A) at frequencies between3.5 Hz and 20kHz with an additional suitable microphone. It is an instrument consisting of a first class microphone, coupled to a good amplifier and terminating in a logarithmic indicator calibrated in decibels. A meter of this type lacks certain characteristics of the human ear but is, nevertheless, very useful for measuring the levels of sounds. It should be remembered that it is not equally sensitive at all frequencies. Two sounds at different frequencies may have exactly same sound pressure levels assessed by the ear to be louder than the other due to the varying sensitivity of the ear. A sound meter, however, incorporates "frequency weighted" networks designated to provide a closer relationship between the sensitivity of instrument and the response of human ear.

The meter will be positioned in a desired location with no obstruction from the sound source and the reading will be taken. A sound level meter attached to calibrator is shown in Fig. 13.6(a). A *calibrator* produces a reference sound and when attached to the microphone calibrates the meter. The vertical dashed line in the figure indicates the point where the calibrator will be attached to the meter.

Most sound levels meters contain a microphone, amplifiers, weighting networks, and a display reading in decibels [Fig. 13.6 (b)], Sound pressure coming from a source reaches the microphone, where a pressure transducer such as a diaphragm mimics he pressure pattern of the source and converts the signal to a small current of electricity. The electrical signal thus produced will be then amplified in a preamplifier.

Sound pressure that the meter receives is not the same as what the ear would perceive. For this reason an electronic circuitry called a *weighting network* is built into the meter so as to produce a readout that closely resembles a human response. A series of three internationally accepted weighting scales has been adopted. These are the A, B, and C weighting networks. The *A network* approximates human response to low-intensity sounds, the *B network* approximates the response to medium-intensity sounds, and the *C network* approximates the human response to high-intensity sounds. Figure 13.6(c) shows the plot for the three networks.

The networks circuitry electronically subtracts the actual sound pressure level of a particular frequency that reaches the microphone according to in which network the actual pressure level reaching the meter is subtracted by 9 and the difference is shown as the readout of the meter. Depending on whether the network used is A, B, or C, readouts are in dB(A), dB(B), or dB(C). As shown in the figure, at low frequencies, the A network subtracts the incoming signal severely; the C network less; and the B network in between. Since the readings are not actually the true sound pressure levels, readouts from the A, B and C networks are called *sound levels*, not sound pressure levels. This is why the meters are called *sound level meters* and not *sound pressure level meters*.

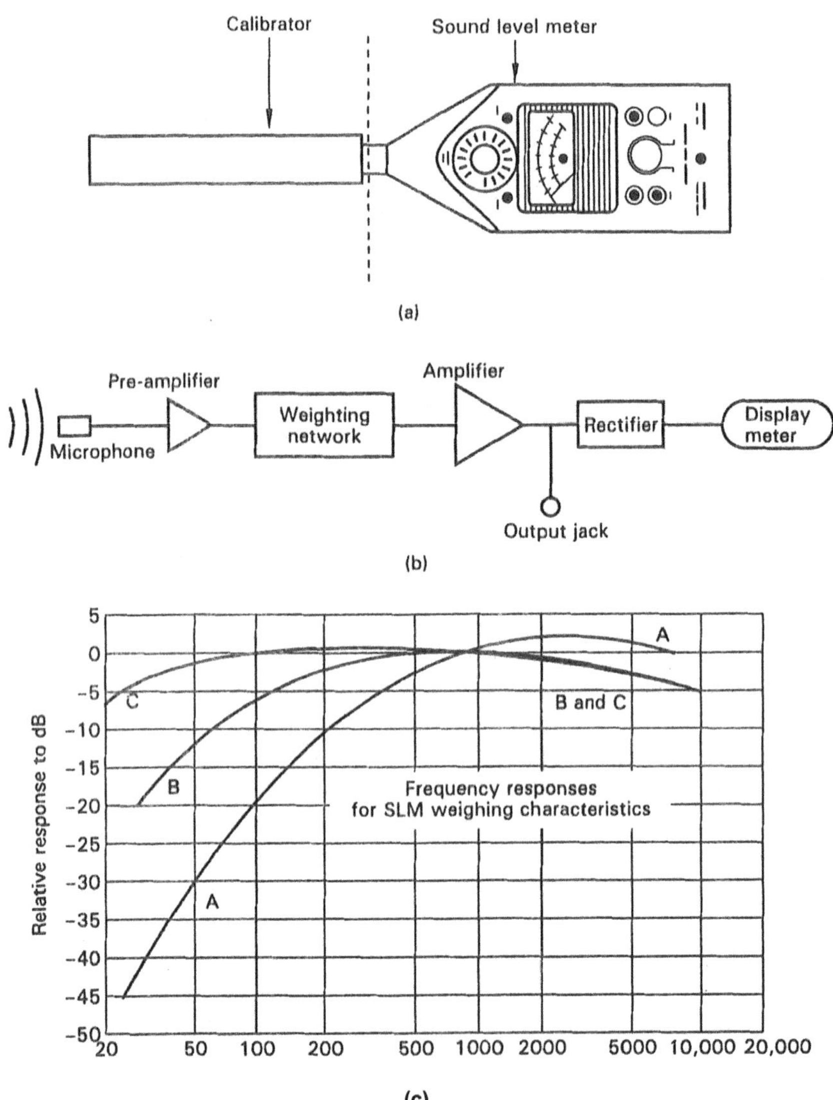

Fig. 13.6(a) *Sound level meter;* **(b)** *Main parts of sound level meter;* **(c)** *Weighting networks plots [3]*

The A network is more commonly used. Another network, the *D-weighting network* has also been recommended as closely resembling the human response to noise at airports.

13.3 Effects of Noise on People

13.3.1 Normal hearing–Frequency range and sensitivity

The ear of the healthy, adult male responds to sound waves in the frequency range 20–16,000 Hz (Figure 13.7). Young children and women often have the capacity to respond to frequencies up to 20,000 Hz. The speech zone lies in the frequency range of 500–2,000 Hz. The ear is most sensitive in the frequency range 2,000–5,000 Hz. The smallest perception of sound pressure in this frequency range is 20 μPa. A sound pressure of 20 μPa at 1,000Hz in air corresponds to a 1.0 nm displacement of the air molecules. The thermal motion of the air molecules corresponds to a sound pressure of about 1μPa. If the ear were much more sensitive, you would hear the air molecules crashing against your ear like waves on the beach.

13.3.2 Repeated interference with sleep

Noise can awaken people from sleep and it can keep them awake. Frequent awakening or awakening for long periods can be very disruptive even if not awakened by noise, a person's sleep pattern can be significantly disturbed, and a reduced feeling of well-being can result the next day. Frequent and prolonged sleep disturbances can result in physical, mental or emotional illness.

13.3.3 Effects on communication

External sounds can interfere with conversations and use of the telephone as well as the enjoyment of radios and television programmes. It can thus affect the efficiency of offices, [Fig. 13.8(a)] schools and other place where communication is of vital importance. The maximum acceptable level of noise under such conditions is 55dB.70 dB is considered very noisy and causes serious interference with verbal communication. Figure 13.8(b) presents how the quality of speech communication is affected by different background sounds and talker-to-listener distances.

13.3.4 Physiological responses

Physiological responses to noise exposure include [8] (Table 13.1)

1. A vascular response characterised by peripheral vaso-constriction, changes in heart rate and blood pressure.
2. Various glandular charges such as increased output of adrenaline evidenced by chemical charges in blood.

3. Slow, deep breathing.
4. A change in the electrical resistance of skin with changes in activity of the sweat glands.
5. Brief changes in skeleton muscle tension.

Constant noise may cause blood vessels to contract, skin to become pale, muscles to contract and adrenaline to be shot into the bloodstream. The adrenaline is responsible for both excitatory and inhibitory responses in living beings. This is the reason factory workers develop abnormal heart-beat rates and suffer from insomnia, nervousness and impaired motor co-ordination. Considering these points U.S. Government has kept 90 decibels as a health hazard for an 8-hr-day working environment. It has been proved that high noise is bad, particularly for those suffering from high tension diabetes. Noise also produces startling effects on babies and they may even develop a fear psychosis as a result of sharp and sudden noise.

13.3.5 Epidemiological studies

Several researchers have conducted field studies testing industrial workers, collating their health records in an attempt to overcome the limitations of duration and realism in laboratory studies. The physiological problems amongst two groups of workers are classified as from *very noisy industries* and *less noisy industries* respectively. The difference between the two groups indicates a higher incidence of problems in the high-noise group than in the low-noise group. Noise level, however, is not isolated in such a study as the cause. The two groups are from two separate industries; the high- noise group from light industries such as textile industries. Between these industries there were numerous other differences which could have effects on health, such as heat, physical workload, anxiety, and the type of people.

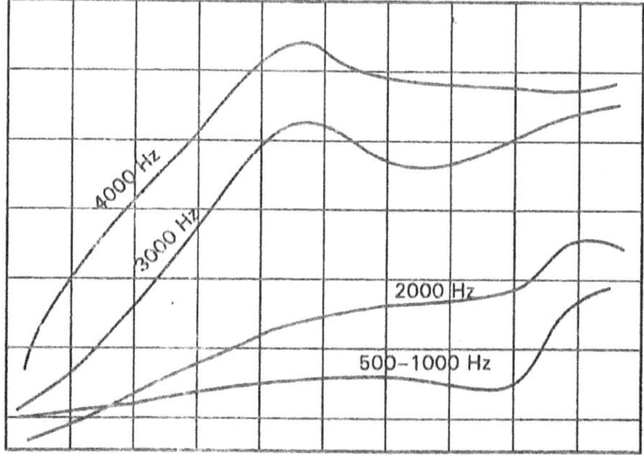

Fig. 13.7(a) *Hearing impairment with years of exposure [4]*

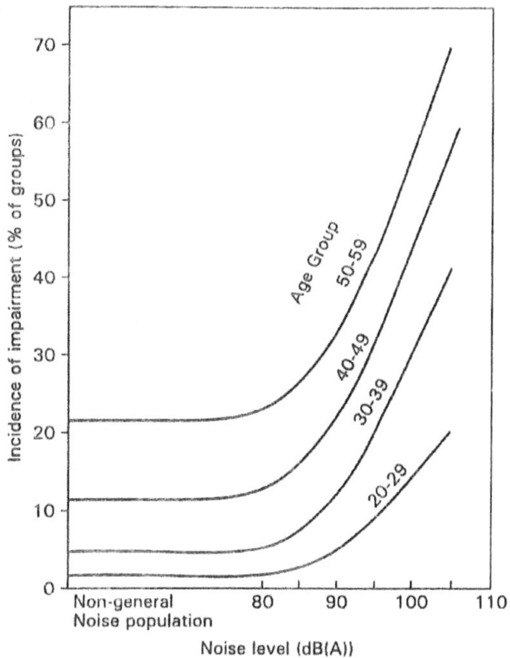

Fig. 13.7(b) *Incidence of hearing impairment (average hearing threshold level in excess of 15 db ASA at 500, 10000 and 2,000 cps) [5]*

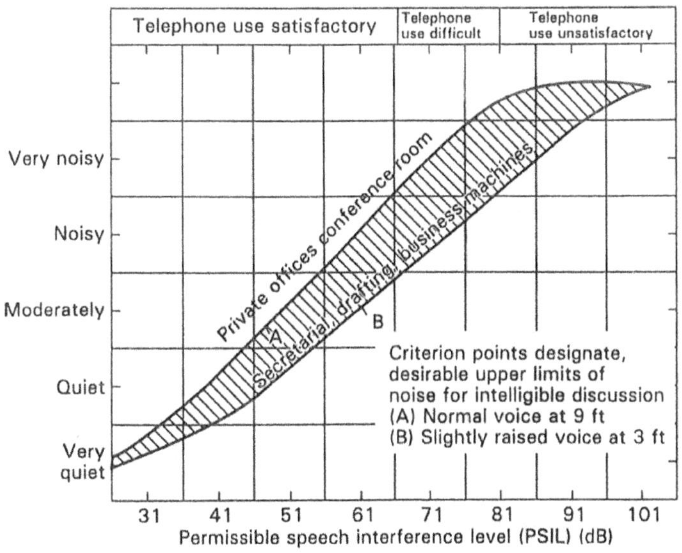

Fig. 13.8(a) *Rating chart for office noises [6]*

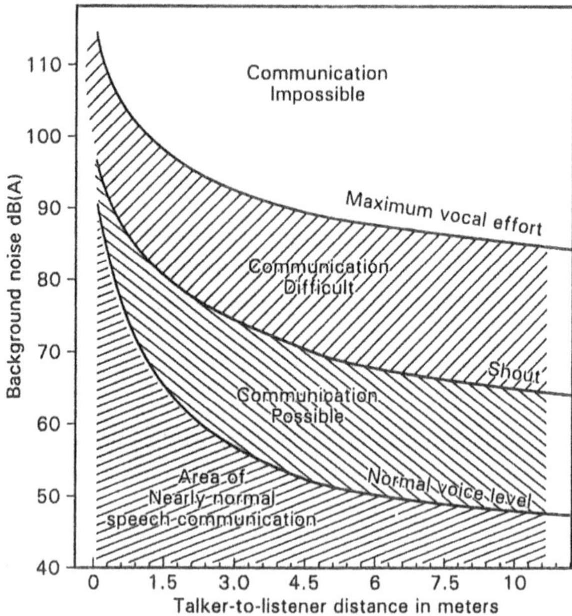

Fig. 13.8(b) *Quality of speech communication with distance (in relation to the A-weighted sound level of noise, dB(A) and the distance between the talker and the listener) [7]*

Table 13.1 *Effects of high intensity noise on human beings*

Noise (dB)	Effects observed
0	Threshold of change in pulse rate
150	Significance change in pulse rate
110	Stimulation of reception in skin
120	Pain threshold
130–135	Nausea, vomiting, dizziness, with touch and muscle sense
140	Pain in ear, prolonged exposure causing insanity
140	Extreme limit of human noise tolerance
150	Prolonged exposure causing burning of skin
160	Minor permanent damage if prolong
190	Major permanent damage in a short time

13.3.6 Noise rating system

Any noise rating system must take frequency into account, must be able to differentiate between day-time and night-time noise, but must be capable of describing the cumulative noise exposure. These requirements can be met by a statistical system only.

13.3.6.1 L_N and L_{eq} systems for noise rating

L_N system: L_N is a statistical measure that indicates how frequently a particular sound level is exceeded. For example, L_{40} = 72 dB(A), indicates that 72 dB(A) is exceeded 40% of the measuring time. A plot of L_N against N = 1%, 10%, 20% etc. is shown in Fig. 13.9. Figure 13.9 gives the cumulative distribution curve or the probability distribution curve which will show how often the noise levels fall into certain class intervals. It can be seen from the figure that 22% of the time the measured noise levels ranged between 70 and 72 dB(A); for 17% of the time they ranged between 72 to 74 dB(A); and so on. By adding the percentages given in successive class intervals from right to left, we can arrive at a corresponding L_N where N is the sum of the percentages and L is the lower limit of the leftmost class interval added, thus-L_{40}:

$$L(2 + 7 + 14 + 17) = 72\text{dB(A)}$$

The L_{eq} system is a method for describing fluctuating noise levels in which the equivalent continuous equal energy level, L_{eq} concept can be used for describing the constant noise level, over a given time, and is given by Equation (13.25).

$$L_{eq} = 10\log\frac{1}{t}\int_{0}^{t}10^{L(t)/10}\,dt \qquad \text{......(13.25)}$$

where t = the time over which L_{eq} is determined; $L(t)$ = the time varying noise level in dB(A). As there is no well-defined relationship between $L(t)$ and time a number of discrete samples of $L(t)$ have to be taken. L_{eq} is given by

$$L_{eq} = 10\log\sum_{i=1}^{i=n}10^{L_i/10}\,t_1 \qquad \text{......(13.26)}$$

where n = the total number of samples taken; L_i = the noise level in dB(A) of the ith sample: t_1 = fraction of total sample time.

Problem 13.4

A noise level of 95 dB(A) exists for ten minutes and is followed by a reduced noise level of 45 dB(A) for 40 min. Estimate equivalent continuous equal energy level for the 90 min period for a five-minute sampling interval.

Solution: If the sampling interval is 10 min, then the total number of samples (n) is 9, and the fraction of total sample time (t_1) for each sample is 1/9 = 0.101. Then we can calculate

$$\sum_{t=1}^{2} = 10^{95/10}(0.10) + 10^{45/10}(0.10)$$

$$= (3.1 \times 10^{8}) + (3.1 \times 10^{3}) = 3.1 \times 10^{11} = 114.9 = 115 \text{ dB(A)}.$$

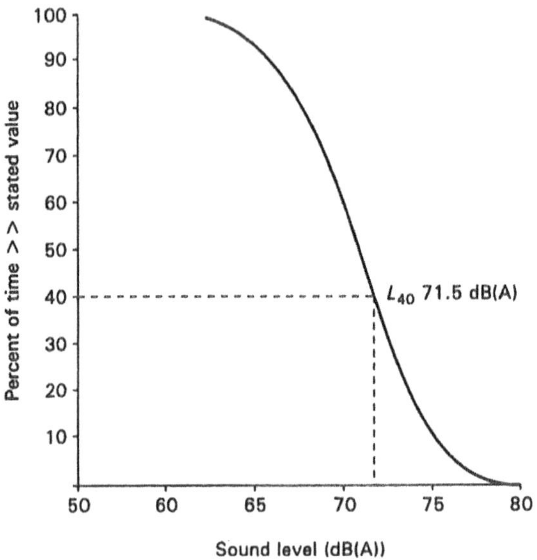

Fig. 13.9(a) *Cumulative distribution curve*

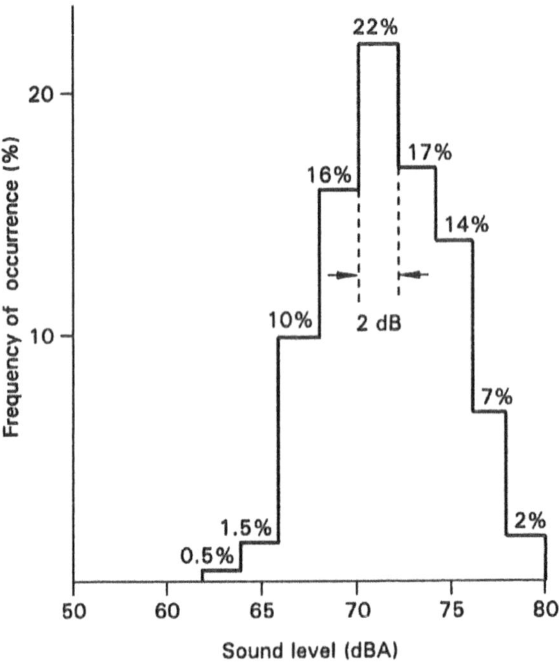

Fig. 13.9(b) *Probability distribution plot*

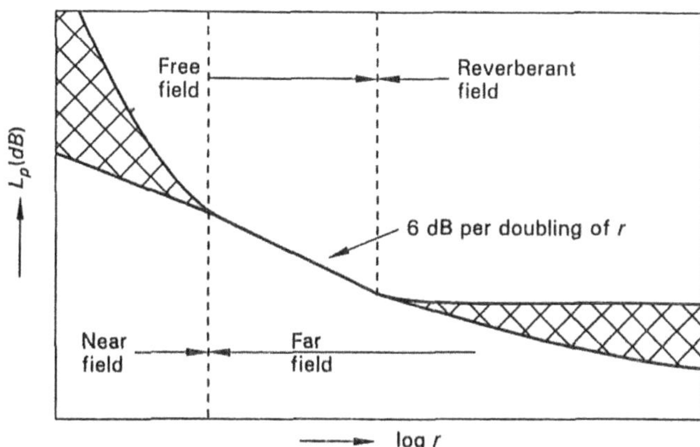

Fig. 13.10 *Variation of sound pressure level in an enclosure along the radius r from a noise source*

13.4 Propagation of Noise in Ambient Air

13.4.1 Variation of character of wave radiation with varying distance from a sound source

The character of the wave radiation from a noise source will vary with distance from the source (Fig. 13.10). At locations close to the source, the near field, the particle velocity is not in phase with the sound pressure. In this area, L_p fluctuates with distance and does not follow the inverse square law (Fig. 13.11). When the particle velocity and sound pressure are in phase, the location of the sound measurement is said to be in the far field. If the sound source is in free space, that is there are no reflecting surfaces, then measurements in the far field are also free field measurements. If the sound source is in a highly reflective space, for example, a room with steel walls, ceiling, and floor then measurements in the far field are also reverberant field measurements. The shaded area in the far field of Fig. 13.10 shows that L_p does not follow the inverse square law in the reverberant field.

The sound intensity I at a distance from a point noise source radiating uniformly in all directions in the surroundings is $P_w/4\pi r^2$, where $4\pi r^2$ is the spherical area receiving the noise. If the source is not a point source, or if the source is not radiating uniformly in all directions, a factor Q, called the directivity, may be factored in. Hence, in general, I at a distance r from the source may be written as

$$I = Q\frac{P_w}{4\pi r^2} \qquad \text{......(13.27)}$$

where $4\pi r^2$ is now a spherical area equivalent to the area receiving the noise. The area receiving the noise may also be designated generally as the area S, whether spherical or nonspherical. With power P_w the general intensity at the distance r of S is

$$I = \frac{P_w}{S} \qquad \text{......(13.28)}$$

Noise dose D is

$$D = \Sigma \frac{C_1}{T_c} \qquad \text{......(13.28a)}$$

where C is the total exposure time of a given steady state noise level and T_c is allowable time for exposure.

Dividing both sides of above equation by 10^{-12} W/m^2, taking the "decibel" logarithm, performing the necessary algebra, and using the definition of the respective decibels, the following equation is obtained:

$$L_p \approx L_I = L_w + 10\log\frac{Q}{r^2} - 11 \qquad \text{......(13.29)}$$

where

L_p = sound pressure level, dB = 10 log $P_w/10^{-12}$ = 20μPa

L_w = sound power level, dB = 10 log I/10^{-12} = 10^{-12} W

r = distance between source and receiver, m

20 log r = decibel transform = 10 log r^2

11 = decibel transform = 10 log (4π) = 10.99

This equation is the general representation of the *inverse square law* (Fig. 13.11).

From Equations (13.27) and (13.28) we have

$$Q = \frac{4\pi r^2}{S} \qquad \text{......(13.30)}$$

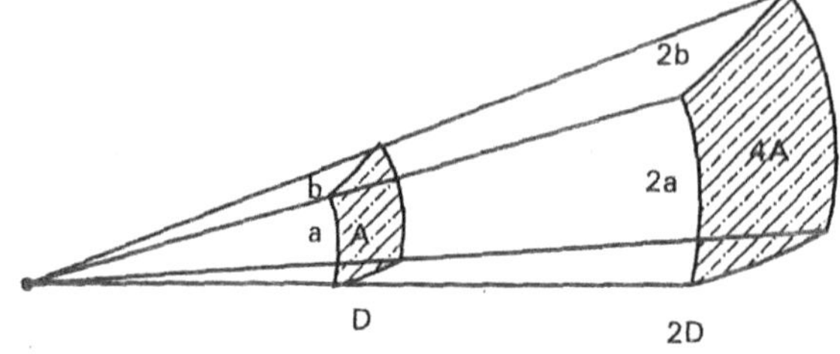

Fig. 13.11 *Illustration of inverse square law*

13.4.2 Directivity

Most real sources do not radiate sound uniformly in all directions. If you were to measure the sound pressure level in a given frequency band at a fixed distance from a real source, you would find different levels for different directions. If you plotted these data in polar coordinates, you would obtain the directivity pattern of the source.

The directivity factor is the numerical measure of the directivity of a sound source. The logarithm form of directivity factor Q is called as directivity index DT_θ.

For a spherical source it is defined as follows:

$$DI_\theta = L_{p\theta} - L_{ps} \quad \text{......(13.31)}$$

where

$L_{p\theta}$ = sound pressure level measured at distance r and angle θ^o from a directive source radiating power W into anechoic space dB

L_{ps} = sound pressure level measured at distance r from a non-directive source radiating power W into anechoic space, dB

For a source located on or near a hard, flat surface, the directivity index take the following form.

$$DI_\theta = L_{p\theta} - L_{ps} + 3 \quad \text{......(13.32)}$$

The 3 dB is made because the measurement is made over a hemisphere instead of a sphere. That is, the intensity at a radius r, is twice as large if a source radiates into a hemisphere rather than the ideal sphere we have used up to this point. Each directivity index is applicable only to the angle at which $L_{p\theta}$ was measured and only for the frequency at which it was measured.

As the directivity pattern does not change its shape regardless of the distance from the source we can apply the inverse square law to directive sources simply by adding the directivity index.

$$L_{p\theta} = L_w + DI_\theta - 20\log r - 11 - A_e \quad \text{......(13.33)}$$

The values of $L_{p\theta}$ are at the distance r.

Problem 13.5

Determine the value of Q for the following placement of a point source: (a) radiating uniformly in all directions, (b) in a corner, and (c) on a flat surface.

Solution:

Let r be the distance.

$$Q = \frac{4\pi r^2}{S} \quad \text{......(13.34)}$$

(a) $S = 4\pi r^2$; $\quad Q = \dfrac{4\pi r^2}{4\pi r^2} = 1$

(b) $S = \dfrac{1}{8}(4\pi r^2)$; $\quad Q = \dfrac{4\pi r^2}{\frac{1}{8}(4\pi r^2)} = 8$

(c) $S=\frac{1}{2}(4\pi r^2)$; $Q=\frac{4\pi r^2}{\frac{1}{2}(4\pi r^2)}=2$

13.4.3 Effects of atmospheric conditions on noise transmission

Sound energy is absorbed in isotropic air by molecular excitation and relaxation of oxygen molecules and, at very low temperatures, by heat conduction and viscosity in the air. Molecular excitation is a complex formation of the frequency of noise, humidity and temperature. In general, we may say that as the humidity decreases, sound absorption increases. As the temperature increases to about 10–20 °C (depending upon the noise frequency), absorption increases. Above 25 °C, absorption decreases. Sound absorption is higher at higher frequencies.

The vertical temperature profile greatly alters the propagation paths of sound. If a superadiabatic lapse rate exists, sound rays bend upward and noise shadow zones are formed. If an inversion exists, sound rays are bent back toward the ground. This results in an increase in the sound level. These effects are negligible for short distances but may exceed 10 dB at distances over 800 m.

13.4.4 Basic point source model for noise transmission outdoors

A directional source having a sound power level L_w will produce a sound pressure level at a receiver, which can be estimated by Equation (13.32) which is the basic point source equation (modified inverse square law if the last term A_e is moved).

The A_e term has units of dB and is the excess attenuation beyond wave divergence which is caused by environmental conditions.

The A_e term may be further divided into six terms as follows:

A_{e1} = effect of the difference in values of the density and temperature of ambient air when the ambient temperature and barometric pressure differ appreciably from values that make pc = 400, for example, 38.9 °C and 101.325 kPa expressed in dB:

A_{e2} = attenuation by absorption in the air, dB.

A_{e3} = attenuation by rain, sleet, snow, or fog, dB

A_{e4} = attenuation by barriers, dB

A_{e5} = attenuation by grass, shrubbery, and trees, dB

A_{e6}=attenuation and fluctuation owing to wind and temperature gradients, to atmospheric turbulence, and to the characteristics of the ground, dB.

13.4.5 Indoor noise propagation

The propagation of noise in an indoor space depends on the acoustic properties of the walls, ceiling, floors, the materials present inside the space, and the reverberation effect as the sound waves bounce from one surface to the other. It is intuitive that propagation would somehow follow the inverse square law where in modified form

$$L_p = L_w + 10\log\frac{Q}{r^2} - 11 \qquad \text{......(13.35)}$$

This equation may also be written as

$$L_p = L_w + 10\log\frac{Q}{4\pi\, r^2} \qquad \text{......(13.36)}$$

But $Q = 4\pi r^2/S$. Substituting in Equation (13.36), we have

$$L_p = L_w + 10\log\frac{1}{S} \qquad \text{......(13.37)}$$

As applied in outdoor noise transmission, S is the area that receives the sound coming from the source. To apply Equation (13.37) for indoor transmission, S is defined incorporating indoor receiving parameters as

$$\frac{1}{S} = \frac{1}{4\pi\, r^2} + \frac{4}{R} \qquad \text{......(13.38)}$$

where R is called the *room constant*, $4\pi r^2$ is the equivalent spherical area, r (distance from noise source to a given point). The term containing R gives the reverberating effect as sound bounces back and forth inside the room. Expressed in terms of total surface area receiving the sound (S_T), R is expressed as

$$R = \frac{S_T\bar{\alpha}}{1-\bar{\alpha}} \qquad \text{......(13.39)}$$

The *sound absorption* coefficient $\bar{\alpha}$ is the fraction of the incident noise that is absorbed by the surface and $1 - \bar{\alpha}$ is the fraction reflected given by

$$\bar{\alpha} = \frac{\Sigma\alpha_i S_i}{S_T} \qquad \text{......(13.40)}$$

where S is the area with absorption coefficient α and i is the index for the various surfaces. Table 13.2 shows some values of α.

L_p is now written as

$$L_p = L_w + 10\log\left(\frac{1}{4\pi r^2} + \frac{4}{R}\right) \qquad \text{......(13.41)}$$

which is the equation for indoor space sound propagation.

Table 13.2 *Absorption coefficient α for some building materials*

Material	Octave band centre frequency (Hz)					
	125	250	500	1000	2000	4000
Marble or glazed tile	0.01	0.01	0.01	0.01	0.02	0.02
Plaster, gypsum or lime, smooth finish on tile or brick	0.01	0.02	0.02	0.03	0.04	0.05
Plaster, gypsum or lime Rough finish on lath	0.13	0.10	0.05	0.05	0.04	0.03
Smooth finish	0.15	0.10	0.05	0.04	0.04	0.03
Plywood paneling, 1 cm thick	0.30	0.20	0.20	0.09	0.10	0.11
Brick, unglazed	0.02	0.02	0.03	0.04	0.05	0.06
Brick, unglazed, painted	0.01	0.01	0.02	0.02	0.02	0.02
Carpet, heavy On concrete	0.01	0.05	0.15	0.40	0.60	0.70
On 40 mm hairfelt or foam rubber	0.07	0.25	0.60	0.70	0.7	0.74
With impermeable latex backing on 40 mm hairfelt or foam rubber	0.07	0.30	0.40	0.35	0.50	0.65
Concrete block Coarse	0.35	0.45	0.30	0.30	0.40	0.25
Painted	0.10	0.05	0.06	0.07	0.09	0.08
Glass Large panes of heavy plate glass	0.15	0.05	0.04	0.03	0.02	0.02
Ordinary window glass	0.40	0.30	0.20	0.15	0.07	0.04
Gypsum board	0.30	0.10	0.05	0.04	0.07	0.09
Floors Concrete	0.01	0.01	0.02	0.02	0.02	0.02
Asphalt, rubber, or tile on concrete	0.02	0.03	0.03	0.03	0.03	0.02
Wood	0.15	0.15	0.10	0.06	0.06	0.07

13.5 Noise sources

13.5.1 Traffic noise

Problems due to noise pollution are chiefly confined to urban areas. In the rural areas, noise is not yet a health problem as the noise levels are normally low. Most of the urban area problems are due to movement of various types of vehicles. Every moving vehicle produces vibrations because of the construction of that vehicle, age of the vehicle and the way it is handled. Automobile, the result of the invention of the internal combustion engine, is prominent in producing noise.

The noise production from an automobile can be examined starting from the tires onwards. If the tread of the tire is broad, the area touching the ground will be more and there will be more noise production. If there is not enough tire pressure, then also there will be more noise due to higher friction. So one of the factors to be kept in mind is the tire pressure. Less pressure increases petrol consumption due to higher frictional drag. The noise from the tires also increases with the speed of the car [9] as depicted in the graph (Fig. 13.12).

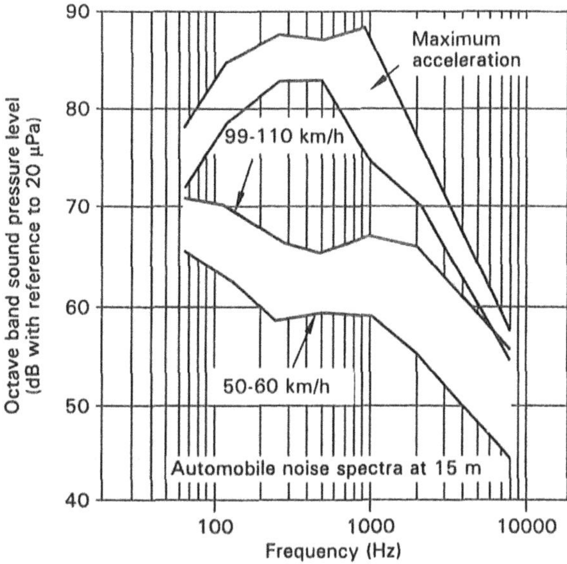

Fig. 13.12 *Noise of tyres increasing with speed*

More prominent is the exhaust pipe noise. The compressed exhaust gases from the cylinders enter the exhaust manifold and enter the 'silencer' where the gases expand and noise level also drops down considerably. Depending on the design of the silencer, the gases and noise also have to take a devious route, thus reducing the noise. Much of the noise produced is curtailed at the silencer level.

The connecting rod from the gear box to the rear wheels, the transmission mechanism produces noise due to the coupling mechanisms which have to be free-moving for efficiency, but produce noise. By proper maintenance, servicing with oil and grease this noise can be reduced.

13.5.2 Highway vehicle noise

Exhaust noise is the predominant source for normal operation below about 55 km/h (Fig. 13.13) for most automobiles while the tire noise becomes the dominant noise source at speeds above 80km/h. The contribution of automobiles to the noise environment is significant because of the large number of vehicles in operation.

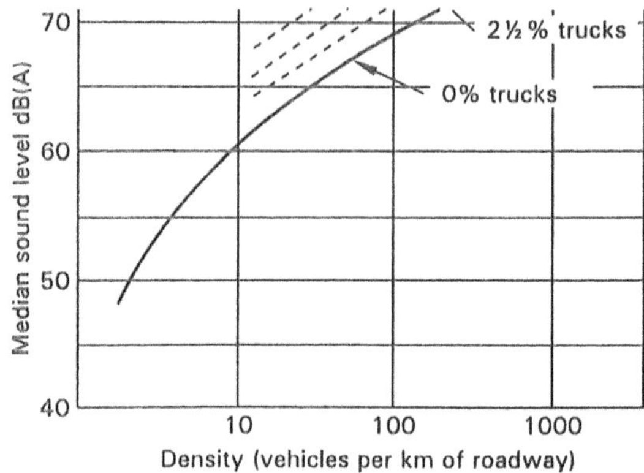

Fig. 13.13 *Median noise level estimates of mixed traffic at 80 kmph*

Diesel trucks are 8 to 10 dB noisier than petrol-powered ones. At speeds above 80 km/h, tire noise source of the truck predominates (figure 13.14). The "crossbar" tread is the noisiest.

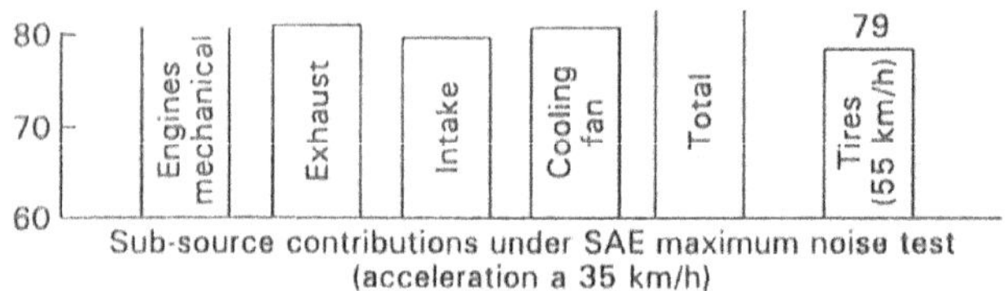

Fig. 13.14 *Typical example of diesel truck noise*

Table 13.3 *Summary of noise characteristics of internal combustion engines*

Source	A weighted noise energy (kWh/d)[a]	Typical A weighted noise level at 15.2m [dB(A)]	8 h exposure level [dB(A)][b] Average	Maximum	Typical exposure time (h)
Lawn mowers	63	74	74	82	1.5
Garden tractors	63	78	N/A	N/A	N/A
Chain saws	40	82	85	95	1
Snow blowers	40	84	61	75	1
Lawn mowers	16	78	67	75	½
Model aircraft	12	78	70[c]	79[c]	¼
Leaf blowers	3.2	76	67	75	¼
Generators	0.8	71	-	-	-
Tillers	0.4	70	72	80	1

(a) Based on estimates of the total number of units in operation per day; (b) Equivalent level for evaluation of relative hearing damage risk; (c) During engine trimming operation.

Internal combustion engines due to their wide usage are important sources of noise (Table 13.3).

13.5.3 Prediction of traffic noise levels

The methodology developed under National Cooperative Highway Research Program of USA (NCHRP174) [10] is a widely adaptable methodology for the prediction of highway noise as it is very simple with a high success rate. This methodology involves a four-step prediction procedure.

The first step is known as a short method in which a gross quick and gross prediction of the noise levels is made, the potential problem areas are identified and no problem areas in terms of noise levels are eliminated to make the assessment process simple. Because of the complexity involved in the assessment of true highway noise levels in the short method the prediction will be always on the higher side.

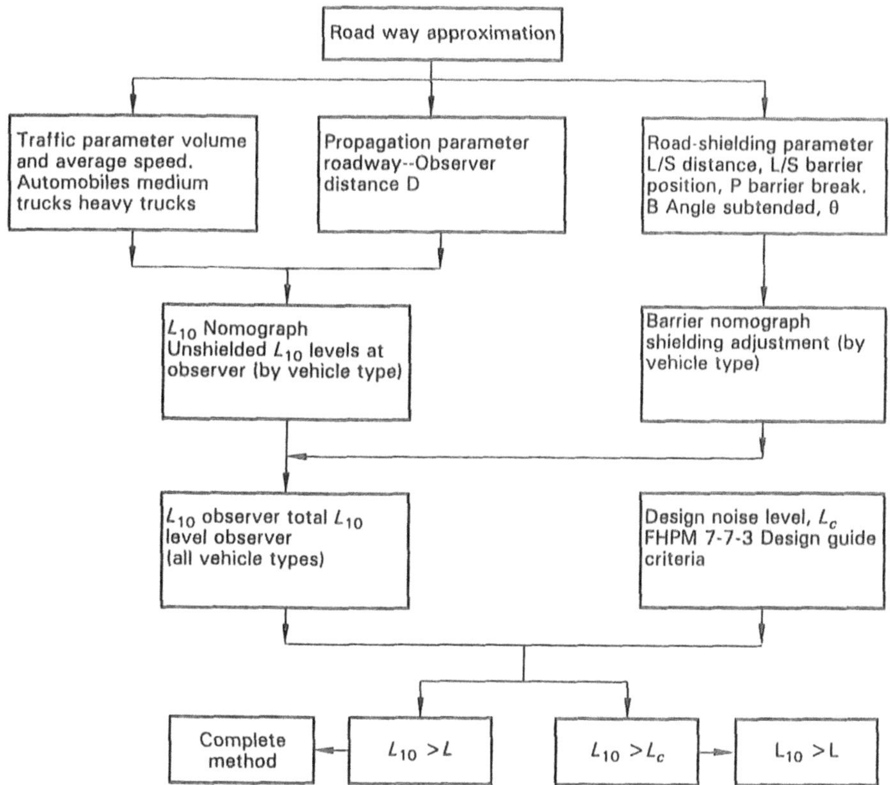

Fig. 13.15 *Flow diagram of methodology for applying NCHRP 174 method for estimation of L_{10} from traffic*

In the short method prediction can be carried out using two nomographs (a nomograph is a graph that provides the solution to an equation or a series of equations containing three or more variables) for traffic and roadway parametric data. As the short method is based on a number of assumptions and approximations, it should be further refined for arriving at final conclusions.

The flow diagram presenting the methodology of short methods is given in Fig. 13.15. In this method the roadway will be approximated by one infinite element with constant traffic parameters and roadway characteristics.

The first step in this method involves defining an infinite straight-line approximation to the real highway configuration omitting on-ramps, off-ramps, and interchange ramps.

After selecting an approximate roadway, the following parameters must be estimated or computed (a) the traffic parameters, which include the speed and volume of each class of vehicles; (b) the propagation characteristics which describe the location of the receiver relative to the roadway; and (c) the roadway shielding parameters, which describe the shielding provided by the roadway, if any.

Then the traffic and propagation parameters have to be combined in the L_{10} nomograph to determine for each type of source, the observer unshielded L_{10} level at the observer.

The final result should be then compared to the criteria level, L_c, at the observer to define a "no problem" or "potential problem" condition. After identification of the potential problem area, the observer location in question should be evaluated using the complete method.

The second step of methodology is termed as NCHRP 174 "complete method" which utilizes a fairly large computer program to refine the predictions made in the first step.

The third step is in this procedure is termed as "selection of a noise control design".

The fourth step is termed as "check the design operation" in which the second step will be repeated and refined to arrive at the final conclusions.

13.5.4 Sources of noise from industry and controlling measures

No environmental factor has caused as much confusion regarding its effect on worker efficiency and worker health as industrial noise. Noise in industry originates from processes causing impact, vibration or reciprocation movements, friction and turbulence in air or gas streams.

Noise emission standards have only an indirect control over the noise radiated by a machine. They state maximum permissible sound levels in work places; acceptable day- and night-time levels in residential, commercial and industrial areas and maximum permissible noise crossing industrial and construction site boundaries. The measured noise in such cases may be produced by a single machine or by a combination of many kinds of machinery.

The maximum permissible sound level at a worker's ears and the time of exposure are not related directly to the noise produced by any one machine but depend upon the total noise in the area, where the workers are located with respect to the machine and other factors. For this reason, noise emission standards or their intent must be confined with product noise emission regulations.

The maximum permissible sound levels crossing industrial sites into residential and commercial zones may be stated in terms of overall A-weighted sound pressure levels. It has been reported that high intensities, high frequencies and intermittent nature of noise are the factors of annoyance for workers [8]. Such a situation not only brings about physical and physiological damage but also impairs worker efficiency giving rise to low production and causing dissatisfaction. Community response to industrial noise is seen in setting up acceptable limits for community areas but this is difficult to establish precisely because of the variety and complexity of the different factors involved.

Basic noise levels for industrial zone should not exceed 55 dB at night and 65 dB during the day time. Noise contributes to development of cardiovascular problems like heart diseases and high blood pressure. Workers exposed to high noise levels have more circulatory problems, cardiac disturbances, neuro-sensory and motor impairment and even more social conflicts at home and at work.

13.6 Noise Control Measures

13.6.1 Noise control

There are five general methods of controlling noise: enclosing the noise source, enclosing the noise receiver, putting a barrier between the noise source and the receiver, controlling the noise generator, and regulating controls by pressure studies.

13.6.2 Enclosures

To enclose a noisy machine in a separate room or box than to quieten it by altering its design, operation, or parts is one of the practical and economical options. The walls of the enclosure should be massive and airtight to contain the sound. Absorbent lining on the interior surfaces of the enclosure will reduce the reverberant build-up of noise within it. Structural contact between the noise source and the enclosure must be avoided, or else the source vibration will be transmitted to the enclosure walls and thus, short–circuit the isolation.

13.6.3 Screen the receiver

In most of the industries, noise is an essential part of the machines used and if a worker is continuously exposed to sound hazards for long, he may suffer from annoyance, loss of efficiency and damage to hearing. Employees should be given ear muffs or ear plugs to protect them from such losses.

13.6.4 Ear protection

Molded and pliable earplugs, cup type protectors, and helmets are commercially available as hearing protectors. Such devices may provide noise reductions ranging from 15 to 35 dB. Ear plugs are effective only if they are properly fitted by medical personnel [11].

These devices should be used as alternatives other methods when failed to lower the noise level to acceptable limits.

13.6.5 Control of noise during transmission

Devices can be used in the transmission path to block or reduce the flow of sound energy before it reaches the ears. This can be done in several ways.

(a) Absorb the sound along the path
(b) Deflect the sound in some other direction by placing a reflecting barrier in its path
(c) Contain the sound by placing the source –insulating box or enclosure.

13.6.6 Separation

Recently in 1997, some researchers in United States reported total reduction of noise transmission from the source by employing a vibrator that produces similar frequencies in the opposite path nullifying the noise level completely. If experiments in this area are successful many industries as well as vehicles may be able to reduce noise levels considerably.

The absorptive capacity of the atmosphere can be made use of as a simple, economical method of reducing the noise level. Air absorbs high-frequency sounds more effectively than it absorbs low-frequency sounds. If we can double the distance from the point source, we will succeed in lowering the sound pressure level by 6 dB.

13.6.7 Absorbing materials

Noise, like light, will bounce from one hard surface to another. In noise control work, this is called *reverberation*. Sound absorbing materials are rated either by their **Sabin absorption coefficients** (α_{SAB}) at 125,500, 1,000, 2,000 and 4,000 Hz or by a single number rating called the **noise reduction coefficient (NRC).**

The ratio of the sound energy incident on one surface of a panel to the energy radiated from the opposite surface is called the **sound transmission loss (TL)**. The actual energy loss is partially reflected and partially absorbed.

Sound absorbing materials such as acoustical tiles, carpets, and drapes placed on ceilings, floors or walls can reduce the noise level in most rooms by about 5–10 dB for high frequency sounds, but only by 2 or 3 dB for low frequency sounds.

Noise transmitted through ducts, pipes, or electrical channels can be reduced effectively by lining the inside surfaces of such passageways with sound absorbing materials. A comparable degree of noise reduction for the lower frequency sounds is considerably more difficult to achieve because it usually requires at least a doubling of the thickness and/or length of acoustical treatment.

13.6.8 Barriers and panels

Placing barriers, screens, or deflectors in the noise path can be an effective way of reducing noise transmission, provided that the barriers are large enough in size, and depending upon whether the noise is high frequency or low frequency. High frequency noise is reduced more effectively than low frequency noise. The effectiveness of a barrier depends on its location, its height and its length.

13.6.9 Control of industrial noise source by design changes in Equipment

13.6.9.1 Reduce impact factors

Many machines and items of equipment are designed with parts that strike forcefully against other parts, producing noise. Often this striking action or impact is essential to the machine function. Several steps can be taken to reduce noise from impact forces. The particular remedy to be applied will be determined by the nature of the machine in question.

Some of the suggested design modifications are as follows:

1. The design may be modified by reducing the weight, size, or height of fall of the impacting mass, changing the metal, or inserting noise absorbers if possible at the places of noise production.
2. Anchoring the machine to the floor with noise absorbing supports.
3. Whenever practical, one of the impact heads or surfaces can be made of non-metallic material to reduce resonance.
4. Substitute the application of a small impact force over a long period for a large force over short period to achieve the same result.
5. Smooth out acceleration of moving parts by applying accelerating forces gradually. Avoid high, jerky acceleration or jerky motion.
6. Minimize overshoot, backlash and loose play in cams, followers, gears, linkages and other parts

13.6.9.2 Control of noise from industry (Figure 13.16)

13.6.9.2.1 Acoustic design criteria

One of the first decisions that should be made before starting a noise reduction programme on a machine is to establish an acoustic design goal.

The attenuation requirement for a majority of industrial noise problems is usually in the range 15–25 dB. This order of attenuation on machines can only be attained by complete enclosure if only insulation principles have been used. Attention to detail in the design, construction and erection of enclosures is extremely important as any gaps or leaks severely limit the insulation potential.

13.6.9.3 Vibration isolation

A vibration machine can be isolated from the surrounding structure by supporting it on resistant mountings. This may be associated as a simple mass-spring.

13.6.9.4 Miscellaneous methods

Impact and vibration noises are considerably reduced by the mass, careful design of shape and arrangement of parts and machines so that resonance is avoided. Nevertheless certain machines will remain inherently noisy. These need to be surrounded with absorbent or insulating screens. Noise caused by gas streams can be attenuated or even eliminated by the use of suitable ducts and by correct design and positioning of inlets and outlets.

13.6.9.5 Internal design changes

Centrifugal compressor noise can be controlled to a limited extent by internal design changes, but large, high speed high horsepower machines require additional external sound control when they must meet low noise criteria. Noise reduction can be accomplished by selecting the proper combination of blades, and by controlling clearances between rotating and stationary vanes.

13.6.9.6 Maintenance for low frictional resistance

Lowering the friction between rotating, sliding, or moving parts in mechanical systems results in smoother operation and lower noise output. Proper maintenance of the machinery by oiling and greasing the moving parts reduces noise considerably.

13.6.9.7 Minimize noise leakage

In many cases, machine cabinets can be made into rather effective sound proof enclosures through simple design changes and the application of some sound absorbing treatment.

13.6.9.8 Minimizing vibrational surface area

In general the larger the vibrating part or surface, the greater the noise output. The rule of thumb for quiet machine design is to minimize the effective radiating surface areas of the parts without impairing their operation or structural strength. This can be done by making parts smaller, removing excess material, or cutting openings, slots, or perforations in the parts.

13.6.9.9 Isolate and dampen vibrating elements

Energy transmission between the source and surfaces that radiate the energy can be solved by isolation. Dissipate or attenuate the energy somewhere in the structure by damping.

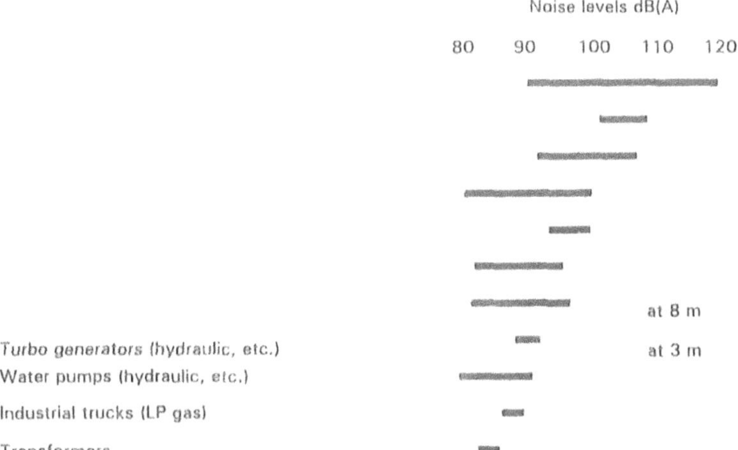

Fig. 13.16(a) *Construction equipment noise ranges*

Noise level (dB(A)) at 15 m

60 70 80 90 100 110

Equipment powered by internal combustion engines

Earth moving: Compactors (Rollers), Front loaders, Backhoes, Tractors, Scrapers, graders, Pavers, Trucks

Handling materials: Concrete mixers, Concrete pumps, Cranes (movables), Cranes (derrick)

Stationery: Pumps, Generators, Compressors

Impact equipment: Pneumatic wrenches, Jackhammers and rock drills, Impact pile drivers (peaks)

Other: Vibrator, Saws

Fig. 13.16(b) *Range of industrial machinery, equipment, and process noise level of operator's position*

13.6.9.10 Provide mufflers/silencers

Mufflers or silencers can be used as acoustical filters and are used when fluid flow noise is to be reduced. The devices can be classified into two fundamental groups: adsorptive mufflers and reactive mufflers.

13.6.9.11 Balance rotating parts

One of the main source of machinery noise is structural vibration caused by the rotation of poorly balanced parts, such as fans, flywheels, pulleys, cams, and shafts. Measures used to correct this condition involve the addition of counterweights to the rotating unit or the removal of some weight from the unit.

13.6.9.12 Reduce frictional resistance

A well designed machine that has been poorly maintained can become a serious source of noise. General cleaning and lubrication of all rotating, sliding or meshing parts at contact points should go a long way towards fixing the problem.

13.6.9.13 Damping materials for noise reduction

The application of any material that reduces or restrains the vibrational motion of that body will decrease its noise output. Three types of damping materials are available:

(a) Liquid mastics which are applied with a spray gun and harden into relatively solid materials, the most common being automobile "undercoating".

(b) Pads of rubber, felt, plastic foam, leaded vinyls, adhesive tapes, or fibrous blankets, which are glued to the vibrating surface.

(c) Sheet metal, visco-elastic laminates or composites, which are bonded to the vibrating surface.

13.6.9.14 Seal noise leaks

Small holes in an otherwise noise tight structure can reduce the effectiveness of the noise control measures. If the designed transmission loss of an acoustical enclosure is 40 dB, an opening that comprises only 0.1% of the surface area will reduce the effectiveness of the enclosure by 10 dB.

Because automobile tire noise increases in relation to pavement roughness indicate that maintenance of the pavement surface is essential to keep noise at minimum levels. Normal road wear can yield noise increases of the order of 6 dB (A).

When exposure to intense noise fields is required and none of the measures are practical then the following two techniques are commonly employed to protect persons from effects of noise.

13.6.9.15 Alter work schedule

Limit the amount of continuous exposure to high noise levels. In terms of hearing protection, it is preferable to schedule an intensely noisy operation for a short interval of time each day over a period of several days rather than a continuous eight-hour run for a day or two. Inherently noisy operations such as street repair, municipal trash collection, factory operation, and aircraft traffic should be curtailed at night and early morning hours to avoid disturbing the sleep of the community.

13.6.9.16 Set a limit by legislation

The following environmental noise levels will be satisfactory for most people:

Close to hospitals, old peoples' homes, etc.

35 dB at night, 45 dB during day, peak noise 55 dB

Residential zones

45 dB at night, 55 dB during day, peak noise 70 dB.

Commercial area

60dB average, peak noise 75 dB

Industrial area

65 dB average, peak noise 80 dB.

Table 13.4 *Noise criteria levels to protect public health and welfare*

Effect	Level	Area
Hearing loss	$L_{eq}(24) \le 70$dB	All areas
Outdoor activity	$L_{dn} \le 55$ dB	Outdoor in residential areas and farms, and other outdoor areas where people spend widely varying amounts of time and other places in which quiet is a basis for use
	$L_{eq}(24) \le 55$ dB	Outdoor areas where people spend limited amount of time, such as schoolyards, and playgrounds, etc.
Indoor activity Inference and annoyance	$L_{dn} \le 55$ dB	Indoor residential areas
	$L_{eq}(24) \le 45$ dB	Other indoor areas with human activities, such as schools

13.6.10 Noise control criteria

Table 13.4 shows noise criteria in terms of NC (noise criteria) levels for various activities, while Table 13.5 and 13.6 shows speech interference levels for various degrees of communication effectiveness.

The U.S. Federal Highway Administration (FHA) developed the standards shown in Table 13.7 for new construction. These levels are those that would be expected to yield no complaints.

Table 13.5 *Acceptable NC levels for various activities*

Activity	Suggested range of noise criteria, NC
Acceptable working conditions with minimum speech interference:	
Industrial areas, garages, laundries	NC-45 to NC -55
Moderately fair listening conditions:	
Lobbies, cafeterias, drafting rooms, business, machine areas	NC-40 to NC -45
Fair listening conditions required:	
Large offices, restaurants, retail shops, and stores, etc.	NC-25 to NC-40
Good listening conditions required:	NC-20 to NC-25
Private offices, school classrooms, small conference rooms, libraries, TV listening	
Very good listening conditions required:	
Auditorium, theaters	NC-15 to NC-20
Large meeting and conference rooms	NC-15 to NC-20
Excellent listening conditions required:	
Concrete halls, recording studios, etc.	NC-15 to NC-20
Sleeping, resting, and relaxing:	
Suburban and rural: homes, apartments, hotels, hospitals, etc.	NC-15 to NC-20
Urban: homes, apartments, hotels, hospitals, etc.	NC-15 to NC-20

Table 13.6 *Speech interference levels for various degrees of conversation effectiveness*

Telephone communication	
PSH (dB)	**Telephone intelligibility**
> 865	Very poor
< 865	Good to very poor

Face to face communication

Vocal effort	PSIL (dB)	Distance from speaker to listener	Communication intelligibility
Maximum vocal effort	100	30 cm	Difficult
	100 and above	30 cm	Impossible
Short	90	30 cm	Possible
Very loud voice	70	30 cm	Possible
	60	1.5 cm	Possible
	50	3.0 cm	Possible
	40	4.5 cm	Possible

Table 13.7 *FHA noise standards for highways*

Land use Category	Design level Leq	(dB(A)) L10	Description of land use category
A	57	60	Tracts of lands in which serenity and quiet are of extraordinary significance and serve an important public need, and where the preservation of those qualities is essential if the area is to continue to serve its intended purpose. Such areas could include amphitheaters, particular parks or portions of parks, or open spaces that are dedicated or recognized by appropriate local officials for activities requiring special qualities of serenity and quiet
B	67	70	Residences, motels, hotels, public meeting rooms, schools, churches, libraries, hospitals, picnic areas, recreational areas, playgrounds, active sports, areas and parks
C	72	75	Developed land properties, or activities not included in categoriesA and B above
D	Unlimited	Unlimited	Undeveloped lands
E	52	55	Public meeting rooms, schools, churches, libraries, hospitals, and similar buildings

13.6.11 Indian Govt. noise pollution and control rules

Govt. of India formulated noise pollution control rules through Ministry of Environment & Forest for management of noise pollution

The Noise Pollution (Regulation and Control) Rules, 2000.

MINISTRY OF ENVIRONMENT & FORESTS NOTIFICATION

Ambient Air Quality Standards In Respect of Noise for Different Areas/Zones

1. The ambient air quality standards in respect of Noise or different areas/zones shall be such as specified in the Schedule annexed to these rules.
2. "The State Government [shall categorize] the areas into industrial, commercial, residential or silence areas/zones the purpose or implementation of noise standards for different areas.
3. 'The State Government shall take measures for abatement of noise including noise emanating from vehicular movements, "[blowing of horns, bursting of sound emitting fire crackers, use of loud speak'ers or public address system and sound producing instruments] and ensure that the existing noise levels do not exceed the ambient air quality standards specified under these rules.
4. All development authorities, local bodies and other concerned authorities while planning developmental activity or carrying out functions relating to town and country planning shall take into consideration all aspects of noise pollution as a parameter of quality of life to avoid noise menace and to achieve the objective of maintaining the ambient air quality standards in respect of noise

5. An area comprising not less than 100 meters around hospitals, educational institutions and courts may be declared as silence area/zone for the purpose of these rules.
6. Responsibility As To Enforcement of Noise Pollution Control Measures.-
   (a) The noise levels in any area/zone shall not exceed the ambient air quality standards in respect of noise as specified in the Schedule.
   (b) The authority shall be responsible for the enforcement of noise pollution control measures and the due compliance of the ambient air quality standards in respect of noise.
   (c) The respective State Pollution Control Boards or Pollution Control Committees in consultation with the Central Pollution Control Board shall collect. compile and publish technical and statistical data relating to noise pollution and measures devised for its effective prevention. control and abatement. I
7. Restriction for the use of loud speakers/public address systems and public address systems
   (a) A loud speaker or a public address system shall not be used except after obtaining written permission from the authority.
   (b) A loud speaker or a public address system or any sound producing instrument or a musical instrument or a sound amplifier shall not be used at night time except in closed premises for communication within. like auditoria, conference rooms, community halls, banquet halls or during a public emergency.
   (c) Notwithstanding anything contained in sub-rule (2), the State Government may subject to such terms and conditions as are necessary to reduce noise pollution, permit use of loud speakers or [public address systems and the like during night hours] (between 10.00 p.m. to 12.00 midnight) on or during any' cultural or religious festive occasion of a limited duration not exceeding fifteen days in all during a calendar year.] [The Concerned State Government shall generally specify in advance, the number and particulars of the days on which such exemption would be operative].
   (d) The noise level at the boundary of the public place, where loudspeaker or public address system or any other noise source is being used shall not exceed 10 dB(A) above the ambient noise standards for the area or 75 dB(A) whichever is lower.
   (e) The peripheral noise level of a privately owned sound system or a sound producing instrument shall not, at the boundary of the private place, exceed by more than 5 dB(A) the ambient noise standards specified for the area in which it is used l.
8. Restrictions for the use of Horns, Loud Speakers, Public address systems Sound Producing Instruments and Bursting of Fire Crackers

(a) No horn shall be used in silence zones or during night tinie in residential areas except during a public emergency.
(b) Sound emitting fire crackers shall not be burst in silence zone or during night time.
(c) Sound emitting construction equipments shall not be used or operated during night time in residential areas and silence zones. I

9. Consequences of Any Violation in Silence Zone/Area: Whoever, in any place covered under the silence zone/area commits any of the following offence, he shall be liable for penalty under the provisions of the Act:
(a) who ever, plays any music or uses any sound amplifiers,
(b) whoever, beats a drum or tom-tom or blows a horn either musical or pressure, or trumpet or beats or sounds any instrument,
(c) whoever, exhibits any mimetic, musical or other performances of a nature to attract crowds,
(d) whoever, bursts sound emitting fire crackers; or
(e) who ever, uses a loud speaker or a public address system.

10. Complaints to Be Made to the Authority: A person may, if the noise level exceeds the ambient noise standards by 10 dB(A) or more given in the corresponding columns against any area/zone [or,
(a) if there is a violation of any provision of these rules regarding restrictions imposed during night time], make a complaint to the authority.
(b) The authority shall act on the complaint and take action against the violator in accordance with the provisions of these rules and any other law in force.

11. Power to Prohibit Etc. Continuance of Music Sound or Noise:
(a) If the authority is satisfied from the report of an officer in-charge of a police station or other information received by him [including from the complainant] that it is necessary to do so in order to prevent annoyance, disturbance, discomfort or injury or risk person who dwell or occupy property on the vicinity, he may, by a written order issue such directions as he may consider necessary to any person for preventing, prohibiting, controlling or regulating:

The incidence or continuance in or upon, any premises of
(i) Any vocal or instrumental music,
(ii) sounds caused by playing, beating, clashing, blowing or use in any manner whatsoever of any instrument including loudspeakers, [public address systems, horn, construction equipment, appliance or apparatus] or contrivance which is capable of producing or reproducing sound,
(iii) sound caused by bursting of sound emitting fire crackers, or

(b) The carrying on in or upon, any premises of any trade, a vocation or operation or process resulting in or attended with noise.

12. The authority empowered under sub-rule (1) may, either on its own motion, or on the application of any person aggrieved by an order made under sub-rule (1 either rescind, modify or alter any such order provided that before any such application is disposed of, the said authority shall afford to the applicant [3] [and to the original complainant, as the case may bel an opportunity of appearing before it either in person or by a person representing him and showing cause against the order and shall. If it rejects any such application either wholly or in part, record its reasons for such rejection.

Ambient Air Quality Standards in respect of Noise

Area code	Category of Area/Zone	Limits in dB(A) Leq*	
		Night Time	Day Time
(A)	Industrial area	75	70
(B)	Commercial area	65	55
(C)	Residential area	55	45
(D)	Silence Zone	50	40

Note:
1. Day time shall mean from 6.00 a.m. to 10.00 p.m.
2. Night time shall mean from 10.00 p.m. to 6.00 a.m.
3. Silence zone is an area comprising not less than 100 metres around hospitals, educational institutions, courts, religious places or any other area which is declared as such by the competent authority.
4. Mixed categories of areas may be declared as one of the four above mentioned categories by the competent authority.

*dB(A) Leq denotes the time weighted average of the level of sound in decibels on scale A which is relatable to human hearing. A "decibel" is a unit in which noise is measured.

"A", in dB(A) Leq, denotes the frequency weighting in the measurement of noise and corresponds to frequency response characteristics of the human ear.

Leq: It is an energy mean of the noise level over a specific period

Questions

Short Answer Questions

1. How does noise differ from sound?
2. Differentiate between air and noise pollution.
3. Write a short note on characteristics of sound.
4. Deduce the relationships among intensity, pressure and power levels.
5. What do you understand about noise rating systems?
6. Explain the effects of atmospheric conditions on noise transmission.
7. Write a brief note on indoor noise propagation.

Essay Questions

1. Explain in detail about Sound pressure, power and intensity.
2. Describe about sound measuring instrumentation.
3. Explain the effects of noise on human beings in detail.
4. Bring out the phenomena of noise propagation in ambient air.
5. Discuss the noise control measures carried out in an industry scale.
6. List out and explain the design changes in equipment to control noise in industrial application.
7. Explain the different Government Noise Pollution and Control Rules.

References

1. Council of Environmental Quality. USA Report, 1997.
2. Transportation Noise Pollution Control and Abatement. 1970. Langley Research Center.
3. P.Cunn, 1977. Environmental noise pollution. Wiley, New York.
4. L. L. Beranek and L. N. Miller, 1967. The anatomy of noise. Machine design.
5. J. C. Ratcliff, 1970. Guidelines for noise exposure control sound and vibration.
6. A.P. G. Peterson and E. E. Groves, 1972. Handbook of noise measurement. General Radio Co.
7. USEPA report. 1972. Effect of noise on people.
8. K. D. Kryter, 1970. Effects of noise on man. Academic Press.
9. USEPA, 1971.Transportationnoise. US Govt. Printing Office, Washington DC.
10. National Cooperative Highway Research Program of USA (NCHRP 174). 1976.
11. National Bureau of Standards Handbook. 1976.

Suggested References

1. C. Michael Hogan and Gary L. Latshaw, "The relationship between highway planning and urban noise", Proceedings of the ASCE, Urban Transportation, May 21–23, 1973, Chicago, Illinois. By American Society of Civil Engineers. Urban Transportation Division.
2. Hoffmann, Barbara; Moebus, Susanne; Stang, Andreas; Beck, Eva-Maria; Dragano, Nico; Möhlenkamp, Stephan; Schmermund, Axel; Memmesheimer, Michael; Mann, Klaus (2006-11-01). "Residence close to high traffic and prevalence of coronary heart disease". European Heart Journal. 27 (22): 2696–2702. doi:10.1093/eurheartj/ehl278. ISSN 0195-668X. PMID 17003049.
3. “Results and Discussion - Effects - Noise Effect on Wildlife - Noise - Environment – FHWA”. www.fhwa.dot.gov. Retrieved 2015-12-21.

4. Rosen and P. Olin, Hearing Loss and Coronary Heart Disease, Archives of Otolaryngology, 82:236 (1965).
5. J.M. Field, Effect of personal and situational variables upon noise annoyance in residential areas, Journal of the Acoustical Society of America, 93: 2753-2763 (1993).
6. "Noise Pollution". World Health Organisation. 2009.
7. Jefferson, Catrice. "Noise Pollution". U.S. Environmental Protection Agency.2013.

CHAPTER 14

Prediction and Assessment of Impacts on the Air Environment

14.1 Air Quality Impact Assessment

Environmental impact assessment (EIA) of air pollution diffusion from different kinds of hazardous sources can provide support information for decision making of emergency plan preparation and approval for any new development activity. For any EIA, Air quality impact assessment (AQIA) is an important technique for determining the relative contribution to ground level pollutant concentrations of specific current or future source emissions at receptor sites. The principal activities in AQIA are air quality modelling and monitoring techniques. A six step methodology for evaluating the impacts of any human activity on the air environment, is generally adopted, for which it is necessary to acquire basic information on the following aspects of air pollution.

14.1.1 Basic information on air pollution

Basic information on air pollution sources is required and the effects of specific air pollutants have to be examined. Though these aspects are discussed in detail in Chapter 1, these are summarized in the following pages for continuity of discussion.

14.1.2 Air pollution sources

The sources of air pollution can be classified based on the type of activity, the frequency of occurrence and spatial distribution, and the types of emissions and can be delineated as arising from natural sources or man-made sources. Sources like wind blown dust, volcanic eruptions, lightning-generated forest fires and biological activity etc. can be termed as natural sources while transportation vehicles, industrial processes, power plants, construction activities, and military training activities, etc. can be termed as man-made sources.

Based on the number and spatial distribution, air pollution sources can be classified as single or point sources (stationary), area or multiple sources (stationary or mobile), and line sources. Pollutant emissions from industrial process stacks, as well as stacks for different fuel combustion processes are typical examples of point sources while vehicular traffic, fugitive dust emissions from resource or material stockpiles or construction, or military training activities over large geographical areas, etc. are examples of area sources. Figure 14.1 presents various types of sources which can be used for analyzing air pollutant sources in a given geographical area.

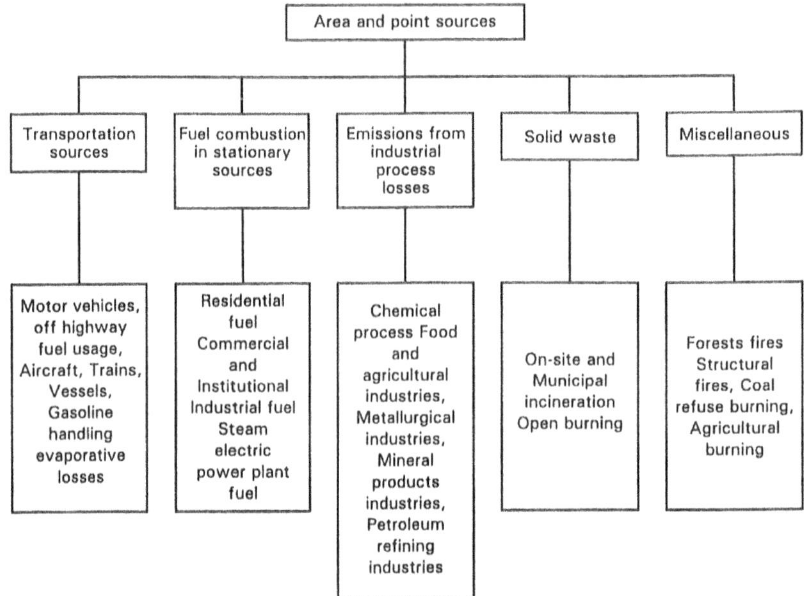

Fig. 14.1 *Types of air pollution sources [I]*

Based on the preliminary regulations and standards prescribed by various statutory and regulating agencies, the impacts of various activities on the air quality have to be examined.

14.1.3 Air pollution effects (human health, material damage, climatic and aesthetic effects)

The air pollution effects can be grouped into:

14.1.3.1 Effects on human and animal health

Include eye irritation, headaches, and aggravation of respiratory difficulties. Plants and crops have been subjected to abnormal growth patterns, leaf discoloration or spotting and death. Animals such as cattle have been subjected to the undesirable consequences of atmospheric fluorides.

14.13.2 Materials damage

Examples are property devaluation because of odours, deterioration of materials such as concrete statuary, and discoloration of painted surfaces on cars, buildings, and bridge structures.

14.1.3.3 Climatic changes

Greenhouse effects, ozone depletion, etc.

Aesthetic effects: Include reduction in visibility, discoloration of air, photochemical smog-related traffic disruptions at airports, and the general nuisance aspects of odors and dust.

14.2 Generalized Approach for Assessment of the Impact of Air Pollution

To evaluate the impacts on air quality by any project activity, a six-step or six activity model is proposed for planning and assessment. It is highly flexible and can be modified for adoption to different types of pollutants and for addressing specific problems in various types of locations.

14.2.1 The following are six main steps in the proposed model for EIA study on air environment

1. Evaluation and identification of sources and quantity of air pollutant emissions of different phases of the proposed activity like construction, operation and development;
2. Detailed evaluation of the project area for the existing ambient air quality, emission Inventory, and meteorological data;
3. Enumeration of appropriate laws, regulations, or criteria to be followed for maintaining ambient air quality and/or pollutant emission standards;
4. Carrying out impact assessment studies adopting mass balances, dispersion calculations, comprehensive mathematical models, and/or qualitative predictions based on case studies and professional judgement;
5. Assessment of significance of anticipated beneficial and detrimental impacts; and
6. Development of appropriate mitigation measures for the adverse impacts.

14.2.2 Step 1: Evaluation and identification of sources and quantity of air pollutant emissions from different phases of the proposed activity

In the first step one has to examine what types and quantities of air pollutants are likely to be emitted during the construction and/or operational phases of the proposed project. The typical air pollution sources of some projects are given in Table 14.1. One can use the emission factor information based on the project type or activity. An emission factor is the average rate at which a pollutant is released into the atmosphere as a result of some activity, such as combustion or industrial production, divided by the level of that activity [2], Emission factors relate the types and quantities of pollutants emitted to indicators such as production capacity, quality of fuel burned, or vehicle-km traveled by an automobile. The emission factors for different air pollutants from various types of refuse incinerators and automobiles are presented in Table 14.2 and 14.3 respectively.

Table 14.1 *Air pollution sources of some activities*

Activity	Air pollution sources
Solid waste disposal	Refuse incineration open burning, sewage sludge incineration
Combustion activity	Coal combustion as fuel oil combustion, Natural gas combustion,
Industrial mineral products production	Wood waste combustion in boiling lignite carbon Coal cleaning, Sand and gravel process, Stone quarrying process
Chemical industrial process	(a) Boiler coal burning (b) Chemical reactor gaseous effluents (c) Waste water treatment plants (d) Acidification and neutralization reaction. (e) Solvent evaporation (f) Solvent evaporation in reaction process

Table 14.2 *Emission factors for refuse incinerators without controls-emission factor rating*

Incinerator type	Particulates kg/MT	Sulphur oxides[b] kg/MT	Carbon Monoxide kg/MT	Hydrocarbons[c] kg/MT	Nitrogen oxides[d] kg/MT
Municipal					
Multiple chamber uncontrolled	15	0.75	17.5	0.75	1
With settling chamber and water-spray system	7	0.75	17.5	0.75	1
Industrial-commercial					
Multiple chamber	3.5	0.75	5	1.5	1.5
Single chamber	7.5	0.75	10	7.5	1
Controlled air	0.7	0.75	–ve	–ve	5
Flue-fed	15	0.25	10	7.5	1.5
Flue-fed (modified)	3	0.25	5	1.5	5
Domestic single chamber					
Without primary burner	17.5	0.25	150	50	0.5
With primary burner	3.5	0.25	–ve	1	1
Pathological	4	–ve	–ve	–ve	1.5

(a) average factors given based on EPA procedures for incinerator stack testing. Using high side of particulate, HC and CO emission ranges when operation is intermittent and combustion conditions are poor

(b) Expressed as SO_2: (c) Expressed as methane: (d) Expressed as NO_2

Source: Adapted from US Environmental Protection Agency, 1973. pp. 2.1-3. [3]

Table 14.3 *Air pollution emission factors for passenger vehicles*

Transpost mode	Co_2 per km	Organic compounds	CO	No_x Grams per kilometer	So_x
Truck (petrol)					
Single occupancy	5.29	11.28	96.86	7.23	0.811
Average occupancy	2.85	5.92	49.91	3.80	0.42
Car					
Single occupancy	3.95	9.06	71.81	5.67	0.49
Average occupancy	2.39	5.32	42.25	3.35	0.21
Vehicle ride share					
Three-person car pool	1.30	3.03	23.95	1.90	0.17
Four –person car pool	0.987	2.25	17.95	1.41	0.10
Nine-person car pool	0.599	1.26	10.75	0.81	0.10
Bus (diesel)					
Transit	1.37	0.88	4.26	6.41	N.A.
Rail intercity					
Diesel	1.51	3.95	2.11	3.17	1.79
Electric	0.917	neg	0.17	3.88	7.30
Commuter (diesel)	1.869	3.66	5.07	14.46	2.22
Transit (electric)	1.30	neg	0.21	5.22	10.19
Aircraft	2.01	1.76	1.83	3.80	0.28
Bicycle	0	0	0	0	0
Walk	0	0	0	0	0

14.2.3 Step 2: Detailed evaluation of existing ambient air quality meteorological conditions and natural quality existing in the project area

14.2.3.1 Compilation air quality information

All the information on the existing air quality should be compiled, particularly for the pollutants likely to be emitted from various project activities as identified in Step 1. To utilize this information appropriately, one must carefully describe the characteristics of each sampling site, including any unique factors about the site, such as surrounding land usage, height of the sampling device above the ground surface, and the type and calibration history of the sampling equipment. Graphical presentation of air quality information may be useful, particularly if there appear to be trends, either upwards or downward, in the air quality levels of any of the air pollutants.

The collected raw data should be compared in terms of the existing air data and presented in accordance with the averaging times in pertinent ambient air quality standards. Evaluation of annual average concentrations, along with the pertinent statistical distributions should be done. Data distributions may be required for arriving at 8 h or 24 h averaging data.

If the baseline air quality in the proposed project or activity area is not available, data from nearby areas with similar characteristics in terms of land usage and climatological features be utilized.

Baseline ambient-air quality data will be useful for assessing whether the air quality before the project activity exceeds attains or does not comply with relevant standards. Some relevant toxic air pollutants also have to be examined in addition to normal air pollutants. Greater significance should be given to, and greater attention should be given to, those pollutants which do not meet or barely meet the allowable ambient air concentrations. If one or more pollutants are in a non-attainment area some additional air quality management methods may be needed.

14.2.3.2 Procurement or development of emission inventory

For evaluating the potential air quality impacts of a proposed project or activity, it is necessary to identify the study area (potential area or region of influence) associated with the air pollution emissions. The delineation of the study area can be made using the boundaries of the land associated with the project, or the delineation can include larger area by considering the atmospheric dispersion patterns within the vicinity of the proposed project or activity. So an emission inventory gives the overall scenario of air pollutants emissions in the existing meteorological conditions of the project area.

If the appropriate emission inventory suitable for the study area is not available, a comprehensive emission inventory of the study areas has to be prepared taking into account the following points.

(a) All pollutants and sources of emissions in the project area are to be classified and considered.

(b) 1nformation on emission factors for each of the identified pollutants and sources have to be collected with average values.

(c) Daily quality and quantity of materials handled, processed, or burned, or other unit production information, of the individual identified sources have to be estimated.

(d) The rate at which each pollutant is emitted to the atmosphere and annual averages have to be estimated.

(e) Specific pollutant emissions from each of the identified sources have to be added UP.

14.2.3.4 Key meteorological data

Categorization of meteorological parameters which hinder dispersion of pollutants emitted to the atmosphere have to be made on the following lines.

(a) Meteorological data which describe the general air pollution dispersion characteristics of the study area.
(b) Meteorological data useful to describe the atmospheric dispersion of air pollutants from a project activity quantitatively.
(c) Meteorological data necessary for air pollution dispersion modeling.

The general atmospheric dispersion conditions will provide a fundamental understanding of atmospheric transport. More importantly, limiting times, months, or seasons can be identified during this process and this information can be used in construction - phase planning and operational – phase decision making. Data indicative of the general characteristics of the area with regard to air pollution dispersion include mixing height, inversion height and mean annual wind speeds. Wind rose information can be used to qualitatively disclose the atmospheric dispersion of air pollutants from an activity.

14.2.3.5 Baseline monitoring

To establish the concentration of specific pollutants, ambient air monitoring has to be carried out to verify the experienced changes in air quality concentrations for those pollutants which are identified as potential problems in the earlier steps.

14.2.4 Step 3: Examination of appropriate air quality emission regulation laws and air quality standards to be maintained as per local, state and central government notifications

The primary sources of information on air quality standards, criteria, and policies of the basic local, state, and central government agencies like state PCBs and central PCBs which have the statutory authority to maintain the air resources have to be collected. Documentation of this information will allow the determination of the significance of air quality impacts incurred during projects or activities and will aid in deciding between alternative actions or in assessing the need for mitigation measures for a given alternative. Specific air quality management policies or requirements may be in existence for particular areas, and the particular requirements of such policies have to be ascertained.

14.2.5 Step 4: Carrying out impact assessment using mass-balance, mathematic modelling and other qualitative approaches

Impact assessment can be carried out using a number of scientific methods like mass balances, ambient air dispersion models, and plume dispersion models.

14.2.5.1 Mass balance approach

With reference to the existing air environment inventory of the project area, air pollutant emissions from the construction and/or operational phase of a project have to be considered for which an inventory based on mass balance approach has to be prepared.

The preparation of emission inventory for a proposed project or activity involves the following steps

1. In the project area the pollutants likely to be emitted from different phases of the proposed project or activity like both the construction and the operational phases have to be classified and the sources have to be identified.
2. Information on the emission factors for each of the identified sources for each pollutant and their annual averages and aggregates have to be completed.
3. Procuring the specific unit-production information based upon the source and its type which may be either number of kilometres of vehicle traveled or tons of coal consumed or the extent of the area under construction, which multiplied with the relevant emission factor gives the overall mass balance value.
4. Computation of the rate at which each pollutant is emitted into the atmosphere, with this rate typically being extended to an annual basis. Systematic comparison of the emission from the proposed project or activity with the existing emission inventory for the area has to be carried out on an annual basis which can be computed by first assessing the rate at which each pollutant is emitted into the atmosphere from different project activities.
5. The pollutant emissions from each of the identified source categories associated with the proposed project or activity are to be added to arrive at the total value.

14.2.5.2 A meso scale impact calculation

A meso scale impact calculation has to be carried out based on the expected increase in the existing emission inventory for one or more pollutants as a result of the construction and or operational phase of the proposed project or activity using the relationship

Percentage increase in inventory

$$= \frac{\text{Project} - \text{activity emmission inventory information (100)}}{\text{Existing emission inverntory information}} \quad .....(14.1)$$

The increase in the percentage inventory for each pollutant for each project activity has to be assessed and the total percentage increase has to be arrived at by adding all the individual values in the inventory.

The percentage change in the current inventory for one or more air pollutants can be used to interpret and assess the impact on the basis of following criteria.

(a) The existing air quality for the pollutants of interest.
(b) The quantity of emission and magnitude of the percentage change.
(c) The time period of the expected percentage change.
(d) The potential for visibility reduction.
(e) Any local, sensitive receptors to damage from the pollutants.

Further in addition to the above factors expected emissions from the proposed activity with reference to applicable emission standards have to be examined, Though it is expected that the proposed activities will be in compliance with pertinent emission standards, the extent of compliance involved in the proposed activity should be thoroughly examined.

14.2.5.3 Box model approaches

The box model is a simple atmospheric dispersion model, which can be adopted to compute the ground level concentrations of specific air pollutants of concern emitted from the project activity [4]. The details are discussed in Chapter 5. As discussed earlier with a box model it is assumed that the pollutants emitted to the atmosphere are uniformly mixed in a volume, or “box” of air of fixed dimensions [5]. In this model the downwind, crosswind and vertical dimensions of the box and the time period over which pollutant emissions to be considered must be established. It is also assumed that the emissions, wind speed, and characteristics of air available for dilution will not vary over time, i.e. the box is under steady state conditions [6]. Further in this model it is also assumed that discharges mix completely and instantaneously with the air available for dilution and the released material is chemically stable and remains in the air.

The average concentrations of the pollutant using the box model can be mathematically expressed as

$$C = \frac{Qt}{xyz} \qquad .....(14.2)$$

where

C = average concentration of gas or particulate <20 μm in size, throughout the box, including at ground level, μg/m^3

Q = release rate of gas, or particulates <20 μm in size, from source type(s), μg/s

t = time period over which assumption of uniform mixing in box is valid, s (typical period, 1 h)

x = downwind dimension of box, m

y = crosswind dimension of box, m

z = vertical dimension of box, m

Adopting mass balance approach the important pollutants of interest can be examined for their ground level concentrations. One method of establishing the dimensions' of the box and the time period for the emissions is to use the data on limiting meteorological conditions corresponding to the worst case conditions.

The data from the box model can be examined and evaluated on a pollutant to pollutant basis, with reference to existing ambient air quality and corresponding standards, comparing existing pollutant concentrations and the concentration from the proposed project activity.

For more precise estimation of the dispersion of pollutants from a point source, the Following air emission models are used.

14.2.5.4 Air quality dispersion models

Based on the types of sources [elevated point (stacks), ground level point, ground level area, or line], pollutant type (gases or particulates), averaging times (short-term. 24-h, monthly, or annual), and atmospheric reactions (deposition, photochemical smog formation, or acid rain formation) the air pollutant dispersion models can be classified as:

(a) Manual calculation (or calculator) models and

(b) Computer models

14.2.5.4.1 Manual calculation models

There are three basic manual calculation models: (a) Pas quill-Gifford (b) Ground level point source model and (c) Area sources model which are widely used for calculating, the impact on air quality of projects or activities. They are useful for calculating short term, average concentrations of air pollutants at specific locations.

14.2.5.4.1.1 Pasquill model as modified by Gifford

Elevated point source model. This is one of the important manual calculation models (discussed in detail in Chapter 5) and can be indirectly used to analyse the impact on air quality of single, elevated point sources [7]. It is particularly suitable for various industrial gaseous emissions as many industries like chemical plants, heat and steam generation facilities have elevated stacks.

$$C_{x,y,0} = \frac{Q}{\pi\sigma_y\sigma_z u}\exp\left[-\left(\frac{H^2}{2\sigma_z^2}+\frac{y^2}{2\sigma_y^2}\right)\right] \qquad .....(14.3)$$

where $C_{x,y,0}$ = ground level concentration of gas, or particulates <20 μm in size, at a distance x, in m downwind from source, and distance y in m crosswind (90° from wind direction) from source, in μg/m^3

Q = release rate of gas, or particulates <20 μm in size, from elevated point source, μg/s

σ_y = horizontal dispersion coefficient which represents amount of plume spreading in crosswind direction from source, and under a given atmospheric stability condition, m

α_z = vertical, dispersion coefficient which represents amount of plume spreading in vertical direction at a distance x downwind from source, and under a given atmospheric stability condition, m

u= mean wind speed, m/sec

H = effective stack height (actual physical height plus any rise of plume as it leaves the stack), m; plume rise is a result of momentum effect caused by vertical velocity.

14.4.6.4.1.2 Ground level point source model

This is another manual calculation model that can be used for ground level point sources. The ground level concentration is given by:

$$C_{x,y,0} = \frac{Q}{\pi\sigma_y\sigma_z u}\exp - \left[\frac{y^2}{2\sigma_y^2}\right] \qquad .....(14.4)$$

14.2.5.4.1.3 Area source model

The other air pollution sources associated with many projects and activities are area sources (examples include air pollutants from agricultural operations, open burning, wind erosion, and pesticide applications) and line sources (examples include unpaved roads and vehicular traffic) for which the following equation can be used for the calculation of the ground level concentrations.

$$C_{x,0,0} = \frac{Q}{\pi(\sigma_y^2 + \sigma_{y0}^2)^{1/2}\sigma_z u} \qquad .....(14.5)$$

where $C_{x,y,0}$ = ground level concentration of gas, or particulates less than 20 μm in size, directly downwind and at a distance x in m downwind from the source, $\mu g/m^3$

σ_{y0} = one-fourth of emission width of area or line source along axis which coincides with wind direction, m

14.2.5.4.1.4 Computer models

A number of computer-based air quality simulation models which area a part of the U.S.EPA's UNAMAP (Users Network for Applied Modeling of Air Pollution) programme are available.

Using these models not only the impacts of the project or activity of the project on air quality can be assessed but also various modifications of the proposed project or activity can be evaluated to assess the effectiveness of the efforts to minimize the impacts of the project activity. Based on both the technical capabilities of models and model-related managerial issues such as economic considerations, and necessary training and experience of model users, the appropriate model can be selected for a given condition.

A PC compatible **SCREEN model** was developed by the USEPA to facilitate an initial evaluation of the air quality impacts of stationary sources [8]. It can be used in a **"Screening mode"** to calculate ground level concentrations under limited dispersion conditions.

A number of models specific for computation of fugitive-dust emissions, emissions at military airbases, heavy gases evaporation from toxic spills and hazardous waste sites are also available. The **Fugitive-Dust Model** (FDM) is a computerised air quality model specifically designed for computing concentration and deposition impacts from fugitive dust sources [9]. Emissions for each source are apportioned by the user into a series of particle size classes. A gravitational setting velocity and a deposition velocity are calculated by FDM for each class.

A user friendly model, **Emissions and Dispersion Modeling System" (EDMS)** was developed by Segal:

(a) To compute emission inventory of all sources at an airport or airbase.

(b) The concentrations produced by these sources at four airport locations.

Atmospheric dispersion of denser than air vapors can be computed by a package called **SLAB** [10]which can be applied to ground level area sources, an elevated horizontal jet, and an instantaneous volume source. Subsequent dispersion of the released vapor is modeled as either a steady-state plume, a transient puff, or a combination of the two, depending on the duration of release.

A model was developed for estimating evaporation which can occur from a toxic-chemical spill. The model is based on calculating the temperature of the spilled chemical Pool based on the net energy input into the pool from all possible sources. The pool temperature is then used to calculate the evaporation rate under steady state conditions.

A model to address site specific area source emissions hazardous wastes sites also was developed [1 1].

Table 14.4 *Selected USEPA-recommended plune models and appropriate applications*

Model	Averaging period	Source type	Terrain	Land use
SCREEN	Hourly, daily	Point, area	Simple, complex	Rural, urban
ISCST2	Hourly to annual	Point, area, volume	Simple	Rural urban
ISCLT2	Monthly, seasonal annual	Point, area, volume	Simple	Rural, urban
MPTER	Hourly to annual	Point	Simple	Rural
COMPLEX 1	Hourly to annual	Point	Complex	Rural
SHORTZ	Hourly to annual	Point, area	Complex	Urban
LONGZ	Seasonal to annual	Point, area	Complex	Urban

14.2.6 Step 5: Assessment of significance of impacts of predicted beneficial and detrimental effects

Evaluation of the significance of anticipated changes related to the proposed project should be carried out through public meetings and public participation programs. Professional judgement is required based on the percentage changes from base line conditions in terms of air pollutant emissions levels and /or exposed human population, or the PSI. These changes should be considered during both the construction and operational phases of a project.

For certain type of projects or air-pollutant prediction methods, there are numerical standards or criteria which can be used as a basis for interpretation. A final impact significance can be assessed based on the specific effects of the types of air pollutants from a proposed project-activity and identification of sensitive receptors in the study area [12, 13].

14.2.7 Step 6: Development of appropriate mitigation or remediation plans for reducing adverse Impacts

Remediation or mitigation measures involve project-activity design or operational Features that can be used to minimize the magnitude of the impact on air quality. Mainly the design should be revised to reduce the air pollutants expected to be emitted from the project. The revised project or activity can then be reassessed to determine whether other remediation or mitigation measures will help in eliminating or sufficiently minimizing the deleterious impact on air quality.

The following are some examples when mitigation measures can help in minimizing detrimental effects of air pollution.

1. Regulatory control on the practice of open burning of agricultural crop residues; like delineation of specified times for burning, and the establishment of distance requirements between residences and open burning areas. This will help in reducing air pollutant concentrations in sensitive areas [5].
2. Development of vegetation cover and watering or use of wind breaks and chemical stabilizers Watering is the most common method for temporary dust control(nearly 50%).Planting of rapid growing vegetation in construction areas will reduce dust generation [14].
3. For reducing air pollutant emissions from unpaved roads, paving the surface, treating with penetration chemicals, working soil-stabilization chemicals into the roadbed, watering, and traffic control regulations are some of the remediation methods [15]:
4. In the case of open waste piles and staging areas, dry surface impoundments, landfills, land treatment systems, and waste stabilization measures are adopted as Fugitive dust control techniques [15].
5. For controlling or reducing airborne pesticide residues resulting from spraying, pesticide and other materials, low pressure spray nozzles can be used to minimize the generation of fine particles. Further by spraying the pesticide in periods of low wind velocities its dispersion can be reduced.
6. Several alternative fuels, including methanol, ethanol, compressed natural gas liquefied petroleum gas, electricity, and reformulated gasoline [16] are recommended by USEPA for reducing air pollution from various combustion process in automobiles and processes.

7. Exhaust emission control technologies for petrol- and diesel-fueled cars, trucks, and buses are available to minimize pollutants from the transportation sector.
8. A number of control systems like cyclones, scrubbers, fabric filters, and electrostatic precipitators for control of particulate emissions are now available. Flue-gas desulphurization and carbon adsorption can be used for SO_2 control. Thermal incineration, flares, carbon adsorption, absorption, condensers, fabric filters, electrostatic precipitators, and venturi scrubbers are some of the control systems using for reducing pollutant levels from various point source air pollutants [17, 18].

14.3 Detailed Methodology for Air Pollution Dispersion Modeling Approaches for assessment of Air Quality Impacts (AQI) for a Specific Study Area (19-91)

14.3.1 Various key steps in air pollution dispersion modeling approach for air quality impact assessment for a specific study area

Air Quality Impact assessment methodology using Air Pollution Dispersion modeling approach involves the following 5 Key Steps

1. Input data collection
2. Dispersion modelling
3. Processing dispersion model output data
4. Interpretation of dispersion modelling results
5. Preparation of an impact assessment report

14.3.2 Input data collection

The first stage in the Air Quality impact assessment is the collection of all the information required to complete the dispersion modelling. This includes development of (a) air emissions inventory; (b) compilation of meteorological data; (c). back ground air quality data; and (d). terrain data.

14.3.2.1 Emissions inventory

The emissions inventory is the foundation of the air quality impact assessment. Developing a sound emissions inventory should be a priority task and requires the collation of a significant amount of data. A thorough air emissions inventory for a study area identifies all sources of air pollution, the air pollutants emitted from each source, and estimates the emission concentration and rate of all air pollutants emitted.

The details of each step of the development of an emissions inventory are given below

14.3.2.1.1 Identify all sources of air pollution and potential emissions

A thorough understanding of the premises is essential in developing an emissions inventory. Undertaking a site visit of the existing premises or a detailed review of the engineering drawings for the proposed premises is necessary to identify all sources of air pollution and gain an understanding of the process and industry. This knowledge can be supplemented with a literature review on the industry and its most prevalent air pollution issues. For all sources of air pollution at a premises, identify the following:

- release type and source release parameters
- location (in metres relative to fixed origin, elevation and discharge geometry)
- potential air pollutants emitted.

Release type

Source configuration may be one of the following types:

Point sources

For a point source, emissions emanate from a very small opening such as a stack or vent. Stacks usually emit hot gases forcefully into the atmosphere at a fixed height above ground level.

Tall point sources: The term 'tall' point source usually refers to sources that protrude out of the surface boundary layer (e.g. over 30 to 50 m tall).

Wake-affected point sources: Where nearby buildings interfere with the trajectory and growth of the plume, the source is called a wake-affected point source. A point source is wake-affected if stack height is less than or equal to 2.5 times the height of buildings located within a distance of 5L (where L is the lesser of the height or width of the building) from each release point.

Wake-free point sources: Wake-free point sources are more than 2.5 times the height of the largest nearby building, so that surrounding buildings do not influence the stack top airflow.

Area sources

An area source has a more realistic two-dimensional structure but only a limited vertical extent. It is a source with a large surface area such as a liquid surface (pond, lagoon) or a landfill surface.

Line sources

A line source is a special case of a long, thin area source. In practice, these sources are taken to be at ground level and thin. A line source becomes an area source if the breadth exceeds 20% of the length.

Volume sources

A volume source is an essentially three-dimensional structure. Usually there are a sufficient number of emission points to consider a uniform emission rate over the full source structure. They are diffuse sources, such as emissions from within a building.

Source release parameters: For proposed study area this information is obtained from the engineering drawings and plans.

Point: stack height, stack diameter, temperature, discharge velocity, moisture, pressure, carbon dioxide and oxygen concentration

Diffuse area: surface area, side length and release height

Diffuse volume: side length, release height, and initial horizontal and vertical plume spread (σ_y and σ_z).

Estimation of emission rates

There are a number of methods that can be used to estimate the emission rate from each source. The EPA's preferred methods are direct measurement for existing sources and manufacturers' design specifications for proposed sources. Emission factors are generally used when there is no other information available or when emissions can reasonably be demonstrated to be negligible.

Manufacturers' design specifications or performance guarantees

Manufacturers' design specifications or performance guarantees can be used to estimate the emission rate of air pollutants from proposed sources. Such specifications provide a reliable means of determining the upper limit to the emission rate or concentration of air pollutants for sources that are maintained and operated in a proper and efficient manner.

Post-commissioning testing may be required to establish that sources comply with the manufacturers' design specifications and/or performance guarantees.

Direct measurement

For sources where manufacturers' design specifications or performance guarantees are unknown, emission rates and source release parameters should normally be established from the results of source emission sampling and analysis.

Emission factors

An emission factor is usually an equation that relates the quantity of a pollutant released to process throughput. These factors are generally averages of all available data of acceptable quality, and are generally assumed to be representative of long-term averages for all facilities in the source category. As stated above, emission factors are generally used when there is no other information available or when emissions can reasonably be demonstrated to be negligible. Some databases of emission factors include: • US EPA's AP-42 Emissions Factors (www.epa.gov/ttn/chief)

- National Pollutant Inventory Emissions Estimation Technique Manuals (www.npi.gov.au/handbooks/approved_handbooks/sector-manuals.html).

Calculate emission concentration for point sources

The concentration of a pollutant emitted from a source is calculated using equation :

$$Ci = ERi_{FR} \quad .....(14.6)$$

where:

C_i = the concentration of pollutant i emitted from a source in mg/m^3

ER_i = the rate pollutant i is emitted from the source in mg/s

FR = the gaseous volumetric flow rate in m^3/s

The inventory should contain two emission concentrations:

1. Actual concentration of a pollutant emitted from a source (mg/Am^3) calculated using the actual gaseous volumetric flow rate (Am^3/s) and measured emission rate in Equation 14.6
2. Concentration of a pollutant emitted from a source corrected to the reference conditions as specified in the Regulation. This is calculated using the gaseous volumetric flow rate corrected to normal conditions (dry, 273K, 101.3 kPa) and the measured emission rate in Equation 14.6.

Assess compliance with the Regulation

The inventory must be used to demonstrate compliance with the Regulation. All sources of air emissions must comply with the requirements of the Regulation. If a source does not comply, the emissions inventory must be revised to reflect the implementation of new technology and/or pollution control equipment that will comply with the Regulation.

Presentation of emissions inventory

The results of the emissions inventory must be presented to include the following information:

- all release parameters of stack and fugitive sources (e.g. temperature, exit velocity, stack dimensions, flow rate, moisture content, pressure, carbon dioxide and oxygen concentration)
- pollutant emission concentrations and a comparison against the relevant requirements of the Regulation.

A suggested format for summarising and presenting the results of the emissions inventory in the impact assessment report is provided in Tables 14.5 and 14.6. The additional data that should be included in the impact assessment report for complex mixtures of odour and hydrogen sulfide is included in Table 14.7.

Table 14.5 *Stack source release parameters*

Source	Release type	Stack height (m)	Exit temp (C)	Exit diameter (m)	Exit velocity (m/s)	Oxygen conc. (%)	Moisture content (%)	Flow rate (Am/s)	Flow rare (Nm)
Boiler No. 1	Wake-affected	20	150	4	15	10	15	188.5	103.4

14.3.2.2 Compilation of meteorological data

The meteorological data used in the dispersion model is of fundamental importance as it drives the transport and dispersion of the air pollutants in the atmosphere. The most critical parameters are wind direction, which determines the initial direction of transport of pollutants from their sources; wind speed, which dilutes the plume in the direction of transport and determines the travel time from source to receptor; and atmospheric turbulence, which indicates the dispersive ability of the atmosphere.

Table 14.6 *Stack emission concentrations and regulation limits*

Pollutant	Emission rate (g/s)	Emission (mg/Am concentration)	Corrected emission concentration 3 at stack (mg/Nm reference conditions)	Regulation emission concentration 3 at limit (mg/Nm stack reference conditions)
Sulfur dioxide	40	212.2	N/A	N/A
Solid particles	2	10.6	31.6	100
Oxides of nitrogen	15	79.6	237.4	350

Table 14.7 *Peak odour emission rates*

Source	Source type	Odour emission rate (OUm 3/s)	Stability class	Peak odour emission rate (OUm 3/s)	
				Near-field	Far-feild
Lagoon No.1	Area	20,000	A, B, C, D	50,000	46,000
			E , F	46,000	38,000

14.3.2.2.1 Minimum data requirements

The meteorological data used in the dispersion modelling is one factor that determines the level of assessment.

Level 1 impact assessments are conducted using 'synthetic' worst-case meteorological data. Table 14.8lists the wind speed and stability class combinations that need to be included in the synthetic worst-case meteorological data file.

Level 2 impact assessments are conducted using at least one year of site-specific meteorological data. The meteorological data must be 90% complete in order to be acceptable for use in Level 2 impact assessments (i.e. for one year, there can be no more than 876 hours of data missing). If site-specific meteorological data are not available for a Level 2 impact assessment, at least one year of site-representative meteorological data must be used. The site representative data should be:

- preferably collected at a meteorological monitoring station. Where measured data is unavailable or of insufficient quality for dispersion modelling purposes, a meteorological data file may be generated using a prognostic meteorological model such as TAPM.
- correlated against a longer-duration site-representative meteorological database of at least five years (preferably five consecutive years) to be deemed acceptable. It must be clearly established that the data adequately describes the expected meteorological patterns at the site under investigation (e.g. wind speed, wind direction, ambient temperature, atmospheric stability class, inversion conditions and katabatic drift).

14.3.2.2.2 Preparation of Level 1 meteorological data

Wind speed and stability class

Gaussian plume dispersion models use stability categories as indicators of atmospheric turbulence and the dispersive properties of the atmosphere. Based on the work of Pasquill and Gifford, seven stability categories have been defined: A – very unstable; B – unstable; C – slightly unstable; D – neutral; E – slightly stable; F – stable; and G – very stable conditions. In most dispersion models, stability classes F and G are combined into one class, F.

The stability class at any given time depends on:

- static stability (vertical temperature profile of the atmosphere, i.e. migrating high and low air-pressure masses)
- convective or thermal turbulence (caused by the rising of air heated at ground level)
- mechanical turbulence (a function of wind speed and surface roughness, i.e. wind flow over rough terrain, trees or buildings).

Table 14.8 lists the minimum wind speed and stability class combinations that must be included in a Level 1 meteorological data file.

Table 14.8 *Wind speed and stability class combinations for a Level 1 data file*

	Wind speed (m/s)																		
Stability class	**0.5**	**1**	**1.5**	**2**	**2.5**	**3**	**3.5**	**4**	**4.5**	**5**	**6**	**7**	**8**	**10**	**12**	**14**	**16**	**18**	**20**
A	*	*	*	*	*	*													
B	*	*	*	*	*	*	*	*	*	*									
C	*	*	*	*	*	*	*	*	*	*	*	*	*	*					
D	*	*	*	*	*	*	*	*	*	*	*	*	*	*	*	*	*	*	*
E	*	*	*	*	*	*	*	*	*	*									
F	*	*	*	*	*	*													

Ambient temperature

For Level 1 impact assessments, the maximum and minimum ambient temperatures that are representative of the site must be included in the Level 1 meteorological data file to account for the range in possible plume rise. Higher ambient temperatures will result in the lowest plume rise and hence the largest impacts.

Mixing height

For Level 1 impact assessments, the mixing height for neutral and unstable conditions (classes A–D) can be calculated using an estimate of the mechanically driven mixing height. The mechanical mixing height, h, can be calculated as follows:

Mechanical mixing height for stability classes A–D

$$h = 0.3 \times u^* / f \ldots\ldots \qquad \ldots\ldots(14.7)$$

where

h = mixing height (m) u* = friction velocity (m/s) f = coriolis parameter

For Level 1 impact assessments, the mixing height, h, for stable conditions (classes *E* and *F*) can either be set at an unlimited value (e.g. 5000 m) or calculated as follows:

Mechanical mixing height for stability classes E and F

$$h = 0.4 \times (u^* L / f)^{0.5} \ldots\ldots \qquad \ldots\ldots(14.8)$$

where

h = mixing height (m); u^* = friction velocity (m/s);

L = Monin–Obukhov length (m) F = coriolis parameter

Monin–Obukhov length

The Monin–Obukhov length, L, characterises the stability of the surface layer. The surface layer is defined as the layer above the ground in which the vertical variation of heat and momentum flux is negligible. The surface layer is typically 10% the height of the mixed layer. The parameter, L, can be calculated using the linear approximation to Golder's plot (Golder 1972) as follows:

Monin–Obukhov length

$$1/L = X + Y \times \log_{10}(Z_o) \ldots\ldots\ldots\ldots\ldots \qquad \ldots\ldots(14.9)$$

where

L = Monin–Obukhov length (m)

X & Y = parameters dependent on the Pasquill–Gifford stability class (Table 14.9) Z_o = surface roughness height (m) (Table 14.10)

Table 14.9 *Parameterisation of Golder's plot*

	Pasquill–Gifford stability class					
Parameter	**A**	**B**	**C**	**D**	**E**	**F**
X	–0.096	–0.037	–0.002	0.000	0.004	0.035
Y	0.029	0.025	0.018	0.000	–0.018	–0.0365

In Equation 14.9

- the value of Z_o is the surface roughness height, unless the surface roughness height is outside the range Z_{omin} to Z_{omax} presented in Table 14.10
- if the surface roughness height < Z_{omin} use the value of Z_{omin} for $\mathbf{Z_o}$• if the surface roughness > Z_{omax} use the value of Z_{omax} for Z_o.

Table 14.10 *Upper and lower limits for surface roughness heights for each Pasquill– Gifford stability class*

	Pasquill–Gifford stability class					
Parameter	**A**	**B**	**C**	**D**	**E**	**F**
Zomin	0.001	0.001	0.001	0.001	0.001	0.001
Zomax	18.0	30.0	1.25	50.0	1.6	9.0

Table 14.11 presents typical values of surface roughness height for various land uses.

Table 14.11 *Typical values of surface roughness height for various land-use categories (AUSPLUME version 6.0)*

Land-use category	Roughness height Z_o (m)	Land-use category	Roughness height Z_o (m)
Hill	2.0	High-rise	1.0
Industrial area	0.8	Commercial	0.8
Forest	0.8	Residential	0.4
Rolling rural	0.4	Flat rural	0.1
Flat desert	0.01	Water	0.0001

Surface friction velocity

The surface friction velocity, u^*, is a measure of mechanical turbulence and is directly related to the surface roughness. The parameter, u^*, can be calculated using the procedure given in (Businger and Fleagle 1980; McRae 1981).

Coriolis parameter

The coriol is parameter accounts for variation in wind direction with height (wind shear) at different latitudes and can be calculated in accordance with well-established techniques. The coriolis parameter, *f*, can be calculated as follows:

$$f = 2\Omega \sin(\Psi)$$

where

Ω = Earth's rotation rate ($2\Pi/86400$ or 7.29×10^{-5} rad·s^{-1}) Π = pi or 3.1416 radians (rad)

86400 = number of seconds in the day (s/day)

Ψ = latitude in radians (rad)

Table 14.12 an example of typical mixing heights for a location with a similar latitude to Sydney (34°) and in a rural location (surface roughness height of 0.3 m) to be included in the Level 1 meteorological data file.

Table 14.12 *Typical mixing heights for a rural location (km)*

	Wind speed (m/s)																		
Stability class	**0.5**	**1**	**1.5**	**2**	**2.5**	**3**	**3.5**	**4**	**4.5**	**5**	**6**	**7**	**8**	**10**	**12**	**14**	**16**	**18**	**20**
A	0.2	0.4	0.6	0.8	1.0	1.2													
B	0.2	0.4	0.5	0.7	0.9	1.0	1.2	1.4	1.6	1.8									
C	0.2	0.3	0.5	0.6	0.8	0.9	1.1	1.2	1.4	1.5	1.8	2.1	2.4	3.1					
D	0.2	0.3	0.4	0.6	0.7	0.8	1.0	1.1	1.3	1.4	1.7	2.0	2.2	2.8	3.3	3.9	4.5	5.0	5.0
E	5.0	5.0	5.0	5.0	5.0	5.0	5.0	5.0	5.0	5.0									
F	5.0	5.0	5.0	5.0	5.0	5.0	5.0												

Table 14.13 lists an example of typical mixing heights for a location with a similar latitude to Sydney (34°) and in an urban location (surface roughness height of 1.0 m) to be included in the Level 1 meteorological data file.

Table 14.13 *Typical mixing heights for an urban location (km)*

	Wind speed (m/s)																		
Stability class	**0.5**	**1**	**1.5**	**2**	**2.5**	**3**	**3.5**	**4**	**4.5**	**5**	**6**	**7**	**8**	**10**	**12**	**14**	**16**	**18**	**20**
A	0.3	0.6	1.0	1.3	1.6	2.0													
B	0.3	0.5	0.8	1.1	1.4	1.7	1.9	2.2	2.5	2.7									
C	0.2	0.4	0.7	0.9	1.1	1.3	1.5	1.8	2.0	2.2	2.6	3.1	3.5	4.4					
D	0.2	0.4	0.6	0.8	1.1	1.3	1.5	1.7	1.9	2.1	2.6	2.9	3.4	4.3	5.0	5.0	5.0	5.0	5.0
E	5.0	5.0	5.0	5.0	5.0	5.0	5.0	5.0	5.0	5.0									
F	5.0	5.0	5.0	5.0	5.0	5.0	5.0												

14.3.2.2.3 Preparation of Level 2 meteorological data

For guidance on processing meteorological data for dispersion modelling purposes, the USEPA guide (USEPA 2000) and USEPA processor (USEPA 1996) should be used.

Stability class

In order of preference, the following methods outlined in USEPA (2000) should be used to determine stability class:

Turner's 1964 method: This requires information on solar altitude or zenith angle, cloud cover, cloud ceiling height and wind speed. Solar altitude can easily be calculated, but cloud cover and ceiling height are generally determined through visual observations.

Solar radiation–delta temperature method: This retains the basic structure and rationale of Turner's 1964 method but eliminates the need for observations of cloud cover and ceiling height. The method uses the surface-layer wind speed (measured at 10 m) in combination with measurements of total solar radiation during the day and a low-level vertical temperature difference (i.e. at 2 m and 10 m) at night.

Sigma theta method (the standard deviation of the horizontal wind direction fluctuation): All modern meteorological data loggers include software to determine sigma theta.

For Level 2 impact assessments, hourly stability class should be estimated using the USEPA meteorological pre-processor for regulatory models (USEPA 1996) or a processor that includes similar techniques.

Mixing height

Mixing height is the depth through which pollutants released to the atmosphere are typically mixed by dispersive processes. Dispersion of pollutants in the lower atmosphere is greatly aided by the convective and turbulent mixing that takes place. Mixing height determines the vertical extent of dispersion for releases occurring below that height. Releases occurring above that height are assumed to have no ground-level impact (with the exception of fumigation episodes). Therefore, the greater the vertical extent of the mixed layer, the larger the volume available to dilute pollutant emissions.

Morning and afternoon mixing heights are estimated using vertical temperature profiles (otherwise known as 'upper air data') and surface temperature measurements. For Level 2 impact assessments, hourly mixing heights should be estimated from the twice-daily mixing height values, sunrise and sunset times, and hourly stability categories by using the USEPA meteorological pre-processor for regulatory models (USEPA 1996) or a processor which includes similar techniques.

14.3.2.3 Background air quality, terrain, sensitive receptors and building wake effects

14.3.2.3.1 Background air quality data

Including background concentrations of pollutants in the assessment enables the total impact of the proposal (i.e. impact of emissions on existing air quality) to be assessed. The background concentrations of air pollutants are ideally obtained from ambient monitoring data collected at the proposed site. As this is extremely rare, data is typically obtained from a monitoring site as close as possible to the proposed location where the sources of air pollution resemble the existing sources at the proposal site.

Accounting for background concentrations

For impact assessments of sulfur dioxide (SO_2), nitrogen dioxide (NO_2), ozone (O_3), PM_{10}, total suspended particulates (TSP), deposited dust, lead (Pb), carbon monoxide (CO) and hydrogen fluoride (HF), the existing background concentrations of the pollutants in the study area should be assessed.

Level 1 assessments

- Obtain ambient monitoring data that includes at least one year of continuous measurements.
- Determine the maximum background concentration of the pollutant being assessed for each relevant averaging period.

Level 2 assessments

- Obtain ambient monitoring data that includes at least one year of continuous measurements and is contemporaneous with the meteorological data used in the dispersion modelling.
- At each receptor, add each individual dispersion model prediction to the corresponding measured background concentration (e.g. add the first hourly average dispersion model prediction to the first hourly average background concentration) to obtain hourly predictions of total impact.

14.3.2.4 Terrain and sensitive receptors

The dispersion modelling input file requires information regarding the surrounding terrain and sensitive receptors. Terrain and receptor files to be developed which include the location and height in metres relative to a fixed origin. The location of any particularly sensitive receptors (and likely future sensitive receptors) such as residences, schools and hospitals can also be specifically included in the receptor file.

Building wake effects: PRIME is the EPA's preferred building wake algorithm. AUSPLUME v. 6.0 includes the PRIME building wake algorithm and the Building Profile Input Program (BPIP) for entering the location and dimension of buildings. The location and dimensions of buildings located within a distance of 5L (where *L* is the lesser of the height or width of the building) from each release point for buildings with a height greater than 0.4 times the stack height should be entered in BPIP.

14.3.3 Dispersion modelling

14.3.3.1 Dispersion models

Dispersion models provide the ability to mathematically simulate atmospheric conditions and behaviour. They are used to calculate spatial and temporal sets of concentrations and particle deposition due to emissions from various sources. Dispersion models can be used to determine the affected zone around an emitter by producing results that can be compared against impact assessment criteria.

Dispersion models can provide concentration or deposition estimates over an almost unlimited grid of user-specified locations, and can be used to evaluate both existing and forecast emission scenarios. In this capacity, air dispersion modelling is a useful tool in assessing the air quality impacts associated with existing and proposed emission sources. The results of the dispersion modelling analysis can be used to develop control strategies that should ensure compliance with the assessment criteria. Dispersion models can also be used to estimate the cumulative impacts of various industries that are located close to one another.

Dispersion models are widely used by environmental regulators in Australia, New Zealand, the United States, the United Kingdom and Europe, and industry well understands their limitations. The results have been shown, through numerous model evaluation studies, to be sufficiently robust to be relied on to calculate concentration limits for point-source stack emissions.

14.4 Approved dispersion Models for Air Quality Assessment for Typical Urban Areas

AUSPLUME v. 6.0 is the approved dispersion model for use in most simple, near field applications in NSW, where coastal effects and complex terrain are of no concern.

AUSPLUME is a Gaussian plume model, based on the assumption that cross-sections through elevated plumes from point sources of pollution have a Gaussian (or normal) distribution of concentration. AUSPLUME is also a steady-state model, which assumes the atmosphere is in a state of uniform flow, and wind velocity is a function of height alone and does not vary with direction. The mathematical basis of AUSPLUME is the Victorian EPA's Plume Calculation Procedure (EPA Victoria 1985), which itself is an extension of the ISC model (USEPA 1995). AUSPLUME v. 6.0 is specifically *not* approved for use in the following applications:

- **complex terrain, non-steady-state conditions:** AUSPLUME is a steady-state model and is unable to adjust the winds to reflect the effects of terrain. The straight-line trajectory assumption of the plume model is unable to handle the curved flow associated with terrain-induced deflection of channelling. AUSPLUME should not be used for terrain where the height of any receptor exceeds the lowest release height.
- **buoyant line plumes** (e.g. discharges from the roof vents of aluminium smelters)
- **coastal effects such as fumigation:** AUSPLUME is unable to consider large changes in meteorological conditions which can occur over short distances across a coastline.
- **high frequency of stable calm night-time conditions:** Pollutants can accumulate under such conditions or will flow downwind with the drainage flow. AUSPLUME has no memory of the previous hour's weather conditions as each hour is treated independently of the next and material is carried away instantaneously, to the edge of the grid, even if only light winds are prevailing.
- **high frequency of calm conditions:** AUSPLUME cannot handle calm conditions because of the inverse wind speed dependence plume equation. AUSPLUME assumes a minimum wind speed, which shoots the plume out to the edge of the grid even though the plume may not have moved at all.

- **inversion break-up fumigation conditions**: There are also other situations where another dispersion model may be more scientifically sound than AUSPLUME. In these instances, CALPUFF or TAPM (Section 6.3) may be appropriate. The two key factors that should be considered in evaluating whether to use a conventional plume model, such as AUSPLUME, or a more sophisticated approach are:
  1. Is the steady-state assumption in the plume model valid?
  2. Do the technical parameterisations in the plume model adequately treat the situation to be modelled?

14.4.1 Advanced dispersion models for specialist application

In circumstances where the AUSPLUME dispersion model is not approved or suitable for use, other dispersion models may be appropriate. Guidance on choosing appropriate alternative dispersion models can be found in the USEPA publication *Guideline on Air Quality Models* (USEPA 1999). CALPUFF and TAPM are the most commonly used alternative dispersion models for regulatory dispersion modelling applications in NSW. However only experienced and appropriately trained professionals should use them.

CALPUFF

CALPUFF is a multi-layer, multi-species, non-steady-state Gaussian puff dispersion model that is able to simulate the effects of time- and space-varying meteorological conditions on pollutant transport. This enables the model to account for a variety of effects such as spatial variability of meteorological conditions, causality effects, dry deposition and dispersion over a variety of spatially varying land surfaces, plume fumigation, low wind speed dispersion, pollutant transformation and wet removal. CALPUFF has various algorithms for parameterising dispersion processes, including the use of turbulence-based dispersion coefficients derived from similarity theory or observations.

CALPUFF has been accepted by the USEPA as a guideline model to be used in regulatory applications involving the long-range transport of pollutants (> 50 km). It can also be used on a case-by-case basis in situations involving complex flow and non-steady-state cases up to 50 kilometres from the source.

CALPUFF v. 5.7, CALMET v. 5.5 and CALPOST v. 5.4 or later should be used.

TAPM

The Air Pollution Model (TAPM) was developed by the CSIRO to simulate three dimensional meteorology and pollution dispersion in areas where meteorological data are sparse or non-existent. The modelling system contains a number of dispersion modules. These include a particle/puff dispersion model for dispersion from point, line, area and volume sources, and a three-dimensional grid-point model for urban air pollution studies. The dispersion models allow for plume rise and building wake effects, wet and dry deposition and photochemistry for urban air shed applications. TAPM v. 2.0 or later should be used.

Dispersion modelling methodology using AUSPLUME

Unless otherwise stated, the default options specified in the *Technical Users Manual* (EPA Victoria 2000) must be used with AUSPLUME. These options are the most appropriate for most impact assessment applications.

Terrain effects

- Use the Egan half-height approach to account for terrain effects.

Building wake effects

- All building dimensions must be entered in 10-degree increments. Use the USEPA utility program BPIP (USEPA 1995) within AUSPLUME to calculate the 36 wind-direction dependent building dimensions.
- Use the PRIME method to account for building wake effects.
- The USEPA's guidance document on good engineering practice (USEPA 1985) must be taken into account when designing new stacks to avoid building wake effects.

Horizontal dispersion curves

- For stacks < 100 m high, use sigma theta values or Pas quill–Gifford curves.
- For stacks > 100 m high, use Briggs rural curves.

Vertical dispersion curves

- For stacks < 100 m high, use Pas quill–Gifford curves.
- For stacks > 100 m high, use Briggs rural curves.

'Enhance plume spreads for buoyancy'

- Enable this option for both the horizontal and vertical dimensions. **'Adjust Pas quill–Gifford formulae for roughness height'** • Use this option.

Plume rise parameters

- Use the AUSPLUME defaults.

Wind speed categories

- Use the AUSPLUME defaults.

Wind profile exponents

- Use Irwin rural wind profile exponents for rural areas.
- Use Irwin urban wind profile exponents for urban areas.

Dispersion modelling methodology using CALPUFF

CALPUFF includes an option to automatically set all the options to the USEPA default values. These include:

- Wind speed profile: ISC Rural
- Transitional plume rise modelled
- Stacktip downwash
- Partial plume penetration

- Dispersion curves: Pasquill–Gifford or dispersion coefficients using turbulence-based micro-meteorology
- No adjustment of dispersion curves for roughness
- Terrain: partial plume adjustment method

14.4.2 Processing dispersion model output data

14.4.2.1 Calculate peak concentrations

The evaluation of odour impacts requires the estimation of short or peak concentrations on the time scale of less than one second. Dispersion model predictions are typically valid for averaging periods of one hour and longer. Dispersion models, such as AUSPLUME, therefore need to be supplemented to accurately simulate atmospheric dispersion of odours and the instantaneous perception of odours by the human nose.

The prediction of peak concentrations from estimates of ensemble means can be obtained from a ratio between extreme short-term concentration and longer-term averages. Properly defined peak-to-mean ratios depend upon the type of source, atmospheric stability and distance downwind. Table14. 15 shows the EPA-recommended factors for estimating peak concentrations for different source types, stabilities and distances as developed by Kate stone Scientific (1995 and 1998).

The P/M60 ratios in Table 14.15 are for an idealised situation for one source in flat terrain where the receptor is located along the centerline of the single plume. The ratios do not consider fluctuations away from the centre line, terrain influences or plume interaction from multiple sources.

The EPA requires peak ground-level concentrations to be calculated for the following pollutants:

- hydrogen sulfide
- complex mixtures of odour.

A screening level assessment of peak GLCS can be undertaken by applying the ratios in Table 14.14 to multiple sources at a premises. These ratios can be applied to the emission rates entered into the dispersion model as follows:

1. Determine the source type, stability class and if the receptors are near-field or farfield or both.
2. Select the appropriate P/M60 ratios from Table 14.15
3. For wake-affected point sources, determine the meteorological conditions (i.e. wind speed and stability class) under which the source is wake-affected and wake-free.
4. Apply P/M60 ratios to odour and hydrogen sulfide emission rates so they vary with wind speed and stability class.

Table 14.14 *Factors for estimating peak concentrations in flat terrain (Kate stone Scientific 1995 and 1998)*

Source type	Pasquill–Gifford stability class	Near-field P/M60*	Far-field P/M60*
Area	A, B, C, D	2.5	2.3
	E, F	2.3	1.9
Line	A–F	6	6
Surface wake-free point	A, B, C	12	4
	D, E, F	25	7
Tall wake-free point	A, B, C	17	3
	D, E, F	35	6
Wake-affected point	A–F	2.3	2.3
Volume	A–F	2.3	2.3

* Ratio of peak 1-second average concentrations to mean 1-hour average concentrations

Availability of dispersion modelling software and guidance documents

Windows-based **AUSPLUME** v. 6.0 can be purchased by writing to: Environment Protection Authority of Victoria, 27 Francis Street, Melbourne Victoria 3000

The **BPIP PRIME** (BPIPPRM) user's manual and software can be electronically downloaded, free of charge, from the USEPA website at

www.epa.gov/scram001/tt22.htm#bpipprm

The **CALPUFF** dispersion modelling package and guidance documents can be electronically downloaded, free of charge, from the Earth Tech Incorporated website at

earthtec.vwh.net/download/download.htm The **TAPM** software can be purchased by writing to:

Dr Peter Hurley, CSIRO Atmospheric Research PMB 1, Aspendale Victoria 3195

Guidance documents and information about TAPM can be obtained from the CSIRO website at www.dar.csiro.au/tapm/

Other dispersion modelling software and guidance documents can be electronically downloaded free of charge from the USEPA website at www.epa.gov/ttn/scram/

14.4.2.2 Interpretation of dispersion modelling results

The primary purpose of an air quality impact assessment is to determine whether emissions from a premises will comply with the appropriate environmental outcomes.

The assessment criteria outlined below reflect the environmental outcomes adopted by the EPA.

To ensure the environmental outcomes are achieved, emissions from a premises must be assessed against the assessment criteria. The cumulative impact of emissions from several facilities also needs to be considered. Impacts of sulfur dioxide (SO_2), nitrogen dioxide (NO_2), ozone (O_3), Lead (Pb), particles (PM_{10}), total suspended particulates (TSP), deposited dust, carbon monoxide (CO) and hydrogen fluoride (HF) must be combined with existing background levels before comparison with the relevant impact assessment criteria. (Table 14.15)

Table 14.15 *Impact assessment criteria for SO_2, NO_2, O_3, Pb, PM_{10}, TSP, deposited dust, CO and HF*

Pollutant	Averaging period	Concentration		Source
		pphm	**μg/m³**	
Sulfur dioxide (SO_2)	10 minutes	25	712	NGMRC (1996)
	1 hour	20	570	NEPC (1998)
	24 hours	8	228	NEPC (1998)
	Annual	2	60	NEPC (1998)
Nitrogen dioxide (NO_2)	1 hour	12	246	NEPC (1998)
	Annual	3	62	NEPC (1998)
Photochemical oxidants (as ozone)	1 hour	10	214	NEPC (1998)
	4 hours	8	171	NEPC (1998)
Lead	Annual	–	0.5	NEPC (1998)
PM_{10}	24 hours	-	50	NEPC (1998)
	Annual	-	30	EPA (1998)
Total suspended particulates (TSP)	Annual	-	90	NHMRC (1996)
		g/m²/month [a]	**g/m²/month [b]**	
Deposited dust [c]	Annual	2	4	NERDDC (1988)
		ppm	**mg/m³**	
Carbon monoxide (CO)	15 minutes	87	100	WHD (2000)
	1 hour	25	30	WHO (2000)
	8 hours	9	10	NEPC (1998)
		μg/m³ [d]	**μg/m³ [e]**	
Hydrogen fluoride	90 days	0.5	0.25	ANGECC (1990)
	30 days	0.84	0.4	ANZECC (1990)
	7 days	1.7	0.8	ANZECC (1990)
	24 hours	2.9	1.5	ANZECC (1990)

[a]Maximum increase in deposited dust level. [b]Maximum total deposited dust level. [c]Dust is assessed as insoluble solids as defined by AS 3580.10.1–1991 (AM-19). [d]General land use, which includes all areas other than specialised land use. [e]Specialised land use, which includes all areas with vegetation sensitive to fluoride, such as grape vines and stone fruits.

14.4.2.3 Presentation of Air quality Impact assessment results

The results of an impact assessment should be presented as follows:

1. Concentration, hazard index and/or risk contours (isopleths) to define potential affected zones
2. Concentration, hazard index and/or risk predictions in tabular form for each of the following:
   (a) existing and likely future sensitive receptors
   (b) maximum exposed off-site receptor
   (c) maximum outside the boundary of the premises.

14.4.2.4 What if impact assessment criteria are exceeded?

If the EPA's impact assessment criteria are exceeded, the dispersion modelling must be revised to include various pollution control strategies until compliance is achieved. To determine incremental increases in the cost of air pollution abatement, a sensitivity analysis can be carried out by varying:

- source release parameters
- separation distance
- level of management practice.

The results can be used to select the most cost-effective and environmentally effective control strategy.

14.4.3 Application of GIS for presentation of Air Quality Impact

Environmental impact assessment (EIA) of air pollution diffusion of different air pollutants on GIS platform can provide a decision support system with enhanced micro scale spatial air pollution analysis for sustainable environmental planning

The Geographic Information System (GIS), as a product of the combination traditional sciences (geography, cartography and surveying) with modern science and technology (computer science, etc.), is gradually evolving into an integrated modern discipline for processing spatial information. GIS is an effective managing tool for the geographical information resources, and its powerful databases and ability of spatial analysis have found more and more applications in the fields of environmental impact assessment.

Arc GIS Engine is a collection of GIS components and developer resources that can be embedded, allowing the developers to add dynamic mapping and GIS capabilities to existing applications or build new custom mapping applications.

New Software's are being developed in which Arc Engine is used to deploy GIS data and maps, and visualization of atmospheric pollution diffusion using application programming interfaces (APIs) for COM and NET (Figure 14.2).

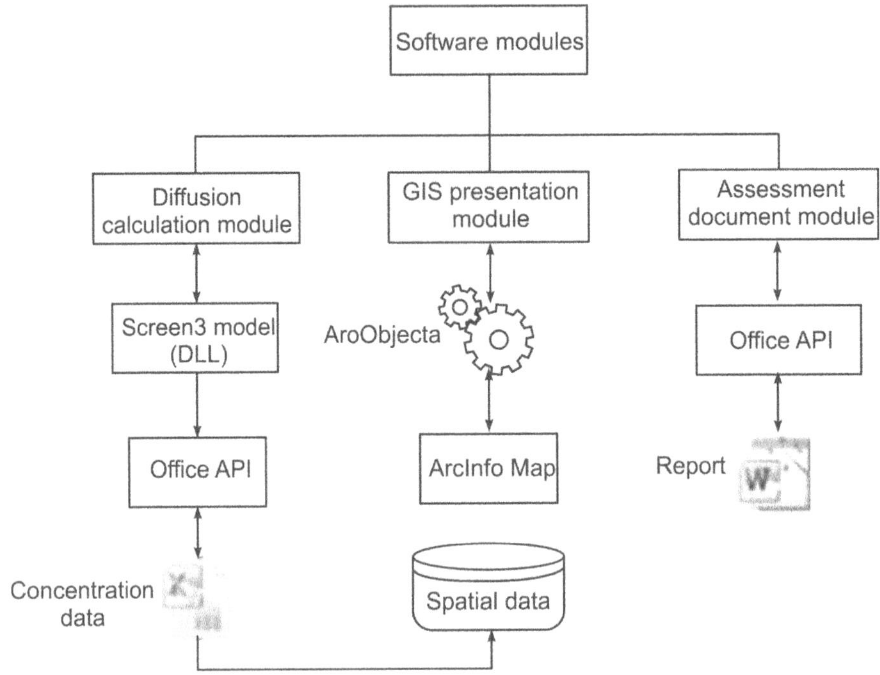

Fig. 14.2 *Integrated Air quality Models and GIS Software Architecture*

Spatial data will be the base of maps, and it stores the necessary space information of the system and provides data support for spatial analysis and graphics operation. The spatial data is managed by the internal layer database of the GIS platform. The features of real world can be represented as different layers in a map. A new map layer will be built to load the concentration information into an existing map. Then the embedded interpolation algorithm of the Arc GIS engine will be used to constructing a raster layer and the interpolation results would be used to create a concentration contour lines layer. The raster values represent the concentrations of the regional area that one could use different colors to differ the pollution situation by using the embedded function in Arc GIS engine to render the raster layer. Finally, the raster layer and the contour layer will be loaded into the map, so that the distribution of the pollutant is shown on the map with good visualization effect.

The inherently spatial nature of transportation-related air quality analysis makes the geographic information system (GIS) particularly well suited to enhancing microscale air quality analysis. GIS provides several features ideal for the type of analysis necessary to determine transportation-related air quality impacts. It is an excellent storage, manipulation, modeling, and mapping tool for spatial data. Spatial information such as street coordinates and accompanying attributes can be exported and manipulated as input to air quality models such as CALINE3 and CAL3QHC. Output from air quality models in the form of pollution concentrations at specified receptor locations can be input to GIS for hot-spot identification, estimation of contributions of off-road mobile sources, and impact analysis. GIS tools applied to air quality analysis include contour generation, classification, thematic analysis, point-in-polygon analysis, and polygon overlay. Several case studies demonstrating these capabilities using TRANSCAD, a transportation-based GIS package, are presented for microscale air quality analysis.

14.4.3.1 Case study of surface ozone over Mississippi (USA) Gulf Coast

The spatial distribution of simulated ozone, SO_2 and NO_2 at 1000 CST 9 June'06 representing daytime is shown in Figure 14.3. The precursors SO_2 and NO_2 have higher concentrations in the northern and northeastern Mississippi and northwestern Alabama. The peak ozone concentration at this time is about 45-55 ppbv and it is spread over northern and southeast Louisiana, and northwestern and southern parts of Mississippi. The concentration patterns of NO_2 and SO_2 species indicate large plume like peaks due to sources like coal fired power plants, manufacturing and other fuel combustion units distributed in the northern, central parts and along the Gulf coast in the domain.

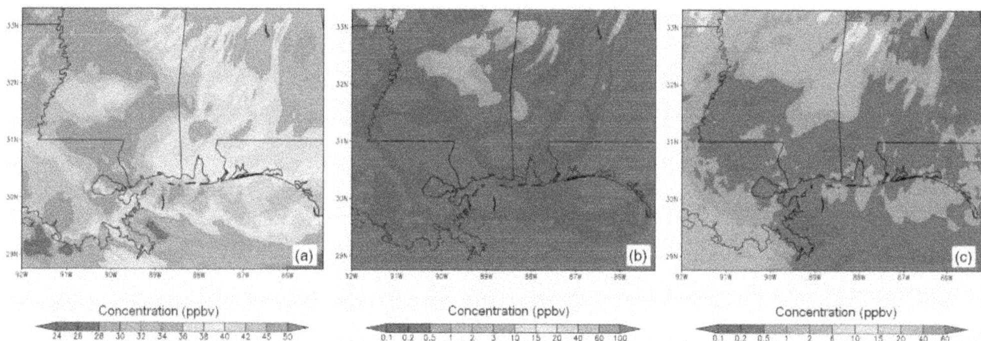

Fig. 14.3 *Spatial distribution of concentrations of simulated a) Ozone b) SO_2 and c) NO_2 from the model lowest layer at 16 UTC 9 June. The concentration units are in ppbv*

14.4.3.2 Presentation of Air quality impacts of a Transportation corridor using integrated air quality dispersion model SCREEN3 and ArcGIS modules

Figure 14.4 presents the output of an integrated air quality dispersion model SCREEN3 and ArcGIS modules for a typical transportation corridor showing spatial presentation of the source information, the pollutant concentration distribution of the plume and a concentration curve at the plume centerline

The parameters of the emission source are as follow:

Source type = POINT

Emission rate (g/S) =100.0

Stack height (m) = 100.0 ζ Stack inner diameter (m) = 2.5

Stack exit velocity (m/s) = 25.0

Stack gas temperature (K) = 450.0

Ambient air temperature (K) = 293.0
Receptor height (m) = 0.0
Urban/Rural option = Urban
Wind direction = Northeastern

The model examines a range of stability classes and wind speeds to identify the "worst case" meteorological conditions intended to estimate maximum ground-level concentrations.

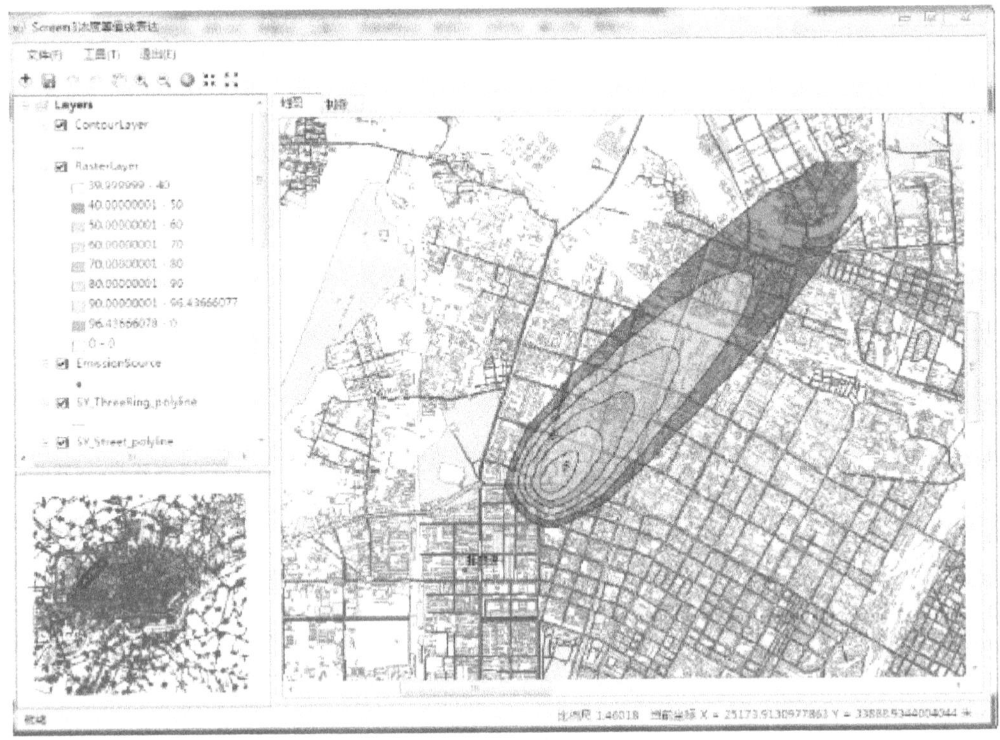

Fig. 14.4 *Contour Representation of air pollutant concentration of a plume arising from a transportation corridor*

Questions

1. Discuss effects of Air pollution on health, materials damage, climatic change, aesthetic effects.
2. You have been assigned to carry out Environment impact assessment on Air pollution monitoring. Discuss in detail the steps required to carryout EIA.
3. Describe Pasquill model.
4. Describe various sources of Air pollution and potential emissions.

5. Discuss in detail various parameters to study Air pollution.
6. Describe Dispersion modelling.
7. Write a short note on dispersion modelling using AUSPLUME.
8. Discuss the output data processing of Dispersion model
9. Explain in detail the dispersion of air quality impact assessment result
10. Discuss the role of GIS in Air quality impact assessment

References

1. USEPA. Air quality atlas. EPA 4001 K – 92 - 002. Office of Air Quality Planning and standards. Research Triangle Park, NC. May 1992.
2. USEPA. Research Triangle Park, NC Appr. 1993. 8-1. Compilation of air pollutant emission factors. 2nd Edition. AP- 42.
3. USEPA 1973.2.1-3.
4. USEPA 1972. Guidelines on air quality models. EPA 450/2 –78 –027. Research Triangle Park, NC Appr. 1928. 16-27. A-12-A-24.
5. L.W. Canter. 1985. Environmental impacts of agricultural production activities. Lewis Chelsea Publishers, MI. 169-209.
6. L Ortolano, 1985. Estimating air quality impacts. Environmental impact assessment review. 5:9-35. March.
7. D. B. Turner, 1979. Atmospheric dispersion modelling a critical review. Journal of the air pollution control assessment. 29(5):502-5 19.
8. R. W. Brole, 1988. Screening procedures for estimating the air quality impact of stationary sources. EPA 450/4 – 88 - 010 USEPA. Research Triangle Park, NC. August.
9. P. Carry, 1990. Fugitive Dust Model (FDM) for micro computers.EPA SW / DK 90 – 94 USEPA Seattle.
10. D. L. Ermark, 1988, SLAB. Denser than air atmospheric dispersion model UCRL-99882, Lawrence Livermore National Laboratory. Livermore, Calif. October.
11. J. S. Touma, 1989. Review and evaluation of a new source dispersion algorithms for emission sources of superfuel sites. EPA 450/4 - 89 - 020 USEPA. Research Triangle Park, NC. November.
12. J. R. Barker and D. T. Tingey, (Eds), 1991. Air pollution effects on biodiversity. Van Nostrand Reinhold, New York.
13. D. A. Cataldo, 1990. Evaluation, actualization of mechanisms controlling facilities effects of arising smokes (transport, transformations, rate and terrestrial ecological effects of brass obstructions).AD – A 227 BY / Y/ WEP. Battle Pacific Northwest Laboratory. Richland Wash. August.

14. U.S. Army construction engineering Research Lab. Environmental review guide from USA Rev. 5 volumes. 1989.
15. C. Conherd, P. Engle Client, G. E. Muleski, J. S. Kimmey and K. D. Resbury, 1990. Control of fugitive and hazardous dust. Arryes Data Corporation Park, Ridge NJ.
16. USEPA 1998b.
17. A. J. Buonicose and W. T. Davis, 1992. Air pollution engineering manual. Van Nostrand Reinhold New York.
18. M. K. Sink. 1991. Handbook: control technologies for hazardous air pollutants. EPA 625/6-9-041, US. Environmental Protection Agency, Cincinnati.
19. Earth Tech 2000, A User's Guide for the CALPUFF Dispersion Model (Version 5), Earth Tech Incorporated, Long Beach CA, USA.
20. Earth Tech 2000, A User's Guide for the CALMET Meteorological Model (Version 5), Earth Tech Incorporated, Long Beach CA, USA.
21. EPA Victoria 1985, Plume Calculation Procedure: An Approved Procedure under Schedule E of State Environment Protection Policy (The Air Environment), Publication 210, Environment Protection Authority of Victoria, Melbourne.
22. EPA Victoria 1986. The AUSPLUME Gaussian Plume Dispersion Model, First Edition, Publication 264, Environment Protection Authority of Victoria, Melbourne.
23. EPA Victoria 1999, AUSPLUME Gaussian Plume Dispersion Model: Technical User Manual, Publication 671, Environment Protection Authority of Victoria, Melbourne.
24. EPA Victoria 2000, AUSPLUME Gaussian Plume Dispersion Model: Technical User Manual, Environment Protection Authority of Victoria, Melbourne.
25. USEPA 1999, Guideline on Air Quality Models, 40 CFR, Chapter I, Part 51, Appendix W, United States Environmental Protection Agency, Washington DC, USA.
26. DEC 2005, Approved Methods for the Sampling and Analysis of Air Pollutants in New South Wales, Department of Environment and Conservation NSW, Sydney.
27. EPA 2001, Draft Policy: Assessment and Management of Odour from Stationary Sources in NSW, NSW Environment Protection Authority, Sydney.
28. EPA 2001, Technical Notes – Draft Policy: Assessment and Management of Odour from Stationary Sources in NSW, NSW Environment Protection Authority, Sydney.
29. EPA Victoria 1985, Plume Calculation Procedure: An Approved Procedure under Schedule E of State Environment Protection Policy (The Air Environment), Publication 210, Environment Protection Authority of Victoria, Melbourne.
30. EPA Victoria 1990, The AUSPLUME Gaussian Plume Dispersion Model, First Edition, Publication 264, Environment Protection Authority of Victoria, Melbourne. EPA Victoria 1999, AUSPLUME Gaussian Plume Dispersion Model: Technical User Manual, Publication 671, Environment Protection Authority of Victoria, Melbourne.

31. EPA Victoria 2000, AUSPLUME Gaussian Plume Dispersion Model: Technical User Manual, Environment Protection Authority of Victoria, Melbourne.
32. Department of Environment and Heritage, National Pollutant Inventory Emissions Estimation Technique Manuals (www.npi.gov.au/handbooks/approved handbooks/sector-manuals html)
33. USEPA 1995, AP 42, Fifth Edition Compilation of Air Pollutant Emission Factors, Volume
34. 1: Stationary Point and Area Sources (www.epa.gov/ttn/chief/ap42/index.html)
35. Businger, J. and Fleagle, R.1980, An Introduction to Atmospheric Physics, Academic Press, New York.
36. Golder, D. 1972, 'Relations among stability parameters in the surface layer', Boundary Layer Meteorology 3, 47–58.
37. Hurley, P. 2005a, The Air Pollution Model (TAPM) Version 3, Part 1: Technical Description, CSIRO Atmospheric Research Technical Paper No. 71, CSIRO Division of Atmospheric Research, Melbourne.
38. Hurley, P. 2005b, The Air Pollution Model (TAPM) Version 3: User Manual, CSIRO Atmospheric Research Technical Paper No. 31, CSIRO Division of Atmospheric Research, Melbourne.
39. Hurley, P., Physick, W., Luhar, A. and Edwards, M. 2005, The Air Pollution Model (TAPM) Version 3, Part 2: Summary of Some Verification Studies, CSIRO Atmospheric Research Technical Paper No. 72, CSIRO Division of Atmospheric Research, Melbourne.
40. McRae, G.J. 1981, Mathematical Modelling of Photochemical Air Pollution, Chapter 4, 'Turbulent Diffusion Coefficients', PhD Thesis, California Institute of Technology, Los Angeles.
41. Standards Australia 1987a, AS 2922–1987, Ambient Air – Guide for the Siting of Sampling Units, Standards Australia, Sydney.
42. Standards Australia 1987b, AS 2923–1987, Ambient Air – Guide for the Measurement of Horizontal Wind for Air Quality Applications, Standards Australia, Sydney.
43. USEPA 1996, Meteorological Processor for Regulatory Model (MPRM), User's Guide, EPA454/B-96-002, United States Environmental Protection Agency, Washington DC, USA.
44. USEPA 2000, Meteorological Monitoring Guidance for Regulatory Modelling Applications, EPA-450/R-99-005, United States Environmental Protection Agency, Washington DC, USA.
45. Zanetti, P. 1990, Air Pollution Modelling: Theories, Computational Methods and Available Software, Van Nostrand Reinhold, New York.
46. DEC 4th Quarter 2003–present, Quarterly Air Quality Monitoring Report Part A: DEC Data, Department of Environment and Conservation NSW, Sydney.
47. DEC 4th Quarter 2003–present, Quarterly Air Quality Monitoring Report Part B: Industry Data, Department of Environment and Conservation NSW, Sydney.
48. EPA 1992–3rd Quarter 2003, Quarterly Air Quality Monitoring Report Part A: EPA Data, NSW Environment Protection Authority, Sydney.

49. EPA 1992–3rd Quarter 2003, Quarterly Air Quality Monitoring Report Part B: Industry Data, NSW Environment Protection Authority, Sydney.
50. EPA Victoria 2000, AUSPLUME Gaussian Plume Dispersion Model: Technical User Manual, Environment Protection Authority of Victoria, Melbourne.
51. Schulman, L.L., Strimaitis, D.G. and Scire, J.S. 1998, Development and Evaluation of the PRIME Plume Rise and Building Downwash Model, Preprint Volume for the Tenth Joint Conference on Applications of Air Pollution Meteorology with A&WMA, American Meteorological Society, Phoenix, Arizona.
52. SPCC 1975–1991, Quarterly Air Quality Monitoring Reports, State Pollution Control Commission, Sydney.
53. USEPA 1995, User's Guide to the Building Profile Input Program, Revised February 1995, PA-454/R-93-038, United States Environmental Protection Agency, Washington DC, USA.
54. Hurley, P., Physick, W., Luhar, A. and Edwards, M. 2005, The Air Pollution Model (TAPM) Version 3, Part 2: Summary of Some Verification Studies, CSIRO Atmospheric Research Technical Paper No. 72, CSIRO Division of Atmospheric Research, Melbourne.
55. Katestone Scientific 1995, The Evaluation of Peak-to-Mean Ratios for Odour Assessments, volumes I and II, Katestone Scientific Pty Ltd, Brisbane.
56. Katestone Scientific 1998, Peak-to-Mean Concentration Ratios for Odour Assessments, Katestone Scientific Pty Ltd, Brisbane.
57. Schulman, L.L., Strimaitis, D.G. and Scire, J.S. 1998, Development and Evaluation of the PRIME Plume Rise and Building Downwash Model, Preprint Volume for the Tenth Joint Conference on Applications of Air Pollution Meteorology with A&WMA, American Meteorological Society, Phoenix, Arizona, 1998.
58. USEPA 1985, Guideline for Determination of Good Engineering Practice Stack Height (Technical Support Document for the Stack Height Regulations), Revised, EPA-450/4-80023R, United States Environmental Protection Agency, Washington DC, USA.
59. USEPA 1995, User's Guide to the Building Profile Input Program, Revised February 1995, EPA-454/R-93-038, United States Environmental Protection Agency, Washington DC, USA.
60. USEPA 1995, User's Guide for the Industrial Source Complex (ISC3) Dispersion Models, Volume 1 and Volume 2, US EPA Office of Air Quality Planning and Standards Emissions, Monitoring, and Analysis Division Research Triangle Park, North Carolina.
61. USEPA 1997, Addendum to ISC3 User's Guide, The PRIME Plume Rise and Building Downwash Model, United States Environmental Protection Agency, Washington DC, USA.
62. USEPA 1999, Guideline on Air Quality Models, 40 CFR, Chapter I, Part 51, Appendix W, United States Environmental Protection Agency, Washington DC, USA.
63. ANZECC 1990, National Goals for Fluoride in Ambient Air and Forage, Australian and New Zealand Environment and Conservation Council, Canberra.

64. AWT 2001, Literature Review – Australian and Overseas Odour Threshold Data and Ambient Air Quality Criteria for Hydrogen Sulphide, A Report to the NSW EPA, Australian Water Technologies, Sydney.
65. Carson, P. and Round, J. 1989, Feedlot Odours, Department of Primary Industry, Brisbane.
66. CARB 2003a, Hot Spots Analysis and Reporting Program (HARP), California Environmental
67. Protection Agency, Air Resources Board, www.arb.ca.gov/toxics/harp/harp.htm CARB 2003b, Hot Spots Analysis and Reporting Program, User Guide Version 1.0,
68. California Environmental Protection Agency, Air Resources Board, Sacramento. en Health 2001, Exposure Scenarios and Exposure Settings, March 2001,
69. Department of Health and Ageing and en Health Council, Canberra. enHealth 2002, Environmental Health Risk Assessment: Guidelines for Assessing Human Health Risks from Environmental Hazards, June 2002, Department of Health and Ageing and enHealth Council, Canberra. enHealth 2003,
70. Australian Exposure Assessment Handbook, Consultation Draft, December 2003, Department of Health and Ageing and en Health Council, Canberra.
71. EPA 1998, Action for Air: The NSW Government's 25-Year Air Quality Management Plan, NSW Environment Protection Authority, Sydney.
72. EPA 2001, Draft Policy: Assessment and Management of Odour from Stationary Sources in NSW, NSW Environment Protection Authority, Sydney.
73. EPA 2001, Technical Notes – Draft Policy: Assessment and Management of Odour from Stationary Sources in NSW, NSW Environment Protection Authority, Sydney.
74. EPA Victoria 1985, Plume Calculation Procedure: An Approved Procedure Under Schedule E of State Environment Protection Policy (The Air Environment), Publication 210, Environment Protection Authority of Victoria, Melbourne.
75. NEPC 1998, Ambient Air – National Environment Protection Measure for Ambient Air Quality, National Environment Protection Council, Canberra.
76. NERDDC 1988, Air Pollution from Surface Coal Mining: Measurement, Modelling and Community Perception, Project No. 921, National Energy Research Development and Demonstration Council, Canberra.
77. NHMRC 1996, Ambient Air Quality Goals Recommended by the National Health and Medical Research Council, National Health and Medical Research Council, Canberra.
78. OEHHA 1994, Benzo[a] pyrene as a Toxic Air Contaminant, Executive Summary, California Air Resources Board and the Office of Environmental Health Hazard Assessment, Sacramento.
79. OEHHA 2003, The Air Toxics Hot Spots Program Guidance Manual for Preparation of Health Risk Assessments, Office of Environmental Health Hazard Assessment, California Environmental Protection Agency, Sacramento.

80. Streeton, J.A. 1990, Air Pollution, Health Effects and Air Quality Objectives in Victoria, Environment Protection Authority of Victoria, Melbourne.
81. SPCC 1983, Air Pollution from Coal Mining and Related Developments, State Pollution Control Commission, Sydney.
82. Victorian Government Gazette, Friday 21 December 2001, 'State Environment Protection Policy (Air Quality Management)', No. S 240, Government of Victoria, Melbourne.
83. WHO 2000, WHO Air Quality Guidelines for Europe, 2nd Edition, World Health Organization, Geneva.
84. Anjaneyulu Yerramilli, Venkata Srinivas C, Venkata Bhaskar Rao Dodla, LaToya Myles, William R Pendergrass, Christoph A Vogel, Hari Prasad Dasari, Francis Tuluri, Julius M Baham; Robert Hughes, Chuck Patrick, John Young, Shelton Swanier; 2010, A simulation study of surface Ozone pollution in the Central Gulf Coastal region using WRF/Chem; Atmospheric Environment; ATMENV-D-10-00784; Elsevier Publications.
85. Software for Environmental Impact Assessment of Air Pollution Dispersion Based on ArcGIS; Shengjun Zhonga*, Legang Zhoua, Zhufang Wangb; Procedia Environmental Sciences 10 (2011) 2792 – 2797 (Elsievier).

APPENDIX

Reference Tables

Table A.1 *Dimensional prefixes*

Symbol	Prefix	Multiple
T	tera units	10^{12}
G	giga units	10^{9}
M	mega units	10^{6}
k	kilo units	10^{3}
h	hector units	10^{2}
Da	deca units	10^{1}
	units	10^{0}
d	deci units	10^{-1}
c	centi units	10^{-2}
m	milli units	10^{-3}
μ	micro units	10^{-6}
n	nano units	10^{-9}
p	pico units	10^{-12}
f	femto units	10^{-15}
a	atto units	10^{-18}

Table A.2 *The Greek alphabet*

α	Alpha	ρ	Rho
β	Beta	σ	Sigma
γ	Gamma	τ	Tau
δ	Delta	υ	Upsilon
ε	Epsilon	ϕ	Phi
ζ	Zeta	χ	Chi
η	Eta	ψ	Psi
θ	Theta	*w*	Omega
ι	Iota		
κ	Kappa		
λ	Lambda		
m	Mu		
ν	Nu		
ξ	Xi		
o	Omicron		
p	Pi		

Table A.3 *Conversion factors*

Length

$12\frac{\text{in.}}{\text{ft}}$ $3\frac{\text{ft}}{\text{yd}}$ $5280\frac{\text{ft}}{\text{mile}}$ $1.15\frac{\text{mile}}{\text{nautical mile}}$

$24\frac{\text{cm}}{\text{in.}}$ $3.28\frac{\text{ft}}{\text{m}}$ $10^6\frac{\text{microns}}{\text{m}}$ $10^4\frac{\text{angstroms}}{\text{micron}}$

Area

$144\frac{\text{in.}^2}{\text{ft}^2}$ $6.45\frac{\text{cm}^2}{\text{in.}^2}$ $43{,}560\frac{\text{ft}^2}{\text{acre}}$ $640\frac{\text{acres}}{\text{mi}^2}$

Volume

$1728\frac{\text{in.}^3}{\text{ft}^3}$ $35.5\frac{\text{ft}^3}{\text{m}^3}$ $7.48\frac{\text{gal}}{\text{ft}^3}$ $3.78\frac{\text{literes}}{\text{gal}}$

$8\frac{\text{pt}}{\text{gal}}$ $28.3\frac{\text{leters}}{\text{ft}^3}$ $29.6\frac{\text{ml}}{\text{oz}}$ $42\frac{\text{gal}}{\text{bbl}}\text{petroleum}$

Force

$2000\frac{\text{lb}_r}{\text{ton}}$ $2.2\frac{\text{lb}f}{\text{kg}}$ $16\frac{\text{oz}}{\text{lb}f}$ $7000\frac{\text{grains}}{\text{lb}_f}$

$4.45\times10^5\frac{\text{dynes}}{\text{lb}_f}$ $4.45\frac{\text{Newtons}}{\text{lb}_f}$ $10^5\frac{\text{dynes}}{\text{lb}_m}$

Mass

$454\frac{\text{g}}{\text{lb}_m}$ $2.20\frac{\text{lb}_m}{\text{kg}}$ $28.4\frac{\text{q}}{\text{oz}}$ $7000\frac{\text{grains}}{\text{lb}_m}$

$2000\frac{\text{lb}_m}{\text{ton}}$ $2240\frac{\text{lb}_m}{\text{kg}}$ $284\frac{\text{q}}{\text{oz}}$ $7000\frac{\text{grains}}{\text{lb}_m}$

Energy and power

$33{,}000\frac{\text{ft}-\text{lb}_f/\text{min}}{\text{hp}}$ $2544\frac{\text{bthu}}{\text{hp}}$ $1.34\frac{\text{hp}}{\text{kW}}$ $252\frac{\text{cal}}{\text{Btu}}$

$778\frac{\text{ft}-\text{lb}_f}{\text{Btu}}$ $4.19\frac{\text{Jouls}}{\text{cal}}$ $1.8\frac{\text{Btu.lb}_f}{\text{cal}/\text{g}}$ $1\frac{\text{Joule}/\text{sec}}{\text{W}}$

Dinsity

$0.016\frac{\text{g}/\text{cm}^2}{\text{lb}_m/\text{ft}^3}$ $8.34\frac{\text{lb}_m/\text{gal}}{\text{g}/\text{cm}^3}$

Temperature

$^\circ\text{K} = ^\circ\text{C}+273$ $^\circ\text{R} = ^\circ\text{F}+460$ $^\circ\text{C} = 5/9(^\circ\text{F}-32)$

$^\circ\text{F} = 1.8^\circ\text{C}+32$ $1.8\frac{\Delta^\circ\text{F}}{\Delta^\circ\text{C}}$ $1.8\frac{\Delta^\circ\text{R}}{\Delta^\circ\text{F}}$

Viscosity

$1\frac{\text{Poise}}{\text{g}/\text{cm}-\text{sec}}$ $1\frac{\text{cP}}{6.72\times10^{-4}\text{lb}_m/\text{ft}-\text{sec}}$

Source: E.I Shaheen, "Basic Practice of Chemical enginnering", Second edition, International Institute of Technology, Inc., Joplin, MO, 1983.

Table A.4 *Physical constants*

Constant	Magnitude	Units
Air		
Density (70°F & 29.92 in. Hg)	0.075	lbm/ft^3
(0°C & 760 mm Hg)	1.29	g/litre
Viscosity (70°F& 29.92 in Hg)	4×10^{-7}	lbf-sec/ft^2
(20°C & 760 mm Hg)	0.018	cP
Avogadro's number	6.023×10^{23}	molecules/g-mole
Boltzmann's constant	1.380×10^{-23}	crgs/°C
Electronic charge	1.602×10^{-19}	Coulombs
Electronic mass	9.106×10^{-28}	g
Faraday's constant	96,493	coul/equivalent
Gas constant	1.987	cal/g-mole°C
	1.987	btu/lbm mole °F
	82.057	cm3-atm/g-mole °K
	8.314	Joules/g-mole °K
Gravitational acceleration	980.665	cm/sec^2
Planck's constant	6.624×10^{27}	erg sec
Speed of		
light (in vacuum)	2.998×10^{10}	cm/sec
sound (20°C, 1 atm, air)	344	m/sec
Standand volume of gas	22.4	litre/g-mole
(0°C & 760 mm Hg)		
Water		
density (68°F)	62.4	lbm/fr^3
	8.33	lbm/gal
density (4°C)	1.0	g/ml
viscosity (20°C)	1.0	cP

Table A.5 *Properties of the elements*

Element	Symbol	Atomic Number	Atomic Weight*	Melting Piont, °C	Density of Solid, g/cc
Actinium	Ac	89	(227)	1050	
Aluminum	Al	13	26.982	660	2.70
Americium	Am	95	(243)	11.7	
Antimony	Sb	51	121.75	630.5	6.88
Argon	Ar	18	39.948	–189.4	1.67
Arsenic	As	33	74.922	817	5.72
Astatine	At	85	(210)	300	
Barium	Ba	56	137.34	714	3.5
Berkelium	13k	97	(247)		
Beryllium	Be	4	9.012	1277	1.85
Bismuth	Bi	83	208.980	271	9.80
Boron	B	5	10.811	2030	2.34
Bromine	Br	35	79.909		7.2
Cadmium	Cd	48	112.40	321	8.65
Calcium	Ca	20	40.08	850	1.55
Californium	Cf	98	(249)		

Table A.5 ***Contd...***

Carbon	c	6	12.011	3727	3.51
Cerium	Ce	58	140.12	804	6.77
Cesium	CS	55	132.905	28.7	1.90
Chlorine	Cl	17	35.453	–101	1.9
Chromium	Cr	24	51.996	1875	7.19
Cobalt	Co	27	58.933	1495	8.85
Copper	Cu	29	63.54	1083	8.96
Curium	CM	96	(247)		
Dysprosium	Dy	66	162.50	1407	8.55
Einsteinium	Es	99	(254)		
Erbium	Er	68	167.26	1497	9.15
Europium	Eu	63	151.96	826	5.24
Fermium	Fm	100	(253)		
Fluorine	F	9	18.998	–220	1.3
Francium	Fr	87	(223)	27	
Gadolinium	Gd	64	157.25	1312	7.86
Gallium	Ga	31	69.72	29.8	5.91
Germanium	Ge	32	72	937	5.32
Gold	Au	79	59	1063	19.32
Hafnium	Hf	72	196.967	2222	13.1
Helium	He	2	4.003	–269.7	
Holmium	Ho	67	164.930	1461.	6.79
Hydrogen	H	1	1.008		259,2
Iodine	I	53	126.904	113.7	4.94
Iridium	Ir	77	192.2	2454	22.4
Iron	Fe	26	55,847	1537	7.87
Krypton	Kr	36	83.80	–157.3	3.0
Lanthanum	La	57	138.91	920	6.19
Lawrencium	Lw	103	(257)		
Lead	Pb	82	207.19	327.3	11,34
Lithium	Li	3	6.939	180,5	0,53
Lutetium	Lu	71	174.97	1650	9.85
Magnesium	Mg	12	24.312	650	1.74
Manganese	Mn	25	54.938	1245	7.43
Mendelevium	Mv	1,01	(256)		
Mercury	Hg	80	200.59	–38.4	14.19
Molybdenum	Mo	42	95.94	2610	10.2
Neodymium	Nd	60	144.24	1019	7.0
Neon	Ne	10	20.183	–248.7	1.45
Neptunium	Np	93	(237)	637	19.5
Nickel	Ni	28	58.71	1453	8.9
Niobium	Nb	41	92.906	2415	8.6
Nitrogen	N	7	14.007	–210	1.03
Nobelium	No	102	(253)		
Osmium	OS	76	190.2	2706	22.57
Oxygen	0	8	15.999	–218.8	1.43
Palladium	Pd	46	106.4	1552	12.02
Phosphorus (white)	P	15	30.974	44.2	1.83
Platinum	Pt	78	195.09	1769	21.45

Table A.5 ***Contd...***

Plutonium	Pu	94	(242)	640	19.3
Polonium	Po	84	(210)	254	9.2
Potassium	K	19	39.102	63.7	0.86
Praseodymium	Pr	59	140.907	919	6.77
Promethium	Pm	61	(147)	1027	
Protoactinium	Pa	91	(231)	1230	15.4
Radium	Ra	88	(226)	700	5.0.
Radon	Rn	86	(222)	–71	4
Rhenium	Re	75	186.2	3180	21.04
Rhodium	Rh	45	102.905	1966	12.44
Rubidium	Rb	37	85,47	38.8	1.53
Ruthenium	Ru	44	101.07	2500	12.2
Samarium	Sm	62	150.32	1072	7.49
Scandium	Sc	21	44.956	1539	2.99
Selenium	Se	34	78.96	217	4.79
Silicon	Si	14	28.086	1410	2.34
Strontium	Sr	38	87.62	768	2.60
Sulfur (yellow)	S	16	32.604	119	2.07
Silver	Ag	47	107.807	960.8	10.5
Sodium	Na	11	22.990	97.8	0.97
Tantalum	Ta	73	180.948	2996	16.6
Technetium	Tc	43	(99)	2200	11.5
Tellurium	Te	52	127.60	450	6.24
Terbium	Tb	65	158.924	1356	8.25
Thalliurn	Tl	81	204.37	303	11.85
Thor-um	Th	90	232.038	1750	11.66
Thulium	Tm	69	168.934	1545	9.31
Tin	Sn	50	118.69	231.9	7.30
Titanium	Ti	22	47.90	1668	4.51
Tungsten	W	74	183.85	3410	19.3
Uranium	U	92	238.03	1132	19.05
Vanadium	v	23	50.942	1900	6.1
Xenon	Xe	54	131.30	–112	3.6
Ytterbium	Yb	70	173.04	824	6.96
Yttrium	y	39	88.905	1509	4.47
Zinc	Zn	30	65.37	419.5	7.13

*Atomic weights base on carbon 12 rounded off to three decimal places. Atomic weights in parentheses are for the most stable isotopes. a melting point at 28 atm prssue. B Sublimes. C Diamond.

Source: E.I, Shaheen, "Basic Practice of Chemical Engineering", Second Edition, International Institute of Technology, Inc., Joplin, MO, 1983.

Index

D

E

F

Q

R

S

T

U

V

W

X

Z

www.ingramcontent.com/pod-product-compliance
Ingram Content Group UK Ltd.
Pitfield, Milton Keynes, MK11 3LW, UK
UKHW061535190726
13845UKWH00005B/111